LINEAR AND INTEGER OPTIMIZATION

Theory and Practice

Third Edition

Advances in Applied Mathematics

Series Editor: Daniel Zwillinger

Published Titles

Green's Functions with Applications, Second Edition *Dean G. Duffy*

Linear and Integer Optimization: Theory and Practice, Third Edition
Gerard Sierksma and Yori Zwols

Markov Processes *James R. Kirkwood*

Pocket Book of Integrals and Mathematical Formulas, 5th Edition
Ronald J. Tallarida Stochastic Partial Differential Equations,
Second Edition *Pao-Liu Chow*

Advances in Applied Mathematics

LINEAR AND INTEGER OPTIMIZATION

Theory and Practice

Third Edition

Gerard Sierksma

University of Groningen, The Netherlands

Yori Zwols

Google, London, United Kingdom

CRC Press
Taylor & Francis Group
Boca Raton London New York

CRC Press is an imprint of the
Taylor & Francis Group, an **informa** business

A CHAPMAN & HALL BOOK

All code in this book is subject to the MIT open source license. See http://opensource.org/licenses/MIT.

CRC Press
Taylor & Francis Group
6000 Broken Sound Parkway NW, Suite 300
Boca Raton, FL 33487-2742

© 2015 by Taylor & Francis Group, LLC
CRC Press is an imprint of Taylor & Francis Group, an Informa business

No claim to original U.S. Government works

ISBN 13: 978-1-4987-1016-9 (hbk)

Visit the Taylor & Francis Web site at
http://www.taylorandfrancis.com

and the CRC Press Web site at
http://www.crcpress.com

To Rita and Rouba

Contents

II   Linear optimization practice: advanced techniques

Appendices

List of Figures

List of Figures

List of Tables

Preface

The past twenty years showed an explosive growth of applications of mathematical algorithms. New and improved algorithms and theories appeared, mainly in practice-oriented scientific journals. Linear and integer optimization algorithms have established themselves among the most-used techniques for quantitative decision support systems.

Many universities all over the world used the first two editions of this book for their linear optimization courses. The main reason of choosing this textbook was the fact that this book shows the strong and clear relationships between theory and practice.

Linear optimization, also commonly called *linear programming*, can be described as the process of transforming a real-life decision problem into a mathematical model, together with the process of designing algorithms with which the mathematical model can be analyzed and solved, resulting in a proposal that may support the procedure of solving the real-life problem. The mathematical model consists of an objective function that has to be maximized or minimized, and a finite number of constraints, where both the objective function and the constraints are linear in a finite number of decision variables. If the decision variables are restricted to be integer-valued, the linear optimization model is called an integer (linear) optimization model.

Linear and integer optimization are among the most widely and successfully used decision tools in the quantitative analysis of practical problems where rational decisions have to be made. They form a main branch of management science and operations research, and they are indispensable for the understanding of nonlinear optimization. Besides the fundamental role that linear and integer optimization play in economic optimization problems, they are also of great importance in strategic planning problems, in the analysis of algorithms, in combinatorial problems, and in many other subjects. In addition to its many practical applications, linear optimization has beautiful mathematical aspects. It blends algorithmic and algebraic concepts with geometric ones.

New in the third edition

New developments and insights have resulted in a rearrangement of the second edition, while maintaining its original flavor and didactic approach. Considerable effort has been made to modernize the content. Besides the fresh new layout and completely redesigned figures, we added new modern examples and applications of linear optimization.

Where the second edition of the book had a disk with a linear optimization package and an interior path algorithm solver, the internet has made this redundant. Solving a linear optimization model is now easier than ever, with several packages available on the internet which readily solve even moderately large linear optimization models. We made the book more practical by adding computer code in the form of models in the GNU Mathematical Programming Language (GMPL). These computer models and the corresponding data files are available for download from the official website, at `http://www.lio.yoriz.co.uk/`. Moreover, these models can be readily solved using the online solver provided on the website.

Organization

This book starts at an elementary level. All concepts are introduced by means of simple prototype examples which are extended and generalized to more realistic settings. The reader is challenged to understand the concepts and phenomena by means of mathematical arguments. So besides the insight into the practical use of linear and integer optimization, the reader obtains a thorough knowledge of its theoretical background. With the growing need for very specific techniques, the theoretical knowledge has become more and more practically useful. It is very often not possible to apply standard techniques in practical situations. Practical problems demand specific adaptations of standard models, which are efficiently solvable only with a thorough mathematical understanding of the techniques.

The book consists of two parts. Part I covers the theory of linear and integer optimization. It deals with basic topics such as Dantzig's simplex algorithm, duality, sensitivity analysis, integer optimization models, and network models, as well as more advanced topics such as interior point algorithms, the branch-and-bound algorithm, cutting planes, and complexity. Part II of the book covers case studies and more advanced techniques such as column generation, multiobjective optimization, and game theory.

All chapters contain an extensive number of examples and exercises. The book contains five appendices, a list of symbols, an author index, and a subject index. The literature list at the end of the book contains the relevant literature usually from after 1990.

Examples, computer exercises, and advanced material are marked with icons in the margin:

 Example    Computer exercise   Advanced material

For course lecturers

The book can be used as a textbook for advanced undergraduate students and graduate students in the fields – among others – of operations research and management science, mathematics, computer science, and industrial engineering. It can be used in a one-quarter course on linear and integer optimization where the emphasis is on both the practical and the mathematical aspects. Since all theorems in this book can be considered as summaries of preceding explanations, the rigorous mathematical proofs can be omitted in undergraduate courses. Such a course may include the chapters 1 through 5, and 7 through 9, without all rigorous mathematical discussions and proofs, and may emphasize Part II on model building and practical case studies. The book is written in such a way that the rigorous mathematical parts are easily recognizable (they are printed in small type and marked with a vertical line), and their omission does not affect the readability of the text. The theoretical level of the book requires only very elementary set theory, trigonometry, and calculus. The more advanced mathematics is explained in the text and in appendices.

For an advanced one-quarter course, the elementary parts can be treated quickly, so that there is enough time to cover most of the material, including Chapter 6 on Karmarkar's algorithm. The flow diagram in Figure 0.1 can be used to structure a course. The diagram outlines which sections form the basic material of the book, and which sections are considered advanced material. It also shows the case-study chapters in Part II that are relevant for each chapter in Part I.

Overview of Part I

In Chapter 1, the reader is introduced to linear optimization. The basic concepts of linear optimization are explained, along with examples of linear optimization models. Chapter 2 introduces the mathematical theory needed to study linear optimization models. The geometry and the algebra, and the relationship between them, are explored in this chapter. In Chapter 3, Dantzig's simplex algorithm for solving linear optimization problems is developed. Since many current practical problems, such as large-scale crew scheduling problems, may be highly degenerate, we pay attention to this important phenomenon. For instance, its relationship with multiple optimal solutions and with shadow prices is discussed in detail. This discussion is also indispensable for understanding the output of linear optimization computer software. Chapter 4 deals with the crucial concepts of duality and optimality, and Chapter 5 offers an extensive account to the theory and practical use of sensitivity analysis. In Chapter 6, we discuss the interior path version of Karmarkar's interior point algorithm for solving linear optimization problems. Among all versions of Karmarkar's algorithm, the interior path algorithm is one of the most accessible and elegant. The algorithm determines optimal solutions by following the so-called interior path through the (relative) interior of the feasible region towards an optimal solution. Chapter 7 deals with integer linear optimization, and discusses several solution techniques such as the branch-and-bound algorithm, and Gomory's cutting-plane algorithm. We also discuss algorithms for mixed-integer linear optimization models. Chapter 8 can be seen as an extension of Chapter 7; it discusses

Chapter	Techniques discussed
10. Designing a reservoir for irrigation	Practical modeling
11. Classifying documents by language	Machine learning
12. Production planning; a single product case	Dynamic optimization
13. Production of coffee machines	Dynamic optimization
14. Conflicting objectives: producing versus importing	Multiobjective optimization, goal optimization, fuzzy optimization
15. Coalition forming and profit distribution	Game theory
16. Minimizing trimloss when cutting cardboard	Column generation
17. Off-shore helicopter routing	Combinatorial optimization, column generation, approximation algorithms
18. The catering service problem	Network modeling, network simplex algorithm

Table 0.1: Techniques discussed in the chapters of Part II.

the network simplex algorithm, with an application to the transshipment problem. It also presents the maximum flow/minimum cut theorem as a special case of linear optimization duality. Chapter 9 deals with computational complexity issues such as polynomial solvability and NP-completeness. With the use of complexity theory, mathematical decision problems can be partitioned into 'easy' and 'hard' problems.

Overview of Part II

The chapters in Part II of this book discuss a number of (more or less) real-life case studies. These case studies reflect both the problem-analyzing and the problem-solving ability of linear and integer optimization. We have written them in order to illustrate several advanced modeling techniques, such as network modeling, game theory, and machine learning, as well as specific solution techniques such as column generation and multiobjective optimization. The specific techniques discussed in each chapter are listed in Table 0.1.

Acknowledgments

We are grateful to a few people who helped and supported us with this book — to Vašek Chvátal, Peter van Dam, Shane Legg, Cees Roos, Gert Tijssen, and Theophane Weber, to the LaTeX community at `tex.stackexchange.com`, and to our families without whose support this book would not have been written.

Groningen and London, January 2015   *Gerard Sierksma* and *Yori Zwols*

Applications   Basic material   Advanced material

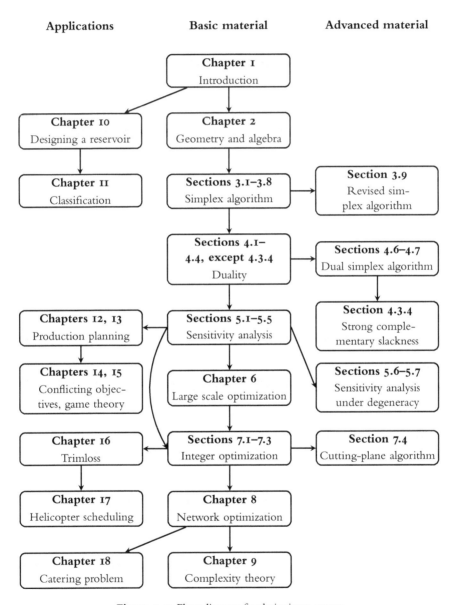

Figure 0.1: Flow diagram for designing a course.

CHAPTER I

Basic concepts of linear optimization

Overview

In 1827, the French mathematician JEAN-BAPTISTE JOSEPH FOURIER (1768–1830) published a method for solving systems of linear inequalities. This publication is usually seen as the first account on linear optimization. In 1939, the Russian mathematician LEONID V. KANTOROVICH (1912–1986) gave linear optimization formulations of resource allocation problems. Around the same time, the Dutch economist TJALLING C. KOOPMANS (1910–1985) formulated linear optimization models for problems arising in classical, Walrasian (LÉON WALRAS, 1834–1910), economics. In 1975, both KANTOROVICH and KOOPMANS received the Nobel Prize in economic sciences for their work. During World War II, linear optimization models were designed and solved for military planning problems. In 1947, GEORGE B. DANTZIG (1914–2005) invented, what he called, the simplex algorithm. The discovery of the simplex algorithm coincided with the rise of the computer, making it possible to computerize the calculations, and to use the method for solving large-scale real life problems. Since then, linear optimization has developed rapidly, both in theory and in application. At the end of the 1960's, the first software packages appeared on the market. Nowadays linear optimization problems with millions of variables and constraints can readily be solved.

Linear optimization is presently used in almost all industrial and academic areas of quantitative decision making. For an extensive – but not exhaustive – list of fields of applications of linear optimization, we refer to Section 1.6 and the case studies in Chapters 10–11. Moreover, the theory behind linear optimization forms the basis for more advanced nonlinear optimization.

In this chapter, the basic concepts of linear optimization are discussed. We start with a simple example of a so-called *linear optimization model* (abbreviated to LO-model) containing two decision variables. An optimal solution of the model is determined by means of the 'graphical method'. This simple example is used as a warming up exercise for more realistic cases, and the general form of an LO-model. We present a few LO-models that illustrate the use of linear optimization, and that introduce some standard modeling techniques. We also describe how to use an online linear optimization package to solve an LO-model.

1.1 The company Dovetail

We start this chapter with a prototype problem that will be used throughout the book. The problem setting is as follows. The company Dovetail produces two kinds of matches: long and short ones. The company makes a profit of 3 (×\$1,000) for every 100,000 boxes of long matches, and 2 (×\$1,000) for every 100,000 boxes of short matches. The company has one machine that can produce both long and short matches, with a total of at most 9 (×100,000) boxes per year. For the production of matches the company needs wood and boxes: three cubic meters of wood are needed for 100,000 boxes of long matches, and one cubic meter of wood is needed for 100,000 boxes of short matches. The company has 18 cubic meters of wood available for the next year. Moreover, Dovetail has 7 (×100,000) boxes for long matches, and 6 (×100,000) for short matches available at its production site. The company wants to maximize its profit in the next year. It is assumed that Dovetail can sell any amount it produces.

1.1.1 Formulating Model Dovetail

In order to write the production problem that company Dovetail faces in mathematical terms, we introduce the *decision variables* x_1 and x_2:

$$x_1 = \text{the number of boxes } (\times 100{,}000) \text{ of long matches to be made the next year,}$$
$$x_2 = \text{the number of boxes } (\times 100{,}000) \text{ of short matches to be made the next year.}$$

The company makes a profit of 3 (×\$1,000) for every 100,000 boxes of long matches, which means that for x_1 (×100,000) boxes of long matches, the profit is $3x_1$ (×\$1,000). Similarly, for x_2 (×100,000) boxes of short matches the profit is $2x_2$ (×\$1,000). Since Dovetail aims at maximizing its profit, and it is assumed that Dovetail can sell its full production, the *objective* of Dovetail is:

$$\text{maximize } 3x_1 + 2x_2.$$

The function $3x_1 + 2x_2$ is called the *objective function* of the problem. It is a function of the decision variables x_1 and x_2. If we only consider the objective function, it is obvious that the production of matches should be taken as high as possible. However, the company also has to take into account a number of *constraints*. First, the machine capacity is 9 (×100,000) boxes per year. This yields the constraint:

$$x_1 + x_2 \leq 9. \tag{1.1}$$

Second, the limited amount of wood yields the constraint:

$$3x_1 + x_2 \leq 18. \tag{1.2}$$

Third, the numbers of available boxes for long and short matches is restricted, which means that x_1 and x_2 have to satisfy:

$$x_1 \leq 7, \tag{1.3}$$

$$\text{and } x_2 \leq 6. \tag{1.4}$$

The inequalities $(1.1) - (1.4)$ are called *technology constraints*. Finally, we assume that only nonnegative amounts can be produced, i.e.,

$$x_1, x_2 \geq 0.$$

The inequalities $x_1 \geq 0$ and $x_2 \geq 0$ are called *nonnegativity constraints*. Taking together the six expressions formulated above, we obtain Model Dovetail:

Model Dovetail.

$$
\begin{array}{llrl}
\max & 3x_1 + 2x_2 & & \\
\text{s.t.} & x_1 + x_2 & \leq 9 & \quad (1.1) \\
& 3x_1 + x_2 & \leq 18 & \quad (1.2) \\
& x_1 & \leq 7 & \quad (1.3) \\
& x_2 & \leq 6 & \quad (1.4) \\
& x_1, x_2 \geq 0. & &
\end{array}
$$

In this model 's.t.' means 'subject to'. Model Dovetail is an example of a *linear optimization model*. We will abbreviate 'linear optimization model' as 'LO-model'. The term 'linear' refers to the fact that the objective function and the constraints are linear functions of the decision variables x_1 and x_2. In the next section we will determine an *optimal solution* (also called *optimal point*) of Model Dovetail, which means that we will determine values of x_1 and x_2 satisfying the constraints of the model, and such that the value of the objective function is maximum for these values.

LO-models are often called 'LP-models', where 'LP' stands for *linear programming*. The word 'programming' in this context is an old-fashioned word for optimization, and has nothing to do with the modern meaning of programming (as in 'computer programming'). We therefore prefer to use the word 'optimization' to avoid confusion.

1.1.2 The graphical solution method

Model Dovetail has two decision variables, which allows us to determine a solution graphically. To that end, we first draw the constraints in a rectangular coordinate system. We start with the nonnegativity constraints (see Figure 1.1). In Figure 1.1, the values of the decision variables x_1 and x_2 are nonnegative in the shaded area above the line $x_1 = 0$ and to the right of the line $x_2 = 0$. Next, we draw the line $x_1 + x_2 = 9$ corresponding to constraint (1.1) and determine on which side of this line the values of the decision variables satisfying

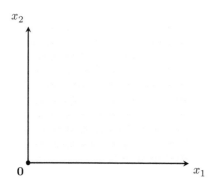

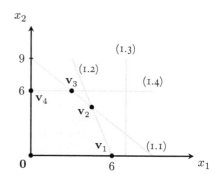

Figure 1.1: Nonnegativity constraints.   **Figure 1.2:** The feasible region of Model Dovetail.

$x_1 + x_2 \leq 9$ are located. Figure 1.2 is obtained by doing this for all constraints. We end up with the region $\mathbf{0}\mathbf{v}_1\mathbf{v}_2\mathbf{v}_3\mathbf{v}_4$, which is called the *feasible region* of the model; it contains the points $\begin{bmatrix} x_1 \\ x_2 \end{bmatrix}$ that satisfy the constraints of the model. The points $\mathbf{0}$, $\mathbf{v}_1$, $\mathbf{v}_2$, $\mathbf{v}_3$, and $\mathbf{v}_4$ are called the *vertices* of the feasible region. It can easily be calculated that:

$$\mathbf{v}_1 = \begin{bmatrix} 6 \\ 0 \end{bmatrix}, \mathbf{v}_2 = \begin{bmatrix} 4\frac{1}{2} \\ 4\frac{1}{2} \end{bmatrix}, \mathbf{v}_3 = \begin{bmatrix} 3 \\ 6 \end{bmatrix}, \text{ and } \mathbf{v}_4 = \begin{bmatrix} 0 \\ 6 \end{bmatrix}.$$

In Figure 1.2, we also see that constraint (1.3) can be deleted without changing the feasible region. Such a constraint is called *redundant with respect to the feasible region*. On the other hand, there are reasons for keeping this constraint in the model. For example, when the right hand side of constraint (1.2) is sufficiently increased (thereby moving the line in Figure 1.2 corresponding to (1.3) to the right), constraint (1.3) becomes nonredundant again. See Chapter 5.

Next, we determine the points in the feasible region that attain the maximum value of the objective function. To that end, we draw in Figure 1.2 a number of so-called level lines. A *level line* is a line for which all points $\begin{bmatrix} x_1 \\ x_2 \end{bmatrix}$ on it have the same value of the objective function. In Figure 1.3, five level lines are drawn, namely $3x_1 + 2x_2 = 0, 6, 12, 18$, and 24. The arrows in Figure 1.3 point in the direction of increasing values of the objective function $3x_1 + 2x_2$. These arrows are in fact perpendicular to the level lines.

In order to find an optimal solution using Figure 1.3, we start with a level line corresponding to a small objective value (e.g., 6) and then (virtually) 'move' it in the direction of the arrows, so that the values of the objective function increase. We stop moving the level line when it reaches the boundary of the feasible region, so that moving the level line any further would mean that no point of it would lie in the region $\mathbf{0}\mathbf{v}_1\mathbf{v}_2\mathbf{v}_3\mathbf{v}_4$. This happens for the level line $3x_1 + 2x_2 = 4\frac{1}{2}$. This level line intersects the feasible region at exactly one point, namely $\begin{bmatrix} 4\frac{1}{2} \\ 4\frac{1}{2} \end{bmatrix}$. Hence, the optimal solution is $x_1^* = 4\frac{1}{2}, x_2^* = 4\frac{1}{2}$, and the optimal objective value

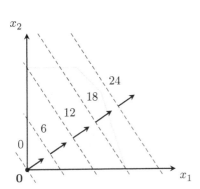
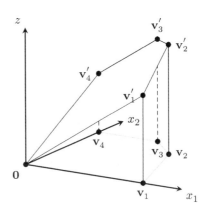

Figure 1.3: Level lines and the feasible region.   **Figure 1.4:** Three-dimensional picture.

is $22\frac{1}{2}$. Note that the optimal point is a vertex of the feasible region. This fact plays a crucial role in linear optimization; see Section 2.1.2. Also note that this is the only optimal point.

In Figure 1.4, the same model is depicted in three-dimensional space. The values of $z = 3x_1 + 2x_2$ on the region $\mathbf{0}\mathbf{v}_1\mathbf{v}_2\mathbf{v}_3\mathbf{v}_4$ form the region $\mathbf{0}\mathbf{v}_1'\mathbf{v}_2'\mathbf{v}_3'\mathbf{v}_4'$. From Figure 1.4 it is obvious that the point $\mathbf{v}_2$ with coordinate values $x_1 = 4\frac{1}{2}$ and $x_2 = 4\frac{1}{2}$ is the optimal solution. At $\mathbf{v}_2$, the value of the objective function is $z^* = 22\frac{1}{2}$, which means that the maximum profit is \$22,500. This profit is achieved by producing $450,000$ boxes of long matches and $450,000$ boxes of short matches.

1.2 Definition of an LO-model

In this section we define the standard form of an LO-model. Actually, the literature on linear optimization contains a wide range of definitions. They all consist of an objective together with a number of constraints. All such standard forms are equivalent; see Section 1.3.

1.2.1 The standard form of an LO-model

We first formulate the various parts of a general LO-model. It is always assumed that for any positive integer d, $\mathbb{R}^d$ is a Euclidean vector space of dimension d; see Appendix B. Any LO-model consists of the following four parts:

Decision variables. An LO-model contains real-valued decision variables, denoted by $x_1, \ldots, x_n$, where n is a finite positive integer. In Model Dovetail the decision variables are x_1 and x_2.

Objective function. The objective function $c_1x_1 + \ldots + c_nx_n$ is a linear function of the n decision variables. The constants $c_1, \ldots, c_n$ are real numbers that are called the *objective coefficients*. Depending on whether the objective function has to be maximized or minimized, the *objective* of the model is written as:

$$\max\ c_1 x_1 + \ldots + c_n x_n, \qquad \text{or} \qquad \min\ c_1 x_1 + \ldots + c_n x_n,$$

respectively. In the case of Model Dovetail the objective is $\max 3x_1 + 2x_2$ and the objective function is $3x_1 + 2x_2$. The value of the objective function at a point $\mathbf{x}$ is called the *objective value at* $\mathbf{x}$.

Technology constraints. A technology constraint of an LO-model is either a '$\leq$', a '$\geq$', or an '$=$' expression of the form:

$$a_{i1} x_1 + \ldots + a_{in} x_n \quad (\leq, \geq, =) \quad b_i,$$

where $(\leq, \geq, =)$ means that either the sign '$\leq$', or '$\geq$', or '$=$' holds. The entry a_{ij} is the coefficient of the j'th decision variable x_j in the i'th technology constraint. Let m be the number of technology constraints. All (left hand sides of the) technology constraints are linear functions of the decision variables $x_1, \ldots, x_n$.

Nonnegativity and nonpositivity constraints. A *nonnegativity constraint* of an LO-model is an inequality of the form $x_i \geq 0$; similarly, a *nonpositivity constraint* is of the form $x_i \leq 0$. It may also happen that a variable x_i is not restricted by a nonnegativity constraint or a nonpositivity constraint. In that case, we say that x_i is a *free* or *unrestricted* variable. Although nonnegativity and nonpositivity constraints can be written in the form of a technology constraint, we will usually write them down separately.

For $i \in \{1, \ldots, m\}$ and $j \in \{1, \ldots, n\}$, the real-valued entries a_{ij}, b_i, and c_j are called the *parameters* of the model. The technology constraints, nonnegativity and nonpositivity constraints together are referred to as the *constraints* (or *restrictions*) of the model.

A vector $\mathbf{x} \in \mathbb{R}^n$ that satisfies all constraints is called a *feasible point* or *feasible solution* of the model. The set of all feasible points is called the *feasible region* of the model. An LO-model is called *feasible* if its feasible region is nonempty; otherwise, it is called *infeasible*. An *optimal solution* of a maximizing (minimizing) LO-model is a point in the feasible region with maximum (minimum) objective value, i.e., a point such that there is no other point with a larger (smaller) objective value. Note that there may be more than one optimal solution, or none at all. The objective value at an optimal solution is called the *optimal objective value*.

Let $\mathbf{x} \in \mathbb{R}^n$. A constraint is called *binding* at the point $\mathbf{x}$ if it holds with equality at $\mathbf{x}$. For example, in Figure 1.2, the constraints (1.1) and (1.1) are binding at the point $\mathbf{v}_2$, and the other constraints are not binding. A constraint is called *violated* at the point $\mathbf{x}$ if it does not hold at $\mathbf{x}$. So, if one or more constraints are violated at $\mathbf{x}$, then $\mathbf{x}$ does not lie in the feasible region.

A maximizing LO-model with only '$\leq$' technology constraints and nonnegativity constraints can be written as follows:

$$
\begin{aligned}
\max \quad & c_1 x_1 + \quad \ldots + \quad c_n x_n \\
\text{s.t.} \quad & a_{11} x_1 + \quad \ldots + \quad a_{1n} x_n \leq b_1 \\
& \quad \vdots \qquad\qquad\quad \vdots \qquad \vdots \\
& a_{m1} x_1 + \quad \ldots + \quad a_{mn} x_n \leq b_m \\
& x_1, \ldots, x_n \geq 0.
\end{aligned}
$$

Using the summation sign '$\sum$', this can also be written as:

$$
\begin{aligned}
\max \quad & \sum_{j=1}^{n} c_j x_j \\
\text{s.t.} \quad & \sum_{j=1}^{n} a_{ij} x_j \leq b_i \qquad\qquad \text{for } i = 1, \ldots, m \\
& x_1, \ldots, x_n \geq 0.
\end{aligned}
$$

In terms of matrices there is an even shorter notation. The superscript '$\top$' transposes a row vector into a column vector, and an (m, n) matrix into an (n, m) matrix $(m, n \geq 1)$. Let

$$
\mathbf{c} = \begin{bmatrix} c_1 & \ldots & c_n \end{bmatrix}^{\top} \in \mathbb{R}^n, \mathbf{b} = \begin{bmatrix} b_1 & \ldots & b_m \end{bmatrix}^{\top} \in \mathbb{R}^m,
$$

$$
\mathbf{x} = \begin{bmatrix} x_1 & \ldots & x_n \end{bmatrix}^{\top} \in \mathbb{R}^n, \text{ and } \mathbf{A} = \begin{bmatrix} a_{11} & \ldots & a_{1n} \\ \vdots & & \vdots \\ a_{m1} & \ldots & a_{mn} \end{bmatrix} \in \mathbb{R}^{m \times n}.
$$

The matrix $\mathbf{A}$ is called the *technology matrix* (or *coefficients matrix*), $\mathbf{c}$ is the *objective vector*, and $\mathbf{b}$ is the *right hand side vector* of the model. The LO-model can now be written as:

$$
\max\{ \mathbf{c}^{\top} \mathbf{x} \mid \mathbf{A} \mathbf{x} \leq \mathbf{b}, \mathbf{x} \geq \mathbf{0} \},
$$

where $\mathbf{0} \in \mathbb{R}^n$ is the n-dimensional all-zero vector. We call this form the *standard form* of an LO-model (see also Section 1.3). It is a *maximizing* model with '$\leq$' technology constraints, and nonnegativity constraints. The feasible region F of the standard LO-model satisfies:

$$
F = \{ \mathbf{x} \in \mathbb{R}^n \mid \mathbf{A} \mathbf{x} \leq \mathbf{b}, \mathbf{x} \geq \mathbf{0} \}.
$$

In the case of Model Dovetail, we have that:

$$
\mathbf{c} = \begin{bmatrix} 3 & 2 \end{bmatrix}^{\top}, \mathbf{b} = \begin{bmatrix} 9 & 18 & 7 & 6 \end{bmatrix}^{\top}, \mathbf{x} = \begin{bmatrix} x_1 & x_2 \end{bmatrix}^{\top}, \text{ and } \mathbf{A} = \begin{bmatrix} 1 & 1 \\ 3 & 1 \\ 1 & 0 \\ 0 & 1 \end{bmatrix}.
$$

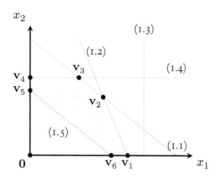

Figure 1.5: The feasible region of Model Dovetail*.

So Model Dovetail in standard form reads:

$$\max\left\{\begin{bmatrix} 3 & 2 \end{bmatrix}\begin{bmatrix} x_1 \\ x_2 \end{bmatrix} \,\middle|\, \begin{bmatrix} 1 & 1 \\ 3 & 1 \\ 1 & 0 \\ 0 & 1 \end{bmatrix}\begin{bmatrix} x_1 \\ x_2 \end{bmatrix} \leq \begin{bmatrix} 9 \\ 18 \\ 7 \\ 6 \end{bmatrix}, \begin{bmatrix} x_1 \\ x_2 \end{bmatrix} \geq \begin{bmatrix} 0 \\ 0 \end{bmatrix}\right\}.$$

Suppose that we want to add the following additional constraint to Model Dovetail. The manager of Dovetail has an agreement with retailers to deliver a total of at least 500,000 boxes of matches next year. Using our decision variables, this yields the new constraint:

$$x_1 + x_2 \geq 5. \tag{1.5}$$

Instead of a '$\leq$' sign, this inequality contains a '$\geq$' sign. With this additional constraint, Figure 1.3 changes into Figure 1.5. From Figure 1.5 one can graphically derive that the optimal solution $(x_1^* = x_2^* = 4\frac{1}{2})$ is not affected by adding the new constraint (1.5).

Including constraint (1.5) in Model Dovetail, yields Model Dovetail*:

Model Dovetail*.

$$\begin{aligned}
\max \quad & 3x_1 + 2x_2 \\
\text{s.t.} \quad & x_1 + x_2 \leq 9 & (1.1) \\
& 3x_1 + x_2 \leq 18 & (1.2) \\
& x_1 \quad\;\; \leq 7 & (1.3) \\
& \quad\; x_2 \leq 6 & (1.4) \\
& x_1 + x_2 \geq 5 & (1.5) \\
& x_1, x_2 \geq 0.
\end{aligned}$$

In order to write this model in the standard form, the '$\geq$' constraint has to be transformed into a '$\leq$' constraint. This can be done by multiplying both sides of it by -1. Hence, $x_1 + x_2 \geq 5$ then becomes $-x_1 - x_2 \leq -5$. Therefore, the standard form of Model

Dovetail* is:

$$
\max\left\{ [3 \ 2]\begin{bmatrix} x_1 \\ x_2 \end{bmatrix} \;\middle|\; \begin{bmatrix} 1 & 1 \\ 3 & 1 \\ 1 & 0 \\ 0 & 1 \\ -1 & -1 \end{bmatrix}\begin{bmatrix} x_1 \\ x_2 \end{bmatrix} \leq \begin{bmatrix} 9 \\ 18 \\ 7 \\ 6 \\ -5 \end{bmatrix}, \begin{bmatrix} x_1 \\ x_2 \end{bmatrix} \geq \begin{bmatrix} 0 \\ 0 \end{bmatrix} \right\}.
$$

Similarly, a constraint with '=' can be put into standard form by replacing it by two '$\leq$' constraints. For instance, $3x_1 - 8x_2 = 11$ can be replaced by $3x_1 - 8x_2 \leq 11$ and $-3x_1 + 8x_2 \leq -11$. Also, the minimizing LO-model $\min\{\mathbf{c}^\mathsf{T}\mathbf{x} \mid \mathbf{Ax} \leq \mathbf{b}, \mathbf{x} \geq \mathbf{0}\}$ can be written in standard form, since

$$
\min\{\mathbf{c}^\mathsf{T}\mathbf{x} \mid \mathbf{Ax} \leq \mathbf{b}, \mathbf{x} \geq \mathbf{0}\} = -\max\{-\mathbf{c}^\mathsf{T}\mathbf{x} \mid \mathbf{Ax} \leq \mathbf{b}, \mathbf{x} \geq \mathbf{0}\};
$$

see also Section 1.3.

1.2.2 Slack variables and binding constraints

Model Dovetail in Section 1.1 contains the following machine capacity constraint:

$$
x_1 + x_2 \leq 9. \tag{1.1}
$$

This constraint expresses the fact that the machine can produce at most 9 ($\times 100{,}000$) boxes per year. We may wonder whether there is excess machine capacity (*overcapacity*) in the case of the optimal solution. For that purpose, we introduce an additional nonnegative variable x_3 in the following way:

$$
x_1 + x_2 + x_3 = 9.
$$

The variable x_3 is called the *slack variable* of constraint (1.1). Its optimal value, called the *slack*, measures the unused capacity of the machine. By requiring that x_3 is nonnegative, we can avoid the situation that $x_1 + x_2 > 9$, which would mean that the machine capacity is exceeded and the constraint $x_1 + x_2 \leq 9$ is violated. If, at the optimal solution, the value of x_3 is zero, then the machine capacity is completely used. In that case, the constraint is binding at the optimal solution.

Introducing slack variables for all constraints of Model Dovetail, we obtain the following model:

Model Dovetail with slack variables.

$$
\begin{aligned}
\max \quad & 3x_1 + 2x_2 \\
\text{s.t.} \quad & x_1 + x_2 + x_3 && = 9 \\
& 3x_1 + x_2 + x_4 && = 18 \\
& x_1 + x_5 && = 7 \\
& x_2 + x_6 && = 6 \\
& x_1, x_2, x_3, x_4, x_5, x_6 \geq 0.
\end{aligned}
$$

In this model, x_3, x_4, x_5, and x_6 are the nonnegative slack variables of the constraints (1.1), (1.2), (1.3), and (1.4), respectively. The number of slack variables is therefore equal to the number of inequality constraints of the model. In matrix notation the model becomes:

$$
\max\left\{ \begin{bmatrix} 3 & 2 \end{bmatrix} \begin{bmatrix} x_1 \\ x_2 \end{bmatrix} \,\middle|\, \begin{bmatrix} 1 & 1 & 1 & 0 & 0 & 0 \\ 3 & 1 & 0 & 1 & 0 & 0 \\ 1 & 0 & 0 & 0 & 1 & 0 \\ 0 & 1 & 0 & 0 & 0 & 1 \end{bmatrix} \begin{bmatrix} x_1 \\ x_2 \\ x_3 \\ x_4 \\ x_5 \\ x_6 \end{bmatrix} = \begin{bmatrix} 9 \\ 18 \\ 7 \\ 6 \end{bmatrix}, \begin{bmatrix} x_1 \\ x_2 \\ x_3 \\ x_4 \\ x_5 \\ x_6 \end{bmatrix} \geq \begin{bmatrix} 0 \\ 0 \\ 0 \\ 0 \\ 0 \\ 0 \end{bmatrix} \right\}.
$$

If $\mathbf{I}_m$ denotes the identity matrix with m rows and m columns ($m \geq 1$), then the general form of an LO-model with slack variables can be written as:

$$
\max\left\{ \mathbf{c}^\mathsf{T}\mathbf{x} \,\middle|\, \begin{bmatrix} \mathbf{A} & \mathbf{I}_m \end{bmatrix} \begin{bmatrix} \mathbf{x} \\ \mathbf{x}_s \end{bmatrix} = \mathbf{b}, \mathbf{x} \geq \mathbf{0} \right\},
$$

with $\mathbf{x} \in \mathbb{R}^n$, $\mathbf{x}_s \in \mathbb{R}^m$, $\mathbf{c} \in \mathbb{R}^n$, $\mathbf{b} \in \mathbb{R}^m$, $\mathbf{A} \in \mathbb{R}^{m \times n}$, and $\mathbf{I}_m \in \mathbb{R}^{m \times m}$. Note that the value of $\mathbf{x}_s$ (the vector of slack variables) satisfies $\mathbf{x}_s = \mathbf{b} - \mathbf{A}\mathbf{x}$ and is therefore completely determined by the value of $\mathbf{x}$. This means that if the values of the entries of the vector $\mathbf{x}$ are given, then the values of the entries of $\mathbf{x}_s$ are fixed.

1.2.3 Types of optimal solutions and feasible regions

It may happen that an LO-model has more than one optimal solution. For instance, if we replace the objective of Model Dovetail by

$$
\max x_1 + x_2,
$$

then all points on the line segment $\mathbf{v}_2\mathbf{v}_3$ (see Figure 1.2) have the same optimal objective value, namely 9, and therefore all points on the line segment $\mathbf{v}_2\mathbf{v}_3$ are optimal. In this case, we say that there are *multiple optimal solutions*; see also Section 3.7 and Section 5.6.1. The feasible region has two optimal vertices, namely $\mathbf{v}_2$ and $\mathbf{v}_3$.

Three types of feasible regions can be distinguished, namely:

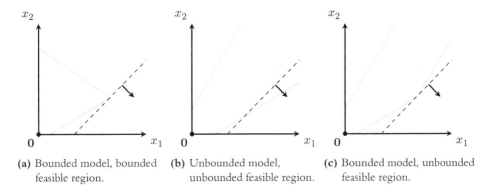

(a) Bounded model, bounded feasible region.

(b) Unbounded model, unbounded feasible region.

(c) Bounded model, unbounded feasible region.

Figure 1.6: Bounded and unbounded feasible regions.

Feasible region bounded and nonempty. A feasible region is called *bounded* if all decision variables are bounded on the feasible region (i.e., no decision variable can take on arbitrarily large values on the feasible region). An example is drawn in Figure 1.6(a). If the feasible region is bounded, then the objective values are also bounded on the feasible region and hence an optimal solution exists. Note that the feasible region of Model Dovetail is bounded; see Figure 1.2.

Feasible region unbounded. A nonempty feasible region is called *unbounded* if it is not bounded; i.e., at least one of the decision variables can take on arbitrarily large values on the feasible region. Examples of an unbounded feasible region are shown in Figure 1.6(b) and Figure 1.6(c). Whether an optimal solution exists depends on the objective function. For example, in the case of Figure 1.6(b) an optimal solution does not exist. Indeed, the objective function takes on arbitrarily large values on the feasible region. Therefore, the model has no optimal solution. An LO-model with an objective function that takes on arbitrarily large values is called *unbounded*; it is called *bounded* otherwise. On the other hand, in Figure 1.6(c), an optimal solution does exist. Hence, this is an example of an LO-model with an unbounded feasible region, but with a (unique) optimal solution.

Feasible region empty. In this case we have that $F = \emptyset$ and the LO-model is called *infeasible*. For example, if an LO-model contains the (contradictory) constraints $x_1 \geq 6$ and $x_1 \leq 3$, then its feasible region is empty. If an LO-model is infeasible, then it has no feasible points and, in particular, no optimal solution. If $F \neq \emptyset$, then the LO-model is called *feasible*.

So, an LO-model either has an optimal solution, or it is infeasible, or it is unbounded. Note that an unbounded LO-model necessarily has an unbounded feasible region, but the converse is not true. In fact, Figure 1.6(c) shows an LO-model that is bounded, although it has an unbounded feasible region.

1.3 Alternatives of the standard LO-model

In the previous sections we mainly discussed LO-models of the form

$$\max\{\mathbf{c}^\mathsf{T}\mathbf{x} \mid \mathbf{A}\mathbf{x} \leq \mathbf{b}, \mathbf{x} \geq \mathbf{0}\},$$

with $\mathbf{A} \in \mathbb{R}^{m \times n}$. We call this form the *standard form* of an LO-model. The standard form is characterized by a maximizing objective, '$\leq$' technology constraints, and nonnegativity constraints.

In general, many different forms may be encountered, for instance with both '$\geq$' and '$\leq$' technology constraints, and both nonnegativity ($x_i \geq 0$) and nonpositivity constraints ($x_i \leq 0$). All these forms can be reduced to the standard form $\max\{\mathbf{c}^\mathsf{T}\mathbf{x} \mid \mathbf{A}\mathbf{x} \leq \mathbf{b}, \mathbf{x} \geq \mathbf{0}\}$.

The following rules can be applied to transform a nonstandard LO-model into a standard model:

A minimizing model is transformed into a maximizing model by using the fact that minimizing a function is equivalent to maximizing minus that function. So, the objective of the form 'min $\mathbf{c}^\mathsf{T}\mathbf{x}$' is equivalent to the objective '$-\max(-\mathbf{c})^\mathsf{T}\mathbf{x}$'. For example, 'min $x_1 + x_2$' is equivalent to '$-\max -x_1 - x_2$'.

A '$\geq$' constraint is transformed into a '$\leq$' constraint by multiplying both sides of the inequality by -1 and reversing the inequality sign. For example, $x_1 - 3x_2 \geq 5$ is equivalent to $-x_1 + 3x_2 \leq -5$.

A '$=$' constraint of the form '$\mathbf{a}^\mathsf{T}\mathbf{x} = b$' can be written as '$\mathbf{a}^\mathsf{T}\mathbf{x} \leq b$ and $\mathbf{a}^\mathsf{T}\mathbf{x} \geq b$'. The second inequality in this expression is then transformed into a '$\leq$' constraint (see the previous item). For example, the constraint '$2x_1 + x_2 = 3$' is equivalent to '$2x_1 + x_2 \leq 3$ and $-2x_1 - x_2 \leq -3$'.

A nonpositivity constraint is transformed into a nonnegativity constraint by replacing the corresponding variable by its negative. For example, the nonpositivity constraint '$x_1 \leq 0$' is transformed into '$x_1' \geq 0$' by substituting $x_1 = -x_1'$.

A free variable is replaced by the difference of two new nonnegative variables. For example, the expression 'x_1 free' is replaced by '$x_1' \geq 0$, $x_1'' \geq 0$', and substituting $x_1 = x_1' - x_1''$.

The following two examples illustrate these rules.

Example 1.3.1. *Consider the nonstandard LO-model:*

$$
\begin{aligned}
\min \quad & -5x_1 + 3x_2 & \text{(1.7)} \\
\text{s.t.} \quad & x_1 + 2x_2 \geq 10 \\
& x_1 - x_2 \leq 6 \\
& x_1 + x_2 = 12 \\
& x_1, x_2 \geq 0.
\end{aligned}
$$

In addition to being a minimizing model, the model has a '$\geq$' constraint and a '$=$' constraint. By applying the above rules, the following equivalent standard form LO-model is found:

$$
\begin{aligned}
-\max \quad & 5x_1 - 3x_2 \\
\text{s.t.} \quad & -x_1 - 2x_2 \leq -10 \\
& x_1 - x_2 \leq 6 \\
& x_1 + x_2 \leq 12 \\
& -x_1 - x_2 \leq -12 \\
& x_1, x_2 \geq 0.
\end{aligned}
$$

The (unique) optimal solution of this model reads: $x_1^* = 9$, $x_2^* = 3$. This is also the optimal solution of the nonstandard model (1.7).

Example 1.3.2. *Consider the nonstandard LO-model:*

$$
\begin{aligned}
\max \quad & 5x_1 + x_2 \\
\text{s.t.} \quad & 4x_1 + 2x_2 \leq 8 \\
& x_1 + x_2 \leq 12 \\
& x_1 - x_2 \leq 5 \\
& x_1 \geq 0, \ x_2 \text{ free.}
\end{aligned} \tag{1.8}
$$

This model has a maximizing objective, and only '$\leq$' constraints. However, the variable x_2 is unrestricted in sign. Applying the above rules, the following equivalent standard form LO-model is found:

$$
\begin{aligned}
\max \quad & 5x_1 + x_2' - x_2'' \\
\text{s.t.} \quad & 4x_1 + 2x_2' - 2x_2'' \leq 8 \\
& x_1 + x_2' - x_2'' \leq 12 \\
& x_1 - x_2' + x_2'' \leq 5 \\
& x_1, x_2', x_2'' \geq 0.
\end{aligned} \tag{1.9}
$$

The point $\hat{\mathbf{x}} = \begin{bmatrix} x_1^* & (x_2')^* & (x_2'')^* \end{bmatrix}^\mathsf{T} = \begin{bmatrix} 3 & 0 & 2 \end{bmatrix}^\mathsf{T}$ is an optimal solution of this model. Hence, the point

$$
\mathbf{x}^* = \begin{bmatrix} x_1^* \\ x_2^* \end{bmatrix} = \begin{bmatrix} x_1^* \\ (x_2')^* - (x_2'')^* \end{bmatrix} = \begin{bmatrix} 3 \\ -2 \end{bmatrix}
$$

is an optimal solution of model (1.8). Note that $\hat{\mathbf{x}}' = \begin{bmatrix} 3 & 10 & 12 \end{bmatrix}^\mathsf{T}$ is another optimal solution of (1.9) (why?), corresponding to the same optimal solution $\mathbf{x}^*$ of model (1.8). In fact, the reader may verify that every point in the set

$$
\left\{ \begin{bmatrix} 3 \\ \alpha \\ 2 + \alpha \end{bmatrix} \in \mathbb{R}^3 \ \middle|\ \alpha \geq 0 \right\}
$$

is an optimal solution of (1.9) that corresponds to the optimal solution $\mathbf{x}^*$ of model (1.8).

We have listed six possible general nonstandard models below. Any method for solving one of the models (I)–(VI) can be used to solve the others, because they are all equivalent. The matrix $\mathbf{A}$ in (III) and (VI) is assumed to be of full row rank (i.e., $\text{rank}(\mathbf{A}) = m$; see Appendix B and Section 3.8). The alternative formulations are:

(I) $\max\{\mathbf{c}^\mathsf{T}\mathbf{x} \mid \mathbf{A}\mathbf{x} \leq \mathbf{b}, \mathbf{x} \geq \mathbf{0}\}$;   (IV) $\min\{\mathbf{c}^\mathsf{T}\mathbf{x} \mid \mathbf{A}\mathbf{x} \geq \mathbf{b}, \mathbf{x} \geq \mathbf{0}\}$;
   (standard form)   (standard dual form; see Chapter 4)

(II) $\max\{\mathbf{c}^\mathsf{T}\mathbf{x} \mid \mathbf{A}\mathbf{x} \leq \mathbf{b}\}$;   (V) $\min\{\mathbf{c}^\mathsf{T}\mathbf{x} \mid \mathbf{A}\mathbf{x} \geq \mathbf{b}\}$;

(III) $\max\{\mathbf{c}^\mathsf{T}\mathbf{x} \mid \mathbf{A}\mathbf{x} = \mathbf{b}, \mathbf{x} \geq \mathbf{0}\}$;   (VI) $\min\{\mathbf{c}^\mathsf{T}\mathbf{x} \mid \mathbf{A}\mathbf{x} = \mathbf{b}, \mathbf{x} \geq \mathbf{0}\}$.

Formulation (I) can be reduced to (II) by writing:

$$\max\{\mathbf{c}^\mathsf{T}\mathbf{x} \mid \mathbf{A}\mathbf{x} \leq \mathbf{b}, \mathbf{x} \geq \mathbf{0}\} = \max\left\{\mathbf{c}^\mathsf{T}\mathbf{x} \;\middle|\; \begin{bmatrix} \mathbf{A} \\ -\mathbf{I}_n \end{bmatrix}\mathbf{x} \leq \begin{bmatrix} \mathbf{b} \\ \mathbf{0} \end{bmatrix}\right\}.$$

The notation $\begin{bmatrix} \mathbf{A} \\ -\mathbf{I}_n \end{bmatrix}$ and $\begin{bmatrix} \mathbf{b} \\ \mathbf{0} \end{bmatrix}$ is explained in Appendix B. Formulation (II) can be reduced to (I) as follows. Each vector $\mathbf{x}$ can be written as $\mathbf{x} = \mathbf{x}' - \mathbf{x}''$ with $\mathbf{x}', \mathbf{x}'' \geq \mathbf{0}$. Hence,

$$\max\{\mathbf{c}^\mathsf{T}\mathbf{x} \mid \mathbf{A}\mathbf{x} \leq \mathbf{b}\} = \max\{\mathbf{c}^\mathsf{T}(\mathbf{x}' - \mathbf{x}'') \mid \mathbf{A}(\mathbf{x}' - \mathbf{x}'') \leq \mathbf{b}, \; \mathbf{x}', \mathbf{x}'' \geq \mathbf{0}\}$$
$$= \max\left\{\begin{bmatrix} \mathbf{c}^\mathsf{T} & -\mathbf{c}^\mathsf{T} \end{bmatrix}\begin{bmatrix} \mathbf{x}' \\ \mathbf{x}'' \end{bmatrix} \;\middle|\; \begin{bmatrix} \mathbf{A} & -\mathbf{A} \end{bmatrix}\begin{bmatrix} \mathbf{x}' \\ \mathbf{x}'' \end{bmatrix} \leq \mathbf{b}, \; \mathbf{x}' \geq \mathbf{0}, \; \mathbf{x}'' \geq \mathbf{0}\right\},$$

and this has the form (I). The reduction of (I) to (III) follows by introducing slack variables in (I). Formulation (III) can be reduced to (I) by noticing that the constraints $\mathbf{A}\mathbf{x} = \mathbf{b}$ can be written as the two constraints $\mathbf{A}\mathbf{x} \leq \mathbf{b}$ and $\mathbf{A}\mathbf{x} \geq \mathbf{b}$. Multiplying the former by -1 on both sides yields $-\mathbf{A}\mathbf{x} \leq -\mathbf{b}$. Therefore, (III) is equivalent to:

$$\max\left\{\mathbf{c}^\mathsf{T}\mathbf{x} \;\middle|\; \begin{bmatrix} \mathbf{A} \\ -\mathbf{A} \end{bmatrix}\mathbf{x} \leq \begin{bmatrix} \mathbf{b} \\ -\mathbf{b} \end{bmatrix}\right\}.$$

The disadvantage of this transformation is that the model becomes considerably larger. In Section 3.8, we will see an alternative, more economical, reduction of (III) to the standard form. Similarly, (IV), (V), and (VI) are equivalent. Finally, (III) and (VI) are equivalent because

$$\min\{\mathbf{c}^\mathsf{T}\mathbf{x} \mid \mathbf{A}\mathbf{x} = \mathbf{b}, \mathbf{x} \geq \mathbf{0}\} = -\max\{(-\mathbf{c})^\mathsf{T}\mathbf{x} \mid \mathbf{A}\mathbf{x} = \mathbf{b}, \mathbf{x} \geq \mathbf{0}\}.$$

1.4 Solving LO-models using a computer package

Throughout this book, we will illustrate the ideas in each chapter by means of computer examples. There are numerous computer packages for solving linear optimization models. Roughly two types of linear optimization packages can be distinguished:

Linear optimization solvers. A linear optimization solver takes input values for the cost vector, the technology matrix, and the right hand side vector of the LO-model that needs to be solved. The purpose of the package is to find an optimal solution to that LO-model. As described in Section 1.2.3, it is not always possible to find an optimal solution because the LO-model may be infeasible or unbounded. Another reason why a solver might fail is because there may be a limit on the amount of time that is used to find a solution, or because of round-off errors due to the fact that computer algorithms generally do calculations with limited precision using so-called *floating-point numbers* (see also the discussion in Section 3.6.1). Examples of linear optimization solvers are CPLEX and the `linprog` function in MATLAB.

Linear optimization modeling languages. Although any LO-model can be cast in a form that may serve as an input for a linear optimization solver (see Section 1.3), it is often times rather tedious to write out the full cost vector, the technology matrix, and the right hand side vector. This is, for example, the case when we have variables $x_1, \ldots, x_{100}$ and we want the objective of the LO-model to maximize $\sum_{i=1}^{100} x_i$. Instead of having to type a hundred times the entry 1 in a vector, we would prefer to just tell the computer to take the sum of these variables. For this purpose, there are a few programming languages available that allow the user to write LO-models in a more compact way. The purpose of the linear optimization programming package is to construct the cost vector, the technology matrix, and the right hand side vector from an LO-model written in that language. Examples of such linear optimization programming languages are GNU MathProg (also known as GMPL), AMPL, GAMS, and AIMMS.

We will demonstrate the usage of a linear optimization package by using the online solver provided on the website of this book. The online solver is able to solve models that are written in the GNU MathProg language. To solve Model Dovetail using the online solver, the following steps need to be taken:

Start the online solver from the book website: `http://www.lio.yoriz.co.uk/`.

In the editor, type the following code (without the line numbers).

```
1     x1 >= 0;
2     x2 >= 0;
3
4        z:      3*x1 + 2*x2;
5
6           c11:    x1 +   x2 <=  9;
7           c12: 3*x1 +   x2 <= 18;
8           c13:    x1        <=  7;
9           c14:          x2 <=  6;
10
11    ;
```

Listing 1.1: Model Dovetail as a GNU MathProg model.

This is the representation of Model Dovetail in the MathProg language. Since Model Dovetail is a relatively simple model, its representation is straightforward. For more details on how to use the MathProg language, see Appendix F.

Press the 'Solve' button to solve the model. In this step, a few things happen. The program first transforms the model from the previous step into a cost vector, technology matrix, and right hand side vector in some standard form of the LO-model (this standard form depends on the solver and need not correspond to the standard form we use in this book). Then, the program takes this standard form model and solves it. Finally, the solution of the standard-form LO-model is translated back into the language of the original model.

Press the 'Solution' button to view the solution. Among many things that might sound unfamiliar for now, the message states that the solver found an optimal solution with objective value 22.5. Also, the optimal values of x_1 and x_2 are listed. As should be expected, they coincide with the solution we found earlier in Section 1.1.2 by applying the graphical solution method.

1.5   Linearizing nonlinear functions

The definition of an LO-model states that the objective function and the (left hand sides of the) constraints have to be linear functions of the decision variables. In some cases, however, it is possible to rewrite a model with nonlinearities so that the result is an LO-model. We describe a few of them in this section.

1.5.1   Ratio constraints

Suppose that the company Dovetail wants to make at least 75% of its profit from long matches. At the optimal solution in the current formulation (Model Dovetail), the profit from long matches is $(4\frac{1}{2} \times 3 \ =) \ 13\frac{1}{2}$ ($\times$\$1,000) and the profit from short matches is $(4\frac{1}{2} \times 2 \ =) \ 9$ ($\times$\$1,000). Thus, currently, the company gets $(13\frac{1}{2}/22\frac{1}{2} \ =) \ 60\%$ of its profit from long matches. In order to ensure that at least 75% of the profit comes from the production of long matches, we need to add the following constraint:

$$\frac{3x_1}{3x_1 + 2x_2} \geq \tfrac{3}{4}. \tag{1.10}$$

The left hand side of this constraint is clearly nonlinear, and so the constraint – in its current form – cannot be used in an LO-model. However, multiplying both sides by $3x_1 + 2x_2$ (and using the fact that $x_1 \geq 0$ and $x_2 \geq 0$), we obtain the formulation:

$$3x_1 \geq \tfrac{3}{4}(3x_1 + 2x_2), \quad \text{or, equivalently,} \quad \tfrac{3}{4}x_1 - \tfrac{3}{2}x_2 \geq 0,$$

which is linear. Note that the left hand side of (1.10) is not defined if x_1 and x_2 both have value zero; the expression $\tfrac{3}{4}x_1 - \tfrac{3}{2}x_2 \geq 0$ does not have this problem. Adding the constraint to Model Dovetail yields the optimal solution $x_1^* = 5\frac{1}{7}$, $x_2^* = 2\frac{4}{7}$ with corresponding objective value $20\frac{4}{7}$. The profit from long matches is $15\frac{3}{7}$ ($\times$\$1,000), which is exactly 75% of the total profit, as required.

1.5.2 Objective functions with absolute value terms

Consider the following minimization model in which we have an absolute value in the objective function:

$$
\begin{aligned}
\min \ & |x_1| + x_2 \\
\text{s.t.} \ \ & 3x_1 + 2x_2 \geq 1 \\
& x_1 \text{ free}, x_2 \geq 0.
\end{aligned}
\tag{1.11}
$$

The model is not an LO-model, because of the absolute value in the objective function, which makes the objective function nonlinear. Model (1.11) can be written as an LO-model by introducing two new nonnegative variables u_1 and u_2 that will have the following relationship with x_1:

$$
u_1 = \begin{cases} x_1 & \text{if } x_1 \geq 0 \\ 0 & \text{otherwise}, \end{cases} \quad \text{and} \quad u_2 = \begin{cases} -x_1 & \text{if } x_1 \leq 0 \\ 0 & \text{otherwise}. \end{cases}
\tag{1.12}
$$

A more compact way of writing (1.12) is:

$$
u_1 = \max\{0, x_1\} \text{ and } u_2 = \max\{0, -x_1\}.
$$

Note that we have that $x_1 = u_1 - u_2$ and $|x_1| = u_1 + u_2$. Because u_1 and u_2 should never be simultaneously nonzero, they need to satisfy the relationship $u_1 u_2 = 0$. By adding the constraint $u_1 u_2 = 0$ and substituting $u_1 - u_2$ for x_1 and $u_1 + u_2$ for $|x_1|$ in (1.11), we obtain the following optimization model:

$$
\begin{aligned}
\min \ & u_1 + u_2 + x_2 \\
\text{s.t.} \ \ & 3u_1 - 3u_2 + 2x_2 \geq 1 \\
& u_1 u_2 \qquad\qquad = 0 \\
& x_2, x_3, u_2 \geq 0.
\end{aligned}
\tag{1.13}
$$

This model is still nonlinear. However, the constraint $u_1 u_2 = 0$ can be left out. That is, we claim that it suffices to solve the following optimization model (which is an LO-model):

$$
\begin{aligned}
\min \ & u_1 + u_2 + x_2 \\
\text{s.t.} \ \ & 3u_1 - 3u_2 + 2x_2 \geq 1 \\
& u_1, u_2, x_2 \geq 0.
\end{aligned}
\tag{1.14}
$$

To see that the constraint $u_1 u_2 = 0$ may be left out, we will show that it is automatically satisfied at any optimal solution of (1.14). Let $\mathbf{x}^* = \begin{bmatrix} u_1^* \ u_2^* \ x_2^* \end{bmatrix}^\mathsf{T}$ be an optimal solution of (1.14). Suppose for a contradiction that $u_1^* u_2^* \neq 0$. This implies that both $u_1^* > 0$ and $u_2^* > 0$. Let $\varepsilon = \min\{u_1^*, u_2^*\} > 0$, and consider $\hat{\mathbf{x}} = \begin{bmatrix} \hat{u}_1 \ \hat{u}_2 \ \hat{x}_2 \end{bmatrix}^\mathsf{T} = \begin{bmatrix} u_1^* - \varepsilon \ u_2^* - \varepsilon \ x_2^* \end{bmatrix}^\mathsf{T}$. It is easy to verify that $\hat{\mathbf{x}}$ a feasible solution of (1.14), and that the corresponding objective value $\hat{z}$ satisfies $\hat{z} = u_1^* - \varepsilon + u_2^* - \varepsilon + x_2^* < u_1^* + u_2^* + x_2^*$. Thus, we have constructed a feasible solution $\hat{\mathbf{x}}$ of (1.14), the objective value of which is smaller than the objective value of $\mathbf{x}^*$, contradicting the fact that $\mathbf{x}^*$ is an optimal solution of (1.14). Hence, the constraint $u_1 u_2 = 0$ is automatically satisfied by any optimal solution of (1.14) and, hence, any optimal solution of (1.14) is also an optimal solution of (1.13). We leave it to the reader to show that

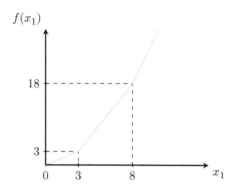

Figure 1.7: Piecewise linear function $f(x_1)$.

if $\mathbf{x}^* = \begin{bmatrix} u_1^* & u_2^* & x_2^* \end{bmatrix}^{\mathsf{T}}$ is an optimal solution of (1.14), then $\begin{bmatrix} x_1^* & x_2^* \end{bmatrix}^{\mathsf{T}}$ with $x_1^* = u_1^* - u_2^*$ is an optimal solution of (1.11); see Exercise 1.8.9.

Model (1.14) can be solved using the graphical method as follows. Recall that at least one of u_1 and u_2 has to have value zero in the optimal solution. So we can distinguish two cases: either $u_1 = 0$, or $u_2 = 0$. These two cases give rise to two different LO-models:

$$
\begin{array}{lll}
\min & u_2 + x_2 & \\
\text{s.t.} & -3u_2 + 2x_2 \geq 1 & \text{and} \\
& u_2, x_2 \geq 0,
\end{array}
\qquad
\begin{array}{ll}
\min & u_1 + x_2 \\
\text{s.t.} & 3u_1 + 2x_2 \geq 1 \qquad (1.15) \\
& u_1, x_2 \geq 0.
\end{array}
$$

Since both LO-models have two decision variables, they can be solved using the graphical method. The optimal solutions are $\begin{bmatrix} 0 & \frac{1}{2} \end{bmatrix}^{\mathsf{T}}$ (with optimal objective value $\frac{1}{2}$) and $\begin{bmatrix} \frac{1}{3} & 0 \end{bmatrix}^{\mathsf{T}}$ (with optimal objective value $\frac{1}{3}$), respectively. The optimal solution of (1.14) is found by choosing the solution among these two that has the smallest objective value. This gives $u_1^* = \frac{1}{3}$, $u_2^* = 0$, and $x_2^* = 0$, with optimal objective value $\frac{1}{3}$. The corresponding optimal solution of (1.11) satisfies $x_1^* = u_1^* - u_2^* = \frac{1}{3}$ and $x_2^* = 0$.

It is important to realize that it is not true that every <u>feasible</u> solution of (1.13) corresponds to a feasible solution of (1.11). For example, the vector $\begin{bmatrix} u_1 & u_2 & x_2 \end{bmatrix}^{\mathsf{T}} = \begin{bmatrix} 2 & 1 & 0 \end{bmatrix}^{\mathsf{T}}$ is a feasible solution of (1.13) with objective value 3. However, the corresponding vector $\begin{bmatrix} x_1 & x_2 \end{bmatrix}^{\mathsf{T}} = \begin{bmatrix} 1 & 0 \end{bmatrix}^{\mathsf{T}}$ in (1.11) has objective value 1. The reason for this mismatch is the fact that u_1 and u_2 are simultaneously nonzero. Recall that this never happens at an optimal feasible solution of (1.13).

This method also works for a maximizing objective in which the absolute value appears with a negative coefficient. However, it does not work for a maximizing objective in which the absolute value appears in the objective function with a positive coefficient. The reader is asked to verify this in Exercise 1.8.10.

1.5.3 Convex piecewise linear functions

The example from the previous section can be generalized to so-called *convex piecewise linear functions*. For instance, consider the following optimization model:

$$\begin{aligned} \min \quad & f(x_1) + 4x_2 \\ \text{s.t.} \quad & x_1 + x_2 \geq 10 \\ & x_1, x_2 \geq 0, \end{aligned} \tag{1.16}$$

where

$$f(x_1) = \begin{cases} x_1 & \text{if } 0 \leq x_1 \leq 3 \\ 3 + 3(x_1 - 3) & \text{if } 3 < x_1 \leq 8 \\ 18 + 5(x_1 - 8) & \text{if } x_1 > 18. \end{cases}$$

Figure 1.7 shows the graph of $f(x_1)$. The function $f(x_1)$ is called *piecewise linear*, because it is linear on each of the intervals $[0, 3]$, $(3, 8]$, and $(8, \infty)$ separately. Like in the previous subsection, model (1.16) can be solved by an alternative formulation. We start by introducing three new nonnegative decision variables u_1, u_2, and u_3. They will have the following relationship with x_1:

$$u_1 = \begin{cases} x_1 & \text{if } 0 \leq x_1 \leq 3 \\ 3 & \text{if } x_1 > 3, \end{cases} \quad u_2 = \begin{cases} 0 & \text{if } 0 \leq x_1 \leq 3 \\ x_1 - 3 & \text{if } 3 < x_1 \leq 8 \\ 5 & \text{if } x_1 > 8, \end{cases} \quad u_3 = \begin{cases} 0 & \text{if } x_1 \leq 8 \\ x_1 - 8 & \text{if } x_1 > 8. \end{cases}$$

Crucially, we have that:

$$u_1 + u_2 + u_3 = x_1 \quad \text{and} \quad u_1 + 3u_2 + 5u_3 = f(x_1). \tag{1.17}$$

To see this, we need to consider three different cases.

$0 \leq x_1 \leq 3$. Then $u_1 = x_1$, $u_2 = 0$, and $u_3 = 0$. Hence, $u_1 + u_2 + u_3 = x_1$ and $u_1 + 3u_2 + 5u_3 = x_1 = f(x_1)$.

$3 < x_1 \leq 8$. Then $u_1 = 3$, $u_2 = x_1 - 3$, and $u_3 = 0$. Hence, $u_1 + u_2 + u_3 = 3 + x_1 - 3 + 0 = x_1$ and $u_1 + 3u_2 + 5u_3 = 3 + 3(x_1 - 3) = f(x_1)$.

$x_1 > 8$. Then $u_1 = 3$, $u_2 = 5$, $u_3 = x_1 - 8$, and $f(x_1) = 18 + 5(x_1 - 8)$. Hence, $u_1 + u_2 + u_3 = 3 + 5 + x_1 - 8 = x_1$ and $u_1 + 3u_2 + 5u_3 = 3 + 15 + 5(x_1 - 8) = 18 + 5(x_1 - 8) = f(x_1)$.

In each of the cases, it is clear that (1.17) holds. Notice that u_1, u_2, u_3 satisfy the following inequalities:

$$0 \leq u_1 \leq 3, \quad 0 \leq u_2 \leq 5, \quad u_3 \geq 0.$$

Moreover, we have the following equations (this follows from the definition of u_1, u_2, u_3):

$$u_2(3 - u_1) = 0 \quad \text{and} \quad u_3(5 - u_2) = 0.$$

The first equation states that either $u_2 = 0$, or $u_1 = 3$, or both. Informally, this says that u_2 has a positive value only if u_1 is at its highest possible value 3. The second equation states that either $u_3 = 0$, or $u_2 = 5$, or both. This says that u_3 has a positive value only if u_3 is at its highest possible value 5. The two equations together imply that u_3 has a positive value only if both u_1 and u_2 are at their respective highest possible values (namely, $u_1 = 3$ and $u_2 = 5$).

Adding these equations to (1.16) and substituting the expressions of (1.17), we obtain the following (nonlinear) optimization model:

$$
\begin{aligned}
\min\ & u_1 + 3u_2 + 5u_3 + 4x_2 \\
\text{s.t.}\ & u_1 + u_2 + u_3 + x_2 > 10 \\
& u_1 \leq 3 \\
& u_2 \leq 5 \\
& (3 - u_1)u_2 = 0 \\
& (5 - u_2)u_3 = 0 \\
& u_1, u_2, u_3, x_2 \geq 0.
\end{aligned}
\tag{1.18}
$$

As in Section 1.5.2, it turns out that the nonlinear constraints $(3 - u_1)u_2 = 0$ and $(5 - u_2)u_3 = 0$ may be omitted. To see this, let $\mathbf{x}^* = \begin{bmatrix} u_1^* & u_2^* & u_3^* & x_2^* \end{bmatrix}^\mathsf{T}$ be an optimal solution of (1.18), and let z^* be the corresponding optimal objective value. Suppose for a contradiction that $u_2^*(3 - u_1^*) \neq 0$. Because $u_2^* \geq 0$ and $u_1^* \leq 3$, this implies that $u_1^* < 3$ and $u_2^* > 0$. Let $\varepsilon = \min\{3 - u_1^*, u_2^*\} > 0$, and define $\hat{u}_1 = u_1^* + \varepsilon$, $\hat{u}_2 = u_2^* - \varepsilon$, $\hat{u}_3 = u_3^*$, and $\hat{x}_2 = x_2^*$. It is straightforward to check that the vector $\hat{\mathbf{x}} = \begin{bmatrix} \hat{u}_1 & \hat{u}_2 & \hat{u}_3 & \hat{x}_2 \end{bmatrix}^\mathsf{T}$ is a feasible solution of (1.18). The objective value corresponding to $\hat{\mathbf{x}}$ satisfies:

$$
\begin{aligned}
\hat{u}_1 + 3\hat{u}_2 + 5\hat{u}_3 + 4\hat{x}_2 &= (u_1^* + \varepsilon) + 3(u_2^* - \varepsilon) + 5u_3^* + 4x_2^* \\
&= u_1^* + 3u_2^* + 5u_3^* + 4x_2^* - 2\varepsilon = z^* - 2\varepsilon < z^*,
\end{aligned}
$$

contrary to the fact that $\mathbf{x}^*$ is an optimal solution of (1.18). Therefore, $(3 - u_1^*)u_2^* = 0$ is satisfied for any optimal solution $\begin{bmatrix} u_1^* & u_2^* & u_3^* & x_2^* \end{bmatrix}^\mathsf{T}$ of (1.18), and hence the constraint $(3 - u_1)u_2 = 0$ can be omitted. Similarly, the constraint $(5 - u_2)u_3 = 0$ can be omitted. This means that model (1.16) may be solved by solving the following LO-model:

$$
\begin{aligned}
\min\ & u_1 + 3u_2 + 5u_3 + 4x_2 \\
\text{s.t.}\ & u_1 + u_2 + u_3 + x_2 \geq 10 \\
& u_1 \leq 3 \\
& u_2 \leq 5 \\
& u_1, u_2, u_3, x_2 \geq 0.
\end{aligned}
\tag{1.19}
$$

An optimal solution of (1.19) can be found using the graphical method, similarly to the discussion in Section 1.5.2. The cases to consider are: (1) $u_2 = u_3 = 0$, (2) $u_1 = 3$ and $u_3 = 0$, and (3) $u_1 = 3$ and $u_2 = 5$. An optimal solution turns out to be $u_1^* = 3$, $u_2^* = 5$, $u_3^* = 0$, and $x_2^* = 2$. This means that an optimal solution of (1.16) is $x_1^* = 8$ and $x_2^* = 2$.

The piecewise linear function $f(x_1)$ has a special property: it is a *convex function* (see Appendix D). This is a necessary condition for the method described in this section to work.

In Exercise 1.8.12, the reader is given a model with a nonconvex piecewise linear objective function and is asked to show that the technique does not work in that case.

1.6 Examples of linear optimization models

Linear and integer linear optimization are used in a wide range of subjects. ROBERT E. BIXBY (born 1945) has collected the following impressive list of applications.

Transportation – airlines: fleet assignment; crew scheduling; personnel scheduling; yield management; fuel allocation; passenger mix; booking control; maintenance scheduling; load balancing, freight packing; airport traffic planning; gate scheduling; upset recovery and management.

Transportation – others: vehicle routing; freight vehicle scheduling and assignment; depot and warehouse location; freight vehicle packing; public transportation system operation; rental car fleet management.

Financial: portfolio selection and optimization; cash management; synthetic option development; lease analysis; capital budgeting and rationing; bank financial planning; accounting allocations; securities industry surveillance; audit staff planning; assets liabilities management; unit costing; financial valuation; bank shift scheduling; consumer credit delinquency management; check clearing systems; municipal bond bidding; stock exchange operations; debt financing optimization.

Process industries (chemical manufacturing, refining): plant scheduling and logistics; capacity expansion planning; pipeline transportation planning; gasoline and chemical blending.

Manufacturing: product mix planning; blending; manufacturing scheduling; inventory management; job scheduling; personnel scheduling; maintenance scheduling and planning; steel production scheduling; blast furnace burdening in the steel industry.

Coal industry: coal sourcing and transportation logistics; coal blending; mining operation management.

Forestry: Forest land management; forest valuation models; planting and harvesting models.

Agriculture: production planning; farm land management; agriculture pricing models; crop and product mix decision models; product distribution.

Oil and gas exploration and production: oil and gas production scheduling; natural gas transportation planning.

Public utilities and natural resources: electric power distribution; power generator scheduling; power tariff rate determination; natural gas distribution planning; natural gas pipeline transportation; water resource management; alternative water supply evaluation; water reservoir management; public water transportation models; mining excavation models.

Food processing: food blending; recipe optimization; food transportation logistics; food manufacturing logistics and scheduling.

Communications and computing: circuit board (VLSI) layout; logical circuit design; magnetic field design; complex computer graphics; curve fitting; virtual reality systems; computer system capacity planning; office automation; multiprocessor scheduling; telecommunications scheduling; telephone operator scheduling; telemarketing site selection.

Health care: hospital staff scheduling; hospital layout; health cost reimbursement; ambulance scheduling; radiation exposure models.

Pulp and paper industry: inventory planning; trimloss minimization; waste water recycling; transportation planning.

Textile industry: pattern layout and cutting optimization; production scheduling.

Government and military: post office scheduling and planning; military logistics; target assignment; missile detection; manpower deployment.

Miscellaneous applications: advertising mix/media scheduling; sales region definition; pollution control models; sales force deployment.

The current section contains a number of linear optimization models that illustrate the wide range of applications from real world problems. They also illustrate the variety and the complexity of the modeling process. See also Chapters 10–11 for more real world applications.

1.6.1 The diet problem

A doctor prescribes to a patient exact amounts of daily vitamin A and vitamin C intake. Specifically, the patient should choose her diet so as to consume exactly 3 milligrams of vitamin A and exactly 75 milligrams of vitamin C. The patient considers eating three kinds of food, which contain different amounts of vitamins and have different prices. She wants to determine how much of each food she should buy in order to minimize her total expenses, while making sure to ingest the prescribed amounts of vitamins. Let x_i be the amount of food i that she should buy ($i = 1, 2, 3$). Each unit of food 1 contains 1 milligram of vitamin A and 30 milligrams of vitamin C, each unit of food 2 contains 2 milligrams of vitamin A and 10 milligrams of vitamin C, and each unit of food 3 contains 2 milligrams of vitamin A and 20 milligrams of vitamin C. The unit cost of food 1, 2, and 3 is \$40, \$100, and \$150 per week, respectively. This problem can be formulated as follows:

$$
\begin{aligned}
\min \quad & 40x_1 + 100x_2 + 150x_3 \\
\text{s.t.} \quad & x_1 + 2x_2 + 2x_3 = 3 \\
& 30x_1 + 10x_2 + 20x_3 = 75 \\
& x_1, x_2, x_3 \geq 0.
\end{aligned}
$$

Finding an optimal solution of this LO-model can be done by elimination as follows. First, subtracting ten times the first constraint from the second one yields:

$$20x_1 - 10x_2 = 45, \quad \text{or, equivalently,} \quad x_2 = 2x_1 - 4.5.$$

Similarly, subtracting five times the first constraint from the second one yields:

$$25x_1 + 10x_3 = 60, \quad \text{or, equivalently,} \quad x_3 = 6 - 2.5x_1.$$

Substituting these expressions for x_2 and x_3 into the original model, we obtain:

$$\begin{array}{ll} \min & 40x_1 + 100(2x_1 - 4.5) + 150(6 - 2.5x_1) \\ \text{s.t.} & 2x_1 - 4.5 \geq 0 \\ & 6 - 2.5x_1 \geq 0 \\ & x_1 \geq 0. \end{array}$$

Hence, the model becomes:

$$\begin{array}{ll} \min & -135x_1 + 450 \\ \text{s.t.} & x_1 \geq 2.25 \\ & x_1 \leq 2.4 \\ & x_1 \geq 0. \end{array}$$

Since the objective coefficient of x_1 is negative, the optimal solution is found by choosing x_1 as large as possible, i.e., $x_1^* = 2.4$. Thus, the optimal solution is $x_1^* = 2.4$, $x_2^* = 2 \times 2.4 - 4.5 = 0.3$, $x_3^* = 6 - 2.5 \times 2.4 = 0$, and $z^* = -135 \times 2.4 + 450 = 126$, which means that she should buy 2.4 units of food 1, 0.3 units of food 2, and none of food 3. The total cost of this diet is \$126 per week.

An interesting phenomenon appears when one of the right hand side values is changed. Suppose that the doctor's prescription was to take 85 milligrams of vitamin C instead of 75 milligrams. So, the corresponding LO-model is:

$$\begin{array}{ll} \min & 40x_1 + 100x_2 + 150x_3 \\ \text{s.t.} & x_1 + 2x_2 + 2x_3 = 3 \\ & 30x_1 + 10x_2 + 20x_3 = 85 \\ & x_1, x_2, x_3 \geq 0. \end{array}$$

The optimal solution of this new model is $x_1^* = 2.8$, $x_2^* = 0.1$, $x_3^* = 0$, and $z^* = 122$. Observe that the corresponding diet is \$4 cheaper than the original diet. Hence the patient gets more vitamins for less money. This is the so-called *more-for-less paradox*. It is of course only a seeming paradox, because the fact that 'an exact amount' of the vitamins is prescribed can in practice be relaxed to 'at least the amount'. Replacing the equality signs in the original LO-model by '$\geq$' signs gives the optimal solution $x_1^* = 3$, $x_2^* = 0$, $x_3^* = 0$, and $z^* = 120$. The corresponding vitamin A and C intake is 3 and 90 milligrams, respectively. This 'paradox' therefore only tells us that we have to be careful when formulating LO-models, and with the interpretation of the solution.

We conclude with the model code for the diet problem:

```
1    x1 >= 0;
2    x2 >= 0;
3    x3 >= 0;
4
5        z:
6    40 * x1 + 100 * x2 + 150 * x3;
7
8            vitaminA:
9    x1 + 2 * x2 + 2 * x3 = 3;
10
11           vitaminC:
12   30 * x1 + 10 * x2 + 20 * x3 = 75;
13
14   ;
```

Listing 1.2: The diet problem.

1.6.2 Estimation by regression

Suppose that we want to estimate a person's federal income tax based on a certain personal profile. Suppose that this profile consists of values for a number of well-defined attributes. For example, we could use the number of semesters spent in college, the income, the age, or the total value of the real estate owned by the person. In order to carry out the estimation, we collect data of say $(n =)$ 11 persons. Let m be the number of attributes, labeled $1, \ldots, m$. We label the persons in the sample as $i = 1, \ldots, n$. For $i = 1, \ldots, n$, define $\mathbf{a}_i = \begin{bmatrix} a_{i1} & \ldots & a_{im} \end{bmatrix}^\mathsf{T}$ with a_{ij} the value of attribute j of person i. Moreover, let b_i be the amount of the federal income tax to be paid by person i. It is assumed that the values of the eleven vectors $\mathbf{a}_1 = \begin{bmatrix} a_{11} & a_{12} \end{bmatrix}^\mathsf{T}, \ldots, \mathbf{a}_{11} = \begin{bmatrix} a_{11,1} & a_{11,2} \end{bmatrix}^\mathsf{T}$, together with the values of $b_1, \ldots, b_{11}$ are known. Table 1.1 lists an example data set for eleven persons of whom we collected the values of two attributes, a_1 and a_2. In Figure 1.8, we have plotted these data points.

The question now is: how can we use this data set in order to estimate the federal income tax b of any given person (who is not in the original data set) based on the values of a given profile vector $\mathbf{a}$? To that end, we construct a graph 'through' the data points $\begin{bmatrix} a_{11} & \ldots & a_{1m} & b \end{bmatrix}^\mathsf{T}, \ldots,$ $\begin{bmatrix} a_{n1} & \ldots & a_{nm} & b \end{bmatrix}^\mathsf{T}$ in such a way that the total distance between these points and this graph is as small as possible. Obviously, how small or large this total distance is depends on the shape of the graph. In Figure 1.8, the shape of the graph is a plane in three-dimensional space. In practice, we may take either a convex or a concave graph (see Appendix D). However, when the data points do not form an apparent 'shape', we may choose a hyperplane. This hyperplane is constructed in such a way that the sum of the deviations of the n data points $\begin{bmatrix} a_{i1} & \ldots & a_{im} & b_i \end{bmatrix}^\mathsf{T}$ $(i = 1, \ldots, n)$ from this hyperplane is as small as possible. The general form of such a hyperplane $H(\mathbf{u}, v)$ is:

$$H(\mathbf{u}, v) = \left\{ \begin{bmatrix} a_1 & \ldots & a_m & b \end{bmatrix}^\mathsf{T} \in \mathbb{R}^{m+1} \;\middle|\; b = \mathbf{u}^\mathsf{T}\mathbf{a} + v \right\},$$

(see also Section 2.1.1) with variables $\mathbf{a}$ ($\in \mathbb{R}^m$) and b ($\in \mathbb{R}$). The values of the parameters $\mathbf{u}$ ($\in \mathbb{R}^m$) and v ($\in \mathbb{R}$) need to be determined such that the total deviation between the n points $\begin{bmatrix} a_{i1} & \ldots & a_{im} & b_i \end{bmatrix}^\mathsf{T}$ (for $i = 1, \ldots, n$) and the hyperplane is as small as possible. As the deviation of the data points from the hyperplane, we use the 'vertical' distance. That is, for each $i = 1, \ldots, n$, we take as the distance between the hyperplane H and the point $\begin{bmatrix} a_{i1} & \ldots & a_{im} & b_i \end{bmatrix}^\mathsf{T}$:

$$|\mathbf{u}^\mathsf{T}\mathbf{a}_i + v - b_i|.$$

In order to minimize the total deviation, we may solve the following LO-model:

$$\begin{aligned} \min \; & y_1 + \ldots + y_n \\ \text{s.t.} \; & -y_i \le \mathbf{a}_i^\mathsf{T}\mathbf{u} + v - b_i \le y_i && \text{for } i = 1, \ldots, n \\ & y_i \ge 0 && \text{for } i = 1, \ldots, n. \end{aligned}$$

In this LO-model, the variables are the entries of $\mathbf{y}$, the entries of $\mathbf{u}$, and v. The values of $\mathbf{a}_i$ and b_i ($i = 1, \ldots, n$) are given. The 'average' hyperplane 'through' the data set reads:

$$H(\mathbf{u}^*, v^*) = \left\{ \begin{bmatrix} a_1 & \ldots & a_m & b \end{bmatrix}^\mathsf{T} \in \mathbb{R}^{m+1} \;\middle|\; b = (\mathbf{u}^*)^\mathsf{T}\mathbf{x} + v^* \right\},$$

where $\mathbf{u}^*$ and v^* are optimal values for $\mathbf{u}$ and v, respectively. Given this hyperplane, we may now estimate the income tax to be paid by a person that is not in our data set, based on the person's profile. In particular, for a person with given profile $\hat{\mathbf{a}}$, the estimated income tax to be paid is $\hat{b} = (\mathbf{u}^*)^\mathsf{T}\hat{\mathbf{a}} + v^*$. The optimal solution obviously satisfies $y_i^* = |\mathbf{a}_i^\mathsf{T}\mathbf{u}^* + v^* - b_i|$, with $\mathbf{u}^*$ and v^* the optimal values, so that the optimal value of y_i^* measures the deviation of data point $\begin{bmatrix} a_{i1} & \ldots & a_{im} & b \end{bmatrix}$ from the hyperplane.

Person i	b_i	a_{i1}	a_{i2}
1	4,585	2	10
2	7,865	9	10
3	3,379	7	2
4	6,203	3	6
5	2,466	1	9
6	3,248	7	5
7	4,972	6	7
8	3,437	9	4
9	3,845	1	4
10	3,878	9	2
11	5,674	5	9

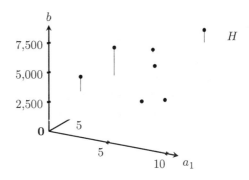

Table 1.1: Federal income tax data for eleven persons. The second column contains the amount of income tax paid, and the third and fourth columns contain the profile of each person.

Figure 1.8: Plot of eleven persons' federal income tax b as a function of two attributes a_1 and a_2. The dots are the data points. The lines show the distance from each data point to the hyperplane H.

For the example with the data given in Table 1.1, the optimal solution turns out to be:

$$u_1^* = 249.5, \ u_2^* = 368.5, \ v^* = 895.5, \ y_1^* = 494.5, \ y_2^* = 1039, \ y_3^* = 0, \ y_4^* = 2348,$$
$$y_5^* = 1995.5, \ y_6^* = 1236.5, \ y_7^* = 0, \ y_8^* = 1178, \ y_9^* = 1226, \ y_{10}^* = 0, \ y_{11}^* = 214.5.$$

Thus, the best hyperplane H through the points is given by:

$$H = \left\{ \begin{bmatrix} a_1 & a_2 & b \end{bmatrix}^{\mathsf{T}} \in \mathbb{R}^3 \ \middle| \ b = 249.5a_1 + 368.5a_2 + 895.5 \right\}.$$

Moreover, for a person with profile $\hat{\mathbf{a}} = \begin{bmatrix} \hat{a}_1 & \hat{a}_2 \end{bmatrix}$, the estimated income tax to be paid is $\hat{b} = 249.5\hat{a}_1 + 368.5\hat{a}_2 + 895.5$. This means that the second attribute has more influence on the estimate than the first attribute. Also, from the values of $y_1^*, \ldots, y_{11}^*$, we see that the distances from the data points corresponding to persons 5 and 6 are largest, meaning that the amount of income tax that they had to pay deviates quite a bit from the amount predicted by our model. For example, for person 5, the estimated income tax is $(249.5 \times 1 + 368.5 \times 9 + 895.5 =)$ \$4,461.50, whereas this person's actual income tax was \$2,466.00.

The model described above is a special kind of (linear) *regression model*. Regression models are used widely in statistics. One of the most common such models is the so-called *least squares* regression model, which differs from the model above in that the distance between the data points and the hyperplane is not measured by the absolute value of the deviation, but by the square of the deviation. Thus, in a least squares regression model, the objective is to minimize the sum of the squared deviations. In addition to different choices of the distance function, it is also possible to apply 'better' graphs, such as convex or concave functions. Such models are usually nonlinear optimization models and hence these topics lie outside of the scope of this book.

1.6.3 Team formation in sports

Another application of linear optimization is the formation of sports teams. In this section we describe a simple LO-model for choosing a soccer line-up from a given set of players. In soccer, a line-up consists of a choice of eleven players (from a potentially larger set) that are assigned to eleven positions. The positions depend on the system of play. For example, in a so-called 1-4-3-3 system, the team consists of a goal keeper, four defenders, three midfielders, and three forward players.

Let N denote the number of players that we can choose from, and let M denote the number of positions in the field (i.e., $M = 11$ in the case of soccer). We assume that $N \geq M$, i.e., there are enough players available to play a game. The players that are not lined up will not play. For each $i = 1, \ldots, N$ and $j = 1, \ldots, M$, define the decision variable x_{ij}, with the following meaning:

$$x_{ij} = \begin{cases} 1 & \text{if player } i \text{ is assigned to position } j \\ 0 & \text{otherwise.} \end{cases}$$

Since not every assignment of 0's and 1's to the x_{ij}'s represents a legal line-up, we need to impose some restrictions on the values of the x_{ij}'s. For one, we need to make sure that exactly one player is assigned to each position. This is captured by the following set of M constraints:

$$\sum_{i=1}^{N} x_{ij} = 1 \qquad \text{for } j = 1, \ldots, M.$$

These constraints state that, for each position j, the sum of $x_{1j}, \ldots, x_{Nj}$ equals 1. Since each of the x_{ij}'s should equal either 0 or 1, this implies that exactly one of $x_{1j}, \ldots, x_{Nj}$ will have value 1. Moreover, although not every player has to be lined up, we require that no player is lined up on two different positions. This is achieved by the following set of N constraints:

$$\sum_{j=1}^{M} x_{ij} \leq 1 \qquad \text{for } i = 1, \ldots, N.$$

These constraints state that, for each player i, at most one of the $x_{i1}, \ldots, x_{iM}$ has value 1.

In order to formulate an optimization model, we also need to add an objective function which measures how good a given line-up is. To do so, we introduce the parameter c_{ij} ($i = 1, \ldots, N$ and $j = 1, \ldots, M$), which measures how well player i fits on position j. A question that arises is of course: how to determine the values of the c_{ij}'s? We will come back to this later. With the values of the c_{ij}'s at hand, we can now write an objective. Let us say that if player i is assigned to position j, this player contributes c_{ij} to the objective function. Thus, the objective is to maximize the sum of the c_{ij}'s, over all pairs i, j, such that player i is assigned to position j. The corresponding objective function can be written as a linear function of the x_{ij}'s. The objective then reads:

$$\max \sum_{i=1}^{N} \sum_{j=1}^{M} c_{ij} x_{ij}.$$

Since $x_{ij} = 1$ if and only if player i is assigned to position j, the term $c_{ij} x_{ij}$ equals c_{ij} if player i is assigned to position j, and 0 otherwise, as required.

Combining the constraints and the objective function, we have the following optimization model:

$$\max \sum_{i=1}^{N} \sum_{j=1}^{M} c_{ij} x_{ij}$$

$$\text{s.t.} \quad \sum_{i=1}^{N} x_{ij} = 1 \quad \text{for } j = 1, \ldots, M$$

$$\sum_{j=1}^{M} x_{ij} \leq 1 \quad \text{for } i = 1, \ldots, N$$

$$x_{ij} \in \{0, 1\} \quad \text{for } i = 1, \ldots, N \text{ and } j = 1, \ldots, M.$$

This optimization model, in its current form, is not an LO-model, because its variables are restricted to have integer values and this type of constraint is not allowed in an LO-model.

Actually, the model is a so-called *integer linear optimization model* (abbreviated as *ILO-model*). We will see in Chapter 7 that, in general, integer linear optimization models are hard to solve. If we want to write the model as an LO-model, we will have to drop the constraint that the decision variables be integer-valued. So let us consider the following optimization model instead:

$$
\begin{aligned}
\max \quad & \sum_{i=1}^{N}\sum_{j=1}^{M} c_{ij}x_{ij} \\
\text{s.t.} \quad & \sum_{i=1}^{N} x_{ij} = 1 \quad \text{for all } j = 1, \ldots, M \\
& \sum_{j=1}^{M} x_{ij} \leq 1 \quad \text{for all } i = 1, \ldots, N \\
& 0 \leq x_{ij} \leq 1 \quad \text{for all } i = 1, \ldots, N \text{ and } j = 1, \ldots, M.
\end{aligned}
\tag{1.20}
$$

This model is an LO-model. But recall from Model Dovetail that, in general, an LO-model may have a solution whose coordinate values are fractional. Therefore, by finding an optimal solution of (1.20), we run the risk of finding an optimal solution that has a fractional value for x_{ij} for some i and j. Since it does not make sense to put only part of a player in the field, this is clearly not desirable (although one could interpret half a player as a player who is playing only half of the time). However, for this particular LO-model something surprising happens: as it turns out, (1.20) always has an optimal solution for which all x_{ij}'s are integer-valued, i.e., they are either 0 or 1. The reason for this is quite subtle and will be described in Chapter 8. But it does mean that (1.20) correctly models the team formation problem.

We promised to come back to the determination of the values of the c_{ij}'s. One way is to just make educated guesses. For example, we could let the values of c_{ij} run from 0 to 5, where 0 means that the player is completely unfit for the position, and 5 means that the player is perfect for the position.

A more systematic approach is the following. We can think of forming a team as a matter of economic supply and demand of qualities. Let us define a list of 'qualities' that play a role in soccer. The positions demand certain qualities, and the players supply these qualities. The list of qualities could include, for example, endurance, speed, balance, agility, strength, inventiveness, confidence, left leg skills, right leg skills. Let the qualities be labeled $1, \ldots, Q$. For each position j, let d_{jq} be the 'amount' of quality q demanded on position j and, for each player i, let s_{iq} be the 'amount' of quality q supplied by player i. We measure these numbers all on the same scale from 0 to 5. Now, for all i and j, we can define how well player i fits on position j by, for example, calculating the average squared deviation of player i's supplied qualities compared to the qualities demanded for position j:

$$
c_{ij} = -\sum_{q=1}^{Q}(s_{iq} - d_{jq})^2.
$$

The negative sign is present because we are maximizing the sum of the c_{ij}'s, so a player i that has exactly the same qualities as demanded by a position j has $c_{ij} = 0$, whereas any deviation from the demanded qualities will give a negative number.

Observe that c_{ij} is a nonlinear function of the s_{iq}'s and d_{jq}'s. This, however, does not contradict the definition of an LO-model, because the c_{ij}'s are not decision variables of the model; they show up as parameters of the model, in particular as the objective coefficients.

The current definition of c_{ij} assigns the same value to a positive deviation and a negative deviation. Even worse: suppose that there are two players, i and i', say, who supply exactly the same qualities, i.e., $s_{iq} = s_{i'q}$, except for some quality q, for which we have that $s_{iq} = d_{jq} - 1$ and $s_{i'q} = d_{jq} + 2$ for position j. Then, although player i' is clearly better for position j than player i is, we have that $c_{ij} > c_{i'j}$, which does not make sense. So, a better definition for c_{ij} should not have that property. For example, the function:

$$c_{ij} = -\sum_{q=1}^{Q} \left(\min(0, s_{iq} - d_{jq}) \right)^2$$

does not have this property. It is left to the reader to check this assertion.

1.6.4 Data envelopment analysis

In this section, we describe *data envelopment analysis* (DEA), an increasingly popular management tool for comparing and ranking the relative efficiencies of so-called *decision making units* (DMUs), such as banks, branches of banks, hospitals, hospital departments, universities, or individuals. After its introduction in 1978, data envelopment analysis has become a major tool for performance evaluation and benchmarking, especially for cases where there are complex relationships between multiple inputs and multiple outputs. The interested reader is referred to Cooper *et al.* (2011). An important part of DEA concerns the concept of efficiency. The *efficiency* of a DMU is roughly defined as the extent to which the inputs for that DMU are used to produce its outputs, i.e.,

$$\text{efficiency} = \frac{\text{output}}{\text{input}}.$$

The efficiency is defined in such a way that an efficiency of 1 means that the DMU makes optimal use of its inputs. So, the efficiency can never exceed 1. This measure of efficiency, however, may not be adequate because there are usually multiple inputs and outputs related to, for example, different resources, activities, and environmental factors. It is not immediately apparent how to compare these (possibly very different) inputs and outputs with each other. DEA takes a *data driven* approach. We will discuss here the basic ideas behind DEA, and use an example to show how relative efficiencies can be determined and how targets for relatively inefficient DMUs can be set.

As an example, consider the data presented in Table 1.2. We have ten universities that are the DMUs. All ten universities have the same inputs, namely real estate and wages. They also produce the same outputs, namely economics, business, and mathematics graduates. The universities differ, however, in the amounts of the inputs they use, and the amounts of the outputs they produce. The first two columns of Table 1.2 contain the input amounts for each university, and the last three columns contain the output amounts.

Given these data, comparing universities 1 and 2 is straightforward. University 1 has strictly larger outputs than university 2, while using less inputs than university 2. Hence, university 1 is more efficient than university 2. But how should universities 3 and 4 be compared? Among these two, university 3 has the larger number of graduated economics students, and university 4 has the larger number of graduated business and mathematics students. So, the outputs of universities 3 and 4 are hard to compare. Similarly, the inputs of universities 3 and 5 are hard to compare. In addition, it is hard to compare universities 9 and 10 to any of the other eight universities, because they are roughly twice as large as the other universities.

In DEA, the different inputs and outputs are compared with each other by assigning weights to each of them. However, choosing weights for the several inputs and outputs is generally hard and rather arbitrary. For example, it may happen that different DMUs have organized their operations differently, so that the output weights should be chosen differently. The key idea of DEA is that each DMU is allowed to choose its own set of weights. Of course, each DMU will then choose a set of weights that is most favorable for their efficiency assessment. However, DMUs are not allowed to 'cheat': a DMU should choose the weights in such a way that the efficiency of the other DMUs is restricted to at most 1. It is assumed that all DMUs convert (more or less) the same set of inputs into the same set of outputs: only the weights of the inputs and outputs may differ among the DMUs. DEA can be formulated as a linear optimization model as follows.

| DMU | Inputs | | Outputs | | |
	Real estate	Wages	Economics graduates	Business graduates	Math graduates
1	72	81	77	73	78
2	73	82	73	70	69
3	70	59	72	67	80
4	87	83	69	74	84
5	53	64	57	65	65
6	71	85	78	72	73
7	65	68	81	71	69
8	59	62	64	66	56
9	134	186	150	168	172
10	134	140	134	130	130

Table 1.2: Inputs and outputs for ten universities.

Let the DMUs be labeled $(k =) 1, \ldots, N$. For each $k = 1, \ldots, N$, the *relative efficiency* (or *efficiency rate*) $RE(k)$ of DMU k is defined as:

$$RE(k) = \frac{\text{weighted sum of output values of DMU } k}{\text{weighted sum of input values of DMU } k},$$

where $0 \leq RE(k) \leq 1$. DMU k is called *relatively efficient* if $RE(k) = 1$, and *relatively inefficient* otherwise. Note that if DMU k is relatively efficient, then its total input value equals its total output value. In other words, a relative efficiency rate $RE(k)$ of DMU k means that DMU k is able to produce its outputs with a $100 RE(k)$ percent use of its inputs. In order to make this definition more precise, we introduce the following notation.

Let $m \, (\geq 1)$ be the number of inputs, and $n \, (\geq 1)$ the number of outputs. For each $i = 1, \ldots, m$ and $j = 1, \ldots, n$, define:

$$
\begin{aligned}
x_i \quad &= \text{ the weight of input } i; \\
y_j \quad &= \text{ the weight of output } j; \\
u_{ik} \quad &= \text{ the (positive) amount of input } i \text{ to DMU } k; \\
v_{jk} \quad &= \text{ the (positive) amount of output } j \text{ to DMU } k.
\end{aligned}
$$

These definitions suggest that the same set of weights, the x_i's and the y_i's, are used for all DMUs. As described above, this is not the case. In DEA, each DMU is allowed to adopt its own set of weights, namely in such a way that its own relative efficiency is maximized. Hence, for each DMU, the objective should be to determine a set of input and output weights that yields the highest efficiency for that DMU in comparison to the other DMUs. The optimization model for DMU $k \, (= 1, \ldots, N)$ can then be formulated as follows:

$$
\begin{aligned}
RE^*(k) = \max \quad & \frac{v_{1k}y_1 + \ldots + v_{nk}y_n}{u_{1k}x_1 + \ldots + u_{mk}x_m} \\
\text{s.t.} \quad & \frac{v_{1r}y_1 + \ldots + v_{nr}y_n}{u_{1r}x_1 + \ldots + u_{mr}x_m} \leq 1 \ \text{ for } r = 1, \ldots, N \qquad (M_1(k)) \\
& x_1, \ldots, x_m, y_1, \ldots, y_n \geq \varepsilon.
\end{aligned}
$$

The decision variables in this model are constrained to be at least equal to some small positive number ε, so as to avoid any input or output becoming completely ignored in determining the efficiencies. We choose $\varepsilon = 0.00001$. Recall that in the above model, for each k, the weight values are chosen so as to maximize the efficiency of DMU k. Also note that we have to solve N such models, namely one for each DMU.

Before we further elaborate on model $(M_1(k))$, we first show how it can be converted into a linear model. Model $(M_1(k))$ is a so-called *fractional linear model*, i.e., the numerator and the denominator of the fractions are all linear in the decision variables. Since all denominators are positive, this fractional model can easily be converted into a common LO-model. In order to do so, first note that when maximizing a fraction, it is only the relative value of the numerator and the denominator that are of interest and not the individual values. Therefore, we can set the value of the denominator equal to a constant value (say, 1), and then maximize the numerator. For each $k = 1, \ldots, N$, we obtain the following LO-model:

$$
\begin{aligned}
RE^*(k) = \max\ & v_{1k}y_1 + \ldots + v_{nk}y_n \\
\text{s.t.}\quad & u_{1k}x_1 + \ldots + u_{mk}x_m = 1 \\
& v_{1r}y_1 + \ldots + v_{nr}y_n - u_{1r}x_1 - \ldots - u_{mr}x_m \leq 0 \qquad (M_2(k)) \\
& \qquad\qquad\qquad\qquad\qquad\qquad\qquad \text{for } r = 1, \ldots, N \\
& x_1, \ldots, x_m, y_1, \ldots, y_n \geq \varepsilon.
\end{aligned}
$$

Obviously, if $RE^*(k) = 1$, then DMU k is relatively efficient, and if $RE^*(k) < 1$, then there is a DMU that is more efficient than DMU k. Actually, there is always one that is relatively efficient, because at least one of the '$\leq$'-constraints of $(M_2(k))$ is satisfied with equality in the optimal solution of $(M_2(k))$; see Exercise 1.8.17. If $RE^*(k) < 1$, then the set of all relatively efficient DMUs is called the *peer group* $PG(k)$ of the inefficient DMU k. To be precise, for any DMU k with $RE^*(k) < 1$, we have that:

$$
PG(k) = \{p \in \{1, \ldots, N\} \mid RE^*(p) = 1\}.
$$

The peer group contains DMUs that can be set as a target for the improvement of the relative efficiency of DMU k.

Consider again the data of Table 1.2. Since there are ten universities, there are ten models to be solved. For example, for $k = 1$, the LO-model $M_2(1)$ reads:

$$
\begin{aligned}
\max\quad & 72y_1 + 81y_2 \\
\text{s.t.}\quad & 77x_1 + 73x_2 + 78x_3 = 1 \\
& 72y_1 + 81y_2 - 77x_1 - 73x_2 - 78x_3 \leq 0 \\
& 73y_1 + 82y_2 - 73x_1 - 70x_2 - 69x_3 \leq 0 \\
& 70y_1 + 59y_2 - 72x_1 - 67x_2 - 80x_3 \leq 0 \\
& 87y_1 + 83y_2 - 69x_1 - 74x_2 - 84x_3 \leq 0 \\
& 53y_1 + 64y_2 - 57x_1 - 65x_2 - 65x_3 \leq 0 \\
& 71y_1 + 85y_2 - 78x_1 - 72x_2 - 73x_3 \leq 0 \\
& 65y_1 + 68y_2 - 81x_1 - 71x_2 - 69x_3 \leq 0 \\
& 59y_1 + 62y_2 - 64x_1 - 66x_2 - 56x_3 \leq 0 \\
& 67y_1 + 93y_2 - 75x_1 - 84x_2 - 86x_3 \leq 0 \\
& 67y_1 + 70y_2 - 67x_1 - 65x_2 - 65x_3 \leq 0 \\
& x_1, x_2, x_3, y_1, y_2 \geq 0.00001.
\end{aligned}
$$

Listing 1.3 contains the optimization model written as a GMPL model.

```
1    INPUT;
2    OUTPUT;
3
4       N >= 1;
5       u{1..N, INPUT};      # input values
6       v{1..N, OUTPUT};     # output values
7
```

```
8        K;                      # the DMU to assess
9        eps > 0;
10
11   x{i    INPUT} >= eps;
12   y{j    OUTPUT} >= eps;
13
14        objective:
15     {j    OUTPUT} y[j] * v[K, j];
16
17          this_dmu:
18     {i    INPUT} x[i] * u[K, i] = 1;
19
20          other_dmus{k    1..N}:
21     {j    OUTPUT} y[j] * v[k, j] <=    {i    INPUT} x[i] * u[k, i];
22
23   ;
24
25     eps := 0.00001;
26     K := 1;
27     N := 10;
28
29   INPUT  := realestate wages;
30   OUTPUT := economics business mathematics;
31
32     u: realestate wages :=
33   1 72 81
34   2 73 82
35   3 70 59
36   4 87 83
37   5 53 64
38   6 71 85
39   7 65 68
40   8 59 62
41   9 134 186
42   10 134 140;
43
44     v: economics business mathematics :=
45   1 77 73 78
46   2 73 70 69
47   3 72 67 80
48   4 69 74 84
49   5 57 65 65
50   6 78 72 73
51   7 81 71 69
52   8 64 66 56
53   9 150 168 172
54   10 134 130 130;
55
56   ;
```

Listing 1.3: DEA model.

The optimal solution reads: $y_1^* = 0.01092, y_2^* = 0.00264, x_1^* = 0.00505, x_2^* = 0.00001,$ $x_3^* = 0.00694$, with optimal objective value 0.932. Hence, the relative efficiency of uni-

| DMU | Input weights | | Output weights | | | Efficiency | |
	Real estate	Wages	Economics graduates	Business graduates	Math graduates	Efficiency	Peer set
1	10.92	2.64	5.05	0.01	6.94	93.2%	3, 7, 9
2	10.13	3.18	4.16	1.71	6.03	83.9%	3, 5, 7, 9
3	0.01	16.94	0.01	0.01	12.48	100.0%	
4	8.68	2.95	0.01	1.30	8.68	82.5%	3, 5, 9
5	5.46	11.11	0.01	14.89	0.49	100.0%	
6	14.07	0.01	7.59	0.01	4.34	91.0%	7, 9
7	1.58	13.20	12.33	0.01	0.01	100.0%	
8	5.54	10.85	1.86	13.34	0.01	100.0%	
9	11.79	2.26	4.24	8.11	0.01	100.0%	
10	5.60	8.93	1.27	11.70	0.54	88.1%	3, 5, 7, 8

Table 1.3: DEA results for the data in Table 1.2. The weights have been multiplied by 1,000.

versity 1 is:

$$RE^*(1) = \frac{72y_1^* + 81y_2^*}{77x_1^* + 73x_2^* + 78x_3^*} = z^* = 0.932.$$

We have listed the optimal solutions, the relative efficiencies, and the peer sets in Table 1.3. It turns out that universities 3, 5, 7, 8, and 9 are relatively efficient, and that the other universities are relatively inefficient. Note that universities 9 and 10 are about twice as large as the other universities. Recall that the DEA approach ignores the scale of a DMU and only considers the relative sizes of the inputs and the outputs.

1.6.5 Portfolio selection; profit versus risk

An investor considers investing $10,000 in stocks during the next month. There are n ($n \geq 1$) different stocks that can be bought. The investor wants to buy a portfolio of stocks at the beginning of the month, and sell them at the end of the month, without making any changes to the portfolio during that time. Since the investor wants to make as much profit as possible, a portfolio should be selected that has the highest possible total selling price at the end of the month.

Let $i = 1, \ldots, n$. Define R_i to be the *rate of return* of stock i, i.e.:

$$R_i = \frac{V_i^1}{V_i^0},$$

where V_i^0 is the current value of stock i, and V_i^1 is the value of stock i in one month. This means that if the investor decides to invest $1 in stock i, then this investment will be worth $\$R_i$ at the end of the month. The main difficulty with portfolio selection, however, is that the rate of return is not known in advance, i.e., it is *uncertain*. This means that we cannot know in advance how much any given portfolio will be worth at the end of the month.

| Scenario s | \multicolumn{5}{c}{Stock i} |
	1	2	3	4	5
1	−4.23	−1.58	0.20	5.50	2.14
2	8.30	0.78	−0.34	5.10	2.48
3	6.43	1.62	1.19	−2.90	4.62
4	0.35	3.98	2.14	−0.19	−2.72
5	1.85	0.61	1.60	−3.30	−0.58
6	−6.10	1.79	0.61	2.39	−0.24
μ_i	1.10	1.20	0.90	1.10	0.95
ρ_i	4.43	1.27	0.74	3.23	2.13

Table 1.4: The values of R_i^s of each stock i in each scenario s, along with the expected rate of return μ_i, and the mean absolute deviation ρ_i. All numbers are in percentages.

One way to deal with this uncertainty is to assume that, although we do not know the exact value of R_i in advance, we know a number of possible scenarios that may happen. For example, we might define three scenarios, describing a bad outcome, an average outcome, and a good outcome. Let S be the number of scenarios. We assume that all scenarios are equally likely to happen. For each i, let R_i^s be the rate of return of stock i in scenario s. Table 1.4 lists an example of values of R_i^s for $(n =)$ 5 stocks and $(S =)$ 6 scenarios. For example, in scenario 1, the value of stock 1 decreases by 4.23% at the end of the month. So, in this scenario, an investment of $1 in stock 1 will be worth $(1 - 0.0423 =)$ $0.9567 at the end of the month. On the other hand, in scenario 2, this investment will be worth $(1 + 0.083 =)$ $1.083. Since we do not know in advance which scenario will actually happen, we need to base our decision upon all possible scenarios. One way to do this is to consider the so-called *expected rate of return* of stock i, which is denoted and defined by:

$$\mu_i = \frac{1}{S} \sum_{s=1}^{S} R_i^s.$$

The expected rate of return is the average rate of return, where the average is taken over all possible scenarios. It gives an indication of the rate of return at the end of the month. Since the value of R_i^s is assumed to be known, the value of μ_i is known as well. Hence, it seems reasonable to select a portfolio of stocks that maximizes the total expected rate of return. However, there is usually a trade-off between the expected rate of return of a stock and the associated *risk*. A low-risk stock is a stock that has a rate of return that is close to its expected value in each scenario. Putting money on a bank account (or in a term-deposit) is an example of such a low-risk investment: the interest rate r (expressed as a fraction, e.g., 2% corresponds to 0.02) is set in advance, and the investor knows that if x is invested $(x \geq 0)$, then after one month, this amount will have grown to $(1 + r)x$. In this case, the rate of return is the same for each scenario. On the other hand, the rate of return of a high-risk stock varies considerably among the different scenarios. Stock 1 in Table 1.4 is an example of a high-risk stock.

Risk may be measured in many ways. One of them is the *mean absolute deviation*. The mean absolute deviation of stock i is denoted and defined as:

$$\rho_i = \frac{1}{S} \sum_{s=1}^{S} |R_i^s - \mu_i|.$$

So, ρ_i measures the average deviation of the rate of return of stock i compared to the expected rate of return of stock i.

For each $i = 1, \ldots, n$, define the following decision variable:

$x_i =$ the fraction of the \$10,000 to be invested in stock i.

Since the investor wants to invest the full \$10,000, the x_i's should satisfy the constraint:

$$x_1 + \ldots + x_n = 1.$$

If the investor simply wants the highest expected rate of return, then one may choose the stock i that has the largest value of μ_i and set $x_i = 1$. However, this strategy is risky: it is better to diversify. So, the problem facing the investor is a so-called *multiobjective* optimization problem: on the one hand, the expected rate of return should be as large as possible; on the other hand, the risk should be as small as possible. Since stocks with a high expected rate of return usually also have a high risk, there is a trade-off between these two objectives; see also Chapter 14. If \$$x_i$ is invested in stock i, then it is straightforward to check that the expected rate of return μ of the portfolio as a whole satisfies:

$$\mu = \frac{1}{S} \sum_{s=1}^{S} \sum_{i=1}^{n} R_i^s x_i = \sum_{i=1}^{n} \mu_i x_i. \tag{1.21}$$

The mean absolute deviation ρ of a portfolio in which \$$x_i$ is invested in stock i satisfies:

$$\rho = \frac{1}{S} \sum_{s=1}^{S} \left| \sum_{i=1}^{n} R_i^s x_i - \mu \right| = \frac{1}{S} \sum_{s=1}^{S} \left| \sum_{i=1}^{n} (R_i^s - \mu_i) x_i \right|.$$

To solve the multiobjective optimization problem, we introduce a positive weight parameter λ which measures how much importance we attach to maximizing the expected rate of return of the portfolio, relative to minimizing the risk of the portfolio. We choose the following objective:

$$\max \ \lambda(\text{expected rate of return}) - (\text{mean absolute deviation}).$$

Thus, the objective is $\max(\lambda\mu - \rho)$. The resulting optimization problem is:

$$\max \lambda \sum_{i=1}^{n} \mu_i x_i - \frac{1}{S} \sum_{s=1}^{S} \left| \sum_{i=1}^{n} (R_i^s - \mu_i) x_i \right|$$

$$\text{s.t.} \sum_{i=1}^{n} x_i = 1 \tag{1.22}$$

$$x_1, \ldots, x_n \geq 0.$$

Although this is not an LO-model because of the absolute value operations, it can be turned into one using the technique of Section 1.5.2. To do so, we introduce, for each $s = 1, \ldots, S$, the decision variable u_s, and define u_s to be equal to the expression $\sum_{i=1}^{n} (R_i^s - \mu_i) x_i$ inside the absolute value bars in (1.22). Note that u_s measures the rate of return of the portfolio in scenario s. Next, we write $u_s = u_s^+ - u_s^-$ and $|u_s| = u_s^+ + u_s^-$. In Section 1.5.2, this procedure is explained in full detail. This results in the following LO-model:

$$\max \lambda \sum_{i=1}^{n} \mu_i x_i - \frac{1}{S} \sum_{s=1}^{S} (u_s^+ + u_s^-)$$

$$\text{s.t.} \sum_{i=1}^{n} x_i = 1 \tag{1.23}$$

$$u_s^+ - u_s^- = \sum_{i=1}^{n} (R_i^s - \mu_i) x_i \quad \text{for } s = 1, \ldots, S$$

$$x_1, \ldots, x_n, u_s^+, u_s^- \geq 0 \quad \text{for } s = 1, \ldots, S.$$

Consider again the data in Table 1.4. As a validation step, we first choose the value of λ very large, say $\lambda = 1000$. The purpose of this validation step is to check the correctness of the model by comparing the result of the model to what we expect to see. When the value of λ is very large, the model tries to maximize the expected rate of return, and does not care much about minimizing the risk. As described above, in this case we should choose $x_j^* = 1$ where j is the stock with the largest expected rate of return, and $x_i^* = 0$ for $i \neq j$. Thus, when solving the model with a large value of λ, we expect to see exactly this solution. After solving the model with $\lambda = 1000$ using a computer package, it turns out that this is indeed an optimal solution. On the other hand, when choosing $\lambda = 0$, the model tries to minimize the total risk. The optimal solution then becomes:

$$x_1^* = 0.036, \ x_2^* = 0, \ x_3^* = 0.761, \ x_4^* = 0.151, \ x_5^* = 0.052,$$
$$u_4^+ = 0.0053, \ u_1^+ = u_2^+ = u_3^+ = u_5^+ = u_6^+ = 0,$$
$$u_5^- = 0.0018, \ u_6^- = 0.0035, \ u_1^- = u_2^- = u_3^- = u_4^- = 0.$$

The corresponding expected rate of return is 0.94%, and the corresponding average absolute deviation of the portfolio is 0.177%. This means that if the investor invests, at the beginning of the month, \$360 in stock 1, \$7,610 in stock 3, \$1,510 in stock 4, and \$520 in stock 5, then the investor may expect a 0.94% profit at the end of the month, i.e., the investor may expect that the total value of the stocks will be worth \$10,094. Note, however, that the

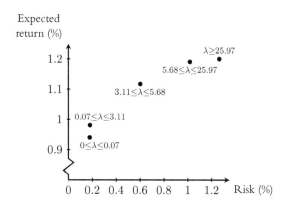

Figure 1.9: Return versus risk in the portfolio selection problem.

actual value of this portfolio at the end of the month is not known in advance. The actual value can only be observed at the end of the month and may deviate significantly from the expected value. For example, if scenario 6 happens to occur, then the rate of return of the portfolio will be $(0.94 - 0.35 =)\ 0.59\%$; hence, the portfolio will be worth only $10,059. On the other hand, if scenario 4 occurs, the portfolio will be worth $10,147, significantly more than the expected value.

By setting the value of λ to a strictly positive number, the expected return and the risk are balanced. As an example, we choose $\lambda = 5$. In that case, the following is an optimal solution:

$$x_1^* = 0,\ x_2^* = 0.511,\ x_3^* = 0,\ x_4^* = 0.262,\ x_5^* = 0.227,$$
$$u_2^+ = 0.0118,\ u_4^+ = 0.0025,\ u_6^+ = 0.0037,\ u_1^+ = u_3^+ = u_5^+ = 0,$$
$$u_5^- = 0.018,\ u_1^- = u_2^- = u_3^- = u_4^- = u_6^- = 0,$$

with an expected return of 1.12% (about 30% higher than the minimum-risk portfolio) and an average absolute deviation of 0.6% (more than three times as much as the minimum-risk portfolio).

In general, the choice of the value of λ depends on the investor. If the investor is very risk averse, a small value of λ should be chosen; if the investor is risk-seeking, a large value should be chosen. Since it is not clear beforehand what exact value should be chosen, we can solve the model for various values of λ. Figure 1.9 shows the expected return and the risk for different optimal solutions with varying values of λ. This figure illustrates the fact that as λ increases (i.e., as the investor becomes increasingly risk-seeking), both the expected return and the risk increase. The investor can now choose the portfolio that provides the combination between expected return and risk that meets the investors' preferences best. The process in which we analyze how optimal values depend on the choice of a parameter (in this case, λ) is called *sensitivity analysis*, which is the main topic of Chapter 5.

1.7 Building and implementing mathematical models

In practical situations, designing mathematical models is only one part of a larger decision-making process. In such settings, designing a mathematical model is therefore not a goal by itself. The mathematical models form a tool to help make a suitable and informed decision. In this section, we discuss mathematical models in the broader setting of the decision-making process.

This decision-making process usually involves a team of people from various disciplines, e.g., business managers and subject experts. Among the subject experts, there may be one or more operations research experts. The objective of the operations research expert is to design useful mathematical models that are appropriate for the real-life problem at hand. Since the expert needs to work closely with the other team members, the job requires not only the appropriate technical and mathematical knowledge, but also social skills.

Complicated practical problems can be investigated by using models that reflect the relevant features of the problem, in such a way that properties of the model can be related to the real (actual) situation. One of the main decisions to be made by the operations research expert is which features of the problem to take into account, and which ones to leave out. Clearly, the more features are taken into account, the more realistic the model is. On the other hand, taking more features into account usually complicates the interpretation of the model results, in addition to making the model harder to solve. Choosing which features to take into account usually means finding a good balance between making the model sufficiently realistic on the one hand, and keeping the model as simple as possible on the other hand. There are no general rules to decide which features are appropriate, and which ones are not. Therefore, mathematical modeling can be considered as an art.

Throughout the decision process, the following questions need to be asked. What is the problem? Is this really the problem? Can we solve this problem? Is mathematical optimization an appropriate tool to solve the problem, or should we perhaps resort to a different mathematical (or nonmathematical) technique? Once it is clear that mathematical optimization is appropriate, other questions need to be answered. What data do we need to solve the problem? Once we have found a solution, the question then arises whether or not this is a useful solution. How will the solution be implemented in practice? By 'solving a problem' we mean the successful application of a solution to a practical problem, where the solution is a result from a mathematical model.

The following steps provide some guidelines for the decision-making process; usually these steps do not have to be made explicit during the actual model building and implementing process. Usually, there is a certain budget available for the completion of the decision process. During the whole process, the budget has to be carefully monitored.

Algorithm 1.7.1. (*Nine steps of the decision process*)

Step 1: Problem recognition;

Step 2: Problem definition;

Step 3: Observation and problem analysis;

Step 4: Designing a conceptual model;

Step 5: Formulating a mathematical model;

Step 6: Solving the mathematical model;

Step 7: Taking a decision;

Step 8: Implementing the decision;

Step 9: Evaluation.

We will discuss each of the nine steps in more detail.

Step 1. Problem recognition.
The recognition of a problem precedes the actual procedure. It may start when a manager in a business does not feel quite happy about the current situation, but does not exactly know what is wrong. More often than not, the issue simmers for a while, before the manager decides to take action. For instance, a planner discovers that orders are delivered late, monitors the situation for a couple of weeks, and, when the situation does not change, informs the logistics department.

Step 2. Problem definition.
The next step in the decision-making process concerns the definition of the problem. For example, the definition of the problem is: 'The warehouse is too small', or 'The orders are not manufactured before their due dates'. This step is usually carried out by a small team of experts. This team discusses and reconsiders the problem definition, and – if necessary – improves it. An important aspect is setting the goals of the decision-making process. A provisional goal could be to improve the insight and understanding of the complex situation of the problem.

Step 3. Observation and problem analysis.
Since a problem usually occurs in large (sub)systems, it is necessary to observe and analyze the problem in the context of the company as a whole. It is important to communicate with other experts within and outside the organization. Relevant data need to be collected and analyzed to understand all aspects of the problem. It may happen that it is necessary to return to Step 2 in order to revise the problem definition. At the end of Step 3 the structure of the problem should become clear.

Step 4. Designing a conceptual model; validation.
During Step 3, the relevant information has become available. It is now clear which aspects and parameters are of importance and should be taken into account for further considerations. However, this information may not be enough to solve the problem. The conceptual model reflects the relationships between the various aspects or parameters of the problem, and the objective that has to be reached. Usually, not all aspects can be included. If vital details are omitted, then the model will not be realistic. On the other

hand, if too many details are included, then attention will be distracted from the crucial factors and the results may become more difficult to interpret. The trade-off between the size of the model (or, more precisely, the time to solve the model) and its practical value must also be taken into account. Conceptual models have to be *validated*, which means investigating whether or not the model is an accurate representation of the practical situation. The conceptual model (which usually does not contain mathematical notions) can be discussed with experts inside and outside the company. In case the model does not describe the original practical situation accurately enough, the conceptual model may need to be changed, or one has to return to Step 3. At the end of Step 4, the project team may even decide to change the problem definition, and to return to Step 2.

Step 5. Designing a mathematical model; verification.
In this step the conceptual relationships between the various parameters are translated into mathematical relationships. This is not always a straightforward procedure. It may be difficult to find the right type of mathematical model, and to formulate the appropriate specification without an overload of variables and constraints. If significant problems occur, it may be useful to return to Step 3, and to change the conceptual model in such a way that the mathematical translation is less difficult. If the project team decides to use a linear optimization model, then the relationships between the relevant parameters should be linear. This is in many situations a reasonable restriction. Instead of solving the whole problem immediately, it is usually better to start with smaller subproblems; this may lead to new insights and to a reformulation of the conceptual model. The mathematical model has to be *verified*, which means determining whether or not the model performs as intended, and whether or not the computer running time of the solution technique that is used is acceptable (see also Chapter 9). This usually includes checking the (in)equalities and the objective of the model. Another useful technique is to compute solutions of the model for a number of extreme worst-case instances, or for data for which realizations are already known. Sensitivity analysis on a number of relevant parameters may also provide insight into the accuracy and robustness of the model. These activities may lead to a return to Step 3 or Step 4.

Step 6. Solving the mathematical model.
Sometimes a solution of the mathematical model can be found by hand, without using a computer. In any case, computer solution techniques should only be used after the mathematical model has been carefully analyzed. This analysis should include looking for simplifications which may speed up the processing procedures, analyze the models outputs thoroughly, and carry out some sensitivity analysis. Sensitivity analysis may, for instance, reveal a strong dependence between the optimal solution and a certain parameter, in which case the conceptual model needs to be consulted in order to clarify this situation. Moreover, sensitivity analysis may provide alternative (sub)optimal solutions which can be used to satisfy constraints that could not be included in the mathematical model.

Step 7. Taking a decision.
As soon as Step 6 has been finished, a decision can be proposed. Usually, the project team

formulates a number of alternative proposals together with the corresponding costs and other business implications. In cooperation with the management of the business, it is decided which solution will be implemented or, if implementation is not possible, what should be done otherwise. Usually, this last situation means returning to either Step 4, or Step 3, or even Step 2.

Step 8. Implementing the decision.
Implementing the decision requires careful attention. The expectations that are set during the decision process have to be realized. Often a pilot-project can be started to detect any potential implementation problems at an early stage. The employees that have to work with the new situation, need to be convinced of the effectiveness and the practicability of the proposed changes. Sometimes, courses need to be organized to teach employees how to deal with the new situation. One possibility to prevent problems with the new implementation is to communicate during the decision process with those involved in the actual decision and execution situation.

Step 9. Evaluation.
During the evaluation period, the final checks are made. Questions to be answered are: Does everything work as intended? Is the problem solved? Was the decision process organized well? Finally, the objectives of the company are compared to the results of the implemented project.

1.8 Exercises

Exercise 1.8.1.   In Section 1.2.1, we defined the standard form of an LO-model. Write each of the following LO-models in the standard form $\max\{\mathbf{c}^\mathsf{T}\mathbf{x} \mid \mathbf{A}\mathbf{x} \le \mathbf{b}, \mathbf{x} \ge \mathbf{0}\}$.

(a) $\min \quad 25x_1 + \quad 17x_2$
s.t. $0.21x_1 + 0.55x_2 \ge 3$
$\quad\ \, 0.50x_1 + 0.30x_2 \ge 7$
$\quad\ \, 0.55x_1 + 0.10x_2 \ge 5$
$\quad\quad\quad x_1, x_2 \text{ free}$

(b) $\min \quad 12x + |5y|$
s.t. $\quad x + \ 2y \ge 4$
$\quad\ \, 5x + \ 6y \le 7$
$\quad\ \, 8x + \ 9y = 5$
$\quad\quad x, y \text{ free}$

(c) $\min 3x_1 + \quad x_3$
s.t. $x_1 + \ x_2 + \ x_3 + x_4 = 10$
$\quad x_1 - 2x_2 + 2x_3 \quad\quad = 6$
$\quad 0 \le x_1 \le 4, 0 \le x_2 \le 4$
$\quad 0 \le x_3 \le 4, 0 \le x_4 \le 12$

(d) $\min |x_1| + |x_2| + |x_3|$
s.t. $x_1 - 2x_2 \quad\quad = 3$
$\quad\quad - \ x_2 + \ x_3 \le 1$
$\quad x_1, x_2, x_3 \text{ free}$

Exercise 1.8.2.   Show that an LO-model of the form

$$\max\{\mathbf{c}_1^\mathsf{T}\mathbf{x}_1 + \mathbf{c}_2^\mathsf{T}\mathbf{x}_2 \mid \mathbf{A}_1\mathbf{x}_1 = \mathbf{b}_1, \ \mathbf{A}_2\mathbf{x}_2 = \mathbf{b}_2, \ \mathbf{x}_1 \ge \mathbf{0}, \mathbf{x}_2 \ge \mathbf{0}\}$$

with $\mathbf{c}_i \in \mathbb{R}^{n_i}$, $\mathbf{x}_i \in \mathbb{R}^{n_i}$, $\mathbf{A}_i \in \mathbb{R}^{m_i \times n_i}$, $\mathbf{b}_i \in \mathbb{R}^{m_i}$ $(i = 1, 2)$, can be solved by solving the two models

$$\max\{\mathbf{c}_1^\mathsf{T}\mathbf{x}_1 \mid \mathbf{A}_1\mathbf{x}_1 = \mathbf{b}_1, \ \mathbf{x}_1 \geq \mathbf{0}\} \quad \text{and} \quad \max\{\mathbf{c}_2^\mathsf{T}\mathbf{x}_2 \mid \mathbf{A}_2\mathbf{x}_2 = \mathbf{b}_2, \ \mathbf{x}_2 \geq \mathbf{0}\}.$$

Given an optimal solution of these two, what is the optimal solution for the original model, and what is the corresponding optimal objective value?

Exercise 1.8.3. Solve the following LO-models by using the graphical solution method.

(a) $\max \quad x_1 + 2x_2$
 s.t. $-x_1 + 2x_2 \leq 4$
        $3x_1 + x_2 \leq 9$
        $x_1 + 4x_2 \geq 4$
        $x_1, x_2 \geq 0$

(b) $\min \ 11x_1 + 2x_2$
 s.t. $x_1 + x_2 \leq 4$
        $15x_1 - 2x_2 \geq 0$
        $5x_1 + x_2 \geq 5$
        $2x_1 + x_2 \geq 3$
        $x_1, x_2 \geq 0$

(c) $\max \ 2x_1 + 3x_2 - 2x_3 + 3x_4$
 s.t. $x_1 + x_2 \leq 6$
        $2x_3 + 3x_4 \leq 12$
        $-x_3 + x_4 \geq -2$
        $2x_1 - x_2 \leq 4$
        $2x_3 - x_4 \geq -1$
        $0 \leq x_2 \leq 2, \ x_1, x_3, x_4 \geq 0$

(d) $\max \ 3x_1 + 7x_2 + 4x_3 - 3x_4$
 s.t. $x_1 + 2x_2 \leq 6$
        $x_3 + x_4 \leq 6$
        $4x_1 + 5x_2 \leq 20$
        $-2x_1 + x_2 \leq 1$
        $2x_3 - x_4 \leq 4$
        $x_3 \geq 2$
        $1 \leq x_2 \leq 2, \ x_1, x_4 \geq 0$

Exercise 1.8.4. The LO-models mentioned in this chapter all have '$\leq$', '$\geq$', and '$=$' constraints, but no '$<$' or '$>$' constraints. The reason for this is the fact that models with such constraints may not have an optimal solution, even if the feasible region is bounded. Show this by constructing a bounded model with '$<$' and/or '$>$' constraints, and argue that the constructed model does not have an optimal solution.

Exercise 1.8.5. Consider the constraints of Model Dovetail in Section 1.1. Determine the optimal vertices in the case of the following objectives:

(a) $\max 2x_1 + x_2$

(b) $\max x_1 + 2x_2$

(c) $\max \frac{3}{2}x_1 + \frac{1}{2}x_2$

Exercise 1.8.6. In Section 1.1.2 the following optimal solution of Model Dovetail is found:

$$x_1^* = 4\tfrac{1}{2}, \ x_2^* = 4\tfrac{1}{2}, \ x_3^* = 0, \ x_4^* = 0, \ x_5^* = 2\tfrac{1}{2}, \ x_6^* = 1\tfrac{1}{2}.$$

What is the relationship between the optimal values of the slack variables x_3, x_4, x_5, x_6 and the constraints (1.1), (1.2), (1.3), (1.4)?

Exercise 1.8.7.   Let $\alpha, \beta \in \mathbb{R}$. Consider the LO-model:

$$
\begin{aligned}
\min \quad & x_1 + x_2 \\
\text{s.t.} \quad & \alpha x_1 + \beta x_2 \geq 1 \\
& x_1 \geq 0, \ x_2 \ \text{free}.
\end{aligned}
$$

Determine necessary and sufficient conditions for α and β such that the model

(a)   is infeasible,

(b)   has an optimal solution,

(c)   is feasible, but unbounded,

(d)   has multiple optimal solutions.

Exercise 1.8.8.   Solve the following LO-model with the graphical solution method.

$$
\begin{aligned}
\max \quad & 2x_1 - 3x_2 - 4x_3 \\
\text{s.t.} \quad & -x_1 - 5x_2 + x_3 \qquad = 4 \\
& -x_1 - 3x_2 \qquad + x_4 = 2 \\
& x_1, x_2, x_3, x_4 \geq 0.
\end{aligned}
$$

Exercise 1.8.9.   Show that if $\mathbf{x}^* = \begin{bmatrix} u_1^* & u_2^* & x_2^* \end{bmatrix}^\mathsf{T}$ is an optimal solution of (1.14), then $\begin{bmatrix} x_1^* & x_2^* \end{bmatrix}^\mathsf{T}$ with $x_1^* = u_1^* - u_2^*$ is an optimal solution of (1.11).

Exercise 1.8.10.   The method of Section 1.5.2 to write a model with an absolute value as an LO-model does not work when the objective is to maximize a function in which an absolute value appears with a positive coefficient. Show this by considering the optimization model $\max\{|x| \mid -5 \leq x \leq 5\}$.

Exercise 1.8.11.   Show that any convex piecewise linear function is continuous.

Exercise 1.8.12.   The method of Section 1.5.3 to write a model with a piecewise linear function as an LO-model fails if the piecewise linear function is not convex. Consider again model (1.16), but with the following function f:

$$
f(x_1) = \begin{cases} x_1 & \text{if } 0 \leq x_1 \leq 3 \\ 3 + 3(x_1 - 3) & \text{if } 3 < x_1 \leq 8 \\ 18 + (x_1 - 8) & \text{if } x_1 > 18. \end{cases}
$$

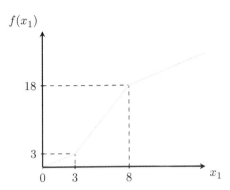

Figure 1.10: Nonconvex piecewise linear function $f(x_1)$.

This function is depicted in Figure 1.10. What goes wrong when the model is solved using the solution technique of Section 1.5.3?

Exercise 1.8.13. Consider the team formation model in Section 1.6.3. Take any $p \in \{1, \ldots, N\}$. For each of the following requirements, modify the model to take into account the additional requirement:

(a) Due to an injury, player p cannot be part of the line-up.

(b) Player p has to play on position 1.

(c) Due to contractual obligations, player p has to be part of the line-up, but the model should determine at which position.

(d) Let $A \subseteq \{1, \ldots, N\}$ and $B \subseteq \{1, \ldots, M\}$. The players in set A can only play on positions in set B.

Exercise 1.8.14. Consider the data envelopment analysis example described in Section 1.6.4. Give an example that explains that if a DMU is relatively efficient according to the DEA solution, then this does not necessarily mean that it is 'inherently efficient'. (Formulate your own definition of 'inherently efficient'.) Also explain why DMUs that are relatively inefficient in the DEA approach, are always 'inherently inefficient'.

Exercise 1.8.15. One of the drawbacks of the data envelopment analysis approach is that all DMUs may turn out to be relatively efficient. Construct an example with three different DMUs, each with three inputs and three outputs, and such that the DEA approach leads to three relatively efficient DMUs.

Exercise 1.8.16. Construct an example with at least three DMUs, three inputs and three outputs, and such that all but one input and one output have weights equal to ε in the optimal solution of model $(M_2(k))$.

Exercise 1.8.17. Consider model $(M_2(k))$. Show that there is always at least one DMU that is relatively efficient. (Hint: show that if all '$\leq$'-constraints of $(M_2(k))$ are satisfied with strict inequality in the optimal solution of $(M_2(k))$, then a 'more optimal' solution of $(M_2(k))$ exist, and hence the optimal solution was not optimal after all.)

Exercise 1.8.18. Consider the following data for three baseball players.

Players	Bats	Hits	Home runs
Joe	123	39	7
John	79	22	3
James	194	35	5

We want to analyze the efficiency of these baseball players using data envelopment analysis.

(a) What are the decision making units, and what are the inputs and outputs?

(b) Use DEA to give an efficiency ranking of the three players.

Exercise 1.8.19. Consider again the portfolio selection problem of Section 1.6.5. Using the current definition of risk, negative deviations from the expected rate of return have the same weight as positive deviations. Usually, however, the owner of the portfolio is more worried about negative deviations than positive ones.

(a) Suppose that we give the negative deviations a weight α and the positive deviations a weight β, i.e., the risk of the portfolio is:

$$\rho = \frac{1}{S} \sum_{s=1}^{S} |f(R_i^s - \mu_i)|,$$

where $f(x) = -\alpha x$ if $x < 0$ and $f(x) = \beta x$ if $x \geq 0$. Formulate an LO-model that solves the portfolio optimization problem for this definition of risk.

(b) What conditions should be imposed on the values of α and β?

Exercise 1.8.20. When plotting data from a repeated experiment in order to test the validity of a supposed linear relationship between two variables x_1 and x_2, the data points are usually not located exactly on a straight line. There are several ways to find the 'best fitting line' through the plotted points. One of these is the method of *least absolute deviation regression*. In least absolute deviation regression it is assumed that the exact underlying relationship is a straight line, say $x_2 = \alpha x_1 + \beta$. Let $\{[a_1\ b_1], \ldots, [a_n\ b_n]\}$ be the data set. For each $i = 1, \ldots, n$, the absolute error, r_i, is defined by $r_i = a_i - b_i \alpha - \beta$. The problem is to determine values of α and β such that the sum of the absolute errors, $\sum_{i=1}^{n} |r_i|$, is as small as possible.

(a) Formulate the method of least absolute deviation regression as an LO-model.

Employee	Age	Salary (×$1,000)
1	31	20
2	57	50
3	37	30
4	46	38
5	32	21
6	49	34
7	36	26
8	51	40
9	53	38
10	55	42
11	43	26
12	28	21
13	34	23
14	62	40
15	31	25
16	58	43
17	47	35
18	65	42
19	28	25
20	43	33

Table 1.5: Ages and salaries of twenty employees of a company.

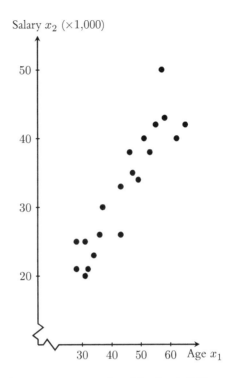

Figure 1.11: Scatter plot of the data in Table 1.5.

(b) Determine the least absolute deviation regression line $x_2 = \alpha x_1 + \beta$ for the following data set: $\{[1\ 1], [2\ 2], [3\ 4]\}$.

(c) Table 1.5 contains the ages and salaries of twenty employees of a company. Figure 1.11 shows a scatter plot of the same data set. Determine the least absolute deviation regression line $x_2 = \alpha x_1 + \beta$ for the data set, where x_1 is the age, and x_2 is the salary of the employee.

(d) Modify the model so that instead of minimizing the sum of the absolute deviations, the maximum of the absolute deviations is minimized. Solve the model for the data set in Table 1.5.

Exercise 1.8.21. The new budget airline CheapNSafe in Europe wants to promote its brand name. For this promotion, the company has a budget of €95,000. CheapNSafe hired a marketing company to make a television advertisement. It is now considering buying 30-second advertising slots on two different television channels: ABC and XYZ. The number of people that are reached by the advertisements clearly depends on how many slots are bought. Fortunately, the marketing departments of ABC and XYZ have estimates of how many people are reached depending on the number of slots bought. ABC reports that the first twenty slots on their television channel reach 20,000 people per slot; the next twenty slots reach another 10,000 each; any additional slots do not reach any additional people. So, for example, thirty slots reach $(20 \times 20{,}000 + 10 \times 10{,}000 =)\ 500{,}000$ people. One slot on ABC costs €2,000. On the other hand, XYZ reports that the first fifteen slots on XYZ reach 30,000 people each; the next thirty slots reach another 15,000 each; any additional slots do not reach any additional people. One slot on XYZ costs €3,000.

Suppose that CheapNSafe's current objective is to maximize the number of people reached by television advertising.

(a) Draw the two piecewise linear functions that correspond to this model. Are the functions convex or concave?

(b) Assume that people either watch ABC, or XYZ, but not both. Design an LO-model to solve CheapNSafe's advertising problem.

(c) Solve the LO-model using a computer package. What is the optimal advertising mix, and how many people does this mix reach?

(d) XYZ has decided to give a quantity discount for its slots. If a customer buys more than fifteen slots, the slots in excess of fifteen cost only €2,500 each. How can the model be changed in order to incorporate this discount? Does the optimal solution change?

Part I

Linear optimization theory
Basic techniques

CHAPTER 2

Geometry and algebra of feasible regions

Overview

In Chapter 1, we introduced the basic concepts of linear optimization. In particular, we showed how to solve a simple LO-model using the graphical method. We also observed the importance of the vertices of the feasible region. In the two-dimensional case (in which case the graphical method works), it was intuitively clear what was meant by a vertex, but since the feasible region of an LO-model is in general a subset of n-dimensional Euclidean space, $\mathbb{R}^n$ ($n \geq 1$), we will need to properly define what is meant by a vertex. In the first part of the current chapter we will formalize the concept of vertex, along with other concepts underlying linear optimization, and develop a geometric perspective of the feasible region of an LO-model.

In the second part of this chapter, we will take a different viewpoint, namely the *algebraic* perspective of the feasible region. The algebraic description of the feasible region forms the basic building block for one of the most widely used methods to solve LO-models, namely the *simplex algorithm*, which will be described in Chapter 3. We will also describe the connection between the geometric and the algebraic viewpoints.

2.1 The geometry of feasible regions

In this section we will look at the feasible region of an LO-model from a geometric perspective. As was pointed out in Section 1.1.2, vertices of the feasible region play a crucial role in linear optimization. In this section, we will define formally what is meant by a vertex, and we will prove the *optimal vertex theorem* (Theorem 2.1.5) which states that if a standard LO-model has an optimal solution, then it has an optimal solution that is a vertex of the feasible region.

2.1.1 Hyperplanes and halfspaces

A point on a line, a line in a plane, and a plane in a space are examples of 'hyperplanes' that divide the space in which they lie into two halfspaces. Let $n \geq 1$. A *hyperplane* in $\mathbb{R}^n$ is an $(n-1)$-dimensional set of points $\mathbf{x} = \begin{bmatrix} x_1 & \dots & x_n \end{bmatrix}^\top$ in $\mathbb{R}^n$ satisfying a linear equation of the form $a_1 x_1 + \dots + a_n x_n = b$ where $a_1, \dots, a_n, b$ are real numbers and not all a_i's are equal to zero. Any hyperplane H in $\mathbb{R}^n$ can be written as:

$$
H = \left\{ \begin{bmatrix} x_1 \\ \vdots \\ x_n \end{bmatrix} \in \mathbb{R}^n \;\middle|\; a_1 x_1 + \dots + a_n x_n = b \right\} = \left\{ \mathbf{x} \in \mathbb{R}^n \;\middle|\; \mathbf{a}^\top \mathbf{x} = b \right\},
$$

where $a_1, \dots, a_n, b$ are real numbers, $\mathbf{x} = \begin{bmatrix} x_1 & \dots & x_n \end{bmatrix}^\top$, and $\mathbf{a} = \begin{bmatrix} a_1 & \dots & a_n \end{bmatrix}^\top$ with $\mathbf{a} \neq \mathbf{0}$.

Example 2.1.1.   *The set $\left\{ \begin{bmatrix} x_1 & x_2 \end{bmatrix}^\top \in \mathbb{R}^2 \mid 3x_1 + 2x_2 = 8 \right\}$ is a hyperplane in $\mathbb{R}^2$, namely a line in the plane. The hyperplane $\left\{ \begin{bmatrix} x_1 & x_2 & x_3 \end{bmatrix}^\top \in \mathbb{R}^3 \mid -2x_1 = 3 \right\}$ is parallel to the coordinate plane $\mathbf{0}x_2 x_3$, because the coefficients of x_2 and x_3 are zero, meaning that x_2 and x_3 can take on any arbitrary value.*

Every hyperplane $H = \left\{ \mathbf{x} \in \mathbb{R}^n \mid \mathbf{a}^\top \mathbf{x} = b \right\}$ with $\mathbf{a} \neq \mathbf{0}$ divides $\mathbb{R}^n$ into two *halfspaces*, namely:

$$H^+ = \left\{ \mathbf{x} \in \mathbb{R}^n \mid \mathbf{a}^\top \mathbf{x} \leq b \right\}, \text{ and}$$
$$H^- = \left\{ \mathbf{x} \in \mathbb{R}^n \mid \mathbf{a}^\top \mathbf{x} \geq b \right\}.$$

Note that $H^+ \cap H^- = H$, $H^+ \cup H^- = \mathbb{R}^n$, and H is the boundary (see Appendix D.1) of both H^+ and H^-. Also note that there is some ambiguity in the definition of H^+ and H^-, because the equation $\mathbf{a}^\top \mathbf{x} = b$ is equivalent to $-\mathbf{a}^\top \mathbf{x} = -b$.

The vector $\mathbf{a}$ is called the *normal vector* of the hyperplane H. It is perpendicular to H, meaning that $\mathbf{a}$ is perpendicular to every line in the hyperplane (see Appendix B). So, to prove that $\mathbf{a}$ is indeed perpendicular to H, consider any line L in H. We will prove that $\mathbf{a}$ is perpendicular to L. Take any two distinct points $\mathbf{x}'$ and $\mathbf{x}''$ on the line L. The vector $\mathbf{x}' - \mathbf{x}''$ is the direction of L; see Figure 2.1. Because $\mathbf{x}' \in H$ and $\mathbf{x}'' \in H$, we have that $\mathbf{a}^\top \mathbf{x}' = b$ and $\mathbf{a}^\top \mathbf{x}'' = b$, and hence that $\mathbf{a}^\top (\mathbf{x}' - \mathbf{x}'') = \mathbf{a}^\top \mathbf{x}' - \mathbf{a}^\top \mathbf{x}'' = b - b = 0$. It follows that $\mathbf{a}$ is perpendicular to $\mathbf{x}' - \mathbf{x}''$.

The collection of hyperplanes $H_1, \dots, H_k$ is called *(linearly) independent* if and only if the corresponding collection of normal vectors is linearly independent; it is called *(linearly) dependent* otherwise. (The concepts of 'linear independent' and 'linear dependent' collections of vectors are explained in Appendix B.)

Example 2.1.2.   *The hyperplanes $\left\{ \begin{bmatrix} x_1 & x_2 \end{bmatrix}^\top \in \mathbb{R}^2 \mid x_1 = 0 \right\}$ and $\left\{ \begin{bmatrix} x_1 & x_2 \end{bmatrix}^\top \in \mathbb{R}^2 \mid x_2 = 0 \right\}$ are linearly independent (to be precise: they form a linearly independent collection of hyperplanes), because the normal vectors $\begin{bmatrix} 1 & 0 \end{bmatrix}^\top$ and $\begin{bmatrix} 0 & 1 \end{bmatrix}^\top$ are linearly independent (to be precise: they form*

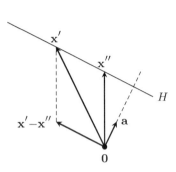

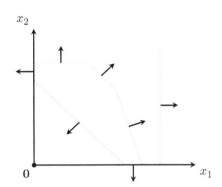

Figure 2.1: The vector $\mathbf{a}$ is a normal of
$$H = \left\{ \mathbf{x} \in \mathbb{R}^n \,\middle|\, \mathbf{a}^\mathsf{T}\mathbf{x} = b \right\}.$$

Figure 2.2: Normal vectors pointing out of the feasible region of Model Dovetail*.

a linearly independent collection of vectors). On the other hand, the hyperplanes $\{[x_1 \ x_2]^\mathsf{T} \in \mathbb{R}^2 \mid x_1 = 0\}$ and $\{[x_1 \ x_2]^\mathsf{T} \in \mathbb{R}^2 \mid 2x_1 = 5\}$ are not linearly independent, because the normal vectors $[1 \ 0]^\mathsf{T}$ and $[2 \ 0]^\mathsf{T}$ are not linearly independent.

Let $m, n \geq 0$, let $\mathbf{A}$ be an (m, n) matrix with entries a_{ij} for $i = 1, \ldots, m$ and $j = 1, \ldots, n$, let $\mathbf{b} = [b_1 \ \ldots \ b_m]^\mathsf{T}$. Let $F = \{\mathbf{x} \in \mathbb{R}^n \mid \mathbf{A}\mathbf{x} \leq \mathbf{b}, \mathbf{x} \geq \mathbf{0}\}$ be the feasible region of the LO-model $\max\{\mathbf{c}^\mathsf{T}\mathbf{x} \mid \mathbf{A}\mathbf{x} \leq \mathbf{b}, \mathbf{x} \geq \mathbf{0}\}$. The feasible region F is the intersection of $m + n$ halfspaces, as follows. Each point $\mathbf{x} = [x_1 \ \ldots \ x_n]^\mathsf{T} \in F$ has to satisfy, simultaneously, the following constraints:

$$
\begin{aligned}
a_{11}x_1 + \ldots + a_{1n}x_n &\leq b_1 \\
&\vdots \\
a_{m1}x_1 + \ldots + a_{mn}x_n &\leq b_m \\
-x_1 &\leq 0 \\
&\ddots \\
& \\
-x_n &\leq 0.
\end{aligned}
$$

Defining

$$H_i^+ = \left\{ \begin{bmatrix} x_1 \\ \vdots \\ x_n \end{bmatrix} \in \mathbb{R}^n \,\middle|\, a_{i1}x_1 + \ldots + a_{in}x_n \leq b_i \right\} \qquad \text{for } i = 1, 2, \ldots, m,$$

$$\text{and } H_{m+j}^+ = \left\{ \begin{bmatrix} x_1 \\ \vdots \\ x_n \end{bmatrix} \in \mathbb{R}^n \,\middle|\, -x_j \leq 0 \right\} \qquad \text{for } j = 1, 2, \ldots, n,$$

it follows that $F = H_1^+ \cap \cdots \cap H_{m+n}^+$. The normal vectors of the corresponding hyperplanes $H_1, \ldots, H_{m+n}$ are, respectively:

$$\mathbf{a}_1 = \begin{bmatrix} a_{11} \\ \vdots \\ a_{1n} \end{bmatrix}, \ldots, \mathbf{a}_m = \begin{bmatrix} a_{m1} \\ \vdots \\ a_{mn} \end{bmatrix}, \mathbf{a}_{m+1} = \begin{bmatrix} -1 \\ 0 \\ \vdots \\ 0 \end{bmatrix}, \ldots, \mathbf{a}_{m+n} = \begin{bmatrix} 0 \\ \vdots \\ 0 \\ -1 \end{bmatrix},$$

and they are precisely the rows of $\begin{bmatrix} \mathbf{A} \\ \mathbf{I}_n \end{bmatrix}$. Notice that any halfspace is convex. Hence, F is the intersection of $m + n$ convex sets, namely $H_1^+, \ldots, H_{m+n}^+$, and therefore F itself is convex (see Appendix D). Observe also that, for $i = 1, \ldots, m$, the hyperplane H_i corresponding to the halfspace H_i^+ consists of all points in $\mathbb{R}^n$ for which the i'th technology constraint holds with equality. Similarly, for $j = 1, \ldots, n$, the hyperplane H_{m+j} corresponding to the halfspace H_{m+j}^+ consists of all points in $\mathbb{R}^n$ for which the j'th nonnegativity constraint $-x_j \leq 0$ holds with equality, i.e., $H_{m+j} = \left\{ \begin{bmatrix} x_1 & \cdots & x_n \end{bmatrix}^\top \mid x_j = 0 \right\}$.

Example 2.1.3. *In the case of Model Dovetail* (Section 1.2.1), the halfspaces are:*

$$H_1^+ = \left\{ \begin{bmatrix} x_1 & x_2 \end{bmatrix}^\top \mid x_1 + x_2 \leq 9 \right\}, H_2^+ = \left\{ \begin{bmatrix} x_1 & x_2 \end{bmatrix}^\top \mid 3x_1 + x_2 \leq 18 \right\},$$
$$H_3^+ = \left\{ \begin{bmatrix} x_1 & x_2 \end{bmatrix}^\top \mid x_1 \leq 7 \right\}, H_4^+ = \left\{ \begin{bmatrix} x_1 & x_2 \end{bmatrix}^\top \mid x_2 \leq 6 \right\},$$
$$H_5^+ = \left\{ \begin{bmatrix} x_1 & x_2 \end{bmatrix}^\top \mid -x_1 - x_2 \leq -5 \right\}, H_6^+ = \left\{ \begin{bmatrix} x_1 & x_2 \end{bmatrix}^\top \mid -x_1 \leq 0 \right\},$$
$$H_7^+ = \left\{ \begin{bmatrix} x_1 & x_2 \end{bmatrix}^\top \mid -x_2 \leq 0 \right\}.$$

and $F = H_1^+ \cap H_2^+ \cap H_3^+ \cap H_4^+ \cap H_5^+ \cap H_6^+ \cap H_7^+$. The normal vectors are:

$$\begin{bmatrix} 1 \\ 1 \end{bmatrix}, \begin{bmatrix} 3 \\ 1 \end{bmatrix}, \begin{bmatrix} 1 \\ 0 \end{bmatrix}, \begin{bmatrix} 0 \\ 1 \end{bmatrix}, \begin{bmatrix} -1 \\ -1 \end{bmatrix}, \begin{bmatrix} -1 \\ 0 \end{bmatrix}, \begin{bmatrix} 0 \\ -1 \end{bmatrix}.$$

If we draw each of the normal vectors in the previous example as an arrow from some point on the corresponding hyperplane, then each vector 'points out of the feasible region F'; see Figure 2.2. Formally, we say that, for $l = 1, \ldots, m + n$, the vector $\mathbf{a}_l$ 'points out of the feasible region F' if, for every point $\mathbf{x}^0 \in H_l$, it holds that $\left\{ \mathbf{x}^0 + \lambda \mathbf{a}_l \mid \lambda > 0 \right\} \subset H_l^- \backslash H_l$. To see that, for each $l = 1, \ldots, m+n$, the vector $\mathbf{a}_l$ indeed points out of the feasible region F, let $\lambda > 0$. We have that $\mathbf{a}_l^\top (\mathbf{x}^0 + \lambda \mathbf{a}_l) = \mathbf{a}_l^\top \mathbf{x}^0 + \lambda \mathbf{a}_l^\top \mathbf{a}_l = b_l + \lambda \|\mathbf{a}_l\|^2 > b_l$, where we have used the fact that $\|\mathbf{a}_l\| > 0$, which in turn follows from the fact that $\mathbf{a}_l \neq \mathbf{0}$ (see Appendix B). Thus, $\mathbf{x}^0 + \lambda \mathbf{a}_l \in H_l^-$ and $\mathbf{x}^0 + \lambda \mathbf{a}_l \notin H_l^+$, and hence $\mathbf{a}_l$ points out of the feasible region. Notice that, since $F \subset H_l^+$, it follows that F and $\mathbf{x}^0 + \lambda \mathbf{a}_l$ are on different sides of H_l for all $\lambda > 0$.

2.1.2 Vertices and extreme directions of the feasible region

Consider the standard LO-model:

$$\max\{\mathbf{c}^\mathsf{T}\mathbf{x} \mid \mathbf{A}\mathbf{x} \le \mathbf{b}, \mathbf{x} \ge \mathbf{0}\}.$$

We will first define the concept 'vertex' more precisely. Let F be the feasible region of the LO-model and let $H_1, \ldots, H_{m+n}$ be the hyperplanes corresponding to its constraints. There are three equivalent ways of defining a vertex (or *extreme point*):

> The vector $\mathbf{x}^0 \in F$ is called a *vertex* of F if and only if there are n independent hyperplanes in the collection $\{H_1, \ldots, H_{m+n}\}$ that intersect at $\mathbf{x}^0$.

> The vector $\mathbf{x}^0 \in F$ is called a *vertex* of F if and only if there is a hyperplane H (not necessarily one of $H_1, \ldots, H_{m+n}$) with corresponding halfspace H^+ such that $F \subseteq H^+$ and $F \cap H = \{\mathbf{x}^0\}$.

> The vector $\mathbf{x}^0 \in F$ is called a *vertex* of F if and only if there are no two distinct $\mathbf{x}', \mathbf{x}'' \in F$ such that $\mathbf{x}^0 = \lambda\mathbf{x}' + (1 - \lambda)\mathbf{x}''$ for some $\lambda \in (0, 1)$.

The first definition is in terms of hyperplanes corresponding to F. The second definition is in terms of a halfspace that contains F. Note that if there is a hyperplane $H = \{\mathbf{x} \in \mathbb{R}^n \mid \mathbf{a}^\mathsf{T}\mathbf{x} = b\}$ with $F \subseteq H^+$, then there is also 'another' hyperplane $\tilde{H}$ with $F \subseteq \tilde{H}^-$, namely $\tilde{H} = \{\mathbf{x} \in \mathbb{R}^n \mid (-\mathbf{a})^\mathsf{T}\mathbf{x} = (-b)\}$. The third definition states that a point is an extreme point of F if and only if it cannot be written as a convex combination of two other points of F. In Appendix D it is shown that the three definitions are equivalent.

Example 2.1.4. *Consider again the feasible region F of Model Dovetail. In Figure 2.3, this is the region $\mathbf{0}\mathbf{v}_1\mathbf{v}_2\mathbf{v}_3\mathbf{v}_4$. The halfspaces that determine F are:*

$$H_1^+ = \left\{ \begin{bmatrix} x_1 & x_2 \end{bmatrix}^\mathsf{T} \mid x_1 + x_2 \le 9 \right\}, H_2^+ = \left\{ \begin{bmatrix} x_1 & x_2 \end{bmatrix}^\mathsf{T} \mid 3x_1 + x_2 \le 18 \right\},$$
$$H_3^+ = \left\{ \begin{bmatrix} x_1 & x_2 \end{bmatrix}^\mathsf{T} \mid x_1 \le 7 \right\}, H_4^+ = \left\{ \begin{bmatrix} x_1 & x_2 \end{bmatrix}^\mathsf{T} \mid x_2 \le 6 \right\},$$
$$H_5^+ = \left\{ \begin{bmatrix} x_1 & x_2 \end{bmatrix}^\mathsf{T} \mid -x_1 \le 0 \right\}, H_6^+ = \left\{ \begin{bmatrix} x_1 & x_2 \end{bmatrix}^\mathsf{T} \mid -x_2 \le 0 \right\}.$$

Note that $H_1^+, H_2^+, H_3^+, H_4^+$ correspond to the technology constraints of the model, and H_5^+ and H_6^+ to the nonnegativity constraints. The corresponding hyperplanes are:

$$H_1 = \left\{ \begin{bmatrix} x_1 & x_2 \end{bmatrix}^\mathsf{T} \mid x_1 + x_2 = 9 \right\}, H_2 = \left\{ \begin{bmatrix} x_1 & x_2 \end{bmatrix}^\mathsf{T} \mid 3x_1 + x_2 = 18 \right\},$$
$$H_3 = \left\{ \begin{bmatrix} x_1 & x_2 \end{bmatrix}^\mathsf{T} \mid x_1 = 7 \right\}, H_4 = \left\{ \begin{bmatrix} x_1 & x_2 \end{bmatrix}^\mathsf{T} \mid x_2 = 6 \right\},$$
$$H_5 = \left\{ \begin{bmatrix} x_1 & x_2 \end{bmatrix}^\mathsf{T} \mid -x_1 = 0 \right\}, H_6 = \left\{ \begin{bmatrix} x_1 & x_2 \end{bmatrix}^\mathsf{T} \mid -x_2 = 0 \right\}.$$

To illustrate the first definition of the concept of a vertex, $\mathbf{v}_1$ is the unique point in the intersection of H_2 and H_6, vertex $\mathbf{v}_3$ is the unique point in the intersection of H_1 and H_4, and $\mathbf{0}$ is the unique point in the intersection of H_5 and H_6. Therefore, the first definition of a vertex implies that $\mathbf{v}_1$, $\mathbf{v}_3$, and $\mathbf{0}$ are vertices of F (and, similarly, $\mathbf{v}_2$ and $\mathbf{v}_4$ are vertices of F).

To illustrate the second definition, consider $\mathbf{v}_3$. The dotted line in the figure shows the hyperplane $H = \left\{ \begin{bmatrix} x_1 & x_2 \end{bmatrix}^\mathsf{T} \mid x_1 + 3x_2 = 21 \right\}$. Graphically, it is clear that $\mathbf{v}_3$ is the unique point in the

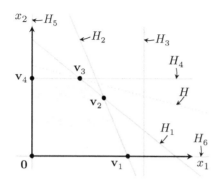

Figure 2.3: Vertices of the feasible region of Model Dovetail.

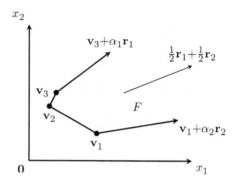

Figure 2.4: Extreme points and directions of F.

intersection of H and F, and the corresponding halfspace $H^+ = \left\{ \begin{bmatrix} x_1 & x_2 \end{bmatrix}^\top \mid x_1 + 3x_2 \leq 21 \right\}$ satisfies $F \subset H^+$. Therefore, the second definition implies that $\mathbf{v}_3$ is a vertex. Notice that H is not among the hyperplanes $H_1, \ldots, H_6$. In fact, for $i = 1, \ldots, 6$, the intersection $F \cap H_i$ does not consist of a single point (e.g., $H_1 \cap F$ contains the line segment $\mathbf{v}_2 \mathbf{v}_3$, and $H_3 \cap F$ contains no point at all). Therefore, we could not even have chosen H to be one of the $H_1, \ldots, H_6$. Constructing similar hyperplanes for the other vertices is left to the reader.

Finally, to illustrate the third definition, we observe that none of the points $\mathbf{0}, \mathbf{v}_1, \ldots, \mathbf{v}_4$ can be written as a convex combination of other points in F. For example, $\mathbf{v}_1$ cannot be written as a convex combination of some other two points in F. This is because, if it could be written as a convex combination of two other points, $\mathbf{x}, \mathbf{x}' \in F$, say, then $\mathbf{v}_1$ would lie on the line segment connecting $\mathbf{x}$ and $\mathbf{x}'$, which is clearly impossible. It is left to the reader to convince oneself that every point in F except $\mathbf{0}, \mathbf{v}_1, \ldots, \mathbf{v}_4$ can be written as a convex combination of two other points in F.

In the case of Model Dovetail, the vertices describe the full feasible region in the sense that the feasible region is the convex hull (see Appendix D) of these vertices; see Figure 2.3. This is, however, not true in the case of an unbounded feasible region. Consider for example the feasible region drawn in Figure 2.4. This feasible region is unbounded and can therefore not be written as the convex hull of (finitely many) vertices (see Exercise 2.3.9). To allow for the possibility of an unbounded feasible region in an LO-model, we need another concept to describe the feasible region of an arbitrary LO-model. We say that a vector $\mathbf{r} \in \mathbb{R}^n$ is a *direction of unboundedness* (or, *direction*) of the feasible region F if $\mathbf{r} \neq \mathbf{0}$ and there exists a point $\mathbf{x} \in F$ such that $\mathbf{x} + \alpha \mathbf{r} \in F$ for all $\alpha \geq 0$. For example, in Figure 2.4, the vectors $\mathbf{r}_1$ and $\mathbf{r}_2$ are directions of the feasible region. Notice that if F is bounded, then it has no directions of unboundedness; see Exercise 2.3.4. Observe that if $\mathbf{r}$ is a direction of the feasible region, then so is $\alpha \mathbf{r}$ with $\alpha > 0$. Because the vectors $\mathbf{r}$ and $\alpha \mathbf{r}$ (with $\alpha > 0$) point in the same direction, they are essentially the same, and we think of them as the same direction.

Informally, the definition of a direction of F states that $\mathbf{r}$ is a direction if there exists some point $\mathbf{x} \in F$ so that, if we move away from $\mathbf{x}$ in the direction $\mathbf{r}$, we stay inside F. The following theorem states that for a direction $\mathbf{r}$ it holds that, if we move away from *any* point in the direction $\mathbf{r}$, we stay inside F.

Theorem 2.1.1.
Let $F \subset \mathbb{R}^n$ be the feasible region of a standard LO-model. Then, a vector $\mathbf{r} \in \mathbb{R}^n$ is a direction of F if and only if for every $\mathbf{x} \in F$ it holds that $\mathbf{x} + \alpha \mathbf{r} \in F$ for all $\alpha \geq 0$.

Proof. Write $F = \left\{ \mathbf{x} \in \mathbb{R}^n \ \middle| \ \mathbf{a}_i^\mathsf{T} \mathbf{x} \leq b_i, i = 1, \ldots, m+n \right\}$. Note that for $i = m+1, \ldots, m+n$, $\mathbf{a}_i$ is a unit vector and $b_i = 0$. The 'if' part of the statement is trivial. For the 'only if' part, let $\mathbf{r}$ be a direction of F. We will first prove that $\mathbf{a}_i^\mathsf{T} \mathbf{r} \leq 0$ for $i = 1, \ldots, m+n$. Suppose for a contradiction that $\mathbf{a}_i^\mathsf{T} \mathbf{r} > 0$ for some $i \in \{1, \ldots, m+n\}$. By the definition of a direction, there exists a point $\hat{\mathbf{x}} \in F$ such that $\hat{\mathbf{x}} + \alpha \mathbf{r} \in F$ for all $\alpha \geq 0$. Because $\mathbf{a}_i^\mathsf{T} \mathbf{r} > 0$, we may choose α large enough so that $\mathbf{a}_i^\mathsf{T} (\hat{\mathbf{x}} + \alpha \mathbf{r}) = \mathbf{a}_i^\mathsf{T} \hat{\mathbf{x}} + \alpha \mathbf{a}_i^\mathsf{T} \mathbf{r} > b_i$. But this implies that $\hat{\mathbf{x}} + \alpha \mathbf{r} \notin F$, which is a contradiction.

Hence, $\mathbf{a}_i^\mathsf{T} \mathbf{r} \leq 0$ for $i = 1, \ldots, m+n$. Now let $\mathbf{x}$ be any point of F and let $\alpha \geq 0$. We have that:
$$\mathbf{a}_i^\mathsf{T} (\mathbf{x} + \alpha \mathbf{r}) = \mathbf{a}_i^\mathsf{T} \mathbf{x} + \alpha \mathbf{a}_i^\mathsf{T} \mathbf{r} \leq \mathbf{a}_i^\mathsf{T} \mathbf{x} \leq b_i,$$
for $i = 1, \ldots, m+n$. Hence, $\mathbf{x} + \alpha \mathbf{r} \in F$, as required. $\qquad \square$

If $\mathbf{r}_1$ and $\mathbf{r}_2$ are directions of the feasible region, then so is any convex combination of them. To see this, let $\lambda \in [0,1]$. We will show that $\mathbf{r} = \lambda \mathbf{r}_1 + (1-\lambda) \mathbf{r}_2$ is a direction of the feasible region as well. By the definition of a direction, there exist points $\mathbf{x}_1, \mathbf{x}_2 \in F$ such that $\mathbf{x}_1 + \alpha \mathbf{r}_1 \in F$ and $\mathbf{x}_2 + \alpha \mathbf{r}_2 \in F$ for every $\alpha \geq 0$. Now let $\alpha \geq 0$ and let $\mathbf{x} = \lambda \mathbf{x}_1 + (1-\lambda) \mathbf{x}_2$. It follows that:
$$\mathbf{x} + \alpha \mathbf{r} = \lambda \mathbf{x}_1 + (1-\lambda) \mathbf{x}_2 + \alpha(\lambda \mathbf{r}_1 + (1-\lambda)\mathbf{r}_2) = \lambda(\mathbf{x}_1 + \alpha \mathbf{r}_1) + (1-\lambda)(\mathbf{x}_2 + \alpha \mathbf{r}_2).$$

Therefore, $\mathbf{x} + \alpha \mathbf{r}$ is a convex combination of two points in F, namely $\mathbf{x}_1 + \alpha \mathbf{r}_1$ and $\mathbf{x}_2 + \alpha \mathbf{r}_2$. Hence, because F is convex, $\mathbf{x} + \alpha \mathbf{r} \in F$. Since this holds for all $\alpha \geq 0$, it follows that $\mathbf{r}$ is a direction of F.

Since we know that the set of directions of F is a convex set, it makes sense to define extreme directions: a direction $\mathbf{r}$ of the feasible region F of an LO-model is called an *extreme direction* if $\mathbf{r}$ cannot be written as a convex combination of two other directions of F. For example, in Figure 2.4, the extreme directions of the feasible region are all vectors of the form either $\alpha \mathbf{r}_1$ or $\alpha \mathbf{r}_2$ with $\alpha > 0$; the vector $\frac{1}{2}\mathbf{r}_1 + \frac{1}{2}\mathbf{r}_2$, also in the figure, is a direction but not an extreme direction.

The following theorem states that any feasible region can be described in terms of its vertices and extreme directions. Note that by 'the extreme directions' of the feasible region F, we

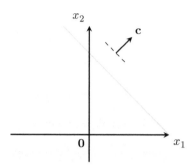

Figure 2.5: The feasible region of a nonstandard LO-model without vertices.

really mean all extreme directions up to scaling: if $\mathbf{r}$ is an extreme direction of F, then $\alpha\mathbf{r}$ is listed for exactly one value of $\alpha > 0$. So, the extreme directions of the feasible region in Figure 2.4 are $\mathbf{r}_1$ and $\mathbf{r}_2$.

> **Theorem 2.1.2.**
> Let $F \subset \mathbb{R}^n$ be the feasible region of a standard LO-model. Let $\mathbf{v}_1, \ldots, \mathbf{v}_k$ be the vertices of F and let $\mathbf{r}_1, \ldots, \mathbf{r}_l$ be its extreme directions. Let $\mathbf{x} \in \mathbb{R}^n$. Then $\mathbf{x} \in F$ if and only if $\mathbf{x}$ can be written as:
>
> $$\mathbf{x} = \lambda_1\mathbf{v}_1 + \ldots + \lambda_k\mathbf{v}_k + \mu_1\mathbf{r}_1 + \ldots + \mu_l\mathbf{r}_l,$$
>
> with $\lambda_1, \ldots, \lambda_k, \mu_1, \ldots, \mu_l \geq 0$ and $\lambda_1 + \ldots + \lambda_k = 1$.

The requirement that the LO-model is in standard form is necessary. Figure 2.5 shows the unbounded feasible region of the LO-model $\max\{x_1 + x_2 \mid x_1 + x_2 \leq 5\}$. This LO-model is not in standard form. The extreme directions of the feasible region are $\mathbf{r}_1 = \begin{bmatrix} 1 & -1 \end{bmatrix}^\mathsf{T}$ and $\mathbf{r}_2 = \begin{bmatrix} -1 & 1 \end{bmatrix}^\mathsf{T}$, but the feasible region has no vertex. Clearly, no point on the line $\{x_1 + x_2 \mid x_1 + x_2 = 5\}$ can be written as a linear combination of $\mathbf{r}_1$ and $\mathbf{r}_2$.

In Section 2.1.3, we will prove Theorem 2.1.2 for the case when F is bounded. In that case, F has no extreme directions, and hence the theorem states that a point $\mathbf{x} \in \mathbb{R}^n$ is a point of F if and only if $\mathbf{x}$ can be written as a convex combination of the vertices of F. In other words, if F is bounded, then it is the convex hull of its vertices. The interested reader is referred to, e.g., Bertsimas and Tsitsiklis (1997) for a proof for the general case.

2.1.3   Faces of the feasible region

Consider the region:

$$F = \big\{\mathbf{x} \in \mathbb{R}^n \mid \mathbf{a}_i^\mathsf{T}\mathbf{x} \leq b_i, \ i = 1, \ldots, m+n\big\}.$$

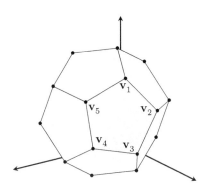

Figure 2.6: Three-dimensional feasible region.

For any set $J \subseteq \{1, \ldots, m+n\}$, let F_J be the subset of the region F in which all '$\leq$' constraints with indices in J are replaced by '$=$' constraints. That is, define:

$$F_J = \left\{ \mathbf{x} \in \mathbb{R}^n \mid \mathbf{a}_j^\mathsf{T} \mathbf{x} = b_j \text{ for } j \in J, \ \mathbf{a}_i^\mathsf{T} \mathbf{x} \leq b_i \text{ for } i \in \bar{J} \right\},$$

where $\bar{J} = \{1, \ldots, m+n\} \setminus J$ denotes the complement of J. If $F_J \neq \emptyset$, then F_J is called a *face* of F. Clearly, we have that $F = F_\emptyset$, and hence F is a face of itself (provided that $F \neq \emptyset$). An equivalent way of expressing F_J is:

$$F_J = F \cap \left(\bigcap_{j \in J} H_j \right),$$

where $H_1, \ldots, H_{m+n}$ denote the hyperplanes of which $\mathbf{a}_1, \ldots, \mathbf{a}_{m+n}$ are the normal vectors. The reader is asked to prove this in Exercise 2.3.10. From this expression, it follows immediately that if $J' \subseteq J$, then $F_J \subseteq F_{J'}$.

Example 2.1.5. *Consider again Figure 2.3. Let $J_1 = \{1\}$. F_{J_1} is the set of all points in F for which the inequality (1.2) of Model Dovetail holds with equality. That is, $F_{J_1} = F \cap H_1$. Graphically, it should be clear that the face F_{J_1} is exactly the line segment $\mathbf{v}_2\mathbf{v}_3$. Similarly, let $J_2 = \{2\}$. The face F_{J_2} is exactly the line segment between $\mathbf{v}_1$ and $\mathbf{v}_2$. Next, let $J_3 = \{1, 2\}$. Then $F_{J_3} = F \cap H_1 \cap H_2$, which means that the face F_{J_3} consists of exactly one point, namely the point $\mathbf{v}_2$ in $H_1 \cap H_2$. Note that $\mathbf{v}_2$ is a vertex. Finally, let $J_4 = \{1, 2, 3\}$. Then, $F_{J_4} = \emptyset$, and hence F_{J_4} is not a face of F. The reader may check that the faces of F are: F itself, the line segments $\mathbf{0}\mathbf{v}_1$, $\mathbf{v}_1\mathbf{v}_2$, $\mathbf{v}_2\mathbf{v}_3$, $\mathbf{v}_3\mathbf{v}_4$, $\mathbf{v}_4\mathbf{0}$, and the points $\mathbf{0}$, $\mathbf{v}_1$, $\mathbf{v}_2$, $\mathbf{v}_3$, and $\mathbf{v}_4$. Notice that vertices of a face are also vertices of F; see Theorem D.3.3.*

Example 2.1.6. *Consider the feasible region F in $\mathbb{R}^3$ drawn in Figure 2.6. The faces of F are: F itself, the twelve pentagon-shaped 'sides' (e.g., the pentagon $\mathbf{v}_1\mathbf{v}_2\mathbf{v}_3\mathbf{v}_4\mathbf{v}_5\mathbf{v}_1$), the thirty 'edges' (e.g., the line segments $\mathbf{v}_1\mathbf{v}_2$ and $\mathbf{v}_2\mathbf{v}_3$), and the twenty vertices.*

The above examples suggest that if F_J is a singleton, then F_J is a vertex. The reader is asked to prove this fact in Exercise 2.3.11.

Note that faces may in turn have faces themselves. As it turns out, these 'faces of faces' of F are faces of F too. For example, in Figure 2.6, the vertices $\mathbf{v}_1, \ldots, \mathbf{v}_5$, and the line segments $\mathbf{v}_1\mathbf{v}_2, \ldots, \mathbf{v}_4\mathbf{v}_5, \mathbf{v}_5\mathbf{v}_1$ are all faces of both F and the pentagonal face $\mathbf{v}_1\mathbf{v}_2\mathbf{v}_3\mathbf{v}_4\mathbf{v}_5$. This fact is proved in general in Theorem D.3.3 in Appendix D.

2.1.4 The optimal vertex theorem

We observed in Section 1.1.2 that the unique optimal solution of Model Dovetail is a vertex of its feasible region. In Theorem 2.1.5, we will show that this holds in general, i.e., whenever a standard LO-model has an optimal solution, it has an optimal solution that corresponds to a vertex of its feasible region. As is the case in the statement of Theorem 2.1.2, the qualification 'standard' is necessary. Indeed, Figure 2.5 shows an example of a nonstandard LO-model which has an optimal solution, but no optimal solution that is a vertex.

Theorem 2.1.3 states that, in general, if a (standard or nonstandard) LO-model has an optimal solution, then it holds that the set of all optimal solutions is a face of the feasible region. In the case of Figure 2.6, this should be intuitively clear. For the LO-model depicted in Figure 2.5, the set of optimal points is $\{\mathbf{x} \in \mathbb{R}^2 \mid x_1 + x_2 = 5\}$, which is indeed a face of the feasible region.

Theorem 2.1.3.
Consider any LO-model with feasible region

$$F = \{\mathbf{x} \in \mathbb{R}^n \mid \mathbf{a}_i^\mathsf{T}\mathbf{x} \le b_i, i = 1, \ldots, m+n\},$$

and objective $\max \mathbf{c}^\mathsf{T}\mathbf{x}$. Suppose that the LO-model has an optimal solution. Then, the set of optimal solutions is a face of the feasible region of the model.

Proof. Let $Z \subseteq F$ be the set of optimal solutions of the model, and note that by assumption $Z \ne \emptyset$. Let F_J be a face that contains Z and let J be as large as possible with that property. Note that $Z \subseteq F_J$. So, in order to prove that $Z = F_J$, it suffices to prove that $F_J \subseteq Z$.

Take any $i \in \bar{J}$. We claim that there exists a point $\hat{\mathbf{x}}^i \in Z$ such that $\mathbf{a}_i^\mathsf{T}\hat{\mathbf{x}}^i < b_i$. Assume to the contrary that $\mathbf{a}_i^\mathsf{T}\mathbf{x} = b_i$ for every $\mathbf{x} \in Z$. Then, $Z \subseteq F_{J \cup \{i\}}$, and hence we could have chosen $J \cup \{i\}$ instead of J. This contradicts the fact that J is largest.

For each $i \in \bar{J}$, let $\hat{\mathbf{x}}^i \in Z$ be such that $\mathbf{a}_i^\mathsf{T}\hat{\mathbf{x}}^i < b_i$, and define $\mathbf{x}^* = \frac{1}{|\bar{J}|}\sum_{i \in \bar{J}}\hat{\mathbf{x}}^i$. Because Z is convex, we have that $\mathbf{x}^* \in Z$ (why?). Moreover, it is straightforward to check that $\mathbf{a}_i^\mathsf{T}\mathbf{x}^* < b_i$ for each $i \in \bar{J}$, i.e., $\mathbf{x}^*$ lies in the relative interior (see Appendix D) of F_J.

Let $\hat{\mathbf{x}}$ be any point in F_J and let $\mathbf{d} = \hat{\mathbf{x}} - \mathbf{x}^*$. Since, for small enough $\varepsilon > 0$, we have that $\mathbf{x}^* + \varepsilon\mathbf{d} \in F_J \subseteq F$ and $\mathbf{x}^* - \varepsilon\mathbf{d} \in F_J \subseteq F$, we also have that

$$\mathbf{c}^\mathsf{T}(\mathbf{x}^* + \varepsilon\mathbf{d}) \le \mathbf{c}^\mathsf{T}\mathbf{x}^* \text{ and } \mathbf{c}^\mathsf{T}(\mathbf{x}^* - \varepsilon\mathbf{d}) \le \mathbf{c}^\mathsf{T}\mathbf{x}^*.$$

Hence, we have that $\mathbf{c}^\mathsf{T}\mathbf{d} = 0$, and so:

$$\mathbf{c}^\mathsf{T}\hat{\mathbf{x}} = \mathbf{c}^\mathsf{T}\mathbf{x}^* + \mathbf{c}^\mathsf{T}(\hat{\mathbf{x}} - \mathbf{x}^*) = \mathbf{c}^\mathsf{T}\mathbf{x}^* + \mathbf{c}^\mathsf{T}\mathbf{d} = \mathbf{c}^\mathsf{T}\mathbf{x}^*.$$

Therefore, $\hat{\mathbf{x}}$ is an optimal solution. Since $\hat{\mathbf{x}} \in F_J$ was chosen arbitrarily, this implies that $F_J \subseteq Z$. $\qquad\square$

We have seen in Figure 2.5 that the set of optimal solutions may be a face that contains no vertex. In that case, there is no optimal vertex. This situation, however, cannot happen in the case of a standard LO-model. Informally, this can be seen as follows. The feasible region of a standard LO-model can be iteratively constructed by starting with the set $\{\mathbf{x} \in \mathbb{R}^n \mid \mathbf{x} \geq \mathbf{0}\}$ and intersecting it by one halfspace at a time. The crucial observations are that the initial region $\{\mathbf{x} \in \mathbb{R}^n \mid \mathbf{x} \geq \mathbf{0}\}$ has the property that every face of it contains a vertex and that, no matter how we intersect it with halfspaces, it is not possible to destroy this property.

So, in order to find a vertex of the feasible region of a standard LO-model, we should consider any face of the feasible region, and then take a smallest face contained in the former face. The latter face should be a vertex. The following theorem proves that this is in fact the case.

Theorem 2.1.4.
The feasible region of any standard LO-model has at least one vertex.

Proof. Let F_J be a face of F with J as large as possible. Let I ($\subseteq \bar{J}$) be as small as possible such that the following equation holds:

$$F_J(I) = \left\{\mathbf{x} \in \mathbb{R}^n \;\middle|\; \mathbf{a}_j^\mathsf{T}\mathbf{x} = b_j \text{ for } j \in J, \;\; \mathbf{a}_i^\mathsf{T}\mathbf{x} \leq b_i \text{ for } i \in I\right\}.$$

Suppose for a contradiction that $I \neq \emptyset$. Let $i \in I$. If $F_J(I \setminus \{i\}) = F_J(I)$, then we should have chosen $I \setminus \{i\}$ instead of I. Hence, since $F_J(I) \subseteq F_J(I \setminus \{i\})$, it follows that there exists $\mathbf{x}^1 \in F_J(I \setminus \{i\})$ but $\mathbf{x}^1 \notin F_J(I)$. So, $\mathbf{x}^1$ satisfies $\mathbf{a}_i^\mathsf{T}\mathbf{x}^1 > b_i$.

Let $\mathbf{x}^2$ be any point in $F_J(I)$. It satisfies $\mathbf{a}_i^\mathsf{T}\mathbf{x}^2 \leq b_i$. Since $\mathbf{x}^1$ and $\mathbf{x}^2$ are both in $F_J(I \setminus \{i\})$, it follows from the convexity of this set that the line segment $\mathbf{x}^1\mathbf{x}^2$ lies in $F_J(I \setminus \{i\})$, and hence there exists a point $\hat{\mathbf{x}} \in F_J(I)$ on the line segment $\mathbf{x}^1\mathbf{x}^2$ that satisfies $\mathbf{a}_i^\mathsf{T}\hat{\mathbf{x}} = b_i$. But this means that the set $F_{J \cup \{i\}}$ is nonempty, contrary to the choice of J.

So, we have proved that in fact $F_J = \left\{\mathbf{x} \in \mathbb{R}^n \;\middle|\; \mathbf{a}_j^\mathsf{T}\mathbf{x} = b_j \text{ for } j \in J\right\}$. Hence, F_J is the intersection of finitely many hyperplanes. If F_J is not a singleton, then it contains a line, and hence F contains a line. But this contradicts the fact that $\mathbf{x} \geq \mathbf{0}$. This proves the theorem. $\quad\square$

With these tools at hand, we can prove the optimal vertex theorem.

Theorem 2.1.5. (*Optimal vertex theorem*)
A standard LO-model has an optimal solution if and only if its feasible region contains an optimal vertex.

Proof. Let Z be the set of optimal solutions of the model $\max\left\{\mathbf{c}^\mathsf{T}\mathbf{x} \mid \mathbf{A}\mathbf{x} \leq \mathbf{b}, \mathbf{x} \geq \mathbf{0}\right\}$, i.e.,

$$Z = \left\{\mathbf{x} \in \mathbb{R}^n \mid \mathbf{A}\mathbf{x} \leq \mathbf{b}, \mathbf{x} \geq \mathbf{0}, \mathbf{c}^\mathsf{T}\mathbf{x} = z^*\right\},$$

where z^* is the optimal objective value. The set Z is itself the feasible region of the standard LO-model

$$\max\left\{\mathbf{c}^\mathsf{T}\mathbf{x} \mid \mathbf{A}\mathbf{x} \leq \mathbf{b}, \mathbf{c}^\mathsf{T}\mathbf{x} \leq z^*, -\mathbf{c}^\mathsf{T}\mathbf{x} \leq -z^*, \mathbf{x} \geq \mathbf{0}\right\}.$$

Hence, by Theorem 2.1.4, Z contains a vertex $\mathbf{x}^*$, which is an optimal solution of the original LO-model $\max\left\{\mathbf{c}^\mathsf{T}\mathbf{x} \mid \mathbf{A}\mathbf{x} = \mathbf{b}, \mathbf{x} \geq \mathbf{0}\right\}$. By Theorem 2.1.3, Z is a face of the feasible region F. Since $\mathbf{x}^*$ is a vertex of a face of F, $\mathbf{x}^*$ is also a vertex of F. Hence, $\mathbf{x}^*$ is an optimal vertex.

The converse assertion of the theorem is obvious. $\qquad\qquad\square$

2.1.5 Polyhedra and polytopes

Consider again the feasible region F of an LO-model. Recall that F is the intersection of halfspaces $H_1, \ldots, H_{m+n}$; see Section 2.1.1. If, in addition, F is bounded, then F has no directions of unboundedness, and hence Theorem 2.1.2 shows that any point in F can be written as a convex combination of the vertices of F. In other words, if F is bounded, then it is the *convex hull* of its finitely many vertices. This means that any bounded feasible region can be viewed in two complementary ways: it is the intersection of a collection of hyperplanes, while at the same time it is the convex hull of a set of points.

For example, in the case of Figure 2.6, it should be clear that the feasible region depicted in the figure can be seen as the intersection of twelve hyperplanes (one for each face), but also as the convex hull of its twenty vertices.

This observation leads to the following two definitions. A set $F \subseteq \mathbb{R}^n$ is called a *polyhedron* (plural: *polyhedra*) if it is the intersection of finitely many hyperplanes. A set $F \subseteq \mathbb{R}^n$ is called a *polytope* if it is the convex hull of finitely many points. In this language, the feasible region in Figure 2.6 is both a polyhedron and a polytope.

An important theorem in the theory of polyhedra and polytopes is the so-called *Weyl-Minkowski theorem*[1], which asserts that bounded polyhedra and polytopes are in fact the same objects.

Theorem 2.1.6. (*Weyl-Minkowski theorem*)
A set $F \subseteq \mathbb{R}^n$ is a polytope if and only if F is a bounded polyhedron.

---

[1]Named after the German mathematician, theoretical physicist and philosopher HERMANN K. H. WEYL (1885–1955) and the German mathematician HERMANN MINKOWSKI (1864–1909).

While the statement of Theorem 2.1.6 may be intuitively and graphically clear, it requires careful work to actually prove the theorem; see the proof of Theorem 2.1.2 in Appendix D, which shows the 'if' direction of Theorem 2.1.6. Note that not every polyhedron is a polytope: since a polytope is necessarily bounded, any unbounded polyhedron is not a polytope.

The important difference between polyhedra and polytopes lies in the information by which they are described, namely hyperplanes in the case of polyhedra, and vertices in the case of polytopes. The collection of hyperplanes that describe a bounded polyhedron is often called its *H-representation*, and the collection of vertices that describe a polytope is called its *V-representation*. The Weyl-Minkowski theorem shows that any bounded polyhedron has both an H-representation and a V-representation. However, it is computationally very hard to determine the V-representation from the H-representation and vice versa. The interested reader is referred to Grünbaum (2003).

2.2 Algebra of feasible regions; feasible basic solutions

The geometric concepts introduced in Section 2.1 were used to gain some insight in the various aspects of feasible regions. However, an algebraic description is more suitable for computer calculations. The current section describes the various algebraic aspects of feasible regions and its relationship with the corresponding geometric concepts.

2.2.1 Notation; row and column indices

We first introduce some useful notation that is used throughout the book. Let $\mathbf{A}$ be any (m, n)-matrix. Suppose that every row of $\mathbf{A}$ is associated with a unique integer called the *index* of the row, and the same holds for every column of $\mathbf{A}$. Let R be the set of row indices, and C the set of column indices. For example, the columns may have indices $1, \ldots, n$ and the rows $1, \ldots, m$. When $\mathbf{A}$ is the technology matrix of a standard LO-model, the columns of $\mathbf{A}$ are associated with the decision variables of the model, and the column indices are taken to be the subscripts of the corresponding decision variables.

Let $I \subseteq R$ and $J \subseteq C$. Define:

$$(\mathbf{A})_{I,J} = \text{the } (|I|, |J|)\text{-submatrix of } \mathbf{A} \text{ with row indices in } I, \text{ and column indices in } J.$$

When taking a submatrix, the indices of the rows and columns are preserved, i.e., if a row has index $i \in R$ in $\mathbf{A}$, then it also has index i in $(\mathbf{A})_{I,J}$ (provided that $i \in I$), and similarly for column indices. An equivalent way to define $(\mathbf{A})_{I,J}$ is: $(\mathbf{A})_{I,J}$ is the matrix obtained from $\mathbf{A}$ by deleting all rows whose index is not in I and all columns whose index is not in J. Define the following short hand notation:

$$(\mathbf{A})_{I,\star} = (\mathbf{A})_{I,C}, \text{ i.e., the } (|I|, n)\text{-submatrix of } \mathbf{A} \text{ with row indices in } I,$$
$$(\mathbf{A})_{\star,J} = (\mathbf{A})_{R,J}, \text{ i.e., the } (m, |J|)\text{-submatrix of } \mathbf{A} \text{ with column indices in } J.$$

Define the *complement* of $I \subseteq C$ by $\bar{I} = C \setminus I$. Similarly, define the complement of J by $\bar{J} = R \setminus J$. So, the complement of a set is defined with respect to the larger set (e.g., $\{1, \ldots, n\}$) of which the set is a subset. Usually, it is clear from the context what is meant by this larger set. Finally, for any integer k, define $I + k = \{i + k \mid i \in I\}$. Similarly, define $k + I = I + k$ and $I - k = I + (-k)$.

Example 2.2.1.   *Consider the matrix*

$$\mathbf{A} = \begin{matrix} & \begin{matrix} 1 & 2 & 3 & 4 \end{matrix} \\ \begin{matrix} 1 \\ 2 \\ 3 \end{matrix} & \begin{bmatrix} 2 & 3 & 5 & 0 \\ 0 & 1 & 6 & 1 \\ 8 & 2 & 1 & 3 \end{bmatrix} \end{matrix}.$$

The row indices are written left of each row, and the column indices above each column. Let $I = \{1, 3\} \subset \{1, 2, 3\}$ and $J = \{3\} \subset \{1, 2, 3, 4\}$. Then,

$$(\mathbf{A})_{I,\star} = \begin{bmatrix} 2 & 3 & 5 & 0 \\ 8 & 2 & 1 & 3 \end{bmatrix}, (\mathbf{A})_{\star,J} = \begin{bmatrix} 5 \\ 6 \\ 1 \end{bmatrix}, \text{ and } (\mathbf{A})_{I,J} = \begin{bmatrix} 5 \\ 1 \end{bmatrix}.$$

Furthermore, $\bar{I} = \{2\}$, $\bar{J} = \{1, 2, 4\}$, and $2 + \bar{J} = \{3, 4, 6\}$.

Whenever this does not cause any confusion, we omit the parentheses in the notation $(\mathbf{A})_{\star,J}$, $(\mathbf{A})_{I,\star}$, and $(\mathbf{A})_{I,J}$, and we write $\mathbf{A}_{\star,J}$, $\mathbf{A}_{I,\star}$, and $\mathbf{A}_{I,J}$, respectively, instead. Define $(\mathbf{A})_{i,\star} = (\mathbf{A})_{\{i\},\star}$; $(\mathbf{A})_{\star,j}$ and $(\mathbf{A})_{i,j}$ are defined analogously. The same notation also applies to vectors, but we will have only one subscript. For example, for a vector $\mathbf{x} \in \mathbb{R}^n$ with entry indices R, $(\mathbf{x})_I$ is the subvector of $\mathbf{x}$ containing the entries with indices in I ($I \subseteq R$). Note that, for $i \in R$, we have that $(\mathbf{x})_{\{i\}} = x_i$.

2.2.2 Feasible basic solutions

Let $m \geq 1$ and $n \geq 1$. Consider the system of equalities:

$$\begin{bmatrix} \mathbf{A} & \mathbf{I}_m \end{bmatrix} \begin{bmatrix} \mathbf{x} \\ \mathbf{x}_s \end{bmatrix} = \mathbf{b},$$

with $\mathbf{x} \in \mathbb{R}^n$, $\mathbf{x}_s \in \mathbb{R}^m$, $\mathbf{b} \in \mathbb{R}^m$, $\mathbf{A} \in \mathbb{R}^{m \times n}$, $\mathbf{I}_m \in \mathbb{R}^{m \times m}$, $\mathbf{x} \geq \mathbf{0}$, and $\mathbf{x}_s \geq \mathbf{0}$. The i'th column of the matrix $\begin{bmatrix} \mathbf{A} & \mathbf{I}_m \end{bmatrix}$ is multiplied by the i'th variable in $\begin{bmatrix} \mathbf{x} \\ \mathbf{x}_s \end{bmatrix}$ ($i = 1, \ldots, n + m$). So, each column of $\begin{bmatrix} \mathbf{A} & \mathbf{I}_m \end{bmatrix}$ corresponds to a decision variable or a slack variable. Let the index of each column be the subscript of the corresponding decision or slack variable. So the columns have indices $1, \ldots, n + m$. The rank (see Appendix B) of $\begin{bmatrix} \mathbf{A} & \mathbf{I}_m \end{bmatrix}$ is m, because this matrix contains the submatrix $\mathbf{I}_m$ which has rank m. Suppose that, from the $n + m$ columns of $\begin{bmatrix} \mathbf{A} & \mathbf{I}_m \end{bmatrix}$, we select a subset of m linearly independent columns (see Appendix B). Let $\mathbf{B}$ be the (m, m)-submatrix determined by these columns, and let $\mathbf{N}$ be the (m, n)-submatrix consisting of the remaining columns. That is, define:

$$\mathbf{B} = \begin{bmatrix} \mathbf{A} & \mathbf{I}_m \end{bmatrix}_{\star, BI}, \text{ and } \mathbf{N} = \begin{bmatrix} \mathbf{A} & \mathbf{I}_m \end{bmatrix}_{\star, NI},$$

where the columns of $\begin{bmatrix} \mathbf{A} & \mathbf{I}_m \end{bmatrix}$ have indices $1, \ldots, n + m$, and:

$$
\begin{array}{ll}
BI &= \text{ the set of column indices of } \mathbf{B} \text{ in } \{1, \ldots, n + m\}, \text{ and} \\
NI &= \{1, \ldots, n + m\} \setminus BI \text{ (so, } NI = \overline{BI}).
\end{array}
$$

Recall that the column indices are preserved when taking submatrices, so the columns of $\mathbf{B}$ have indices in BI and the columns of $\mathbf{N}$ have indices in NI. We will write:

$$\begin{bmatrix} \mathbf{A} & \mathbf{I}_m \end{bmatrix} \equiv \begin{bmatrix} \mathbf{B} & \mathbf{N} \end{bmatrix},$$

where the symbol $\equiv$ means 'equality up to an appropriate permutation of the rows and columns'. See Appendix B.2. The vector $\mathbf{x}_{BI}$ is the subvector of $\begin{bmatrix} \mathbf{x} \\ \mathbf{x}_s \end{bmatrix}$ consisting of all variables with indices corresponding to the columns of $\mathbf{B}$; the entries of $\mathbf{x}_{BI}$ are called the *basic variables*. Similarly, the vector $\mathbf{x}_{NI}$ is the subvector of $\begin{bmatrix} \mathbf{x} \\ \mathbf{x}_s \end{bmatrix}$ consisting of all variables that have indices in NI; the entries of $\mathbf{x}_{NI}$ are called the *nonbasic variables*. So, we have that:

$$\mathbf{x}_{BI} = \begin{bmatrix} \mathbf{x} \\ \mathbf{x}_s \end{bmatrix}_{BI} \text{ and } \mathbf{x}_{NI} = \begin{bmatrix} \mathbf{x} \\ \mathbf{x}_s \end{bmatrix}_{NI}.$$

Moreover, because $BI \cup NI = \{1, \ldots, n + m\}$ and $BI \cap NI = \emptyset$, it holds that:

$$\begin{bmatrix} \mathbf{x} \\ \mathbf{x}_s \end{bmatrix} \equiv \begin{bmatrix} \mathbf{x}_{BI} \\ \mathbf{x}_{NI} \end{bmatrix}.$$

The matrix $\mathbf{B}$ is square and consists of linearly independent columns and is therefore invertible (see Appendix B). Any such invertible (m, m)-submatrix $\mathbf{B}$ of $\begin{bmatrix} \mathbf{A} & \mathbf{I}_m \end{bmatrix}$ is called a *basis matrix*. The reason for this name is the fact that the columns of $\mathbf{B}$ form a basis for the space $\mathbb{R}^m$; see Appendix B.

The system of equations $\begin{bmatrix} \mathbf{A} & \mathbf{I}_m \end{bmatrix} \begin{bmatrix} \mathbf{x} \\ \mathbf{x}_s \end{bmatrix} = \mathbf{b}$ is equivalent to $\begin{bmatrix} \mathbf{B} & \mathbf{N} \end{bmatrix} \begin{bmatrix} \mathbf{x}_{BI} \\ \mathbf{x}_{NI} \end{bmatrix} = \mathbf{b}$ (meaning that both systems have the same set of solutions). Hence, $\mathbf{B}\mathbf{x}_{BI} + \mathbf{N}\mathbf{x}_{NI} = \mathbf{b}$ and, since $\mathbf{B}$ is invertible, it follows that:

$$\mathbf{x}_{BI} = \mathbf{B}^{-1}\mathbf{b} - \mathbf{B}^{-1}\mathbf{N}\mathbf{x}_{NI}. \tag{2.1}$$

So it is possible to express the basic variables in terms of the corresponding nonbasic variables, meaning that any choice of values for the nonbasic variables fixes the values of the corresponding basic variables. In fact, since the matrix $\mathbf{B}$ is invertible, there is a unique way of expressing the basic variables in terms of the nonbasic variables.

Example 2.2.2. *Consider Model Dovetail. Let $\mathbf{B}$ consist of the columns 1, 3, 5, and 6 of the matrix $\begin{bmatrix} \mathbf{A} & \mathbf{I}_4 \end{bmatrix}$. We have that $BI = \{1, 3, 5, 6\}$, $NI = \{2, 4\}$, so that $\mathbf{x}_{BI} =$*

$\begin{bmatrix} x_1 & x_3 & x_5 & x_6 \end{bmatrix}^\mathsf{T}$, $\mathbf{x}_{NI} = \begin{bmatrix} x_2 & x_4 \end{bmatrix}^\mathsf{T}$, and:

$$
\begin{bmatrix} \mathbf{A} & \mathbf{I}_4 \end{bmatrix} = \begin{matrix} & \begin{smallmatrix} x_1 & x_2 & x_3 & x_4 & x_5 & x_6 \end{smallmatrix} \\ \begin{bmatrix} 1 & 1 & 1 & 0 & 0 & 0 \\ 3 & 1 & 0 & 1 & 0 & 0 \\ 1 & 0 & 0 & 0 & 1 & 0 \\ 0 & 1 & 0 & 0 & 0 & 1 \end{bmatrix} \end{matrix}, \mathbf{B} = \begin{matrix} & \begin{smallmatrix} x_1 & x_3 & x_5 & x_6 \end{smallmatrix} \\ \begin{bmatrix} 1 & 1 & 0 & 0 \\ 3 & 0 & 0 & 0 \\ 1 & 0 & 1 & 0 \\ 0 & 0 & 0 & 1 \end{bmatrix} \end{matrix}, \text{ and } \mathbf{N} = \begin{matrix} & \begin{smallmatrix} x_2 & x_4 \end{smallmatrix} \\ \begin{bmatrix} 1 & 0 \\ 1 & 1 \\ 0 & 0 \\ 1 & 0 \end{bmatrix} \end{matrix}.
$$

One can easily check that $\mathbf{B}$ *is invertible, and that:*

$$
\begin{aligned}
x_1 &= 6 - \tfrac{1}{3}x_2 - \tfrac{1}{3}x_4, \\
x_3 &= 3 - \tfrac{2}{3}x_2 + \tfrac{1}{3}x_4, \\
x_5 &= 1 + \tfrac{1}{3}x_2 + \tfrac{1}{3}x_4, \\
x_6 &= 6 - x_2.
\end{aligned}
$$

The values of the basic variables x_1, x_3, x_5, x_6 *are thus completely determined by the values of the nonbasic variables* x_2 *and* x_4.

Let $\mathbf{B}$ be a basis matrix. If $\begin{bmatrix} \mathbf{x} \\ \mathbf{x}_s \end{bmatrix} \equiv \begin{bmatrix} \mathbf{x}_{BI} \\ \mathbf{x}_{NI} \end{bmatrix}$ with $\mathbf{x}_{BI} = \mathbf{B}^{-1}\mathbf{b}$ and $\mathbf{x}_{NI} = \mathbf{0}$, then the vector $\mathbf{x}$ is called the *basic solution* with respect to the basis matrix $\mathbf{B}$. If $\mathbf{B}^{-1}\mathbf{b} \geq \mathbf{0}$, then $\begin{bmatrix} \mathbf{x} \\ \mathbf{x}_s \end{bmatrix}$ is a solution of $\begin{bmatrix} \mathbf{A} & \mathbf{I}_m \end{bmatrix} \begin{bmatrix} \mathbf{x} \\ \mathbf{x}_s \end{bmatrix} = \mathbf{b}$ satisfying $\mathbf{x} \geq \mathbf{0}$ and $\mathbf{x}_s \geq \mathbf{0}$, and hence $\mathbf{x}$ is a point in the feasible region $F = \{\mathbf{x} \in \mathbb{R}^n \mid \mathbf{A}\mathbf{x} \leq \mathbf{b}, \mathbf{x} \geq \mathbf{0}\}$. Therefore, if $\mathbf{B}^{-1}\mathbf{b} \geq \mathbf{0}$, then $\mathbf{x}$ is called a *feasible basic solution*. If, on the other hand, $\mathbf{B}^{-1}\mathbf{b} \geq \mathbf{0}$ does not hold (i.e., at least one entry of $\mathbf{B}^{-1}\mathbf{b}$ is negative), then $\mathbf{x}$ is called an *infeasible basic solution*. If at least one entry of $\mathbf{x}_{BI}$ has value zero, then $\mathbf{x}$ is called a *degenerate basic solution*, and otherwise $\mathbf{x}$ is called a *nondegenerate basic solution*; see also Section 2.2.4.

In general, there are $\binom{n+m}{m}$ candidates for feasible basic solutions, because there are $\binom{n+m}{m}$ possibilities to select m columns out of the $n+m$ columns of $\begin{bmatrix} \mathbf{A} & \mathbf{I}_m \end{bmatrix}$. One of the choices is $\mathbf{I}_m$. Note that not all such choices give rise to a feasible basic solution, e.g., $\mathbf{B}^{-1}\mathbf{b}$ may not satisfy $\mathbf{B}^{-1}\mathbf{b} \geq \mathbf{0}$, or $\mathbf{B}$ may not be invertible. In the former case, the choice gives rise to an infeasible basic solution, and in the latter case the choice does not give rise to a basic solution at all.

For $\mathbf{b} \geq \mathbf{0}$, the system of equalities $\begin{bmatrix} \mathbf{A} & \mathbf{I}_m \end{bmatrix} \begin{bmatrix} \mathbf{x} \\ \mathbf{x}_s \end{bmatrix} = \mathbf{b}$ with $\begin{bmatrix} \mathbf{x} \\ \mathbf{x}_s \end{bmatrix} \geq \mathbf{0}$ has at least one feasible basic solution, namely the basic solution corresponding to the basis matrix $\mathbf{I}_m$, i.e., $\mathbf{x} = \mathbf{0}$, $\mathbf{x}_s = \mathbf{b}$. The choice $\mathbf{B} = \mathbf{I}_m$ implies that $\mathbf{x}_{BI} = \mathbf{x}_s$. This solution corresponds to the vertex $\mathbf{0}$ of the feasible region. If, however, $\mathbf{b} \not\geq \mathbf{0}$, then this solution is an infeasible basic solution.

There is a close relationship between feasible basic solutions and vertices of the feasible region. In fact, each feasible basic solution of an LO-model corresponds to a vertex of the feasible region; see Theorem 2.2.2. Also, it may happen that a vertex corresponds to multiple feasible basic solutions, in which case these feasible basic solutions are degenerate; see Section 2.2.4.

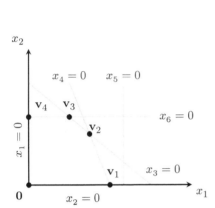

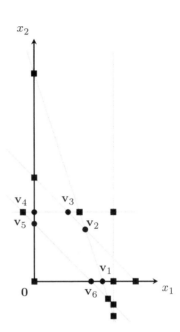

Figure 2.7: Zero-variable representation.   **Figure 2.8:** Feasible (●) and infeasible (■) basic solutions.

Example 2.2.3.   *Each equation associated with a technology constraint of Model Dovetail corresponds to a slack variable with value zero, namely (see Figure 2.7):*

$$x_1 + x_2 = 9 \text{ corresponds to } x_3 = 0,$$
$$3x_1 + x_2 = 18 \text{ corresponds to } x_4 = 0,$$
$$x_1 = 7 \text{ corresponds to } x_5 = 0,$$
$$\text{and} \qquad x_2 = 6 \text{ corresponds to } x_6 = 0.$$

We have seen that each feasible basic solution corresponds to a $(4,4)$ basis matrix $\mathbf{B}$ with nonnegative $\mathbf{B}^{-1}\mathbf{b}$. There are $\binom{6}{4} = 15$ possible ways to choose such a matrix $\mathbf{B}$ in $\begin{bmatrix} \mathbf{A} & \mathbf{I}_4 \end{bmatrix}$; only five of them correspond to a vertex of the feasible region in the following way. Each vertex of the feasible region (see Figure 2.7) is determined by setting the values of two nonbasic variables equal to 0, as follows:

$$\mathbf{0} \; : x_1 = 0, x_2 = 0 \text{ with basic variables } x_3, x_4, x_5, x_6;$$
$$\mathbf{v}_1 : x_2 = 0, x_4 = 0 \text{ with basic variables } x_1, x_3, x_5, x_6;$$
$$\mathbf{v}_2 : x_3 = 0, x_4 = 0 \text{ with basic variables } x_1, x_2, x_5, x_6;$$
$$\mathbf{v}_3 : x_3 = 0, x_6 = 0 \text{ with basic variables } x_1, x_2, x_4, x_5;$$
$$\mathbf{v}_4 : x_1 = 0, x_6 = 0 \text{ with basic variables } x_2, x_3, x_4, x_5.$$

For instance, $\mathbf{v}_2$ is the intersection of the lines corresponding to $x_1 + x_2 = 9$ and $3x_1 + x_2 = 18$, which in turn correspond to the slack variables x_3 and x_4, respectively. Note that vertices that are adjacent on the boundary of the feasible region differ in precisely one nonbasic variable. For instance, the adjacent vertices $\mathbf{v}_1$ and $\mathbf{v}_2$ share the zero-valued variable x_4 but differ in the remaining nonbasic variable. On the other hand, the nonadjacent vertices $\mathbf{v}_1$ and $\mathbf{v}_4$ differ in two nonbasic variables. These

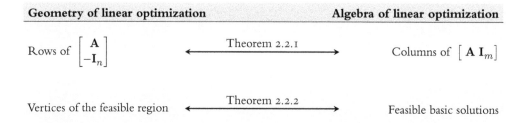

Figure 2.9: Geometry-algebra relationships.

observations hold in general and are crucial in the simplex algorithm that we will develop in Chapter 3. In Section 2.2.5 we will deal with adjacency in more detail.

In the case of vertex $\mathbf{0}$, the basis matrix $\mathbf{B}$ consists of the last four columns of the matrix $\begin{bmatrix} \mathbf{A} & \mathbf{I}_m \end{bmatrix}$. In the case of vertex $\mathbf{v}_1$, the matrix $\mathbf{B}$ consists of the columns 1, 3, 5, 6, because $x_2 = 0$ and $x_4 = 0$.

Not every pair $x_i = 0$, $x_j = 0$ with $i \neq j$, determines a vertex of the feasible region. For instance, the pair $x_5 = 0$, $x_6 = 0$ corresponds to the basis matrix

$$\mathbf{B} = \begin{bmatrix} \overset{x_1}{1} & \overset{x_2}{1} & \overset{x_3}{1} & \overset{x_4}{0} \\ 3 & 1 & 0 & 1 \\ 1 & 0 & 0 & 0 \\ 0 & 1 & 0 & 0 \end{bmatrix},$$

with $BI = \{1, 2, 3, 4\}$ and $NI = \{5, 6\}$. However, $\mathbf{B}^{-1}\mathbf{b} = \begin{bmatrix} 7 & 6 & -4 & -9 \end{bmatrix}^{\mathsf{T}}$ is not nonnegative, and the point $\begin{bmatrix} 7 & 6 \end{bmatrix}^{\mathsf{T}}$ is not in the feasible region. So, $\begin{bmatrix} 7 & 6 & -4 & -9 & 0 & 0 \end{bmatrix}^{\mathsf{T}}$ is an infeasible basic solution. Figure 2.8 shows all basic solutions, feasible as well as infeasible, of Model Dovetail*. In this figure the feasible basic solutions are shown as circles, and the infeasible ones are shown as squares. It is left to the reader to calculate all basic solutions.

2.2.3   Relationship between feasible basic solutions and vertices

The discussion in the previous section suggests that there a strong relationship between the feasible basic solutions of an LO-model and the vertices of its feasible region. The current section explores this relationship in more detail. Figure 2.9 gives an overview of the relationships found in this section.

We have seen in Section 2.2.2 that feasible basic solutions are determined by choosing m columns from the matrix $\begin{bmatrix} \mathbf{A} & \mathbf{I}_m \end{bmatrix}$. Recall that $BI \subseteq \{1, \ldots, n + m\}$ with $|BI| = m$ is the set of indices of these columns. The columns of $\begin{bmatrix} \mathbf{A} & \mathbf{I}_m \end{bmatrix}$ corresponding to the indices in BI form the basis matrix $\mathbf{B}$. Whether or not a specific choice of BI corresponds to a basic solution depends on whether or not the corresponding basis matrix $\mathbf{B}$ is invertible.

Vertices of the feasible region, on the other hand, are determined by selecting n linearly independent hyperplanes from the hyperplanes $H_1, \ldots, H_{m+n}$ (see Section 2.1.1). The normal vectors of $H_1, \ldots, H_{m+n}$ are precisely the rows of the matrix $\begin{bmatrix} \mathbf{A} \\ -\mathbf{I}_n \end{bmatrix}$. This means that, effectively, a vertex is determined by selecting n linearly independent rows from the latter matrix.

Suppose now that we have a feasible basic solution $\begin{bmatrix} \hat{\mathbf{x}} \\ \hat{\mathbf{x}}_s \end{bmatrix}$ with index sets BI and NI. Let $I = BI \cap \{1, \ldots, n\}$ and let $J = (BI \cap \{n+1, \ldots, n+m\}) - n$. That is, I is the set of indices in BI corresponding to decision variables, and $n + J$ is the set of indices in BI corresponding to slack variables. Note that we have that $I \subseteq \{1, \ldots, n\}$, $J \subseteq \{1, \ldots, m\}$, and

$$ BI = I \cup (n + J), \text{ and } NI = \bar{I} \cup (n + \bar{J}). $$

Recall that, at this feasible basic solution, we have that $\hat{\mathbf{x}}_{NI} = \mathbf{0}$. This implies that:

For $i \in \bar{I}$, x_i is a <u>decision</u> variable with value zero. Thus, the constraint $x_i \geq 0$ is binding (i.e., it holds with equality) at the feasible basic solution. This constraint corresponds to the hyperplane H_{m+i}.

For $j \in \bar{J}$, x_{n+j} is a <u>slack</u> variable with value zero. Thus, the j'th technology constraint is binding at the feasible basic solution. This constraint corresponds to the hyperplane H_j.

This means that the indices in NI provide us with $(|NI| =)$ n hyperplanes that contain $\hat{\mathbf{x}}$. The indices of these hyperplanes are: $m + i$ for $i \in \bar{I}$, and j for $j \in \bar{J}$. Define:

$$ BI^c = \bar{J} \cup (m + \bar{I}), \text{ and } NI^c = J \cup (m + I). $$

Note that $|NI^c| = |BI| = m$ and $|BI^c| = |NI| = n$. The set BI^c is called the *complementary dual set* of BI, and NI^c is called the complementary dual set of NI. The superscript 'c' refers to the fact that the set is the complementary dual set. Note that the complementary dual set BI^c is not the same as the complement $\overline{BI}$ of BI. The $(|BI^c| =)$ n hyperplanes with indices in BI^c contain $\hat{\mathbf{x}}$. They correspond to n constraints (possibly including nonnegativity constraints) that are binding at $\hat{\mathbf{x}}$. See also Section 4.3.1.

Example 2.2.4. *Consider Model Dovetail. We have that $n = 2$ and $m = 4$. Take $BI = \{1, 3, 5, 6\}$ and $NI = \{2, 4\}$. We saw in Example 2.2.2 that this choice of BI and NI corresponds to a feasible basic solution $\begin{bmatrix} \hat{\mathbf{x}}_{BI} \\ \hat{\mathbf{x}}_{NI} \end{bmatrix}$ with $\hat{\mathbf{x}}_{BI} = \mathbf{B}^{-1}\mathbf{b} \geq \mathbf{0}$ and $\hat{\mathbf{x}}_{NI} = \mathbf{0}$, where $\mathbf{B}$ is the basis matrix corresponding to BI. To see that this feasible basic solution corresponds to a vertex of the feasible region, recall that a vertex is a point in the feasible region for which n $(= 2)$ linearly independent hyperplanes hold with equality. The choice $NI = \{2, 4\}$ means setting $x_2 = 0$ and $x_4 = 0$. That is, the constraints $x_2 \geq 0$ and $3x_1 + x_2 \leq 18$ hold with equality. From Figure 2.3, it is clear that this corresponds to the vertex $\mathbf{v}_1$. In terms of the hyperplanes defined in Example 2.1.4, $x_2 = 0$ and $x_4 = 0$ correspond to the hyperplanes H_6 and H_2, respectively. So, the vertex corresponding to this particular choice of BI and NI is the unique point in the intersection of H_2 and H_6, i.e., hyperplanes H_2 and H_6 are binding at the feasible basic solution.*

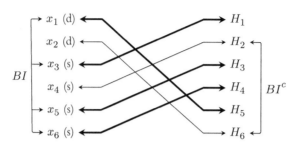

Figure 2.10: The set BI and its complementary dual set BI^c for Model Dovetail. The 'd' and 's' in parentheses mean 'decision variable' and 'slack variable', respectively.

The set of binding hyperplanes can be derived using the above notation as follows. We have that $I = \{1\}$ (corresponding to the basic decision variable x_1). Moreover, $n + J = \{3, 5, 6\}$ (corresponding to the basic slack variables x_3, x_5, and x_6), so that $J = \{1, 3, 4\}$. It follows that $\bar{I} = \{2\}$, and $\bar{J} = \{2\}$. Therefore, the complementary dual sets are $BI^c = \bar{J} \cup (m + \bar{I}) = \{2, 6\}$ and $NI^c = J \cup (m + I) = \{1, 3, 4, 5\}$. We have that:

Since $\bar{I} = \{2\}$, the decision variable x_2 has value zero at the feasible basic solution $\begin{bmatrix} \hat{\mathbf{x}}_{BI} \\ \hat{\mathbf{x}}_{NI} \end{bmatrix}$. The equation $x_2 = 0$ corresponds to the hyperplane $(H_{4+2} =) H_6$.

Since $\bar{J} = \{2\}$, the slack variable $(x_{2+2} =) x_4$ has value zero. The equation $x_4 = 0$ corresponds to the hyperplane H_2.

Thus, the hyperplanes with indices in BI^c are binding. Figure 2.10 shows the correspondence between the variables with indices in BI and the hyperplanes with indices in BI^c. The set BI^c is the complement of the set of indices of hyperplanes that correspond to the variables in BI.

The question that remains is: how can we be sure that the n hyperplanes are linearly independent? In other words, do the rows of $\begin{bmatrix} \mathbf{A} \\ -\mathbf{I}_n \end{bmatrix}$ with indices in BI^c form an invertible matrix? The following theorem, Theorem 2.2.1, shows the relationship between the invertibility of the (m, m)-submatrix of $\begin{bmatrix} \mathbf{A} & \mathbf{I}_m \end{bmatrix}$ corresponding to the columns in BI on the one hand, and the invertibility of the (n, n)-submatrix of $\begin{bmatrix} \mathbf{A} \\ -\mathbf{I}_n \end{bmatrix}$ corresponding to the rows in BI^c on the other hand.

Theorem 2.2.1. (*Row-column complementarity theorem*)
Let $m, n \geq 1$ and let $\mathbf{A}$ be any (m, n)-matrix. Let $BI \subseteq \{1, \ldots, n + m\}$ with $|BI| = m$, and let BI^c be defined as above. The following assertions are equivalent:

(i) $\begin{bmatrix} \mathbf{A} & \mathbf{I}_m \end{bmatrix}_{\star, BI}$ is an invertible (m, m)-submatrix of $\begin{bmatrix} \mathbf{A} & \mathbf{I}_m \end{bmatrix}$;

(ii) $\begin{bmatrix} \mathbf{A} \\ -\mathbf{I}_n \end{bmatrix}_{BI^c, \star}$ is an invertible (n, n)-submatrix of $\begin{bmatrix} \mathbf{A} \\ -\mathbf{I}_n \end{bmatrix}$.

The matrices

$$\mathbf{B} = \begin{bmatrix} \mathbf{A} & \mathbf{I}_m \end{bmatrix}_{\star,BI} \text{ and } \overline{\mathbf{B}}^\mathsf{T} = \begin{bmatrix} \mathbf{A} \\ -\mathbf{I}_n \end{bmatrix}_{BI^c,\star}$$

are called *complementary dual matrices*. The term 'dual' will become clear in Chapter 4. Note that the columns of $\mathbf{B}$ are columns from the matrix $\begin{bmatrix} \mathbf{A} & \mathbf{I}_m \end{bmatrix}$, and the rows of the matrix $\overline{\mathbf{B}}^\mathsf{T}$ are rows from the matrix $\begin{bmatrix} \mathbf{A} \\ -\mathbf{I}_n \end{bmatrix}$. Note also that $\overline{\mathbf{B}} = \begin{bmatrix} \mathbf{A}^\mathsf{T} & -\mathbf{I}_n \end{bmatrix}_{\star,BI^c}$.

Before we give a proof of Theorem 2.2.1, we illustrate the theorem by means of an example.

Example 2.2.5. *Consider again the technology matrix $\mathbf{A}$ of Model Dovetail; see also Figure 2.3. We have that $n = 2$, $m = 4$, and:*

$$\begin{bmatrix} \mathbf{A} & \mathbf{I}_4 \end{bmatrix} = \begin{array}{c} \begin{array}{cccccc} x_1 & x_2 & x_3 & x_4 & x_5 & x_6 \end{array} \\ \begin{bmatrix} 1 & 1 & 1 & 0 & 0 & 0 \\ 3 & 1 & 0 & 1 & 0 & 0 \\ 1 & 0 & 0 & 0 & 1 & 0 \\ 0 & 1 & 0 & 0 & 0 & 1 \end{bmatrix} \end{array}, \text{ and } \begin{bmatrix} \mathbf{A} \\ -\mathbf{I}_2 \end{bmatrix} = \begin{array}{c} H_1 \\ H_2 \\ H_3 \\ H_4 \\ H_5 \\ H_6 \end{array} \begin{bmatrix} 1 & 1 \\ 3 & 1 \\ 1 & 0 \\ 0 & 1 \\ -1 & 0 \\ 0 & -1 \end{bmatrix}.$$

Note that the columns of $\begin{bmatrix} \mathbf{A} & \mathbf{I}_4 \end{bmatrix}$ correspond to the decision variables and the slack variables, and the rows correspond to the technology constraints of the model with slack variables. For example, the second column corresponds to x_2, and the first row corresponds to the first technology constraint (i.e., $x_1 + x_2 + x_3 = 9$). Similarly, the columns of $\begin{bmatrix} \mathbf{A} \\ -\mathbf{I}_2 \end{bmatrix}$ correspond to the decision variables of the model, and the rows correspond to the hyperplanes that define the feasible region. For example, the third row corresponds to hyperplane H_3.

Take $BI = \{1, 3, 5, 6\}$ and $NI = \{2, 4\}$. We saw in Example 2.2.2 that this choice of BI and NI corresponds to a feasible basic solution, and we saw in Example 2.2.4 that this feasible basic solution in turn corresponds to a vertex of the feasible region, namely the unique point in the intersection of H_2 and H_6. It is easy to see that H_2 and H_6 are linearly independent.

Using the concept of complementary dual matrices, we obtain the following. To see how the invertibility of the basis matrix $\mathbf{B}$ implies that the H_i with $i \in BI^c$ form a collection of linearly independent hyperplanes, we introduce the following 'matrix':

$$\begin{bmatrix} \mathbf{A} & \mathbf{I}_m \\ -\mathbf{I}_n \end{bmatrix} = \begin{array}{c} \overbrace{\begin{array}{cccccc} x_1 & x_2 & x_3 & x_4 & x_5 & x_6 \end{array}}^{BI} \\ \left[\begin{array}{cc|cccc} 1 & 1 & 1 & 0 & 0 & 0 \\ 3 & 1 & 0 & 1 & 0 & 0 \\ 1 & 0 & 0 & 0 & 1 & 0 \\ 0 & 1 & 0 & 0 & 0 & 1 \\ \hline -1 & 0 & & & & \\ 0 & -1 & & & & \end{array}\right] \begin{array}{l} H_1 \\ H_2 \leftarrow \\ H_3 \\ H_4 \\ H_5 \\ H_6 \leftarrow \end{array} \end{array} BI^c$$

The corresponding basis matrix $\mathbf{B}$ *consists of the vertically shaded entries. Let* $\overline{\mathbf{B}}$ *be the transpose of the matrix consisting of the horizontally shaded entries. So, we have that:*

$$\mathbf{B} = \begin{bmatrix} \mathbf{A} & \vdots & \mathbf{I}_4 \end{bmatrix}_{\star,BI} = \begin{matrix} \begin{smallmatrix} x_1 & x_3 & x_5 & x_6 \end{smallmatrix} \\ \begin{bmatrix} 1 & 1 & 0 & 0 \\ 3 & 0 & 0 & 0 \\ 1 & 0 & 1 & 0 \\ 0 & 0 & 0 & 1 \end{bmatrix} \end{matrix}, \ and \ \overline{\mathbf{B}}^\mathsf{T} = \begin{bmatrix} \mathbf{A} \\ \cdots \\ -\mathbf{I}_2 \end{bmatrix}_{BI^c,\star} = \begin{matrix} H_2 \\ H_6 \end{matrix} \begin{bmatrix} 3 & 1 \\ 0 & -1 \end{bmatrix}.$$

The columns of the matrix $\overline{\mathbf{B}}$ *are the normal vectors corresponding to* H_2 *and* H_6. *Therefore,* H_2 *and* H_6 *are linearly independent if and only if the matrix* $\overline{\mathbf{B}}$ *is invertible. Applying Gaussian elimination (see Appendix B), we find that:*

$$\mathbf{B} \sim \begin{bmatrix} \mathbf{3} & 0 & 0 & 0 \\ 0 & 1 & 0 & 0 \\ 0 & 0 & 1 & 0 \\ 0 & 0 & 0 & 1 \end{bmatrix} \ and \ \overline{\mathbf{B}}^\mathsf{T} \sim \begin{bmatrix} \mathbf{3} & 0 \\ 0 & -1 \end{bmatrix}.$$

So, both $\mathbf{B}$ *and* $\overline{\mathbf{B}}$ *are equivalent to a block diagonal matrix (see Appendix B.3) which has* $\begin{bmatrix} 3 \end{bmatrix}$ *(marked in bold in the matrices above) and the negative of an identity matrix as its blocks. Since the matrices* $\begin{bmatrix} 3 \end{bmatrix}$ *and* $\begin{bmatrix} -1 \end{bmatrix}$ *are trivially invertible, it follows that the matrices* $\mathbf{B}$ *and* $\overline{\mathbf{B}}$ *are invertible as well; see also Appendix B.7.*

We now prove Theorem 2.2.1.

Proof of Theorem 2.2.1. Take any $BI \subseteq \{1,\ldots,m+n\}$ with $|BI| = m$. Let $NI = \{1,\ldots,m+n\} \setminus BI$. Define $I = BI \cap \{1,\ldots,n\}$ and $J = (BI \cap \{n+1,\ldots,n+m\}) - n$. We have that:

$$BI = I \cup (n+J), NI = \bar{I} \cup (n+\bar{J}), BI^c = \bar{J} \cup (m+\bar{I}), NI^c = J \cup (m+I).$$

Note that $|I|+|J| = m$, $|\bar{I}|+|\bar{J}| = n$, $|\bar{I}| = n-|I| = |J|$, and $|\bar{J}| = m-|J| = |I|$. Consider the 'matrix' $\begin{bmatrix} \mathbf{A} & \vdots & \mathbf{I}_m \\ \cdots \\ -\mathbf{I}_n & \vdots \end{bmatrix}$. Using the partition of $\{1,\ldots,n\}$ into I and $\bar{I}$, and the partition of $\{1,\ldots,m\}$ into J and $\bar{J}$, we find that:

$$\begin{bmatrix} \mathbf{A} & \vdots & \mathbf{I}_m \\ \cdots \\ -\mathbf{I}_n & \vdots \end{bmatrix} \equiv \begin{matrix} \overbrace{\phantom{\mathbf{A}_{\bar{J},I} \quad \mathbf{A}_{\bar{J},\bar{I}}}}^{1,\ldots,n} \quad \overbrace{\phantom{(\mathbf{I}_m)_{\bar{J},\bar{J}} \quad (\mathbf{I}_m)_{\bar{J},J}}}^{1,\ldots,m} \\ \left[\begin{array}{cc|cc} \mathbf{A}_{\bar{J},I} & \mathbf{A}_{\bar{J},\bar{I}} & (\mathbf{I}_m)_{\bar{J},\bar{J}} & (\mathbf{I}_m)_{\bar{J},J} \\ \mathbf{A}_{J,I} & \mathbf{A}_{J,\bar{I}} & (\mathbf{I}_m)_{J,\bar{J}} & (\mathbf{I}_m)_{J,J} \\ \hline (-\mathbf{I}_n)_{I,I} & (-\mathbf{I}_n)_{I,\bar{I}} & & \\ (-\mathbf{I}_n)_{\bar{I},I} & (-\mathbf{I}_n)_{\bar{I},\bar{I}} & & \end{array} \right] \begin{matrix} {\Big\}}{\scriptstyle 1,\ldots,m} \\ \\ {\Big\}}{\scriptstyle 1,\ldots,n.} \end{matrix} \end{matrix}$$

In this matrix, the row and column indices are given by the lists at the curly braces. Note that $(\mathbf{I}_m)_{\bar{J},J}$, $(\mathbf{I}_m)_{J,\bar{J}}$, $(-\mathbf{I}_n)_{I,\bar{I}}$, and $(-\mathbf{I}_n)_{\bar{I},I}$ are all-zero matrices. Using this fact, and applying Gaussian elimination, where the 1's of $(\mathbf{I}_m)_{J,J}$ and the -1's of $(-\mathbf{I}_n)_{\bar{I},\bar{I}}$ are the pivot entries, we obtain that:

$$\begin{bmatrix} \mathbf{A} & \mathbf{I}_m \\ \hline -\mathbf{I}_n & \end{bmatrix} \sim \begin{bmatrix} \mathbf{A}_{\bar{J},I} & \mathbf{0} & (\mathbf{I}_m)_{\bar{J},\bar{J}} & \mathbf{0} \\ \mathbf{0} & \mathbf{0} & \mathbf{0} & (\mathbf{I}_m)_{J,J} \\ (-\mathbf{I}_n)_{I,I} & \mathbf{0} & & \\ \mathbf{0} & (-\mathbf{I}_n)_{\bar{I},\bar{I}} & & \end{bmatrix}.$$

We will show that (i) $\implies$ (ii). The proof of (ii) $\implies$ (i) is left to the reader (Exercise 2.3.8). We have that:

$$\begin{bmatrix} \mathbf{A} & \mathbf{I}_m \end{bmatrix}_{\star,BI} = \begin{bmatrix} \mathbf{A}_{\star,I} & (\mathbf{I}_m)_{\star,J} \end{bmatrix} \equiv \begin{bmatrix} \mathbf{A}_{\bar{J},I} & (\mathbf{I}_m)_{\bar{J},J} \\ \mathbf{A}_{J,I} & (\mathbf{I}_m)_{J,J} \end{bmatrix} \sim \begin{bmatrix} \mathbf{A}_{\bar{J},I} & \mathbf{0} \\ \mathbf{0} & (\mathbf{I}_m)_{J,J} \end{bmatrix}.$$

The latter matrix is a block diagonal matrix (see Appendix B.3), which is invertible if and only if its blocks, i.e., the square matrices $\mathbf{A}_{\bar{J},I}$ and $(\mathbf{I}_m)_{J,J}$, are invertible (see Appendix B.7). Hence, because $(\mathbf{I}_m)_{J,J}$ is invertible and $\begin{bmatrix} \mathbf{A}_{\star,I} & (\mathbf{I}_m)_{\star,J} \end{bmatrix}$ is invertible by assumption, it follows that $\mathbf{A}_{\bar{J},I}$ is also invertible. This implies that $\begin{bmatrix} \mathbf{A}_{\bar{J},I} & \mathbf{0} \\ \mathbf{0} & (-\mathbf{I}_m)_{\bar{I},\bar{I}} \end{bmatrix}$ is invertible as well, and since this matrix is equivalent to $\begin{bmatrix} \mathbf{A}_{\bar{J},\star} \\ (-\mathbf{I}_n)_{\bar{I},\star} \end{bmatrix}$, it follows that the latter is invertible. The latter matrix is exactly $\begin{bmatrix} \mathbf{A} \\ -\mathbf{I}_n \end{bmatrix}_{BI^c,\star}$. This proves that (i) $\implies$ (ii). $\qquad\square$

So, we have established that choosing m linearly independent columns from $\begin{bmatrix} \mathbf{A} & \mathbf{I}_m \end{bmatrix}$ is equivalent to choosing n linearly independent rows from $\begin{bmatrix} \mathbf{A} \\ -\mathbf{I}_n \end{bmatrix}$. To summarize, we have the following equivalences:

$$\text{Basis matrix } \mathbf{B} \iff m \text{ independent columns from } \begin{bmatrix} \mathbf{A} & \mathbf{I}_m \end{bmatrix} \text{ (with indices in } BI\text{)}$$
$$\iff n \text{ independent rows from } \begin{bmatrix} \mathbf{A} \\ -\mathbf{I}_n \end{bmatrix} \text{ (with indices in } BI^c\text{)}$$
$$\iff n \text{ independent hyperplanes } (H_i \text{ with } i \in BI^c).$$

We can now use this insight to show that feasible basic solutions indeed correspond to vertices of the feasible region, and vice versa.

Theorem 2.2.2. (*Relationship between feasible basic solutions and vertices*)
Consider the standard LO-model $\max\{\mathbf{c}^\mathsf{T}\mathbf{x} \mid \mathbf{A}\mathbf{x} \le \mathbf{b}, \mathbf{x} \ge \mathbf{0}\}$, with $\mathbf{A} \in \mathbb{R}^{m \times n}$ and $\mathbf{b} \in \mathbb{R}^m$. Every feasible basic solution of the LO-model corresponds to a vertex of the feasible region $F = \{\mathbf{x} \in \mathbb{R}^n \mid \mathbf{A}\mathbf{x} \le \mathbf{b}, \mathbf{x} \ge \mathbf{0}\}$, and vice versa.

Proof of Theorem 2.2.2. To prove that every feasible basic solution corresponds to a vertex, suppose first that $\begin{bmatrix} \hat{\mathbf{x}}_{BI} \\ \hat{\mathbf{x}}_{NI} \end{bmatrix}$ is a feasible basic solution with respect to the invertible (m,m)-submatrix $\mathbf{B}$ of $\begin{bmatrix} \mathbf{A} & \mathbf{I}_m \end{bmatrix}$; i.e., $\hat{\mathbf{x}}_{BI} = \mathbf{B}^{-1}\mathbf{b} \ge \mathbf{0}$, and $\hat{\mathbf{x}}_{NI} = \mathbf{0}$. Let $\begin{bmatrix} \hat{\mathbf{x}} \\ \hat{\mathbf{x}}_s \end{bmatrix} \equiv \begin{bmatrix} \hat{\mathbf{x}}_{BI} \\ \hat{\mathbf{x}}_{NI} \end{bmatrix}$; the indices of $\hat{\mathbf{x}}$ are $1, \ldots, n$ and the indices of $\hat{\mathbf{x}}_s$ are $n+1, \ldots, n+m$. We will show that $\hat{\mathbf{x}}$ is a vertex of F. To that end, we look for n independent hyperplanes $H_1, \ldots, H_n$ that intersect at $\hat{\mathbf{x}}$.

Let $I = BI \cap \{1, \ldots, n\}$ and let $J = (BI \cap \{n+1, \ldots, n+m\}) - n$. (So, I contains the indices of the basic decision variables, and $n+J$ contains the indices of the basic slack variables.) Clearly, $BI = I \cup (n+J)$, $NI = \bar{I} \cup (n+\bar{J})$, $|I| + |J| = m$, and $|\bar{I}| + |\bar{J}| = n$. Hence, $\mathbf{B} = \left[\mathbf{A}_{\star, I} \; (\mathbf{I}_m)_{\star, J}\right]$. According to Theorem 2.2.1, $\overline{\mathbf{B}}^\mathsf{T} = \begin{bmatrix} \mathbf{A} \\ -\mathbf{I}_n \end{bmatrix}_{BI^c, \star} = \begin{bmatrix} \mathbf{A}_{\bar{J}, \star} \\ (-\mathbf{I}_n)_{\bar{I}, \star} \end{bmatrix}$ is an invertible (n, n)-submatrix of $\begin{bmatrix} \mathbf{A} \\ -\mathbf{I}_n \end{bmatrix}$, and it consists of the rows of $\mathbf{A}$ with indices in $\bar{J}$, together with the rows of $-\mathbf{I}_n$ with indices in $\bar{I}$. Define the n hyperplanes:

$$
\begin{aligned}
H_i &= \left\{ \begin{bmatrix} x_1 & \cdots & x_n \end{bmatrix}^\mathsf{T} \in \mathbb{R}^n \;\middle|\; a_{i1}x_1 + \ldots + a_{in}x_n = b_i \right\} && \text{for } i \in \bar{J}, \\
\text{and } H_{m+j} &= \left\{ \begin{bmatrix} x_1 & \cdots & x_n \end{bmatrix}^\mathsf{T} \in \mathbb{R}^n \;\middle|\; x_j = 0 \right\} && \text{for } j \in \bar{I}.
\end{aligned}
$$

The normal vectors of the H_j's and the H_{m+i}'s are precisely the rows of $\begin{bmatrix} \mathbf{A}_{\bar{J}, \star} \\ (-\mathbf{I}_n)_{\bar{I}, \star} \end{bmatrix}$, which is $\overline{\mathbf{B}}^\mathsf{T}$, so that these normals are linearly independent. Hence, the hyperplanes H_i with $i \in \bar{J}$ together with the hyperplanes H_{m+j} with $j \in \bar{I}$ form an independent set of $|\bar{I}| + |\bar{J}| = n$ hyperplanes.

It remains to show that $\hat{\mathbf{x}}$ is in the intersection of these n hyperplanes. We will show that:

$$
\begin{aligned}
a_{i1}\hat{x}_1 + \ldots + a_{in}\hat{x}_n &= b_i && \text{for } i \in \bar{J}, \\
\text{and } \hat{x}_j &= 0 && \text{for } j \in \bar{I}.
\end{aligned}
\tag{2.2}
$$

The fact that $\hat{x}_j = 0$ for each $j \in \bar{I}$ follows from the fact that $\bar{I} \subseteq NI$. The expression '$a_{i1}\hat{x}_1 + \ldots + a_{in}\hat{x}_n = b_i$ for $i \in \bar{J}$' can be written in matrix form as:

$$
\mathbf{A}_{\bar{J}, \star} \hat{\mathbf{x}} = \mathbf{b}_{\bar{J}},
\tag{2.3}
$$

where $\mathbf{A}_{\bar{J}, \star}$ consists of the rows of $\mathbf{A}$ with indices in $\bar{J}$, and $\mathbf{b}_{\bar{J}}$ consists of the corresponding entries of $\mathbf{b}$ in $\bar{J}$. In order to prove (2.3), first notice that $\begin{bmatrix} \mathbf{A} & \mathbf{I}_m \end{bmatrix} \begin{bmatrix} \hat{\mathbf{x}} \\ \hat{\mathbf{x}}_s \end{bmatrix} = \mathbf{b}$. Partitioning $\mathbf{I}_m$ according to J and $\bar{J}$ yields:

$$
\begin{bmatrix} \mathbf{A} & (\mathbf{I}_m)_{\star, J} & (\mathbf{I}_m)_{\star, \bar{J}} \end{bmatrix} \begin{bmatrix} \hat{\mathbf{x}} \\ (\hat{\mathbf{x}}_s)_J \\ (\hat{\mathbf{x}}_s)_{\bar{J}} \end{bmatrix} = \mathbf{b}.
$$

This is equivalent to:

$$
\mathbf{A}\hat{\mathbf{x}} + (\mathbf{I}_m)_{\star, J}(\hat{\mathbf{x}}_s)_J + (\mathbf{I}_m)_{\star, \bar{J}}(\hat{\mathbf{x}}_s)_{\bar{J}} = \mathbf{b}.
$$

Taking in this expression only the rows with indices in $\bar{J}$, we find that:

$$
\mathbf{A}_{\bar{J}, \star}\hat{\mathbf{x}} + (\mathbf{I}_m)_{\bar{J}, J}(\hat{\mathbf{x}}_s)_J + (\mathbf{I}_m)_{\bar{J}, \bar{J}}(\hat{\mathbf{x}}_s)_{\bar{J}} = \mathbf{b}_{\bar{J}}.
$$

Since $(\mathbf{I}_m)_{J, \bar{J}}$ is an all-zeroes matrix and $(\hat{\mathbf{x}}_s)_{\bar{J}} = \mathbf{0}$ because $\bar{J} \subseteq NI$, it follows that $\mathbf{A}_{\bar{J}, \star}\hat{\mathbf{x}} = \mathbf{b}_{\bar{J}}$. This proves that (2.3) holds, which means that (2.2) holds. Hence, $\hat{\mathbf{x}}$ indeed corresponds to a vertex of F.

For the converse, let $\hat{\mathbf{x}}$ be a vertex of F. Hence, $\mathbf{A}\hat{\mathbf{x}} \leq \mathbf{b}$, $\hat{\mathbf{x}} \geq \mathbf{0}$, and there are n independent hyperplanes $H_{i_1}, \ldots, H_{i_n}$ that intersect at $\hat{\mathbf{x}}$. The normal vectors of $H_{i_1}, \ldots H_{i_n}$ are rows of the matrix $\begin{bmatrix} \mathbf{A} \\ -\mathbf{I}_n \end{bmatrix}$. Hence, the hyperplanes $H_{i_1}, \ldots, H_{i_n}$ correspond to an invertible (n, n)-submatrix $\overline{\mathbf{B}}^\mathsf{T}$ of $\begin{bmatrix} \mathbf{A} \\ -\mathbf{I}_n \end{bmatrix}$. Let $I \subseteq \{1, \ldots, n\}$, $J \subseteq \{1, \ldots, m\}$, with $|\bar{I}| + |\bar{J}| = n$, be such that $\overline{\mathbf{B}}^\mathsf{T} = \begin{bmatrix} \mathbf{A}_{\bar{J},\star} \\ (-\mathbf{I}_n)_{\bar{I},\star} \end{bmatrix}$. Therefore, $\mathbf{A}_{\bar{J},\star}\hat{\mathbf{x}} = \mathbf{b}_{\bar{J}}$ and $\hat{\mathbf{x}}_{\bar{I}} = \mathbf{0}$. According to Theorem 2.2.1, the matrix $\mathbf{B} = \begin{bmatrix} \mathbf{A}_{\star,I} & (\mathbf{I}_m)_{\star,J} \end{bmatrix}$ is an invertible (m, m)-submatrix of $\begin{bmatrix} \mathbf{A} & \mathbf{I}_m \end{bmatrix}$. Define $\hat{\mathbf{x}}_s = \mathbf{b} - \mathbf{A}\hat{\mathbf{x}}$, and let $\begin{bmatrix} \hat{\mathbf{x}} \\ \hat{\mathbf{x}}_s \end{bmatrix} \equiv \begin{bmatrix} \hat{\mathbf{x}}_{BI} \\ \hat{\mathbf{x}}_{NI} \end{bmatrix}$ with $BI = I \cup (n + J)$ and $NI = \bar{I} \cup (n + \bar{J})$. Since $\mathbf{A}_{\star,\bar{J}}\hat{\mathbf{x}} = \mathbf{b}_{\bar{J}}$, it follows that the corresponding vector of slack variables satisfies $(\hat{\mathbf{x}}_s)_{\bar{J}} = \mathbf{0}$, so that

$$\hat{\mathbf{x}}_{NI} = \begin{bmatrix} \hat{\mathbf{x}} \\ \hat{\mathbf{x}}_s \end{bmatrix}_{NI} = \begin{bmatrix} \hat{\mathbf{x}}_{\bar{I}} \\ (\hat{\mathbf{x}}_s)_{\bar{J}} \end{bmatrix} = \begin{bmatrix} \mathbf{0} \\ \mathbf{0} \end{bmatrix}.$$

Since $\hat{\mathbf{x}} \geq \mathbf{0}$, we also have that $\hat{\mathbf{x}}_{BI} = \mathbf{B}^{-1}\mathbf{b} \geq \mathbf{0}$, and hence $\begin{bmatrix} \hat{\mathbf{x}}_{BI} \\ \hat{\mathbf{x}}_{NI} \end{bmatrix}$ is in fact a feasible basic solution with respect to $\mathbf{B}$. $\square$

From Theorem 2.2.2, it follows that every vertex of the feasible region corresponds to at least one feasible basic solution, while a feasible basic solution corresponds to precisely one vertex.

2.2.4   Degeneracy and feasible basic solutions

In Section 2.1.1 we saw that a vertex of a feasible region $F \subseteq \mathbb{R}^n$ is determined by n (≥ 1) linearly independent hyperplanes. In general, however, if the feasible region F is the intersection of a number of halfspaces, then a vertex of F may lie in more than n hyperplanes corresponding to these halfspaces. In that case, we speak of a 'degenerate vertex'. Formally, let $\mathbf{v}$ be a vertex of $F = \bigcap_{i=1}^{m+n} H_i^+$. If $\mathbf{v}$ is in at least $n + 1$ hyperplanes from the set $\{H_1, \ldots, H_{m+n}\}$, then $\mathbf{v}$ is called a *degenerate vertex* of F. A vertex that is not degenerate is called a *nondegenerate vertex*.

As an example, Figure 2.11 depicts the feasible region of Model Dovetail, except that the constraint $x_1 \leq 7$ has been replaced by the constraint $x_1 \leq 6$, which is redundant with respect to the feasible region. The vertex $\mathbf{v}_1 = \begin{bmatrix} 6 & 0 \end{bmatrix}^\mathsf{T}$ is the intersection of three hyperplanes, whereas any two of them already determine $\mathbf{v}_1$. See also Figure 1.2.

However, degeneracy does not necessarily imply that there is a constraint that is redundant with respect to the feasible region. For instance, if F is a pyramid in $\mathbb{R}^3$ with a square basis (see Figure 2.12), then the top vertex $\mathbf{v}$ is the intersection of four hyperplanes, whereas three hyperplanes already determine $\mathbf{v}$. Interestingly, if $n \leq 2$, then degeneracy does imply that there is a constraint that is redundant with respect to the feasible region. The reader is asked to show this in Exercise 2.3.13.

The vertices $\mathbf{v}$ in Figure 2.12 and $\mathbf{v}_1$ in Figure 2.11 are degenerate vertices.

Recall that a feasible basic solution corresponding to a basis matrix $\mathbf{B}$ is degenerate if $\mathbf{B}^{-1}\mathbf{b}$ has at least one entry with value zero. In Theorem 2.2.3, we establish a one-to-one rela-

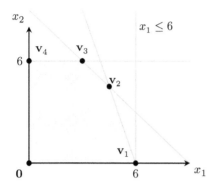

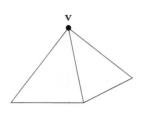

Figure 2.11: Degenerate vertex and redundant hyperplane.

Figure 2.12: Pyramid with a degenerate top.

tionship between degenerate vertices and degenerate feasible basic solutions, and between uniqueness of feasible basic solutions and nondegeneracy.

Theorem 2.2.3. (*Degenerate vertices and degenerate feasible basic solutions*)
Let $\mathbf{A} \in \mathbb{R}^{m \times n}$, $\mathbf{b} \in \mathbb{R}^m$, and let $F = \{\mathbf{x} \in \mathbb{R}^n \mid \mathbf{Ax} \leq \mathbf{b}, \mathbf{x} \geq \mathbf{0}\}$ be nonempty. The following statements are equivalent:

(i) $\mathbf{v}$ is a degenerate vertex.

(ii) $\mathbf{v}$ corresponds to multiple feasible basic solutions.

(iii)   The feasible basic solutions corresponding to $\mathbf{v}$ are degenerate.

Proof of Theorem 2.2.3. (i) $\implies$ (ii). Suppose that $\mathbf{v}$ is degenerate. Let $H_1, \ldots, H_{m+n}$ be the hyperplanes corresponding to the rows of the matrix $\begin{bmatrix} \mathbf{A} \\ -\mathbf{I}_n \end{bmatrix}$. Since $\mathbf{v}$ is a vertex, there exist n linearly independent hyperplanes among $H_1, \ldots, H_{m+n}$ that contain $\mathbf{v}$. Let $NI \subseteq \{1, \ldots, n + m\}$ be the set of indices of these n hyperplanes. For $i = 1, \ldots, m + n$, let $\mathbf{a}_i$ ($\in \mathbb{R}^n$) be the normal vector corresponding to hyperplane H_i. Since $\mathbf{v}$ is degenerate, there exists another hyperplane, say H_p with normal vector $\mathbf{a}_p$, that contains $\mathbf{v}$. By Theorem B.1.1 applied to the vectors $\mathbf{a}_i$ ($i \in NI$) and $\mathbf{a}_p$, there exists $NI' \subseteq NI \cup \{p\}$ such that the n vectors $\mathbf{a}_i$ ($i \in NI'$) are linearly independent. Let $BI = \{1, \ldots, n + m\} \setminus NI$ and $BI' = \{1, \ldots, n + m\} \setminus NI'$. It follows from Theorem 2.2.1 that BI and BI' are the index sets of the basic variables of two distinct basic solutions, and it follows from the construction of BI and BI' that these basic solutions are in fact feasible basic solutions that correspond to $\mathbf{v}$.

(ii) $\implies$ (iii). Consider any feasible basic solution $\begin{bmatrix} \mathbf{x}_{BI} \\ \mathbf{x}_{NI} \end{bmatrix} \equiv \begin{bmatrix} \mathbf{x} \\ \mathbf{x}_s \end{bmatrix}$ corresponding to $\mathbf{v}$. Let NI' be the index set of nonbasic variables of a different feasible basic solution that corresponds to $\mathbf{v}$. Because $\begin{bmatrix} \mathbf{x} \\ \mathbf{x}_s \end{bmatrix}_{NI} = \mathbf{0}$ and $\begin{bmatrix} \mathbf{x} \\ \mathbf{x}_s \end{bmatrix}_{NI'} = \mathbf{0}$, we have that $\begin{bmatrix} \mathbf{x} \\ \mathbf{x}_s \end{bmatrix}$ has at least $m + 1$ entries with value zero, and hence at most $n-1$ entries with nonzero value. It follows that $\mathbf{x}_{BI}$ has at least one entry with value zero, which means that this feasible basic solution is degenerate.

(iii) $\implies$ (i). Let $\begin{bmatrix} \mathbf{x}_{BI} \\ \mathbf{0} \end{bmatrix}$ be a degenerate feasible basic solution with respect to the invertible (m,m)-submatrix $\mathbf{B}$ of $\begin{bmatrix} \mathbf{A} & \mathbf{I}_m \end{bmatrix}$. Hence $\mathbf{x}_{BI}$ contains at least one zero entry, say x_k with $k \in BI$. Let $\mathbf{v}$ be the vertex of F corresponding to $\begin{bmatrix} \mathbf{x}_{BI} \\ \mathbf{0} \end{bmatrix}$ (see Theorem 2.2.2) determined by the n equalities $x_i = 0$ with $i \in NI$. Besides the n hyperplanes corresponding to $x_i = 0$ with $i \in NI$, the equation $x_k = 0$ ($k \in BI$) corresponds to a hyperplane that also contains $\mathbf{v}$. Hence, there are at least $n+1$ hyperplanes that contain $\mathbf{v}$, and this implies that $\mathbf{v}$ is degenerate.

$\square$

By the definition of a degenerate feasible basic solution, it holds that the feasible basic solutions corresponding to a particular vertex $\mathbf{v}$ are either all degenerate, or all nondegenerate. In fact, Theorem 2.2.3 shows that in the latter case, there is a unique nondegenerate feasible basic solutions. So, for any vertex $\mathbf{v}$ of the feasible region of a standard LO-model, there are two possibilities:

1. The vertex $\mathbf{v}$ is degenerate. Then, $\mathbf{v}$ corresponds to multiple feasible basic solutions and they are all degenerate.

2. The vertex $\mathbf{v}$ is nondegenerate. Then, $\mathbf{v}$ corresponds to a unique feasible basic solution, and this feasible basic solution is nondegenerate.

Degeneracy plays an important role in the development of the simplex algorithm in Chapter 3; see also Section 3.5.1. In the following example, we illustrate the first possibility, i.e., a degenerate vertex with degenerate feasible basic solutions.

Example 2.2.6. *Consider again the degenerate vertex $\mathbf{v}_1 = \begin{bmatrix} 6 & 0 \end{bmatrix}^\mathsf{T}$ in Figure 2.11. Introducing the slack variables x_3, x_4, x_5, and x_6 for the constraints (1.1), (1.2), (1.3), and (1.4), respectively, we find three different feasible basic solutions associated with $\mathbf{v}_1$, namely the feasible basic solutions with as nonbasic variables the pairs $\{x_2, x_4\}$, $\{x_2, x_5\}$, and $\{x_4, x_5\}$. Since these three feasible basic solutions correspond to the same vertex, it follows from Theorem 2.2.3 that the vertex $\mathbf{v}_1$ is degenerate and so are the three feasible basic solutions corresponding to it.*

2.2.5 Adjacency

In Chapter 3, we will see that, using the so-called simplex algorithm, an optimal solution can be found by starting at vertex $\mathbf{0}$ of the feasible region (provided that the point $\mathbf{0}$ is in fact a feasible solution) and then proceeding along the boundary of the feasible region towards an optimal vertex. In fact, an optimal vertex is reached by 'jumping' from vertex to vertex along the boundary of the feasible region, i.e., from feasible basic solution to feasible basic solution, in such a way that in each step one nonbasic variable changes into a basic variable and thus one basic variable becomes a nonbasic variable.

Let $n, m \geq 1$, and let $F = \bigcap_{i=1}^{m+n} H_i^+$ be nonempty where $H_1, \dots, H_{m+n}$ are hyperplanes in $\mathbb{R}^n$. Two distinct vertices $\mathbf{u}$ and $\mathbf{v}$ of F are called *adjacent* if and only if there are two index sets $I_1, I_2 \subseteq \{1, \ldots, m + n\}$ that differ in exactly one index, such that $\{\mathbf{u}\} = \bigcap_{i \in I_1} H_i$ and $\{\mathbf{v}\} = \bigcap_{i \in I_2} H_i$. Two adjacent vertices $\mathbf{u}$ and $\mathbf{v}$ therefore 'share' $n - 1$ independent hyperplanes, say $H_1, \ldots, H_{n-1}$; the remaining two hyperplanes, say H_n and H_{n+1}, determine the locations of the vertices $\mathbf{u}$ and $\mathbf{v}$ on the 'line' $H_1 \cap \ldots \cap H_{n-1}$. (Why is this intersection one-dimensional?) Hence, $\{\mathbf{u}\} = H_1 \cap \ldots \cap H_{n-1} \cap H_n$ and $\{\mathbf{v}\} = H_1 \cap \ldots \cap H_{n-1} \cap H_{n+1}$, whereas both $\{H_1, \ldots, H_{n-1}, H_n\}$ and $\{H_1, \ldots, H_{n-1}, H_{n+1}\}$ are independent sets of hyperplanes.

In Appendix D it is shown that the definition of adjacency of vertices given above is equivalent to the following one. Two distinct vertices $\mathbf{u}$ and $\mathbf{v}$ are adjacent in F if and only if there exists a hyperplane H, with a corresponding halfspace H^+, such that $F \subset H^+$ and $F \cap H = \mathrm{conv}\{\mathbf{u}, \mathbf{v}\}$. When comparing the definitions of 'vertex' and 'adjacent vertices' (see Section 2.1.2), one may observe that $\mathrm{conv}\{\mathbf{u}, \mathbf{v}\}$ is a face of F; see Section 2.1.3.

Example 2.2.7.   *Consider again Model Dovetail (where $n = 2$), and Figure 1.2. It is clear from the figure that the vertices $\mathbf{v}_2$ and $\mathbf{v}_3$ are adjacent in F. The hyperplanes H_1, H_2, H_4 (see Section 2.1.1) determine $\mathbf{v}_2$ and $\mathbf{v}_3$, namely $\{\mathbf{v}_2\} = H_1 \cap H_2$ and $\{\mathbf{v}_3\} = H_1 \cap H_4$. See also Exercise 2.3.2.*

We are now ready to formulate the 'vertex adjacency' theorem. We will use the expression 'two feasible basic solutions differ in one index'. By this, we mean that the corresponding index sets BI_1 and BI_2 differ in precisely one index, i.e., $|BI_1 \setminus BI_2| = |BI_2 \setminus BI_1| = 1$.

Theorem 2.2.4. (*Adjacent vertices and feasible basic solutions*)
Let $\mathbf{A} \in \mathbb{R}^{m \times n}$, $\mathbf{b} \in \mathbb{R}^m$, and let $F = \{\mathbf{x} \in \mathbb{R}^n \mid \mathbf{A}\mathbf{x} \leq \mathbf{b}, \mathbf{x} \geq \mathbf{0}\}$ be nonempty. If two vertices $\mathbf{u}$ and $\mathbf{v}$ of F are adjacent, then there exist two feasible basic solutions corresponding to $\mathbf{u}$ and $\mathbf{v}$, respectively, that differ in one index. Conversely, two feasible basic solutions that differ in one index correspond to either two adjacent vertices of F, or to one degenerate vertex of F.

Proof of Theorem 2.2.4. Let $\mathbf{u}$ and $\mathbf{v}$ be two adjacent vertices of the nonempty feasible region $F = \{\mathbf{x} \in \mathbb{R}^n \mid \mathbf{A}\mathbf{x} \leq \mathbf{b}, \mathbf{x} \geq \mathbf{0}\}$. According to the definition of adjacency, there are $n + 1$ hyperplanes $H_1, \ldots, H_n, H_{n+1}$ corresponding to the rows of $\begin{bmatrix} \mathbf{A} \\ -\mathbf{I}_n \end{bmatrix}$, such that $\{\mathbf{u}\} = H_1 \cap \ldots \cap H_{n-1} \cap H_n$ and $\{\mathbf{v}\} = H_1 \cap \ldots \cap H_{n-1} \cap H_{n+1}$, and both $\{H_1, \ldots, H_{n-1}, H_n\}$ and $\{H_1, \ldots, H_{n-1}, H_{n+1}\}$ are independent sets of hyperplanes. Let $\overline{\mathbf{B}}_1$ and $\overline{\mathbf{B}}_2$ be the (n, n)-submatrices of $\begin{bmatrix} \mathbf{A} \\ -\mathbf{I}_n \end{bmatrix}$ with $\overline{\mathbf{B}}_1$ consisting of the rows of $\begin{bmatrix} \mathbf{A} \\ -\mathbf{I}_n \end{bmatrix}$ corresponding to $H_1, \ldots, H_{n-1}, H_n$, and $\overline{\mathbf{B}}_2$ consisting of the rows of $\begin{bmatrix} \mathbf{A} \\ -\mathbf{I}_n \end{bmatrix}$ corresponding to $H_1, \ldots, H_{n-1}, H_{n+1}$. Hence, $\overline{\mathbf{B}}_1$ and $\overline{\mathbf{B}}_2$ are invertible. The complementary dual matrices $\mathbf{B}_1$ and $\mathbf{B}_2$ of $\overline{\mathbf{B}}_1$ and $\overline{\mathbf{B}}_2$, respectively, are invertible submatrices of $\begin{bmatrix} \mathbf{A} & \mathbf{I}_m \end{bmatrix}$ according to Theorem 2.2.1.

One can easily check that $\overline{\mathbf{B}}_1$ and $\overline{\mathbf{B}}_2$ differ in precisely one row and therefore that $\mathbf{B}_1$ and $\mathbf{B}_2$ differ in precisely one column. Let BI_1 consist of the indices corresponding to the columns of $\mathbf{B}_1$ in $\begin{bmatrix} \mathbf{A} & \mathbf{I}_m \end{bmatrix}$, and BI_2 of the indices corresponding to the columns of $\mathbf{B}_2$ in $\begin{bmatrix} \mathbf{A} & \mathbf{I}_m \end{bmatrix}$. Then $\begin{bmatrix} \mathbf{x}_{BI_1} \\ \mathbf{0} \end{bmatrix}$ and $\begin{bmatrix} \mathbf{x}_{BI_2} \\ \mathbf{0} \end{bmatrix}$ are feasible basic solutions with respect to $\mathbf{B}_1$ and $\mathbf{B}_2$, respectively, that differ in precisely one index.

Conversely, let $\begin{bmatrix} \mathbf{x}_{BI_1} \\ \mathbf{0} \end{bmatrix}$ and $\begin{bmatrix} \mathbf{x}_{BI_2} \\ \mathbf{0} \end{bmatrix}$ be feasible basic solutions corresponding to the invertible (m, m)-submatrices $\mathbf{B}_1$ and $\mathbf{B}_2$, respectively, of $\begin{bmatrix} \mathbf{A} & \mathbf{I}_m \end{bmatrix}$ that differ in precisely one index. The sets of hyperplanes $\{H_1, \ldots, H_{n-1}, H_n\}$ and $\{H_1, \ldots, H_{n-1}, H_{n+1}\}$, corresponding to the complementary dual matrices $\overline{\mathbf{B}}_1$ and $\overline{\mathbf{B}}_2$ of $\mathbf{B}_1$ and $\mathbf{B}_2$, respectively, are both independent (see Theorem 2.2.1), and define therefore a vertex $\mathbf{u}$ and a vertex $\mathbf{v}$ of F, respectively. If $\mathbf{u} \neq \mathbf{v}$, then $\mathbf{u}$ and $\mathbf{v}$ are adjacent. If $\mathbf{u} = \mathbf{v}$, then $\mathbf{u} \, (= \mathbf{v})$ is a degenerate vertex of F, because all $n + 1$ hyperplanes $H_1, \ldots, H_{n-1}, H_n, H_{n+1}$ contain $\mathbf{u} \, (= \mathbf{v})$. $\qquad \square$

The phrase "there exists" in Theorem 2.2.4 is important. Not every pair of feasible basic solutions corresponding to adjacent vertices is also a pair of adjacent feasible basic solutions. Also, in the case of degeneracy, adjacent feasible basic solutions may correspond to the same vertex, and hence they do not correspond to adjacent vertices.

2.3 Exercises

Exercise 2.3.1.

(a) Give an example in $\mathbb{R}^2$ of an LO-model in which the feasible region has a degenerate vertex that has no redundant binding constraints.

(b) Give an example in $\mathbb{R}^3$ of an LO-model in which the optimal solution is degenerate without redundant binding technology constraints and all nonnegativity constraints are redundant.

Exercise 2.3.2.   Consider the feasible region

$$F = \left\{ \begin{bmatrix} x_1 & x_2 & x_3 \end{bmatrix}^{\mathsf{T}} \in \mathbb{R}^3 \; \middle| \; x_1 + x_2 + x_3 \leq 1, \; x_1 \leq 1, \; x_1, x_2, x_3 \geq 0 \right\}.$$

(a) Draw F and identify its vertices and their adjacencies.

(b) For each vertex of F, identify the hyperplanes that determine it.

(c) For each pair of adjacent vertices of F, check that they are adjacent using the definition of adjacency in terms of hyperplanes.

Exercise 2.3.3.   Consider the feasible region

$$F = \left\{ \begin{bmatrix} x_1 & x_2 & x_3 \end{bmatrix}^{\mathsf{T}} \in \mathbb{R}^3 \; \middle| \; x_1 + x_2 + x_3 \geq 1, \; x_1, x_2 \geq 0, \; 0 \leq x_3 \leq 1 \right\}.$$

(a)   Determine all vertices and directions of unboundedness of F.

(b)   Write F in the form $F = P + C$ ($= \{x + y \mid x \in P, y \in C\}$), where P is a polytope and C a polyhedral cone (see Appendix D) of $\mathbb{R}^3$.

Exercise 2.3.4.   Show that any LO-model with a bounded feasible region has no directions of unboundedness.

Exercise 2.3.5.   Consider the following assertion. If every vertex of the feasible region of a feasible LO-model is nondegenerate, then the optimal solution is unique. If this assertion is true, give a proof; otherwise, give a counterexample.

Exercise 2.3.6.   Show that the feasible region of the following LO-model is unbounded, and that it has multiple optimal solutions.

$$
\begin{aligned}
\max \quad & -2x_1 + x_2 + x_3 \\
\text{s.t.} \quad & x_1 - x_2 - x_3 \geq -2 \\
& x_1 - x_2 + x_3 \leq 2 \\
& x_1, x_2 \geq 0,\ x_3 \text{ free.}
\end{aligned}
$$

Determine all vertices (extreme points) and extreme directions of the feasible region.

Exercise 2.3.7.   Consider a feasible standard LO-model. Assume that this model has an unbounded 'optimal solution' in such a way that the objective function can be made arbitrarily large by increasing the value of some decision variable, x_k, say, while keeping the values of the other variables fixed. Let $\alpha > 0$, and consider the original model extended with the constraint $x_k \leq \alpha$. Assuming that this new model has a (bounded) optimal solution, show that it has an optimal solution $\mathbf{x}^*$ for which $x_k^* = \alpha$.

Exercise 2.3.8.   Prove that (ii) implies (i) in the proof of Theorem 2.2.1.

Exercise 2.3.9.   Prove that an unbounded feasible region cannot be written as the convex hull of finitely many points.

Exercise 2.3.10.   Prove that the $F_J = F \cap \left(\bigcap_{j \in J} H_j \right)$ in Section 2.1.3 holds.

Exercise 2.3.11.   Let $F \subseteq \mathbb{R}^n$ be the feasible region of a standard LO-model with m constraints (including any nonnegativity and nonpositivity constraints) and let F_J be a face of F with $J \subseteq \{1, \ldots, m\}$. Show that if F_J contains exactly one point, then this point is a vertex of F.

Exercise 2.3.12.   Consider Model Dovetail. Using the method in Example D.3.1, construct values of $\lambda_1, \ldots, \lambda_5 \geq 0$ with $\sum_{i=1}^{5} \lambda_i$ so that $\mathbf{x}^0 = \lambda \mathbf{v}_1 + \ldots + \lambda_4 \mathbf{v}_4 + \lambda_5 \mathbf{0}$. Is this choice of values of λ_i's unique? Explain your answer.

Exercise 2.3.13.   For $n = 1, 2$, show that if the feasible region $F \subseteq \mathbb{R}^n$ of an LO-model has a degenerate vertex, then there is a redundant constraint.

CHAPTER 3

Dantzig's simplex algorithm

Overview

The simplex algorithm that is described in this chapter was invented in 1947 by GEORGE B. DANTZIG (1914–2005). In 1963, his famous book *Linear Programming and Extensions* was published by Princeton University Press, Princeton, New Jersey. Since that time, the implementations of the algorithm have improved drastically. Linear optimization models with millions of variables and constraints can nowadays readily be solved by the simplex algorithm using modern computers and sophisticated implementations. The basic idea of Dantzig's simplex algorithm for solving LO-models is to manipulate the columns of the technology matrix of the model in such a way that after a finite number of steps an optimal solution is achieved. These steps correspond to jumps from vertex to vertex along the edges of the feasible region of the LO-model, while these jumps in turn correspond to manipulations of the rows of the technology matrix. This relationship between manipulating columns and manipulating rows was explained in Chapter 2.

3.1 From vertex to vertex to an optimal solution

In this chapter we will introduce a step-by-step procedure (an *algorithm*) for solving linear optimization models. This procedure is the so-called *simplex algorithm*. Before we explain the general form of the simplex algorithm in Section 3.3, we discuss its working by means of Model Dovetail; see Section 1.1. The objective function of this model is $3x_1 + 2x_2$. We found in Section 1.1.2 that the objective function attains its maximum value at a vertex of the feasible region, namely vertex $\mathbf{v}_2 = \begin{bmatrix} 4\frac{1}{2} & 4\frac{1}{2} \end{bmatrix}^\mathsf{T}$; see Figure 3.1.

In Section 2.2.2, we saw that each vertex of the feasible region is determined by setting two variables (the nonbasic variables) to zero in Model Dovetail with slack variables (Section 1.2.2), and that the basic variables can be expressed in terms of the nonbasic variables. We also saw that there are fifteen possible ways to select a $(4, 4)$-submatrix in the technology

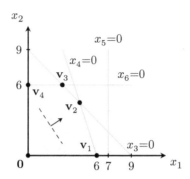

Figure 3.1: The feasible region of Model Dovetail.

matrix $\begin{bmatrix} \mathbf{A} & \mathbf{I}_4 \end{bmatrix}$ of Model Dovetail with slack variables and that, therefore, there are fifteen candidate choices of basic variables that determine a basic solution. Some of these candidates lead to feasible basic solutions, and some of them may lead to infeasible basic solutions. The feasible basic solutions correspond to vertices of the feasible region. Therefore, in the light of the optimal vertex theorem (Theorem 2.1.5), we could in principle find an optimal solution by determining, for each of these fifteen choices, whether they correspond to a feasible basic solution, and then choosing among them the one that has the largest objective value. This approach, however, has two major drawbacks. First, in general, the number of feasible basic solutions can be extremely large: for a standard LO-model with n variables and m constraints, the number of candidates is $\binom{m+n}{m}$. Therefore, checking all possibilities is very inefficient; see also Chapter 9. As we will see, the simplex algorithm is a procedure that considers vertices in a more systematic way. A second drawback is that considering all feasible basic solutions does not provide a way to detect that the LO-model at hand has an unbounded solution. The simplex algorithm does not have this restriction.

The idea behind the simplex algorithm is surprisingly simple: it starts at a vertex of the feasible region, if possible $\mathbf{0}$, and then continues by choosing a vertex adjacent to the starting vertex. In each iteration the simplex algorithm finds, except possibly in the case of degeneracy, a vertex adjacent to the previous one but with a larger objective value. The algorithm stops when the objective value cannot be improved by choosing a new adjacent vertex.

Recall Model Dovetail with slack variables:

$$
\begin{aligned}
\max \quad & 3x_1 + 2x_2 && \text{(M1)}\\
\text{s.t.} \quad & x_1 + x_2 + x_3 && = 9\\
& 3x_1 + x_2 + x_4 && = 18\\
& x_1 + x_5 && = 7\\
& x_2 + x_6 && = 6\\
& x_1, x_2, x_3, x_4, x_5, x_6 \geq 0.
\end{aligned}
$$

At the origin vertex $\mathbf{0}$ of this model, x_1 and x_2 are the nonbasic variables, and the slack variables x_3, x_4, x_5, and x_6 are the basic variables. The objective function $3x_1 + 2x_2$

has value 0 at that vertex. The coefficients of the objective function corresponding to the nonbasic variables x_1 and x_2 (3 and 2, respectively) are both positive. Therefore, we can increase the objective value by increasing the value of one of the nonbasic variables. Suppose that we choose to increase the value of x_1. (It is left to the reader to repeat the calculations when choosing x_2 instead of x_1.) The value of x_2 remains zero. In terms of Figure 3.1, this means that we start at the origin $\mathbf{0}$, and we move along the horizontal axis in the direction of vertex $\mathbf{v}_1$. An obvious question is: by how much can the value of x_1 be increased without 'leaving' the feasible region, i.e., without violating the constraints? It is graphically clear from Figure 3.1 that we can move as far as vertex $\mathbf{v}_1$, but no further.

To calculate algebraically by how much the value of x_1 may be increased, we move to the realm of feasible basic solutions. We started at the feasible basic solution for which x_1 and x_2 are the nonbasic variables, corresponding to the origin vertex $\mathbf{0}$. Since we are increasing the value of x_1, the variable x_1 has to become a basic variable. We say that x_1 *enters* the set of basic variables, or that x_1 is the *entering nonbasic variable*. Recall that we want to jump from one feasible basic solution to another. Recall also that any given feasible basic solution corresponds to a set of exactly m ($= 4$) basic variables. Since we are jumping to a new vertex and x_1 becomes a basic variable, this means that we have to determine a variable that leaves the set of basic variables. Moreover, because the value of x_2 remains zero, the variable x_2 remains nonbasic. Taking $x_2 = 0$ in the constraints of (M1), and using the fact that the slack variables have to have nonnegative values, we obtain the inequalities:

$$x_1 \leq 9, \ 3x_1 \leq 18, \ \text{and} \ x_1 \leq 7.$$

So, we may increase the value of x_1 to at most $\min\{9/1, 18/3, 7/1\} = \min\{9, 6, 7\} = 6$. Any larger value will yield a point outside the feasible region. Determining by how much the value of the entering variable can be increased involves finding a minimum ratio (in this case among $9/1$, $18/3$, and $7/1$). This process is called the *minimum-ratio test*.

Figure 3.1 illustrates the situation geometrically: starting at the origin vertex $\mathbf{0}$, we move along the x_1-axis as the value of x_1 increases. As we move 'to the right', we consecutively hit the lines corresponding to $x_4 = 0$ (at $x_1 = 6$), $x_5 = 0$ (at $x_1 = 7$), and $x_3 = 0$ (at $x_1 = 9$). Clearly, we should stop increasing the value of x_1 when x_1 has value 6, because if the value of x_1 is increased more than that, we will leave the feasible region. Observe that the values $6, 7, 9$ in parentheses correspond to the values in the minimum-ratio test.

Because we want to end up at a vertex, we set x_1 to the largest possible value, i.e., we set $x_1 = 6$, and we leave $x_2 = 0$, as it was before. This implies that $x_3 = 3$, $x_4 = 0$, $x_5 = 1$, $x_6 = 6$. Since we stopped increasing the value of x_1 when we hit the line $x_4 = 0$, we necessarily have that $x_4 = 0$ at the new vertex. Thus, we can choose x_4 as the new nonbasic variable. We say that x_4 *leaves* the set of basic variables, or that x_4 is the *leaving basic variable*. So we are now at the feasible basic solution with nonbasic variables x_2 and x_4. This feasible basic solution corresponds to vertex $\mathbf{v}_1$, i.e., the vertex determined by the lines $x_2 = 0$ and $x_4 = 0$.

We now rewrite (M1) by means of the constraint:

$$3x_1 + x_2 + x_4 = 18.$$

The reason for choosing this particular constraint (instead of one of the other constraints) is that this is the only constraint of (M1) that contains the leaving variable x_4. Solving for x_1, we find that:

$$x_1 = \tfrac{1}{3}(18 - x_2 - x_4). \tag{3.1}$$

Substituting (3.1) into the objective $\max 3x_1 + 2x_2$, the new objective becomes:

$$\max 18 + x_2 - x_4.$$

We can rewrite the remaining constraints in a similar way by substituting (3.1) into them. After straightforward calculations and reordering the constraints, the model becomes:

$$
\begin{array}{llll}
\min & x_2 - x_4 & + 18 & \text{(M2)} \\
\text{s.t.} & \tfrac{1}{3}x_2 + \tfrac{1}{3}x_4 + x_1 & = 6 \\
& \tfrac{2}{3}x_2 - \tfrac{1}{3}x_4 + x_3 & = 3 \\
& -\tfrac{1}{3}x_2 - \tfrac{1}{3}x_4 + x_5 & = 1 \\
& x_2 + x_6 & = 6 \\
& x_2, x_4, x_1, x_3, x_5, x_6 \geq 0.
\end{array}
$$

This model is equivalent to (M1) in the sense that both models have the same feasible region and the same objective; this will be explained in more detail in Section 3.2. Notice also that we have written the model in such a way that the current basic variables x_1, x_3, x_5, and x_6 play the role of slack variables. In particular, they appear in exactly one equation of (M2), and they do so with coefficient 1. Figure 3.2 shows the feasible region of (M2), when x_1, x_3, x_5, and x_6 are interpreted as slack variables. Observe that, although the shape of the feasible region has changed compared to the one shown in Figure 3.1, the vertices and the adjacencies between the vertices are preserved. Notice also that the objective vector has rotated accordingly and still points towards the optimal vertex $\mathbf{v}_2$.

Informally speaking, it is as if we are looking at the feasible region with a camera, and we move the camera position from $\mathbf{0}$ to $\mathbf{v}_1$, so that we are now looking at the feasible region from the point $\mathbf{v}_1$. We therefore say that the feasible region depicted in Figure 3.2 is the feasible region of Model Dovetail from the perspective of the vertex $\mathbf{v}_1$. The vertex $\mathbf{v}_1$ is now the origin of the feasible region. That is, it corresponds to setting $x_2 = 0$ and $x_4 = 0$. Substituting these equations into the constraints and the objective function of (M2) implies that $x_1 = 6$, $x_3 = 3$, $x_5 = 1$, and $x_6 = 6$, and the current corresponding objective value is 18.

Comparing the objective function of (M1) (namely, $3x_1 + 2x_2$) with the objective function of (M2), (namely, $18 + x_2 - x_4$), we see that in the first case the coefficients of the nonbasic variables x_1 and x_2 are both positive, and in the second case the coefficient of x_2 is positive

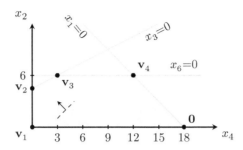

Figure 3.2: The feasible region from the perspective of vertex $\mathbf{v}_1$.

and that of x_4 is negative. The constant in the objective function is 0 in the first case, and 18 in the second case.

Next, since the objective coefficient of x_2 in (M2) is positive, we choose x_2 to enter the set of basic variables. This means that we will increase the value of x_2, while x_4 remains a nonbasic variable and, therefore, the value of x_4 remains zero. In terms of Figure 3.2, this means that we are moving upward along the x_2-axis, starting at $\mathbf{v}_1$, in the direction of $\mathbf{v}_2$. Taking $x_4 = 0$ in the constraints of (M2), we find the following inequalities:

$$\tfrac{1}{3}x_2 \le 6, \ \tfrac{2}{3}x_2 \le 3, \ -\tfrac{1}{3}x_2 \le 1, \text{ and } x_2 \le 6.$$

The third inequality is, because of the negative coefficient in front of x_2, not restrictive when determining a maximum value for x_2. Hence, according to the minimum-ratio test, the value of x_2 can be increased to at most:

$$\min\{3/\tfrac{2}{3}, 6/\tfrac{1}{3}, \star, 6/1\} = \min\{4\tfrac{1}{2}, 18, \star, 6\} = 4\tfrac{1}{2},$$

where '$\star$' signifies the nonrestrictive third inequality. Taking $x_2 = 4\tfrac{1}{2}$ and $x_4 = 0$, we find that $x_1 = 4\tfrac{1}{2}$, $x_3 = 0$, $x_5 = 2\tfrac{1}{2}$, and $x_6 = 1\tfrac{1}{2}$. Geometrically, we see from Figure 3.2 that, as we move from $\mathbf{v}_1$ towards $\mathbf{v}_2$, the first constraint that is hit is the one corresponding to the line $x_3 = 0$. Therefore, x_3 is the new nonbasic variable, i.e., it leaves the set of basic variables. So, currently, the basic variables are x_1, x_2, x_5, and x_6.

We again rewrite the model by using the unique constraint of (M2) that contains the leaving variable x_3, namely:

$$\tfrac{2}{3}x_2 - \tfrac{1}{3}x_4 + x_3 = 3.$$

Solving for x_2, we find that:

$$x_2 = 4\tfrac{1}{2} + \tfrac{1}{2}x_4 - \tfrac{3}{2}x_3.$$

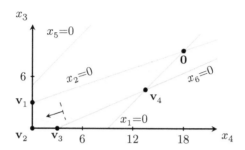

Figure 3.3: The feasible region from the perspective of vertex $\mathbf{v}_2$.

Again, after substituting this expression into the objective function and the constraints of (M2), we find the following equivalent LO-model:

$$
\begin{aligned}
\max \quad & -1\tfrac{1}{2}x_3 - \tfrac{1}{2}x_4 && + 22\tfrac{1}{2} && \text{(M3)} \\
\text{s.t.} \quad & -\tfrac{1}{2}x_3 + \tfrac{1}{2}x_4 + x_1 && = 4\tfrac{1}{2} \\
& 1\tfrac{1}{2}x_3 - \tfrac{1}{2}x_4 \quad + x_2 && = 4\tfrac{1}{2} \\
& \tfrac{1}{2}x_3 - \tfrac{1}{2}x_4 \quad\quad + x_5 && = 2\tfrac{1}{2} \\
& -1\tfrac{1}{2}x_3 + \tfrac{1}{2}x_4 \quad\quad\quad + x_6 && = 1\tfrac{1}{2} \\
& x_3, x_4, x_1, x_2, x_5, x_6 \geq 0.
\end{aligned}
$$

Now it is clear that the objective value cannot be improved anymore, because the coefficients of both x_3 and x_4 are negative: increasing the value of x_3 or x_4 will decrease the objective value. For $x_3 = 0$, $x_4 = 0$ (corresponding to vertex $\mathbf{v}_2$), the objective value is $22\tfrac{1}{2}$, which is therefore the optimal value. The values of the basic variables x_1, x_2, x_5, and x_6 are $4\tfrac{1}{2}$, $4\tfrac{1}{2}$, $2\tfrac{1}{2}$, and $1\tfrac{1}{2}$, respectively. Figure 3.3 shows the feasible region of (M3), when x_1, x_2, x_5, and x_6 are interpreted as slack variables.

Summarizing the above procedure, we have taken the following steps (see Figure 3.4):

1. Start at $\mathbf{0}$ ($x_1 = 0, x_2 = 0$); the objective value is 0.

2. Go to vertex $\mathbf{v}_1$ ($x_2 = 0, x_4 = 0$); the objective value is 18.

3. Go to vertex $\mathbf{v}_2$ ($x_3 = 0, x_4 = 0$); the objective value is $22\tfrac{1}{2}$.

4. Stop, because the coefficients of x_3 and x_4 in the objective function are negative.

In each step, we produced a new feasible basic solution, and we rewrote the original LO-model in such a way that it was straightforward to decide whether the current feasible basic solution was optimal and, if not, how to determine the next step. This idea of rewriting the model in an equivalent form is one of the key features of the simplex algorithm.

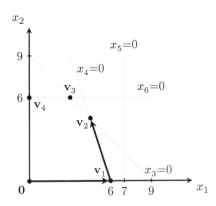

Figure 3.4: Path taken by the simplex algorithm when applied to Model Dovetail.

3.2 LO-model reformulation

As mentioned in Section 3.1, reformulating the LO-model is a crucial ingredient of the simplex algorithm. Recall the solution method explained in Section 3.1. During the steps of this solution method, we produced three intermediate LO-models (M1), (M2), and (M3). We saw that these models had a special structure: in each of these LO-models, the basic variables appear in exactly one equation, and they do so with coefficient 1. Additionally, in each of the LO-models, the basic variables do not appear in the objective function. To see how this works in general, consider the standard LO-model with slack variables:

$$\max \ \mathbf{c}^\mathsf{T}\mathbf{x}$$
$$\text{s.t.} \quad \mathbf{A}\mathbf{x} + \mathbf{x}_s = \mathbf{b} \tag{3.2}$$
$$\mathbf{x} \geq \mathbf{0}, \mathbf{x}_s \geq \mathbf{0}.$$

Now consider a feasible basic solution. Let $BI \subseteq \{1, \ldots, n+m\}$ be the set of indices of the corresponding basic variables, and let $NI = \{1, \ldots, n+m\} \setminus BI$ be the set of indices of the nonbasic variables. Note that $BI \cap NI = \emptyset$ and $BI \cup NI = \{1, \ldots, n+m\}$. Define:

$$\mathbf{B} = \begin{bmatrix} \mathbf{A} & \mathbf{I}_m \end{bmatrix}_{\star,BI}, \text{ and } \mathbf{N} = \begin{bmatrix} \mathbf{A} & \mathbf{I}_m \end{bmatrix}_{\star,NI}.$$

Following the discussion in Section 2.2.2 and in particular equation (2.1), we have that, given any feasible basic solution with a corresponding basis matrix, the basic variables may be uniquely expressed in terms of the nonbasic variables. In particular, we have that:

$$\mathbf{x}_{BI} = \mathbf{B}^{-1}\mathbf{b} - \mathbf{B}^{-1}\mathbf{N}\mathbf{x}_{NI}. \tag{3.3}$$

Recall that, because $\mathbf{B}$ corresponds to a feasible basic solution, we have that $\mathbf{B}^{-1}\mathbf{b} \geq \mathbf{0}$. The fact that the basic variables can be expressed uniquely in terms of the nonbasic variables means that we may rewrite model (3.2) into an equivalent LO-model. By 'equivalent', we mean that the two LO-models have the same set of feasible solutions, and the same objective value for each feasible solution. First, using (3.3), we may rewrite the technology constraints of (3.2) as follows:

$$\mathbf{Ax} + \mathbf{I}_m\mathbf{x}_s = \mathbf{b} \iff \begin{bmatrix} \mathbf{A} & \mathbf{I}_m \end{bmatrix} \begin{bmatrix} \mathbf{x} \\ \mathbf{x}_s \end{bmatrix} = \mathbf{b}$$

$$\iff \begin{bmatrix} \mathbf{B} & \mathbf{N} \end{bmatrix} \begin{bmatrix} \mathbf{x}_{BI} \\ \mathbf{x}_{NI} \end{bmatrix} = \mathbf{b}$$

$$\iff \mathbf{Bx}_{BI} + \mathbf{Nx}_{NI} = \mathbf{b} \tag{3.4}$$

$$\iff \mathbf{B}^{-1}\mathbf{Bx}_{BI} + \mathbf{B}^{-1}\mathbf{Nx}_{NI} = \mathbf{B}^{-1}\mathbf{b}$$

$$\iff \mathbf{I}_m\mathbf{x}_{BI} + \mathbf{B}^{-1}\mathbf{Nx}_{NI} = \mathbf{B}^{-1}\mathbf{b}$$

$$\iff \begin{bmatrix} \mathbf{I}_m & \mathbf{B}^{-1}\mathbf{N} \end{bmatrix} \begin{bmatrix} \mathbf{x}_{BI} \\ \mathbf{x}_{NI} \end{bmatrix} = \mathbf{B}^{-1}\mathbf{b}.$$

Here, we have used the fact that $\mathbf{B}$ is an invertible matrix. Next, again using (3.3), we may rewrite the objective function as:

$$\mathbf{c}^\mathsf{T}\mathbf{x} = \mathbf{c}_{BI}^\mathsf{T}\mathbf{x}_{BI} + \mathbf{c}_{NI}^\mathsf{T}\mathbf{x}_{NI} = \mathbf{c}_{BI}^\mathsf{T}\left(\mathbf{B}^{-1}\mathbf{b} - \mathbf{B}^{-1}\mathbf{Nx}_{NI}\right) + \mathbf{c}_{NI}^\mathsf{T}\mathbf{x}_{NI}$$
$$= \mathbf{c}_{BI}^\mathsf{T}\mathbf{B}^{-1}\mathbf{b} + \left(\mathbf{c}_{NI}^\mathsf{T} - \mathbf{c}_{BI}^\mathsf{T}\mathbf{B}^{-1}\mathbf{N}\right)\mathbf{x}_{NI}. \tag{3.5}$$

Therefore, we have that (3.2) is in fact equivalent to:

$$\begin{aligned} \max \quad & \mathbf{c}_{BI}^\mathsf{T}\mathbf{B}^{-1}\mathbf{b} + \left(\mathbf{c}_{NI}^\mathsf{T} - \mathbf{c}_{BI}^\mathsf{T}\mathbf{B}^{-1}\mathbf{N}\right)\mathbf{x}_{NI} \\ \text{s.t.} \quad & \mathbf{I}_m\mathbf{x}_{BI} + \mathbf{B}^{-1}\mathbf{Nx}_{NI} = \mathbf{B}^{-1}\mathbf{b} \\ & \mathbf{x}_{BI} \geq \mathbf{0}, \mathbf{x}_{NI} \geq \mathbf{0}. \end{aligned} \tag{3.6}$$

It should be stressed once again that the equivalence of (3.2) and (3.6) means that both models have the same set of feasible solutions, and the objective value of each feasible solution is the same in both models. Hence, the two models have the same optimal solution(s). This can be seen as follows. The expression in (3.4) shows that any feasible solution of (3.2) (i.e., any vector $\begin{bmatrix} \mathbf{x} \\ \mathbf{x}_s \end{bmatrix}$ satisfying $\mathbf{Ax} + \mathbf{I}_m\mathbf{x}_s = \mathbf{b}$) is also a feasible solution of (3.6), and vice versa. Moreover, (3.5) shows that the objective functions of the two models coincide.

The term $\mathbf{c}_{BI}^\mathsf{T}\mathbf{B}^{-1}\mathbf{b}$ in the objective function of (3.6) is a constant (i.e., it does not depend on the decision variables), and the variables in $\mathbf{x}_{BI}$ do not appear in the objective function. Moreover, because $\mathbf{I}_m$ is an identity matrix, each basic variable in $\mathbf{x}_{BI}$ appears in exactly one constraint, and it does so with coefficient 1. That is, (3.6) is a standard LO-model with slack variables, where the slack variables are the variables in $\mathbf{x}_{BI}$. We can think of (3.6) as the formulation of (3.2) from the perspective of the feasible basic solution at hand.

The current feasible basic solution corresponding to (3.6) can be easily found by setting $\mathbf{x}_{NI} = \mathbf{0}$; the constraints of (3.6) then state that $\mathbf{x}_{BI} = \mathbf{B}^{-1}\mathbf{b}$. Since $\mathbf{B}^{-1}\mathbf{b} \geq \mathbf{0}$, we have that $\begin{bmatrix} \mathbf{x}_{BI} \\ \mathbf{x}_{NI} \end{bmatrix}$ is in fact a feasible solution of (3.6) (and, hence, also of (3.2)).

In virtue of (3.3), each constraint of (3.6) corresponds to a unique basic variable. Therefore, for each constraint of (3.6), we sometimes add the corresponding basic variable as a row label.

Example 3.2.1.   *Consider again* (M2). *We have that the basic variables corresponding to the feasible basic solution are x_1, x_3, x_5, x_6. That is, $BI = \{1, 3, 5, 6\}$ and $NI = \{2, 4\}$. Hence, we have that:*

$$
\mathbf{B} = \begin{array}{c} \\ \\ \\ \\ \end{array}\!\!\begin{bmatrix} \overset{x_1}{1} & \overset{x_3}{1} & \overset{x_5}{0} & \overset{x_6}{0} \\ 3 & 0 & 0 & 0 \\ 1 & 0 & 1 & 0 \\ 0 & 0 & 0 & 1 \end{bmatrix}, \mathbf{N} = \begin{bmatrix} \overset{x_2}{1} & \overset{x_4}{0} \\ 1 & 1 \\ 0 & 0 \\ 1 & 0 \end{bmatrix}, \text{ and } \mathbf{B}^{-1} = \begin{bmatrix} 0 & \frac{1}{3} & 0 & 0 \\ 1 & -\frac{1}{3} & 0 & 0 \\ 0 & -\frac{1}{3} & 1 & 0 \\ 0 & 0 & 0 & 1 \end{bmatrix}.
$$

It is left to the reader to check that:

$$
\mathbf{c}_{BI}^{\mathsf{T}} \mathbf{B}^{-1} \mathbf{b} = 18, \quad \mathbf{c}_{NI}^{\mathsf{T}} - \mathbf{c}_{BI}^{\mathsf{T}} \mathbf{B}^{-1} \mathbf{N} = \begin{bmatrix} \overset{x_2}{1} & \overset{x_4}{-1} \end{bmatrix},
$$

$$
\mathbf{B}^{-1}\mathbf{N} = \begin{array}{c} x_1 \\ x_3 \\ x_5 \\ x_6 \end{array}\!\!\begin{bmatrix} \overset{x_2}{\frac{1}{3}} & \overset{x_4}{\frac{1}{3}} \\ \frac{2}{3} & -\frac{1}{3} \\ -\frac{1}{3} & -\frac{1}{3} \\ 1 & 0 \end{bmatrix}, \quad \mathbf{B}^{-1}\mathbf{b} = \begin{array}{c} x_1 \\ x_3 \\ x_5 \\ x_6 \end{array}\!\!\begin{bmatrix} 6 \\ 3 \\ 1 \\ 6 \end{bmatrix}.
$$

From these expressions, it can be seen that the LO-model formulation (M2) *is of the form* (3.6). *That is,* (M3) *is equivalent to Model Dovetail, but it is written from the perspective of the feasible basic solution with basic variables x_1, x_3, x_5, x_6, so that:*

$$
\mathbf{B}^{-1}\begin{bmatrix}\mathbf{B} & \mathbf{N}\end{bmatrix} = \begin{bmatrix}\mathbf{I}_m & \mathbf{B}^{-1}\mathbf{N}\end{bmatrix} = \begin{array}{c} x_1 \\ x_3 \\ x_5 \\ x_6 \end{array}\!\!\begin{bmatrix} \overset{x_1}{1} & \overset{x_3}{0} & \overset{x_5}{0} & \overset{x_6}{0} & \overset{x_2}{\frac{1}{3}} & \overset{x_4}{\frac{1}{3}} \\ 0 & 1 & 0 & 0 & \frac{2}{3} & -\frac{1}{3} \\ 0 & 0 & 1 & 0 & -\frac{1}{3} & -\frac{1}{3} \\ 0 & 0 & 0 & 1 & 1 & 0 \end{bmatrix}.
$$

Also, the current objective coefficients of the first and second nonbasic variables are 1 and -1, respectively. It is left to the reader to check that the LO-model formulations (M1) *and* (M3) *are also of the form* (3.6).

3.3   The simplex algorithm

In this section, we formalize the solution technique described in Section 3.1, and formulate the simplex algorithm in general. The algorithm solves LO-models of the standard form $\max\{\mathbf{c}^{\mathsf{T}}\mathbf{x} \mid \mathbf{A}\mathbf{x} \leq \mathbf{b}, \mathbf{x} \geq \mathbf{0}\}$. Roughly, it consists of two key steps:

1. (*Initialization*) Start at an initial feasible basic solution.

2. (*Iteration step*) Assume that we have a feasible basic solution. Use this feasible basic solution to either:

   (a) establish that an optimal solution has been found and stop, or

   (b) establish that the model has an unbounded solution and stop, or

(c) find a variable that enters the set of basic variables and a variable that leaves the set of basic variables, and construct a new feasible basic solution.

Because of Theorem 2.2.2 and Theorem 2.2.4, the procedure of 'jumping' from vertex to vertex along the boundary of the feasible region is a routine algebraic matter of manipulating systems of linear equations $\mathbf{Ax} + \mathbf{x}_s = \mathbf{b}$ with $\mathbf{x} \geq \mathbf{0}$ and $\mathbf{x}_s \geq \mathbf{0}$. Jumping to an adjacent vertex means changing the corresponding feasible basic solution by exchanging a basic variable with a suitable nonbasic variable.

3.3.1   Initialization; finding an initial feasible basic solution

To start the simplex algorithm, an initial feasible basic solution is needed. In many cases, the origin $\mathbf{0}$ may be used for this purpose. To do this, we choose as the initial basic variables the slack variables of the model. This means that $BI = \{n + 1, \ldots, n + m\}$, and $NI = \{1, \ldots, n\}$. Because the columns of $\begin{bmatrix} \mathbf{A} & \mathbf{I}_m \end{bmatrix}$ corresponding to the slack variables form the identity matrix $\mathbf{I}_m$, we have that $\mathbf{B} = \mathbf{I}_m$, and $\mathbf{N} = \mathbf{A}$. The corresponding basic solution is $\begin{bmatrix} \mathbf{x}_{BI} \\ \mathbf{x}_{NI} \end{bmatrix}$ with $\mathbf{x}_{BI} = \mathbf{B}^{-1}\mathbf{b} = \mathbf{b}$ and $\mathbf{x}_{NI} = \mathbf{0}$. As long as $\mathbf{b} \geq \mathbf{0}$, this basic solution is feasible. If $\mathbf{b} \not\geq \mathbf{0}$, other methods have to be used to find an initial feasible basic solution. In Section 3.6, two methods are described that deal with finding an initial feasible basic solution in general.

3.3.2   The iteration step; exchanging variables

Suppose that we have a current feasible basic solution, i.e., index sets BI and NI, and the corresponding matrices $\mathbf{B}$ and $\mathbf{N}$. Write $BI = \{BI_1, \ldots, BI_m\}$ and $NI = \{NI_1, \ldots, NI_n\}$ with $BI_1 < BI_2 < \ldots < BI_m$, and $NI_1 < NI_2 < \ldots < NI_n$. From this, we can readily calculate the vector $\mathbf{c}_{NI}^\mathsf{T} - \mathbf{c}_{BI}^\mathsf{T} \mathbf{B}^{-1} \mathbf{N}$, containing the current objective coefficients for the nonbasic variables. The current feasible basic solution is $\begin{bmatrix} \mathbf{x}_{BI} \\ \mathbf{x}_{NI} \end{bmatrix}$ with $\mathbf{x}_{BI} = \mathbf{B}^{-1}\mathbf{b}$ and $\mathbf{x}_{NI} = \mathbf{0}$.

If all current objective coefficients are nonpositive (to be precise, all entries of $\mathbf{c}_{NI}^\mathsf{T} - \mathbf{c}_{BI}^\mathsf{T} \mathbf{B}^{-1} \mathbf{N}$ have a nonpositive value), then it is clear from (3.6) that setting $\mathbf{x}_{NI} = \mathbf{0}$ results in an optimal solution of (3.6), and hence of the original LO-model. This is the situation that we encountered in model (M3) in Section 3.1. Because an optimal solution has been found, we may stop at this point.

Now suppose that one of the nonbasic variables has a positive current objective coefficient. Suppose this variable is x_{NI_α}, with $\alpha \in \{1, \ldots, n\}$. This is the variable that will enter the set of basic variables. Since the current objective coefficient corresponding to x_{NI_α} is positive, we can increase the objective value by increasing the value of x_{NI_α}, while keeping the values of all other nonbasic variables at zero. To determine by how much the value of

x_{NI_α} can be increased, we substitute $x_{NI_i} = 0$ for $i \neq \alpha$ in (3.3).

$$\mathbf{x}_{BI} = \mathbf{B}^{-1}\mathbf{b} - \left[(\mathbf{B}^{-1}\mathbf{N})_{\star,1} \quad \cdots \quad (\mathbf{B}^{-1}\mathbf{N})_{\star,n} \right] \begin{bmatrix} x_{NI_1} \\ \vdots \\ x_{NI_n} \end{bmatrix}$$

$$= \mathbf{B}^{-1}\mathbf{b} - (\mathbf{B}^{-1}\mathbf{N})_{\star,\alpha} x_{NI_\alpha}.$$

Note that $(\mathbf{B}^{-1}\mathbf{N})_{\star,\alpha}$ is the column of $\mathbf{B}^{-1}\mathbf{N}$ that corresponds to the nonbasic variable x_{NI_α}. The above expression represents a system of m linear equations. In fact, it is instructive to explicitly write out these equations:

$$(\mathbf{x}_{BI})_j = x_{BI_j} = (\mathbf{B}^{-1}\mathbf{b})_j - (\mathbf{B}^{-1}\mathbf{N})_{j,\alpha} x_{NI_\alpha} \qquad \text{for } j = 1, \ldots, m. \qquad (3.7)$$

Using the fact that $x_{BI_j} \geq 0$ for $j \in \{1, \ldots, m\}$, we obtain the following inequalities for x_{NI_α}:

$$(\mathbf{B}^{-1}\mathbf{N})_{j,\alpha} x_{NI_\alpha} \leq (\mathbf{B}^{-1}\mathbf{b})_j \qquad \text{for } j = 1, \ldots, m. \qquad (3.8)$$

Among these inequalities, the ones with $(\mathbf{B}^{-1}\mathbf{N})_{j,\alpha} \leq 0$ do not provide an upper bound for the value of x_{NI_α}, i.e., they form no restriction when increasing the value of x_{NI_α}. So, since we are interested in determining by how much the value of x_{NI_α} can be increased without violating any of these inequalities, we can compactly write the relevant inequalities as follows:

$$x_{NI_\alpha} \leq \min \left\{ \frac{(\mathbf{B}^{-1}\mathbf{b})_j}{(\mathbf{B}^{-1}\mathbf{N})_{j,\alpha}} \;\middle|\; (\mathbf{B}^{-1}\mathbf{N})_{j,\alpha} > 0, j = 1, \ldots, m \right\}. \qquad (3.9)$$

Let δ be the value of the right hand side of this inequality. So, $\mathbf{x}_{NI_\alpha}$ can be increased to at most δ. Let β be such that $\frac{(\mathbf{B}^{-1}\mathbf{b})_\beta}{(\mathbf{B}^{-1}\mathbf{N})_{\beta,\alpha}} = \delta$, assuming for now that β exists. This is the situation that we encountered in models (M1) and (M2) in Section 3.1. The case where β does not exist will be discussed later in this section. Set $x_{NI_\alpha} := \delta$, while keeping the values of all other nonbasic variables at zero, i.e., $x_{NI_i} = 0$ for $i \neq \alpha$. Now, (3.7) provides us with the new values for the variables with indices in BI, namely:

$$\mathbf{x}_{BI} := \mathbf{B}^{-1}\mathbf{b} - \delta(\mathbf{B}^{-1}\mathbf{N})_{\star,\alpha}.$$

By the choice of δ, we again have that $\mathbf{x}_{BI} \geq \mathbf{0}$. Moreover, we have that the equation of (3.7) with $j = \beta$ (corresponding to the basic variable x_{BI_β}) becomes:

$$x_{BI_\beta} = (\mathbf{B}^{-1}\mathbf{b})_\beta - \delta(\mathbf{B}^{-1}\mathbf{N})_{\beta,\alpha} = (\mathbf{B}^{-1}\mathbf{b})_\beta - (\mathbf{B}^{-1}\mathbf{N})_{\beta,\alpha} \frac{(\mathbf{B}^{-1}\mathbf{b})_\beta}{(\mathbf{B}^{-1}\mathbf{N})_{\beta,\alpha}} = 0.$$

So, the new value of x_{BI_β} is zero. We can therefore remove this variable from the set of basic variables, and insert it into the set of nonbasic variables. In other words, x_{BI_β} leaves the set of basic variables. Thus, we now have arrived at a new feasible basic solution. The corresponding new set of basic variable indices is $BI := (BI \setminus \{BI_\beta\}) \cup \{NI_\alpha\}$, and the corresponding new set of nonbasic variable indices is $NI := (NI \setminus \{NI_\alpha\}) \cup \{BI_\beta\}$. Since we have arrived at a new feasible basic solution, we may now repeat the iteration step.

Example 3.3.1. *To illustrate, consider again* (M2). *At the current feasible basic solution we have that $BI = \{1, 3, 5, 6\}$ and $NI = \{2, 4\}$. Hence, $BI_1 = 1$, $BI_2 = 3$, $BI_3 = 5$, $BI_4 = 6$, $NI_1 = 2$, and $NI_2 = 4$.*

$$\mathbf{B}^{-1}\mathbf{N} = \begin{array}{c} x_1 \\ x_3 \\ x_5 \\ x_6 \end{array} \begin{bmatrix} \frac{1}{3} & \frac{1}{3} \\ \frac{2}{3} & -\frac{1}{3} \\ -\frac{1}{3} & -\frac{1}{3} \\ 1 & 0 \end{bmatrix} \begin{array}{c} {\scriptstyle x_2} \quad {\scriptstyle x_4} \end{array} \quad and \ \mathbf{B}^{-1}\mathbf{b} = \begin{array}{c} x_1 \\ x_3 \\ x_5 \\ x_6 \end{array} \begin{bmatrix} 6 \\ 3 \\ 1 \\ 6 \end{bmatrix}.$$

We saw at the end of the previous section that the first and second nonbasic variables (x_2 and x_4, respectively) have current objective coefficients 1 and -1, respectively. Since x_2 has a positive current objective coefficient, we let this variable enter the set of basic variables. Thus, $\alpha = 1$ (because $NI_1 = 2$). Therefore, we have that (see Example 3.2.1):

$$(\mathbf{B}^{-1}\mathbf{N})_{\star,\alpha} = (\mathbf{B}^{-1}\mathbf{N})_{\star,1} = \begin{bmatrix} \frac{1}{3} \\ \frac{2}{3} \\ -\frac{1}{3} \\ 1 \end{bmatrix}.$$

This means that (3.9) becomes:

$$x_{NI_1} = x_2 \leq \min\left\{ \frac{6}{1/3}, \frac{3}{2/3}, \star, \frac{6}{1} \right\} = \min\{18, 4\tfrac{1}{2}, \star, 6\} = 4\tfrac{1}{2} =: \delta,$$

where '$\star$' indicates an entry that satisfies $(\mathbf{B}^{-1}\mathbf{N})_{j,\alpha} \leq 0$. The minimum is attained by the second entry, i.e., $\beta = 2$. The corresponding basic variable, $x_{BI_2} = x_3$, leaves the set of basic variables. Now we set $x_{NI_1} = x_2 := 4\tfrac{1}{2}$, while leaving the other nonbasic variable (i.e., x_4) at value zero. The new values of the basic variables with indices in BI are given by:

$$\mathbf{x}_{BI} = \begin{bmatrix} x_1 \\ x_3 \\ x_5 \\ x_6 \end{bmatrix} = \begin{bmatrix} 6 \\ 3 \\ 1 \\ 6 \end{bmatrix} - \begin{bmatrix} \frac{1}{3} \\ \frac{2}{3} \\ -\frac{1}{3} \\ 1 \end{bmatrix} \times 4\tfrac{1}{2} = \begin{bmatrix} 4\tfrac{1}{2} \\ 0 \\ 2\tfrac{1}{2} \\ 1\tfrac{1}{2} \end{bmatrix}.$$

Observe that indeed $x_3 = 0$. The new solution satisfies $x_1 = x_2 = 4\tfrac{1}{2}$, which means that we have arrived at vertex $\mathbf{v}_2$. The corresponding new sets of basic and nonbasic variable indices are now given by (recall that $\alpha = 1$ and $\beta = 2$):

$$BI := (\{1, 3, 5, 6\} \setminus \{BI_\beta\}) \cup \{NI_\alpha\} = (\{1, 3, 5, 6\} \setminus \{3\}) \cup \{2\} = \{1, 2, 5, 6\},$$
$$NI := (\{1, 3, 5, 6\} \setminus \{NI_\alpha\}) \cup \{BI_\beta\} = (\{2, 4\} \setminus \{2\}) \cup \{3\} = \{3, 4\}.$$

Recall that we assumed that β exists. An important question is: what happens if β does not exist? This happens if and only if $(\mathbf{B}^{-1}\mathbf{N})_{j,\alpha} \leq 0$ for each $j \in \{1,\ldots,m\}$ in (3.9). This means that none of the inequalities of (3.8) are restrictive and, as a result, we may increase the value of x_{NI_α} by any arbitrary amount. Since increasing the value of x_{NI_α} results in an increase of the objective value (recall that the current value of the objective coefficient of x_{NI_α} is positive), this means that the LO-model has an unbounded solution; see also Section 3.5.3.

3.3.3 Formulation of the simplex algorithm

The simplex algorithm can now be formulated in the following way. As an input, the algorithm takes the values of the technology matrix, the right hand side values, and the objective coefficients, together with an initial feasible basic solution. If no such feasible basic solution exists, then the model is infeasible. We will discuss how to find an initial feasible basic solution and how to determine infeasibility in Section 3.3.1.

> **Algorithm 3.3.1.** (*Dantzig's simplex algorithm*)
>
> **Input:** Values of $\mathbf{A}$, $\mathbf{b}$, and $\mathbf{c}$ in the LO-model $\max\{\mathbf{c}^\mathsf{T}\mathbf{x} \mid \mathbf{A}\mathbf{x} \leq \mathbf{b}, \mathbf{x} \geq \mathbf{0}\}$, and an initial feasible basic solution for the LO-model with slack variables (see Section 3.6).
>
> **Output:** Either an optimal solution of the model, or the message 'the model is unbounded'.
>
> **Step 1:** *Constructing the basis matrix.* Let BI and NI be the sets of the indices of the basic and nonbasic variables, respectively, corresponding to the current feasible basic solution. Let $\mathbf{B}$ consist of the columns of $\begin{bmatrix} \mathbf{A} & \mathbf{I}_m \end{bmatrix}$ with indices in BI, and let $\mathbf{N}$ consist of the columns of $\begin{bmatrix} \mathbf{A} & \mathbf{I}_m \end{bmatrix}$ with indices in NI. The current solution is $\mathbf{x}_{BI} = \mathbf{B}^{-1}\mathbf{b}$, $\mathbf{x}_{NI} = \mathbf{0}$, with corresponding current objective value $z = \mathbf{c}_{BI}^\mathsf{T}\mathbf{x}_{BI}$. Go to Step 2.
>
> **Step 2:** *Optimality test; choosing an entering variable.* The (row) vector of current objective coefficients for the nonbasic variables is $\mathbf{c}_{NI}^\mathsf{T} - \mathbf{c}_{BI}^\mathsf{T}\mathbf{B}^{-1}\mathbf{N}$. If each entry in this vector is nonpositive then stop: an optimal solution has been reached. Otherwise, select a positive entry in $\mathbf{c}_{NI}^\mathsf{T} - \mathbf{c}_{BI}^\mathsf{T}\mathbf{B}^{-1}\mathbf{N}$; say $\alpha \in \{1,\ldots,n\}$ is the index of that objective coefficient (i.e., it corresponds to variable x_{NI_α}). Then x_{NI_α} will become a basic variable. Go to Step 3.
>
> **Step 3:** *Minimum-ratio test; choosing a leaving variable.* If $(\mathbf{B}^{-1}\mathbf{N})_{j,\alpha} \leq 0$ for each $j \in \{1,\ldots,m\}$, then stop with the message 'the model is unbounded'. Otherwise, determine an index $\beta \in \{1,\ldots,m\}$ such that:
>
> $$\frac{(\mathbf{B}^{-1}\mathbf{b})_\beta}{(\mathbf{B}^{-1}\mathbf{N})_{\beta,\alpha}} = \min\left\{ \frac{(\mathbf{B}^{-1}\mathbf{b})_j}{(\mathbf{B}^{-1}\mathbf{N})_{j,\alpha}} \;\middle|\; (\mathbf{B}^{-1}\mathbf{N})_{j,\alpha} > 0, j = 1,\ldots,m \right\}.$$

Go to Step 4.

Step 4: *Exchanging.* Set $BI := (BI \setminus \{BI_\beta\}) \cup \{NI_\alpha\}$ and $NI := (NI \setminus \{NI_\alpha\}) \cup \{BI_\beta\}$. Return to Step 1.

The computational process of the Steps 2–4 is called *pivoting*. Determining values for α and β together with carrying out one pivot operation is called an *iteration*. The choice of the variable x_{NI_α} that enters the current set of basic variables in Step 2 is motivated by the fact that we want to increase the current objective value. If we deal with a *minimizing* LO-model (see Section 1.3), then we should look for a current objective coefficient which is negative.

The choice of the leaving variable in Step 3 is based on the requirement that all variables must remain nonnegative, keeping the current solution feasible. The leaving variable, determined by means of the minimum-ratio test in Step 3, is a current basic variable whose nonnegativity imposes the most restrictive upper bound on the increase of the value of the entering variable.

Example 3.3.2.   *The simplex algorithm as described above will now be applied to Model Dovetail (see Section 1.1).*

Initialization. The initial feasible solution is $x_1 = x_2 = 0, x_3 = 9, x_4 = 18, x_5 = 7, x_6 = 6$, and $BI = \{3,4,5,6\}$, $NI = \{1,2\}$.

Iteration 1. The current value of the objective coefficient of x_1 is 3, which is positive. Therefore, x_1 will enter the set of basic variables. In order to find out which variable should leave this set, we perform the minimum-ratio test, namely: $\min\{9/1, 18/3, 7/1\} = 18/3$. Hence, $\alpha = 2$, and therefore $(x_{NI_2} =) \ x_4$ leaves the set of basic variables, so that $BI := \{1,3,5,6\}$ and $NI := \{2,4\}$. Rewriting the model from the perspective of the new feasible basic solution, we obtain model (M2) of Section 3.1.

Iteration 2. Now the only positive objective coefficient is the one corresponding to x_2, so that x_2 will enter the set of basic variables. The minimum-ratio test shows that x_3 leaves the set. Hence, $BI := \{1,2,5,6\}$, and $NI := \{3,4\}$. Rewriting the model, we obtain model (M3) of Section 3.1.

Iteration 3. Since none of the current objective coefficients has a positive value at this point, we have reached an optimal solution.

3.4 Simplex tableaus

In every iteration step of the simplex algorithm that is described in the previous section, we need to determine the current formulation of the LO-model. This formulation can in principle be calculated by constructing the matrices $\mathbf{B}$ and $\mathbf{N}$, and deriving the current formulation using the expressions in (3.6). These expressions, however, involve the inverse matrix $\mathbf{B}^{-1}$, which means that in order to obtain the current formulation, an (m,m)-

matrix needs to be inverted. This takes a significant amount of time because computing the inverse of a matrix is time-consuming[1].

There is a useful technique for carrying out the calculations required for the simplex algorithm without inverting the matrix $\mathbf{B}$ in each iteration step. This technique makes use of so-called *simplex tableaus*. A simplex tableau is a partitioned matrix that contains all relevant information for the current feasible basic solution. A simplex tableau can schematically be depicted in the following partitioned and permuted form (see also Example 3.4.1 below):

m columns (BI)	n columns (NI)	one column	
$\mathbf{0}^\mathsf{T}$	$\mathbf{c}_{NI}^\mathsf{T} - \mathbf{c}_{BI}^\mathsf{T}\mathbf{B}^{-1}\mathbf{N}$	$-\mathbf{c}_{BI}^\mathsf{T}\mathbf{B}^{-1}\mathbf{b}$	} one row
$\mathbf{I}_m$	$\mathbf{B}^{-1}\mathbf{N}$	$\mathbf{B}^{-1}\mathbf{b}$	} m rows

A simplex tableau has $m+1$ rows and $m+n+1$ columns. The top row, which is referred to as 'row 0', contains the current objective coefficients for each of the variables, and (the negative of) the current objective value. The remaining rows, which are referred to as 'row 1' through 'row m', contain the coefficients and right hand sides of the constraints in the formulation of the LO-model from the perspective of the current feasible basic solution. Each column, except for the rightmost one, corresponds to a decision variable or a slack variable of the LO-model. The objective value, when included in a simplex tableau, is always multiplied by -1. The reason for this is that it simplifies the calculations when using simplex tableaus.

It needs to be stressed that the above tableau refers to a permutation of an actual simplex tableau. The columns of actual tableaus are always ordered according to the indices of the variables, i.e., according to the ordering $1, \ldots, n, n+1, \ldots, n+m$. On the other hand, the basic variables corresponding to the rows of a simplex tableau are not necessarily ordered in increasing order of their subscripts; see Example 3.4.1. For each $j = 1, \ldots, m$, let $BI_j \in BI$ be the index of the basic variable corresponding to row j. Note that, unlike in Section 3.3.3, it is <u>not necessarily true</u> that $BI_1 < BI_2 < \ldots < BI_m$. To summarize, we have the following correspondence between the rows and columns of a simplex tableau and the variables of the LO-model:

> Each column is associated with either a decision variable or a slack variable. To be precise, column i ($\in \{1, \ldots, n + m\}$) is associated with variable x_i. This association remains the same throughout the execution of the simplex algorithm.

---

[1] VOLKER STRASSEN (born 1936) discovered an algorithm that computes the inverse of an (m, m)-matrix in $O(m^{2.808})$ running time; see Strassen (1969). This has recently been improved to $O(m^{2.373})$ by VIRGINIA VASSILEVSKA WILLIAMS; see Williams (2012). See also Chapter 9.

Each row (except row 0) is associated with a current basic variable. To be precise, row j ($\in \{1, \ldots, m\}$) is associated with x_{BI_j}. Due to the fact that the set BI changes during the execution of the simplex algorithm, this association changes in each iteration step.

To avoid confusion about which column corresponds to which variable, and which row corresponds to which basic variable, it is useful to write the name of the variable above the corresponding column. We use arrows to mark which variables are the basic variables. Moreover, since each row (except the top row) corresponds to a basic variable, we write the name of each basic variable next to the corresponding row. The following example illustrates this convention.

Example 3.4.1.   *In the case of Model Dovetail, when performing the simplex algorithm using simplex tableaus, the following simplex tableau may occur (see Example 3.4.2). It corresponds to the feasible basic solution represented by* (M2).

x_1	x_2	x_3	x_4	x_5	x_6	$-z$	
0	1	0	-1	0	0	-18	
0	$\frac{2}{3}$	1	$-\frac{1}{3}$	0	0	3	x_3
1	$\frac{1}{3}$	0	$\frac{1}{3}$	0	0	6	x_1
0	$-\frac{1}{3}$	0	$-\frac{1}{3}$	1	0	1	x_5
0	1	0	0	0	1	6	x_6

In this tableau, the first, third, fifth, and sixth column correspond to the basic variables x_1, x_3, x_5, and x_6, respectively, and the second and fourth column correspond to the nonbasic variables x_2 and x_4, respectively. The seventh, rightmost, column corresponds to the (negative of the) current objective value and the current right hand side values. Note that the identity matrix in this tableau is in a row-permuted form: the first and second rows are switched. The rows correspond to the basic variables in the order x_3, x_1, x_5, and x_6. So, $BI_1 = 3$, $BI_2 = 1$, $BI_3 = 5$, and $BI_4 = 6$.

Several interesting facts about the current feasible basic solution can immediately be read off a simplex tableau:

The current objective value. This is the negative of the top-right entry of the tableau. In the example above, it is $-(-18) = 18$.

The current objective coefficients. These are the values in row 0. In the example above, the current objective coefficients for the variables $x_1, \ldots, x_6$ are $0, 1, 0, -1, 0, 0$, respectively. The current objective coefficients of the basic variables are zero. Note also that the vector $\mathbf{c}_{NI}^\mathsf{T} - \mathbf{c}_{BI}^\mathsf{T} \mathbf{B}^{-1} \mathbf{N}$ consists of the entries of the top row of the simplex tableau that correspond to the nonbasic variables. So, in the example above, we have that $\mathbf{c}_{NI}^\mathsf{T} - \mathbf{c}_{BI}^\mathsf{T} \mathbf{B}^{-1} \mathbf{N} = \begin{bmatrix} 1 & -1 \end{bmatrix}$.

The current basic variables associated with the constraints. The basic variable corresponding to the k'th row is x_{BI_k} ($k \in \{1, \ldots, m\}$). This fact can also be determined from the (row-permuted) identity matrix in the tableau as follows. Observe that, for each basic variable

x_{BI_k}, the column with index BI_k is equal to the unit vector $\mathbf{e}_k$ ($\in \mathbb{R}^m$). This means that, for each constraint, the basic variable associated with it can be found by looking at the occurrence of a 1 in the column corresponding to this basic variable. For instance, suppose that, in the example above, we want to determine BI_1. We then need to look for the column (corresponding to a basic variable) that is equal to $\mathbf{e}_1$. This is the column with index 3, so that $BI_1 = 3$. Similarly, we can determine that $BI_2 = 1$, $BI_3 = 5$, and $BI_4 = 6$.

The values of the current basic variables. In the example above, the first constraint in the above simplex tableau reads $\frac{2}{3}x_2 + x_3 - \frac{1}{3}x_4 = 3$. It corresponds to the basic variable x_3. Setting the values of the nonbasic variables x_2 and x_4 to zero, we obtain $x_3 = 3$. Thus, the current value of the basic variable x_3 is 3. Similarly, the current values of the other basic variables satisfy: $x_1 = 6$, $x_5 = 1$, and $x_6 = 6$. Thus, this simplex tableau corresponds to the point $\begin{bmatrix} 6 & 0 \end{bmatrix}^\mathsf{T}$ of the feasible region (which is vertex $\mathbf{v}_1$).

Observe also that during the execution of the simplex algorithm, the current basic solution is always a <u>feasible</u> basic solution. This means that the current basis matrix $\mathbf{B}$ satisfies $\mathbf{B}^{-1}\mathbf{b} \geq \mathbf{0}$, which implies that the entries in the rightmost column, except for the top-right entry, are always nonnegative.

3.4.1 The initial simplex tableau

The simplex algorithm can be performed completely by using simplex tableaus. The first step in doing this is to write down an initial simplex tableau. This is done by deciding on an initial set of basic variables (see also Section 3.6), and putting the coefficients of the standard LO-model in a simplex tableau. Note however that in a simplex tableau:

the objective coefficients of the basic variables (in row 0) should be zero,

the columns corresponding to the basic variables (excluding row 0) should form an identity matrix (usually in permuted form), and

the right hand side values should be nonnegative.

If these three conditions are not met, then the coefficients need to be manipulated by means of Gaussian elimination (see Section 3.4.2) to ensure that the first two conditions indeed hold. This is particularly important when the big-M or two-phase procedure is used to find an initial feasible basic solution; see Section 3.6. If, after ensuring that the first two conditions are satisfied, the third condition does not hold, then the basic solution corresponding to the chosen set of basic variables is infeasible, and therefore a different set of basic variables should be chosen (if it exists).

3.4.2 Pivoting using simplex tableaus

Recall that each iteration of the simplex algorithm consists of the following steps:

Determine a nonbasic variable that enters the set of basic variables. If no such variable exists, we stop because an optimal solution has been found.

Determine a constraint that is the most restrictive for the value of the entering variable, i.e., a constraint that attains the minimum in the minimum-ratio test. The basic variable that is associated with this constraint is the variable that leaves the set of basic variables. If no such constraint exists, then we may stop because the LO-model is unbounded.

In order to determine a nonbasic variable that enters the set of basic variables, the simplex algorithm chooses a nonbasic variable with a positive current objective coefficient. In terms of simplex tableaus, this is an almost trivial exercise: all that needs to be done is look for a positive entry in row 0 of the simplex tableau. Recall that the current objective coefficient of any basic variable is zero, so that it suffices to look for a positive entry among the objective coefficients of the nonbasic variables. If such a positive entry is found, say in the column with index NI_α with $\alpha \in \{1, \ldots, n\}$, then x_{NI_α} is the variable that enters the set of basic variables.

When the value of α has been determined, a basic variable that leaves the set of basic variables needs to be determined. Recall that this is done by means of the minimum-ratio test, which means determining a row index β such that:

$$
\frac{(\mathbf{B}^{-1}\mathbf{b})_\beta}{(\mathbf{B}^{-1}\mathbf{N})_{\beta,\alpha}} = \min\left\{ \frac{(\mathbf{B}^{-1}\mathbf{b})_j}{(\mathbf{B}^{-1}\mathbf{N})_{j,\alpha}} \; \middle| \; (\mathbf{B}^{-1}\mathbf{N})_{j,\alpha} > 0, j = 1, \ldots, m \right\}.
$$

The beauty of the simplex-tableau approach is that all these quantities can immediately be read off the tableau. Indeed, the entries of $\mathbf{B}^{-1}\mathbf{b}$ are the entries in the rightmost column of the simplex tableau, and the entries of $(\mathbf{B}^{-1}\mathbf{N})_{\star,\alpha}$ are the entries in the column with index NI_α of the simplex tableau (both excluding row 0). Thus, the minimum-ratio test can be performed directly from the simplex tableau, and the leaving basic variable x_{BI_β} can readily be determined.

What remains is to update the simplex tableau in such a way that we can perform the next iteration step. This is done using Gaussian elimination (so, not by explicitly calculating the matrix $\mathbf{B}^{-1}$). Let x_{NI_α} be the variable that enters the current set of basic variables, and let x_{BI_β} be the variable that leaves this set. The (β, NI_α)'th entry of the simplex tableau is called the *pivot entry*. Recall that a crucial property of a simplex tableau is that, for every basic variable, it holds that the column corresponding to it contains exactly one nonzero entry, which has value 1. To restore this property, observe that all columns corresponding to basic variables before the pivot are already in the correct form. Only the column corresponding to the variable x_{NI_α} that enters the set of basic variables needs to be transformed using Gaussian elimination. In particular, we consecutively perform the following elementary row operations on the simplex tableau:

Divide row β by the value of the (β, NI_α)'th entry of the simplex tableau (the pivot entry).

For each $k = 0, \ldots, m$, $k \neq \beta$, subtract row β, multiplied by the (k, NI_α)'th entry, from row k. Notice that this includes row 0.

This procedure guarantees that, in the new simplex tableau, the column corresponding to variable x_{NI_α} becomes the unit vector $\mathbf{e}_\beta$. It is left to the reader to check that the columns corresponding to the other basic variables have the same property, and that in fact the columns of the simplex tableau corresponding to the basic variables form an identity matrix (when ordered appropriately). Thus, the resulting tableau is a simplex tableau corresponding to the new feasible basic solution, and we can continue to the next iteration. (Explain why the columns with indices in $BI \setminus \{BI_\beta\}$, which are unit vectors, remain unchanged.)

We illustrate the tableau-version of the simplex algorithm in the following example.

Example 3.4.2. *We calculate the optimal solution of Model Dovetail by means of simplex tableaus.*

Initialization. At the initial feasible basic solution, we have that $BI = \{3,4,5,6\}$, and $NI = \{1,2\}$. The corresponding initial simplex tableau reads as follows:

x_1	x_2	x_3	x_4	x_5	x_6	$-z$	
3	2	0	0	0	0	0	
1	1	1	0	0	0	9	x_3
3	1	0	1	0	0	18	x_4
1	0	0	0	1	0	7	x_5
0	1	0	0	0	1	6	x_6

The columns corresponding to the basic variables form the identity matrix, and the current objective coefficients of the basic variables are 0. So this tableau is indeed a simplex tableau. The values of the coefficients in the tableau correspond to the coefficients of (M1). The first row (row 0) of the tableau states that the objective function is $3x_1 + 2x_3$, and the current objective value is $(-0 =) 0$. The second row (row 1) states that $x_1 + x_2 + x_3 = 9$, which is exactly the first constraint of (M1).

Iteration 1. There are two nonbasic variables with a positive current objective coefficient, namely x_1 and x_2. We choose x_1 as the variable to enter the set of basic variables, so $\alpha = 1$ (because $NI_1 = 1$). We leave it to the reader to carry out the calculations when x_2 is chosen. The minimum-ratio test gives:

$$\delta = \min\left\{\tfrac{9}{1}, \tfrac{18}{3}, \tfrac{7}{1}, \star\right\}.$$

The numerators in the above expression are found in the rightmost column of the current simplex tableau, and the denominators in the column corresponding to x_{NI_α} ($= x_1$). We obtain that $\delta = \tfrac{18}{3}$ and, hence, $\beta = 2$. So $(x_{BI_2} =) x_4$ is the leaving variable, and the pivot entry is the entry of the simplex tableau in row 2 and the column 1.

x_1	x_2	x_3	x_4	x_5	x_6	$-z$	
3	2	0	0	0	0	0	
1	1	1	0	0	0	9	x_3
③	1	0	1	0	0	18	x_4
1	0	0	0	1	0	7	x_5
0	1	0	0	0	1	6	x_6

The marked entry is the pivot entry. We set $BI := (\{3,4,5,6\} \setminus \{4\}) \cup \{1\} = \{3,1,5,6\}$ and $NI := (\{1,2\} \setminus \{1\}) \cup \{4\} = \{2,4\}$, and obtain the following 'tableau'. This tableau is not a simplex tableau yet; we need to apply Gaussian elimination to obtain a simplex tableau. We perform the following elementary row operations:

(a) *Divide row 2 (corresponding to x_4) by 3.*

(b) *Subtract 3 times the (new) third row from row 0.*

(c) *Subtract 1 times the (new) third row from row 1.*

(d) *Subtract 1 times the (new) third row from row 3.*

(e) *Leave row 4 as is.*

Iteration 2. *The resulting new simplex tableau is:*

x_1	x_2	x_3	x_4	x_5	x_6	$-z$	
0	1	0	-1	0	0	-18	
0	$\boxed{\frac{2}{3}}$	1	$-\frac{1}{3}$	0	0	3	x_3
1	$\frac{1}{3}$	0	$\frac{1}{3}$	0	0	6	x_1
0	$-\frac{1}{3}$	0	$-\frac{1}{3}$	1	0	1	x_5
0	1	0	0	0	1	6	x_6

The feasible basic solution that corresponds to this simplex tableau satisfies $x_1 = 6$, $x_2 = 0$, $x_3 = 3$, $x_4 = 0$, $x_5 = 1$, $x_6 = 6$, and the corresponding current objective value is 18. The entry in row 0 corresponding to x_2 is the only positive entry, so $NI_\alpha = 2$ (hence, $\alpha = 1$). Therefore, $(x_{NI_1} =) x_2$ is the entering variable. The minimum-ratio test yields:

$$\delta = \min\left\{ \tfrac{3}{2/3}, \tfrac{6}{1/3}, \star, \tfrac{6}{1} \right\} = \min\left\{ 4\tfrac{1}{2}, \tfrac{18}{1}, \star, \tfrac{6}{1} \right\} = 4\tfrac{1}{2}.$$

Hence, $\beta = 1$, and therefore $(x_{BI_\beta} =) x_3$ is the leaving variable. The pivot entry is marked in the simplex tableau above. After Gaussian elimination, we continue to the next iteration.

Iteration 3. $BI := \{1,2,5,6\}$, $NI := \{3,4\}$.

x_1	x_2	x_3	x_4	x_5	x_6	$-z$	
0	0	$-1\frac{1}{2}$	$-\frac{1}{2}$	0	0	$-22\frac{1}{2}$	
0	1	$1\frac{1}{2}$	$-\frac{1}{2}$	0	0	$4\frac{1}{2}$	x_2
1	0	$-\frac{1}{2}$	$\frac{1}{2}$	0	0	$4\frac{1}{2}$	x_1
0	0	$\frac{1}{2}$	$-\frac{1}{2}$	1	0	$2\frac{1}{2}$	x_5
0	0	$-1\frac{1}{2}$	$\frac{1}{2}$	0	1	$1\frac{1}{2}$	x_6

This tableau corresponds to an optimal feasible basic solution, because all objective coefficients are nonpositive. Hence, $x_1^* = 4\frac{1}{2}$, $x_2^* = 4\frac{1}{2}$, $x_3^* = x_4^* = 0$, $x_5^* = 2\frac{1}{2}$, $x_6^* = 1\frac{1}{2}$, and $z^* = 22\frac{1}{2}$.

We now formulate the simplex algorithm using simplex tableaus.

Algorithm 3.4.1. (*Dantzig's simplex algorithm using simplex tableaus*)

Input: Values of $\mathbf{A}$, $\mathbf{b}$, and $\mathbf{c}$ in the LO-model $\max\{\mathbf{c}^\mathsf{T}\mathbf{x} \mid \mathbf{A}\mathbf{x} \leq \mathbf{b}, \mathbf{x} \geq \mathbf{0}\}$, and an initial feasible basic solution for the LO-model with slack variables (see Section 3.6).

Output: Either an optimal solution of the model, or the message 'the model is unbounded'.

Step 1: *Initialization.* Let $BI = \{BI_1, \ldots, BI_m\}$ and $NI = \{NI_1, \ldots, NI_n\}$ be the sets of the indices of the basic and nonbasic variables corresponding to the initial feasible basic solution. Use Gaussian elimination to ensure that the tableau is in the correct form, i.e., the objective coefficients corresponding to the basic variables are zero, and the columns corresponding to the basic variables form a (row-permuted) identity matrix.

Step 2: *Optimality test; choosing an entering variable.* The entries in row 0 with indices in NI contain the current objective coefficients for the nonbasic variables (i.e., they form the vector $\mathbf{c}_{NI}^\mathsf{T} - \mathbf{c}_{BI}^\mathsf{T}\mathbf{B}^{-1}\mathbf{N}$). If each entry is nonpositive, then stop: an optimal solution has been reached. Otherwise, select a positive entry, say with index NI_α ($\alpha \in \{1, \ldots, n\}$). Then, x_{NI_α} is the entering variable. Go to Step 3.

Step 3: *Minimum-ratio test; choosing a leaving variable.* If the entries of column NI_α of the tableau are all nonpositive, then stop with the message 'the model is unbounded'. Otherwise, determine an index $\beta \in \{1, \ldots, m\}$ such that:

$$\frac{(\mathbf{B}^{-1}\mathbf{b})_\beta}{(\mathbf{B}^{-1}\mathbf{N})_{\beta,\alpha}} = \min\left\{ \frac{(\mathbf{B}^{-1}\mathbf{b})_j}{(\mathbf{B}^{-1}\mathbf{N})_{j,\alpha}} \;\middle|\; (\mathbf{B}^{-1}\mathbf{N})_{j,\alpha} > 0, j = 1, \ldots, m \right\}.$$

Determine the current value of BI_β by determining the column (with index in BI) that is equal to $\mathbf{e}_\beta$. The variable x_{BI_β} is the leaving variable. Go to Step 4.

Step 4: *Exchanging.* Set $BI := (BI \setminus \{BI_\beta\}) \cup \{NI_\alpha\}$ and $NI := (NI \setminus \{NI_\alpha\}) \cup \{BI_\beta\}$. Apply Gaussian elimination to ensure that the simplex tableau is in the correct form:

Divide row β by the value of the (β, NI_α)'th entry of the simplex tableau (the pivot entry).

For each $k = 0, \ldots, m$, $k \neq \beta$, subtract row β, multiplied by the (k, NI_α)'th entry, from row k.

Return to Step 2.

3.4.3　Why the identity matrix of a simplex tableau is row–permuted

In each of the simplex tableaus in Example 3.4.2, except for the initial one, the rows are not ordered according to the subscripts of the basic variables corresponding to them. As a result, the identity matrix $\mathbf{I}_4$ appears in row-permuted form in them. In fact, the values of $BI_1, \ldots, BI_m$ determine how the matrix $\mathbf{I}_4$ is permuted: when the rows are rearranged according to increasing values of BI_k ($k \in \{1, \ldots, m\}$), the identity matrix becomes 'nonpermuted' again.

The reason for this permutation is the following. Recall that each row is associated with a basic variable, and this association changes in each iteration. For instance, in the iterations of Example 3.4.2, the basic variables corresponding to the rows of simplex tableaus and the exchanges that happen can be depicted as follows:

	Iteration 1		Iteration 2		Iteration 3
$x_{BI_1} =$	x_3		x_3	$\leftrightarrow$	x_2
$x_{BI_2} =$	x_4	$\leftrightarrow$	x_1		x_1
$x_{BI_3} =$	x_5		x_5		x_5
$x_{BI_4} =$	x_6		x_6		x_6

The arrows indicate the exchanged basic variables. (The reader may verify that, if in Iteration 1 we had chosen x_2 as the entering variable, the ordering of the optimal tableau would have been x_1, x_6, x_5, x_2. This represents the same feasible basic solution as the one found in the above example, but its rows are permuted.)

In general, in each iteration, the association of the row β containing the pivot element changes from x_{BI_β} to x_{NI_α}, while the other associations remain unchanged. As can be seen from the figure above, this operation does not preserve the ordering of the subscripts of the basic variables associated with the rows. Since the values of $BI_1, \ldots, BI_m$ are directly related to the identity matrix in the tableau, this means that the identity matrix appears in row-permuted form.

Recall that $\mathbf{B} = \begin{bmatrix} \mathbf{A} & \mathbf{I}_m \end{bmatrix}_{\star,BI}$. Specifically, the columns of $\mathbf{B}$ are arranged in increasing order of the subscripts of the corresponding basic variables. We therefore refer to such a basis matrix as an *ordered basis matrix*. The actual simplex tableau can be found by arranging the columns of the basis matrix in the order $BI_1, \ldots, BI_m$. Recall that the indices $BI_1, \ldots, BI_m$ are not necessarily in increasing order in the corresponding simplex tableau. Hence, the ordering of the columns is in general different from the ordering of the columns in an ordered basis matrix, and we therefore refer to this matrix as an *unordered basis matrix*. That is, an unordered basis matrix has the form:

$$\mathbf{B}' = \begin{bmatrix} \begin{bmatrix} \mathbf{A} & \mathbf{I}_m \end{bmatrix}_{\star,BI_1} & \cdots & \begin{bmatrix} \mathbf{A} & \mathbf{I}_m \end{bmatrix}_{\star,BI_m} \end{bmatrix},$$

where $BI_1, \ldots, BI_m$ are not necessarily in increasing order. Row 1 through row m of the simplex tableau are then given by:

$$\begin{bmatrix} (\mathbf{B}')^{-1}\begin{bmatrix} \mathbf{A} & \mathbf{I}_m \end{bmatrix} & (\mathbf{B}')^{-1}\mathbf{b} \end{bmatrix}.$$

The reader may check that this matrix is equal to $\begin{bmatrix} \mathbf{B}^{-1}\begin{bmatrix} \mathbf{A} & \mathbf{I}_m \end{bmatrix} & \mathbf{B}^{-1}\mathbf{b} \end{bmatrix}$ up to a row permutation. Note that the ordering of the columns of the basis matrix only affects the ordering of the rows of the simplex tableau; it does not affect the feasible basic solution corresponding to it.

Example 3.4.3. *Consider the simplex tableaus in iteration 2 of Example 3.4.2. For this simplex tableau, we have that $BI_1 = 3$, $BI_2 = 1$, $BI_3 = 5$, and $BI_4 = 6$. The unordered basis matrix corresponding to this tableau is:*

$$\mathbf{B}' = \begin{bmatrix} \overset{x_3}{1} & \overset{x_1}{1} & \overset{x_5}{0} & \overset{x_6}{0} \\ 0 & 3 & 0 & 0 \\ 0 & 1 & 1 & 0 \\ 0 & 0 & 0 & 1 \end{bmatrix}.$$

Compare this to the (ordered) basis matrix $\mathbf{B}$ of Example 3.2.1. The reader may verify that:

$$(\mathbf{B}')^{-1}\begin{bmatrix} \mathbf{A} & \mathbf{I}_m \end{bmatrix} = \begin{bmatrix} \overset{x_1}{0} & \overset{x_2}{\frac{2}{3}} & \overset{x_3}{1} & \overset{x_4}{-\frac{1}{3}} & \overset{x_5}{0} & \overset{x_6}{0} \\ 1 & \frac{1}{3} & 0 & \frac{1}{3} & 0 & 0 \\ 0 & -\frac{1}{3} & 0 & -\frac{1}{3} & 1 & 0 \\ 0 & 1 & 0 & 0 & 0 & 1 \end{bmatrix}, \text{ and } (\mathbf{B}')^{-1}\mathbf{b} = \begin{bmatrix} 3 \\ 6 \\ 1 \\ 6 \end{bmatrix}.$$

This is exactly the order in which the coefficients appear in the simplex tableau of iteration 2 of Example 3.4.2.

Recall that, in each iteration of the simplex algorithm, exactly one of the values of $BI_1, \ldots, BI_m$ (namely, BI_β) is changed. In a computer implementation, it is therefore more efficient to store BI as an (ordered) vector rather than an (unordered) set. When storing BI as a vector, in each iteration, only the β'th entry needs to be updated, i.e., we set $BI_\beta := NI_\alpha$. Similarly, it is more efficient to work with unordered basis matrices rather than the usual ordered basis matrices. When working with ordered basis matrices, in each iteration, the new basis matrix needs to be constructed from scratch by taking the columns of $\begin{bmatrix} \mathbf{A} & \mathbf{I}_m \end{bmatrix}$ with indices in BI. In contrast, when working with unordered basis matrices, obtaining the new basis matrix after pivoting is a matter of replacing the β'th column of the current unordered basis matrix. This is one of the key ideas behind the revised simplex algorithm, to be discussed in Section 3.9.

3.5 Discussion of the simplex algorithm

In the previous two sections, we described two equivalent versions of the simplex algorithm, namely Algorithm 3.3.1 and Algorithm 3.4.1. In this section, we discuss a number of points that need clarification.

▷ In Section 3.3.1, we discussed using the origin as the initial feasible basic solution to start the simplex algorithm. What to do if the origin is not in the feasible region, so that we cannot start the simplex algorithm here? We postpone this discussion to Section 3.6, where two methods are described that deal with this question.

▷ In any iteration step, we look for a nonbasic variable to enter the basis, i.e., one that has a positive current objective coefficient. What to do if more than one current objective coefficient is positive? Which one should we choose?

▷ Similarly, in the minimum-ratio test, we look for a leaving variable. What if there are multiple candidates for the leaving variable? What if there are none?

▷ The value of δ defined in (3.9) determines how much the value of the entering variable is increased. If one of the current right hand side values equals 0, it may happen that $\delta = 0$. In that case, the value of the entering variable cannot be increased. This situation happens in the case of degeneracy. We will see the possible consequences of such a situation in Section 3.5.5.

▷ The simplex algorithm works by moving from one feasible basic solution to another feasible basic solution. We know from Theorem 2.1.5 that if an LO-model has an optimal solution, then it has an optimal vertex. But how do we know that, if there is indeed an optimal vertex, then there is also a feasible basic solution for which all current objective coefficients are nonnegative?

▷ How do we know that we will not, after a number of iterations, return to a feasible basic solution that has already been visited? In that case, the simplex algorithm does not terminate. This phenomenon is called *cycling*. In Section 3.5.5, we will show that this is a real possibility in the case of degeneracy. In Section 3.5.6, we will describe a method to prevent cycling.

These points (except for the first one, which is discussed in Section 3.6) will all be discussed in this section. At the end of the section, we will prove that the simplex algorithm works correctly for any given input. Before we do so, we introduce a useful tool, the simplex adjacency graph.

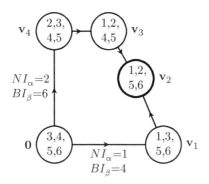

Figure 3.5: Simplex adjacency graph corresponding to Model Dovetail. Vertex $\mathbf{v}_2$ is optimal.

3.5.1 Simplex adjacency graphs

In this section, we introduce the so-called *simplex adjacency graph* of an LO-model. Simplex adjacency graphs form an insightful tool for thinking about the several ways of choosing entering and leaving variables (or, equivalently, choosing a pivot entry in a simplex tableau). A *simplex adjacency graph* is a directed graph (see Appendix C) that expresses these possibilities. The nodes of the simplex adjacency graph are the index sets BI corresponding to the feasible basic solutions, and there is an arc from node $BI(i)$ to node $BI(j)$ if the simplex tableau corresponding to $BI(j)$ can be obtained from the simplex tableau corresponding to $BI(i)$ by means of exactly one pivot step (i.e., there is one iteration between the corresponding tableaus). Note that simplex adjacency graphs do not have much practical value for solving LO-models, because, in order to construct the nodes of a simplex adjacency graph, one would have to determine all feasible basic solutions of the LO-model, of which there may be exponentially many (see Section 9.2). We only use them to gain insight into the workings of the simplex algorithm.

The simplex adjacency graph for Model Dovetail is depicted in Figure 3.5. The nodes correspond to the feasible basic solutions of the feasible region, and the arrows indicate a possible iteration step of the simplex algorithm. Note that, in this figure, all feasible basic solutions are nondegenerate, and hence each node corresponds to a unique vertex. It can be seen from the figure that there are two ways to get from $\mathbf{0}$ to the optimal solution $\mathbf{v}_2$. We mentioned in Example 3.4.2 that, in the first iteration, there are two possible choices for the leaving variable x_{NI_α}, namely x_1 and x_2. The simplex adjacency graph shows that if we choose $NI_\alpha = 1$ in the first iteration, then the simplex algorithm will take the path $\mathbf{0} \to \mathbf{v}_1 \to \mathbf{v}_2$. If, on the other hand, we choose $NI_\alpha = 2$ in the first iteration, then the algorithm will take the path $\mathbf{0} \to \mathbf{v}_4 \to \mathbf{v}_3 \to \mathbf{v}_2$. It can also be seen from the simplex adjacency graph that $\mathbf{0}$ is the only feasible basic solution for which we have multiple choices for NI_α and/or BI_β.

The simplex adjacency graph becomes more interesting in the presence of degeneracy. The following example shows the simplex adjacency graph in the case of degeneracy.

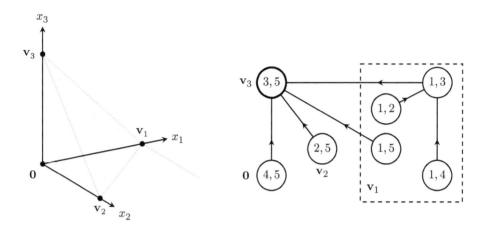

Figure 3.6: The feasible region of model (3.10).

Figure 3.7: Simplex adjacency graph corresponding to model (3.10). Vertex $\mathbf{v}_3$ is optimal.

Example 3.5.1.   *Consider the following LO-model:*

$$
\begin{aligned}
\max \quad & x_3 && (3.10)\\
\text{s.t.} \quad & x_1 + x_2 + x_3 \le 1 \\
& x_1 \le 1 \\
& x_1, x_2, x_3 \ge 0.
\end{aligned}
$$

The feasible region of the model is depicted in Figure 3.6. Notice that vertex $\mathbf{v}_1$ is degenerate. The simplex adjacency graph for model (3.10) is depicted in Figure 3.7. The nodes with $BI = \{1,2\}$, $BI = \{1,3\}$, $BI = \{1,4\}$, and $BI = \{1,5\}$ correspond to vertex $\mathbf{v}_1$ of the feasible region. This vertex actually corresponds to four feasible bases index sets; each two of these index sets differ in precisely one component. Let

$$
\begin{aligned}
H_1 &= \left\{ \begin{bmatrix} x_1 & x_2 & x_3 \end{bmatrix}^{\mathsf{T}} \;\middle|\; x_1 + x_2 + x_3 \le 1 \right\}, \\
H_2 &= \left\{ \begin{bmatrix} x_1 & x_2 & x_3 \end{bmatrix}^{\mathsf{T}} \;\middle|\; x_1 \le 1 \right\}, \\
H_3 &= \left\{ \begin{bmatrix} x_1 & x_2 & x_3 \end{bmatrix}^{\mathsf{T}} \;\middle|\; -x_1 \le 0 \right\}, \\
H_4 &= \left\{ \begin{bmatrix} x_1 & x_2 & x_3 \end{bmatrix}^{\mathsf{T}} \;\middle|\; -x_2 \le 0 \right\}, \\
H_5 &= \left\{ \begin{bmatrix} x_1 & x_2 & x_3 \end{bmatrix}^{\mathsf{T}} \;\middle|\; -x_3 \le 0 \right\}.
\end{aligned}
$$

Using Theorem 2.2.1, it then follows that:

$$
\begin{aligned}
\mathbf{0} &= H_1 \cap H_4 \cap H_5 \text{ corresponds to } BI = \{4,5\}, \\
\mathbf{v}_2 &= H_1 \cap H_2 \cap H_4 \text{ corresponds to } BI = \{2,5\}, \\
\mathbf{v}_3 &= H_2 \cap H_4 \cap H_5 \text{ corresponds to } BI = \{3,5\}, \\
\mathbf{v}_1 &= H_1 \cap H_2 \cap H_3 \text{ corresponds to } BI = \{1,2\}, \\
\mathbf{v}_1 &= H_2 \cap H_3 \cap H_5 \text{ corresponds to } BI = \{1,3\}, \\
\mathbf{v}_1 &= H_1 \cap H_3 \cap H_5 \text{ corresponds to } BI = \{1,4\}, \\
\mathbf{v}_1 &= H_1 \cap H_2 \cap H_5 \text{ corresponds to } BI = \{1,5\}.
\end{aligned}
$$

Here, H_1 and H_2 are the hyperplanes associated with the two constraints, and $H_3, \ldots, H_5$ are the hyperplanes associated with the three nonnegativity constraints.

The first three coordinate entries of $\mathbf{0}$, $\mathbf{v}_1$, $\mathbf{v}_2$, $\mathbf{v}_3$ are the values of the decision variables x_1, x_2, x_3, and the remaining two are the values of the slack variables x_4, x_5; x_4 refers to $x_1 + x_2 + x_3 \leq 1$, and x_5 to $x_1 \leq 1$. The various simplex tableaus are:

$\{4,5\}$:

x_1	x_2	x_3	x_4	x_5	$-z$	
0	0	1	0	0	0	
1	1	1	1	0	1	x_4
1	0	0	0	1	1	x_5

$\{2,5\}$:

x_1	x_2	x_3	x_4	x_5	$-z$	
0	0	1	0	0	0	
1	1	1	1	0	1	x_2
1	0	0	0	1	1	x_5

$\{3,5\}$:

x_1	x_2	x_3	x_4	x_5	$-z$	
-1	-1	0	-1	0	-1	
1	1	1	1	0	1	x_3
1	0	0	0	1	1	x_5

$\{1,2\}$:

x_1	x_2	x_3	x_4	x_5	$-z$	
0	0	1	0	0	0	
1	0	0	0	1	1	x_1
0	1	1	1	-1	0	x_2

$\{1,3\}$:

x_1	x_2	x_3	x_4	x_5	$-z$	
0	-1	0	-1	1	0	
1	0	0	0	1	1	x_1
0	1	1	1	-1	0	x_3

$\{1,4\}$:

x_1	x_2	x_3	x_4	x_5	$-z$	
0	0	1	0	0	0	
1	0	0	0	1	1	x_1
0	1	1	1	-1	0	x_4

$\{1,5\}$:

x_1	x_2	x_3	x_4	x_5	$-z$	
0	0	1	0	0	0	
1	1	1	1	0	1	x_1
0	-1	-1	-1	1	1	x_5

The different pivot steps between the above tableaus correspond to the arcs in the graph depicted in Figure 3.7.

Note also that the feasible basic solutions corresponding to the index sets $\{1,2\}$, $\{1,3\}$, $\{1,4\}$, and $\{1,5\}$ are degenerate. This can be seen from the fact that in each of the corresponding simplex tableaus, one of the right-hand side values in the tableau is 0.

Simplex adjacency graphs have the following interesting properties:

If there is an arc from node $BI(i)$ to $BI(j)$, then these two sets correspond to feasible basic solutions that are adjacent in the feasible region. In particular, $BI(i)$ and $BI(j)$ differ in exactly one index, i.e., $|BI(i) \setminus BI(j)| = |BI(j) \setminus BI(i)| = 1$.

If there is an arc from node $BI(i)$ to $BI(j)$, then the objective values $z(i)$ and $z(j)$ corresponding to $BI(i)$ and $BI(j)$, respectively, satisfy $z(j) \geq z(i)$. Moreover, if $z(i) = z(j)$, then $BI(i)$ and $BI(j)$ refer to the same (degenerate) vertex.

An optimal node is one that has only arcs pointing into that node. Note that a node with no outgoing arcs is not necessarily optimal, because the problem may be unbounded. See for instance Example 3.5.4.

3.5.2 Optimality test and degeneracy

In the language of the simplex adjacency graph, the simplex algorithm jumps from one node to another one, along the arcs of the graph. The algorithm stops when the current objective coefficients are nonpositive. The following theorem shows that, as soon as the current objective coefficients are nonpositive, the current feasible basic solution corresponds to an optimal vertex of the feasible region.

Theorem 3.5.1.
Consider the LO-model $\max\{\mathbf{c}^\mathsf{T}\mathbf{x} \mid \mathbf{A}\mathbf{x} \leq \mathbf{b}, \mathbf{x} \geq \mathbf{0}\}$. Let $\mathbf{B}$ be a basis matrix in $\begin{bmatrix} \mathbf{A} & \mathbf{I}_m \end{bmatrix}$ corresponding to the feasible basic solution $\begin{bmatrix} \hat{\mathbf{x}} \\ \hat{\mathbf{x}}_s \end{bmatrix} \equiv \begin{bmatrix} \hat{\mathbf{x}}_{BI} \\ \hat{\mathbf{x}}_{NI} \end{bmatrix}$ with $\hat{\mathbf{x}}_{BI} = \mathbf{B}^{-1}\mathbf{b} \geq \mathbf{0}$ and $\hat{\mathbf{x}}_{NI} = \mathbf{0}$. If $\mathbf{c}_{NI}^\mathsf{T} - \mathbf{c}_{BI}^\mathsf{T}\mathbf{B}^{-1}\mathbf{N} \leq \mathbf{0}$, then $\hat{\mathbf{x}}$ is an optimal vertex of the LO-model.

Proof. To prove the theorem, it suffices to prove two facts: (1) $\hat{\mathbf{x}}$ is a feasible solution of (3.2), and (2) no feasible solution $\mathbf{x}'$ of (3.2) satisfies $\mathbf{c}^\mathsf{T}\mathbf{x}' > \mathbf{c}^\mathsf{T}\hat{\mathbf{x}}$.

Fact (1) is easily checked by substituting $\mathbf{x}_{BI} = \mathbf{B}^{-1}\mathbf{b}$ and $\mathbf{x}_{NI} = \mathbf{0}$ into (3.6). Fact (2) can be seen as follows. Let $\mathbf{x}'$ be any feasible solution of (3.2). Then, in particular $\mathbf{x}'_{BI} \geq \mathbf{0}$ and $\mathbf{x}'_{NI} \geq \mathbf{0}$. Hence, by (3.5), we have that:
$$\mathbf{c}^\mathsf{T}\mathbf{x}' = \mathbf{c}_{BI}^\mathsf{T}\mathbf{B}^{-1}\mathbf{b} + (\mathbf{c}_{NI}^\mathsf{T} - \mathbf{c}_{BI}^\mathsf{T}\mathbf{B}^{-1}\mathbf{N})\mathbf{x}'_{NI} \leq \mathbf{c}_{BI}^\mathsf{T}\mathbf{B}^{-1}\mathbf{b} = \mathbf{c}^\mathsf{T}\hat{\mathbf{x}},$$
which proves the theorem. $\square$

Theorem 3.5.1 states that a basic solution with respect to the basis matrix $\mathbf{B}$ is optimal if and only if
$$\mathbf{B}^{-1}\mathbf{b} \geq \mathbf{0}, \text{ and } \mathbf{c}_{NI}^\mathsf{T} - \mathbf{c}_{BI}^\mathsf{T}\mathbf{B}^{-1}\mathbf{N} \leq \mathbf{0}.$$

The first condition is called the *feasibility condition*, and the second condition is called the *optimality condition*.

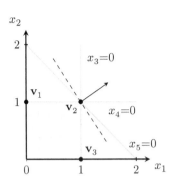

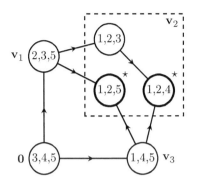

Figure 3.8: The feasible region of model (3.11). **Figure 3.9:** Simplex adjacency graph corresponding to model (3.11). Vertex $\mathbf{v}_2$ is optimal.

Recall that, in the case of degeneracy, there are multiple feasible basic solutions that correspond to the same optimal vertex. The simplex algorithm looks for a feasible basic solution for which the objective coefficients are nonpositive. A question is: is it possible that a feasible basic solution corresponds to an optimal vertex, while one or more of the corresponding objective coefficients are positive? If the simplex algorithm encounters such a feasible basic solution, it will not conclude that it has found an optimal solution, and it will continue with the next iteration instead. This is in fact a real possibility.

We say that the feasible basic solution corresponding to basis matrix $\mathbf{B}$ is an *optimal feasible basic solution* if $\mathbf{c}_{NI}^\mathsf{T} - \mathbf{c}_{BI}^\mathsf{T}\mathbf{B}^{-1}\mathbf{N} \leq \mathbf{0}$. The matrix $\mathbf{B}$ is then called an *optimal basis matrix*. An optimal vertex may correspond to some optimal feasible basic solutions as well as some nonoptimal feasible basic solutions. The following example shows that it may happen that not all feasible basic solutions that correspond to an optimal vertex are also optimal feasible basic solutions.

Example 3.5.2. *Consider the following LO-model.*

$$
\begin{aligned}
\max\quad & 3x_1 + 2x_2 && \text{(3.11)}\\
\text{s.t.}\quad & x_1 && \leq 1\\
& x_2 \leq 1\\
& x_1 + x_2 \leq 2\\
& x_1, x_2 \geq 0.
\end{aligned}
$$

It is easy to see that the unique optimal solution of this model satisfies $x_1^* = x_2^* = 1$. The feasible region of (3.11) is depicted in Figure 3.8, and the corresponding simplex adjacency graph is depicted in Figure 3.9. The simplex tableaus corresponding to the optimal vertex $\mathbf{v}_2$ are:

x_1	x_2	x_3	x_4	x_5	$-z$
0	0	0	1	-3	-5
1	0	0	-1	1	1
0	1	0	1	0	1
0	0	1	1	-1	0

x_1	x_2	x_3	x_4	x_5	$-z$
0	0	-1	0	-2	-5
1	0	1	0	0	1
0	1	-1	0	1	1
0	0	1	1	-1	0

x_1	x_2	x_3	x_4	x_5	$-z$
0	0	-3	-2	0	-5
1	0	1	0	0	1
0	1	0	1	0	1
0	0	-1	-1	1	0

Informally, when looking at the feasible region from the perspective of the feasible basic solution with $BI = \{1, 2, 3\}$ (see Section 3.2), the lines $x_4 = 0$ and $x_5 = 0$, intersecting at $\mathbf{v}_2$, become the axes. During an iteration step, the simplex algorithm 'tries' to move away from $\mathbf{v}_2$, along the line $x_5 = 0$, in the direction of the point $\begin{bmatrix} 2 & 0 \end{bmatrix}^{\mathsf{T}}$, because doing so seemingly increases the objective value (because of the angle between the line $x_5 = 0$ and the level lines of the objective function). However, when doing so, the line $x_3 = 0$ is immediately hit, and so in fact a new feasible basic solution is reached without moving to a new vertex. Clearly, the feasible basic solution corresponding to $BI = \{1, 2, 3\}$ is not optimal.

It is left to the reader to draw the feasible region of (3.11) from the perspective of the feasible basic solution with $BI = \{1, 2, 3\}$, i.e., the feasible region of the equivalent model

$$\max\{x_4 - 3x_5 + 5 \mid -x_4 + x_5 \leq 1,\ x_4 \leq 1,\ x_4 - x_5 \leq 0,\ x_4 \geq 0,\ x_5 \geq 0\}.$$

This discussion begs the following question. The simplex algorithm terminates when it either finds an optimal feasible basic solution, or it determines that the LO-model has an unbounded solution. If the LO-model has an optimal solution, how can we be sure that an optimal feasible <u>basic</u> solution even exists? We know from Theorem 2.1.5 that if the given LO-model has an optimal solution, then it has an optimal vertex. But could it happen that none of the feasible basic solutions corresponding to some optimal vertex are optimal? If that is the case, then it is conceivable that the simplex algorithm cycles between the various feasible basic solutions corresponding to that optimal vertex, without ever terminating.

As we will see in Section 3.5.5, this 'cycling' is a real possibility. The good news, however, is that there are several ways to circumvent this cycling behavior. Once this issue has been settled, we will be able to deduce that, in fact, if a standard LO-model has an optimal vertex, then there is at least one optimal feasible basic solution.

3.5.3   Unboundedness

In step 3, the simplex algorithm (see Section 3.3) tries to determine a leaving variable x_{BI_β}. The algorithm terminates with the message 'the model is unbounded' if such x_{BI_β} cannot be determined. The justification for this conclusion is the following theorem.

Theorem 3.5.2. (*Unboundedness*)

Let $\mathbf{B}$ be a basis matrix corresponding to the feasible basic solution $\begin{bmatrix} \mathbf{x}_{BI} \\ \mathbf{x}_{NI} \end{bmatrix}$, with $\mathbf{x}_{BI} = \mathbf{B}^{-1}\mathbf{b} \geq \mathbf{0}$ and $\mathbf{x}_{NI} = \mathbf{0}$, of the LO-model

$$\max\{\mathbf{c}^\mathsf{T}\mathbf{x} \mid \mathbf{Ax} \leq \mathbf{b}, \mathbf{x} \geq \mathbf{0}\}.$$

Assume that there exists an index $\alpha \in \{1, \ldots, n\}$ that satisfies $(\mathbf{c}_{NI}^\mathsf{T} - \mathbf{c}_{BI}^\mathsf{T}\mathbf{B}^{-1}\mathbf{N})_\alpha > 0$ and $(\mathbf{B}^{-1}\mathbf{N})_{\star,\alpha} \leq \mathbf{0}$. Then, the halfline L defined by:

$$L = \left\{ \mathbf{x} \in \mathbb{R}^n \,\middle|\, \begin{bmatrix} \mathbf{x} \\ \mathbf{x}_s \end{bmatrix} \equiv \begin{bmatrix} \mathbf{x}_{BI} \\ \mathbf{x}_{NI} \end{bmatrix} = \begin{bmatrix} \mathbf{B}^{-1}\mathbf{b} \\ \mathbf{0} \end{bmatrix} + \lambda \begin{bmatrix} -(\mathbf{B}^{-1}\mathbf{N})_{\star,\alpha} \\ \mathbf{e}_\alpha \end{bmatrix}, \lambda \geq 0 \right\}$$

is contained in the feasible region and the model is unbounded. In the above expression, $\mathbf{0} \in \mathbb{R}^n$ and $\mathbf{e}_\alpha$ is the α'th unit vector in $\mathbb{R}^n$.

Proof of Theorem 3.5.2. Let α be as in the statement of the theorem. For every $\lambda \geq 0$, define $\mathbf{x}(\lambda) \equiv \begin{bmatrix} \mathbf{x}_{BI}(\lambda) \\ \mathbf{x}_{NI}(\lambda) \end{bmatrix}$ by $\mathbf{x}_{BI}(\lambda) = \mathbf{B}^{-1}\mathbf{b} - \lambda(\mathbf{B}^{-1}\mathbf{N})_{\star,\alpha}$ and $\mathbf{x}_{NI}(\lambda) = \lambda\mathbf{e}_\alpha$. Note that $(\mathbf{x}_{NI}(\lambda))_\alpha = (\lambda\mathbf{e}_\alpha)_\alpha$ implies that $x_{NI_\alpha}(\lambda) = \lambda$.

Let $\lambda \geq 0$. Since $\mathbf{x}_{BI} = \mathbf{B}^{-1}\mathbf{b} \geq \mathbf{0}$ and $(\mathbf{B}^{-1}\mathbf{N})_{\star,\alpha} \leq \mathbf{0}$, it follows that $\mathbf{x}_{BI}(\lambda) \geq \mathbf{0}$ and $\mathbf{x}_{NI}(\lambda) \geq \mathbf{0}$. In order to show that $\mathbf{x}(\lambda)$ lies in the feasible region, we additionally need to prove that $\mathbf{Bx}_{BI}(\lambda) + \mathbf{Nx}_{NI}(\lambda) = \mathbf{b}$. This equation holds because

$$\mathbf{x}_{BI}(\lambda) + \mathbf{B}^{-1}\mathbf{Nx}_{NI}(\lambda) = \mathbf{B}^{-1}\mathbf{b} - \lambda(\mathbf{B}^{-1}\mathbf{N})_{\star,\alpha} + (\mathbf{B}^{-1}\mathbf{N})\lambda\mathbf{e}_\alpha = \mathbf{B}^{-1}\mathbf{b}.$$

Hence, L lies in the feasible region. To see that the model is unbounded, note that

$$\mathbf{c}^\mathsf{T}\mathbf{x}(\lambda) = \mathbf{c}_{BI}^\mathsf{T}\mathbf{x}_{BI}(\lambda) + \mathbf{c}_{NI}^\mathsf{T}\mathbf{x}_{NI}(\lambda) = \mathbf{c}_{BI}^\mathsf{T}\mathbf{B}^{-1}\mathbf{b} - \lambda\mathbf{c}_{BI}^\mathsf{T}(\mathbf{B}^{-1}\mathbf{N})_{\star,\alpha} + \lambda\mathbf{c}_{NI}^\mathsf{T}\mathbf{e}_\alpha$$
$$= \mathbf{c}_{BI}^\mathsf{T}\mathbf{B}^{-1}\mathbf{b} + \lambda(\mathbf{c}_{NI}^\mathsf{T} - \mathbf{c}_{BI}^\mathsf{T}\mathbf{B}^{-1}\mathbf{N})_\alpha.$$

Thus, since $(\mathbf{c}_{NI}^\mathsf{T} - \mathbf{c}_{BI}^\mathsf{T}\mathbf{B}^{-1}\mathbf{N})_\alpha > 0$, the objective value grows unboundedly along the halfline L. $\qquad\square$

The following example illustrates Theorem 3.5.2.

Example 3.5.3. *Consider the LO-model:*

$$\begin{array}{rl} \max & x_2 \\ \text{s.t.} & x_1 - x_2 \leq 1 \\ & -x_1 + 2x_2 \leq 1 \\ & x_1, x_2 \geq 0. \end{array}$$

After one iteration of the simplex algorithm, the following simplex tableau occurs:

x_1	x_2	x_3	x_4		
$\frac{3}{2}$	0	0	$-\frac{1}{2}$	$-\frac{1}{2}$	
$-\frac{1}{2}$	0	1	$\frac{1}{2}$	$\frac{3}{2}$	x_3
$-\frac{3}{2}$	1	0	$\frac{1}{2}$	$\frac{1}{2}$	x_2

The current objective coefficient of the nonbasic variable x_1 is positive, so $\alpha = 1$ ($NI_1 = 1$). The minimum-ratio test does not give a value for β, because the entries of the column corresponding to x_1 are all negative. By Theorem 3.5.2, the line

$$L = \left\{ \mathbf{x} \in \mathbb{R}^4 \;\middle|\; \mathbf{x} \equiv \begin{bmatrix} \mathbf{x}_{BI} \\ \mathbf{x}_{NI} \end{bmatrix} = \begin{bmatrix} x_3 \\ x_2 \\ x_1 \\ x_4 \end{bmatrix} = \begin{bmatrix} \frac{3}{2} \\ \frac{1}{2} \\ 0 \\ 0 \end{bmatrix} + \lambda \begin{bmatrix} \frac{1}{2} \\ \frac{3}{2} \\ 1 \\ 0 \end{bmatrix}, \lambda \geq 0 \right\}$$

lies in the feasible region, and the objective function takes on arbitrarily large values on L. In terms of the model without slack variables, this is the line

$$\left\{ \mathbf{x} \in \mathbb{R}^2 \;\middle|\; \mathbf{x} = \begin{bmatrix} x_1 \\ x_2 \end{bmatrix} = \begin{bmatrix} 0 \\ \frac{1}{2} \end{bmatrix} + \lambda \begin{bmatrix} 1 \\ \frac{3}{2} \end{bmatrix}, \lambda \geq 0 \right\}.$$

3.5.4 Pivot rules

Consider again Model Dovetail; see, e.g., Section 3.1. The objective function is $z = 3x_1 + 2x_2$. There are two positive objective coefficients, namely 3 and 2. We chose x_3, x_4, and x_5 as the initial basic variables. In Section 3.1, we chose to bring x_1 into the set of basic variables and to take x_4 out. This led to vertex $\mathbf{v}_1$; see Figure 1.3. On the other hand, we could have brought x_2 into the set of basic variables, which would have led to vertex $\mathbf{v}_4$. Clearly, the choice $\mathbf{0} \to \mathbf{v}_1$ leads faster to the optimal vertex $\mathbf{v}_2$ than the choice $\mathbf{0} \to \mathbf{v}_4$. The fact that there are two possible choices for the entering variable is also illustrated in Figure 3.5: there are two arcs going out of the node corresponding to $\mathbf{0}$ in this figure.

So, there is a certain degree of freedom in the simplex algorithm that needs to be resolved. This is done by a so-called *pivot rule*, which prescribes which variable should enter the set of basic variables when there are multiple nonbasic variables with a positive current objective coefficient, and which variable should leave the set of basic variables in case of a tie in the minimum-ratio test. In terms of the simplex adjacency graph, the pivot rule chooses, at each node, which one of the outgoing arcs should be followed.

Unfortunately, it is not possible to predict beforehand which choice of a positive coefficient in the current objective function is the most efficient, i.e., which choice leads in as few steps as possible to an optimal solution. A commonly used pivot rule is to simply choose the most positive current objective coefficient. In the case of Model Dovetail this should be the objective coefficient 3. If there is still a tie, then one could choose the most positive coefficient corresponding to the variable with the smallest subscript. Choosing the most positive current objective coefficient may seem a good idea, because it allows the algorithm to make the most 'progress' towards an optimal solution. However, when this rule is adopted, the simplex algorithm does not necessarily take the shortest path (in terms of the number of iterations) to the optimal vertex. In fact, we will see in Section 3.5.5 that, in the case of degeneracy, the pivot rule of choosing the most positive current objective coefficient may lead to cycling. Fortunately, there exists a pivot rule that avoids cycling; see Section 3.5.7.

3.5.5 Stalling and cycling

In Theorem 2.2.4, we have seen that it may happen that two adjacent feasible basic solutions correspond to one common (necessarily degenerate) vertex of the feasible region. Vertex $\mathbf{v}_1$ in Figure 3.6 and Figure 3.7 is an example of this phenomenon. In such a situation, a jump to a new feasible basic solution may not lead to an improvement of the objective value, because we may stay at the same vertex. This phenomenon is called *stalling*. Even worse, a sequence of iterations in such a degenerate vertex may finally lead to an already encountered feasible basic solution, and the procedure is caught in a cycle. This phenomenon is called *cycling*. The replacement of a column in the basis matrix by another one (or, equivalently, the replacement of a hyperplane containing the corresponding vertex by another hyperplane containing that vertex) leads to a self-repeating sequence of feasible basic solutions. If that is the case, the algorithm continues indefinitely and never arrives at an optimal feasible basic solution.

Note that as long each vertex of the feasible region corresponds to precisely one feasible basic solution (i.e., there are no degenerate vertices), then there is no problem, because each iteration leads to a new vertex, and an iteration is only carried out when it results in an improvement of the objective value. So, once the objective value has improved, we cannot come back to a feasible basic solution that has been visited before.

The following example, due to VAŠEK CHVÁTAL (born 1946), has a degenerate vertex in which the simplex algorithm cycles; see also Chvátal (1983). The corresponding simplex adjacency graph contains a directed cycle of nodes corresponding to this degenerate vertex.

Example 3.5.4. *Consider the following LO-model, in which $\mathbf{0}$ is a degenerate vertex:*

$$
\begin{aligned}
\max \quad & 10x_1 - 57x_2 - 9x_3 - 24x_4 \\
\text{s.t.} \quad & \tfrac{1}{2}x_1 - 5\tfrac{1}{2}x_2 - 2\tfrac{1}{2}x_3 + 9x_4 \leq 0 \\
& \tfrac{1}{2}x_1 - 1\tfrac{1}{2}x_2 - \tfrac{1}{2}x_3 + x_4 \leq 0 \\
& x_1, x_2, x_3, x_4 \geq 0.
\end{aligned}
$$

This model has an unbounded solution, because for any arbitrarily large $\gamma \geq 0$, the point $\begin{bmatrix} \gamma & 0 & \gamma & 0 \end{bmatrix}^\mathsf{T}$ is feasible and has objective value $(10\gamma - 9\gamma =) \gamma$. We will apply the simplex algorithm to it using the following pivot rules:

(a) *The variable that enters the current set of basic variables will be a nonbasic variable that has the largest current objective coefficient.*

(b) *If there is a tie in the minimum-ratio test, then the 'minimum-ratio row' with the smallest index is chosen.*

Introducing the slack variables x_5 and x_6, the model becomes:

$$\max \; 10x_1 - 57x_2 - 9x_3 - 24x_4$$
$$\text{s.t.} \quad \tfrac{1}{2}x_1 - 5\tfrac{1}{2}x_2 - 2\tfrac{1}{2}x_3 + 9x_4 + x_5 \quad\quad = 0$$
$$\tfrac{1}{2}x_1 - 1\tfrac{1}{2}x_2 - \tfrac{1}{2}x_3 + \quad x_4 \quad\quad + x_6 = 0$$
$$x_1, x_2, x_3, x_4, x_5, x_6 \geq 0.$$

When the simplex algorithm is carried out by means of simplex tableaus and following rules (a) and (b), the following iterations and tableaus occur:

Initialization. $BI = \{5, 6\}$, $NI = \{1, 2, 3, 4\}$.

x_1	x_2	x_3	x_4	x_5	x_6	$-z$	
10	-57	-9	-24	0	0	0	
$\boxed{\tfrac{1}{2}}$	$-5\tfrac{1}{2}$	$-2\tfrac{1}{2}$	9	1	0	0	x_5
$\tfrac{1}{2}$	$-1\tfrac{1}{2}$	$-\tfrac{1}{2}$	1	0	1	0	x_6

Iteration 1. $BI = \{1, 6\}$, $NI = \{2, 3, 4, 5\}$.

x_1	x_2	x_3	x_4	x_5	x_6	$-z$	
0	53	41	-204	-20	0	0	
1	-11	-5	18	2	0	0	x_1
0	$\boxed{4}$	2	-8	-1	1	0	x_6

Iteration 2. $BI = \{1, 2\}$, $NI = \{3, 4, 5, 6\}$.

x_1	x_2	x_3	x_4	x_5	x_6	$-z$	
0	0	$14\tfrac{1}{2}$	-98	$-6\tfrac{3}{4}$	$-13\tfrac{1}{4}$	0	
1	0	$\boxed{\tfrac{1}{2}}$	-4	$-\tfrac{3}{4}$	$2\tfrac{3}{4}$	0	x_1
0	1	$\tfrac{1}{2}$	-2	$-\tfrac{1}{4}$	$\tfrac{1}{4}$	0	x_2

Iteration 3. $BI = \{2, 3\}$, $NI = \{1, 4, 5, 6\}$.

x_1	x_2	x_3	x_4	x_5	x_6	$-z$	
-29	0	0	18	15	-93	0	
2	0	1	-8	$-1\tfrac{1}{2}$	$5\tfrac{1}{2}$	0	x_3
-1	1	0	$\boxed{2}$	$\tfrac{1}{2}$	$-2\tfrac{1}{2}$	0	x_2

Iteration 4. $BI = \{3,4\}$, $NI = \{1,2,5,6\}$.

x_1	x_2	x_3	x_4	x_5	x_6	$-z$	
-20	-9	0	0	$10\frac{1}{2}$	$-70\frac{1}{2}$	0	
-2	4	1	0	$\boxed{\frac{1}{2}}$	$-4\frac{1}{2}$	0	x_3
$-\frac{1}{2}$	$\frac{1}{2}$	0	1	$\frac{1}{4}$	$-1\frac{1}{4}$	0	x_4

Iteration 5. $BI = \{4,5\}$, $NI = \{1,2,3,6\}$.

x_1	x_2	x_3	x_4	x_5	x_6	$-z$	
22	-93	-21	0	0	24	0	
-4	8	2	0	1	-9	0	x_5
$\frac{1}{2}$	$-1\frac{1}{2}$	$-\frac{1}{2}$	1	0	$\boxed{1}$	0	x_4

Iteration 6. $BI = \{5,6\}$, $NI = \{1,2,3,4\}$.

x_1	x_2	x_3	x_4	x_5	x_6	$-z$	
10	-57	-9	-24	0	0	0	
$\frac{1}{2}$	$-5\frac{1}{2}$	$-2\frac{1}{2}$	9	1	0	0	x_5
$\frac{1}{2}$	$-1\frac{1}{2}$	$-\frac{1}{2}$	1	0	1	0	x_6

Thus, after six iterations, we see the initial tableau again, so that the simplex algorithm has made a cycle.

Figure 3.10 shows the simplex adjacency graph of the above LO-model. The graph has fifteen nodes, each labeled with a set BI of basic variable indices. These fifteen nodes represent all feasible basic solutions. They were calculated by simply checking all $\left(\binom{6}{2} =\right)$ 15 possible combinations of indices. The cycle corresponding to the above simplex tableaus is depicted with thick arcs in Figure 3.10.

The LO-model of the example above has an unbounded solution. However, the fact that the simplex algorithm makes a cycle has nothing to do with this fact. In Exercise 3.10.4, the reader is asked to show that when adding the constraint $x_1 \leq 0$ to the model, the simplex algorithm still makes a cycle (when using the same pivot rules), although the model has a (finite) optimal solution. Also note that, in the LO-model of the example above, the simplex algorithm cycles in a nonoptimal vertex of the feasible region. It may happen that the algorithm cycles through feasible basic solutions that correspond to an optimal vertex, so that in fact an optimal solution has been found, but the simplex algorithm fails to conclude that the solution is in fact optimal; see Exercise 3.10.4.

We have seen that cycling is caused by the fact that feasible basic solutions may be degenerate. However, not all degenerate vertices cause the simplex algorithm to cycle; see Exercise 3.10.2.

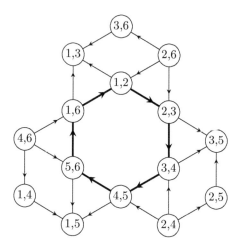

Figure 3.10: Simplex adjacency graph of Chvátal's example. The cycle corresponding to the simplex tableaus is indicated with thick arcs.

3.5.6 Anti-cycling procedures: the perturbation method

There are several procedures that prevent the simplex algorithm from cycling. One of the possibilities is the so-called *perturbation method*. The basic idea of the perturbation method is to replace, for each $i = 1, \ldots, m$, the right hand side b_i of the i'th constraint of the model by $b_i + \varepsilon_i$ with

$$0 < \varepsilon_m \ll \varepsilon_{m-1} \ll \ldots \ll \varepsilon_2 \ll \varepsilon_1 \ll 1.$$

The symbol '$\ll$' means 'much smaller than'. For instance, one may choose a small positive value for ε (e.g., $\varepsilon = 10^{-5}$), and take $\varepsilon_1 = \varepsilon, \varepsilon_2 = \varepsilon^2, \ldots, \varepsilon_m = \varepsilon^m$. The original LO-model then changes into the *perturbed model*:

$$
\begin{array}{rlcrll}
\max & c_1 x_1 & + & \ldots & + & c_n x_n \\
\text{s.t.} & a_{11} x_1 & + & \ldots & + & a_{1n} x_n \le b_1 + \varepsilon^1 \\
\text{s.t.} & a_{21} x_1 & + & \ldots & + & a_{2n} x_n \le b_2 + \varepsilon^2 \\
& \vdots & & & & \vdots \qquad \vdots \\
& a_{m1} x_1 & + & \ldots & + & a_{mn} x_n \le b_m + \varepsilon^m \\
& x_1, \ldots, x_n \ge 0, & & & &
\end{array}
\tag{3.12}
$$

or, in matrix notation:

$$\max\{\mathbf{c}^\mathsf{T}\mathbf{x} \mid \mathbf{A}\mathbf{x} \le \mathbf{b} + \boldsymbol{\epsilon}\},$$

where $\boldsymbol{\epsilon} = \begin{bmatrix} \varepsilon^1 & \ldots & \varepsilon^m \end{bmatrix}^\mathsf{T}$. Informally, this causes the hyperplanes defined by the constraints to 'move' by a little bit, not enough to significantly change the feasible region of the model, but just enough to make sure that no more than n hyperplanes intersect in the same point.

Theorem 3.5.3.

For a small enough value of $\varepsilon > 0$, the perturbed model (3.12) has no degenerate vertices.

Proof of Theorem 3.5.3. Before we prove the theorem, we prove the technical statement $(\star)$ below.

$(\star)$ Let $f(\varepsilon) = a_0 + a_1\varepsilon + a_2\varepsilon^2 + \ldots + a_m\varepsilon^m$, where $a_0, a_1, \ldots, a_m$ are real numbers that are not all equal to zero. There exists $\delta > 0$ such that $f(\varepsilon) \neq 0$ for all $\varepsilon \in (0, \delta)$.

The proof of $(\star)$ is as follows. Let $k \in \{0, 1, \ldots, m\}$ be smallest such that $a_k \neq 0$. For $\varepsilon > 0$, define:

$$g(\varepsilon) = \varepsilon^{-k} f(\varepsilon) = a_k + a_{k+1}\varepsilon + \ldots + a_m\varepsilon^{m-k}.$$

Note that $\lim_{\varepsilon \downarrow 0} g(\varepsilon) = a_k$. Since $g(\varepsilon)$ is continuous (see Appendix D) for $\varepsilon > 0$, this implies that there exists a $\delta > 0$ such that $|g(\varepsilon) - a_k| < \frac{|a_k|}{2}$ for all $\varepsilon \in (0, \delta)$. It follows that $g(\varepsilon) \neq 0$ for all $\varepsilon \in (0, \delta)$, and therefore $f(\varepsilon) \neq 0$ for all $\varepsilon \in (0, \delta)$, as required. This proves $(\star)$.

We may now continue the proof of the theorem. Consider any basic solution (feasible or infeasible), and let $\mathbf{B}$ be the corresponding basis matrix. Take any $k \in \{1, \ldots, m\}$. Define:

$$f(\varepsilon) = (\mathbf{B}^{-1})_{k,\star}(\mathbf{b} + \boldsymbol{\epsilon}) = (\mathbf{B}^{-1})_{k,\star}\mathbf{b} + (\mathbf{B}^{-1})_{k,1}\varepsilon_1 + \ldots + (\mathbf{B}^{-1})_{k,m}\varepsilon_m. \tag{3.13}$$

Since $\mathbf{B}^{-1}$ is nonsingular, it follows that $(\mathbf{B}^{-1})_{k,\star}$ is not the all-zero vector. It follows from $(\star)$ applied to (3.13) that there exists $\delta > 0$ such that $(\mathbf{B}^{-1})_{k,\star}(\mathbf{b} + \boldsymbol{\epsilon}) \neq 0$ for all $\varepsilon \in (0, \delta)$.

This argument gives a strictly positive value of δ for every choice of the basis matrix $\mathbf{B}$ and every choice of $k \in \{1, \ldots, m\}$. Since there are only finitely many choices of $\mathbf{B}$ and k, we may choose the smallest value of δ, say δ^*. Note that $\delta^* > 0$. It follows that $(\mathbf{B}^{-1})_k(\mathbf{b} + \boldsymbol{\epsilon}) \neq 0$ for every basis matrix $\mathbf{B}$ and for every $\varepsilon \in (0, \delta^*)$. $\square$

We now illustrate the perturbation method by elaborating on Chvátal's example.

Example 3.5.5.   *Recall the LO-model from Example 3.5.4. We will apply the perturbation method to it.*

Initialization. *Let $\varepsilon > 0$ with $\varepsilon \ll 1$. The initial simplex tableau becomes:*

x_1	x_2	x_3	x_4	x_5	x_6	$-z$	
10	-57	-9	-24	0	0	0	
$\frac{1}{2}$	$-5\frac{1}{2}$	$-2\frac{1}{2}$	9	1	0	ε	x_5
$\boxed{\frac{1}{2}}$	$-1\frac{1}{2}$	$-\frac{1}{2}$	1	0	1	ε^2	x_6

Iteration 1. *The variable x_1 is the only candidate for entering the set of basic variables. Since $2\varepsilon^2 \ll 2\varepsilon$, it follows that x_6 has to leave the set of basic variables. Hence, $BI := \{1, 5\}$.*

x_1	x_2	x_3	x_4	x_5	x_6	$-z$	
0	-27	1	-44	0	-20	$-20\varepsilon^2$	
0	-4	-2	8	1	-1	$\varepsilon-\varepsilon^2$	x_5
1	-3	-1	2	0	2	$2\varepsilon^2$	x_1

Now, we conclude that the model is unbounded (why?).

Note that applying the perturbation method may render feasible basic solutions of the original LO-model infeasible. In fact, if improperly applied, the perturbation method may turn a feasible LO-model into an infeasible one. In Exercise 3.10.5, the reader is asked to explore such situations.

3.5.7 Anti-cycling procedures: Bland's rule

In Section 3.5.6, we introduced the perturbation method to avoid cycling of the simplex algorithm. In the current section, we describe the *smallest-subscript rule*, also known as *Bland's rule*[2]. Recall that there is a certain degree of freedom in the simplex algorithm when there are multiple choices for the variable that enters the set of basic variables, and the variable that leaves the set of basic variables. The smallest-subscript rule prescribes how to resolve these ties in such a way that cycling does not occur. The pivot rules prescribed by the smallest-subscript rule are:

(a) The entering variable is the nonbasic variable with a positive current objective coefficient that has the smallest index (i.e., which is leftmost in the simplex tableau).

(b) If there is a tie in the minimum-ratio test, then the 'minimum-ratio' row corresponding to the basic variable with the smallest index is chosen.

The proof of the correctness (meaning that no cycling occurs) of the smallest-subscript rule lies outside of the scope of this book. The interested reader is referred to Bertsimas and Tsitsiklis (1997). We illustrate the procedure by applying it to Chvátal's example.

Example 3.5.6. *Recall the LO-model from Example 3.5.4. Applying the simplex algorithm with the smallest subscript rule results in the following simplex tableaus:*

---

[2]Named after the American mathematician and operations researcher ROBERT G. BLAND (born 1948).

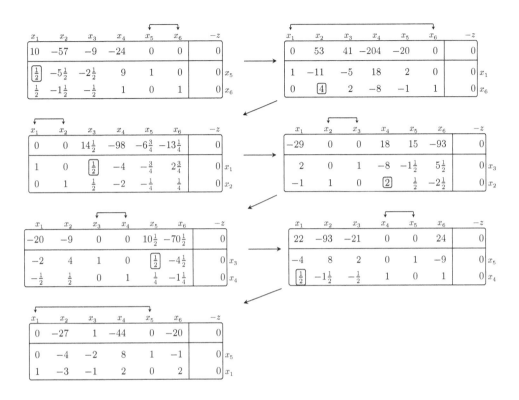

At this point, we conclude that no cycling has occurred, and that the model is unbounded.

3.5.8 Correctness of the simplex algorithm

We now have enough tools to show that, when given a standard LO-model and an initial feasible basic solution, the simplex algorithm terminates in finitely many iterations, and it correctly finds an optimal solution, or establishes that the model is unbounded.

Theorem 3.5.4. (*The simplex algorithm theorem*)
Consider the LO-model $\max\{\mathbf{c}^\mathsf{T}\mathbf{x} \mid \mathbf{A}\mathbf{x} \leq \mathbf{b}, \mathbf{x} \geq \mathbf{0}\}$. Suppose that we have an initial feasible basic solution. Then, the simplex algorithm (Algorithm 3.3.1), with an anti-cycling procedure, terminates in finitely many steps and it correctly returns an optimal feasible basic solution, or determines that the model has an unbounded solution.

Proof. We showed in Section 3.3.2 that if we start with a feasible basic solution in step 1 of Algorithm 3.3.1, then the iteration step either terminates, or it produces a new feasible basic solution. Because an anti-cycling procedure is used, the simplex algorithm does not encounter the same feasible basic solution twice. Since there are at most $\binom{m+n}{m}$ feasible basic solutions, this means that the simplex algorithm takes a finite number of iteration steps. Therefore, the algorithm terminates either in step 2 with an optimal solution, or in step 3 with the message 'the model is unbounded'. In the former case, the current feasible basic solution satisfies $\mathbf{c}_{NI}^{\mathsf{T}} - \mathbf{c}_{BI}^{\mathsf{T}}\mathbf{B}^{-1}\mathbf{N} \leq \mathbf{0}$, and hence it follows from Theorem 3.5.1 that the current feasible basic solution is indeed optimal. In the latter case, an index $\alpha \in \{1, \ldots, n\}$ is found that satisfies $(\mathbf{c}_{NI}^{\mathsf{T}} - \mathbf{c}_{BI}^{\mathsf{T}}\mathbf{B}^{-1}\mathbf{N})_\alpha > 0$ and $(\mathbf{B}^{-1}\mathbf{N})_{\star,\alpha} \leq \mathbf{0}$; hence, it follows from Theorem 3.5.2 that the model is indeed unbounded. $\qquad\square$

Although Theorem 3.5.4 states that the number of steps taken by the simplex algorithm is finite, this does not rule out the possibility that the number of steps may be enormous (exponential). In fact, there are (pathological) LO-models for which the simplex algorithm goes through all (exponentially many) feasible basic solutions before reaching an optimal solution; see Chapter 9. It is still an open question whether there exists a pivot rule such that the simplex algorithm is guaranteed to terminate in a reasonable number of steps for all possible inputs. On the other hand, despite this theoretical problem, the simplex algorithm is extremely efficient in practice.

Now that Theorem 3.5.4 has been proved, we can address the question whether there even exists a feasible basic solution corresponding to an optimal vertex. The answer is affirmative.

Theorem 3.5.5.
Consider a standard LO-model and suppose that it is feasible and bounded. Then there exists an optimal feasible basic solution.

Proof. Suppose the LO-model has an optimal solution. Apply the simplex algorithm to the model, starting from any initial feasible basic solution. Because the LO-model does not have an unbounded solution, it follows from Theorem 3.5.4 that the algorithm returns, in finitely many iterations, an optimal feasible basic solution. Therefore, the model has an optimal feasible basic solution. $\qquad\square$

3.6 Initialization

In order to start the simplex algorithm, an *initial feasible basic solution* is needed. Sometimes, the all-zero solution can be taken. If an initial feasible basic solution is not immediately apparent, there are several possibilities to find one. In Section 1.2.1, we have formulated Model Dovetail* for which $\mathbf{0}$ does not lie in the feasible region. The reason is that this model contains the constraint $x_1 + x_2 \geq 5$, which excludes $\mathbf{0}$ from the feasible region. Indeed, for Model Dovetail*, the choice $BI = \{3, 4, 5, 6, 7\}$ (i.e., the set of indices of the slack variables) corresponds to an infeasible basic solution, namely $\begin{bmatrix} 0 & 0 & 9 & 18 & 7 & 6 & -5 \end{bmatrix}^{\mathsf{T}}$. This solution cannot serve as an initial feasible basic solution for the simplex algorithm.

We will discuss the so-called *big-M procedure* and the *two-phase procedure* for determining an initial feasible basic solution. If all constraints of the primal LO-model are '$\geq$' constraints and all right hand side values are positive, then the all-zero vector is certainly not feasible.

In this case it is sometimes profitable to apply the simplex algorithm to the so-called dual model; see Section 4.6.

3.6.1 The big-M procedure

Let $i = 1, \ldots, m$. Constraints of the form

$$a_{i1}x_1 + \cdots + a_{in}x_n \geq b_i \text{ with } b_i > 0,$$

or of the form

$$a_{i1}x_1 + \cdots + a_{in}x_n \leq b_i \text{ with } b_i < 0$$

exclude the all-zero vector as start feasible basic solution for the simplex algorithm. When applying the big-M procedure, each such constraint i is augmented, together with its slack variable, with a so-called *artificial variable* u_i, and the objective function is augmented with $-Mu_i$, where M is a big positive real number. For big values of M ('big' in relation to the original input data), the simplex algorithm will put highest priority on making the value of the factor Mu_i as small as possible, thereby setting the value of u_i equal to zero. We illustrate the procedure by means of the following example.

Example 3.6.1. *Consider the model:*

$$
\begin{aligned}
\max \quad & 2x_1 + x_2 && (3.14)\\
\text{s.t.} \quad & 2x_1 - x_2 \geq 4\\
& -x_1 + 2x_2 \geq 2\\
& x_1 + x_2 \leq 12\\
& x_1, x_2 \geq 0.
\end{aligned}
$$

The first two constraints both exclude $\mathbf{0}$ from the feasible region. Applying the big-M procedure, these constraints are augmented (in addition to the slack variables) by artificial variables, say u_1 and u_2, respectively. Moreover, the objective function is extended with $-Mu_1 - Mu_2$. So, we obtain the model:

$$
\begin{aligned}
\max \quad & 2x_1 + x_2 && -Mu_1 - Mu_2 && (3.15)\\
\text{s.t.} \quad & 2x_1 - x_2 - x_3 && + u_1 && = 4\\
& -x_1 + 2x_2 \quad - x_4 && + u_2 = 2\\
& x_1 + x_2 \quad\quad + x_5 && = 12\\
& x_1, x_2, x_3, x_4, x_5, u_1, u_2 \geq 0.
\end{aligned}
$$

In this model, we choose the value of M large, say, $M = 10$. An initial feasible basic solution of this model is $u_1 = 4, u_2 = 2, x_5 = 12$, and $x_1 = x_2 = x_3 = x_4 = 0$. In terms of model (3.14), this solution corresponds to the infeasible point $\begin{bmatrix} 0 & 0 \end{bmatrix}^\mathsf{T}$. Since u_1 and u_2 have large negative objective coefficients, the simplex algorithm, applied to the above maximizing LO-model, tries to drive u_1 and u_2 out of the set of basic variables. The obvious initial basic variables are u_1, u_2, and x_5. It is left to the reader to carry out the successive iterations; start with writing this model into a simplex tableau. It is useful to draw a picture of the feasible region and to follow how the simplex algorithm jumps from *infeasible basic solutions to feasible basic solutions, and arrives – if it exists – at an optimal feasible basic solution. To reduce the calculations, as soon as an artificial variable leaves the set of basic variables (and therefore becomes zero), the corresponding column may be deleted. The reader can check that the optimal solution of (3.15) satisfies $x_1^* = 7\frac{1}{3}$, $x_2^* = 4\frac{2}{3}$, $u_1^* = u_2^* = 0$, with optimal objective value $z^* = 19\frac{1}{3}$. The reader should also check that, because $u_1^* = u_2^* = 0$, the optimal solution of (3.14) is obtained by discarding the values of u_1^* and u_2^*. That is, it satisfies $x_1^* = 7\frac{1}{3}$, $x_2^* = 4\frac{2}{3}$, with optimal objective value $z^* = 19\frac{1}{3}$.*

If the simplex algorithm terminates with at least one artificial variable in the optimal 'solution', then either the original LO-model is infeasible, or the value of M was not chosen large enough. The following two examples show what happens in the case of infeasibility of the original model, and in the case when the value of M is not chosen large enough.

Example 3.6.2.   *Consider again model (3.14), but with the additional constraint $x_1 \leq 3$. Let $M = 10$. The optimal solution of this model is $x_1^* = 3$, $x_3^* = 2\frac{1}{2}$, $u_1^* = \frac{1}{2}$, and $u_2^* = 0$. The optimal value of u_1^* is nonzero, which is due to the fact that the LO-model is infeasible.*

The following example shows that, from the fact that some artifical variable has a nonzero optimal value, one cannot immediately draw the conclusion that the original LO-model was infeasible.

Example 3.6.3.   *Consider again model (3.15), but let $M = \frac{1}{10}$. The optimal solution of (3.15) then is $x_1^* = 12$, $x_3^* = 20$, $u_2^* = 14$, and $x_2^* = x_4^* = x_5^* = u_1^* = 0$. Clearly, $\begin{bmatrix} x_1^* & x_2^* \end{bmatrix}^T = \begin{bmatrix} 12 & 0 \end{bmatrix}^T$ is not a feasible solution of (3.14) because the constraint $-x_1 + 2x_2 \geq 2$ is violated.*

This example illustrates the fact that, in order to use the big-M procedure, one should choose a large enough value for M. However, choosing too large a value of M may lead to computational errors. Computers work with so-called *floating-point numbers*, which store only a finite number of digits (to be precise: bits) of a number. Because of this finite precision, a computer rounds off the results of any intermediate calculations. It may happen that adding a small number k to a large number M results in the same number M, which may cause the final result to be inaccurate or plainly incorrect. For example, when using so-called *single-precision* floating-point numbers, only between 6 and 9 significant digits (depending on the exact value of the number) are stored. For instance, if $M = 10^8$, then adding any number k with $|k| \leq 4$ to M results in M again. The following example shows how this phenomenon may have a disastrous effect on the calculations of the simplex algorithm.

Example 3.6.4.   *Consider again Model Dovetail. We choose $M = 10^8$ and use single-precision floating point numbers. It turns out that, given these particular choices, $M + 2$ and $M + 3$ both result in M. This means that the initial simplex tableau after Gaussian elimination becomes:*

x_1	x_2	x_3	x_4	x_5	x_6	x_7	u_1	$-z$
M	M	0	0	0	0	0	$-M$	$5M$
1	1	1	0	0	0	0	0	9
3	1	0	1	0	0	0	0	18
1	0	0	0	1	0	0	0	7
0	1	0	0	0	1	0	0	6
$\boxed{1}$	1	0	0	0	0	-1	1	5

Similarly, $5M - 15$ results in $5M$. Hence, performing one pivot step leads to the tableau:

x_1	x_2	x_3	x_4	x_5	x_6	x_7	u_1	$-z$
0	0	0	0	0	0	0	0	0
0	0	1	0	0	0	1	-1	4
0	-2	0	1	0	0	$\boxed{3}$	-3	3
0	-1	0	0	1	0	1	-1	2
0	1	0	0	0	1	0	0	6
1	1	0	0	0	0	-1	1	5

This 'simplex tableau' states that we have reached an optimal solution. The corresponding solution satisfies $x_1 = 5$, $x_2 = 0$. Since we know that the unique optimal solution of Model Dovetail satisfies $x_1^* = x_2^* = 4\frac{1}{2}$, the solution we find is clearly not optimal. Moreover, the current objective value, as given in the top-right entry of the tableau, is 0. But the actual objective value for $x_1 = 5$, $x_2 = 0$ is 15. So, not only have we ended up with a wrong 'optimal vertex', the reported corresponding objective value is wrong as well.

A better way to apply the big-M procedure is to treat M as a large but unspecified number. This means that the symbol M remains in the simplex tableaus. Whenever two terms need to be compared, M is considered as a number that is larger than any number that is encountered while performing the calculations. So, for example, $2M + 3 > 1.999M + 10000 > 1.999M$. We illustrate this procedure by applying it to Model Dovetail*; see Section 1.2.1.

Example 3.6.5. *Consider Model Dovetail*. Besides the slack variables, an artificial variable u_1 is added to the model with a large negative objective coefficient $-M$. So, we obtain the model:*

$$
\begin{aligned}
\max \quad & 3x_1 + 2x_2 && - Mu_1 \\
\text{s.t.} \quad & x_1 + x_2 + x_3 && = 9 \\
& 3x_1 + x_2 + x_4 && = 18 \\
& x_1 + x_5 && = 7 \\
& x_2 + x_6 && = 6 \\
& x_1 + x_2 - x_7 + u_1 && = 5 \\
& x_1, x_2, x_3, x_4, x_5, x_6, x_7, u_1 \geq 0.
\end{aligned}
$$

Applying the simplex algorithm to this model, the following iterations occur.

Initialization. $BI = \{3, 4, 5, 6, 8\}$, $NI = \{1, 2, 7\}$. *We start by putting the coefficients into the tableau:*

x_1	x_2	x_3	x_4	x_5	x_6	x_7	u_1	$-z$
3	2	0	0	0	0	0	$-M$	0
1	1	1	0	0	0	0	0	9
3	1	0	1	0	0	0	0	18
1	0	0	0	1	0	0	0	7
0	1	0	0	0	1	0	0	6
①1	1	0	0	0	0	-1	1	5

Recall that in a simplex tableau the objective coefficients corresponding to the basic variables have to be zero. So we need to apply Gaussian elimination to obtain a simplex tableau. To do so, we add M times the fifth row to the top row:

x_1	x_2	x_3	x_4	x_5	x_6	x_7	u_1	$-z$
$3+M$	$2+M$	0	0	0	0	$-M$	0	$5M$
1	1	1	0	0	0	0	0	9
3	1	0	1	0	0	0	0	18
1	0	0	0	1	0	0	0	7
0	1	0	0	0	1	0	0	6
①1	1	0	0	0	0	-1	1	5

Iteration 1. $BI = \{1, 3, 4, 5, 6\}$, $NI = \{2, 7, 8\}$.

x_1	x_2	x_3	x_4	x_5	x_6	x_7	u_1	$-z$
0	-1	0	0	0	0	3	$-(3+M)$	-15
0	0	1	0	0	0	1	-1	4
0	-2	0	1	0	0	③3	-3	3
0	-1	0	0	1	0	1	-1	2
0	1	0	0	0	1	0	0	6
1	1	0	0	0	0	-1	1	5

Since $u_1 = 0$, we have arrived at a feasible basic solution of Model Dovetail*. We now delete the column corresponding to u_1.

Iteration 2. $BI = \{1,3,5,6,7\}$, $NI = \{2,4\}$.

x_1	x_2	x_3	x_4	x_5	x_6	x_7	$-z$
0	1	0	-1	0	0	0	-18
0	$\boxed{\frac{2}{3}}$	1	$-\frac{1}{3}$	0	0	0	3
0	$-\frac{2}{3}$	0	$\frac{1}{3}$	0	0	1	1
0	$-\frac{1}{3}$	0	$-\frac{1}{3}$	1	0	0	1
0	1	0	0	0	1	0	6
1	$\frac{1}{3}$	0	$\frac{1}{3}$	0	0	0	6

Iteration 3. $BI = \{1,2,5,6,7\}$, $NI = \{3,4\}$.

x_1	x_2	x_3	x_4	x_5	x_6	x_7	$-z$
0	0	$-\frac{3}{2}$	$-\frac{1}{2}$	0	0	0	$-22\frac{1}{2}$
0	1	$\frac{3}{2}$	$-\frac{1}{2}$	0	0	0	$\frac{9}{2}$
0	0	1	0	0	0	1	4
0	0	$\frac{1}{2}$	$-\frac{1}{2}$	1	0	0	$\frac{5}{2}$
0	0	$-\frac{3}{2}$	$\frac{1}{2}$	0	1	0	$\frac{3}{2}$
1	0	$-\frac{1}{2}$	$\frac{1}{2}$	0	0	0	$\frac{9}{2}$

So we find the optimal solution:

$$x_3^* = x_4^* = 0, \; x_1^* = x_2^* = 4\tfrac{1}{2}, \; x_5^* = 2\tfrac{1}{2}, \; x_6^* = 1\tfrac{1}{2}, \; x_7^* = 4.$$

The simplex algorithm has followed the path $\mathbf{0} \to \mathbf{v}_6 \to \mathbf{v}_1 \to \mathbf{v}_2$ *in Figure 1.5.*

3.6.2 The two-phase procedure

A major drawback of the big-M procedure introduced in the previous section is the fact that the choice of the value of M is not obvious. This could be circumvented by regarding M as a large but unspecified number, but this is complicated to implement in a computer package. The two-phase procedure solves this problem.

The two-phase procedure adds, in the same manner as employed in the big-M procedure, artificial variables to the '$\geq$' constraints with positive right-hand side values, and to the '$\leq$' constraints with negative right-hand side values. But instead of adding each artificial variable to the objective with a large negative coefficient, the objective function (in the first phase) is replaced by minus the sum of all artificial variables. During the first phase, the simplex algorithm then tries to maximize this objective, effectively trying to give all artificial variables the value zero; i.e., the artificial variables are 'driven out' of the set of basic variables, and a feasible basic solution for the first-phase model, and hence for the original model, has been found. When this has happened, the second phase starts by replacing the 'artificial' objective function by the objective function of the original model. The feasible basic solution found in the first phase is then used as a starting point to find an optimal solution of the model.

We will first illustrate the two-phase procedure by means of an example.

Example 3.6.6. *Consider again the model from Example 3.6.1.*

Phase 1. *Introduce artificial variables u_1 and u_2, both nonnegative, and minimize $u_1 + u_2$ (which is equivalent to maximizing $-u_1 - u_2$):*

$$
\begin{array}{lrcll}
\max & & -u_1 - u_2 & \\
\text{s.t.} & 2x_1 - x_2 - x_3 & + u_1 & = 4 \\
& -x_1 + 2x_2 \quad - x_4 & + u_2 & = 2 \\
& x_1 + x_2 \quad + x_5 & & = 12 \\
& x_1, x_2, x_3, x_4, x_5, u_1, u_2 \geq 0. &
\end{array}
$$

An initial feasible basic solution for the simplex algorithm has x_5, u_1, u_2 as the basic variables. It is left to the reader to carry out the different steps that lead to an optimal solution of this model. It turns out that, in an optimal solution of this model, it holds that $u_1 = u_2 = 0$. Deleting u_1 and u_2 from this solution results in a feasible basic solution of the original model.

Phase 2. *The original objective function $z = 2x_1 + x_2$ is now taken into account, and the simplex algorithm is carried out with the optimal feasible basic solution from Phase 1 as the initial feasible basic solution.*

If the optimal solution in Phase 1 contains a nonzero artificial variable u_i, then the original model is infeasible. In case all u_i's are zero in the optimal solution from Phase 1, but one of the u_i's is still in the set of basic variables, then a pivot step can be performed that drives this artificial variable out of the set of basic variables. It is left to the reader to check that this does not change the objective value.

Let us now look at how the two-phase procedure works when it is applied to Model Dovetail*.

Example 3.6.7. *Consider again Model Dovetail*.*

Phase 1. *Only one artificial variable u_1 has to be introduced. We then obtain the model:*

$$
\begin{array}{lrcll}
\max & & -u_1 & \\
\text{s.t.} & x_1 + x_2 + x_3 & & = 9 \\
& 3x_1 + x_2 \quad + x_4 & & = 18 \\
& x_1 \quad + x_5 & & = 7 \\
& x_2 \quad + x_6 & & = 6 \\
& x_1 + x_2 \quad - x_7 + u_1 & & = 5 \\
& x_1, x_2, x_3, x_4, x_5, x_6, x_7, u_1 \geq 0. &
\end{array}
$$

Recall that the objective coefficients of the basic variables in simplex tableaus need to be zero.

Initialization. $BI = \{3, 4, 5, 6, 8\}$, $NI = \{1, 2, 7\}$.

x_1	x_2	x_3	x_4	x_5	x_6	x_7	u_1	$-z$
1	1	0	0	0	0	-1	0	5
1	1	1	0	0	0	0	0	9
3	1	0	1	0	0	0	0	18
1	0	0	0	1	0	0	0	7
0	1	0	0	0	1	0	0	6
1	☐1	0	0	0	0	-1	1	5

We can now bring either x_1 or x_2 into the set of basic variables. Suppose that we select x_2. Then, u_1 leaves the set of basic variables.

Iteration 1. $BI = \{2, 3, 4, 5, 6\}$, $NI = \{1, 7, 8\}$.

x_1	x_2	x_3	x_4	x_5	x_6	x_7	u_1	$-z$
0	0	0	0	0	0	0	-1	0
0	0	1	0	0	0	1	-1	4
2	0	0	1	0	0	1	-1	13
1	0	0	0	1	0	0	0	7
-1	0	0	0	0	1	1	-1	1
1	1	0	0	0	0	-1	1	5

A feasible basic solution has been found, namely $x_1 = x_7 = 0$, $x_2 = 5$, $x_3 = 4$, $x_4 = 13$, $x_5 = 7$, $x_6 = 1$, and $u_1 = 0$.

Phase 2. The simplex algorithm is now applied to the original model. Since u_1 is not a basic variable anymore, we drop the column corresponding to u_1.

Initialization. $BI = \{2, 3, 4, 5, 6\}$, $NI = \{1, 7\}$.

x_1	x_2	x_3	x_4	x_5	x_6	x_7	$-z$
1	0	0	0	0	0	2	-10
0	0	1	0	0	0	1	4
2	0	0	1	0	0	1	13
1	0	0	0	1	0	0	7
-1	0	0	0	0	1	☐1	1
1	1	0	0	0	0	-1	5

Iteration 1. $BI = \{2, 3, 4, 5, 7\}$, $NI = \{1, 6\}$.

x_1	x_2	x_3	x_4	x_5	x_6	x_7	$-z$
3	0	0	0	0	-2	0	-12
①	0	1	0	0	-1	0	3
3	0	0	1	0	-1	0	12
1	0	0	0	1	0	0	7
-1	0	0	0	0	1	1	1
0	1	0	0	0	1	0	6

Iteration 2. $BI = \{1, 2, 4, 5, 7\}$, $NI = \{3, 6\}$.

x_1	x_2	x_3	x_4	x_5	x_6	x_7	$-z$
0	0	-3	0	0	1	0	-21
1	0	1	0	0	-1	0	3
0	0	-3	1	0	②	0	3
0	0	-1	0	1	1	0	4
0	0	1	0	0	0	1	4
0	1	0	0	0	1	0	6

Iteration 3. $BI = \{1, 2, 5, 6, 7\}$, $NI = \{3, 4\}$.

x_1	x_2	x_3	x_4	x_5	x_6	x_7	$-z$
0	0	$-\frac{3}{2}$	$-\frac{1}{2}$	0	0	0	$-22\frac{1}{2}$
1	0	$-\frac{1}{2}$	$\frac{1}{2}$	0	0	0	$\frac{9}{2}$
0	0	$-\frac{3}{2}$	$\frac{1}{2}$	0	1	0	$\frac{3}{2}$
0	0	$\frac{1}{2}$	$-\frac{1}{2}$	1	0	0	$\frac{5}{2}$
0	0	1	0	0	0	1	4
0	1	$\frac{3}{2}$	$-\frac{1}{2}$	0	0	0	$\frac{9}{2}$

Note that this final result is the same as the final tableau of the big-M procedure.

When comparing the big-M procedure with the two-phase procedure, it turns out that, in practice, they both require roughly the same amount of computing time. However, the paths from the infeasible $\mathbf{0}$ to the feasible region can be different for the two methods. For example, in the case of Model Dovetail*, the big-M procedure follows the path $\mathbf{0} \rightarrow \mathbf{v}_6 \rightarrow \mathbf{v}_1 \rightarrow \mathbf{v}_2$, while the two-phase procedure follows the path $\mathbf{0} \rightarrow \mathbf{v}_5 \rightarrow \mathbf{v}_4 \rightarrow \mathbf{v}_3 \rightarrow \mathbf{v}_2$. When we choose to bring x_1 into the set of basic variables in Phase I (Iteration 1), the two methods follow the same path.

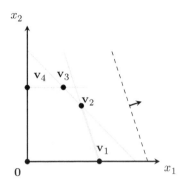

Figure 3.11: Multiple optimal solutions. All points on the thick line are optimal.

3.7 Uniqueness and multiple optimal solutions

It may happen that the optimal solution of an LO-model is not unique, meaning that there are multiple points of the feasible region for which the optimal objective value is attained; see also Section 1.2.3. In general, an LO-model has *multiple optimal solutions* if there is more than one point of the feasible region for which the objective function is optimal.

Consider for instance the LO-model:

$$\begin{aligned}
\max \quad & 3x_1 + x_2 && (3.16)\\
\text{s.t.} \quad & x_1 + x_2 \leq 9\\
& 3x_1 + x_2 \leq 18\\
& x_2 \leq 6\\
& x_1, x_2 \geq 0.
\end{aligned}$$

Figure 3.11 shows the feasible region of this model and one of the level lines of the objective function. The level lines of the objective function are parallel to the line $3x_1 + x_2 = 18$, which is part of the boundary of the feasible region. It can be seen from the figure that the vertices $\mathbf{v}_1$ and $\mathbf{v}_2$ are both optimal. (Using the fact that $\mathbf{v}_1 = \begin{bmatrix} 6 & 0 \end{bmatrix}^\mathsf{T}$ and $\mathbf{v}_2 = \begin{bmatrix} 4\frac{1}{2} & 4\frac{1}{2} \end{bmatrix}^\mathsf{T}$, we see that the objective value is 18 for both points.)

Notice also that the line segment $\mathbf{v}_1\mathbf{v}_2$ is contained in the line $3x_1 + x_2 = 18$. Hence, $\mathbf{v}_1$ and $\mathbf{v}_2$ are not the only optimal points: all points on the line segment $\mathbf{v}_1\mathbf{v}_2$ are optimal. In general, if two or more points (not necessarily vertices) of the feasible region of an LO-model are optimal, then all convex combinations of these points are also optimal. Theorem 3.7.1 makes this precise. Recall that a point $\mathbf{x}$ is a convex combination of the points $\mathbf{w}_1, \ldots, \mathbf{w}_k$ if and only if there exist real numbers $\lambda_1, \ldots, \lambda_k \geq 0$ with $\sum_{i=1}^{k} \lambda_i = 1$ such that $\mathbf{x} = \sum_{i=1}^{k} \lambda_i \mathbf{w}_i$.

Theorem 3.7.1. (*Convex combinations of optimal solutions*)
Let $\mathbf{x}_1^*, \ldots, \mathbf{x}_k^*$ be optimal solutions of the LO-model $\max\{\mathbf{c}^\mathsf{T}\mathbf{x} \mid \mathbf{A}\mathbf{x} \leq \mathbf{b}, \mathbf{x} \geq \mathbf{0}\}$. Every convex combination of $\mathbf{x}_1^*, \ldots, \mathbf{x}_k^*$ is also an optimal solution of the model.

Proof. Let $\mathbf{x}_1^*, \ldots, \mathbf{x}_k^*$ be optimal solutions and let z^* be the corresponding optimal objective value. Let $\mathbf{x}'$ be a convex combination of $\mathbf{x}_1^*, \ldots, \mathbf{x}_k^*$. This means that there exist $\lambda_1, \ldots, \lambda_k$, with $\lambda_i \geq 0$ ($i = 1, \ldots, k$) and $\sum_{i=1}^k \lambda_k = 1$, such that $\mathbf{x}' = \sum_{i=1}^k \lambda_i \mathbf{x}_i^*$. Together with the fact that $\mathbf{x}_i^* \geq \mathbf{0}$ and $\mathbf{A}\mathbf{x}_i^* \leq \mathbf{b}$ for $i = 1, \ldots k$, we have that:

$$\mathbf{x}' = \sum_{i=1}^k \lambda_i \mathbf{x}_i^* \geq \mathbf{0},$$

$$\mathbf{A}\mathbf{x}' = \mathbf{A}\left(\sum_{i=1}^k \lambda_i \mathbf{x}_i^*\right) = \left(\sum_{i=1}^k \lambda_i \mathbf{A}\mathbf{x}_i^*\right) \leq \left(\sum_{i=1}^k \lambda_i \mathbf{b}\right) = \left(\sum_{i=1}^k \lambda_i\right)\mathbf{b} = \mathbf{b}, \text{ and}$$

$$\mathbf{c}^\mathsf{T}\mathbf{x}' = \mathbf{c}^\mathsf{T}\left(\sum_{i=1}^k \lambda_i \mathbf{x}_i^*\right) = \left(\sum_{i=1}^k \lambda_i \mathbf{c}^\mathsf{T}\mathbf{x}_i^*\right) \leq \left(\sum_{i=1}^k \lambda_i z^*\right) = \left(\sum_{i=1}^k \lambda_i\right) z^* = z^*.$$

The first two equations show that $\mathbf{x}'$ is a feasible solution of the model, and the third equation shows that the objective value corresponding to $\mathbf{x}'$ is z^*. Hence, $\mathbf{x}'$ is an optimal solution of the model. $\qquad\square$

A consequence of Theorem 3.7.1 is that every LO-model has either no solution (infeasible), one solution (unique solution), or infinitely many solutions (multiple optimal solutions).

Multiple solutions can be detected using the simplex algorithm as follows. Recall that when the simplex algorithm has found an optimal feasible basic solution of a maximizing LO-model, then all current objective coefficients are nonpositive. If the current objective coefficient corresponding to a nonbasic variable x_i is strictly negative, then increasing the value of x_i leads to a decrease of the objective value, and hence this cannot lead to another optimal solution. If, however, the current objective coefficient corresponding to a nonbasic variable x_i is zero, then increasing the value of x_i (provided that this possible) does not change the objective value and hence leads to another optimal solution. The following example illustrates this.

Example 3.7.1. *Introduce the slack variables x_3, x_4, x_5. After one simplex iteration, model (3.16) changes into:*

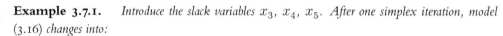

$$
\begin{aligned}
\max \quad & -x_4 && + 18 \\
\text{s.t.} \quad & \tfrac{2}{3}x_2 + x_3 - \tfrac{1}{3}x_4 && = 3 \\
& x_1 + \tfrac{1}{3}x_2 + \tfrac{1}{3}x_4 && = 6 \\
& x_2 + x_5 && = 6 \\
& x_1, x_2, x_3, x_4, x_5 \geq 0.
\end{aligned}
$$

The current basic variables are x_1, x_3, x_5, and the nonbasic variables are x_2 and x_4. However, only the nonbasic variable x_4 has a nonzero objective coefficient. The other nonbasic variable x_2 has zero objective coefficient. Taking $x_2 = x_4 = 0$, we find that $x_1 = 6$, $x_3 = 3$, and $x_5 = 6$. This corresponds to vertex $\mathbf{v}_1$ of the feasible region. The optimal objective value is $z^* = 18$. Since x_2 is not in the current objective function, we do not need to take $x_2 = 0$. In fact, x_2 can take on any *value as long as* $x_2 \geq 0$ *and:*

$$x_3 = 3 - \tfrac{2}{3}x_2 \geq 0$$
$$x_1 = 6 - \tfrac{1}{3}x_2 \geq 0$$
$$x_5 = 6 - x_2 \geq 0.$$

Solving for x_2 *from these three inequalities, gives:*

$$0 \leq x_2 \leq 4\tfrac{1}{2}.$$

Hence, $x_2 = 4\tfrac{1}{2}\lambda$ *is optimal as long as* $0 \leq \lambda \leq 1$. *Using* $x_1 = 6 - \tfrac{1}{3}x_2$, *it follows that* $x_1 = 6 - \tfrac{1}{3}(4\tfrac{1}{2}\lambda) = 6 - 1\tfrac{1}{2}\lambda$. *Letting* $\lambda_1 = 1 - \lambda$ *and* $\lambda_2 = \lambda$, *we find that* $x_1 = 6(\lambda_1 + \lambda_2) - 1\tfrac{1}{2}\lambda_2 = 6\lambda_1 + 4\tfrac{1}{2}\lambda_2$ *and* $x_2 = 4\tfrac{1}{2}\lambda_2$, *so that*

$$\begin{bmatrix} x_1 \\ x_2 \end{bmatrix} = \lambda_1 \begin{bmatrix} 6 \\ 0 \end{bmatrix} + \lambda_2 \begin{bmatrix} 4\tfrac{1}{2} \\ 4\tfrac{1}{2} \end{bmatrix} \text{ with } \lambda_1 + \lambda_2 = 1 \text{ and } \lambda_1, \lambda_2 \geq 0.$$

This is in fact a convex combination of $\begin{bmatrix} 6 & 0 \end{bmatrix}^\mathsf{T}$ *and* $\begin{bmatrix} 4\tfrac{1}{2} & 4\tfrac{1}{2} \end{bmatrix}^\mathsf{T}$; *it corresponds to the line segment* $\mathbf{v}_1\mathbf{v}_2$. *Hence, the line segment* $\mathbf{v}_1\mathbf{v}_2$ *is optimal.*

It may also happen that the set of optimal solutions is unbounded as well. Of course, this is only possible if the feasible region is unbounded. See also Section 4.4.

Example 3.7.2. *Consider the LO-model:*

$$\begin{array}{rl} \max & x_1 - 2x_2 \\ \text{s.t.} & \tfrac{1}{2}x_1 - x_2 \leq 2 \\ & -3x_1 + x_2 \leq 3 \\ & x_1, x_2 \geq 0. \end{array}$$

Figure 3.12 shows the feasible region and a level line of the objective function of this model. Introducing the slack variables x_3 *and* x_4 *and applying the simplex algorithm, one can easily verify that the optimal simplex tableau becomes:*

x_1	x_2	x_3	x_4	$-z$
0	0	-2	0	-4
1	-2	2	0	4
0	-5	6	1	15

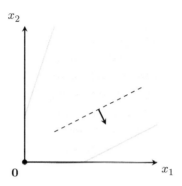

Figure 3.12: Unbounded feasible region.

The model corresponding to this tableau is:

$$\begin{aligned}
\max \quad & & -2x_3 & & +4 \\
\text{s.t.} \quad x_1 - & 2x_2 + & 2x_3 & & = 4 \\
& -5x_2 + & 6x_3 + x_4 & & = 15 \\
x_1, x_2, x_3, x_4 & \geq 0. &&&
\end{aligned}$$

The optimal solution satisfies: $x_1^* = 4$, $x_4^* = 15$, $x_2^* = x_3^* = 0$, *and* $z^* = 4$. *Note that all entries in the column corresponding to x_2 of the optimal simplex tableau are negative, and that the objective coefficient of x_2 is zero. Therefore, instead of taking $x_2 = 0$, we can take $x_2 = \lambda \geq 0$, while staying feasible and without changing the optimal objective value $z^* = 4$. For $x_2 = \lambda \geq 0$, and $x_3 = 0$, we find that $x_1 = 4 + 2\lambda$ and $x_4 = 15 + 5\lambda$. Hence, all vectors*

$$\begin{bmatrix} x_1 \\ x_2 \end{bmatrix} = \begin{bmatrix} 4 \\ 0 \end{bmatrix} + \lambda \begin{bmatrix} 2 \\ 1 \end{bmatrix} \text{ with } \lambda \geq 0$$

are optimal. These vectors form a halfline that starts at $\begin{bmatrix} 4 & 0 \end{bmatrix}^{\mathsf{T}}$ and that has direction vector $\begin{bmatrix} 2 & 1 \end{bmatrix}^{\mathsf{T}}$. On the other hand, the simplex tableau corresponding to the vertex $\begin{bmatrix} 0 & 3 \end{bmatrix}^{\mathsf{T}}$, i.e., to $BI = \{2, 3\}$, reads:

x_1	x_2	x_3	x_4	$-z$
-5	0	0	2	6
$-2\frac{1}{2}$	0	1	1	5
-3	1	0	1	3

The model corresponding to this tableau reads:

$$
\begin{aligned}
\max \quad & -5x_1 && + 2x_4 - 6 \\
\text{s.t.} \quad & -2\tfrac{1}{2}x_1 && + x_3 + x_4 && = 5 \\
& -3x_1 + x_2 && \quad + x_4 && = 3 \\
& x_1, x_2, x_3, x_4 \geq 0.
\end{aligned}
$$

Both the objective coefficient and the matrix coefficients of x_1 are now negative. This means that the halfline with (take $x_1 = \lambda \geq 0$, and $x_3 = x_4 = 0$):

$$
\begin{bmatrix} x_1 \\ x_2 \end{bmatrix} = \begin{bmatrix} 0 \\ 3 \end{bmatrix} + \lambda \begin{bmatrix} 1 \\ 3 \end{bmatrix} \text{ with } \lambda \geq 0
$$

is in the feasible region. The objective value $z = 6 - 5\lambda$ decreases unboundedly along this halfline for $\lambda < \infty$. Hence, the model is unbounded; see also Section 1.1.2.

The above discussions are summarized in the following theorem.

Theorem 3.7.2. (*Multiple optimal solutions*)
Let $\mathbf{B}$ be any basis matrix corresponding to the feasible basic solution $\mathbf{x}_{BI} = \mathbf{B}^{-1}\mathbf{b} \geq \mathbf{0}$, $\mathbf{x}_{NI} = \mathbf{0}$ of the standard LO-model:

$$
\max\{\mathbf{c}^\mathsf{T}\mathbf{x} \mid \mathbf{A}\mathbf{x} \leq \mathbf{b}, \mathbf{x} \geq \mathbf{0}\}.
$$

Assume that there is an index $\alpha \in NI$ with $(\mathbf{c}_{NI}^\mathsf{T} - \mathbf{c}_{BI}^\mathsf{T}\mathbf{B}^{-1}\mathbf{N})_\alpha = 0$. Define:

$$
\delta = \min\left\{ \frac{(\mathbf{B}^{-1}\mathbf{b})_j}{(\mathbf{B}^{-1}\mathbf{N})_{j,\alpha}} \,\middle|\, (\mathbf{B}^{-1}\mathbf{N})_{j,\alpha} > 0, j \in \{1, \ldots, m\} \right\},
$$

where we use the convention that $\min \emptyset = \infty$. Then, every point in the set

$$
L = \left\{ \mathbf{x} \in \mathbb{R}^n \,\middle|\, \begin{bmatrix} \mathbf{x} \\ \mathbf{x}_s \end{bmatrix} \equiv \begin{bmatrix} \mathbf{x}_{BI} \\ \mathbf{x}_{NI} \end{bmatrix} = \begin{bmatrix} \mathbf{B}^{-1}\mathbf{b} \\ \mathbf{0} \end{bmatrix} + \lambda \begin{bmatrix} -(\mathbf{B}^{-1}\mathbf{N})_{\star,\alpha} \\ \mathbf{e}_\alpha \end{bmatrix}, 0 \leq \lambda \leq \delta \right\}
$$

is an optimal solution of the model. In the above expression, $\mathbf{0} \in \mathbb{R}^n$ and $\mathbf{e}_\alpha$ is the α'th unit vector in $\mathbb{R}^n$.

Proof of Theorem 3.7.2. See Exercise 3.10.11. $\qquad\qquad\square$

It is left to the reader to compare this theorem with the results of Example 3.7.2.

Note that, in the case of degeneracy, the value of δ in the statement of Theorem 3.7.2 may be zero. In that case, the set L contains just a single point. This does not imply that the LO-model has a unique solution; see Exercise 3.10.13. If, however, all current objective coefficients of the nonbasic variables are strictly negative, then the LO-model has a unique solution.

Theorem 3.7.3. (*Unique optimal solution*)
Let $\mathbf{B}$ be any basis matrix corresponding to the feasible basic solution $\mathbf{x}_{BI} = \mathbf{B}^{-1}\mathbf{b}$, $\mathbf{x}_{NI} = \mathbf{0}$ of the standard LO-model:

$$\max\{\mathbf{c}^\mathsf{T}\mathbf{x} \mid \mathbf{A}\mathbf{x} \le \mathbf{b}, \mathbf{x} \ge \mathbf{0}\}.$$

Assume that $(\mathbf{c}_{NI}^\mathsf{T} - \mathbf{c}_{BI}^\mathsf{T}\mathbf{B}^{-1}\mathbf{N})_j < 0$ for all $j \in \{1, \dots, n\}$. Then, this feasible basic solution is the unique optimal feasible basic solution of the model, and this feasible basic solution corresponds to the unique optimal solution of the model.

Note that an LO-model may have a unique optimal feasible <u>basic</u> solution, while having multiple optimal points. The reader is asked to construct such a model in Exercise 3.10.14.

Proof of Theorem 3.7.3. Let $\begin{bmatrix} \hat{\mathbf{x}} \\ \hat{\mathbf{x}}_s \end{bmatrix} \equiv \begin{bmatrix} \hat{\mathbf{x}}_{BI} \\ \hat{\mathbf{x}}_{NI} \end{bmatrix}$ with $\hat{\mathbf{x}}_{BI} = \mathbf{B}^{-1}\mathbf{b} \ge \mathbf{0}$ and $\hat{\mathbf{x}}_{NI} = \mathbf{0}$ be the feasible basic solution with respect to $\mathbf{B}$, and let $\hat{z}$ be the corresponding objective value. Consider the model formulation (3.6). Since $(\mathbf{c}_{NI}^\mathsf{T} - \mathbf{c}_{BI}^\mathsf{T}\mathbf{B}^{-1}\mathbf{N})_j < 0$ for all $j \in \{1, \dots, n\}$, it follows that any solution with $\mathbf{x}_{NI} \ne \mathbf{0}$ has a corresponding objective value that is strictly smaller than $\hat{z}$. Hence, $\hat{z}$ is the optimal objective value of the model, and any optimal point $\begin{bmatrix} \mathbf{x}_{BI}^* \\ \mathbf{x}_{NI}^* \end{bmatrix}$ satisfies $\mathbf{x}_{NI}^* = \mathbf{0}$. Therefore, the constraints of (3.6) imply that $\mathbf{x}_{BI}^* = \mathbf{B}^{-1}\mathbf{b}$. Thus, every optimal point of (3.6) satisfies $\mathbf{x}_{BI}^* = \hat{\mathbf{x}}_{BI}$ and $\mathbf{x}_{NI}^* = \hat{\mathbf{x}}_{NI}$. In other words, $\begin{bmatrix} \hat{\mathbf{x}}_{BI} \\ \hat{\mathbf{x}}_{NI} \end{bmatrix}$ is the unique optimal point of (3.6), and hence $\hat{\mathbf{x}}$ is the unique optimal solution of the standard LO-model. $\square$

3.8 Models with equality constraints

The simplex algorithm can also be applied to models with equality constraints, i.e., to LO-models of the form:

$$\max\{\mathbf{c}^\mathsf{T}\mathbf{x} \mid \mathbf{A}\mathbf{x} = \mathbf{b}, \mathbf{x} \ge \mathbf{0}\}. \tag{3.17}$$

See also (III) in Section 1.3. Before we explain how this model may be solved using the simplex algorithm, recall that, in Section 1.3, we described the following transformation from a model of the form (3.17) to a standard LO-model:

$$\max\{\mathbf{c}^\mathsf{T}\mathbf{x} \mid \mathbf{A}\mathbf{x} = \mathbf{b}, \mathbf{x} \ge \mathbf{0}\} = \max\left\{\mathbf{c}^\mathsf{T}\mathbf{x} \,\middle|\, \begin{bmatrix} \mathbf{A} \\ -\mathbf{A} \end{bmatrix}\mathbf{x} \le \begin{bmatrix} \mathbf{b} \\ -\mathbf{b} \end{bmatrix}, \mathbf{x} \ge \mathbf{0}\right\}.$$

The simplex algorithm can certainly be applied directly to this model. However, the disadvantage of this transformation is that the resulting model has twice as many constraints as the original model. Moreover, as Theorem 3.8.1 shows, all feasible basic solutions become degenerate.

Theorem 3.8.1.

All feasible basic solutions of the model $\max\left\{\mathbf{c}^\mathsf{T}\mathbf{x} \,\middle|\, \begin{bmatrix} \mathbf{A} \\ -\mathbf{A} \end{bmatrix}\mathbf{x} \le \begin{bmatrix} \mathbf{b} \\ -\mathbf{b} \end{bmatrix}, \mathbf{x} \ge \mathbf{0}\right\}$ are degenerate.

Proof of Theorem 3.8.1. Let $\tilde{\mathbf{A}} = \begin{bmatrix} \mathbf{A} & \mathbf{I}_m & \mathbf{0} \\ -\mathbf{A} & \mathbf{0} & \mathbf{I}_m \end{bmatrix}$, and take any invertible $(2m, 2m)$-submatrix $\tilde{\mathbf{B}} = \begin{bmatrix} \mathbf{B} & \mathbf{B}' & \mathbf{0} \\ -\mathbf{B} & \mathbf{0} & \mathbf{B}'' \end{bmatrix}$ of $\tilde{\mathbf{A}}$, with $\mathbf{B}$ a submatrix of $\mathbf{A}$, and $\mathbf{B}'$ and $\mathbf{B}''$ submatrices of $\mathbf{I}_m$. Let $\mathbf{x}'$ and $\mathbf{x}''$ be the nonnegative variables corresponding to $\mathbf{B}'$ and $\mathbf{B}''$, respectively. Moreover, let $\mathbf{x}_{BI}$ be the vector of variables corresponding to the column indices of $\mathbf{B}$. Then, the system of equations

$$\tilde{\mathbf{B}}\begin{bmatrix} \mathbf{x}_{BI} \\ \mathbf{x}' \\ \mathbf{x}'' \end{bmatrix} = \begin{bmatrix} \mathbf{b} \\ -\mathbf{b} \end{bmatrix}$$

is equivalent to $\mathbf{B}\mathbf{x}_{BI} + \mathbf{B}'\mathbf{x}' = \mathbf{b}$, $-\mathbf{B}\mathbf{x}_{BI} + \mathbf{B}''\mathbf{x}'' = -\mathbf{b}$. Adding these two expressions gives $\mathbf{B}'\mathbf{x}' + \mathbf{B}''\mathbf{x}'' = \mathbf{0}$. Now recall that $\mathbf{B}'$ is a submatrix of the identity matrix $\mathbf{I}_m$, and each column of $\mathbf{B}'$ is a unit vector. Therefore, $\mathbf{B}'\mathbf{x}'$ is an m-dimensional vector consisting of the entries of $\mathbf{x}'$ and a number of zeroes (to be precise, m minus the number of entries in $\mathbf{x}'$). The same is true for $\mathbf{B}''\mathbf{x}''$. Therefore, since $\mathbf{x}', \mathbf{x}'' \ge \mathbf{0}$, the fact that $\mathbf{B}'\mathbf{x}' + \mathbf{B}''\mathbf{x}'' = \mathbf{0}$ implies that $\mathbf{x}' = \mathbf{x}'' = \mathbf{0}$. Because the variables in $\mathbf{x}_{BI}$, $\mathbf{x}'$, and $\mathbf{x}''$ are exactly the basic variables corresponding to the basis matrix $\tilde{\mathbf{B}}$, it follows that the feasible basic solution $\begin{bmatrix} \mathbf{x}_{BI} \\ \mathbf{x}' \\ \mathbf{x}'' \end{bmatrix}$ is indeed degenerate. $\square$

In order to apply the simplex algorithm to models of the form (3.17), we need to understand the concept of feasible basic solutions for such models. In Section 2.2, we defined feasible basic solutions for LO-models of the standard form $\max\{\mathbf{c}^\mathsf{T}\mathbf{x} \mid \mathbf{A}\mathbf{x} \le \mathbf{b}, \mathbf{x} \ge \mathbf{0}\}$, with $\mathbf{A}$ an (m, n)-matrix. The key component of the definition of a feasible basic solution is the concept of basis matrix. A basis matrix is any <u>invertible</u> (m, m)-submatrix of the matrix $\begin{bmatrix} \mathbf{A} & \mathbf{I}_m \end{bmatrix}$. Recall that the latter matrix originates from the fact that the constraints of the model with slack variables can be written as $\begin{bmatrix} \mathbf{A} & \mathbf{I}_m \end{bmatrix}\begin{bmatrix} \mathbf{x} \\ \mathbf{x}_s \end{bmatrix} = \mathbf{b}$.

Consider now an LO-model of the form (3.17). The concept of 'basis matrix' in such a model is defined in an analogous way: a *basis matrix* for model (3.17) is any <u>invertible</u> (m, m)-submatrix of the matrix $\mathbf{A}$. The corresponding basic solution is then $\begin{bmatrix} \mathbf{x}_{BI} \\ \mathbf{x}_{NI} \end{bmatrix}$ with $\mathbf{x}_{BI} = \mathbf{B}^{-1}\mathbf{b}$ and $\mathbf{x}_{NI} = \mathbf{0}$.

In the case of a model of the form (3.17), it may happen that $\mathbf{A}$ contains no basis matrices at all. This happens when $\mathbf{A}$ is not of full row rank, i.e., if $\mathrm{rank}(\mathbf{A}) < m$. In that case, $\mathbf{A}$ contains no (m, m)-submatrix, and hence the model has no basic solutions at all. This problem does not occur in the case of a model with inequality constraints, because the matrix $\begin{bmatrix} \mathbf{A} & \mathbf{I}_m \end{bmatrix}$ contains the (m, m) identity matrix, which has rank m. (See also Section B.4.)

In practice one may encounter matrices that do not have full row rank. For example, every instance of the transportation problem (see Section 8.2.1) has a technology matrix that does not have full row rank, so that its rows are not linearly independent. In such cases we can apply Gaussian elimination and delete zero rows until either the remaining matrix has full row rank, or this matrix augmented with its right hand side values gives rise to an inconsistent set of equalities. In case of the transportation problem, the deletion of one arbitrary row already leads to a full row rank matrix; see Section 8.2.1. In general, multiple rows may need to be deleted.

The mechanics of the simplex algorithm for models with equality constraints are the same as in the case of models with inequality constraints. The only caveat is that Gaussian elimination is required to turn the initial tableau into a simplex tableau. The following example illustrates this.

Example 3.8.1. *Consider the following LO-model:*

$$\begin{aligned} \max \quad & 40x_1 + 100x_2 + 150x_3 \\ \text{s.t.} \quad & x_1 + 2x_2 + 2x_3 = 3 \\ & 30x_1 + 10x_2 + 20x_3 = 75 \\ & x_1, x_2, x_3 \geq 0. \end{aligned}$$

The technology constraints of this LO-model are equality constraints. Putting the coefficients of the model into a tableau, we obtain the following tableau:

x_1	x_2	x_3	$-z$
40	100	150	0
1	2	2	3
30	10	20	75

This is not a simplex tableau yet, because we have not specified the set of basic variables and the tableau does not contain a $(2, 2)$ identity matrix. Let us choose $BI = \{1, 2\}$. After applying Gaussian elimination, we find the following tableau:

x_1	x_2	x_3	$-z$
0	0	54	-126
1	0	$\frac{2}{5}$	$2\frac{2}{5}$
0	1	$\frac{4}{5}$	$\frac{3}{10}$

This is indeed a simplex tableau: the columns of the technology matrix corresponding to the basic variables form an identity matrix, the current objective coefficients of the basic variables are all zero, and *the right hand side values are nonnegative. Since this is a simplex tableau, the simplex algorithm may be applied to find an optimal feasible basic solution. An optimal solution is in fact found after one iteration. It is left to the reader to check that the optimal objective value of the model is $146\frac{1}{4}$, and $x_1^* = 2\frac{1}{4}$, $x_2^* = 0$, $x_3^* = \frac{3}{8}$ is an optimal solution.*

The initial choice of BI in the above example is in some sense lucky: had we used a different choice of BI, then it could happen that one or more right hand side values turned out being negative. This happens, for instance, for $BI = \{2,3\}$. In that case, we would have found an infeasible initial basic solution, and we could not have applied the simplex algorithm. In general, the big-M procedure (see Section 3.6.1) can be applied to avoid having to guess an initial basis. This is illustrated in the following example.

Example 3.8.2. *The model from Example 3.8.1 may be solved by augmenting it with artificial variables u_1 and u_2 as follows:*

$$
\begin{aligned}
\max \quad & 40x_1 + 100x_2 + 150x_3 - Mu_1 - Mu_2 \\
\text{s.t.} \quad & x_1 + 2x_2 + 2x_3 + u_1 \qquad\qquad = 3 \\
& 30x_1 + 10x_2 + 20x_3 \qquad + u_2 = 75 \\
& x_1, x_2, x_3, u_1, u_2 \geq 0.
\end{aligned}
$$

Now, choosing u_1 and u_2 as the initial basic variables yields an initial feasible basic solution.

3.9 The revised simplex algorithm

In this section, we present a special implementation of the simplex algorithm, called the *revised simplex algorithm*. The advantage of the revised simplex algorithm is that the number of arithmetic operations (i.e., addition, subtraction, multiplication, and division) when running the algorithm is kept to a minimum. This is achieved by the fact that – in contrast to the simplex algorithm – the simplex tableau is not updated in full in every iteration. The computational savings become particularly clear when the number of variables is much larger than the number of constraints, and when the technology matrix contains a lot of zero entries. The revised simplex algorithm is nowadays implemented in almost all commercial computer codes.

3.9.1 Formulating the algorithm

Consider the standard LO-model:

$$
\begin{aligned}
\max \ & \mathbf{c}^\mathsf{T}\mathbf{x} \\
\text{s.t. } & \mathbf{A}\mathbf{x} \leq \mathbf{b} \\
& \mathbf{x} \geq \mathbf{0},
\end{aligned}
$$

where $\mathbf{A}$ is an (m,n)-matrix. For any basis matrix $\mathbf{B}$ in $\begin{bmatrix} \mathbf{A} & \mathbf{I}_m \end{bmatrix}$ we write, as usual, $\begin{bmatrix} \mathbf{A} & \mathbf{I}_m \end{bmatrix} \equiv \begin{bmatrix} \mathbf{B} & \mathbf{N} \end{bmatrix}$. An essential part of any iteration of the standard simplex algorithm is

$$\begin{bmatrix} \mathbf{0}^\mathsf{T} & \mathbf{c}_{NI}^\mathsf{T} - \mathbf{c}_{BI}^\mathsf{T} \mathbf{B}^{-1} \mathbf{N} & \vdots & -\mathbf{c}_{BI}^\mathsf{T} \mathbf{B}^{-1} \mathbf{b} \\ \mathbf{I}_m & \mathbf{B}^{-1} \mathbf{N} & \vdots & \mathbf{B}^{-1} \mathbf{b} \end{bmatrix}$$

Figure 3.13: Simplex tableau.

$$\begin{bmatrix} \mathbf{c}_{BI}^\mathsf{T} \mathbf{B}^{-1} & \vdots & -\mathbf{c}_{BI}^\mathsf{T} \mathbf{B}^{-1} \mathbf{b} \\ \mathbf{B}^{-1} & \vdots & \mathbf{B}^{-1} \mathbf{b} \end{bmatrix}$$

Figure 3.14: Revised simplex tableau.

the updating of the current basic index set BI:

$$BI_{\text{new}} := (BI \setminus \{BI_\beta\}) \cup \{NI_\alpha\},$$

where NI_α is the index of the entering variable, and BI_β is the index of the leaving variable. The index α satisfies:

$$\left(\mathbf{c}_{NI}^\mathsf{T} - \mathbf{c}_{BI}^\mathsf{T} \mathbf{B}^{-1} \mathbf{N} \right)_\alpha > 0;$$

see Section 3.3. If such an α does not exist, then the current basis matrix is optimal and the simplex algorithm terminates. Otherwise, once such an α has been calculated, the leaving basic variable x_{BI_β} is determined by performing the minimum-ratio test. The minimum-ratio test determines an index $\beta \in \{1, \ldots, m\}$ such that:

$$\frac{(\mathbf{B}^{-1}\mathbf{b})_\beta}{(\mathbf{B}^{-1}\mathbf{N})_{\beta,\alpha}} = \min\left\{ \frac{(\mathbf{B}^{-1}\mathbf{b})_j}{(\mathbf{B}^{-1}\mathbf{N})_{j,\alpha}} \;\middle|\; j \in \{1, \ldots, m\}, (\mathbf{B}^{-1}\mathbf{N})_{j,\alpha} > 0 \right\}.$$

Then x_{BI_β} is the leaving basic variable.

It follows from this discussion that at each iteration, we really only need to calculate the current objective coefficients, the values of the current set of basic variables, and the column α of $\mathbf{B}^{-1}\mathbf{N}$. Hence, only the calculation of the inverse $\mathbf{B}^{-1}$ is required. There is no need to update and store the complete simplex tableau (consisting of $m + 1$ rows and $m + n + 1$ columns) at each iteration, the updating of a tableau consisting of $m + 1$ rows and $m + 1$ columns suffices. Figure 3.13 and Figure 3.14 schematically show the standard simplex tableau and the *revised simplex tableau*.

Each column of $\mathbf{B}^{-1}\mathbf{N}$ in the standard simplex tableau can be readily computed from the original input data ($\mathbf{N}$ is a submatrix of $\begin{bmatrix} \mathbf{A} & \mathbf{I}_m \end{bmatrix}$) as soon as $\mathbf{B}^{-1}$ is available. Moreover, row 0 of the standard simplex tableau can easily be calculated from the original input data if the value of $\mathbf{c}_{BI}^\mathsf{T} \mathbf{B}^{-1}$ is known. The revised simplex algorithm, in the case of a maximization model, can now be formulated as follows.

Algorithm 3.9.1. (*Revised simplex algorithm*)

Input: Values of $\mathbf{A}$, $\mathbf{b}$, and $\mathbf{c}$ in the LO-model $\max\{\mathbf{c}^\mathsf{T}\mathbf{x} \mid \mathbf{A}\mathbf{x} \leq \mathbf{b}, \mathbf{x} \geq \mathbf{0}\}$, and an initial feasible basic solution for the LO-model with slack variables (see Section 3.6).

Output: Either an optimal solution of the model, or the message 'the model is unbounded'.

Step 1: *Selection of the entering variable.*
Let $\mathbf{B}$ be the current basis matrix. The matrix $\mathbf{B}^{-1}$ and the vector $\mathbf{c}_{BI}^\mathsf{T}\mathbf{B}^{-1}$ are readily available in the revised simplex tableau, so that $\mathbf{c}_{NI}^\mathsf{T} - \mathbf{c}_{BI}^\mathsf{T}\mathbf{B}^{-1}\mathbf{N}$ may be calculated efficiently. Choose an index $\alpha \in \{1, \ldots, n\}$ that satisfies $\left(\mathbf{c}_{NI}^\mathsf{T} - \mathbf{c}_{BI}^\mathsf{T}\mathbf{B}^{-1}\mathbf{N}\right)_\alpha > 0$. If no such α exists, then stop: the current feasible basic solution is optimal. Otherwise, variable x_{NI_α} is the new basic variable. Go to Step 2.

Step 2: *Selection of the leaving variable.*
Determine all row indices $j \in \{1, \ldots, m\}$ for which $\left(\mathbf{B}^{-1}\mathbf{N}\right)_{j,\alpha} > 0$. If no such indices exist, then output the message 'the model is unbounded' and stop. Apply the minimum-ratio test to determine a leaving variable x_{BI_β} with $\beta \in \{1, \ldots, m\}$. This requires the vector $\mathbf{B}^{-1}\mathbf{b}$, which is readily available in the revised simplex tableau, and column α of $\mathbf{B}^{-1}\mathbf{N}$. Go to Step 3.

Step 3: *Updating the basis matrix.*
Let $BI_\beta := NI_\alpha$, and $NI = \{1, \ldots, n+m\} \setminus \{BI_1, \ldots, BI_m\}$. Replace column β of $\mathbf{B}$ (corresponding to x_{BI_β}) by column α of $\mathbf{N}$ (corresponding to x_{NI_α}), and go to Step 1.

In Section 3.9.3, we will apply the revised simplex algorithm to Model Dovetail.

3.9.2 The product form of the inverse

In Step 3 of the revised simplex algorithm, the new basis matrix $\mathbf{B}_{\text{new}}$ differs from the current basis matrix $\mathbf{B}$ by only one column. This makes it possible to calculate the inverse of $\mathbf{B}_{\text{new}}$ from $\mathbf{B}$ in a very effective manner. This procedure uses the so-called *product form of the inverse*. The procedure can be described in the following way. Recall that the inverse of $\mathbf{B}$ is readily available in the revised simplex tableau. In the next iteration, the column of $\mathbf{B}$, say $\mathbf{B}_{\star,\beta}$ (with $\beta \in \{1, \ldots, m\}$), corresponding to the leaving variable x_{BI_β}, is replaced by the column $\mathbf{N}_{\star,\alpha}$ of $\mathbf{N}$, corresponding to the entering variable x_{NI_α}. So, the new basis matrix is:

$$\mathbf{B}_{\text{new}} = \begin{bmatrix} \mathbf{B}_{\star,1} & \cdots & \mathbf{B}_{\star,\beta-1} & \mathbf{N}_{\star,\alpha} & \mathbf{B}_{\star,\beta+1} & \cdots & \mathbf{B}_{\star,m} \end{bmatrix}.$$

This matrix $\mathbf{B}_{\text{new}}$ can be written as follows:

$$\begin{aligned} \mathbf{B}_{\text{new}} &= \begin{bmatrix} \mathbf{B}\mathbf{e}_1 & \cdots & \mathbf{B}\mathbf{e}_{\beta-1} & \mathbf{N}_{\star,\alpha} & \mathbf{B}\mathbf{e}_{\beta+1} & \cdots & \mathbf{B}\mathbf{e}_m \end{bmatrix} \\ &= \mathbf{B}\begin{bmatrix} \mathbf{e}_1 & \cdots & \mathbf{e}_{\beta-1} & \mathbf{B}^{-1}\mathbf{N}_{\star,\alpha} & \mathbf{e}_{\beta+1} & \cdots & \mathbf{e}_m \end{bmatrix} = \mathbf{B}\mathbf{T}, \end{aligned}$$

where, for $i = 1, \ldots, m$, $\mathbf{e}_i$ is the i'th unit vector in $\mathbb{R}^m$, and $\mathbf{T}$ the identity matrix with the β'th column replaced by $\mathbf{B}^{-1}\mathbf{N}_{\star,\alpha}$ ($= (\mathbf{B}^{-1}\mathbf{N})_{\star,\alpha}$). Because $\det(\mathbf{B}_{\text{new}}) = \det(\mathbf{B})\det(\mathbf{T})$ (see Appendix B.7), and $\mathbf{B}$ and $\mathbf{T}$ are nonsingular square matrices, it follows that $\mathbf{B}_{\text{new}}$ is nonsingular. Its inverse satisfies:

$$\mathbf{B}_{\text{new}}^{-1} = \mathbf{T}^{-1}\mathbf{B}^{-1}.$$

Define $\mathbf{E} = \mathbf{T}^{-1}$. $\mathbf{E}$ is called an *elementary matrix*. Write $\begin{bmatrix} M_1 & \ldots & M_m \end{bmatrix}^\mathsf{T} = \mathbf{B}^{-1}\mathbf{N}_{\star,\alpha}$. It is easy to verify (by verifying that $\mathbf{ET} = \mathbf{I}_m$) that $\mathbf{E}$ satisfies:

$$\mathbf{E} = \begin{bmatrix} 1 & \ldots & M_1 & \ldots & 0 \\ \vdots & & \vdots & & \vdots \\ 0 & \ldots & M_\beta & \ldots & 0 \\ \vdots & & \vdots & & \vdots \\ 0 & \ldots & M_m & \ldots & 1 \end{bmatrix}^{-1} = \begin{bmatrix} 1 & \ldots & -M_1/M_\beta & \ldots & 0 \\ \vdots & & \vdots & & \vdots \\ 0 & \ldots & 1/M_\beta & \ldots & 0 \\ \vdots & & \vdots & & \vdots \\ 0 & \ldots & -M_m/M_\beta & \ldots & 1 \end{bmatrix}.$$

Note that $\mathbf{E}$ also differs from the identity matrix by one column. For this reason, $\mathbf{E}$ can be stored as a single column vector: its β'th column, together with an additional entry, indicating its column position in the identity matrix.

Assume that the initial basis matrix (recall that the revised simplex algorithm can always start with the big-M or the two-phase procedure) is the identity matrix. Denote by $\mathbf{B}_{(i)}$ the new basis matrix at the end of the i'th iteration of the revised simplex algorithm, and $\mathbf{E}_{(i)}$ the elementary matrix of the i'th iteration. Then we can write:

$$\mathbf{B}_{(0)}^{-1} = \mathbf{I}_m,$$
$$\mathbf{B}_{(1)}^{-1} = \mathbf{E}_{(1)}\mathbf{I}_m = \mathbf{E}_{(1)},$$
$$\mathbf{B}_{(2)}^{-1} = \mathbf{E}_{(2)}\mathbf{B}_{(1)}^{-1} = \mathbf{E}_{(2)}\mathbf{E}_{(1)},$$
$$\vdots$$
$$\mathbf{B}_{(i)}^{-1} = \mathbf{E}_{(i)}\mathbf{B}_{(i-1)}^{-1} = \mathbf{E}_{(i)}\mathbf{E}_{(i-1)}\ldots\mathbf{E}_{(1)}.$$

Hence, at each iteration of the revised simplex algorithm, the inverse of the new basis matrix can be written as a product of elementary matrices.

3.9.3 Applying the revised simplex algorithm

In the following example, we apply the revised simplex algorithm, while making use of the product form of the inverse, to Model Dovetail (see Section 1.1).

Example 3.9.1.   *After rewriting the model in standard form with slack variables, we have the following LO-model:*

$$\begin{aligned} \max \quad & 3x_1 + 2x_2 \\ \text{s.t.} \quad & x_1 + x_2 + x_3 && = 9 \\ & 3x_1 + x_2 + x_4 && = 18 \\ & x_1 + x_5 && = 7 \\ & x_2 + x_6 && = 6 \\ & x_1, x_2, x_3, x_4, x_5, x_6 \geq 0. \end{aligned}$$

We write $\begin{bmatrix} \mathbf{A} & \mathbf{I}_4 \end{bmatrix} = \begin{bmatrix} \mathbf{a}_1 & \mathbf{a}_2 & \mathbf{a}_3 & \mathbf{a}_4 & \mathbf{a}_5 & \mathbf{a}_6 \end{bmatrix}$, *so that* $\mathbf{a}_3 = \mathbf{e}_1$, $\mathbf{a}_4 = \mathbf{e}_2$, $\mathbf{a}_5 = \mathbf{e}_3$, *and* $\mathbf{a}_6 = \mathbf{e}_4$.

Initialization. *As the initial basis matrix, we take the columns of* $\begin{bmatrix} \mathbf{A} & \mathbf{I}_4 \end{bmatrix}$ *corresponding to the variables* x_3, x_4, x_5, *and* x_6. *So for the initial solution, it holds that:*

$$
\begin{aligned}
\mathbf{B}_{(0)} &= \begin{bmatrix} \mathbf{a}_3 & \mathbf{a}_4 & \mathbf{a}_5 & \mathbf{a}_6 \end{bmatrix} = \mathbf{I}_4, \\
\mathbf{c}_{BI}^{\mathsf{T}} &= \begin{bmatrix} c_3 & c_4 & c_5 & c_6 \end{bmatrix} = \begin{bmatrix} 0 & 0 & 0 & 0 \end{bmatrix}, \\
\mathbf{x}_{BI}^{\mathsf{T}} &= \begin{bmatrix} x_3 & x_4 & x_5 & x_6 \end{bmatrix} = \begin{bmatrix} 9 & 18 & 7 & 6 \end{bmatrix}, \\
\mathbf{x}_{NI}^{\mathsf{T}} &= \begin{bmatrix} x_1 & x_2 \end{bmatrix} \quad\;\; = \begin{bmatrix} 0 & 0 \end{bmatrix}, \\
z &\qquad\qquad\qquad\;\; = 0, \\
\mathbf{B}_{(0)}^{-1} &\qquad\qquad\qquad\;\; = \mathbf{I}_4.
\end{aligned}
$$

Iteration 1. *The first iteration starts with* $BI = \{3, 4, 5, 6\}$ *and* $\mathbf{B}_{(0)}^{-1} = \mathbf{I}_4$.

Step 1. *Since* $\mathbf{c}_{NI}^{\mathsf{T}} - \mathbf{c}_{BI}^{\mathsf{T}} \mathbf{B}_{(0)}^{-1} \begin{bmatrix} \mathbf{a}_1 & \mathbf{a}_2 \end{bmatrix} = \begin{bmatrix} 3 & 2 \end{bmatrix}$, *and both entries of this vector are positive; we select* x_1 *as the entering variable.*

Step 2. $(\mathbf{B}_{(0)}^{-1} \mathbf{a}_1)^{\mathsf{T}} = \begin{bmatrix} 1 & 3 & 1 & 0 \end{bmatrix}$, *and* $(\mathbf{B}_{(0)}^{-1} \mathbf{b})^{\mathsf{T}} = \begin{bmatrix} 9 & 18 & 7 & 6 \end{bmatrix}$. *The minimum-ratio test yields:* $\min\{9/1, 18/3, 7/1, \star\} = 6 = (\mathbf{B}_{(0)}^{-1} \mathbf{b})_4 / (\mathbf{B}_{(0)}^{-1} \mathbf{a}_1)_4$. *Hence,* x_4 *will be the leaving variable.*

Step 3. $BI := (BI \setminus \{4\}) \cup \{1\} = \{3, 1, 5, 6\}$. *Since*

$$
\mathbf{E}_{(1)} = \begin{bmatrix} 1 & -\frac{1}{3} & 0 & 0 \\ 0 & \frac{1}{3} & 0 & 0 \\ 0 & -\frac{1}{3} & 1 & 0 \\ 0 & 0 & 0 & 1 \end{bmatrix},
$$

it follows that $\mathbf{B}_{(1)}^{-1} = \mathbf{E}_{(1)}$.

Iteration 2. *The second iteration starts with* $BI = \{3, 1, 5, 6\}$ *and* $\mathbf{B}_{(1)}^{-1} = \mathbf{E}_{(1)}$.

Step 1. *Since* $\mathbf{c}_{BI} = \begin{bmatrix} c_3 & c_1 & c_5 & c_6 \end{bmatrix}^{\mathsf{T}}$, *and* $\mathbf{c}_{NI} = \begin{bmatrix} c_4 & c_2 \end{bmatrix}^{\mathsf{T}}$, *it follows that* $\mathbf{c}_{NI}^{\mathsf{T}} - \mathbf{c}_{BI}^{\mathsf{T}} \mathbf{B}_{(1)}^{-1} \begin{bmatrix} \mathbf{a}_4 & \mathbf{a}_2 \end{bmatrix} = \begin{bmatrix} -1 & 1 \end{bmatrix}$. *Only the entry corresponding to* x_2 *is positive, so that* x_2 *is the entering variable.*

Step 2. $(\mathbf{B}_{(1)}^{-1} \mathbf{a}_2)^{\mathsf{T}} = \begin{bmatrix} \frac{2}{3} & \frac{1}{3} & -\frac{1}{3} & 1 \end{bmatrix}$, *and* $(\mathbf{B}_{(1)}^{-1} \mathbf{b})^{\mathsf{T}} = \begin{bmatrix} 3 & 6 & 1 & 6 \end{bmatrix}$. *The minimum-ratio test yields:* $\min\{4\frac{1}{2}, 18, \star, 6\} = 4\frac{1}{2} = (\mathbf{B}_{(1)}^{-1} \mathbf{b})_3 / (\mathbf{B}_{(1)}^{-1} \mathbf{a}_2)_3$. *Hence,* x_3 *will be the leaving variable.*

Step 3. $BI := (BI \setminus \{3\}) \cup \{2\} = \{2, 1, 5, 6\}$. *Note that* $\alpha = 2$ *and* $\beta = 3$. *Hence,*

$$
\mathbf{E}_{(2)} = \begin{bmatrix} \frac{3}{2} & 0 & 0 & 0 \\ -\frac{1}{2} & 1 & 0 & 0 \\ \frac{1}{2} & 0 & 1 & 0 \\ -\frac{3}{2} & 0 & 0 & 1 \end{bmatrix}, \text{ and } \mathbf{B}_{(2)}^{-1} = \mathbf{E}_{(2)} \mathbf{E}_{(1)} = \begin{bmatrix} \frac{3}{2} & -\frac{1}{2} & 0 & 0 \\ -\frac{1}{2} & \frac{1}{2} & 0 & 0 \\ \frac{1}{2} & -\frac{1}{2} & 1 & 0 \\ -\frac{3}{2} & \frac{1}{2} & 0 & 1 \end{bmatrix}.
$$

Iteration 3. Recall that $\mathbf{c}_{BI}^{\mathsf{T}} = \begin{bmatrix} c_2 & c_1 & c_5 & c_6 \end{bmatrix}$, and $\mathbf{c}_{NI}^{\mathsf{T}} = \begin{bmatrix} c_4 & c_3 \end{bmatrix}$.

Step 1. $\mathbf{c}_{NI}^{\mathsf{T}} - \mathbf{c}_{BI}^{\mathsf{T}}\mathbf{B}_{(2)}^{-1}\begin{bmatrix} \mathbf{a}_4 & \mathbf{a}_3 \end{bmatrix} = \begin{bmatrix} -\frac{1}{2} & -1\frac{1}{2} \end{bmatrix}$. *Since none of the entries of this vector is positive, the current basis matrix is optimal. In order to calculate the optimal objective value, we determine* $\mathbf{x}_{BI}$ *as follows:*

$$
\mathbf{x}_{BI} = \mathbf{B}_{(2)}^{-1}\mathbf{b} = \mathbf{E}_2(\mathbf{E}_1\mathbf{b}) = \begin{bmatrix} \frac{3}{2} & 0 & 0 & 0 \\ -\frac{1}{2} & 1 & 0 & 0 \\ \frac{1}{2} & 0 & 1 & 0 \\ -\frac{3}{2} & 0 & 0 & 1 \end{bmatrix}\begin{bmatrix} 3 \\ 6 \\ 1 \\ 6 \end{bmatrix} = \begin{bmatrix} 4\frac{1}{2} \\ 4\frac{1}{2} \\ 2\frac{1}{2} \\ 1\frac{1}{2} \end{bmatrix}.
$$

Note that the values of the entries of $\mathbf{E}_1\mathbf{b}$ *were already calculated in the previous iteration. The optimal objective value is* $22\frac{1}{2}$.

3.10   Exercises

Exercise 3.10.1.   Consider the following LO-model:

$$
\begin{aligned}
\max \quad & x_1 + x_2 \\
\text{s.t.} \quad & x_1 && \leq 2 && (1) \\
& 2x_1 + 3x_2 \leq 5 && && (2) \\
& x_1, x_2 \geq 0.
\end{aligned}
$$

Introduce slack variables x_3 and x_4 for the two constraints. Solve the model in a way as described in Section 3.1. Start the procedure at the vertex $\begin{bmatrix} 0 & 0 \end{bmatrix}^{\mathsf{T}}$, then include x_1 in the set of basic variables while removing x_3, and finally exchange x_4 by x_2.

Exercise 3.10.2.   Show that the feasible region of the LO-model:

$$
\begin{aligned}
\max \quad & 6x_1 + 4x_2 + x_3 - 12 \\
\text{s.t.} \quad & x_1 + x_2 + x_3 && \leq 1 \\
& x_1 && \leq 1 \\
& x_1, x_2, x_3 \geq 0
\end{aligned}
$$

contains a degenerate vertex, but the simplex algorithm does not make a cycle.

Exercise 3.10.3.   Consider the following LO-model:

$$
\begin{aligned}
\max \quad & 3x_1 - 80x_2 + 2x_3 - 24x_4 \\
\text{s.t.} \quad & x_1 - 32x_2 - 4x_3 + 36x_4 \leq 0 \\
& x_1 - 24x_2 - x_3 + 6x_4 \leq 0 \\
& x_1, x_2, x_3, x_4 \geq 0.
\end{aligned}
$$

(a)   Draw the simplex adjacency graph.

(b) Use (a) to determine the variables (including slack variables) that are unbounded.

(c) Show that the simplex algorithm cycles when using the following pivot rule: in case of ties, choose the basic variable with the smallest subscript.

(d) Apply the perturbation method.

(e) Extend the model with the constraint $x_3 \leq 1$, and solve it.

Exercise 3.10.4. Consider the LO-model of Example 3.5.4, with the additional constraint $x_1 \leq 0$.

(a) Argue that, in contrast to the LO-model of Example 3.5.4, the new model has a bounded optimal solution.

(b) Find, by inspection, the optimal feasible basic solution of the model.

(c) Show that the simplex algorithm cycles when the same pivot rules are used as in Example 3.5.4.

Exercise 3.10.5.

(a) Give an example of an LO-model in which a feasible basic solution becomes infeasible when the model is perturbed.

(b) Give an example in which the perturbation method turns a feasible LO-model into an infeasible LO-model.

(c) Argue that, if the perturbation method is applied by adding ε^i to the right hand side of each '$\leq$' constraint i and subtracting ε^i from the right hand side of each '$\geq$', the situation of (b) does not occur.

Exercise 3.10.6. A company produces two commodities. For both commodities three kinds of raw materials are needed. In the table below the used quantities are listed (in kg) for each kind of raw material for each commodity unit.

	Material 1	**Material 2**	**Material 3**
Commodity A	10	5	12
Commodity B	7	7	10

The price of raw material 1 is \$2 per kg, the price of raw material 2 is \$3 per kg, and the price of raw material 3 is \$4 per kg. A maximum of only 20,000 kgs of raw material 1 can be purchased, a maximum of only 15,000 kgs of raw material 2, and a maximum of only 25,000 kgs of raw material 3 can be purchased.

(a) The company wants to maximize the production. Write the problem as a standard LO-model and solve it.

(b) The company also wants to keep the costs for raw material as low as possible, but it wants to produce at least 1,000 kgs of each commodity. Write this problem as a standard LO-model and solve it using the Big-M procedure.

Exercise 3.10.7.   The company Eltro manufactures radios and televisions. The company has recently developed a new type of television. In order to draw attention to this new product, Eltro has embarked on an ambitious TV-advertising campaign and has decided to purchase one-minute commercial spots on two types of TV programs: comedy shows and basketball games. Each comedy commercial is seen by 4 million women and 2 million men. Each basketball commercial is seen by 2 million women and 6 million men. A one-minute comedy commercial costs $50,000 and a one-minute basketball commercial costs $100,000. Eltro would like the commercials to be seen by at least 20 million women and 18 million men. Determine how Eltro can meet its advertising requirements at minimum cost. (Hint: use the two-phase procedure.)

Exercise 3.10.8.   Solve the following LO-model by using the simplex algorithm:

$$
\begin{aligned}
\max \quad & 2x_1 - 3x_2 - 4x_3 \\
\text{s.t.} \quad & -x_1 - 5x_2 + x_3 \qquad = 4 \\
& -x_1 - 3x_2 \qquad + x_4 = 2 \\
& x_1, x_2, x_3, x_4 \geq 0.
\end{aligned}
$$

Exercise 3.10.9.   Use the simplex algorithm to solve the following LO-model: (1) by converting it into a maximization model, (2) by solving it directly as a minimization model. Compare the various iterations of both methods.

$$
\begin{aligned}
\min \quad & 2x_1 - 3x_2 \\
\text{s.t.} \quad & x_1 + x_2 \leq 4 \\
& x_1 - x_2 \leq 6 \\
& x_1, x_2 \geq 0.
\end{aligned}
$$

Exercise 3.10.10.   Dafo Car manufactures three types of cars: small, medium, and large ones. The finishing process during the manufacturing of each type of car requires finishing materials and two types of skilled labor: assembling labor and painting labor. For each type of car, the amount of each resource required to finish ten cars is given in the table below.

Resource	Large car	Medium car	Small car
Finishing materials	800 kgs	600 kgs	100 kgs
Assembling hours	4	2	1.5
Painting hours	2	1.5	0.5

At present, 4,800 kgs metal, 20 assembling hours, and 8 painting hours are available. A small car sells for $2,000, a medium car for $3,000, and a large car for $6,000. Dafo Car expects that demand for large cars and small cars is unlimited, but that at most 5 ($\times 100{,}000$) medium cars can be sold. Since the available resources have already been purchased, Dafo Car wants to maximize its total revenue. Solve this problem.

Exercise 3.10.11. Prove Theorem 3.7.2. (Hint: first read the proof of Theorem 3.5.2.)

Exercise 3.10.12.

(a) Construct an example for which $\delta = \infty$ in Theorem 3.7.2.

(b) Construct an example for which $\delta = 0$ in Theorem 3.7.2.

Exercise 3.10.13. Construct an LO-model and an optimal feasible basic solution for which the following holds: the value of δ in the statement of Theorem 3.7.2 is zero, but the model has multiple optimal solutions.

Exercise 3.10.14. Construct an LO-model that has a unique optimal feasible basic solution, but multiple optimal points.

Exercise 3.10.15. Determine all optimal solutions of the following LO-model; compare the results with those of the model from the beginning of Section 3.7.

$$
\begin{aligned}
\max \quad & x_1 + x_2 \\
\text{s.t.} \quad & x_1 + x_2 \le 9 \\
& 3x_1 + x_2 \le 18 \\
& x_2 \le 6 \\
& x_1, x_2 \ge 0.
\end{aligned}
$$

Exercise 3.10.16. Show that the LO-model

$$
\max\Big\{ x_1 + x_2 + x_3 \ \Big|\ \mathbf{A}^{\mathsf{T}} \begin{bmatrix} x_1 & x_2 & x_3 \end{bmatrix}^{\mathsf{T}} \ge \mathbf{b},\ x_1, x_2, x_3, x_4 \ge 0 \Big\},
$$

with $\mathbf{b} = \begin{bmatrix} 1 & 0 & 0 & 0 \end{bmatrix}^{\mathsf{T}}$ and $\mathbf{A} = \begin{bmatrix} 1 & -1 & 0 & 0 \\ 0 & 1 & -1 & 0 \\ 0 & 0 & 1 & -1 \end{bmatrix}$,

is infeasible. (Hint: apply the two-phase procedure.)

Exercise 3.10.17.   Consider the following LO-model:

$$\begin{array}{ll} \max & x_1 + 2x_2 \\ \text{s.t.} & -x_1 + x_2 \leq 6 \\ & x_1 - 2x_2 \leq 4 \\ & x_1, x_2 \geq 0. \end{array}$$

(a) Determine all vertices and extreme directions of the feasible region of the following LO-model by drawing its feasible region.

(b) Use the simplex algorithm to show that the model is unbounded.

(c) Determine a halfline in the feasible region on which the objective values increase unboundedly.

Exercise 3.10.18.   Consider the following LO-models:

$$\begin{array}{ll} \max & 2x_1 + 3x_2 + 8x_3 + x_4 + 5x_5 \\ \text{s.t.} & 3x_1 + 7x_2 + 12x_3 + 2x_4 + 7x_5 \leq 10 \\ & x_1, x_2, x_3, x_4, x_5 \geq 0, \end{array}$$

and

$$\begin{array}{ll} \max & 2x_1 + x_2 + 4x_3 + 5x_5 + x_6 \\ \text{s.t.} & 3x_1 + 6x_2 + 3x_3 + 2x_4 + 3x_5 + 4x_6 \leq 60 \\ & x_1, x_2, x_3, x_4, x_5, x_6 \geq 0. \end{array}$$

For each model, answer the following questions.

(a) Determine all feasible basic solutions.

(b) Draw the simplex adjacency graph.

(c) When starting the simplex algorithm in the all-zero solution, how many possibilities are there to reach the optimal solution?

Exercise 3.10.19.   Determine the simplex adjacency graphs of the following LO-models.

(a) $\max\{x_1 \mid x_1 + x_2 + x_3 \leq 1, x_1 \leq 1, \ x_1, x_2, x_3 \geq 0\}$.

(b) $\max\{x_2 \mid x_1 + x_2 + x_3 \leq 1, x_1 \leq 1, \ x_1, x_2, x_3 \geq 0\}$.

Exercise 3.10.20.   Let $BI(1)$ and $BI(2)$ be two nodes of a simplex adjacency graph, each corresponding to a degenerate vertex. Suppose that $BI(1)$ and $BI(2)$ differ in precisely one entry. Show, by constructing an example, that $BI(1)$ and $BI(2)$ need not be connected by an arc.

Exercise 3.10.21.

(a) Show that, when applying the simplex algorithm, if a variable leaves the set of basic variables in some iteration, then it cannot enter the set of basic variables in the following iteration.

(b) Argue that it may happen that a variable that leaves the set of basic variables enters the set of basic variable in a later iteration.

Exercise 3.10.22. Consider a standard LO-model.

(a) Show that once the simplex algorithm has reached a feasible basic solution that corresponds to an optimal vertex $\mathbf{v}^*$ of the feasible region, all feasible basic solutions that are subsequently visited correspond to the same vertex $\mathbf{v}^*$.

(b) Show, using (a), that every optimal vertex of the feasible region corresponds to at least one optimal feasible basic solution. (Hint: use a similar argument as in the proof of Theorem 3.5.5.)

Exercise 3.10.23. Consider the following LO-model:

$$
\begin{aligned}
\max \quad & x_1 + 2x_2 \\
\text{s.t.} \quad & 2x_1 - x_2 - x_3 \geq -2 \\
& x_1 - x_2 + x_3 \geq -1 \\
& x_1, x_2, x_3 \geq 0.
\end{aligned}
$$

(a) Use the simplex algorithm to verify that the LO-model has no optimal solution.

(b) Use the final simplex tableau to determine a feasible solution with objective value at least 300.

Exercise 3.10.24. Apply the big-M procedure to show that the following LO-model is infeasible.

$$
\begin{aligned}
\max \quad & 2x_1 + 4x_2 \\
\text{s.t.} \quad & 2x_1 - 3x_2 \geq 2 \\
& -x_1 + x_2 \geq 1 \\
& x_1, x_2 \geq 0.
\end{aligned}
$$

Exercise 3.10.25. Convert the following LO-models into standard form, and determine optimal solutions with the simplex algorithm.

(a) $\max \ -2x_1 + x_2$
$\quad$ s.t. $\quad x_1 + x_2 \leq 4$
$\qquad\qquad x_1 - x_2 \leq 62$
$\qquad\qquad x_1 \geq 0,\ x_2$ free.

(b)   max $3x_1 + x_2 - 4x_3 + 5x_4 + 9x_5$
   s.t. $4x_1 - 5x_2 - 9x_3 + x_4 - 2x_5 \leq 6$
      $x_1 + 3x_2 + 4x_3 - 5x_4 + x_5 \leq 9$
      $x_1 + x_2 - 5x_3 - 7x_4 + 11x_5 \leq 10$
      $x_1, x_2, x_4 \geq 0,\ x_3, x_5$ free.

Exercise 3.10.26.   Consider the following LO-model:

$$\max\ 2x_1 + 3x_2 - x_3 - 12x_4$$
$$\text{s.t.}\ 2x_1 + 9x_2 - x_3 - 9x_4 \geq 0$$
$$\tfrac{1}{3}x_1 + x_2 - \tfrac{1}{3}x_3 - 2x_4 \leq 0$$
$$x_1, x_2, x_3, x_4 \geq 0.$$

(a)   Draw the simplex adjacency graph.

(b)   Apply the simplex algorithm with the pivot rule: Choose the current largest positive objective coefficient for the entering basic variable, while ties in the minimum-ratio test are broken by favoring the row with the lowest index. Show that cycling occurs.

(c)   Formulate a tie breaking rule in the minimum-ratio test that excludes cycling; show that the model is unbounded.

Exercise 3.10.27.   Show that the cycle drawn with thick edges in Figure 3.10 is the only simplex cycle in the graph of Figure 3.10.

Exercise 3.10.28.   Apply the revised simplex algorithm, by using the product form of the inverse, to the following LO-models:

(a)   max $2x_1 + 3x_2 - x_3 + 4x_4 + x_5 - 3x_6$
   s.t. $x_1 - 2x_2 + x_4 + 4x_5 + \tfrac{1}{2}x_6 \leq 10$
      $x_1 + x_2 + 3x_3 + 2x_4 + x_5 - x_6 \leq 16$
      $2x_1 + \tfrac{1}{2}x_2 - x_3 - x_4 + x_5 + 5x_6 \leq 8$
      $x_1, x_2, x_3, x_4, x_5, x_6 \geq 0.$

(b)   max $-3x_1 + x_2 + x_3$
   s.t. $x_1 - 2x_2 + x_3 \leq 11$
      $-4x_1 + x_2 + 2x_3 \geq 3$
      $2x_1 - x_3 = -1$
      $x_1, x_2, x_3 \geq 0.$

CHAPTER 4

Duality, feasibility, and optimality

Overview

The concept of duality plays a key role in the theory of linear optimization. In 1947, JOHN VON NEUMANN (1903–1957) formulated the so-called dual of a linear optimization model. The variables of the dual model correspond to the technology constraints of the original model. The original LO-model is then called the primal model. Actually, the terms primal and dual are only relative: the dual of the dual model is again the primal model. In practical terms, if the primal model deals with quantities, then the dual deals with prices. This relationship will be discussed thoroughly in Chapter 5. In this chapter we will show the relationship with the other important concept of linear optimization, namely optimality. The theory is introduced by means of simple examples. Moreover, the relevant geometrical interpretations are extensively discussed.

We saw in Chapter 3 that any simplex tableau represents a feasible basic solution of the primal LO-model. In this chapter, we will see that any simplex tableau also represents a basic solution of the dual model. In contrast to the primal solution, however, this dual solution satisfies the dual optimality criterion, but it is not necessarily feasible. We will see that the simplex algorithm can be viewed as an algorithm that:

seeks primal optimality, while keeping primal feasibility, and

seeks dual feasibility, while keeping dual optimality.

We will describe an alternative simplex algorithm, namely the so-called dual simplex algorithm, which seeks primal feasibility, while keeping primal optimality. Although we will not show this here, this dual simplex algorithm can be viewed as the primal simplex algorithm applied to the dual problem.

4.1   The companies Dovetail and Salmonnose

In this section we introduce the fictitious company Salmonnose, which can be considered as a 'dual' of company Dovetail that was introduced in Chapter 1. The LO-models of both companies will serve as running examples throughout the text.

4.1.1   Formulating the dual model

The new company, called Salmonnose has been set up in the neighborhood of the company Dovetail, introduced in Section 1.1. Salmonnose is in the business of manufacturing wooden crates. The management team has noticed that both Dovetail and Salmonnose need the same raw material: wood. They are therefore considering buying out Dovetail. Obviously, they want to pay as little as possible for Dovetail. However, if the price quoted is too low, then the owners of company Dovetail will refuse the offer. So, how does Salmonnose decide on the price to offer?

We assume that Salmonnose has access to all information about Dovetail. It knows that the production facility at Dovetail allows it to manufacture up to 9 ($\times 100,000$) boxes a year, and its stock consists of 18 ($\times 100,000$) cubic meters of wood, 7 ($\times 100,000$) boxes for long matches and 6 ($\times 100,000$) boxes for short matches. It also knows that Dovetail can sell whatever it produces, and that it makes a profit of 3 ($\times \$1,000$) per box of long matches and a profit of 2 ($\times \$1,000$) per box of short matches.

In order to find out how much it should offer, Salmonnose decides to estimate how much each of the raw materials and the production facility is worth to Dovetail. Let the prices be as follows.

y_1 = the price per unit (i.e., $100,000$ boxes) of production;
y_2 = the price per unit ($100,000$ m^3) of wood;
y_3 = the price of one unit ($100,000$ boxes) of long matches;
y_4 = the price of one unit ($100,000$ boxes) of short matches.

Consider now the production of long matches. Each unit of long matches requires one unit of production capacity, three units of wood, and one unit of boxes for long matches. Once processed, these raw materials yield a profit of 3 ($\times \$1,000$) for this unit of long matches. This means that the value of this particular combination of raw materials is at least 3 ($\times \$1,000$). Thus, the price that Salmonnose has to pay cannot be less than the profit that is made by selling this unit of long matches. Therefore,

$$y_1 + 3y_2 + y_3 \geq 3. \qquad (4.1)$$

Similarly, considering the production of short matches, Salmonnose will come up with the constraint

$$y_1 + y_2 + y_4 \geq 2. \qquad (4.2)$$

Given these constraints, Salmonnose wants to minimize the amount it pays to the owners of Dovetail in order to buy all its production capacity and in-stock raw materials, i.e., it wants to minimize $9y_1 + 18y_2 + 7y_3 + 6y_4$. Of course, the prices are nonnegative, so that $y_1, y_2, y_3, y_4 \geq 0$. Therefore, the model that Salmonnose solves is the following.

Model Salmonnose.

$$\begin{aligned}
\min \quad & 9y_1 + 18y_2 + 7y_3 + 6y_4 \\
\text{s.t.} \quad & y_1 + 3y_2 + y_3 \qquad \geq 3 & (4.1) \\
& y_1 + y_2 \qquad\quad + y_4 \geq 2 & (4.2) \\
& y_1, y_2, y_3, y_4 \geq 0.
\end{aligned}$$

The optimal solution of this model turns out to be: $y_1^* = 1\frac{1}{2}$, $y_2^* = \frac{1}{2}$, $y_3^* = 0$, $y_4^* = 0$, with the optimal objective value $z^* = 22\frac{1}{2}$; see Section 4.2. This means that 'Salmonnose' has to pay \$22,500 to Dovetail to buy it out. The value $z^* = 22\frac{1}{2}$ should not come as a surprise. Model Dovetail showed that the company was making \$22,500 from its facilities, and so any offer of more than this amount would have been acceptable for their owners. The beauty of Model Salmonnose lies in its interpretations, which are given in the next section.

4.1.2 **Economic interpretation**

The results of Section 4.1.1 can be interpreted in two ways: globally if the price of the whole transaction is considered, and marginally if the price of one additional production unit is considered.

The global interpretation. In Section 4.1.1 it is noted that Salmonnose has to pay $22\frac{1}{2}$ ($\times$\$1,000) for the whole transaction, which includes the rent of the machine during one year and the purchase of the inventory of wood and boxes. The various means of production have their own prices as well. For instance, the price of the rent of the machine is equal to $9 \times 1\frac{1}{2} = 13\frac{1}{2}$ ($\times$\$1,000). It should be mentioned that these prices only hold for the whole transaction.

The fact that the 'long' and 'short' boxes in stock go with it for free ($y_3^* = 0$ and $y_4^* = 0$) does not mean, of course, that they do not represent any value; if the 'long' and 'short' boxes are excluded from the transaction, then the total price may become lower. Actually, the prices of the 'long' and 'short' boxes are discounted in the prices of the other means of production.

The marginal interpretation. The fact that $y_3^* = 0$ means that a change in the inventory of boxes for long matches has no implications for the value of the total transaction. The same holds for the inventory of the 'short' boxes. However, a change in the inventory of wood, say by an (additive) quantity γ (called the *perturbation factor*), has consequences for the total value because the value of the corresponding variable y_2 is larger than 0. For instance, the fact that $y_2^* = \frac{1}{2}$ means that if the amount of wood inventory changes from 18 to $18 + \gamma$, then the optimal profit changes from $22\frac{1}{2}$ to $22\frac{1}{2} + \frac{1}{2}\gamma$ ($\times$ \$1,000). It should be mentioned that this only holds for small values of γ; see Chapter 5.

4.2 Duality and optimality

In this section we show how Model Salmonnose can be derived directly from Model Dovetail. The main purpose of this section is explaining how to use duality to determine optimality of solutions.

4.2.1 Dualizing the standard LO-model

Let z^* be the optimal objective value of Model Dovetail. In order to obtain upper bounds for z^*, we can use the constraints of the model in the following way. For instance, the first constraint can be used to derive that $z^* \leq 27$. Namely, since $3x_1 + 2x_2 \leq 3(x_1 + x_2) \leq 3 \times 9 = 27$ for each $x_1, x_2 \geq 0$, it follows that $z^* = \max(3x_1 + 2x_2) \leq 27$. On the other hand, when combining the second and the fourth constraint in the following way:

$$3x_1 + 2x_2 = (3x_1 + x_2) + (x_2) \leq 18 + 6 = 24,$$

we find that $z^* \leq 24$. Generalizing this idea, we can construct any nonnegative linear combination of all four constraints (with nonnegative weight factors y_1, y_2, y_3, y_4), and obtain the inequality:

$$(x_1 + x_2)y_1 + (3x_1 + x_2)y_2 + (x_1)y_3 + (x_2)y_4 \leq 9y_1 + 18y_2 + 7y_3 + 6y_4,$$

provided that $y_1, y_2, y_3, y_4 \geq 0$. By reordering the terms of the left hand side, this inequality can be written as:

$$(y_1 + 3y_2 + y_3)x_1 + (y_1 + y_2 + y_4)x_2 \leq 9y_1 + 18y_2 + 7y_3 + 6y_4.$$

Clearly, as long as the values of y_1, y_2, y_3, and y_4 are chosen so that $y_1 + 3y_2 + y_3 \geq 3$ and $y_1 + y_2 + y_4 \geq 2$, the left hand side is at least $3x_1 + 2x_2$. In that case, we obtain that:

$$z^* \leq 9y_1 + 18y_2 + 7y_3 + 6y_4. \tag{4.3}$$

Note that, for $y_1 = 3$ and $y_2 = y_3 = y_4 = 0$, we obtain $z^* \leq 27$, and for $y_1 = y_3 = 0$ and $y_2 = y_4 = 1$ we obtain $z^* \leq 24$. Now, we may ask the question: what choice of values for y_1, y_2, y_3, and y_4 provides the best (i.e., lowest) upper bound for z^*? To answer this question, we need to determine values of y_1, y_2, y_3, and y_4 so that the right hand side of (4.3) is as small as possible, taking into account the conditions that we imposed on these values. In other words, the following inequality provides the best upper bound:

$$
\begin{aligned}
z^* \leq \min \quad & 9y_1 + 18y_2 + 7y_3 + 6y_4 \\
\text{s.t.} \quad & y_1 + 3y_2 + y_3 \geq 3 \\
& y_1 + y_2 + y_4 \geq 2 \\
& y_1, y_2, y_3, y_4 \geq 0.
\end{aligned}
$$

Note that the LO-model in the right hand side part is exactly Model Salmonnose. So, an upper bound for the optimal objective value z^* of Model Dovetail is given by the optimal objective value of Model Salmonnose. In Theorem 4.2.4 we will show that z^* is in fact equal to the optimal objective value of Model Salmonnose. In the realm of duality, Model Dovetail is called the *primal model* and Model Salmonnose the *dual model*. The general standard form (see Section 1.2.1) of both models is as follows:

STANDARD PRIMAL MODEL	STANDARD DUAL MODEL
$\max\left\{\mathbf{c}^{\mathsf{T}}\mathbf{x} \mid \mathbf{A}\mathbf{x} \leq \mathbf{b}, \mathbf{x} \geq \mathbf{0}\right\}$	$\min\left\{\mathbf{b}^{\mathsf{T}}\mathbf{y} \mid \mathbf{A}^{\mathsf{T}}\mathbf{y} \geq \mathbf{c}, \mathbf{y} \geq \mathbf{0}\right\}$

with $\mathbf{x} \in \mathbb{R}^{n}, \mathbf{y} \in \mathbb{R}^{m}, \mathbf{c} \in \mathbb{R}^{n}, \mathbf{b} \in \mathbb{R}^{m}, \mathbf{A} \in \mathbb{R}^{m \times n}$.

The following remarkable connections between the primal and the dual model exist.

STANDARD PRIMAL MODEL	STANDARD DUAL MODEL
Maximizing model	Minimizing model
n decision variables	m decision variables
m constraints	n constraints
Right hand side vector is $\mathbf{b}$	Right hand side vector is $\mathbf{c}$
Objective coefficients vector is $\mathbf{c}$	Objective coefficients vector is $\mathbf{b}$
The technology matrix is $\mathbf{A}$	The technology matrix is $\mathbf{A}^{\mathsf{T}}$

The i'th dual constraint corresponds to the primal decision variable x_i $(i = 1, \ldots, n)$.
The dual decision variable y_j corresponds to the j'th primal constraint $(j = 1, \ldots, m)$.
The constraint coefficients of the i'th primal variable are the coefficients in the i'th dual constraint.

The process described above, in which the dual model is derived from a given primal model, is called *dualization*. We will see in the following section that dualizing a standard dual model results in the original standard primal model.

4.2.2 Dualizing nonstandard LO-models

In Section 1.3 we have seen how LO-models that are not in standard form can always be transformed into standard form. This means that nonstandard LO-models can be dualized by first transforming them into standard form, and then applying the dualization rules, as formulated in Section 4.2.1. However, the dualization rules for nonstandard LO-models are actually very simple, and so the detour of transforming them into standard form is superfluous. We will derive the dualization rules for nonstandard LO-models once in the following theorem, and then take them for granted.

Theorem 4.2.1.

Consider the following general LO-model:

$$\max \quad \mathbf{c}_1^\mathsf{T}\mathbf{x}_1 + \mathbf{c}_2^\mathsf{T}\mathbf{x}_2 + \mathbf{c}_3^\mathsf{T}\mathbf{x}_3 \qquad\qquad\text{(GM)}$$
$$\text{s.t.} \quad \mathbf{A}_1\mathbf{x}_1 + \mathbf{A}_2\mathbf{x}_2 + \mathbf{A}_3\mathbf{x}_3 \leq \mathbf{b}_1$$
$$\mathbf{A}_4\mathbf{x}_1 + \mathbf{A}_5\mathbf{x}_2 + \mathbf{A}_6\mathbf{x}_3 \geq \mathbf{b}_2$$
$$\mathbf{A}_7\mathbf{x}_1 + \mathbf{A}_8\mathbf{x}_2 + \mathbf{A}_9\mathbf{x}_3 = \mathbf{b}_3$$
$$\mathbf{x}_1 \geq \mathbf{0}, \mathbf{x}_2 \leq \mathbf{0}, \mathbf{x}_3 \text{ free.}$$

The dual model of model (GM) is:

$$\min \quad \mathbf{b}_1^\mathsf{T}\mathbf{y}_1 + \mathbf{b}_2^\mathsf{T}\mathbf{y}_2 + \mathbf{b}_3^\mathsf{T}\mathbf{y}_3 \qquad\qquad\text{(DGM)}$$
$$\text{s.t.} \quad \mathbf{A}_1^\mathsf{T}\mathbf{y}_1 + \mathbf{A}_4^\mathsf{T}\mathbf{y}_2 + \mathbf{A}_7^\mathsf{T}\mathbf{y}_3 \geq \mathbf{c}_1$$
$$\mathbf{A}_2^\mathsf{T}\mathbf{y}_1 + \mathbf{A}_5^\mathsf{T}\mathbf{y}_2 + \mathbf{A}_8^\mathsf{T}\mathbf{y}_3 \leq \mathbf{c}_2$$
$$\mathbf{A}_3^\mathsf{T}\mathbf{y}_1 + \mathbf{A}_6^\mathsf{T}\mathbf{y}_2 + \mathbf{A}_9^\mathsf{T}\mathbf{y}_3 = \mathbf{c}_3$$
$$\mathbf{y}_1 \geq \mathbf{0}, \mathbf{y}_2 \leq \mathbf{0}, \mathbf{y}_3 \text{ free.}$$

The dimensions of the symbols in model (GM) are omitted. The expression '$\mathbf{x}$ free' means that the entries of the vector $\mathbf{x}$ are not restricted in sign, and so they can be either nonnegative or nonpositive. Often, expressions of the form '$\mathbf{x}$ free' are omitted.

Proof. In order to dualize model (GM), we first transform it into the standard form, i.e., into the form $\max\left\{\mathbf{c}^\mathsf{T}\mathbf{x} \mid \mathbf{A}\mathbf{x} \leq \mathbf{b}, \mathbf{x} \geq \mathbf{0}\right\}$. Handling the constraints is straightforward: each '$\geq$' constraint is turned into a '$\leq$' constraint by multiplying both sides of the inequality by -1, and each '$=$' constraint is turned into two '$\leq$' constraints (see also Section 1.3). This leaves the expressions '$\mathbf{x}_2 \leq \mathbf{0}$' and '$\mathbf{x}_3$ free'. These can be handled as follows:

    Multiply $\mathbf{x}_2 \leq \mathbf{0}$ by -1 (resulting in $-\mathbf{x}_2 \geq \mathbf{0}$) and replace every occurrence of $\mathbf{x}_2$ in the model by $-\mathbf{x}_2$.

    Replace $\mathbf{x}_3$ by $\mathbf{x}_3' - \mathbf{x}_3''$ with $\mathbf{x}_3' \geq \mathbf{0}$ and $\mathbf{x}_3'' \geq \mathbf{0}$.

The standard LO-model (SGM) associated with (GM) becomes:

$$\max \quad \mathbf{c}_1^\mathsf{T}\,\mathbf{x}_1 + (-\mathbf{c}_2^\mathsf{T})\mathbf{x}_2 + \mathbf{c}_3^\mathsf{T}\,\mathbf{x}_3' + (-\mathbf{c}_3^\mathsf{T})\mathbf{x}_3'' \qquad\text{(SGM)}$$
$$\text{s.t.} \quad \mathbf{A}_1\,\mathbf{x}_1 + (-\mathbf{A}_2)\mathbf{x}_2 + \mathbf{A}_3\,\mathbf{x}_3' + (-\mathbf{A}_3)\mathbf{x}_3'' \leq \quad\mathbf{b}_1$$
$$(-\mathbf{A}_4)\mathbf{x}_1 + \mathbf{A}_5\,\mathbf{x}_2 + (-\mathbf{A}_6)\mathbf{x}_3' + \mathbf{A}_6\,\mathbf{x}_3'' \leq -\mathbf{b}_2$$
$$\mathbf{A}_7\,\mathbf{x}_1 + (-\mathbf{A}_8)\mathbf{x}_2 + \mathbf{A}_9\,\mathbf{x}_3' + (-\mathbf{A}_9)\mathbf{x}_3'' \leq \quad\mathbf{b}_3$$
$$(-\mathbf{A}_7)\mathbf{x}_1 + \mathbf{A}_8\,\mathbf{x}_2 + (-\mathbf{A}_9)\mathbf{x}_3' + \mathbf{A}_9\,\mathbf{x}_3'' \leq -\mathbf{b}_3$$
$$\mathbf{x}_1, \mathbf{x}_2, \mathbf{x}_3', \mathbf{x}_3'' \geq \mathbf{0}.$$

Let $\mathbf{y}_1$, $\mathbf{y}_2$, $\mathbf{y}_3'$, and $\mathbf{y}_3''$ be the vectors of the dual decision variables corresponding to the constraints of (SGM) with right hand side vectors $\mathbf{b}_1$, $-\mathbf{b}_2$, $\mathbf{b}_3$, and $-\mathbf{b}_3$, respectively. The dual (DSGM) of (SGM) then is:

$$
\begin{array}{rl}
\min & \mathbf{b}_1^\mathsf{T}\,\mathbf{y}_1 + (-\mathbf{b}_2^\mathsf{T})\mathbf{y}_2 + \quad \mathbf{b}_3^\mathsf{T}\,\mathbf{y}_3' + (-\mathbf{b}_3^\mathsf{T})\mathbf{y}_3'' \qquad \text{(DSGM)} \\
\text{s.t.} & \mathbf{A}_1^\mathsf{T}\,\mathbf{y}_1 + (-\mathbf{A}_4^\mathsf{T})\mathbf{y}_2 + \quad \mathbf{A}_7^\mathsf{T}\,\mathbf{y}_3' + (-\mathbf{A}_7^\mathsf{T})\mathbf{y}_3'' \geq \quad \mathbf{c}_1 \\
& (-\mathbf{A}_2^\mathsf{T})\mathbf{y}_1 + \quad \mathbf{A}_5^\mathsf{T}\,\mathbf{y}_2 + (-\mathbf{A}_8^\mathsf{T})\mathbf{y}_3' + \quad \mathbf{A}_8^\mathsf{T}\,\mathbf{y}_3'' \geq -\mathbf{c}_2 \\
& \mathbf{A}_3^\mathsf{T}\,\mathbf{y}_1 + (-\mathbf{A}_6^\mathsf{T})\mathbf{y}_2 + \quad \mathbf{A}_9^\mathsf{T}\,\mathbf{y}_3' + (-\mathbf{A}_9^\mathsf{T})\mathbf{y}_3'' \geq \quad \mathbf{c}_3 \\
& (-\mathbf{A}_3^\mathsf{T})\mathbf{y}_1 + \quad \mathbf{A}_6^\mathsf{T}\,\mathbf{y}_2 + (-\mathbf{A}_9^\mathsf{T})\mathbf{y}_3' + \quad \mathbf{A}_9^\mathsf{T}\,\mathbf{y}_3'' \geq -\mathbf{c}_3 \\
& \mathbf{y}_1, \mathbf{y}_2, \mathbf{y}_3', \mathbf{y}_3'' \geq \mathbf{0}.
\end{array}
$$

Recall that, before the dualization process, we have applied the following transformations to (GM):

(a) Multiply each '$\geq$' constraint by -1 (resulting in a '$\leq$' constraint).

(b) Replace each '$=$' constraint by a pair of '$\leq$' constraints; namely, replace $\mathbf{A}_7\mathbf{x}_1 + \mathbf{A}_8\mathbf{x}_2 + \mathbf{A}_9\mathbf{x}_3 = \mathbf{b}_3$ by $\mathbf{A}_7\mathbf{x}_1 + \mathbf{A}_8\mathbf{x}_2 + \mathbf{A}_9\mathbf{x}_3 \leq \mathbf{b}_3$ and $-\mathbf{A}_7\mathbf{x}_1 - \mathbf{A}_8\mathbf{x}_2 - \mathbf{A}_9\mathbf{x}_3 \leq -\mathbf{b}_3$. Their corresponding dual variables are $\mathbf{y}_3'$ and $\mathbf{y}_3''$, respectively.

(c) Multiply $\mathbf{x}_2 \leq \mathbf{0}$ by -1; replace $-\mathbf{x}_2$ by $\mathbf{x}_2$.

(d) Replace $\mathbf{x}_3$ by $\mathbf{x}_3' - \mathbf{x}_3''$, and replace '$\mathbf{x}_3$ is free' by '$\mathbf{x}_3', \mathbf{x}_3'' \geq \mathbf{0}$'.

The 'dual versions' of (a)–(d) are:

(a') Multiply $\mathbf{y}_2 \geq \mathbf{0}$ by -1; replace $-\mathbf{y}_2$ by $\mathbf{y}_2$.

(b') Replace $\mathbf{y}_3' - \mathbf{y}_3''$ by $\mathbf{y}_3$ (implying that $\mathbf{y}_3$ is free).

(c') Multiply the second constraint (with right hand side $-\mathbf{c}_2$) by -1.

(d') Replace the last two '$\geq$' constraints (corresponding to $\mathbf{x}_3'$ and $\mathbf{x}_3''$, respectively) by one '$=$' constraint, such that its coefficients are the same as the coefficients of $\mathbf{x}_3$ in the original model.

Applying (a')–(d') to (DSGM) leads to model (DGM), the dual of (GM). $\qquad\square$

It is left to the reader to show that the dual of model (DGM) is the original model (GM); see Exercise 4.8.9. Hence, (GM) and (DGM) are mutually dual. In Table 4.1 we have summarized the primal-dual relationships ($\mathbf{A} \in \mathbb{R}^{m \times n}$, $\mathbf{b} \in \mathbb{R}^m$, $\mathbf{c} \in \mathbb{R}^n$, $i = 1, \ldots, n$, and $j = 1, \ldots, m$).

What can be said about the primal and the dual slack variables? Recall that the standard primal model with slack variables is $\max\{\mathbf{c}^\mathsf{T}\mathbf{x} \mid \mathbf{A}\mathbf{x} + \mathbf{x}_s = \mathbf{b}, \mathbf{x} \geq \mathbf{0}, \mathbf{x}_s \geq \mathbf{0}\}$, and the standard dual model with slack variables is $\min\{\mathbf{b}^\mathsf{T}\mathbf{y} \mid \mathbf{A}^\mathsf{T}\mathbf{y} - \mathbf{y}_s = \mathbf{c}, \mathbf{y} \geq \mathbf{0}, \mathbf{y}_s \geq \mathbf{0}\}$. Note that in both models the slack variables are nonnegative. What would happen if we apply the above formulated dualization rules to $\max\{\mathbf{c}^\mathsf{T}\mathbf{x} \mid \mathbf{A}\mathbf{x} \leq \mathbf{b}, \mathbf{x} \geq \mathbf{0}\}$ with $\mathbf{x} \geq \mathbf{0}$ considered as a technology constraint? That is, consider the following nonstandard LO-

	PRIMAL/DUAL		DUAL/PRIMAL
1.	Maximizing model	1.	Minimizing model
2.	Technology matrix $\mathbf{A}$	2.	Technology matrix $\mathbf{A}^\mathsf{T}$
3.	Right hand side vector $\mathbf{b}$	3.	Objective coefficients vector $\mathbf{b}$
4.	Objective coefficients vector $\mathbf{c}$	4.	Right hand side vector $\mathbf{c}$
5.	j'th constraint '=' type	5.	Decision variable y_j free
6.	j'th constraint '$\leq$' type (slack var. $x_{n+j} \geq 0$)	6.	Decision variable $y_j \geq 0$
7.	j'th constraint '$\geq$' type (slack var. $x_{n+j} \leq 0$)	7.	Decision variable $y_j \leq 0$
8.	Decision variable x_i free	8.	i'th constraint '=' type
9.	Decision variable $x_i \geq 0$	9.	i'th constraint '$\geq$' type (slack var. $y_{m+i} \geq 0$)
10.	Decision variable $x_i \leq 0$	10.	i'th constraint '$\leq$' type (slack var. $y_{m+i} \leq 0$)

Table 4.1: Dualization rules for nonstandard LO-models.

model:

$$\max \mathbf{c}^\mathsf{T}\mathbf{x}$$
$$\text{s.t. } \mathbf{A}\mathbf{x} \leq \mathbf{b}, \mathbf{x} \geq \mathbf{0}$$
$$\mathbf{x} \text{ free.}$$

The technology matrix of this model is $\begin{bmatrix} \mathbf{A} \\ \mathbf{I}_n \end{bmatrix}$. Let $\mathbf{y}_1$ ($\in \mathbb{R}^m$) be the vector of dual variables corresponding to $\mathbf{A}\mathbf{x} \leq \mathbf{b}$, and $\mathbf{y}_2$ ($\in \mathbb{R}^n$) the vector of dual variables corresponding to $\mathbf{x} \geq \mathbf{0}$. Applying the above formulated dualization rules, we obtain the following model:

$$\min \mathbf{b}^\mathsf{T}\mathbf{y}_1 + \mathbf{0}^\mathsf{T}\mathbf{y}_2$$
$$\text{s.t. } \mathbf{A}^\mathsf{T}\mathbf{y}_1 + \mathbf{y}_2 = \mathbf{c}$$
$$\mathbf{y}_1 \geq \mathbf{0}, \mathbf{y}_2 \leq \mathbf{0}.$$

This model is a standard dual model with slack variables, except for the fact that the vector of slack variables $\mathbf{y}_s$ of the original standard dual model satisfies $\mathbf{y}_s = -\mathbf{y}_2$. So, the dual slack variables of a standard LO-model are equal to the negatives of the dual variables corresponding to the 'constraint' $\mathbf{x} \geq \mathbf{0}$. If we want to avoid this difference in sign, we should have defined the standard primal model as $\max\{\mathbf{c}^\mathsf{T}\mathbf{x} \mid \mathbf{A}\mathbf{x} \leq \mathbf{b}, -\mathbf{x} \leq \mathbf{0}\}$; see also Section 5.3.3. On the other hand, some LO-packages only accept nonnegative decision variables, and so − if relevant − we should pay extra attention to the actual signs of the optimal values of the slack variables presented in the output of the package.

Example 4.2.1. *We illustrate the dualization rules using the following LO-model.*

$$\begin{aligned}
\min \quad & 2x_1 + x_2 - x_3 \\
\text{s.t.} \quad & x_1 + x_2 - x_3 = 1 \\
& x_1 - x_2 + x_3 \geq 2 \\
& x_2 + x_3 \leq 3 \\
& x_1 \geq 0,\ x_2 \leq 0,\ x_3 \text{ free.}
\end{aligned}$$

Replacing x_2 by $-x_2$, and x_3 by $x_3' - x_3''$, the '=' constraint by two '$\geq$' constraints with opposite sign, and multiplying the third constraint by -1, we find the following standard LO-model:

$$\begin{aligned}
\min \quad & 2x_1 - x_2 - x_3' + x_3'' \\
\text{s.t.} \quad & x_1 - x_2 - x_3' + x_3'' \geq \ \ 1 \\
& -x_1 + x_2 + x_3' - x_3'' \geq -1 \\
& x_1 + x_2 + x_3' - x_3'' \geq \ \ 2 \\
& x_2 - x_3' + x_3'' \geq -3 \\
& x_1, x_2, x_3', x_3'' \geq 0.
\end{aligned}$$

The dual of this model is:

$$\begin{aligned}
\min \quad & y_1' - y_1'' + 2y_2 - 3y_3 \\
\text{s.t.} \quad & y_1' - y_1'' + \ \ y_2 \qquad \ \ \leq \ \ 2 \\
& -y_1' + y_1'' + \ \ y_2 + \ \ y_3 \leq -1 \\
& -y_1' + y_1'' + \ \ y_2 - \ \ y_3 \leq -1 \\
& y_1' - y_1'' - \ \ y_2 + \ \ y_3 \leq \ \ 1 \\
& y_1', y_1'', y_2, y_3 \geq 0.
\end{aligned}$$

Applying rules (a)–(d) and (a')–(d'), we multiply the second constraint by -1, replace the third and fourth inequalities by one equation, replace $y_1' - y_1'' = y_1$ by 'y_1 free', and replace y_3 by $-y_3$. The dual of the original model then becomes:

$$\begin{aligned}
\min \quad & y_1 + 2y_2 + 3y_3 \\
\text{s.t.} \quad & y_1 + \ \ y_2 \qquad \ \ \leq \ \ 2 \\
& y_1 - \ \ y_2 + \ \ y_3 \geq \ \ 1 \\
& -y_1 + \ \ y_2 + \ \ y_3 = -1 \\
& y_1 \text{ free},\ y_2 \geq 0,\ y_3 \leq 0.
\end{aligned}$$

It is left to the reader to check that this result follows directly from the dualization rules of Table 4.1.

4.2.3 Optimality and optimal dual solutions

When we use the simplex algorithm (see Chapter 3) to solve Model Salmonnose, the optimal objective value turns out to be equal to $22\frac{1}{2}$, which is equal to the optimal primal objective value. This remarkable fact holds in general for feasible LO-models and is expressed in the following theorems. These theorems are formulated for the standard LO-models $\max\{\mathbf{c}^\mathsf{T}\mathbf{x} \mid \mathbf{A}\mathbf{x} \leq \mathbf{b}, \mathbf{x} \geq \mathbf{0}\}$ and $\min\{\mathbf{b}^\mathsf{T}\mathbf{y} \mid \mathbf{A}^\mathsf{T}\mathbf{y} \geq \mathbf{c}, \mathbf{y} \geq \mathbf{0}\}$. However, since non-standard LO-models can easily be transformed into standard models, these theorems can be generalized for nonstandard LO-models as well. Examples of such generalizations can be found in the exercises at the end of this chapter. In Section 5.7, models with equality constraints receive special attention.

We start by proving the so-called *weak duality theorem*, which states that the optimal objective value of a standard LO-model is at most the optimal objective value of its dual model.

> **Theorem 4.2.2.** (*Weak duality theorem*)
> Consider a standard primal LO-model and its dual model. If both models are feasible, then:
>
> $$\max\{\mathbf{c}^\mathsf{T}\mathbf{x} \mid \mathbf{A}\mathbf{x} \leq \mathbf{b}, \mathbf{x} \geq \mathbf{0}\} \leq \min\{\mathbf{b}^\mathsf{T}\mathbf{y} \mid \mathbf{A}^\mathsf{T}\mathbf{y} \geq \mathbf{c}, \mathbf{y} \geq \mathbf{0}\}.$$

Proof. Take any $\hat{\mathbf{x}} \geq \mathbf{0}$ and $\hat{\mathbf{y}} \geq \mathbf{0}$ with $\mathbf{A}\hat{\mathbf{x}} \leq \mathbf{b}$ and $\mathbf{A}^\mathsf{T}\hat{\mathbf{y}} \geq \mathbf{c}$. Then,

$$\mathbf{c}^\mathsf{T}\hat{\mathbf{x}} \leq \left(\mathbf{A}^\mathsf{T}\hat{\mathbf{y}}\right)^\mathsf{T}\hat{\mathbf{x}} = \hat{\mathbf{y}}^\mathsf{T}\mathbf{A}\hat{\mathbf{x}} = \hat{\mathbf{y}}^\mathsf{T}(\mathbf{A}\hat{\mathbf{x}}) \leq \hat{\mathbf{y}}^\mathsf{T}\mathbf{b} = \mathbf{b}^\mathsf{T}\hat{\mathbf{y}}.$$

Hence,

$$\max\left\{\mathbf{c}^\mathsf{T}\mathbf{x} \mid \mathbf{A}\mathbf{x} \leq \mathbf{b}, \mathbf{x} \geq \mathbf{0}\right\} \leq \min\left\{\mathbf{b}^\mathsf{T}\mathbf{y} \mid \mathbf{A}^\mathsf{T}\mathbf{y} \geq \mathbf{c}, \mathbf{y} \geq \mathbf{0}\right\}.$$

$\square$

The following important theorem shows that if a pair of vectors $\hat{\mathbf{x}}$ and $\hat{\mathbf{y}}$ can be found satisfying the constraints of the respective primal and dual models, and such that their objective values are equal, then $\hat{\mathbf{x}}$ and $\hat{\mathbf{y}}$ are both optimal.

> **Theorem 4.2.3.** (*Optimality condition*)
> Let $n, m \geq 1$. Suppose that $\hat{\mathbf{x}}$ $(\in \mathbb{R}^n)$ and $\hat{\mathbf{y}}$ $(\in \mathbb{R}^m)$ satisfy the constraints of a standard primal LO-model and its dual, respectively. If $\mathbf{c}^\mathsf{T}\hat{\mathbf{x}} = \mathbf{b}^\mathsf{T}\hat{\mathbf{y}}$, then $\hat{\mathbf{x}}$ and $\hat{\mathbf{y}}$ are optimal for their respective models.

Proof of Theorem 4.2.3. Take any $\hat{\mathbf{x}}$ and $\hat{\mathbf{y}}$ satisfying the conditions of the theorem. It follows from Theorem 4.2.2 that:

$$\mathbf{c}^\mathsf{T}\hat{\mathbf{x}} \leq \max\left\{\mathbf{c}^\mathsf{T}\mathbf{x} \mid \mathbf{A}\mathbf{x} \leq \mathbf{b}, \mathbf{x} \geq \mathbf{0}\right\}$$
$$\leq \min\left\{\mathbf{b}^\mathsf{T}\mathbf{y} \mid \mathbf{A}^\mathsf{T}\mathbf{y} \geq \mathbf{c}, \mathbf{y} \geq \mathbf{0}\right\} \leq \mathbf{b}^\mathsf{T}\hat{\mathbf{y}} = \mathbf{c}^\mathsf{T}\hat{\mathbf{x}}.$$

Because the leftmost term equals the rightmost term in this expression, we have that the inequalities are in fact equalities. Therefore, we have that:

$$\mathbf{c}^\mathsf{T}\hat{\mathbf{x}} = \max\left\{\mathbf{c}^\mathsf{T}\mathbf{x} \mid \mathbf{A}\mathbf{x} \leq \mathbf{b}, \mathbf{x} \geq \mathbf{0}\right\}, \text{ and } \mathbf{b}^\mathsf{T}\hat{\mathbf{y}} = \min\left\{\mathbf{b}^\mathsf{T}\mathbf{y} \mid \mathbf{A}^\mathsf{T}\mathbf{y} \geq \mathbf{c}, \mathbf{y} \geq \mathbf{0}\right\}.$$

This implies that $\hat{\mathbf{x}}$ is an optimal solution of the primal model, and $\hat{\mathbf{y}}$ is an optimal solution of the dual model.   □

The following *strong duality theorem* implies that the converse of Theorem 4.2.3 is also true, i.e., the optimal objective value of a standard LO-model is equal to the optimal objective value of its dual model.

Theorem 4.2.4. (*Strong duality theorem*)
If a standard LO-model has an optimal solution, then the dual model has an optimal solution as well, and vice versa; the optimal objective values of both the primal and dual model are the same. In particular, if $\mathbf{B}$ is any optimal primal basis matrix, then

$$\mathbf{y}^* = (\mathbf{B}^{-1})^\mathsf{T}\mathbf{c}_{BI}$$

is an optimal dual solution.

Proof of Theorem 4.2.4. Recall that there is always an optimal solution that corresponds to a feasible basic solution; see Theorem 2.1.5 and Theorem 2.2.2. Let $\begin{bmatrix} \mathbf{x}^*_{BI} \\ \mathbf{x}^*_{NI} \end{bmatrix}$ be an optimal feasible basic solution of the primal model, with $\mathbf{x}^*_{BI} = \mathbf{B}^{-1}\mathbf{b}$ and $\mathbf{x}^*_{NI} = \mathbf{0}$. The definition of an optimal feasible basic solution implies that $\mathbf{c}^\mathsf{T}_{NI} - \mathbf{c}^\mathsf{T}_{BI}\mathbf{B}^{-1}\mathbf{N} \leq \mathbf{0}$; see Section 3.5.2. Now, let $\hat{\mathbf{y}} = (\mathbf{B}^{-1})^\mathsf{T}\mathbf{c}_{BI}$. We will show that $\hat{\mathbf{y}}$ is an optimal solution of the dual model. To that end we first show that $\hat{\mathbf{y}}$ is feasible; i.e., that $\hat{\mathbf{y}}$ satisfies $\mathbf{A}^\mathsf{T}\hat{\mathbf{y}} \geq \mathbf{c}$ and $\hat{\mathbf{y}} \geq \mathbf{0}$, which is equivalent to

$$\begin{bmatrix} \mathbf{A}^\mathsf{T} \\ \mathbf{I}_m \end{bmatrix}\hat{\mathbf{y}} \geq \begin{bmatrix} \mathbf{c} \\ \mathbf{0} \end{bmatrix}, \text{ and hence to } \begin{bmatrix} \mathbf{B}^\mathsf{T} \\ \mathbf{N}^\mathsf{T} \end{bmatrix}\hat{\mathbf{y}} \geq \begin{bmatrix} \mathbf{c}_{BI} \\ \mathbf{c}_{NI} \end{bmatrix}.$$

The proof of this latter inequality is as follows:

$$\begin{bmatrix} \mathbf{B}^\mathsf{T} \\ \mathbf{N}^\mathsf{T} \end{bmatrix}\hat{\mathbf{y}} = \begin{bmatrix} \mathbf{B}^\mathsf{T}\hat{\mathbf{y}} \\ \mathbf{N}^\mathsf{T}\hat{\mathbf{y}} \end{bmatrix} = \begin{bmatrix} \mathbf{B}^\mathsf{T}(\mathbf{B}^{-1})^\mathsf{T}\mathbf{c}_{BI} \\ \mathbf{N}^\mathsf{T}(\mathbf{B}^{-1})^\mathsf{T}\mathbf{c}_{BI} \end{bmatrix} \geq \begin{bmatrix} \mathbf{c}_{BI} \\ \mathbf{c}_{NI} \end{bmatrix},$$

where we have used the fact that $\mathbf{c}^\mathsf{T}_{NI} - \mathbf{c}^\mathsf{T}_{BI}\mathbf{B}^{-1}\mathbf{N} \leq \mathbf{0}$. Hence, $\hat{\mathbf{y}}$ is dual feasible. It remains to show that $\hat{\mathbf{y}}$ is optimal. We have that:

$$\mathbf{b}^\mathsf{T}\hat{\mathbf{y}} = \mathbf{b}^\mathsf{T}(\mathbf{B}^{-1})^\mathsf{T}\mathbf{c}_{BI} = (\mathbf{B}^{-1}\mathbf{b})^\mathsf{T}\mathbf{c}_{BI} = (\mathbf{x}^*_{BI})^\mathsf{T}\mathbf{c}_{BI} = \mathbf{c}^\mathsf{T}_{BI}\mathbf{x}^*_{BI} = \mathbf{c}^\mathsf{T}\mathbf{x}^*.$$

According to Theorem 4.2.3, it follows that $\hat{\mathbf{y}} = (\mathbf{B}^{-1})^\mathsf{T}\mathbf{c}_{BI}$ is an optimal solution of the dual model. The 'vice versa' part follows by interchanging the terms primal and dual.   □

Using Theorem 4.2.4, we can immediately calculate an optimal dual solution as soon as an optimal primal basis matrix is at hand.

Example 4.2.2.   *Consider again Model Dovetail. The optimal primal basis matrix, its inverse, and the corresponding vector* $\mathbf{c}_{BI}$ *are*

$$
\mathbf{B} = \begin{array}{cccc} \scriptstyle x_1 & \scriptstyle x_2 & \scriptstyle x_5 & \scriptstyle x_6 \\ \begin{bmatrix} 1 & 1 & 0 & 0 \\ 3 & 1 & 0 & 0 \\ 1 & 0 & 1 & 0 \\ 0 & 1 & 0 & 1 \end{bmatrix} \end{array}, \mathbf{B}^{-1} = \frac{1}{2}\begin{bmatrix} -1 & 1 & 0 & 0 \\ 3 & -1 & 0 & 0 \\ 1 & -1 & 2 & 0 \\ -3 & 1 & 0 & 2 \end{bmatrix}, \text{ and } \mathbf{c}_{BI} = \begin{bmatrix} 3 & 2 & 0 & 0 \end{bmatrix}^{\mathsf{T}}.
$$

Hence, $\begin{bmatrix} y_1 & y_2 & y_3 & y_4 \end{bmatrix} = \mathbf{B}^{-1}\mathbf{c}_{BI} = \begin{bmatrix} 1\frac{1}{2} & \frac{1}{2} & 0 & 0 \end{bmatrix}$ *is an optimal dual solution; see also Section 4.1.1.*

The following general optimality condition holds for nonstandard LO-models; see Exercise 4.8.10. Note that this condition does not use feasible basic solutions.

> **Theorem 4.2.5.** (*Optimality condition for nonstandard LO-models*)
> The vector $\mathbf{x}$ is an optimal solution of the nonstandard LO-model (GM) if
>
> (i) $\mathbf{x}$ is feasible, and
>
> (ii)   there exists a vector $\mathbf{y}$ that is feasible for (DGM) with $\mathbf{c}^{\mathsf{T}}\mathbf{x} = \mathbf{b}^{\mathsf{T}}\mathbf{y}$.

Proof. See Exercise 4.8.10.   □

4.3 Complementary slackness relations

In the previous section, we have established that there is a one-to-one correspondence between the decision variables of a standard primal LO-model and the constraints of the corresponding dual LO-model and, similarly, a correspondence between the constraints of the primal LO-model and the decision variables of the dual LO-model. In the current section, we explore these correspondences in more detail, and relate the optimal values of the primal decision variables with the optimal values of the dual slack variables, and, vice versa, the optimal values of the primal slack variables with the dual decision variables.

4.3.1 Complementary dual variables

In Section 1.2.2 we have explained how an LO-model with '$\leq$' constraint can be transformed into a model with '$=$' constraints by introducing nonnegative slack variables. Introducing the slack variables x_3, x_4, x_5, x_6 for Model Dovetail (the primal model), and y_5, y_6 for Model Salmonnose (the dual model), we obtain:

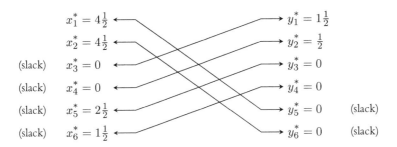

Figure 4.1: Complementary dual variables.

Model Dovetail with slack variables.

$$
\begin{aligned}
\max \quad & 3x_1 + 2x_2 \\
\text{s.t.} \quad & x_1 + x_2 + x_3 && = 9 \\
& 3x_1 + x_2 + x_4 && = 18 \\
& x_1 + x_5 && = 7 \\
& x_2 + x_6 = 6 \\
& x_1, x_2, x_3, x_4, x_5, x_6 \geq 0,
\end{aligned}
$$

and

Model Salmonnose with slack variables.

$$
\begin{aligned}
\min \quad & 9y_1 + 18y_2 + 7y_3 + 6y_4 \\
\text{s.t.} \quad & y_1 + 3y_2 + y_3 - y_5 && = 3 \\
& y_1 + y_2 + y_4 - y_6 = 2 \\
& y_1, y_2, y_3, y_4, y_5, y_6 \geq 0.
\end{aligned}
$$

The optimal solutions of Model Dovetail and Model Salmonnose with slack variables are:

$$
x_1^* = 4\tfrac{1}{2},\ x_2^* = 4\tfrac{1}{2},\ x_3^* = 0,\ x_4^* = 0,\ x_5^* = 2\tfrac{1}{2},\ x_6^* = 1\tfrac{1}{2},
$$

and $\quad y_1^* = 1\tfrac{1}{2},\ y_2^* = \tfrac{1}{2},\ y_3^* = 0,\ y_4^* = 0,\ y_5^* = 0,\ y_6^* = 0,$ respectively.

The first constraint of Model Dovetail with slack variables contains the slack variable x_3 and corresponds to the decision variable y_1 of the dual Model Salmonnose with slack variables; similar relationships hold for the other three constraints. Moreover, the nonnegativity constraint $x_1 \geq 0$ (with decision variable x_1) corresponds to y_5, and $x_2 \geq 0$ to y_6. Hence, we obtain a connection between the decision/slack variables of Model Dovetail and the slack/decision variables of Model Salmonnose, reflected by the arrows in Figure 4.1.

In general, the standard primal and the standard dual model with slack variables can be written as follows.

STANDARD PRIMAL MODEL with slack variables

$$\max\left\{\mathbf{c}^{\mathsf{T}}\mathbf{x} \,\middle|\, \mathbf{A}\mathbf{x} + \mathbf{I}_m\mathbf{x}_s = \mathbf{b}, \ \mathbf{x}, \mathbf{x}_s \geq \mathbf{0}\right\}$$

STANDARD DUAL MODEL with slack variables

$$\min\left\{\mathbf{b}^{\mathsf{T}}\mathbf{y} \,\middle|\, \mathbf{A}^{\mathsf{T}}\mathbf{y} - \mathbf{I}_n\mathbf{y}_s = \mathbf{c}, \ \mathbf{y}, \mathbf{y}_s \geq \mathbf{0}\right\}$$

where $\mathbf{x}_s$ and $\mathbf{y}_s$ are the primal and the dual slack variables, respectively, and $\mathbf{I}_m$ and $\mathbf{I}_n$ are the identity matrices with m and n rows, respectively.

We have seen that, in general, each constraint of a standard LO-model has an associated dual decision variable. Since each decision variable of the primal LO-model corresponds to a dual constraint, this means that each primal decision variable also corresponds to a dual slack variable. Similar reasoning implies that each primal slack variable corresponds to a dual decision variable. In particular, we have the following correspondence between primal and dual variables.

The primal decision variable x_i corresponds to the dual slack variable y_{m+i}, for $i = 1, \ldots, n$.

The primal slack variable x_{n+j} corresponds to the dual decision variable y_j, for $j = 1, \ldots, m$.

The variables x_i and y_{m+i}, as well as x_{n+j} and y_j, are called *complementary dual variables* ($i = 1, \ldots, n$ and $j = 1, \ldots, m$). The corresponding constraints are called *complementary dual constraints*.

Recall that, in Section 2.2.3, we defined the complementary dual set BI^c of BI, and the complementary dual set NI^c of NI. The reader may verify that the vector $\mathbf{y}_{BI^c}$ contains the complementary dual variables of the variables in the vector $\mathbf{x}_{NI}$. Similarly, $\mathbf{y}_{NI^c}$ contains the complementary dual variables of the variables in $\mathbf{x}_{BI}$.

The correspondence between the complementary dual variables is illustrated in Figure 4.2. The matrix in the figure is the technology matrix $\mathbf{A}$ of a standard primal LO-model. From the primal perspective, each column i corresponds to the primal decision variable x_i ($i = 1, \ldots, n$); each row j corresponds to j'th primal constraint, associated with the primal slack variable x_{n+j} ($j = 1, \ldots, m$). From the dual perspective, each row j corresponds to the dual decision variable y_j ($j = 1, \ldots, m$); each column i corresponds to the i'th dual constraint, associated with the dual slack variable y_{m+i} ($i = 1, \ldots, n$).

4.3.2 Complementary slackness

Recall that, for a given feasible (but not necessarily optimal) solution $\hat{\mathbf{x}}$ of an LO-model, a constraint is called *binding* at $\hat{\mathbf{x}}$ if the value of the corresponding slack variable of this constraint is zero at that solution; '$=$' constraints are binding by definition. If a slack variable has a nonzero value at a feasible solution, then the constraint is called *nonbinding* at that solution (see also Section 1.2.2).

$$
\begin{array}{c}
\begin{array}{ccc}
x_1 & x_2 & x_n \\
y_{m+1} & y_{m+2} & y_{m+n} \\
\downarrow & \downarrow & \downarrow
\end{array}
\end{array}
$$

$$
\mathbf{A} = \begin{bmatrix}
a_{11} & a_{12} & \cdots & a_{1n} \\
a_{21} & a_{22} & \cdots & a_{2n} \\
\vdots & \vdots & \ddots & \vdots \\
a_{m1} & a_{m2} & \cdots & a_{mn}
\end{bmatrix}
\begin{array}{ll}
\leftarrow \quad y_1 & x_{n+1} \\
\leftarrow \quad y_2 & x_{n+2} \\
\\
\leftarrow \quad y_m & x_{n+m}
\end{array}
$$

Figure 4.2: Complementary dual variables. The rows of the technology matrix $\mathbf{A}$ correspond to the dual decision variables and the primal slack variables. The columns correspond to the primal decision variables and the dual slack variables.

The following relationships exist between the optimal values of the primal and dual variables (the remarks between parentheses refer to Model Dovetail and Model Salmonnose).

If a primal slack variable has a nonzero value, then the corresponding dual decision variable has value zero. ($x_5^* = 2\frac{1}{2} \Rightarrow y_3^* = 0$, and $x_6^* = 1\frac{1}{2} \Rightarrow y_4^* = 0$).

If a dual decision variable has a nonzero value, then the corresponding primal slack variable has value zero. ($y_1^* = 1\frac{1}{2} \Rightarrow x_3^* = 0$, and $y_2^* = \frac{1}{2} \Rightarrow x_4^* = 0$).

If a dual slack variable has a nonzero value, then the corresponding primal decision variable has value zero. (This does not occur in Model Dovetail and Model Salmonnose, because $y_5^* = 0$, and $y_6^* = 0$).

If a primal decision variable has a nonzero value, then the corresponding dual slack variable has value zero. ($x_1^* = 4\frac{1}{2} \Rightarrow y_5^* = 0$, and $x_2^* = 4\frac{1}{2} \Rightarrow y_6^* = 0$).

In Theorem 4.3.1 it will be shown that these observations are actually true in general. However, the converses of the implications formulated above do not always hold. If, for instance, the value of a primal slack variable is zero, then the value of the corresponding dual variable may be zero as well (see, e.g., Section 5.6).

Note that the observations can be compactly summarized as follows. For each pair of complementary dual variables (i.e., x_i and y_{m+i}, or x_{n+j} and y_j), it holds that at least one of them has value zero. An even more economical way to express this is to say that their product has value zero.

The following theorem gives an optimality criterion when solutions of both the primal and the dual model are known.

Theorem 4.3.1. (*Optimality and complementary slackness*)
Let $n, m \geq 1$. If $\mathbf{x}$ ($\in \mathbb{R}^n$) and $\mathbf{y}$ ($\in \mathbb{R}^m$) satisfy the constraints of the standard primal and the standard dual model, respectively, then the following assertions are equivalent:

(i) $\mathbf{x}$ and $\mathbf{y}$ are optimal (not necessarily feasible basic) solutions of their corresponding models;

(ii) $\mathbf{x}^\mathsf{T}\mathbf{y}_s = 0$ and $\mathbf{x}_s^\mathsf{T}\mathbf{y} = 0$ (with $\mathbf{x}_s$ and $\mathbf{y}_s$ the corresponding slack variables).

Proof of Theorem 4.3.1. The proof of (i) $\Rightarrow$ (ii) is as follows. Using Theorem 4.2.4, it follows that: $\mathbf{x}^\mathsf{T}\mathbf{y}_s = \mathbf{x}^\mathsf{T}(\mathbf{A}^\mathsf{T}\mathbf{y} - \mathbf{c}) = \mathbf{x}^\mathsf{T}(\mathbf{A}^\mathsf{T}\mathbf{y}) - \mathbf{x}^\mathsf{T}\mathbf{c} = (\mathbf{x}^\mathsf{T}\mathbf{A}^\mathsf{T})\mathbf{y} - \mathbf{c}^\mathsf{T}\mathbf{x} = (\mathbf{A}\mathbf{x})^\mathsf{T}\mathbf{y} - \mathbf{c}^\mathsf{T}\mathbf{x} = (\mathbf{b} - \mathbf{x}_s)^\mathsf{T}\mathbf{y} - \mathbf{c}^\mathsf{T}\mathbf{x} = \mathbf{b}^\mathsf{T}\mathbf{y} - \mathbf{x}_s^\mathsf{T}\mathbf{y} - \mathbf{c}^\mathsf{T}\mathbf{x} = -\mathbf{x}_s^\mathsf{T}\mathbf{y}$. Therefore, $\mathbf{x}^\mathsf{T}\mathbf{y}_s = -\mathbf{x}_s^\mathsf{T}\mathbf{y}$, and this implies that $\mathbf{x}^\mathsf{T}\mathbf{y}_s + \mathbf{x}_s^\mathsf{T}\mathbf{y} = 0$. Since $\mathbf{x}, \mathbf{y}, \mathbf{x}_s, \mathbf{y}_s \geq \mathbf{0}$, we have in fact that $\mathbf{x}^\mathsf{T}\mathbf{y}_s = \mathbf{x}_s^\mathsf{T}\mathbf{y} - 0$. The proof of (ii) $\Rightarrow$ (i) is left to the reader. (Hint: use Theorem 4.2.3.) $\square$

The expressions $\mathbf{x}^\mathsf{T}\mathbf{y}_s = 0$ and $\mathbf{x}_s^\mathsf{T}\mathbf{y} = 0$ are called the *complementary slackness relations*. Theorem 4.3.1 states a truly surprising fact: if we have a feasible primal solution $\mathbf{x}^*$ and a feasible dual solution $\mathbf{y}^*$, and they satisfy the complementary slackness relations, then it immediately follows that $\mathbf{x}^*$ and $\mathbf{y}^*$ are optimal solutions (for the standard primal and the standard dual models, respectively).

Example 4.3.1.   *In the case of Model Dovetail and Model Salmonnose, the complementary slackness relations are (see Figure 4.1):*

$$x_1^* \times y_5^* = 0, \qquad x_2^* \times y_6^* = 0, \qquad x_3^* \times y_1^* = 0,$$
$$x_4^* \times y_2^* = 0, \qquad x_5^* \times y_3^* = 0, \qquad x_6^* \times y_4^* = 0.$$

Note that, in the first equation, x_1 is the first decision variable of Model Dovetail, and y_5 is the slack variable of the first constraint of Model Salmonnose. Similarly, in the last equation, x_6 is the slack variable of the last constraint of Model Dovetail, and y_4 is the last decision variable of Model Salmonnose.

The complementary slackness relations can be generalized to nonstandard models. Consider the general LO-model (GM) and the corresponding dual model (DGM). Let $\mathbf{x}_{s1}$ and $\mathbf{x}_{s2}$ be the vectors of the primal slack variables corresponding to the first two sets of constraints of model (GM), and let $\mathbf{y}_{s1}$ and $\mathbf{y}_{s2}$ be the vectors of the dual slack variables corresponding to the first two sets of constraints of model (DGM). So, we have that:

$$\mathbf{x}_{s1} = \mathbf{b}_1 - \mathbf{A}_1\mathbf{x}_1 - \mathbf{A}_2\mathbf{x}_2 - \mathbf{A}_3\mathbf{x}_3,$$
$$\mathbf{x}_{s2} = \mathbf{A}_4\mathbf{x}_1 + \mathbf{A}_5\mathbf{x}_2 + \mathbf{A}_6\mathbf{x}_3 - \mathbf{b}_2,$$
$$\mathbf{y}_{s1} = \mathbf{A}_1^\mathsf{T}\mathbf{y}_1 + \mathbf{A}_2^\mathsf{T}\mathbf{y}_4 + \mathbf{A}_7^\mathsf{T}\mathbf{y}_3 - \mathbf{c}_1, \text{ and}$$
$$\mathbf{y}_{s2} = \mathbf{c}_2 - \mathbf{A}_2^\mathsf{T}\mathbf{y}_1 - \mathbf{A}_5^\mathsf{T}\mathbf{y}_5 - \mathbf{A}_8^\mathsf{T}\mathbf{y}_3.$$

The following theorem summarizes the complementary slackness relations for the general models (GM) and (DGM).

Theorem 4.3.2. *(Optimality and complementary slackness for general LO-models)*
Let $m, n \geq 1$. If

$$\mathbf{x} = \begin{bmatrix} \mathbf{x}_1 \\ \mathbf{x}_2 \\ \mathbf{x}_3 \end{bmatrix} (\in \mathbb{R}^n) \text{ and } \mathbf{y} = \begin{bmatrix} \mathbf{y}_1 \\ \mathbf{y}_2 \\ \mathbf{y}_3 \end{bmatrix} (\in \mathbb{R}^m)$$

satisfy the constraints of the primal model (GM) and the corresponding dual model (DGM), respectively, then the following assertions are equivalent:

(i) $\mathbf{x}$ and $\mathbf{y}$ are optimal (not necessarily feasible basic) solutions of their corresponding models;

(ii) $\mathbf{x}_1^\mathsf{T} \mathbf{y}_{s1} = 0$, $\mathbf{x}_2^\mathsf{T} \mathbf{y}_{s2} = 0$, $\mathbf{x}_{s1}^\mathsf{T} \mathbf{y}_1 = 0$, and $\mathbf{x}_{s2}^\mathsf{T} \mathbf{y}_2 = 0$ (with $\mathbf{x}_{s1}$, $\mathbf{x}_{s2}$, $\mathbf{y}_{s1}$, and $\mathbf{y}_{s2}$ defined as above).

Proof. The proof is left to the reader; see Exercise 4.8.11.   □

4.3.3 Determining the optimality of a given solution

In this section we show how to determine the optimality of a given primal feasible solution by means of the complementary slackness relations without using the actual dual model. Let $\mathbf{x}$ be a feasible solution of a primal LO-model. According to Theorem 4.3.1, $\mathbf{x}$ is optimal if there is a feasible dual solution $\mathbf{y}$ such that $\mathbf{x}^\mathsf{T}\mathbf{y}_s = \mathbf{0}$ and $\mathbf{x}_s^\mathsf{T}\mathbf{y} = \mathbf{0}$. In order to determine this vector $\mathbf{y}$, it is in general not necessary to solve the full dual model. Namely, for each strictly positive entry of $\mathbf{x}$ the corresponding complementary dual variable has value zero. This fact can be used to reduce the dual model. We will illustrate the procedure by means of the following example.

Example 4.3.2. *Consider the following LO-model:*

$$\begin{aligned}
\max \quad & -2x_1 - \tfrac{3}{4}x_2 \\
\text{s.t.} \quad & x_1 + 2x_2 + 2x_3 \geq 5 && \text{(slack } x_4) \\
& 2x_1 - \tfrac{1}{2}x_2 \leq -1 && \text{(slack } x_5) \\
& x_1 + x_2 + x_3 \leq 3 && \text{(slack } x_6) \\
& x_1, x_2, x_3 \geq 0.
\end{aligned}$$

Suppose that we are given the vector $\begin{bmatrix} x_1 & x_2 & x_3 \end{bmatrix}^\mathsf{T} = \begin{bmatrix} 0 & 2 & \tfrac{3}{4} \end{bmatrix}^\mathsf{T}$, and we want to know whether or not this vector is an optimal solution of the LO-model. One can easily check that $x_1 = 0$, $x_2 = 2$, $x_3 = \tfrac{3}{4}$ is a feasible solution, but not a feasible basic solution (why not?). The dual of the above *model reads:*

$$\min \quad 5y_1 - y_2 + 3y_3$$

$$\text{s.t.} \quad y_1 + 2y_2 + y_3 \geq -2 \qquad \text{(slack } y_4)$$

$$2y_1 - \tfrac{1}{2}y_2 + y_3 \geq -\tfrac{3}{4} \qquad \text{(slack } y_5)$$

$$2y_1 \qquad + y_3 \geq 0 \qquad \text{(slack } y_6)$$

$$y_1, y_2, y_3 \geq 0.$$

The pairs of complementary dual variables are (x_1, y_4), (x_2, y_5), (x_3, y_6), (x_4, y_1), (x_5, y_2), *and* (x_6, y_3). *Substituting* $x_1 = 0$, $x_2 = 2$, $x_3 = \tfrac{3}{4}$ *into the constraints of the primal model leads to the following values of the primal slack variables:* $x_4 = \tfrac{1}{2}$, $x_5 = 0$, *and* $x_6 = \tfrac{1}{4}$. *If the given vector is indeed a primal optimal solution, then we should be able to find a corresponding optimal dual solution. Since* x_2, x_3, x_4, *and* x_6 *have nonzero values, it follows that for any optimal dual solution, it must be the case that* $y_5 = y_6 = y_1 = y_3 = 0$. *Hence, the dual model can be reduced to:*

$$\min \quad -y_2$$

$$\text{s.t.} \quad 2y_2 \geq -2$$

$$-\tfrac{1}{2}y_2 = -\tfrac{3}{4}$$

$$y_2 \geq 0.$$

The optimal solution of this model satisfies $y_2 = 1\tfrac{1}{2}$. *The vectors* $\begin{bmatrix} x_1 & x_2 & x_3 & x_4 & x_5 & x_6 \end{bmatrix}^\mathsf{T} = \begin{bmatrix} 0 & 2 & \tfrac{3}{4} & \tfrac{1}{2} & 0 & \tfrac{1}{4} \end{bmatrix}^\mathsf{T}$ *and* $\begin{bmatrix} y_1 & y_2 & y_3 & y_4 & y_5 & y_6 \end{bmatrix}^\mathsf{T} = \begin{bmatrix} 0 & 1\tfrac{1}{2} & 0 & 5 & 0 & 0 \end{bmatrix}^\mathsf{T}$ *now satisfy the complementary slackness relations, and so* $\begin{bmatrix} x_1 & x_2 & x_3 \end{bmatrix}^\mathsf{T} = \begin{bmatrix} 0 & 2 & \tfrac{3}{4} \end{bmatrix}^\mathsf{T}$ *is in fact a (nonbasic) optimal solution.*

This example illustrates how it can be determined whether or not a given solution is optimal, and that the procedure is very effective if the number of nonzero variables (including slack variables) in the given solution is large.

4.3.4 Strong complementary slackness

From Theorem 4.3.1, it follows that a pair of primal and dual feasible solutions, $\mathbf{x}$ and $\mathbf{y}$ respectively, is optimal if and only if

$$x_i y_{m+i} = 0 \quad \text{for } i = 1, \ldots, n,$$
$$\text{and} \quad x_{n+j} y_j = 0 \quad \text{for } j = 1, \ldots, m.$$

So, in the case of optimal primal and dual solutions, the product of two complementary dual variables is always zero. It may even happen that both variables have value zero. In Theorem 4.3.3 it will be shown that there exist optimal primal and dual solutions such that for each pair of complementary dual variables either the value of the primal variable is nonzero or the value of the dual variable is nonzero. In the literature this theorem is known as the *strong complementary slackness theorem.*

Theorem 4.3.3. (*Strong complementary slackness theorem*)
For any standard LO-model that has an optimal solution, there exist an optimal primal solution $\mathbf{x}^* = \begin{bmatrix} x_1^* & \ldots & x_n^* \end{bmatrix}^\mathsf{T}$ and an optimal dual solution $\mathbf{y}^* = \begin{bmatrix} y_1^* & \ldots & y_m^* \end{bmatrix}^\mathsf{T}$ such that every pair of complementary dual variables (x_i^*, y_j^*) has the property that exactly one of x_i^* and y_j^* is nonzero.

Proof of Theorem 4.3.3. Let the standard primal model $\max \left\{ \mathbf{c}^\mathsf{T}\mathbf{x} \mid \mathbf{A}\mathbf{x} \leq \mathbf{0}, \mathbf{x} \geq \mathbf{0} \right\}$ and its dual model $\min \left\{ \mathbf{0}^\mathsf{T}\mathbf{y} \mid \mathbf{A}^\mathsf{T}\mathbf{y} \geq \mathbf{c}, \mathbf{y} \geq \mathbf{0} \right\}$ be denoted by (P) and (D), respectively. We first prove the following weaker statement.

($\star$) For each $i = 1, \ldots, n$, either there exists an optimal solution $\mathbf{x}^* = \begin{bmatrix} x_1^* & \ldots & x_n^* \end{bmatrix}^\mathsf{T}$ of (P) such that $x_i^* > 0$, or there exists an optimal solution $\mathbf{y}^* = \begin{bmatrix} y_1^* & \ldots & y_m^* \end{bmatrix}^\mathsf{T}$ of (D) such that $(\mathbf{A}^\mathsf{T}\mathbf{y}^* - \mathbf{c})_i > 0$.

To prove ($\star$), let z^* be the optimal objective value of (P). Let $i \in \{1, \ldots, n\}$, and consider the LO-model:

$$
\begin{aligned}
\max \quad & \mathbf{e}_i^\mathsf{T}\mathbf{x} \\
\text{s.t.} \quad & \mathbf{A}\mathbf{x} \leq \mathbf{b} \\
& -\mathbf{c}^\mathsf{T}\mathbf{x} \leq -z^* \\
& \mathbf{x} \geq \mathbf{0}.
\end{aligned}
\tag{P'}
$$

(Here, $\mathbf{e}_i$ is the i'th unit vector in $\mathbb{R}^n$.) Model (P') is feasible, because any optimal solution of (P) is a feasible solution of (P'). Let z' be the optimal objective value of (P'). Clearly, $z' \geq 0$. We now distinguish two cases.

Case 1: $z' > 0$. Let $\mathbf{x}^*$ be an optimal solution of (P'). Since $\mathbf{A}\mathbf{x}^* \leq \mathbf{b}$ and $\mathbf{c}^\mathsf{T}\mathbf{x}^* \geq z^* = \mathbf{c}^\mathsf{T}\mathbf{x}$, it follows that $\mathbf{x}^*$ is an optimal solution of (P). Moreover, $x_i^* = \mathbf{e}_i^\mathsf{T}\mathbf{x}^* = z' > 0$. Hence, ($\star$) holds.

Case 2: $z = 0$. Since (P') has optimal objective value 0, it follows from Theorem 4.2.4 that the dual of (P') also has optimal objective value 0:

$$
\begin{aligned}
\min \quad & \mathbf{b}^\mathsf{T}\mathbf{y} - z^*\lambda \\
\text{s.t.} \quad & \mathbf{A}^\mathsf{T}\mathbf{y} - \lambda\mathbf{c} \geq \mathbf{e}_i \\
& \mathbf{y} \geq \mathbf{0}, \lambda \geq 0.
\end{aligned}
\tag{D'}
$$

Let $(\mathbf{y}', \lambda')$ be an optimal solution of (D'). We consider two subcases:

Case 2a: $\lambda' = 0$. Then, we have that $\mathbf{b}^\mathsf{T}\mathbf{y}' = 0$ and $\mathbf{A}^\mathsf{T}\mathbf{y}' \geq \mathbf{e}_i$. Let $\hat{\mathbf{y}}$ be an optimal solution of (D). Then, $\mathbf{y}^* = \hat{\mathbf{y}} + \mathbf{y}'$ satisfies $\mathbf{y}^* = \hat{\mathbf{y}} + \mathbf{y}' \geq \mathbf{0}$, $\mathbf{b}^\mathsf{T}\mathbf{y}^* = \mathbf{b}^\mathsf{T}(\hat{\mathbf{y}} + \mathbf{y}') = \mathbf{b}^\mathsf{T}\hat{\mathbf{y}}$, and $\mathbf{A}^\mathsf{T}\mathbf{y}^* = \mathbf{A}^\mathsf{T}(\hat{\mathbf{y}} + \mathbf{y}') \geq \mathbf{c} + \mathbf{e}_i \geq \mathbf{c}$. Hence, $\mathbf{y}^*$ is an optimal solution of (D), and it satisfies $(\mathbf{A}^\mathsf{T}\mathbf{y}^* - \mathbf{c})_i = (\mathbf{A}^\mathsf{T}(\hat{\mathbf{y}} + \mathbf{y}') - \mathbf{c})_i \geq (\mathbf{e}_i)_i = 1 > 0$, and so ($\star$) holds.

Case 2b: $\lambda' > 0$. Define $\mathbf{y}^* = \mathbf{y}'/\lambda'$. Then, we have that $\mathbf{y}^* \geq \mathbf{0}$, $\mathbf{b}^\mathsf{T}\mathbf{y}^* = (\mathbf{b}^\mathsf{T}\mathbf{y}')/\lambda' = (z^*\lambda')/\lambda' = z^*$, and $\mathbf{A}^\mathsf{T}\mathbf{y}^* - \mathbf{c} = (\mathbf{A}^\mathsf{T}\mathbf{y}' - \lambda'\mathbf{c})/\lambda' \geq \mathbf{e}_i/\lambda'$. Hence, $\mathbf{y}^*$ is an optimal solution of (D), and it satisfies $(\mathbf{A}^\mathsf{T}\mathbf{y}^* - \mathbf{c})_i = (\mathbf{e}_i/\lambda')_i = 1/\lambda' > 0$, and so $(\star)$ holds.

We can now prove the theorem. For $i = 1, \ldots, n$, apply statement $(\star)$ to (P) and (D) to obtain $\mathbf{x}^{(i)}$ and $\mathbf{y}^{(i)}$ such that: (1) $\mathbf{x}^{(i)}$ is an optimal solution of (P), (2) $\mathbf{y}^{(i)}$ is an optimal solution of (P), and (3) either $x_i^{(i)} > 0$ or $(\mathbf{A}^\mathsf{T}\mathbf{y}^{(i)} - \mathbf{c})_i > 0$. Similarly, for $j = 1, \ldots, m$, apply statement $(\star)$ to (D) and (P) (i.e., with the primal and dual models switched) to obtain $\mathbf{x}^{(n+j)}$ and $\mathbf{y}^{(n+j)}$ such that: (1) $\mathbf{x}^{(n+j)}$ is an optimal solution of (P), (2) $\mathbf{x}^{(n\,|\,j)}$ is an optimal solution of (P), and (3) either $y_j^{(n+j)} > 0$ or $(\mathbf{b} - \mathbf{A}\mathbf{x}^{(n+j)})_j > 0$. Define:

$$\mathbf{x}^* = \frac{1}{n+m}\sum_{k=1}^{n+m}\mathbf{x}^{(k)}, \text{ and } \mathbf{y}^* = \frac{1}{n+m}\sum_{k=1}^{n+m}\mathbf{y}^{(k)}.$$

Due to Theorem 3.7.1 and the fact that $\mathbf{x}^*$ is a convex combination of optimal solutions of (P) (why?), we have that $\mathbf{x}^*$ is also a (not necessarily basic) optimal solution of (P). Similarly, $\mathbf{y}^*$ is an optimal solution of (D). Hence, by Theorem 4.3.1, it follows that $\mathbf{x}^*$ and $\mathbf{y}^*$ satisfy the complementary slackness relations.

It remains to show that $\mathbf{x}^*$ and $\mathbf{y}^*$ satisfy the assertions of the theorem. To do so, consider any pair of complementary dual variables (x_i^*, y_{m+i}^*) with $i = 1, \ldots, n$. We have that:

$$x_i^* = \left(\frac{1}{n+m}\sum_{k=1}^{n+m}\mathbf{x}^{(k)}\right)_i = \frac{1}{n+m}\sum_{k=1}^{n+m}x_i^{(k)},$$

and

$$y_{m+i}^* = (\mathbf{A}^\mathsf{T}\mathbf{y}^* - \mathbf{c})_i = \left(\mathbf{A}^\mathsf{T}\left(\frac{1}{n+m}\sum_{k=1}^{n+m}\mathbf{y}^{(k)}\right) - \mathbf{c}\right)_i = \frac{1}{n+m}\sum_{k=1}^{n+m}(\mathbf{A}^\mathsf{T}\mathbf{y}^{(k)} - \mathbf{c})_i.$$

Note that $x_i^{(k)} \geq 0$ and $(\mathbf{A}^\mathsf{T}\mathbf{y}^{(k)} - \mathbf{c})_i \geq 0$ for $k = 1, \ldots, n+m$. Moreover, by the construction of $\mathbf{x}^{(i)}$ and $\mathbf{y}^{(i)}$, we have that either $x_i^{(i)} > 0$ or $(\mathbf{A}^\mathsf{T}\mathbf{y}^{(i)} - \mathbf{c})_i > 0$. Therefore, either $x_i^* > 0$ or $(\mathbf{A}^\mathsf{T}\mathbf{y}^* - \mathbf{c})_i > 0$ (and not both, because $\mathbf{x}^*$ and $\mathbf{y}^*$ satisfy the complementary slackness relations). It is left to the reader that the conclusion also holds for any pair of complementary dual variables (x_{n+j}^*, y_j^*) with $j = 1, \ldots, m$. □

In Theorem 4.3.3, it has been assumed that the LO-model is in standard form. For LO-models containing '=' constraints, we have to be careful when applying the theorem. It may happen that there are multiple optimal primal solutions and that at least one of them is nondegenerate, while the corresponding optimal dual solution is unique and degenerate; see Theorem 5.6.1. If, moreover, the optimal dual variable y corresponding to the '=' constraint (with slack variable $s = 0$) has value 0, then the corresponding complementary slackness relation reads $y \times s = 0 \times 0 = 0$. Since $s = 0$ for all optimal primal solutions, and the optimal dual solution is unique, the strong complementary slackness relation does not hold. For an extended example, see also Exercise 12.7.8.

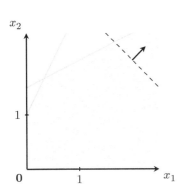

Figure 4.3: Unbounded primal feasible region.

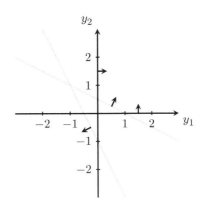

Figure 4.4: Empty dual feasible region.

4.4 Infeasibility and unboundedness; Farkas' lemma

In Theorem 4.2.4 it has been shown that if the primal model has an optimal solution, then the dual model has an optimal solution as well, and the corresponding objective values are the same. Because the dual of the dual of an LO-model is the original model, the converse of this assertion is also true. An LO-model is called *unbounded* if the objective function attains arbitrary large values (in the case of a maximizing model) on the feasible region. Note that in the case of unboundedness, the feasible region has to be unbounded. Also recall that the optimal objective value may be finite even if the feasible region is unbounded; see Section 1.2.3. On the other hand, if the feasible region is empty, then the LO-model is infeasible and has no solution at all.

Example 4.4.1. *Consider the following LO-model:*

$$\begin{aligned}
\max \quad & x_1 + x_2 \\
\text{s.t.} \quad -2x_1 + & x_2 \leq 1 \\
-x_1 + 2x_2 & \leq 3 \\
x_1, x_2 & \geq 0.
\end{aligned}$$

The feasible region of this model is depicted in Figure 4.3. Note that this region is unbounded. For instance, if $x_2 = 1$, then x_1 can have any positive value without violating the constraints of the model. In this figure a level line (with the arrow) of the objective function is drawn. Obviously, the 'maximum' of $x_1 + x_2$ is infinite. So, the model is unbounded, and, therefore, it has no optimal solution. The dual of the above model is:

$$\begin{aligned}
\min \quad & y_1 + 3y_2 \\
\text{s.t.} \quad -2y_1 - & y_2 \geq 1 \\
y_1 + 2y_2 & \geq 1 \\
y_1, y_2 & \geq 0.
\end{aligned}$$

PRIMAL MODEL	DUAL MODEL		
	Optimal	**Infeasible**	**Unbounded**
Optimal	Possible	Impossible	Impossible
Infeasible	Impossible	Possible	Possible
Unbounded	Impossible	Possible	Impossible

Table 4.2: Symmetric duality combinations.

The arrows in Figure 4.4 point into the halfspaces determined by the constraints of this dual model. Clearly, there is no point that simultaneously satisfies all constraints. Hence, the feasible region is empty, and this dual model is infeasible.

In Example 4.4.1, the primal model has an unbounded optimal solution, while its dual model is infeasible. In Theorem 4.4.1 the relationship between unbounded optimal primal solutions and infeasibility of the dual model is formulated.

Theorem 4.4.1. (*Infeasibility and unboundedness*)
If a primal (dual) LO-model is unbounded, then the corresponding dual (primal) model is infeasible.

Proof of Theorem 4.4.1. Recall that for each $\mathbf{x} \geq \mathbf{0}$ and $\mathbf{y} \geq \mathbf{0}$ with $\mathbf{Ax} \leq \mathbf{b}$ and $\mathbf{A}^\mathsf{T}\mathbf{y} \geq \mathbf{c}$, it holds that $\mathbf{c}^\mathsf{T}\mathbf{x} \leq \mathbf{b}^\mathsf{T}\mathbf{y}$ (see the proof of Theorem 4.2.3). Now suppose for a contradiction that the dual model has a feasible solution $\hat{\mathbf{y}}$. Hence, we have that $\mathbf{c}^\mathsf{T}\mathbf{x} \leq \mathbf{b}^\mathsf{T}\hat{\mathbf{y}}$ for every primal feasible solution $\mathbf{x}$. But since $\mathbf{b}^\mathsf{T}\hat{\mathbf{y}} < \infty$, this contradicts the fact that the primal model is unbounded. Therefore, the dual model is infeasible. □

Note that it may also happen that both the primal and the dual model are infeasible. Exercise 4.8.12 presents an LO-model for which this happens.

In Table 4.2, we have summarized the various possibilities when combining the concepts 'there exists an optimal solution' (notation: Optimal), 'infeasibility of the model' (notation: Infeasible), and 'unboundedness' (notation: Unbounded) for the primal or the dual model. The entries 'Possible' and 'Impossible' in Table 4.2 mean that the corresponding combination is possible and impossible, respectively. Note that the table is symmetric due to the fact that the dual model of the dual model of an LO-model is the original LO-model.

In practical situations, feasible regions are usually not unbounded, because the decision variables do not take on arbitrarily large values. On the other hand, empty feasible regions do occur in practice; they can be avoided by sufficiently relaxing the right hand side values (the capacities) of the constraints.

An interesting consequence of linear optimization duality is *Farkas' lemma*[1]. Farkas' lemma is an example of a so-called *theorem of the alternative*. The theorem states two systems of equations and inequalities. Precisely one of the two systems has a solution; the other one is unsolvable (*inconsistent*).

> **Theorem 4.4.2.** (*Farkas' lemma*)
> For any (m, n)-matrix $\mathbf{A}$ and any n-vector $\mathbf{c}$, exactly one of the following two systems has a solution:
>
> (I) $\mathbf{A}\mathbf{x} \leq \mathbf{0}, \mathbf{c}^{\mathsf{T}}\mathbf{x} > 0;$
>
> (II) $\mathbf{A}^{\mathsf{T}}\mathbf{y} = \mathbf{c}, \mathbf{y} \geq \mathbf{0}.$

Proof of Theorem 4.4.2. We will prove that (I) has a solution if and only if (II) has no solution. Consider the primal model (P): $\max\left\{\mathbf{c}^{\mathsf{T}}\mathbf{x} \mid \mathbf{A}\mathbf{x} \leq \mathbf{0}\right\}$, and the corresponding dual model (D): $\min\left\{\mathbf{0}^{\mathsf{T}}\mathbf{y} \mid \mathbf{A}^{\mathsf{T}}\mathbf{y} = \mathbf{c}, \mathbf{y} \geq \mathbf{0}\right\}$. First assume that system (II) has a solution, $\hat{\mathbf{y}}$, say. Since $\mathbf{A}^{\mathsf{T}}\hat{\mathbf{y}} = \mathbf{c}$ and $\hat{\mathbf{y}} \geq \mathbf{0}$, it follows that $\hat{\mathbf{y}}$ is a feasible point of model (D). Moreover, since all feasible points of (D) have objective value 0, $\hat{\mathbf{y}}$ is an optimal solution of (D). Because the optimal objective value of model (D) is 0, model (P) has an optimal solution, and the corresponding optimal objective value is 0; see Exercise 4.8.10(b). Since the maximum objective value on the feasible region of model (P) is 0, it follows that system (I) has no solution.

Now assume that system (II) has no solution. Then, model (D) is infeasible. On the other hand, model (P) is feasible, since $\mathbf{0}$ is in its feasible region. So, model (P) either has an optimal solution, or is unbounded. The former is not the case, because if model (P) had an optimal solution, then model (D) would have been feasible. Hence, model (P) must be unbounded. So, there exist an $\hat{\mathbf{x}}$ with $\mathbf{c}^{\mathsf{T}}\hat{\mathbf{x}} > 0$ and $\mathbf{A}\hat{\mathbf{x}} \leq \mathbf{0}$. Hence, system (I) has a solution.   □

An application of Farkas' lemma is the following. Consider a system of inequalities of the form (I) $\mathbf{A}\mathbf{x} \leq \mathbf{b}, \mathbf{c}^{\mathsf{T}}\mathbf{x} > 0$. Suppose that this system has a solution, and we want to convince someone else that is indeed the case. The most direct way to convince the other person is to present a solution to the system. Any person that questions the claim that the system is solvable can take the presented solution, perform the calculations, and verify that it is indeed a solution of system (I). Such a solution is called a *certificate of solvability* for system (I), because it certifies the statement that system (I) has a solution. But now suppose that the system has no solution. How can we convince the other person of this fact? Farkas' lemma can now be used. If the system indeed has no solution, then, according to the lemma, the system (II) $\mathbf{A}^{\mathsf{T}}\mathbf{y} = \mathbf{c}, \mathbf{y} \geq \mathbf{0}$ must have a solution. So in order to convince the other person that system (I) has no solution, we should present a solution of system (II). Such a solution is called a *certificate of unsolvability* for system (I).

---

[1] Named after the Hungarian mathematician and physicist GYULA FARKAS (1847–1930).

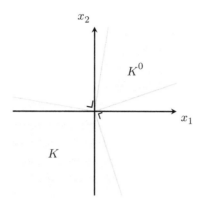

Figure 4.5: Polar cone.

Example 4.4.2. *Consider the system of equations and inequalities:*

$$
\begin{bmatrix} 5 & 4 & 5 & 4 \\ 1 & 4 & 3 & 2 \\ -5 & 4 & 0 & 3 \end{bmatrix} \begin{bmatrix} y_1 \\ y_2 \\ y_3 \\ y_4 \end{bmatrix} = \begin{bmatrix} 5 \\ 2 \\ 2 \end{bmatrix}, \qquad \begin{bmatrix} y_1 \\ y_2 \\ y_3 \\ y_4 \end{bmatrix} \geq \begin{bmatrix} 0 \\ 0 \\ 0 \\ 0 \end{bmatrix} \tag{4.4}
$$

We want to convince the reader that this system has no solution. Note that, looking only at the equality constraints (i.e., ignoring the nonnegativity constraints), it seems reasonable to expect that this system has a solution: the system has more variables than equality constraints, and the matrix has full row rank; therefore the equality constraints have a solution. In fact, it can be checked that the equality constraints have infinitely many solutions.

To see that (4.4) has no solution, we will present a certificate of unsolvability. Let $\mathbf{A}^\top$ be the leftmost $(4, 3)$-matrix, and let $\mathbf{c} = \begin{bmatrix} 5 & 2 & 2 \end{bmatrix}^\top$. Clearly, system (4.4) is of the form $\mathbf{A}^\top \mathbf{y} = \mathbf{c}$, $\mathbf{y} \geq \mathbf{0}$, i.e., form (II) of Theorem 4.4.2. Now, define the vector $\mathbf{x} = \begin{bmatrix} 3 & -9 & 2 \end{bmatrix}^\top$. This is a solution of system (II), because:

$$
\mathbf{Ax} = \begin{bmatrix} 5 & 1 & -5 \\ 4 & 4 & 4 \\ 5 & 3 & 0 \\ 4 & 2 & 3 \end{bmatrix} \begin{bmatrix} 3 \\ -9 \\ 2 \end{bmatrix} = \begin{bmatrix} -4 \\ -16 \\ -12 \\ 0 \end{bmatrix} \leq \begin{bmatrix} 0 \\ 0 \\ 0 \\ 0 \end{bmatrix}, \quad \mathbf{c}^\top \mathbf{x} = \begin{bmatrix} 5 & 2 & 2 \end{bmatrix} \begin{bmatrix} 3 \\ -9 \\ 2 \end{bmatrix} = 1 > 0.
$$

So, the system $\mathbf{Ax} \leq \mathbf{0}$, $\mathbf{c}^\top \mathbf{x} > 0$ has a solution. Therefore, according to Theorem 4.4.2, the system (4.4) has in fact no solution. The vector $\begin{bmatrix} 3 & -9 & 2 \end{bmatrix}^\top$ is the corresponding certificate of unsolvability for the system (4.4). Note that it can be checked that dropping any one of the equality constraints of (4.4) leads to a system that has a solution.

There is also a 'geometric' formulation of Farkas' lemma. A *cone* in $\mathbb{R}^n$ is a subset K of $\mathbb{R}^n$, such that if $\mathbf{x} \in K$ then $\lambda \mathbf{x} \in K$ for each $\lambda \geq 0$; see Appendix D. Let $K =$

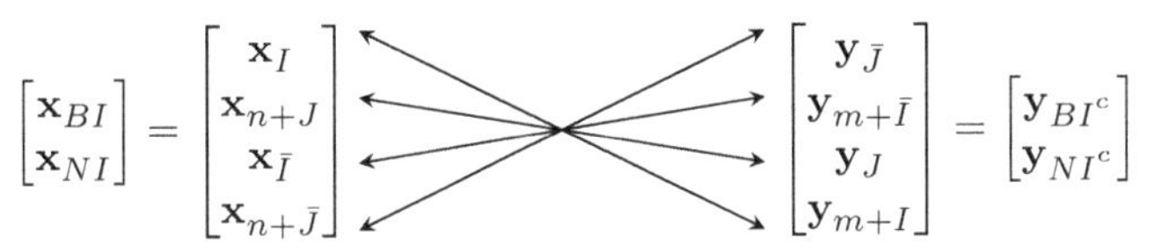

$$\begin{bmatrix} \mathbf{x}_{BI} \\ \mathbf{x}_{NI} \end{bmatrix} = \begin{bmatrix} \mathbf{x}_I \\ \mathbf{x}_{n+J} \\ \mathbf{x}_{\bar{I}} \\ \mathbf{x}_{n+\bar{J}} \end{bmatrix} \qquad \begin{bmatrix} \mathbf{y}_{\bar{J}} \\ \mathbf{y}_{m+\bar{I}} \\ \mathbf{y}_{J} \\ \mathbf{y}_{m+I} \end{bmatrix} = \begin{bmatrix} \mathbf{y}_{BI^c} \\ \mathbf{y}_{NI^c} \end{bmatrix}$$

Figure 4.6: Complementary dual variables.

$\{\mathbf{A}^{\mathsf{T}}\mathbf{y} \mid \mathbf{y} \geq \mathbf{0}\}$. It can easily be seen that K is a convex cone spanned by the row vectors of $\mathbf{A}$ (see Appendix D). Hence, (II) has a solution if and only if $\mathbf{c} \in K$.

The *polar cone* K^0 of a cone K is defined by:

$$K^0 = \{\mathbf{x} \mid \mathbf{d}^{\mathsf{T}}\mathbf{x} \leq \mathbf{0}, \text{ for } \mathbf{d} \in K\}.$$

Hence, for $K = \{\mathbf{A}^{\mathsf{T}}\mathbf{y} \mid \mathbf{y} \geq \mathbf{0}\}$, it follows that:

$$K^0 = \{\mathbf{x} \mid (\mathbf{A}^{\mathsf{T}}\mathbf{y})^{\mathsf{T}}\mathbf{x} \leq \mathbf{0}, \text{ for } \mathbf{y} \geq \mathbf{0}\}$$
$$= \{\mathbf{x} \mid \mathbf{y}^{\mathsf{T}}(\mathbf{Ax}) \leq \mathbf{0}, \text{ for } \mathbf{y} \geq \mathbf{0}\} = \{\mathbf{x} \mid \mathbf{Ax} \leq \mathbf{0}\}.$$

Therefore, (I) has a solution if and only if $\mathbf{c}^{\mathsf{T}}\mathbf{x} > 0$ for some $\mathbf{x} \in K^0$.

Example 4.4.3. *Let* $\mathbf{A} = \begin{bmatrix} -6 & 1 \\ 1 & -3 \end{bmatrix}$. *In Figure 4.5 we have drawn the cone*

$$K = \{\mathbf{A}^{\mathsf{T}}\mathbf{y} \mid \mathbf{y} \geq \mathbf{0}\} = \left\{ \begin{bmatrix} -6y_1 + y_2 \\ y_1 - 3y_2 \end{bmatrix} \middle| y_1, y_2 \geq 0 \right\}$$
$$= \left\{ \begin{bmatrix} -6 \\ 1 \end{bmatrix} y_1 + \begin{bmatrix} 1 \\ -3 \end{bmatrix} y_2 \middle| y_1, y_2 \geq 0 \right\},$$

together with the polar cone of K *which satisfies* $K^0 = \left\{ x_1 \begin{bmatrix} 1 \\ 6 \end{bmatrix} + x_2 \begin{bmatrix} 3 \\ 1 \end{bmatrix} \middle| x_1, x_2 \geq 0 \right\}$. (II)
means that $\mathbf{c} \in K$, *whereas* (I) *means that the vector* $\mathbf{c}$ *makes an acute angle (i.e., an angle of less than* $90°$*) with at least one vector in* K^0.

4.5 Primal and dual feasible basic solutions

We saw in Theorem 4.2.4 how an optimal dual solution can be found once a primal optimal feasible basic solution is found. So far, however, it is not clear which submatrix of $\begin{bmatrix} \mathbf{A}^{\mathsf{T}} & -\mathbf{I}_n \end{bmatrix}$ is a basis matrix that corresponds to a given optimal dual feasible basic solution, and whether or not $\mathbf{y}^* = (\mathbf{B}^{-1})^{\mathsf{T}}\mathbf{c}_{BI}$ (with $\mathbf{B}$ an optimal basis matrix) actually corresponds to a dual feasible basic solution.

In Theorem 4.5.1, it is shown that the complementary dual matrix $\overline{\mathbf{B}}$ of $\mathbf{B}$ (see Section 2.2.3) is an optimal basis matrix in the dual model, if $\mathbf{B}$ is an optimal primal basis (and vice versa). In fact, $\overline{\mathbf{B}}$ is a basis matrix corresponding to $\mathbf{y}^* = (\mathbf{B}^{-1})^\top \mathbf{c}_{BI}$, which means that $\mathbf{y}^*$ corresponds to a vertex of the dual feasible region.

The idea of the proof of the theorem is as follows. Let $\begin{bmatrix} \mathbf{x}_{BI} \\ \mathbf{x}_{NI} \end{bmatrix}$ be the optimal feasible basic solution of the standard primal model with respect to the basis matrix $\mathbf{B}$, and let $\begin{bmatrix} \mathbf{y}_{BI^c} \\ \mathbf{y}_{NI^c} \end{bmatrix}$ be the corresponding optimal solution of the dual model with respect to the complementary dual matrix $\overline{\mathbf{B}}$ of $\mathbf{B}$. Following the notation of Section 2.2.3, define $I = BI \cap \{1, \ldots, n\}$, $J = (BI \cap \{n+1, \ldots, n+m\}) - n$. Since $\mathbf{B}$ and $\overline{\mathbf{B}}$ are complementary dual basis matrices, we have that $BI = I \cup (n+J)$, $NI = \bar{I} \cup (n+\bar{J})$, $BI^c = \bar{J} \cup (m+\bar{I})$, and $NI^c = J \cup (m+I)$. The relationships between these sets are depicted in Figure 4.6. In this figure, the arrows refer to the mutually complementary dual variables. That is, $\mathbf{x}_I$ corresponds to $\mathbf{y}_{m+I}$, $\mathbf{x}_{n+J}$ to $\mathbf{y}_J$, $\mathbf{x}_{\bar{I}}$ to $\mathbf{y}_{m+\bar{I}}$, and $\mathbf{x}_{n+\bar{J}}$ to $\mathbf{y}_{\bar{J}}$. Hence, $\mathbf{x}_{BI}$ corresponds to $\mathbf{y}_{NI^c}$, and $\mathbf{x}_{NI}$ to $\mathbf{y}_{BI^c}$. According to the complementary slackness relations, it follows that both $\mathbf{B}$ is an optimal primal basis matrix and $\overline{\mathbf{B}}$ is an optimal dual basis matrix, if and only if it holds that:

$$\mathbf{x}_{BI}^\top \mathbf{y}_{NI^c} = 0 \text{ and } \mathbf{x}_{NI}^\top \mathbf{y}_{BI^c} = 0.$$

We will verify in the proof that the latter condition in fact holds. Note that primal basic variables correspond to dual nonbasic variables, that primal decision variables correspond to dual slack variables, and that primal slack variables correspond to dual decision variables.

Theorem 4.5.1. (*Optimal primal and dual feasible basic solutions*)
Consider a pair of standard primal and dual LO-models. Let $\mathbf{B}$ and $\overline{\mathbf{B}}$ be complementary dual basis matrices in $\begin{bmatrix} \mathbf{A} & \mathbf{I}_m \end{bmatrix} \equiv \begin{bmatrix} \mathbf{B} & \mathbf{N} \end{bmatrix}$ and $\begin{bmatrix} \mathbf{A}^\top & -\mathbf{I}_n \end{bmatrix} \equiv \begin{bmatrix} \overline{\mathbf{B}} & \overline{\mathbf{N}} \end{bmatrix}$, respectively. If $\mathbf{B}$ and $\overline{\mathbf{B}}$ correspond to feasible basic solutions, then both $\mathbf{B}$ and $\overline{\mathbf{B}}$ are optimal basis matrices, and:

(i) $\mathbf{x}^* = (\overline{\mathbf{B}}^{-1})^\top \mathbf{b}_{BI^c}$ is the vertex of the primal feasible region that corresponds to the primal basis matrix $\mathbf{B}$, and

(ii) $\mathbf{y}^* = (\mathbf{B}^{-1})^\top \mathbf{c}_{BI}$ is the vertex of the dual feasible region that corresponds to the dual basis matrix $\overline{\mathbf{B}}$.

Proof of Theorem 4.5.1. Let $\begin{bmatrix} \hat{\mathbf{x}}_{BI} \\ \hat{\mathbf{x}}_{NI} \end{bmatrix}$ and $\begin{bmatrix} \hat{\mathbf{y}}_{BI^c} \\ \hat{\mathbf{y}}_{NI^c} \end{bmatrix}$ be the feasible basic solutions with $\hat{\mathbf{x}}_{BI} = \mathbf{B}^{-1}\mathbf{b} \geq \mathbf{0}$, $\hat{\mathbf{x}}_{NI} = \mathbf{0}$, and $\hat{\mathbf{y}}_{BI^c} = \overline{\mathbf{B}}^{-1}\mathbf{c} \geq \mathbf{0}$, $\hat{\mathbf{y}}_{NI^c} = \mathbf{0}$ corresponding to $\mathbf{B}$ and $\overline{\mathbf{B}}$. Using the notation of Section 2.2.3, define $I = BI \cap \{1, \ldots, n\}$, $J = (BI \cap \{n+1, \ldots, n+m\}) - n$. We have that (see Theorem 2.2.1):

$$\mathbf{B} = \begin{bmatrix} \mathbf{A} & \mathbf{I}_m \end{bmatrix}_{*,BI} \quad = \begin{bmatrix} \mathbf{A}_{*,I} & (\mathbf{I}_m)_{*,J} \end{bmatrix} \quad \equiv \begin{bmatrix} \mathbf{A}_{\bar{J},I} & \mathbf{0}_{\bar{J},J} \\ \mathbf{A}_{J,I} & (\mathbf{I}_m)_{J,J} \end{bmatrix}$$

$$\overline{\mathbf{B}} = \begin{bmatrix} \mathbf{A}^\top & -\mathbf{I}_n \end{bmatrix}_{*,BI^c} = \begin{bmatrix} (\mathbf{A}^\top)_{*,\bar{J}} & (-\mathbf{I}_n)_{*,\bar{I}} \end{bmatrix} \equiv \begin{bmatrix} (\mathbf{A}^\top)_{I,\bar{J}} & \mathbf{0}_{I,\bar{I}} \\ (\mathbf{A}^\top)_{\bar{I},\bar{J}} & (-\mathbf{I}_n)_{\bar{I},\bar{I}} \end{bmatrix}.$$

Define the $(|I|, |I|)$-matrix $\mathbf{K} = \mathbf{A}_{\bar{J}, I}$. The inverses of $\mathbf{B}$ and $\bar{\mathbf{B}}$ are:

$$\mathbf{B}^{-1} \equiv \begin{bmatrix} \mathbf{K}^{-1} & \mathbf{0}_{I,J} \\ -\mathbf{A}_{J,I}\mathbf{K}^{-1} & (\mathbf{I}_m)_{J,J} \end{bmatrix}, \text{ and } \bar{\mathbf{B}}^{-1} \equiv \begin{bmatrix} (\mathbf{K}^{\mathsf{T}})^{-1} & \mathbf{0}_{\bar{J},\bar{I}} \\ (\mathbf{A}^{\mathsf{T}})_{\bar{I},\bar{J}}(\mathbf{K}^{\mathsf{T}})^{-1} & (-\mathbf{I}_n)_{\bar{I},\bar{I}} \end{bmatrix}.$$

This can be easily checked by showing that $\mathbf{B}\mathbf{B}^{-1} = \mathbf{I}_m$ and $\bar{\mathbf{B}}\bar{\mathbf{B}}^{-1} = \mathbf{I}_n$. (For the proof of the invertibility of $\mathbf{K}$, see the proof of Theorem 2.2.1.) We have that:

$$\hat{\mathbf{x}}_{BI} = \begin{bmatrix} \hat{\mathbf{x}}_I \\ \hat{\mathbf{x}}_{n+J} \end{bmatrix} = \begin{bmatrix} \mathbf{K}^{-1} & \mathbf{0}_{I,J} \\ -\mathbf{A}_{J,I}\mathbf{K}^{-1} & (\mathbf{I}_m)_{J,J} \end{bmatrix} \begin{bmatrix} \mathbf{b}_{\bar{J}} \\ \mathbf{b}_J \end{bmatrix} = \begin{bmatrix} \mathbf{K}^{-1}\mathbf{b}_{\bar{J}} \\ \mathbf{b}_J - \mathbf{A}_{J,I}\mathbf{K}^{-1}\mathbf{b}_{\bar{J}} \end{bmatrix},$$

$$\hat{\mathbf{y}}_{BI^c} = \begin{bmatrix} \hat{\mathbf{y}}_{\bar{J}} \\ \hat{\mathbf{y}}_{m+\bar{I}} \end{bmatrix} = \begin{bmatrix} (\mathbf{K}^{\mathsf{T}})^{-1} & \mathbf{0}_{\bar{J},\bar{I}} \\ (\mathbf{A}^{\mathsf{T}})_{\bar{I},\bar{J}}(\mathbf{K}^{\mathsf{T}})^{-1} & (-\mathbf{I}_n)_{\bar{I},\bar{I}} \end{bmatrix} \begin{bmatrix} \mathbf{c}_I \\ \mathbf{c}_{\bar{I}} \end{bmatrix} = \begin{bmatrix} (\mathbf{K}^{\mathsf{T}})^{-1}\mathbf{c}_I \\ (\mathbf{A}^{\mathsf{T}})_{\bar{I},\bar{J}}(\mathbf{K}^{\mathsf{T}})^{-1}\mathbf{c}_I - \mathbf{c}_{\bar{I}} \end{bmatrix}.$$

Using the fact that $\hat{\mathbf{x}}_{NI} = \begin{bmatrix} \hat{\mathbf{x}}_{\bar{I}} \\ \hat{\mathbf{x}}_{n+\bar{J}} \end{bmatrix} = \mathbf{0}$, and $\hat{\mathbf{y}}_{NI^c} = \begin{bmatrix} \hat{\mathbf{y}}_J \\ \hat{\mathbf{y}}_{m+I} \end{bmatrix} = \mathbf{0}$, it follows that:

$$\begin{aligned}
\hat{\mathbf{x}}_I &= \mathbf{K}^{-1}\mathbf{b}_{\bar{J}} && \text{and} & \hat{\mathbf{y}}_{m+I} &= \mathbf{0} \\
\hat{\mathbf{x}}_{\bar{I}} &= \mathbf{0} && & \hat{\mathbf{y}}_{m+\bar{I}} &= (\mathbf{A}^{\mathsf{T}})_{\bar{I},\bar{J}}\hat{\mathbf{y}}_{\bar{J}} - \mathbf{c}_{\bar{I}} \\
\hat{\mathbf{x}}_{n+J} &= \mathbf{b}_J - \mathbf{A}_{J,I}\hat{\mathbf{x}}_I && & \hat{\mathbf{y}}_J &= \mathbf{0} \\
\hat{\mathbf{x}}_{n+\bar{J}} &= \mathbf{0} && & \hat{\mathbf{y}}_{\bar{J}} &= (\mathbf{K}^{\mathsf{T}})^{-1}\mathbf{c}_I.
\end{aligned}$$

Note that we can recognize the expression for $\hat{\mathbf{x}}_{n+J}$ as the slack of the primal constraints with indices in J. Similarly, we can recognize the expression for $\hat{\mathbf{y}}_{m+\bar{I}}$ as the slack of the dual constraints with indices in $\bar{I}$. Since these primal and dual solutions are feasible (by assumption), and they satisfy the complementary slackness conditions, it follows from Theorem 4.3.2 that $\mathbf{B}$ and $\bar{\mathbf{B}}$ are indeed optimal basis matrices.

For the second part, we have that:

$$\mathbf{x}^* = (\bar{\mathbf{B}}^{-1})^{\mathsf{T}}\mathbf{b}_{BI^c} = \begin{bmatrix} \mathbf{K}^{-1} & \mathbf{K}^{-1}\mathbf{A}_{\bar{J},\bar{I}} \\ \mathbf{0}_{\bar{I},\bar{J}} & (-\mathbf{I}_n)_{\bar{I},\bar{I}} \end{bmatrix} \begin{bmatrix} \mathbf{b}_{\bar{J}} \\ \mathbf{b}_{m+\bar{I}} \end{bmatrix} = \begin{bmatrix} \mathbf{K}^{-1}\mathbf{b}_{\bar{J}} \\ \mathbf{0}_{\bar{I}} \end{bmatrix} = \begin{bmatrix} \hat{\mathbf{x}}_I \\ \hat{\mathbf{x}}_{\bar{I}} \end{bmatrix},$$

$$\mathbf{y}^* = (\mathbf{B}^{-1})^{\mathsf{T}}\mathbf{c}_{BI} = \begin{bmatrix} (\mathbf{K}^{\mathsf{T}})^{-1} & -(\mathbf{A}_{J,I}\mathbf{K}^{-1})^{\mathsf{T}} \\ \mathbf{0}_{J,I} & (\mathbf{I}_m)_{J,J} \end{bmatrix} \begin{bmatrix} \mathbf{c}_I \\ \mathbf{c}_{n+J} \end{bmatrix} = \begin{bmatrix} (\mathbf{K}^{\mathsf{T}})^{-1}\mathbf{c}_I \\ \mathbf{0}_J \end{bmatrix} = \begin{bmatrix} \hat{\mathbf{y}}_{\bar{J}} \\ \hat{\mathbf{y}}_J \end{bmatrix},$$

where we have used the fact that $\mathbf{b}_{m+\bar{I}} = \mathbf{0}$ and $\mathbf{c}_{n+J} = \mathbf{0}$. Therefore, $\mathbf{x}^* = (\bar{\mathbf{B}}^{-1})^{\mathsf{T}}\mathbf{b}_{BI^c}$ is the vertex of the primal feasible region corresponding to $\mathbf{B}$, and $\mathbf{y}^* = (\mathbf{B}^{-1})^{\mathsf{T}}\mathbf{c}_{BI}$ is the vertex of the dual feasible region corresponding to the dual basis matrix $\bar{\mathbf{B}}$. $\qquad\square$

The complementary dual matrix $\bar{\mathbf{B}}$ actually defines a dual (not necessarily feasible) basic solution for any (not necessarily optimal) primal basis matrix $\mathbf{B}$. Moreover, the values of the primal basic variables and the dual basic variables have explicit expressions, which are given in Theorem 4.5.2.

The expressions derived in Theorem 4.5.2 have an interesting interpretation. Let $\mathbf{B}$ and $\bar{\mathbf{B}}$ be complementary dual basis matrices in $\begin{bmatrix} \mathbf{A} & \mathbf{I}_m \end{bmatrix} \equiv \begin{bmatrix} \mathbf{B} & \mathbf{N} \end{bmatrix}$ and $\begin{bmatrix} \mathbf{A}^{\mathsf{T}} & -\mathbf{I}_n \end{bmatrix} \equiv \begin{bmatrix} \bar{\mathbf{B}} & \bar{\mathbf{N}} \end{bmatrix}$, respectively, and let $\begin{bmatrix} \hat{\mathbf{x}} \\ \hat{\mathbf{x}}_s \end{bmatrix} \equiv \begin{bmatrix} \hat{\mathbf{x}}_{BI} \\ \hat{\mathbf{x}}_{NI} \end{bmatrix}$ and $\begin{bmatrix} \hat{\mathbf{y}} \\ \hat{\mathbf{y}}_s \end{bmatrix} \equiv \begin{bmatrix} \hat{\mathbf{y}}_{BI^c} \\ \hat{\mathbf{y}}_{NI^c} \end{bmatrix}$ be the corresponding primal and dual feasible basic solutions. Recall that, by the definition of a basic solution, the primal basic solution is feasible if and only if:

$$\mathbf{B}^{-1}\mathbf{b} \geq \mathbf{0}. \tag{4.5}$$

Similarly, the dual basic solution is feasible if and only if:

$$(\overline{\mathbf{B}}^{-1})^{\mathsf{T}}\mathbf{c} \geq \mathbf{0}. \tag{4.6}$$

Recall also that, by the definition of an optimal basic solution (see Section 3.5.2), the primal basic solution is optimal if and only if:

$$(\mathbf{B}^{-1}\mathbf{N})^{\mathsf{T}}\mathbf{c}_{BI} - \mathbf{c}_{NI} \geq \mathbf{0}. \tag{4.7}$$

Applying this condition to the corresponding dual solution, the dual basic solution is optimal if and only if:

$$(\overline{\mathbf{B}}^{-1}\mathbf{N})^{\mathsf{T}}\mathbf{b}_{BI^c} - \mathbf{b}_{NI^c} \leq \mathbf{0}. \tag{4.8}$$

It follows from Theorem 4.5.2 that the left hand side vectors of (4.5) and (4.8) are in fact equal. Similarly, the left hand side vectors of (4.6) and (4.7) are equal. These two facts have the following important implications:

The primal basic solution satisfies the *primal feasibility condition* (4.5) if and only if the corresponding dual basic solution satisfies the *dual optimality condition* (4.8).

The primal basic solution satisfies the *primal optimality condition* (4.7) if and only if the corresponding dual basic solution satisfies the *dual feasibility condition* (4.6).

Theorem 4.5.2. (*Dual feasibility is primal optimality, and vice versa*)
Consider a pair of standard primal and dual LO-models. Let $\mathbf{B}$ and $\overline{\mathbf{B}}$ be complementary dual basis matrices in $[\mathbf{A}\ \mathbf{I}_m] \equiv [\mathbf{B}\ \mathbf{N}]$ and $[\mathbf{A}^{\mathsf{T}}\ -\mathbf{I}_n] \equiv [\overline{\mathbf{B}}\ \overline{\mathbf{N}}]$, respectively, and let $\begin{bmatrix} \hat{\mathbf{x}} \\ \hat{\mathbf{x}}_s \end{bmatrix} \equiv \begin{bmatrix} \hat{\mathbf{x}}_{BI} \\ \hat{\mathbf{x}}_{NI} \end{bmatrix}$ and $\begin{bmatrix} \hat{\mathbf{y}} \\ \hat{\mathbf{y}}_s \end{bmatrix} \equiv \begin{bmatrix} \hat{\mathbf{y}}_{BI^c} \\ \hat{\mathbf{y}}_{NI^c} \end{bmatrix}$ be the corresponding primal and dual feasible basic solutions. Then,

$$\hat{\mathbf{x}}_{BI} = \mathbf{B}^{-1}\mathbf{b} = \mathbf{b}_{NI^c} - \overline{\mathbf{N}}^{\mathsf{T}}\left(\overline{\mathbf{B}}^{-1}\right)^{\mathsf{T}}\mathbf{b}_{BI^c}, \tag{4.9}$$

$$\hat{\mathbf{y}}_{BI^c} = (\overline{\mathbf{B}}^{-1})^{\mathsf{T}}\mathbf{c} = \mathbf{N}^{\mathsf{T}}(\mathbf{B}^{-1})^{\mathsf{T}}\mathbf{c}_{BI} - \mathbf{c}_{NI}. \tag{4.10}$$

Proof. Note that the first equality in (4.9) follows from the definition of a basic solution, and similarly for (4.10). So, it suffices to prove the second equality in (4.9) and the second equality in (4.10). Using the notation from Theorem 4.5.1, we have that:

$$\mathbf{N} = [\mathbf{A}\ \mathbf{I}_m]_{*,NI} \equiv \begin{bmatrix} \mathbf{A}_{\bar{J},\bar{I}} & (\mathbf{I}_m)_{\bar{J},\bar{J}} \\ \mathbf{A}_{J,\bar{I}} & \mathbf{0}_{J,\bar{J}} \end{bmatrix},$$

$$\overline{\mathbf{N}} = [\mathbf{A}^{\mathsf{T}}\ -\mathbf{I}_n]_{*,NI^c} \equiv \begin{bmatrix} (\mathbf{A}^{\mathsf{T}})_{I,J} & (-\mathbf{I}_n)_{I,I} \\ (\mathbf{A}^{\mathsf{T}})_{\bar{I},J} & \mathbf{0}_{\bar{I},I} \end{bmatrix}.$$

It follows that:

$$\mathbf{b}_{NI^c} - \overline{\mathbf{N}}^{\mathsf{T}}(\overline{\mathbf{B}}^{-1})^{\mathsf{T}}\mathbf{b}_{BI^c} = \mathbf{b}_{NI^c} - \overline{\mathbf{N}}^{\mathsf{T}}\mathbf{x}^*$$

$$\equiv \begin{bmatrix} \mathbf{b}_J \\ \mathbf{b}_{m+I} \end{bmatrix} - \begin{bmatrix} \mathbf{A}_{J,I} & \mathbf{A}_{J,\bar{I}} \\ (-\mathbf{I}_n)_{I,I} & \mathbf{0}_{I,\bar{I}} \end{bmatrix}\begin{bmatrix} \mathbf{x}_I^* \\ \mathbf{0}_{\bar{I}} \end{bmatrix}$$

$$= \begin{bmatrix} \mathbf{b}_J - \mathbf{A}_{J,I}\mathbf{x}_I^* \\ \hat{\mathbf{x}}_I \end{bmatrix} = \begin{bmatrix} \hat{\mathbf{x}}_{n+J} \\ \hat{\mathbf{x}}_I \end{bmatrix} \equiv \hat{\mathbf{x}}_{BI},$$

where we have used the fact that $\hat{\mathbf{x}}_{\bar{I}} = \mathbf{0}$ and $\mathbf{b}_{m+I} = \mathbf{0}$. Similarly,

$$\mathbf{N}^{\mathsf{T}}\left(\mathbf{B}^{-1}\right)^{\mathsf{T}}\mathbf{c}_{BI} - \mathbf{c}_{NI} = \mathbf{N}^{\mathsf{T}}\mathbf{y}^* - \mathbf{c}_{NI}$$

$$\equiv \begin{bmatrix} (\mathbf{A}^{\mathsf{T}})_{\bar{I},\bar{J}} & (\mathbf{A}^{\mathsf{T}})_{\bar{I},J} \\ (\mathbf{I}_m)_{\bar{J},\bar{J}} & \mathbf{0}_{\bar{J},J} \end{bmatrix}\begin{bmatrix} \hat{\mathbf{y}}_{\bar{J}} \\ \mathbf{0}_J \end{bmatrix} - \begin{bmatrix} \mathbf{c}_{\bar{I}} \\ \mathbf{c}_{n+\bar{J}} \end{bmatrix}$$

$$= \begin{bmatrix} (\mathbf{A}^{\mathsf{T}})_{\bar{I},\bar{J}}\hat{\mathbf{y}}_{\bar{J}} - \mathbf{c}_{\bar{I}} \\ \hat{\mathbf{y}}_{\bar{J}} \end{bmatrix} = \begin{bmatrix} \hat{\mathbf{y}}_{m+\bar{I}} \\ \hat{\mathbf{y}}_{\bar{J}} \end{bmatrix} \equiv \hat{\mathbf{y}}_{BI^c},$$

where we have used the fact that $\hat{\mathbf{y}}_J = \mathbf{0}$ and $\mathbf{c}_{n+\bar{J}} = \mathbf{0}$. $\qquad\square$

Note that if $\mathbf{B}$ and $\overline{\mathbf{B}}$ in Theorem 4.5.2 do not correspond to optimal solutions, then $\mathbf{c}^{\mathsf{T}}\hat{\mathbf{x}} \neq \mathbf{b}^{\mathsf{T}}\hat{\mathbf{y}}$.

It follows from Theorem 4.5.2 that, at an optimal solution, we have that:

$$\mathbf{c}_{NI}^{\mathsf{T}} - \mathbf{c}_{BI}^{\mathsf{T}}\mathbf{B}^{-1}\mathbf{N} = -\mathbf{y}_{BI^c}^*,$$

i.e., the optimal dual solution (including the values of both the decision and the slack variables) corresponding to an optimal primal basis matrix $\mathbf{B}$ is precisely the negative of the objective vector (namely, the top row) in the optimal simplex tableau corresponding to $\mathbf{B}$. In the case of Model Dovetail, the objective vector of the optimal simplex tableau is $\begin{bmatrix} 0 & 0 & -1\frac{1}{2} & -\frac{1}{2} & 0 & 0 \end{bmatrix}$; see Section 3.3. According to Theorem 4.5.1, it follows that the corresponding optimal dual solution satisfies: $\begin{bmatrix} y_5^* & y_6^* & y_1^* & y_2^* & y_3^* & y_4^* \end{bmatrix} = \begin{bmatrix} 0 & 0 & 1\frac{1}{2} & \frac{1}{2} & 0 & 0 \end{bmatrix}$.

4.6 Duality and the simplex algorithm

We saw in Chapter 3 that the simplex algorithm operates by moving from feasible basic solution to feasible basic solution. The basic solutions encountered by the simplex algorithm are feasible, but not necessarily optimal.

We also saw (in Section 3.4) that every feasible basic solution corresponds to a simplex tableau, and vice versa. In this section, we will see that any simplex tableau represents not only a primal solution, but also a dual solution. In contrast to the primal solution, however, this dual solution satisfies the dual optimality criterion (4.8), but is not necessarily feasible. We will see that the simplex algorithm can be viewed as an algorithm that:

seeks primal optimality, while keeping primal feasibility, and

seeks dual feasibility, while keeping dual optimality.

In the present section, we consider the geometry of the dual feasible region. In this setting, it is convenient to think of the nonnegativity constraints as technology constraints. So, for a standard primal LO-model with slack variables $\max\{\mathbf{c}^\mathsf{T}\mathbf{x} \mid \mathbf{A}\mathbf{x} + \mathbf{x}_s = \mathbf{b}, \ \mathbf{x}, \mathbf{x}_s \geq \mathbf{0}\}$, we consider the dual model:

$$
\begin{aligned}
\min \ & \mathbf{b}^\mathsf{T}\mathbf{y} \\
\text{s.t.} \ & \begin{bmatrix} \mathbf{A}^\mathsf{T} \\ \mathbf{I}_m \end{bmatrix} \mathbf{y} \geq \begin{bmatrix} \mathbf{c}^\mathsf{T} \\ \mathbf{0}^\mathsf{T} \end{bmatrix}.
\end{aligned}
\tag{4.11}
$$

It will be shown that the objective coefficients in any simplex tableau are the negatives of the slack values of the dual (technology and nonnegativity) constraints. The dual constraints include nonnegativity constraints, and hence the slack values of these dual nonnegativity constraints are actually equal to the values of the dual decision variables. Therefore, the current dual solution can be readily read from the simplex tableau.

4.6.1 Dual solution corresponding to the simplex tableau

Recall from Section 3.4 that the simplex tableau representing the basis matrix $\mathbf{B}$ is, in permuted form, organized as follows:

Theorem 4.2.4 describes how to find a dual optimal solution once a primal optimal basis matrix is known. We saw that, if $\mathbf{B}$ is an optimal primal basis matrix from a standard LO-model, then $\mathbf{y}^* = (\mathbf{B}^{-1})^\mathsf{T}\mathbf{c}_{BI}$ is an optimal dual solution, i.e., it is an optimal solution of the dual model $\min\{\mathbf{b}^\mathsf{T}\mathbf{y} \mid \mathbf{A}^\mathsf{T}\mathbf{y} \geq \mathbf{c}, \mathbf{y} \geq \mathbf{0}\}$. Moreover, this optimal dual solution corresponds to a feasible basic solution of the dual model; see Theorem 4.5.1. We also showed that $\mathbf{y}^* = (\mathbf{B}^{-1})^\mathsf{T}\mathbf{c}_{BI}$ is precisely the dual solution that corresponds to the complementary dual matrix $\overline{\mathbf{B}}$ of $\mathbf{B}$.

It actually makes sense to define a dual solution for any (not necessarily optimal) basis matrix. Let $\mathbf{B}$ be a basis matrix, and let BI be the set of indices of the basic variables. We call $\mathbf{y} = (\mathbf{B}^{-1})^\mathsf{T}\mathbf{c}_{BI}$ the *dual solution* corresponding to the basis matrix $\mathbf{B}$.

The objective value z of the feasible basic solution $\begin{bmatrix} \mathbf{x}_{BI} \\ \mathbf{x}_{NI} \end{bmatrix}$ corresponding to $\mathbf{B}$ is equal to the objective value of the (possibly infeasible) dual solution, because we have that:

$$
\mathbf{b}^\mathsf{T}\mathbf{y} = \mathbf{b}^\mathsf{T}(\mathbf{B}^{-1})^\mathsf{T}\mathbf{c}_{BI} = \mathbf{x}_{BI}^\mathsf{T}\mathbf{c}_{BI} = \mathbf{x}_{BI}^\mathsf{T}\mathbf{c}_{BI} + \mathbf{x}_{NI}^\mathsf{T}\mathbf{c}_{NI} = z.
$$

Row 0 of the simplex tableau contains the negative of the values of the complementary dual variables. This can be seen as follows. Recall that $\mathbf{c}_{BI}^\mathsf{T} - \mathbf{c}_{BI}^\mathsf{T}\mathbf{B}^{-1}\mathbf{B} = \mathbf{0}^\mathsf{T}$, i.e., the value of the objective coefficient of any primal basic variable x_i ($i \in BI$) is zero. Therefore, the row vector of objective coefficients of the simplex tableau satisfies:

$$
\begin{aligned}
\begin{bmatrix} \mathbf{0}^\mathsf{T} & \mathbf{c}_{NI}^\mathsf{T} - \mathbf{c}_{BI}^\mathsf{T}\mathbf{B}^{-1}\mathbf{N} \end{bmatrix} &= \begin{bmatrix} \mathbf{c}_{BI}^\mathsf{T} - \mathbf{c}_{BI}^\mathsf{T}\mathbf{B}^{-1}\mathbf{B} & \mathbf{c}_{NI}^\mathsf{T} - \mathbf{c}_{BI}^\mathsf{T}\mathbf{B}^{-1}\mathbf{N} \end{bmatrix} \\
&= \begin{bmatrix} \mathbf{c}_{BI}^\mathsf{T} & \mathbf{c}_{NI}^\mathsf{T} \end{bmatrix} - \mathbf{c}_{BI}^\mathsf{T}\mathbf{B}^{-1}\begin{bmatrix} \mathbf{B} & \mathbf{N} \end{bmatrix} \\
&\equiv \begin{bmatrix} \mathbf{c}^\mathsf{T} & \mathbf{0}^\mathsf{T} \end{bmatrix} - \mathbf{c}_{BI}^\mathsf{T}\mathbf{B}^{-1}\begin{bmatrix} \mathbf{A} & \mathbf{I}_m \end{bmatrix} \\
&= \begin{bmatrix} \mathbf{c}^\mathsf{T} & \mathbf{0}^\mathsf{T} \end{bmatrix} - \mathbf{y}^\mathsf{T}\begin{bmatrix} \mathbf{A} & \mathbf{I}_m \end{bmatrix} = \begin{bmatrix} \mathbf{c} \\ \mathbf{0} \end{bmatrix} - \begin{bmatrix} \mathbf{A}^\mathsf{T} \\ \mathbf{I}_m \end{bmatrix}.
\end{aligned}
$$

This means that the objective coefficients are the negatives of the slack values corresponding to the constraints of model (4.11). The constraints in model (4.11) include nonnegativity constraints, and hence the slack values of these dual nonnegativity constraints are in fact equal to the values of the dual decision variables.

Let $i \in BI$. Since the value of the objective coefficient of x_i is zero, the complementary dual variable corresponding to the i'th dual constraint has value zero. In other words, the i'th dual constraint is binding. This means that $\mathbf{y}$ is the unique point in the intersection of the hyperplanes associated with the dual constraints with indices in BI. (Why is there a unique point in this intersection?) Due to this fact, we can read the corresponding dual solution from the simplex tableau, as the following example illustrates. In fact, the example shows how an optimal dual basic solution can be determined from an optimal simplex tableau.

Example 4.6.1. *Consider Model Dovetail, and its dual model. The dual model, with the nonnegativity constraints considered as technology constraints, reads:*

$$
\begin{array}{lllll}
\min & 9y_1 + 18y_2 + 7y_3 + 6y_4 & & & \\
\text{s.t.} & y_1 + 3y_2 + y_3 & & \geq 3 & (y_5) \\
& y_1 + y_2 & + y_4 & \geq 2 & (y_6) \\
& y_1 & & \geq 0 & (y_1) \\
& y_2 & & \geq 0 & (y_2) \\
& & y_3 & \geq 0 & (y_3) \\
& & y_4 & \geq 0 & (y_4) \\
\end{array}
$$

y_1, y_2, y_3, y_4 free.

In parentheses, we have written the slack variables corresponding to the several constraints. Consider the (optimal) simplex tableau for $BI = \{2, 1, 5, 6\}$:

x_1	x_2	x_3	x_4	x_5	x_6	$-z$	
0	0	$-1\frac{1}{2}$	$-\frac{1}{2}$	0	0	$-22\frac{1}{2}$	
0	1	$1\frac{1}{2}$	$-\frac{1}{2}$	0	0	$4\frac{1}{2}$	x_2
1	0	$-\frac{1}{2}$	$\frac{1}{2}$	0	0	$4\frac{1}{2}$	x_1
0	0	$\frac{1}{2}$	$-\frac{1}{2}$	1	0	$2\frac{1}{2}$	x_5
0	0	$-1\frac{1}{2}$	$\frac{1}{2}$	0	1	$1\frac{1}{2}$	x_6

The indices of the dual constraints correspond exactly to the indices of the primal variables. Hence, from this tableau, we see that the first, second, fifth, and sixth dual constraints are binding. The third and fourth dual constraints have slack values $1\frac{1}{2}$ and $\frac{1}{2}$, respectively.

Columns 1 and 2 correspond to the dual constraints $y_1 + 3y_2 + y_3 \geq 3$ and $y_1 + y_2 + y_4 \geq 2$. Columns $3, \ldots, 6$ correspond to the dual constraints $y_1 \geq 0$, $\ldots$, $y_4 \geq 0$. The latter provides us with a convenient way to determine the values of the dual variables. For instance, the slack variable for the constraint $y_1 \geq 0$ (represented by the third column) has value $1\frac{1}{2}$. This means that $y_1 = 1\frac{1}{2}$. In contrast, the slack variable for the constraint $y_3 \geq 0$ (represented by the fifth column) has value zero. Hence, we have that $y_3 = 0$. Applying the same arguments to each of the dual constraints, we find that the dual solution $\mathbf{y}$ satisfies:

$$y_1 = 1\tfrac{1}{2}, \ y_2 = \tfrac{1}{2}, \ y_3 = 0, \ y_4 = 0.$$

In fact, because this simplex tableau is optimal, the corresponding (primal) basis matrix is optimal and, hence, Theorem 4.2.4 implies that this dual solution is in fact an optimal solution of the dual model.

4.6.2 The simplex algorithm from the dual perspective

Recall that, in order to find an entering variable, the simplex algorithm tries to find a column with a positive current objective coefficient. Since the current objective coefficients are the negatives of the slack values of the dual constraints (see Section 4.6.1), in terms of the dual model, this corresponds to finding a constraint with a negative slack value. Geometrically, this corresponds to determining a dual constraint that is not satisfied at the current dual solution $\mathbf{y}$. Such a constraint is called a *violated* constraint. So, determining an entering variable x_{NI_α} is equivalent to determining a violated dual constraint with index NI_α.

After choosing the entering variable x_{NI_α}, the simplex algorithm proceeds by choosing a row with corresponding basic variable x_{BI_β} that satisfies the minimum-ratio test. The pivot step that follows includes decreasing the objective coefficient of x_{NI_α} to zero, which means that the slack value of the dual constraint with index NI_α is increased to zero. Geometrically, this means that we move to a dual solution at which the (previously violated) dual constraint with index NI_α becomes binding. At the same time, the slack values of the dual constraints with indices in $BI \setminus \{BI_\beta\}$ are kept at zero. Geometrically, this means that we move from the current dual solution towards the dual constraint with index NI_α, while keeping the dual constraints with indices in $BI \setminus \{BI_\beta\}$ binding. Once all current objective coefficients in the simplex tableau are nonpositive, all slack values are nonnegative, and hence, $\mathbf{y} = (\mathbf{B}^{-1})^{\mathsf{T}}\mathbf{c}_{BI}$ is a feasible dual solution.

To illustrate this, recall Model Dovetail and its dual model, Model Salmonnose. In order to gain insight in the relationship between the iteration steps of the simplex algorithm and the dual model, we will now think of Model Salmonnose as the primal model, and Model Dovetail as the dual model. (Recall that the dual model of the dual model is the original primal model.) We will solve Model Salmonnose using the simplex algorithm. After multiplying the objective function by -1, replacing 'min' by 'max', and adding slack variables y_5 and y_6, Model Salmonnose is equivalent to the following LO-model:

$$
\begin{array}{rl}
\max & -9y_1 - 18y_2 - 7y_3 - 6y_4 \\
\text{s.t.} & y_1 + 3y_2 + y_3 \quad\quad\quad - y_5 \quad\quad = 3 \\
& y_1 + y_2 \quad\quad + y_4 \quad\quad - y_6 = 2 \\
& y_1, y_2, y_3, y_4, y_5, y_6 \geq 0.
\end{array}
$$

Putting this into a 'simplex tableau', we obtain:

y_1	y_2	y_3	y_4	y_5	y_6	$-z$
-9	-18	-7	-6	0	0	0
1	3	1	0	-1	0	3
1	1	0	1	0	-1	2

(Note that this is not yet a simplex tableau because we have not specified which variables are basic.) The columns of this 'simplex tableau' correspond to the constraints of the primal model. The primal constraints corresponding to the dual variables y_5 and y_6 are the nonnegativity constraints $x_1 \geq 0$, $x_2 \geq 0$, respectively, of Model Dovetail.

Finding an initial feasible basic solution is particularly easy in this case, because the columns corresponding to y_3 and y_4 already form the $(2,2)$-identity matrix. Moreover, it can be seen from the model above that choosing y_3 and y_4 as the initial basic variables, thus choosing $y_1 = y_2 = y_5 = y_6 = 0$, leads the initial feasible solution $\mathbf{y} = \begin{bmatrix} 0 & 0 & 3 & 2 & 0 & 0 \end{bmatrix}^{\mathsf{T}}$. So, we may choose $BI = \{3,4\}$ as the initial set of (dual) basic variable indices.

To turn the above 'simplex tableau' into an actual simplex tableau, we apply Gaussian elimination and find the following initial simplex tableau and successive simplex iterations:

Iteration 1. $BI = \{3,4\}$.

y_1	y_2	y_3	y_4	y_5	y_6	$-z$	
4	9	0	0	-7	-6	33	
1	3	1	0	-1	0	3	y_3
$\boxed{1}$	1	0	1	0	-1	2	y_4

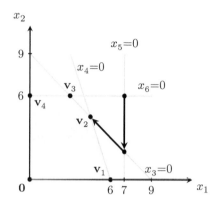

Figure 4.7: Path in the dual problem taken by the simplex algorithm when applied to Model Salmonnose.

The current objective value is 33. The corresponding dual solution (of Model Dovetail) is the unique point in the intersection of the third and fourth constraints of Model Dovetail, i.e., $x_1 = 7$ and $x_2 = 6$, which means that $\mathbf{x} = \begin{bmatrix} 7 & 6 \end{bmatrix}^{\mathsf{T}}$. There are two columns with a positive current objective coefficient. We choose to pivot on y_1; it is left to the reader to carry out the calculations when choosing y_2 instead.

Iteration 2. $BI = \{1, 3\}$.

y_1	y_2	y_3	y_4	y_5	y_6	$-z$	
0	5	0	-4	-7	-2	25	
0	②	1	-1	-1	1	1	y_3
1	1	0	1	0	-1	2	y_1

The current objective value is 25. The corresponding dual solution is the unique point in the intersection of the first and third constraints of Model Dovetail, i.e., $x_1 + x_2 = 9$ and $x_1 = 7$, which implies that $\mathbf{x} = \begin{bmatrix} 7 & 2 \end{bmatrix}^{\mathsf{T}}$.

Iteration 3. $BI = \{1, 2\}$.

y_1	y_2	y_3	y_4	y_5	y_6	$-z$	
0	0	$-2\frac{1}{2}$	$-1\frac{1}{2}$	$-4\frac{1}{2}$	$-4\frac{1}{2}$	$22\frac{1}{2}$	
0	1	$\frac{1}{2}$	$-\frac{1}{2}$	$-\frac{1}{2}$	$\frac{1}{2}$	$\frac{1}{2}$	y_2
1	0	$-\frac{1}{2}$	$1\frac{1}{2}$	$\frac{1}{2}$	$-1\frac{1}{2}$	$1\frac{1}{2}$	y_1

Since all current objective values are nonpositive, we have reached an optimal solution. The optimal objective value is $22\frac{1}{2}$. The corresponding optimal dual solution is the unique point in the intersection of the first and second constraints of Model Dovetail, i.e., $x_1 + x_2 = 9$ and $x_1 + 3x_2 = 18$, which implies that $\mathbf{x} = \begin{bmatrix} 4\frac{1}{2} & 4\frac{1}{2} \end{bmatrix}^{\mathsf{T}}$.

Figure 4.7 depicts the successive dual solutions found by the simplex algorithm. The algorithm started with the (dual infeasible) point $\begin{bmatrix} 7 & 6 \end{bmatrix}^{\mathsf{T}}$, visits the (again infeasible) point

$[7 \ 2]^{\mathsf{T}}$, and finally finds the feasible dual point $\left[4\frac{1}{2} \ 4\frac{1}{2}\right]^{\mathsf{T}}$. So, when viewed from the perspective of the dual model, the simplex algorithm starts with a dual infeasible basic solution (that satisfies the dual optimality criterion), and then performs iteration steps until the dual basic solution becomes feasible.

Recall that the current objective values of y_5 and y_6 are the negatives of the slack variables of the corresponding dual constraints $x_1 \geq 0$, and $x_2 \geq 0$, respectively (see Section 4.6.1). Hence, the dual solutions could also immediately be read from the simplex tableaus by looking at the objective coefficients corresponding to y_5 and y_6. For example, in the initial simplex tableau, we see that these coefficients are -7 and -6, respectively, meaning that $x_1 = 7, x_2 = 6$.

4.7 The dual simplex algorithm

We have seen in the previous section that, when looking from the perspective of the dual problem, the simplex algorithm keeps track of an infeasible dual solution that satisfies the dual optimality criterion. In each iteration step, the algorithm finds a violated constraint and attempts to make this constraint binding. The latter is the key idea of the so-called *dual simplex algorithm*, which we introduce in this section.

The primal simplex algorithm operates on simplex tableaus in which the right hand side values are nonnegative, and works towards making the current objective coefficients in row 0 nonpositive. In contrast, the dual simplex algorithm operates on 'simplex tableaus' in which the current objective coefficients are nonpositive, and works towards making the current right hand side values nonnegative. In each iteration, a violated constraint is selected, and the algorithm chooses an entering and a leaving basic variable so that the constraint becomes binding at the corresponding new basic solution. Before we describe the dual simplex algorithm formally, we give an example.

Example 4.7.1. *Consider the LO-model:*

$$\min \ 2x_1 + 4x_2 + 7x_3$$
$$\text{s.t.} \ \ 2x_1 + \ x_2 + 6x_3 \geq 5$$
$$4x_1 - 6x_2 + 5x_3 \geq 8$$
$$x_1, x_2, x_3 \geq 0.$$

Rewriting this model in standard notation, and introducing the slack variables x_4 and x_5, we obtain:

$$-\max \ -2x_1 - 4x_2 - 7x_3$$
$$\text{s.t.} \ \ -2x_1 - \ x_2 - 6x_3 + x_4 \qquad = -5$$
$$-4x_1 + 6x_2 - 5x_3 \qquad + x_5 = -8$$
$$x_1, x_2, x_3, x_4, x_5 \geq 0.$$

Clearly, $x_1 = x_2 = x_3 = 0, x_4 = -5, x_5 = -8$ is not feasible and can therefore not serve as initial feasible basic solution. However, putting the problem into a simplex tableau, we obtain:

x_1	x_2	x_3	x_4	x_5	$-z$	
-2	-4	-7	0	0	0	
-2	-1	-6	1	0	-5	x_4
-4	6	-5	0	1	-8	x_5

This simplex tableau represents an infeasible basic solution, and this solution satisfies the (primal) optimality criterion. The latter follows from the fact that the current objective coefficients are nonpositive. The dual simplex algorithm now proceeds by finding a violated constraint with index β. At the current solution, both constraints are violated. We choose the first constraint, so that $\beta = 1$ and therefore $(x_{BI_\beta} =) x_4$ is the leaving variable. Next, we choose an entering variable x_{NI_α}, and hence a pivot entry. We choose a value of α such that:

$$\frac{(\mathbf{c}_{NI}^\mathsf{T} - \mathbf{c}_{BI}^\mathsf{T}\mathbf{B}^{-1}\mathbf{N})_\alpha}{(\mathbf{B}^{-1}\mathbf{N})_{\beta,\alpha}} = \min\left\{ \frac{(\mathbf{c}_{NI}^\mathsf{T} - \mathbf{c}_{BI}^\mathsf{T}\mathbf{B}^{-1}\mathbf{N})_j}{(\mathbf{B}^{-1}\mathbf{N})_{\beta,j}} \,\middle|\, \begin{array}{l} (\mathbf{B}^{-1}\mathbf{N})_{\beta,j} < 0, \\ j = 1, \ldots, n \end{array} \right\}. \quad (4.12)$$

In this case, we have that:

$$\min\left\{ \frac{-2}{-2}, \frac{-4}{-1}, \frac{-7}{-6} \right\} = \min\{1, 4, 1\tfrac{1}{6}\} = 1.$$

This means that $\alpha = 1$. After performing a pivot step with the $(\beta, NI_\alpha) = (1, 1)$-entry (with value -2) as the pivot entry, i.e., subtracting row 1 from row 0, subtracting 2 times row 1 from row 2, and dividing row 1 by -2, we find the next simplex tableau:

x_1	x_2	x_3	x_4	x_5	$-z$	
0	-3	-1	-1	0	5	
1	$\frac{1}{2}$	3	$-\frac{1}{2}$	0	$\frac{5}{2}$	x_1
0	8	7	-2	1	2	x_5

Note that in this simplex tableau, the right hand side values are nonnegative. The current objective coefficients are all nonpositive. This means that the current simplex tableau represents an optimal solution. That is, $x_1^* = \frac{5}{2}$, $x_2^* = x_3^* = x_4^* = 0$, $x_5^* = 2$ is an optimal solution.

The expression (4.12) is called the *dual minimum-ratio test*. This test guarantees that, after pivoting on the (β, NI_α)-entry of the simplex tableau, all objective coefficients are nonpositive. To see this, let $\bar{c}_k$ be the current objective coefficient of x_{NI_k} ($k \in \{1, \ldots, n\}$). We need to show that, after the pivot step, the new objective coefficient satisfies:

$$\bar{c}_k' = \bar{c}_k - \frac{\bar{c}_\beta}{(\mathbf{B}^{-1}\mathbf{N})_{\beta,\alpha}}(\mathbf{B}^{-1}\mathbf{N})_{\beta,k} \leq 0. \quad (4.13)$$

Note that $\bar{c}_\beta \leq 0$, and that $(\mathbf{B}^{-1}\mathbf{N})_{\beta,\alpha} < 0$ by the choice of α and β. Hence, $\frac{\bar{c}_\beta}{(\mathbf{B}^{-1}\mathbf{N})_{\beta,\alpha}} \geq 0$. So, if $(\mathbf{B}^{-1}\mathbf{N})_{\beta,k} \geq 0$, then (4.13) holds. If $(\mathbf{B}^{-1}\mathbf{N})_{\beta,k} < 0$, then (4.13) is equivalent to

$$\frac{\bar{c}_k}{(\mathbf{B}^{-1}\mathbf{N})_{\beta,k}} \geq \frac{\bar{c}_\beta}{(\mathbf{B}^{-1}\mathbf{N})_{\beta,\alpha}},$$

which follows immediately from (4.12).

4.7.1 Formulating the dual simplex algorithm

We can now formulate the dual simplex algorithm.

> **Algorithm 4.7.1.** (*The dual simplex algorithm*)
>
> **Input:** Values of $\mathbf{A}$, $\mathbf{b}$, and $\mathbf{c}$ in the LO-model $\max\{\mathbf{c}^\mathsf{T}\mathbf{x} \mid \mathbf{A}\mathbf{x} \leq \mathbf{b}, \mathbf{x} \geq \mathbf{0}\}$, and an initial basic solution for the model with slack variables (see Section 3.6) that satisfies the (primal) optimality condition $(\mathbf{B}^{-1}\mathbf{N})^\mathsf{T}\mathbf{c}_{BI} - \mathbf{c}_{NI} \geq \mathbf{0}$ (see (4.7)).
>
> **Output:** Either an optimal solution of the model, or the message 'no feasible solution'.
>
> **Step 1:** *Constructing the basis matrix.* Let BI and NI be the sets of the indices of the basic and nonbasic variables corresponding to the current basic solution, respectively. Let $\mathbf{B}$ consist of the columns of $[\mathbf{A} \ \mathbf{I}_m]$ with indices in BI, and let $\mathbf{N}$ consist of the columns of $[\mathbf{A} \ \mathbf{I}_m]$ with indices in NI. The current solution is $\mathbf{x}_{BI} = \mathbf{B}^{-1}\mathbf{b}$, $\mathbf{x}_{NI} = \mathbf{0}$, with corresponding current objective value $z = \mathbf{c}_{BI}^\mathsf{T}\mathbf{x}_{BI}$. Go to Step 2.
>
> **Step 2:** *Feasibility test; choosing a leaving variable.* The (column) vector of current right hand side values for the constraints is given by $\mathbf{B}^{-1}\mathbf{b}$. If each entry of this vector is nonnegative, then stop: an optimal solution has been reached. Otherwise, select a negative entry in $\mathbf{B}^{-1}\mathbf{b}$. Let $\beta \in \{1, \ldots, m\}$ be the index of the selected constraint. Then x_{BI_β} is the leaving basic variable. Go to Step 3.
>
> **Step 3:** *Dual minimum-ratio test; choosing an entering variable.* If $(\mathbf{B}^{-1}\mathbf{N})_{\beta,j} \geq 0$ for each $j \in \{1, \ldots, n\}$, then stop with the message 'no feasible solution'. Otherwise, determine an index $\alpha \in \{1, \ldots, n\}$ such that:
>
> $$\frac{(\mathbf{c}_{NI}^\mathsf{T} - \mathbf{c}_{BI}^\mathsf{T}\mathbf{B}^{-1}\mathbf{N})_\alpha}{(\mathbf{B}^{-1}\mathbf{N})_{\beta,\alpha}} = \min\left\{\frac{(\mathbf{c}_{NI}^\mathsf{T} - \mathbf{c}_{BI}^\mathsf{T}\mathbf{B}^{-1}\mathbf{N})_j}{(\mathbf{B}^{-1}\mathbf{N})_{\beta,j}} \ \middle| \ \begin{matrix} (\mathbf{B}^{-1}\mathbf{N})_{\beta,j} < 0, \\ j = 1, \ldots, n \end{matrix} \right\}.$$
>
> Go to Step 4.
>
> **Step 4:** *Exchanging.* Set $BI := (BI \setminus \{BI_\beta\}) \cup \{NI_\alpha\}$ and $NI := (NI \setminus \{NI_\alpha\}) \cup \{BI_\beta\}$. Return to Step 1.

Table 4.3 lists the correspondence between concepts used in the primal and the dual simplex algorithms. Although we will not prove this here, it can be shown that the dual simplex algorithm is just the primal simplex algorithm applied to the dual problem.

The dual simplex algorithm is particularly useful when we have already found an optimal basic solution for an LO-model, and we want to add a new constraint to the model.

4.7.2 Reoptimizing an LO-model after adding a new constraint

It is sometimes necessary to add a new constraint to an LO-model for which an optimal solution has already been found. If the new constraint is satisfied at the previous optimal solution, then this solution is also optimal for the model with the new constraint added. If, however, the new constraint is violated by the previous optimal solution, then this optimal solution is no longer feasible, and hence no longer optimal.

Adding new constraints plays a key role in the cutting plane algorithm, a frequently used solution algorithm for so-called integer optimization problems; see Section 7.4. In the cutting plane algorithm, constraints are added repeatedly, and hence being able to efficiently *reoptimize* is important.

In order to find an optimal solution to the model with the new constraint, we can in principle simply apply the simplex algorithm to the new model, but this is wasteful because the results of the earlier calculations are discarded. It is more efficient to insert the constraint into the optimal simplex tableau and apply the dual simplex algorithm. Usually, the new optimal solution is found in a small number of iterations of the dual simplex algorithm.

The following example illustrates how the dual simplex algorithm can be used when a new constraint is added after a model has been solved.

Example 4.7.2. *Recall the following optimal simplex tableau for Model Dovetail (see Example 3.4.2):*

Primal simplex algorithm	Dual simplex algorithm
Nonnegative right hand side values	Nonpositive current objective values
Attempts to make current objective coefficients nonpositive	Attempts to make the current right hand side values nonnegative
Find a variable whose increase leads to an increase in the objective value; this variable is the entering variable	Find a violated constraint; the basic variable corresponding to this constraint determines the leaving variable
The primal minimum-ratio test determines the leaving variable	The dual minimum-ratio test determines the entering variable

Table 4.3: Correspondence of concepts of the primal simplex algorithm and the dual simplex algorithm.

x_1	x_2	x_3	x_4	x_5	x_6	$-z$	
0	0	$-1\frac{1}{2}$	$-\frac{1}{2}$	0	0	$-22\frac{1}{2}$	
0	1	$1\frac{1}{2}$	$-\frac{1}{2}$	0	0	$4\frac{1}{2}$	x_2
1	0	$-\frac{1}{2}$	$\frac{1}{2}$	0	0	$4\frac{1}{2}$	x_1
0	0	$\frac{1}{2}$	$-\frac{1}{2}$	1	0	$2\frac{1}{2}$	x_5
0	0	$-1\frac{1}{2}$	$\frac{1}{2}$	0	1	$1\frac{1}{2}$	x_6

Matches are, of course, not made of only wood. The match head is composed of several chemicals, such as potassium chlorate, sulfur, and calcium carbonate. Every 100,000 boxes of long matches requires 20 kilograms of this mix of chemicals. The short matches are slightly thinner than the long matches, so every 100,000 boxes of short matches requires only 10 kilograms of the chemicals. Company Dovetail has a separate production line for mixing the chemicals. However, the company can mix at most 130 kilograms of the required chemicals. Hence, we need to impose the additional constraint $20x_1 + 10x_2 \leq 130$, or simply $2x_1 + x_2 \leq 13$. Recall that the optimal solution of Model Dovetail is $\begin{bmatrix} 4\frac{1}{2} & 4\frac{1}{2} \end{bmatrix}^{\mathsf{T}}$. The new constraint is violated at this point, because $2 \times 4\frac{1}{2} + 4\frac{1}{2} = 13\frac{1}{2} \not\leq 13$.

In order to reoptimize without starting all over again, we introduce a new slack variable x_7 and add the new constraint to the simplex tableau:

x_1	x_2	x_3	x_4	x_5	x_6	x_7	$-z$	
0	0	$-1\frac{1}{2}$	$-\frac{1}{2}$	0	0	0	$-22\frac{1}{2}$	
0	1	$1\frac{1}{2}$	$-\frac{1}{2}$	0	0	0	$4\frac{1}{2}$	x_2
1	0	$-\frac{1}{2}$	$\frac{1}{2}$	0	0	0	$4\frac{1}{2}$	x_1
0	0	$\frac{1}{2}$	$-\frac{1}{2}$	1	0	0	$2\frac{1}{2}$	x_5
0	0	$-1\frac{1}{2}$	$\frac{1}{2}$	0	1	0	$1\frac{1}{2}$	x_6
2	1	0	0	0	0	1	13	x_7

This is not a simplex tableau yet, because the columns corresponding to the basic variables do not form an identity matrix. Applying Gaussian elimination to make the first two entries of the last row equal to zero, we find the following simplex tableau:

x_1	x_2	x_3	x_4	x_5	x_6	x_7	$-z$	
0	0	$-1\frac{1}{2}$	$-\frac{1}{2}$	0	0	0	$-22\frac{1}{2}$	
0	1	$1\frac{1}{2}$	$-\frac{1}{2}$	0	0	0	$4\frac{1}{2}$	x_2
1	0	$-\frac{1}{2}$	$\frac{1}{2}$	0	0	0	$4\frac{1}{2}$	x_1
0	0	$\frac{1}{2}$	$-\frac{1}{2}$	1	0	0	$2\frac{1}{2}$	x_5
0	0	$-1\frac{1}{2}$	$\frac{1}{2}$	0	1	0	$1\frac{1}{2}$	x_6
0	0	$-\frac{1}{2}$	$\boxed{-\frac{1}{2}}$	0	0	1	$-\frac{1}{2}$	x_7

This simplex tableau has one violated constraint, and the current objective values are nonpositive. Hence, we can perform an iteration of the dual simplex algorithm. The fifth row corresponds to the violated constraint, so that we choose $\beta = 5$. Note that $BI_\beta = 7$ so that x_7 is the leaving variable. The dual minimum-ratio test gives:

$$\min\left\{\frac{-1\frac{1}{2}}{-\frac{1}{2}}, \frac{-1\frac{1}{2}}{-\frac{1}{2}}\right\} = \min\{3, 1\} = 1.$$

The second entry attains the minimum, so $\alpha = 2$. Therefore, $(x_{NI_2} =) x_4$ is the entering variable. After applying Gaussian elimination, we find the following simplex tableau:

x_1	x_2	x_3	x_4	x_5	x_6	x_7	$-z$	
0	0	-1	0	0	0	-1	-22	
0	1	2	0	0	0	-1	5	x_2
1	0	-1	0	0	0	1	4	x_1
0	0	1	0	1	0	-1	3	x_5
0	0	-2	0	0	1	1	1	x_6
0	0	1	1	0	0	-2	1	x_4

This tableau represents an optimal feasible basic solution. The new optimal objective value is $z^* = 22$, and $x_1^* = 5$, $x_2^* = 4$, $x_3^* = 0$, $x_4^* = 1$, $x_5^* = 3$, $x_6^* = 1$, $x_7^* = 0$ is a new optimal solution.

4.8 Exercises

Exercise 4.8.1. Determine the dual of the following LO-models. Also draw a diagram like Figure 4.1 that illustrates the relationship between the primal variables and the complementary dual variables of the several models.

(a) max $x_1 + 2x_2 - 3x_3$
s.t. $x_1 - 3x_2 \leq 7$
$3x_1 + x_2 + 2x_3 \leq 6$
$-x_1 - 2x_2 - x_3 \geq -5$
$x_1, x_2, x_3 \geq 0.$

(c) max $3x_1 - 5x_2$
s.t. $2x_1 - x_2 \geq 4$
$x_1 - x_2 \geq -3$
$3x_1 - 2x_2 \leq 10$
$x_1, x_2 \geq 0.$

(b) min $x_1 + x_2$
s.t. $2x_1 + 3x_2 + x_3 + x_4 \leq 0$
$-x_1 + x_2 + 2x_3 + x_4 = 6$
$3x_1 + x_2 + 4x_3 + 2x_4 \geq 3$
$x_1 \leq 0, x_2, x_4 \geq 0, x_3$ free.

(d) max $5y_1 + 6y_2 + 3y_3$
s.t. $-y_1 + 2y_2 \leq 1$
$3y_1 - y_2 \geq 2$
$3y_2 + y_3 = 3$
y_1 free, $y_2 \geq 0, y_3 \leq 0.$

Exercise 4.8.2. Use Theorem 4.2.4 in order to decide whether the following LO-models have a finite optimal solution.

(a) max $-9x_1 - 2x_2 + 12x_3$

    s.t. $2x_1 - 2x_2 + 2x_3 \leq 1$

             $-3x_1 + x_2 + x_3 \leq 1$

             $x_1, x_2, x_3 \geq 0.$

(b) max $4x_1 - 3x_2 + 2x_3 - 6x_4$

    s.t. $2x_1 + 4x_2 - x_3 - 2x_4 \leq 1$

             $x_1 - x_2 + 2x_3 - x_4 \leq -1$

             $x_1, x_2, x_3, x_4 \geq 0.$

Exercise 4.8.3. Consider the following LO-model:

$$\begin{aligned}
\max \quad & 2x_1 + 4x_2 + 10x_3 + 15x_4 \\
\text{s.t.} \quad & -x_1 + x_2 + x_3 + 3x_4 \geq 1 \\
& x_1 - x_2 + 2x_3 + x_4 \geq 1 \\
& x_1, x_2, x_3, x_4 \geq 0.
\end{aligned}$$

(a) Solve the model by solving its dual model (using the graphical solution method).

(b) Solve the model by using the complementary slackness relations.

(c) Determine optimal feasible basic solutions of the model and its dual model.

Exercise 4.8.4. Determine, for each value of μ ($\mu \in \mathbb{R}$), the optimal solution of the following model:

$$\begin{aligned}
\max \quad & \mu x_1 - x_2 \\
\text{s.t.} \quad & x_1 + 2x_2 \leq 4 \\
& 6x_1 + 2x_2 \leq 9 \\
& x_1, x_2 \geq 0.
\end{aligned}$$

Exercise 4.8.5. Solve Model Salmonnose as formulated in Section 4.1 (Hint: use either the big-M or the two-phase procedure; see Section 3.6.)

Exercise 4.8.6. Determine the dual of

(a) Model Dovetail* in Section 1.2.1,

(b) Model (1.19) in Section 1.5.3,

(c) the LO-models of Exercise 1.8.3, and

(d) the LO-models (I)–(VI) in Section 1.3.

Exercise 4.8.7. Determine the dual of model (1.20) in Section 1.6.3. (Hint: first write out the model for some small values of N and M, dualize the resulting models, and then generalize.)

Exercise 4.8.8. Solve the dual of the LO-model formulated in Example 3.6.1. Determine the basis matrices corresponding to the optimal primal and dual solutions, and show that they are complementary dual (in the sense of Section 2.2.3).

Exercise 4.8.9.   Show that the dual of model (DGM) is the original model (GM).

Exercise 4.8.10.

(a) Prove the following generalization of Theorem 4.2.3. If $\mathbf{x}$ and $\mathbf{y}$ are feasible solutions of the LO-models (GM) and (DGM), respectively, and $\mathbf{c}^\mathsf{T}\mathbf{x} = \mathbf{b}^\mathsf{T}\mathbf{y}$, then $\mathbf{x}$ and $\mathbf{y}$ are optimal.

(b) Prove the following generalization of Theorem 4.2.4. If the LO-model (GM) has an optimal solution, then the dual model (DGM) has an optimal solution as well, and vice versa; moreover, the optimal objective values of the primal and the dual model are equal.

(c) Prove the optimality criterion for nonstandard LO-models as formulated at the end of Section 4.2.3.

Exercise 4.8.11.   Prove Theorem 4.3.2.

Exercise 4.8.12.   Consider the following LO-model.

$$\begin{array}{rl} \max & 2x_1 - x_2 \\ \text{s.t.} & x_1 - x_2 \leq 1 \\ & x_1 - x_2 \geq 2 \\ & x_1, x_2 \geq 0. \end{array}$$

Show that both this model and the dual of this model are infeasible.

Exercise 4.8.13.   Consider the LO-model:

$$\begin{array}{rll} \max & x_1 + x_2 & \text{(P)} \\ \text{s.t.} & x_1 + x_2 \leq 6 \\ & x_1 \qquad \leq 3 \\ & \qquad x_2 \leq 3 \\ & x_1, x_2 \geq 0. \end{array}$$

(a) Use the graphical method to determine the unique optimal vertex of the model, and all corresponding optimal feasible basic solutions.

(b) Determine the dual of model (P), and determine the optimal feasible basic solutions of this dual model.

(c) Verify that the complementary slackness relations hold for each pair of primal and optimal feasible basic solutions found in (a) and (b). Verify also that the strong complementary slackness relations do not hold.

(d) Construct a pair of primal and dual optimal solutions that satisfy the strong complementary slackness relations.

(a) max $-9x_1 - 2x_2 + 12x_3$
　　s.t. 　　　$2x_1 - 2x_2 + 2x_3 \leq 1$
　　　　　　 $-3x_1 + x_2 + x_3 \leq 1$
　　　　　　 $x_1, x_2, x_3 \geq 0.$

(b) max $4x_1 - 3x_2 + 2x_3 - 6x_4$
　　s.t. 　$2x_1 + 4x_2 - x_3 - 2x_4 \leq 1$
　　　　　 $x_1 - x_2 + 2x_3 - x_4 \leq -1$
　　　　　 $x_1, x_2, x_3, x_4 \geq 0.$

Exercise 4.8.3. Consider the following LO-model:

$$\begin{aligned}
\max \quad & 2x_1 + 4x_2 + 10x_3 + 15x_4 \\
\text{s.t.} \quad & -x_1 + x_2 + x_3 + 3x_4 \geq 1 \\
& x_1 - x_2 + 2x_3 + x_4 \geq 1 \\
& x_1, x_2, x_3, x_4 \geq 0.
\end{aligned}$$

(a) Solve the model by solving its dual model (using the graphical solution method).

(b) Solve the model by using the complementary slackness relations.

(c) Determine optimal feasible basic solutions of the model and its dual model.

Exercise 4.8.4. Determine, for each value of μ ($\mu \in \mathbb{R}$), the optimal solution of the following model:

$$\begin{aligned}
\max \quad & \mu x_1 - x_2 \\
\text{s.t.} \quad & x_1 + 2x_2 \leq 4 \\
& 6x_1 + 2x_2 \leq 9 \\
& x_1, x_2 \geq 0.
\end{aligned}$$

Exercise 4.8.5. Solve Model Salmonnose as formulated in Section 4.1 (Hint: use either the big-M or the two-phase procedure; see Section 3.6.)

Exercise 4.8.6. Determine the dual of

(a) Model Dovetail* in Section 1.2.1,

(b) Model (1.19) in Section 1.5.3,

(c) the LO-models of Exercise 1.8.3, and

(d) the LO-models (I)–(VI) in Section 1.3.

Exercise 4.8.7. Determine the dual of model (1.20) in Section 1.6.3. (Hint: first write out the model for some small values of N and M, dualize the resulting models, and then generalize.)

Exercise 4.8.8. Solve the dual of the LO-model formulated in Example 3.6.1. Determine the basis matrices corresponding to the optimal primal and dual solutions, and show that they are complementary dual (in the sense of Section 2.2.3).

Exercise 4.8.9.   Show that the dual of model (DGM) is the original model (GM).

Exercise 4.8.10.

(a)   Prove the following generalization of Theorem 4.2.3. If $\mathbf{x}$ and $\mathbf{y}$ are feasible solutions of the LO-models (GM) and (DGM), respectively, and $\mathbf{c}^\mathsf{T}\mathbf{x} = \mathbf{b}^\mathsf{T}\mathbf{y}$, then $\mathbf{x}$ and $\mathbf{y}$ are optimal.

(b)   Prove the following generalization of Theorem 4.2.4. If the LO-model (GM) has an optimal solution, then the dual model (DGM) has an optimal solution as well, and vice versa; moreover, the optimal objective values of the primal and the dual model are equal.

(c)   Prove the optimality criterion for nonstandard LO-models as formulated at the end of Section 4.2.3.

Exercise 4.8.11.   Prove Theorem 4.3.2.

Exercise 4.8.12.   Consider the following LO-model.

$$\begin{aligned}
\max\ & 2x_1 - x_2 \\
\text{s.t.}\quad & x_1 - x_2 \le 1 \\
& x_1 - x_2 \ge 2 \\
& x_1, x_2 \ge 0.
\end{aligned}$$

Show that both this model and the dual of this model are infeasible.

Exercise 4.8.13.   Consider the LO-model:

$$\begin{aligned}
\max\ & x_1 + x_2 && \text{(P)} \\
\text{s.t.}\quad & x_1 + x_2 \le 6 \\
& x_1 \qquad\ \le 3 \\
& \qquad x_2 \le 3 \\
& x_1, x_2 \ge 0.
\end{aligned}$$

(a)   Use the graphical method to determine the unique optimal vertex of the model, and all corresponding optimal feasible basic solutions.

(b)   Determine the dual of model (P), and determine the optimal feasible basic solutions of this dual model.

(c)   Verify that the complementary slackness relations hold for each pair of primal and optimal feasible basic solutions found in (a) and (b). Verify also that the strong complementary slackness relations do not hold.

(d)   Construct a pair of primal and dual optimal solutions that satisfy the strong complementary slackness relations.

Exercise 4.8.14. Use Farkas' lemma (see Section 4.4) to determine a certificate of un-solvability for the following systems of inequalities:

(a)
$$x_1 + 2x_2 + 5x_3 - 4x_4 \leq 0,$$
$$x_1 + 4x_2 - 3x_3 + 6x_4 \leq 0,$$
$$2x_1 + 6x_2 + 2x_3 + 2x_4 > 0.$$

(b)
$$3y_1 + 2y_2 - y_3 = 1,$$
$$-5y_1 + 2y_2 + y_3 = -1,$$
$$y_1 - 4y_2 - 3y_3 = 1,$$
$$y_1, y_2, y_3 \geq 0.$$

Exercise 4.8.15. Prove the following variants of Farkas' lemma (see Section 4.4):

(a) For any (m, n)-matrix $\mathbf{A}$ any m-vector $\mathbf{b}$, exactly one of the following systems has a solution: (I) $\mathbf{A}\mathbf{x} = \mathbf{b}, \mathbf{x} \geq \mathbf{0}$, and (II) $\mathbf{A}^\mathsf{T}\mathbf{y} = \mathbf{0}, \mathbf{b}^\mathsf{T}\mathbf{y} < 0$.

(b) For any (m, n)-matrix $\mathbf{A}$, exactly one of the following systems has a solution: (I) $\mathbf{A}\mathbf{x} < \mathbf{0}$, and (II) $\mathbf{A}^\mathsf{T}\mathbf{y} = \mathbf{0}, \mathbf{y} \geq \mathbf{0}, \mathbf{y} \neq \mathbf{0}$. (Hint: consider a primal model with the constraints $\mathbf{A}\mathbf{x} \leq -\varepsilon\mathbf{1}$.)

Exercise 4.8.16. Consider the following LO-model.

$$\max \quad 2x_1 - x_2$$
$$\text{s.t.} \quad x_1 + x_2 \leq 4$$
$$x_1 - x_2 \leq 2$$
$$-x_1 + x_2 \leq 2$$
$$x_1, x_2 \geq 0.$$

(a) Use the (primal) simplex algorithm to determine an optimal simplex tableau for the model.

(b) Use the reoptimization procedure described in Section 4.7.2 to determine an optimal simplex tableau for the model with the additional constraint $x_1 \leq 1$.

(c) Check your answers in (a) and (b) by using the graphical solution method.

CHAPTER 5

Sensitivity analysis

Overview

In most practical situations, the exact values of the model parameters are not known with absolute certainty. For example, model parameters are often obtained by statistical estimation procedures. The analysis of the effects of parameter changes on a given optimal solution of the model is called *sensitivity analysis* or *post-optimality analysis*. Changing the value of a parameter is called a *perturbation* of that parameter. In this chapter we will look at perturbations of several parameters, including objective function coefficients, right hand side coefficient of constraints (including the zeroes of the nonnegativity constraints), and the entries of the technology matrix. In economic terms, while the primal model gives solutions in terms of, e.g., the amount of profit obtainable from the production activity, the dual model gives information on the economic value of the limited resources and capacities. So-called shadow prices are introduced in the case of nondegenerate optimal solutions; in the case of degeneracy, right and left shadow prices are defined.

5.1 Sensitivity of model parameters

The values of the input data used for Model Dovetail and Model Salmonnose (see Section 4.1) are not known with certainty. The quantity of wood that is available for the production of matches in a certain year, might, for instance, turn out to be less than $18,000 \text{ m}^3$, because part of the wood may decompose before production. It is important for the manager of the company to know how sensitive the optimal solution is to changes (*perturbations*) of the amount of available wood, as well as the values of other parameters of the model.

The parameters of the LO-model $\max\{\mathbf{c}^\mathsf{T}\mathbf{x} \mid \mathbf{A}\mathbf{x} \leq \mathbf{b}, \mathbf{x} \geq \mathbf{0}\}$ are the right hand side entries of the constraints (i.e., the entries of the vector $\mathbf{b}$ and the 0 in the nonnegativity constraints $x_i \geq 0$), the objective coefficients (i.e., the entries of the vector $\mathbf{c}$), and the entries of the technology matrix (i.e., the entries of $\mathbf{A}$). Let $\delta \in \mathbb{R}$. By a *perturbation by a factor* δ of a model parameter p, we mean the replacement of p by $p + \delta$ in the model. (So, setting $\delta = 0$ gives the original model.) We will successively discuss perturbations of:

  the entries of the objective coefficients vector (in Section 5.2);

  the entries of the right hand side values of the constraints, include nonnegativity and nonpositivity constraints (in Section 5.3 and Section 5.3.3);

  the entries of the technology matrix (in Section 5.5).

A perturbation by a factor δ of a model parameter will in general influence the optimal solution and the optimal objective value, i.e., another basis matrix may become optimal. However, there is an interval of values of δ for which the optimal basis matrix at hand is optimal. For a given optimal basis matrix, that interval is called the *tolerance interval* (for that basis matrix). The *perturbation function* of a model parameter is the function $z^*(\delta)$, where δ is the perturbation factor, and $z^*(\delta)$ is the optimal objective value of the perturbed model. In the next sections we will determine tolerance intervals and draw perturbation functions.

The general strategy of the analyses in this chapter is as follows. We will always start with a given optimal feasible basic solution (recall that an LO-model may have multiple optimal feasible basic solutions). In Theorem 3.5.1, we established the optimality and feasibility criteria for a feasible basic solution. We found that the basic solution corresponding to the basis matrix $\mathbf{B}$ is optimal if and only if:

$$\mathbf{B}^{-1}\mathbf{b} \geq \mathbf{0}, \qquad\qquad\qquad \text{(feasibility)} \qquad\qquad (5.1)$$
$$\text{and } \mathbf{c}_{NI}^\mathsf{T} - \mathbf{c}_{BI}^\mathsf{T}\mathbf{B}^{-1}\mathbf{N} \leq \mathbf{0}. \qquad \text{(optimality)} \qquad\qquad (5.2)$$

The first condition ensures that the feasible basic solution corresponding to the basis matrix $\mathbf{B}$ is feasible, and the second condition ensures that it is optimal (or, equivalently, that the corresponding dual solution is dual feasible; see Section 4.5). Recall that these inequalities are obtained by writing the model from the perspective of the optimal basic solution (see Section 3.2), and asserting that the right hand side values are nonnegative (the inequalities (5.1)), and the objective coefficients are nonpositive (the inequalities (5.2)).

When the value of some parameter is varied, certain terms in these two expressions change. For example, when perturbing the objective coefficient of a variable that is basic according to the optimal feasible basic solution, the vector $\mathbf{c}_{BI}$ will change. If, however, the right hand side of a technology constraint is varied, then $\mathbf{b}$ will change. For each such a situation, we will analyze how exactly the above optimality and feasibility expressions are affected. This will allow us to determine for which values of the perturbation factor δ the feasible basic solution remains optimal, that is, we will be able to determine the tolerance interval. For values of δ for which the feasible basic solution is no longer optimal, a different feasible basic solution may become optimal, or the problem might become infeasible or unbounded.

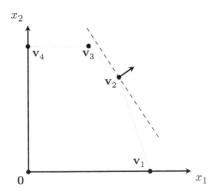

Figure 5.1: Level line $3x_1 + 2x_2 = 22\frac{1}{2}$.

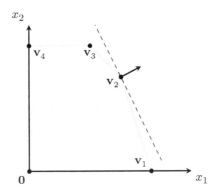

Figure 5.2: Level line $4x_1 + 2x_2 = 27$.

5.2 Perturbing objective coefficients

In this section we discuss the effects of perturbations of the objective coefficients. We distinguish between the coefficients of variables that are basic with respect to a given optimal basis (Section 5.2.1), and variables that are nonbasic with respect to that optimal basis (Section 5.2.2).

5.2.1 Perturbing the objective coefficient of a basic variable

We use our prototype example Model Dovetail for the analysis; see Section 1.1.1. As we have seen in Section 4.1.1, the decision variable x_1 is in the set of basic variables with respect to the optimal feasible basic solution; the corresponding (original) objective coefficient c_1 is 3. We now perturb the objective coefficient 3 by a factor δ. Let $z^*(\delta)$ be the optimal objective value of the perturbed model:

$$
\begin{aligned}
z^*(\delta) = \max \quad & (3+\delta)x_1 + 2x_2 \\
\text{s.t.} \quad & x_1 + \ x_2 \leq \ 9 \\
& 3x_1 + \ x_2 \leq 18 \\
& x_1 \qquad\ \ \leq \ 7 \\
& \qquad\ \ x_2 \leq \ 6 \\
& x_1, x_2 \geq 0.
\end{aligned}
$$

If $\delta = 0$, then the objective function is the same as in the original model (see Figure 5.1). If $\delta = 1$, then the objective function is 'rotated' relative to the original one (see Figure 5.2). In both cases, vertex $\mathbf{v}_2 \ (= \begin{bmatrix} 4\frac{1}{2} & 4\frac{1}{2} \end{bmatrix}^{\mathsf{T}})$ is the optimal solution. However, the optimal objective value is different: for $\delta = 0$, the optimal value is $22\frac{1}{2}$, while for $\delta = 1$, it is $(4 \times 4\frac{1}{2} + 2 \times 4\frac{1}{2} =) 27$. In Figure 5.3, the level lines are drawn for various values of δ. We see that with increasing values of δ, the level line is 'rotated clockwise' around the feasible region, and the optimal solution 'jumps' from vertex $\mathbf{v}_3$ to $\mathbf{v}_2$ to $\mathbf{v}_1$. Also, we see that, as value of δ is increased, starting from -2, vertex $\mathbf{v}_3$ is optimal until $\delta = -1$. At

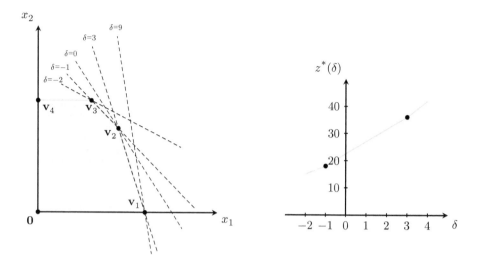

Figure 5.3: Canting level lines.   **Figure 5.4:** Perturbation function for an objective coefficient.

that point, both $\mathbf{v}_3$ and $\mathbf{v}_2$ are optimal (and, in fact, all points on the line segment $\mathbf{v}_2\mathbf{v}_3$ are optimal). Then, as we increase the value of δ slightly beyond -1, $\mathbf{v}_2$ becomes the unique optimal vertex. Vertex $\mathbf{v}_2$ remains optimal up to and including $\delta = 3$, at which point $\mathbf{v}_1$ also becomes optimal (and so do all points on the line segment $\mathbf{v}_1\mathbf{v}_2$). As we increase the value of δ past 3, $\mathbf{v}_1$ remains the unique optimal solution.

In Figure 5.4, the perturbation function $z^*(\delta) = (3 + \delta)x_1^* + 2x_2^*$ is depicted. (Notice that, in the expression for $z^*(\delta)$, the values of x_1^* and x_2^* are the optimal values and, hence, they depend on δ.) We can see that the slope of the graph changes at $\delta = -1$ and at $\delta = 3$. These are the points at which the optimal vertex changes from $\mathbf{v}_3$ to $\mathbf{v}_2$ and from $\mathbf{v}_2$ to $\mathbf{v}_1$, respectively.

Table 5.1 lists the optimal solutions for $\delta = -2, -1, 0, 1, 2, 3, 4$ for Model Dovetail. The second column lists the vertices in which the optimal objective value is attained. The third through the sixth columns contain the objective coefficients and the corresponding optimal values of the decision variables. The seventh and the eighth columns list the changes Δx_1^* and Δx_2^* in x_1^* and x_2^*, respectively. The last column contains the optimal objective value $\mathbf{c}^\mathsf{T}\mathbf{x}^* = c_1 x_1^* + c_2 x_2^*$.

We will now study Table 5.1, Figure 5.3, and Figure 5.4 more carefully. To that end we start at $\delta = 0$, the original situation. At this value of δ, $\mathbf{v}_2$ is the optimal vertex. If δ is relatively small, then $\mathbf{v}_2$ remains the optimal vertex, and the objective value changes by $4\frac{1}{2}\delta$. From Table 5.1 we can conclude that if δ is between -1 and 3, then $\mathbf{v}_2$ is an optimal vertex and the change of the objective value is $4\frac{1}{2}\delta$. If, for instance, the profit (c_1) per box of long matches is not 3 but $2\frac{1}{2}$ (i.e., $\delta = -\frac{1}{2}$), then $\mathbf{v}_2$ is still an optimal vertex, and the objective value changes by $\left(-\frac{1}{2} \times 4\frac{1}{2} =\right) -2\frac{1}{4}$.

δ	optimal vertex	c_1	x_1^*	c_2	x_2^*	Δx_1^*	Δx_2^*	$\mathbf{c}^{\mathsf{T}}\mathbf{x}^*$
-2	$\mathbf{v}_3$	1	3	2	6	$-1\frac{1}{2}$	$1\frac{1}{2}$	15
-1	$\mathbf{v}_3$	2	3	2	6	$-1\frac{1}{2}$	$1\frac{1}{2}$	18
	$\mathbf{v}_2$	2	$4\frac{1}{2}$	2	$4\frac{1}{2}$	0	0	18
0	$\mathbf{v}_2$	3	$4\frac{1}{2}$	2	$4\frac{1}{2}$	0	0	$22\frac{1}{2}$
1	$\mathbf{v}_2$	4	$4\frac{1}{2}$	2	$4\frac{1}{2}$	0	0	27
2	$\mathbf{v}_2$	5	$4\frac{1}{2}$	2	$4\frac{1}{2}$	0	0	$31\frac{1}{2}$
3	$\mathbf{v}_2$	6	$4\frac{1}{2}$	2	$4\frac{1}{2}$	0	0	36
	$\mathbf{v}_1$	6	6	2	0	$1\frac{1}{2}$	$-4\frac{1}{2}$	36
4	$\mathbf{v}_1$	7	6	2	0	$1\frac{1}{2}$	$-4\frac{1}{2}$	42

Table 5.1: Perturbations $\delta = -2, -1, 0, 1, 2, 3, 4$.

Now consider what happens when the value of δ is decreased below -1. If δ decreases from -1 to -2, then the optimal vertex 'jumps' from $\mathbf{v}_2$ to $\mathbf{v}_3$. Similar behavior can be observed when increasing the value of δ to a value greater than 3: if δ increases from 3 to 4, the optimal vertex 'jumps' from $\mathbf{v}_2$ to $\mathbf{v}_1$. It appears that for $-1 \leq \delta \leq 3$, the vertex $\mathbf{v}_2$ is optimal. Thus $-1 \leq \delta \leq 3$ is the tolerance interval for (the feasible basic solution corresponding to) the optimal vertex $\mathbf{v}_2$. For $-2 \leq \delta \leq -1$, $\mathbf{v}_3$ is optimal, and for $3 \leq \delta \leq 4$, $\mathbf{v}_1$ is optimal. For $\delta = -1$ and $\delta = 3$ we have multiple optimal solutions. It is left to the reader to complete the graph of Figure 5.4 for values of δ with $\delta < -2$ and values with $\delta > 4$. Note that the graph of this function consists of a number of line segments connected by kink points. In general, a function is called *piecewise linear* if its graph consists of a number of connected line segments; the points where the slope of the function changes are called *kink points*. In Theorem 5.4.1, it is shown that perturbation functions of objective coefficients are piecewise linear.

We will now show how the tolerance intervals may be calculated without graphing the perturbation function or making a table such as Table 5.1. Model Dovetail, with slack variables and the perturbed objective coefficient c_1, reads as follows (see Section 3.2):

$$
\begin{array}{lrcl}
\max & (3+\delta)x_1 + 2x_2 & & \\
\text{s.t.} & x_1 + x_2 + x_3 & = & 9 \\
& 3x_1 + x_2 + x_4 & = & 18 \\
& x_1 + x_5 & = & 7 \\
& x_2 + x_6 & = & 6 \\
& x_1, x_2, x_3, x_4, x_5, x_6 \geq 0. &
\end{array}
$$

At vertex $\mathbf{v}_2$, the variables x_1, x_2, x_5, and x_6 are basic. The constraints can be rewritten as:

$$\begin{aligned}
1\tfrac{1}{2}x_3 - \tfrac{1}{2}x_4 + x_2 &&&= 4\tfrac{1}{2} \\
-\tfrac{1}{2}x_3 + \tfrac{1}{2}x_4 && + x_1 &= 4\tfrac{1}{2} \\
\tfrac{1}{2}x_3 - \tfrac{1}{2}x_4 &&&+ x_5 &= 2\tfrac{1}{2} \\
-1\tfrac{1}{2}x_3 + \tfrac{1}{2}x_4 &&&&+ x_6 &= 1\tfrac{1}{2}.
\end{aligned}$$

The objective function with perturbation factor δ for the coefficient of x_1 is $(3+\delta)x_1+2x_2$. Using $x_1 = 4\tfrac{1}{2} + \tfrac{1}{2}x_3 - \tfrac{1}{2}x_4$ and $x_2 = 4\tfrac{1}{2} - 1\tfrac{1}{2}x_3 + \tfrac{1}{2}x_4$, the objective function can be expressed in terms of the nonbasic variables x_3 and x_4 as:

$$\begin{aligned}
(3+\delta)x_1 + 2x_2 &= (3+\delta)\big(4\tfrac{1}{2} + \tfrac{1}{2}x_3 - \tfrac{1}{2}x_4\big) + 2\big(4\tfrac{1}{2} - 1\tfrac{1}{2}x_3 + \tfrac{1}{2}x_4\big) \\
&= 22\tfrac{1}{2} + 4\tfrac{1}{2}\delta + \big(\tfrac{1}{2}\delta - 1\tfrac{1}{2}\big)x_3 + \big(-\tfrac{1}{2} - \tfrac{1}{2}\delta\big)x_4.
\end{aligned}$$

Now if $\tfrac{1}{2}\delta - 1\tfrac{1}{2} \leq 0$ and $-\tfrac{1}{2} - \tfrac{1}{2}\delta \leq 0$, then the objective coefficients of both x_3 and x_4 are nonpositive, and, therefore, the feasible basic solution with basic variables x_1, x_2, x_5, and x_6 is still optimal. It is easy to check that the conditions $\tfrac{1}{2}\delta - 1\tfrac{1}{2} \leq 0$ and $-\tfrac{1}{2} - \tfrac{1}{2}\delta \leq 0$ are equivalent to $-1 \leq \delta \leq 3$. So, the tolerance interval is $-1 \leq \delta \leq 3$, and for these values of δ, it holds that:

$$\max\big\{22\tfrac{1}{2} + 4\tfrac{1}{2}\delta + \big(\tfrac{1}{2}\delta - 1\tfrac{1}{2}\big)x_3 + \big(\tfrac{1}{2} - \tfrac{1}{2}\delta\big)x_4\big\} = 22\tfrac{1}{2} + 4\tfrac{1}{2}\delta.$$

In other words, the optimal objective value is attained at $\mathbf{v}_2$ if the profit per box of long matches is between 2 and 6 (with 2 and 6 included).

In general, we have the following situation. Let $\mathbf{B}$ be a basis matrix in $\begin{bmatrix}\mathbf{A} & \mathbf{I}_m\end{bmatrix}$ corresponding to an optimal feasible basic solution. We perturb the entry of $\mathbf{c}_{BI}$ corresponding to the basic variable x_k ($k \in BI$) by δ. Let β be such that $k = BI_\beta$. This means that $\mathbf{c}_{BI}$ is changed into $\mathbf{c}_{BI} + \delta\mathbf{e}_\beta$, where $\mathbf{e}_\beta \in \mathbb{R}^m$ is the β'th unit vector. The feasibility condition (5.1) and the optimality condition (5.2) become:

$$\mathbf{B}^{-1}\mathbf{b} \geq \mathbf{0}, \qquad\qquad \text{(feasibility)}$$

$$\text{and } \mathbf{c}_{NI}^{\mathsf{T}} - \big(\mathbf{c}_{BI} + \delta\mathbf{e}_\beta\big)^{\mathsf{T}}\mathbf{B}^{-1}\mathbf{N} \leq \mathbf{0}. \qquad\qquad \text{(optimality)}$$

Clearly, the first condition is not affected by the changed objective coefficient. This should make sense, because changing the objective function does not change the feasible region. Hence, since the vertex corresponding to $\mathbf{B}$ was feasible before the perturbation, it is still feasible after the perturbation. So, the feasible basic solution corresponding to $\mathbf{B}$ remains optimal as long as the second condition is satisfied. This second condition consists of the n inequalities:

$$\big(\mathbf{c}_{NI}^{\mathsf{T}} - \big(\mathbf{c}_{BI} + \delta\mathbf{e}_\beta\big)^{\mathsf{T}}\mathbf{B}^{-1}\mathbf{N}\big)_j \leq 0 \qquad \text{for } j = 1, \ldots, n. \qquad (5.3)$$

The tolerance interval is the solution set of these n inequalities. In this interval, the feasible basic solution $\begin{bmatrix}\mathbf{x}_{BI}^* \\ \mathbf{x}_{NI}^*\end{bmatrix}$ with $\mathbf{x}_{BI}^* = \mathbf{B}^{-1}\mathbf{b}$ and $\mathbf{x}_{NI}^* = \mathbf{0}$ corresponding to $\mathbf{B}$ remains optimal and, hence, the resulting change Δz in the value of z satisfies (recall that $\mathbf{x}_{NI} = \mathbf{0}$):

$$\Delta z = (\mathbf{c}_{BI} + \delta \mathbf{e}_\beta)^\mathsf{T} \mathbf{B}^{-1} \mathbf{b} - \mathbf{c}_{BI}^\mathsf{T} \mathbf{B}^{-1} \mathbf{b} = \delta \mathbf{e}_\beta^\mathsf{T} \mathbf{B}^{-1} \mathbf{b} = \delta (\mathbf{B}^{-1} \mathbf{b})_\beta.$$

Note that this expression is only valid for values of δ that lie within the tolerance interval. It is left to the reader to derive the tolerance interval $-1 \leq \delta \leq 3$ for the first objective coefficient in Model Dovetail directly from the inequalities in (5.3).

We conclude this section with a remark about the fact that tolerance intervals depend on the choice of the optimal feasible basic solution: different optimal feasible basic solutions may lead to different tolerance intervals. This fact will be illustrated by means of the following example.

Example 5.2.1.   *Consider the LO-model:*

$$
\begin{aligned}
\min \quad & 4x_1 - 4x_2 \\
\text{s.t.} \quad & 2x_1 - x_2 \geq 6 \\
& x_1 - x_2 \geq 2 \\
& x_1 - 2x_2 \geq 0 \\
& x_1, x_2 \geq 0.
\end{aligned}
$$

Let the slack variables of the constraints be x_3, x_4, and x_5, respectively. The objective coefficient of x_1 (which is equal to 4) is perturbed by the factor δ. Using the graphical solution method (see Section 1.1.2), it can easily be shown that there are three optimal feasible basic solutions:

The basic variables are x_1, x_2, x_3. The corresponding optimal simplex tableau yields:

$$
\begin{aligned}
z &= (8 + 4\delta) + (4 + 2\delta)x_4 - \delta x_5 \\
x_1 &= 4 + 2x_4 - x_5 \\
x_2 &= 2 + x_4 - x_5 \\
x_3 &= 0 + 3x_4 - x_5.
\end{aligned}
$$

This solution is optimal as long as $-\delta \geq 0$ and $4 + 2\delta \geq 0$, i.e., $-2 \leq \delta \leq 0$. The perturbation function is $z^*(\delta) = 8 + 4\delta$, for $-2 \leq \delta \leq 0$.

The basic variables are x_1, x_2, x_4. The corresponding optimal simplex tableau yields:

$$
\begin{aligned}
z &= (8 + 4\delta) + \left(\tfrac{4}{3} + \tfrac{2}{3}\delta\right)x_3 + \left(\tfrac{4}{3} - \tfrac{1}{3}\delta\right)x_5 \\
x_1 &= 4 + \tfrac{2}{3}x_3 - \tfrac{1}{3}x_5 \\
x_2 &= 2 + \tfrac{1}{3}x_3 - \tfrac{2}{3}x_5 \\
x_4 &= 0 + \tfrac{1}{3}x_3 + \tfrac{1}{3}x_5.
\end{aligned}
$$

This solution is optimal as long as $\tfrac{4}{3} + \tfrac{2}{3}\delta \geq 0$ and $\tfrac{4}{3} - \tfrac{1}{3}\delta \geq 0$, i.e., $-2 \leq \delta \leq 4$, while $z^*(\delta) = 8 + 4\delta$, for $-2 \leq \delta \leq 4$.

The basic variables are x_1, x_2, x_5. The corresponding optimal simplex tableau yields:

$$
\begin{aligned}
z &= (8+4\delta) + \delta x_3 + (4-\delta)x_4 \\
x_1 &= 4 + x_3 - x_4 \\
x_2 &= 2 + x_3 - 2x_4 \\
x_5 &= 0 - x_3 + 3x_4.
\end{aligned}
$$

This solution is optimal as long as $\delta \geq 0$ and $4 - \delta \geq 0$, i.e., $0 \leq \delta \leq 4$, while $z^*(\delta) = 8 + 4\delta$, for $0 \leq \delta \leq 4$.

The conclusion is that the three optimal feasible basic solutions give rise to different tolerance intervals with respect to the objective coefficient of x_1. It should be clear, however, that the perturbation function is the same in all three cases.

In Exercise 5.8.25, the reader is asked to draw the perturbation function for the above example. Exercise 5.8.26 deals with a situation where a tolerance interval consists of one point.

5.2.2 Perturbing the objective coefficient of a nonbasic variable

Consider again Model Dovetail, but with a different objective function:

$$
\begin{aligned}
\max \quad & 6x_1 + x_2 \\
\text{s.t.} \quad & x_1 + x_2 \leq 9 \\
& 3x_1 + x_2 \leq 18 \\
& x_1 \leq 7 \\
& x_2 \leq 6 \\
& x_1, x_2 \geq 0.
\end{aligned}
$$

(The original objective function cannot be used, since both decision variables are basic at the optimal solution.) The optimal vertex is $\mathbf{v}_1$ with $x_1^* = 6$ and $x_2^* = 0$. At vertex $\mathbf{v}_1$, the basic variables are x_1, x_3, x_5, and x_6. The corresponding optimal objective value is 36. We will consider a perturbation of the objective coefficient of x_2, namely:

$$
\max \quad 6x_1 + (1+\delta)x_2.
$$

Rewriting the model from the perspective of the optimal feasible basic solution (see Section 3.2), we find the following equivalent model:

$$
\begin{aligned}
\max \quad & (\delta - 1)x_2 - 2x_4 + 36 \\
& \tfrac{2}{3}x_2 - \tfrac{1}{3}x_4 + x_3 && = 3 \\
& -\tfrac{1}{3}x_2 + \tfrac{1}{3}x_4 && + x_1 && = 6 \\
& -\tfrac{1}{3}x_2 - \tfrac{1}{3}x_4 && + x_5 && = 1 \\
& x_2 && + x_6 &&= 6.
\end{aligned}
$$

Thus, if $\delta - 1 \leq 0$ (or, equivalently: $\delta \leq 1$), then vertex $\mathbf{v}_1$ is optimal. The tolerance interval is therefore $(-\infty, 1]$. The objective value does not change as long as $\delta \leq 1$, because we have perturbed an objective coefficient of a nonbasic variable (which has value zero).

In general, this can be seen as follows. We perturb the objective coefficient of the nonbasic variable x_k (with $k \in \{1, \ldots, n\}$) by a factor δ. Let α be such that $k = NI_\alpha$. We again consider a matrix $\mathbf{B}$ corresponding to an optimal feasible basic solution. As in Section 5.2.1, we look at the feasibility condition (5.1) and the optimality condition (5.2), which now become:

$$\mathbf{B}^{-1}\mathbf{b} \geq \mathbf{0}, \quad \text{(feasibility)}$$
$$\text{and } (\mathbf{c}_{NI}^\mathsf{T} + \delta\mathbf{e}_\alpha^\mathsf{T}) - \mathbf{c}_{BI}^\mathsf{T}\mathbf{B}^{-1}\mathbf{N} \leq \mathbf{0}. \quad \text{(optimality)}$$

Again, the first condition is unaffected, as should be expected (see the discussion in Section 5.2.1). The second condition is equivalent to:

$$\delta\mathbf{e}_\alpha^\mathsf{T} \leq -(\mathbf{c}_{NI}^\mathsf{T} - \mathbf{c}_{BI}^\mathsf{T}\mathbf{B}^{-1}\mathbf{N}).$$

This condition consists of a set of n inequalities. Only the inequality with index α actually involves δ. Hence, this condition is equivalent to:

$$\delta \leq -(\mathbf{c}_{NI}^\mathsf{T} - \mathbf{c}_{BI}^\mathsf{T}\mathbf{B}^{-1}\mathbf{N})_\alpha.$$

So, the value of $\mathbf{x}_{NI_\alpha}$ ($= x_k$) may be increased by at most the negative of the current objective coefficient of that variable. Recall that, since x_k is a decision variable (i.e., $k \in \{1, \ldots, n\}$), the current objective coefficient is the negative of the value of the complementary dual variable y_{m+k} (see also Figure 4.1 and Figure 4.2). Hence, the feasible basic solution in fact remains optimal as long as:

$$\delta \leq -(\mathbf{c}_{NI}^\mathsf{T} - \mathbf{c}_{BI}^\mathsf{T}\mathbf{B}^{-1}\mathbf{N})_\alpha = y_{m+k}^*. \tag{5.4}$$

Here, y_{m+k}^* is the optimal value of the slack variable of the dual constraint corresponding to x_k. Since we are changing the objective coefficient of a nonbasic variable, and the value of this nonbasic variable remains zero, there is no change in the optimal objective value.

In the model above, the dual constraint corresponding to x_2 is $y_1 + y_2 + y_4 \geq 1$. The optimal dual solution reads: $y_1^* = y_3^* = y_4^* = 0$, and $y_2^* = 2$. Hence, we have that δ should satisfy:

$$\delta \leq (y_1^* + y_2^* + y_4^*) - 1 = 2 - 1 = 1.$$

Therefore, the tolerance interval of optimal feasible solution with respect to a perturbation of the objective coefficient of x_2 is $(-\infty, 1]$.

5.2.3 Determining tolerance intervals from an optimal simplex tableau

The tolerance interval of (the optimal feasible basic solution with respect to a perturbation of) the objective coefficient c_k of the variable x_k ($k \in \{1, \ldots, n\}$) can be read directly from an optimal simplex tableau. We distinguish two cases:

Suppose that x_k is a basic variable, i.e., $k = BI_\beta$ for some $\beta \in \{1, \ldots, m\}$. Rewrite (5.3) in the following way:

$$\left(\mathbf{c}_{NI}^\mathsf{T} - \mathbf{c}_{BI}^\mathsf{T}\mathbf{B}^{-1}\mathbf{N}\right)_j \leq \left(\delta\mathbf{e}_\beta^\mathsf{T}\mathbf{B}^{-1}\mathbf{N}\right)_j \text{ for } j = 1, \ldots, n.$$

Note that $\left(\delta\mathbf{e}_\beta^\mathsf{T}\mathbf{B}^{-1}\mathbf{N}\right)_j = \delta\left(\mathbf{B}^{-1}\mathbf{N}\right)_{\beta,j}$. The tolerance interval of c_k, with lower bound $\delta_{\min}$ and upper bound $\delta_{\max}$, satisfies:

$$\delta_{\min} = \max\left\{\frac{\left(\mathbf{c}_{NI}^\mathsf{T} - \mathbf{c}_{BI}^\mathsf{T}\mathbf{B}^{-1}\mathbf{N}\right)_j}{\left(\mathbf{B}^{-1}\mathbf{N}\right)_{\beta,j}} \,\middle|\, \left(\mathbf{B}^{-1}\mathbf{N}\right)_{\beta,j} < 0, j = 1, \ldots, n\right\}, \text{ and}$$

$$\delta_{\max} = \min\left\{\frac{\left(\mathbf{c}_{NI}^\mathsf{T} - \mathbf{c}_{BI}^\mathsf{T}\mathbf{B}^{-1}\mathbf{N}\right)_j}{\left(\mathbf{B}^{-1}\mathbf{N}\right)_{\beta,j}} \,\middle|\, \left(\mathbf{B}^{-1}\mathbf{N}\right)_{\beta,j} > 0, j = 1, \ldots, n\right\}.$$

Recall that, by definition, $\max(\emptyset) = -\infty$ and $\min(\emptyset) = +\infty$. The numbers in these expressions can be read directly from the optimal simplex tableau, because the term $\mathbf{c}_{NI}^\mathsf{T} - \mathbf{c}_{BI}^\mathsf{T}\mathbf{B}^{-1}\mathbf{N}$ is the row vector of objective coefficients of the nonbasic variables in the optimal simplex tableau corresponding to $\mathbf{B}$, and the term $\left(\mathbf{B}^{-1}\mathbf{N}\right)_{\beta,j}$ is the entry in row β and the column NI_j.

Suppose that x_k is a nonbasic variable, i.e., $k = NI_\alpha$ for some $\alpha \in \{1, \ldots, n\}$. It follows from (5.4) that the tolerance interval is $(-\infty, y_{m+k}^*]$, with y_{m+k}^* the optimal value of the complementary dual variable of x_k. Following the discussion in Section 4.6, the optimal value of the dual slack variable y_{m+k} can be read directly from the optimal simplex tableau.

Example 5.2.2. *Consider the example of Section 5.2.2. The optimal simplex tableau, corresponding to $BI = \{1, 3, 5, 6\}$ and $NI = \{2, 4\}$, is:*

x_1	x_2	x_3	x_4	x_5	x_6	$-z$	
0	-1	0	-2	0	0	-36	
0	$\frac{2}{3}$	1	$-\frac{1}{3}$	0	0	3	x_3
1	$\frac{1}{3}$	0	$\frac{1}{3}$	0	0	6	x_1
0	$-\frac{1}{3}$	0	$-\frac{1}{3}$	1	0	1	x_5
0	1	0	0	0	1	6	x_6

We first determine the tolerance interval for the objective coefficient c_1 of the basic variable x_1. Note that $k = 1$ and $\beta = 2$. The tolerance interval $[\delta_{\min}, \delta_{\max}]$ satisfies:

$$\delta_{\min} = \max\left\{\frac{-1}{\frac{1}{3}}, \frac{-2}{\frac{1}{3}}\right\} = -3, \quad \delta_{\max} = \min\{\star, \star\} = \infty.$$

Thus, if we perturb the objective coefficient c_1 by a factor δ, then the solution in the optimal simplex tableau above remains optimal as long as $\delta \geq -3$. Next, we determine the tolerance interval for the *objective coefficient c_2 of the nonbasic variable x_2. The tolerance interval is $(-\infty, y_6^*]$. Following the discussion in Section 4.6, the optimal dual solution can be read from the optimal simplex tableau, and it reads: $y_1^* = 0$, $y_2^* = 2$, $y_3^* = y_4^* = y_5^* = 0$, and $y_6^* = 1$. So, the tolerance interval for the objective coefficient c_2 is $(-\infty, 1]$.*

5.3 Perturbing right hand side values (nondegenerate case)

In this section we discuss the effects of perturbations of the right hand side values of constraints. We distinguish between perturbations of technology constraints (Section 5.3.2) and perturbations of nonnegativity constraints (Section 5.3.3).

We will study the rate at which the optimal objective value changes as the right hand side value of a constraint is changed. This rate is called the *shadow price* (sometimes also called: *shadow cost*) of that constraint. To be precise, take any constraint (i.e., a technology constraint, a nonnegativity constraint, or a nonpositivity constraint) of an LO-model. Let $z^*(\delta)$ be the optimal objective value of the model in which the right hand side of the selected constraint is perturbed by a sufficiently small factor δ. Then, the shadow price of that constraint is defined as:

$$\text{shadow price of the constraint} = \lim_{\delta \to 0} \frac{z^*(\delta) - z^*(0)}{\delta}. \tag{5.5}$$

In other words, the shadow price of a constraint is the derivative of its perturbation function. The interpretation of the shadow price of a constraint is: if the right hand side value of the constraint is perturbed by a sufficiently small (positive or negative) amount δ, then the optimal objective value increases by δ times the shadow price. The expression 'sufficiently small' means that δ should lie in the tolerance interval. In the case of a 'maximizing' model and $b \geq 0$, the shadow price of the constraint $\mathbf{a}^{\mathsf{T}}\mathbf{x} \leq b$ can be considered as a marginal value that indicates the rate at which the maximum profit z^* increases when a small extra amount of the resource is available in addition to the present amount b.

Recall that any '$\leq$' constraint can be written as a '$\geq$' constraint by multiplying both sides by -1 and reversing the inequality sign. For example, the constraint $\mathbf{a}^{\mathsf{T}}\mathbf{x} \leq b$ is equivalent to the constraint $-\mathbf{a}^{\mathsf{T}}\mathbf{x} \geq -b$. However, increasing the right hand side value of $\mathbf{a}^{\mathsf{T}}\mathbf{x} \leq b$ is equivalent to decreasing the right hand side value of $-\mathbf{a}^{\mathsf{T}}\mathbf{x} \geq -b$. This means that the shadow price of $-\mathbf{a}^{\mathsf{T}}\mathbf{x} \geq -b$ is the negative of the shadow price of $\mathbf{a}^{\mathsf{T}}\mathbf{x} \leq b$. So, the shadow price of a constraint depends on the form of the constraint. Computer packages for linear optimization are often ambiguous about the signs of shadow prices. In the case of nonnegativity constraints in maximizing models, this is caused by the fact that sometimes the computer code uses '$\mathbf{x} \geq \mathbf{0}$', and sometimes '$-\mathbf{x} \leq \mathbf{0}$'.

In general, the correct sign of the shadow price can be easily determined by reasoning. For example, in a maximizing LO-model, we have that:

If <u>increasing</u> the right hand side value results in enlarging the feasible region, then the optimal objective value <u>increases</u> or remains the same when the right hand side value is increased. Hence, the shadow price should be nonnegative.

If <u>increasing</u> the right hand side value results in shrinking the feasible region, then the optimal objective value <u>decreases</u> or remains the same when the right hand side value is increased. Hence, the shadow price should be nonpositive.

For example, consider a perturbation of a constraint of the form $\mathbf{a}^\mathsf{T}\mathbf{x} \leq b$ in a maximizing LO-model. Increasing the right hand side value of this constraint enlarges the feasible region, and hence increases (more precisely: does not decrease) the optimal objective value. Therefore, the shadow price of a '$\leq$' constraint in a maximizing LO-model is nonnegative. In contrast, the shadow price of a constraint of the form $\mathbf{a}^\mathsf{T}\mathbf{x} \geq b$ (such as a nonnegativity constraint) in a maximizing LO-model has a nonpositive shadow price. Similar conclusions hold for minimizing LO-models. For instance, increasing the right hand side value of the constraint $\mathbf{a}^\mathsf{T}\mathbf{x} \leq b$ enlarges the feasible region, and hence decreases (more precisely: does not increase) the optimal objective value in a minimizing LO-model. Therefore, the shadow price of a '$\leq$' constraint in a minimizing LO-model is nonpositive.

The main results of this section are the following. We will consider a standard maximizing primal LO-model $\max\{\mathbf{c}^\mathsf{T}\mathbf{x} \mid \mathbf{A}\mathbf{x} \leq \mathbf{b}, \mathbf{x} \geq \mathbf{0}\}$, and assume that the model has a non-degenerate optimal solution. In Theorem 5.3.3, it is shown that the shadow price of a technology constraint is the optimal value of the corresponding dual decision variable. In Theorem 5.3.4, it is shown that the shadow price of a nonnegativity constraint is the negative of the optimal value of corresponding dual slack variable. So we have the following 'complementary dual' assertions ($i = 1, \ldots, n$ and $j = 1, \ldots, m$):

the shadow price of technology constraint j is equal to the optimal value dual decision variable y_j;

the shadow price of the nonnegativity constraint $x_i \geq 0$ is equal to the negative of the optimal value dual slack variable y_{m+i}.

We will also see that the shadow price of any nonbinding constraint is zero.

5.3.1   Perturbation of nonbinding constraints

The shadow price of a constraint is defined as the rate of increase of the optimal objective value as the right hand side of that constraint is perturbed by a small factor δ. Suppose that we have a technology constraint that is not binding at an optimal point. It should intuitively be clear that, if the right hand side value of this constraint is perturbed by a sufficiently small factor δ, then the original optimal point is still optimal, and hence we do not expect any change in the optimal objective value. This assertion is formulated in the next theorem.

Theorem 5.3.1. (*Shadow prices of nonbinding technology constraints*)
Consider the feasible LO-model $\max\{\mathbf{c}^\mathsf{T}\mathbf{x} \mid \mathbf{Ax} \leq \mathbf{b}, \mathbf{x} \geq \mathbf{0}\}$. The shadow price of any technology constraint that is nonbinding at an optimal solution is zero.

Proof of Theorem 5.3.1. Suppose that the technology constraint $\mathbf{a}_j^\mathsf{T}\mathbf{x} \leq b_j$ is nonbinding at some optimal solution $\mathbf{x}^*$, meaning that $\mathbf{a}_j^\mathsf{T}\mathbf{x}^* < b_j$. This means that there exists a small $\varepsilon > 0$ such that $\mathbf{a}_j^\mathsf{T}\mathbf{x}^* < b_j - \varepsilon$. Consider the perturbed model:

$$z^*(\delta) = \max\left\{\mathbf{c}^\mathsf{T}\mathbf{x} \mid \mathbf{Ax} \leq \mathbf{b} + \delta\mathbf{e}_j, \mathbf{x} \geq \mathbf{0}\right\}, \qquad (5.6)$$

where $\mathbf{e}_j$ is the j'th unit vector in $\mathbb{R}^m$. We will show that, for every δ such that $-\varepsilon \leq \delta \leq \varepsilon$, model (5.6) has the same objective value as the original model, i.e., we will show that $z^*(\delta) = z^*(0)$.

We first show that $z^*(\delta) \geq z^*(0)$. To see this, note that we have that $\mathbf{x}^* \geq \mathbf{0}$ and

$$\mathbf{Ax}^* \leq \mathbf{b} - \varepsilon\mathbf{e}_j \leq \mathbf{b} + \delta\mathbf{e}_j.$$

Therefore, $\mathbf{x}^*$ is a feasible solution of (5.6). Hence, the optimal objective value of (5.6) must be at least as large as the objective value corresponding to $\mathbf{x}^*$. That is, $z^*(\delta) \geq z^*(0)$.

Next, we show that $z^*(\delta) \leq z^*(0)$. Let $\mathbf{x}^{**}$ be an optimal solution of (5.6) and suppose for a contradiction that $\mathbf{c}^\mathsf{T}\mathbf{x}^{**} > \mathbf{c}^\mathsf{T}\mathbf{x}^*$. Let $\lambda = \min\{1, \frac{\varepsilon}{\delta+\varepsilon}\}$. Note that $\lambda \leq \frac{\varepsilon}{\delta+\varepsilon}$ and $0 < \lambda \leq 1$. Define $\hat{\mathbf{x}} = \lambda\mathbf{x}^{**} + (1-\lambda)\mathbf{x}^*$. Then, we have that:

$$\mathbf{c}^\mathsf{T}\hat{\mathbf{x}} = \mathbf{c}^\mathsf{T}(\lambda\mathbf{x}^{**}) + \mathbf{c}^\mathsf{T}((1-\lambda)\mathbf{x}^*) = \lambda(\mathbf{c}^\mathsf{T}\mathbf{x}^{**}) + (1-\lambda)(\mathbf{c}^\mathsf{T}\mathbf{x}^*) > \mathbf{c}^\mathsf{T}\mathbf{x}^*,$$

$$\hat{\mathbf{x}} = \lambda\mathbf{x}^{**} + (1-\lambda)\mathbf{x}^* \geq \lambda\mathbf{0} + (1-\lambda)\mathbf{0} = \mathbf{0}, \text{ and}$$

$$\mathbf{A}\hat{\mathbf{x}} = \lambda\mathbf{Ax}^{**} + (1-\lambda)\mathbf{Ax}^* \leq \lambda(\mathbf{b} + \delta\mathbf{e}_j) + (1-\lambda)(\mathbf{b} - \varepsilon\mathbf{e}_j)$$

$$= \mathbf{b} + \lambda(\delta+\varepsilon)\mathbf{e}_j - \varepsilon\mathbf{e}_j \leq \mathbf{b} + \frac{\varepsilon}{\delta+\varepsilon}(\delta+\varepsilon)\mathbf{e}_j - \varepsilon\mathbf{e}_j = \mathbf{b}.$$

Therefore, $\hat{\mathbf{x}}$ is a feasible solution of the original LO-model. But $\hat{\mathbf{x}}$ has a higher objective value than $\mathbf{x}^*$, contradicting the optimality of $\mathbf{x}^*$. So, $z^*(\delta) \leq z^*(0)$. It follows that $z^*(\delta) = z^*(0)$, and, therefore, the shadow price of the constraint is zero. $\qquad\square$

A fact similar to Theorem 5.3.1 holds for nonnegativity and nonpositivity constraints:

Theorem 5.3.2. (*Shadow prices of nonbinding nonnegativity and nonpositivity constraints*)
For any feasible LO-model, the shadow price of any nonbinding nonnegativity and any nonpositivity constraint is zero.

Proof of Theorem 5.3.2. See Exercise 5.8.3. $\qquad\square$

5.3.2 Perturbation of technology constraints

The right hand side value of the technology constraint $x_1 + x_2 \leq 9$ in Model Dovetail represents the maximum number of boxes ($\times 100{,}000$) that company Dovetail's machine can

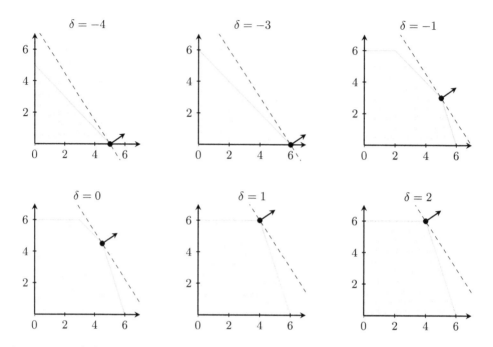

Figure 5.5: Right hand side perturbations. The region marked with solid lines is the perturbed feasible region; the region marked with dotted lines is the unperturbed feasible region.

produce in one year. This number may vary in practice, for instance because the machine is out of order during a certain period, or the capacity is increased because employees work in overtime. Perturbation of the right hand side by a factor δ transforms this constraint into:

$$x_1 + x_2 \leq 9 + \delta.$$

Recall that the objective in Model Dovetail is $\max 3x_1 + 2x_2$. In Figure 5.5, the feasible regions and the corresponding optimal vertices are drawn for $\delta = -4, -3, -1, 0, 1, 2$. The figure shows that, as δ varies, the feasible region changes. In fact, the feasible region becomes larger (to be precise: does not become smaller) as the value of δ is increased. Since the feasible region changes, the optimal vertex also varies as the value of δ is varied.

Table 5.2 lists several optimal solutions for different values of δ. The perturbation function $z^*(\delta)$ of $x_1 + x_2 \leq 9$ is depicted in Figure 5.6. As in Figure 5.4, we again have a piecewise linear function. In Theorem 5.4.2 it is shown that perturbation functions for right hand side parameters are in fact piecewise linear functions.

We are interested in the rate of increase of $z = 3x_1 + 2x_2$ in the neighborhood of the initial situation $\delta = 0$. For $\delta = 0$, this rate of increase is $1\frac{1}{2}$. So, when changing the right hand side by a factor δ, the objective value changes by $1\frac{1}{2}\delta$, i.e., $\Delta z^* = 1\frac{1}{2}\delta$. In Section 4.2, it was calculated that the dual variable y_1 (which corresponds to $x_1 + x_2 \leq 9$) has the optimal value $1\frac{1}{2}$, which is exactly equal to the slope of the function in Figure 5.6 in $\delta = 0$. This fact holds in general and is formulated in Theorem 5.3.3.

δ	$9+\delta$	x_1^*	x_2^*	Δx_1^*	Δx_2^*	$\mathbf{c}^\mathsf{T}\mathbf{x}^*$
-4	5	5	0	$\frac{1}{2}$	$-4\frac{1}{2}$	15
-3	6	6	0	$1\frac{1}{2}$	$-4\frac{1}{2}$	18
-2	7	$5\frac{1}{2}$	$1\frac{1}{2}$	1	-3	$19\frac{1}{2}$
-1	8	5	3	$\frac{1}{2}$	$-1\frac{1}{2}$	21
0	9	$4\frac{1}{2}$	$4\frac{1}{2}$	0	0	$22\frac{1}{2}$
1	10	4	6	$-\frac{1}{2}$	$1\frac{1}{2}$	24
2	11	4	6	$-\frac{1}{2}$	$1\frac{1}{2}$	24

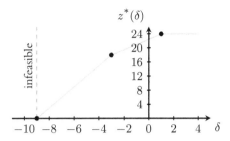

Table 5.2: Perturbations $\delta = -4, -3, -2, -1, 0, 1, 2$.

Figure 5.6: Perturbation function for a technology constraint.

Theorem 5.3.3 expresses the fact that the shadow price of a technology constraint equals the optimal value of the corresponding complementary dual decision variable.

Theorem 5.3.3. (*Shadow prices and dual solutions*)
For any feasible standard LO-model with a nondegenerate optimal solution, the shadow price of any technology constraint exists and is equal to the optimal value of the corresponding dual decision variable.

Proof of Theorem 5.3.3. Consider the feasible standard LO-model with slack variables:

$$\max\left\{\begin{bmatrix}\mathbf{c}^\mathsf{T} & \mathbf{0}^\mathsf{T}\end{bmatrix}\begin{bmatrix}\mathbf{x}\\ \mathbf{x}_s\end{bmatrix} \,\middle|\, \begin{bmatrix}\mathbf{A} & \mathbf{I}_m\end{bmatrix}\begin{bmatrix}\mathbf{x}\\ \mathbf{x}_s\end{bmatrix} = \mathbf{b},\ \mathbf{x}, \mathbf{x}_s \geq \mathbf{0}\right\},$$

with $\mathbf{A} \in \mathbb{R}^{m\times n}$, $\mathbf{I}_m$ the (m,m) identity matrix, $\mathbf{x} \in \mathbb{R}^n$, $\mathbf{x}_s \in \mathbb{R}^m$, and $\mathbf{b} \in \mathbb{R}^m$. Let $\mathbf{B}$ be a basis matrix, corresponding to the optimal solution $\mathbf{x}_{BI}^* = \mathbf{B}^{-1}\mathbf{b} \geq \mathbf{0}$, $\mathbf{x}_{NI}^* = \mathbf{0}$. It follows from Theorem 4.2.4 that $\mathbf{y}^* = (\mathbf{B}^{-1})^\mathsf{T}\mathbf{c}_{BI}$, with $\mathbf{c}_{BI}$ the subvector of $\begin{bmatrix}\mathbf{c}\\\mathbf{0}\end{bmatrix}$ corresponding to $\mathbf{B}$, is an optimal solution of the dual model $\min\left\{\mathbf{b}^\mathsf{T}\mathbf{y} \,\middle|\, \mathbf{A}^\mathsf{T}\mathbf{y} \geq \mathbf{c}, \mathbf{y} \geq \mathbf{0}\right\}$.

Now consider changing the right hand side value of the j'th technology constraint ($j \in \{1,\ldots,m\}$). This means replacing $\mathbf{b}$ by $\mathbf{b} + \delta\mathbf{e}_j$. Following (5.1) and (5.2), the feasible basic solution corresponding to the basis matrix $\mathbf{B}$ remains optimal as long as:

$$\mathbf{B}^{-1}(\mathbf{b} + \delta\mathbf{e}_j) \geq \mathbf{0}.$$

Note that the optimality condition (5.2) is unaffected by the change of $\mathbf{b}$, and therefore does not need to be taken into account. Because of the nondegeneracy, we have that all entries of $\mathbf{B}^{-1}\mathbf{b}$ are strictly positive, and hence $\mathbf{B}$ remains an optimal basis matrix as long as $\delta > 0$ is small enough. (In the case of degeneracy, small changes in the vector $\mathbf{b}$ may give rise to a different optimal basis matrix; see Section 5.6.2).

The perturbation $\delta\mathbf{e}_j$ of $\mathbf{b}$ will influence the optimal solution. Because $\mathbf{B}$ remains an optimal basis matrix, and therefore the same set of technology constraints and nonnegativity constraints (corresponding to NI) remain binding at the optimal vertex, we have that $\mathbf{x}_{NI}^*$ remains equal to

0, but $\mathbf{x}_{BI}^*$ changes from $\mathbf{B}^{-1}\mathbf{b}$ to $\mathbf{B}^{-1}(\mathbf{b}+\delta\mathbf{e}_j)$. The corresponding increase of the objective function satisfies:

$$z^*(\delta) - z^*(0) = \mathbf{c}_{BI}^\mathsf{T}\left(\mathbf{B}^{-1}(\mathbf{b}+\delta\mathbf{e}_j)\right) - \mathbf{c}_{BI}^\mathsf{T}\left(\mathbf{B}^{-1}\mathbf{b}\right) = \mathbf{c}_{BI}^\mathsf{T}\mathbf{B}^{-1}\delta\mathbf{e}_j$$
$$= \left((\mathbf{B}^{-1})^\mathsf{T}\mathbf{c}_{BI}\right)^\mathsf{T}\delta\mathbf{e}_j = (\mathbf{y}^*)^\mathsf{T}\delta\mathbf{e}_j = \delta y_j^*,$$

where $\mathbf{y}^* = \begin{bmatrix} y_1^* & \ldots & y_m^* \end{bmatrix}^\mathsf{T}$ is an optimal dual solution. The equation $z^*(\delta) - z^*(0) = \delta y_j^*$ shows that y_j^* determines the sensitivity of the optimal objective value with respect to perturbations of the vector $\mathbf{b}$. Hence, we have that the shadow price of constraint j is given by:

$$\frac{z^*(\delta) - z^*(0)}{\delta} = \frac{\delta y_j^*}{\delta} = y_j^*.$$

So, y_j^* is the shadow price of constraint j. This proves the theorem. $\qquad\square$

Recall from Section 4.6.1 that the optimal values of the dual variables may be read directly from an optimal simplex tableau of a standard LO-model. Since these values correspond to the shadow prices of the technology constraints, this means that the shadow prices can also be read directly from an optimal simplex tableau.

Example 5.3.1.   *Consider again Model Dovetail (see Section 3.3). The objective vector in the optimal simplex tableau is* $\begin{bmatrix} 0 & 0 & -1\frac{1}{2} & -\frac{1}{2} & 0 & 0 \end{bmatrix}^\mathsf{T}$, *and so the shadow prices of the four technology constraints are* $1\frac{1}{2}, \frac{1}{2}, 0,$ *and* 0, *respectively.*

The shadow price of the constraint $x_1 + x_2 \leq 9$ of Model Dovetail is $1\frac{1}{2}$. This fact holds only in a restricted interval of δ (see Figure 5.6). The tolerance interval of δ, i.e., the (interval of the) values of δ for which $y_1^* = 1\frac{1}{2}$, is determined as follows. In the model with slack variables, the constraint $x_1 + x_2 + x_3 = 9$ turns into the constraint $x_1 + x_2 + x_3 = 9 + \delta$. Rewriting Model Dovetail from the perspective of the optimal feasible basic solution (see Section 3.2) yields:

$$
\begin{aligned}
z^*(\delta) = \max \quad -1\tfrac{1}{2}x_3 &- \tfrac{1}{2}x_4 &&+ 22\tfrac{1}{2} + 1\tfrac{1}{2}\delta \\
\text{s.t.} \quad 1\tfrac{1}{2}x_3 - \tfrac{1}{2}x_4 &+ x_2 && &&= 4\tfrac{1}{2} + 1\tfrac{1}{2}\delta \\
-\tfrac{1}{2}x_3 + \tfrac{1}{2}x_4 && &+ x_1 && &= 4\tfrac{1}{2} - \tfrac{1}{2}\delta \\
\tfrac{1}{2}x_3 - \tfrac{1}{2}x_4 && && + x_5 && = 2\tfrac{1}{2} + \tfrac{1}{2}\delta \\
-1\tfrac{1}{2}x_3 + \tfrac{1}{2}x_4 && && &+ x_6 = 1\tfrac{1}{2} - 1\tfrac{1}{2}\delta \\
x_1, \ldots, x_6 &\geq 0.
\end{aligned}
$$

So, whenever there is one additional unit of machine capacity available for the annual production, the profit increases by $1\frac{1}{2}(\times\$1{,}000)$. At the optimal solution it holds that $x_3 = x_4 = 0$ and $x_1, x_2, x_5, x_6 \geq 0$. Hence,

$$
\begin{aligned}
4\tfrac{1}{2} + 1\tfrac{1}{2}\delta &\geq 0 &&\text{(from } x_2 \geq 0\text{)}, \\
4\tfrac{1}{2} - \tfrac{1}{2}\delta &\geq 0 &&\text{(from } x_1 \geq 0\text{)}, \\
2\tfrac{1}{2} + \tfrac{1}{2}\delta &\geq 0 &&\text{(from } x_5 \geq 0\text{)}, \\
1\tfrac{1}{2} - 1\tfrac{1}{2}\delta &\geq 0 &&\text{(from } x_6 \geq 0\text{)}.
\end{aligned}
$$

As long as these inequalities hold, the optimal solution does not jump to another set of basic variables, i.e., for these values of δ, the optimal feasible basic solutions all correspond to the same basis matrix. From these four inequalities one can easily derive that:

$$-3 \leq \delta \leq 1,$$

and so the tolerance interval is $[-3, 1] = \{\delta \mid -3 \leq \delta \leq 1\}$. Therefore, the shadow price $1\frac{1}{2}$ holds as long as the value of δ satisfies $-3 \leq \delta \leq 1$. In other words, the constraint $x_1 + x_2 \leq 9 + \delta$ has shadow price $1\frac{1}{2}$ in Model Dovetail as long as $-3 \leq \delta \leq 1$.

In the general case, the tolerance interval of the k'th technology constraint ($k = 1, \ldots, m$) with perturbation factor δ is determined by solving the feasibility condition (5.1) and the optimality condition (5.2). Let $\mathbf{B}$ be an optimal basis matrix of the standard LO-model $\max\{\mathbf{c}^\mathsf{T}\mathbf{x} \mid \mathbf{A}\mathbf{x} \leq \mathbf{b}, \mathbf{x} \geq \mathbf{0}\}$. The feasibility condition and optimality condition of the perturbed model are:

$$\mathbf{B}^{-1}(\mathbf{b} + \delta\mathbf{e}_k) \geq \mathbf{0}, \qquad \text{(feasibility)}$$
$$\text{and } \mathbf{c}_{NI}^\mathsf{T} - \mathbf{c}_{BI}^\mathsf{T}\mathbf{B}^{-1}\mathbf{N} \leq \mathbf{0}. \qquad \text{(optimality)}$$

The optimality condition is not affected by the perturbation, and, hence, the current feasible basic solution is optimal as long the following set of m inequalities is satisfied:

$$\mathbf{B}^{-1}\mathbf{b} + \delta\mathbf{B}^{-1}\mathbf{e}_k \geq \mathbf{0}, \tag{5.7}$$

There are two cases:

x_{n+k} is a basic variable. Let β be such that $BI_\beta = n + k$. The term $\mathbf{B}^{-1}\mathbf{b}$ in (5.7) is equal to $\mathbf{x}_{BI}^*$. Recall that the column with index $n + k$ in $\begin{bmatrix} \mathbf{A} & \mathbf{I}_m \end{bmatrix}$ is equal to $\mathbf{e}_k$. Hence, the β'th column of $\mathbf{B}$ is equal to $\mathbf{e}_k$. This implies that $\mathbf{B}\mathbf{e}_\beta = \mathbf{e}_k$, which in turn means that $\mathbf{B}^{-1}\mathbf{e}_k = \mathbf{e}_\beta$. Therefore, (5.7) reduces to $\mathbf{x}_{BI}^* + \delta\mathbf{e}_\beta \geq \mathbf{0}$. This, in turn, reduces to the following inequality:

$$\delta \geq -x_{n+k}^*, \tag{5.8}$$

with x_{n+k}^* the optimal value of the slack variable x_{n+k}. So, the tolerance interval is $[-x_{n+k}^*, \infty)$.

x_{n+k} is a nonbasic variable. Then, the lower bound $\delta_{\min}$ and the upper bound $\delta_{\max}$ of the tolerance interval for δ can be determined by two ratio tests, namely:

$$\delta_{\min} = \max\left\{ -\frac{(\mathbf{B}^{-1}\mathbf{b})_j}{(\mathbf{B}^{-1}\mathbf{e}_k)_j} \;\middle|\; (\mathbf{B}^{-1}\mathbf{e}_k)_j > 0, j = 1, \ldots, m \right\}, \text{ and}$$
$$\delta_{\max} = \min\left\{ -\frac{(\mathbf{B}^{-1}\mathbf{b})_j}{(\mathbf{B}^{-1}\mathbf{e}_k)_j} \;\middle|\; (\mathbf{B}^{-1}\mathbf{e}_k)_j < 0, j = 1, \ldots, m \right\}. \tag{5.9}$$

Recall that, by definition, $\max(\emptyset) = -\infty$ and $\min(\emptyset) = +\infty$. The tolerance interval is $[\delta_{\min}, \delta_{\max}]$.

The tolerance intervals derived above can be obtained directly from the optimal simplex tableau. In the optimal simplex tableau, for $\delta = 0$ and basis matrix $\mathbf{B}$, the vector $\mathbf{B}^{-1}\mathbf{b}$ is the right hand side column, and $\mathbf{B}^{-1}\mathbf{e}_k$ is the column corresponding to the slack variable x_{n+k} of the k'th technology constraint.

Example 5.3.2.   *Consider again the perturbed constraint $x_1 + x_2 \leq 9 + \delta$ of Model Dovetail. The optimal simplex tableau reads:*

x_1	x_2	x_3	x_4	x_5	x_6	z
0	0	$-1\frac{1}{2}$	$-\frac{1}{2}$	0	0	$-22\frac{1}{2}$
0	1	$1\frac{1}{2}$	$-\frac{1}{2}$	0	0	$4\frac{1}{2}$
1	0	$-\frac{1}{2}$	$\frac{1}{2}$	0	0	$4\frac{1}{2}$
0	0	$\frac{1}{2}$	$-\frac{1}{2}$	1	0	$2\frac{1}{2}$
0	0	$-1\frac{1}{2}$	$\frac{1}{2}$	0	1	$1\frac{1}{2}$

Since the first constraint of the model is perturbed, we have that $k = 1$. Hence, the vector $\mathbf{B}^{-1}\mathbf{e}_k$ is the column of the simplex tableau corresponding to x_3, which is a nonbasic variable at the optimal feasible basic solution. Therefore, (5.9) yields that:

$$\delta_{\min} = \max\left\{ -\frac{4\frac{1}{2}}{1\frac{1}{2}}, \star, -\frac{2\frac{1}{2}}{\frac{1}{2}}, \star \right\} = \max\{-3, \star, -5, \star\} = -3,$$

$$\delta_{\max} = \min\left\{ \star, \frac{4\frac{1}{2}}{\frac{1}{2}}, \star, \frac{1\frac{1}{2}}{1\frac{1}{2}} \right\} = \min\{\star, 9, \star, 1\} = 1.$$

Hence, the tolerance interval for the perturbation of the first constraint is $[-3, 1]$. Next, consider a perturbation of the constraint $x_1 \leq 7$. Because this is the third constraint of the model, we have that $k = 3$. The slack variable corresponding to this constraint is x_5. Since $x_5^* = 2\frac{1}{2}$, it follows from (5.8) that the tolerance interval for a perturbation of the constraint $x_1 \leq 7$ is $[-2\frac{1}{2}, \infty)$.

Note that a constraint can be nonbinding at some optimal solutions, and binding at other optimal solutions.

Example 5.3.3.   *We will illustrate case 2 above one more time by considering Model Dovetail. The constraint $x_1 \leq 7$ is nonbinding at the optimal solution, and the corresponding dual variable y_3 has value zero (see Figure 4.1). Perturbing the right hand side by δ, this constraint becomes $x_1 \leq 7 + \delta$, or after adding the slack variable x_5,*

$$x_1 + (x_5 - \delta) = 7.$$

Note that this equation is the same as $x_1 + x_5 = 7$ except that x_5 is replaced by $x_5 - \delta$. We will determine (the interval of) the values of δ for which the constraint $x_1 \leq 7 + \delta$ remains nonbinding, i.e., we will determine the tolerance interval of $x_1 \leq 7$. The constraints of Model Dovetail rewritten

using the basic variables x_1, x_2, x_5, and x_6 are:

$$\begin{aligned}
1\tfrac{1}{2}x_3 - \tfrac{1}{2}x_4 + x_2 \qquad\qquad\qquad &= 4\tfrac{1}{2}\\
-\tfrac{1}{2}x_3 + \tfrac{1}{2}x_4 \qquad + x_1 \qquad\qquad &= 4\tfrac{1}{2}\\
\tfrac{1}{2}x_3 - \tfrac{1}{2}x_4 \qquad\qquad + (x_5 - \delta) \qquad &= 2\tfrac{1}{2}\\
-1\tfrac{1}{2}x_3 + \tfrac{1}{2}x_4 \qquad\qquad\qquad\quad + x_6 &= 1\tfrac{1}{2}.
\end{aligned}$$

The objective function in terms of x_3 and x_4 reads:

$$z = 22\tfrac{1}{2} - 1\tfrac{1}{2}x_3 - \tfrac{1}{2}x_4.$$

Because $x_3 = 0$ and $x_4 = 0$ at the optimal solution, we find that:

$$x_2 = 4\tfrac{1}{2}, \quad x_1 = 4\tfrac{1}{2}, \quad x_5 = 2\tfrac{1}{2} + \delta, \quad x_6 = 1\tfrac{1}{2}.$$

Moreover, $x_5 \geq 0$ implies that $\delta \geq -2\tfrac{1}{2}$. So the tolerance interval is $\{\delta \mid -2\tfrac{1}{2} \leq \delta\}$. The shadow price of the constraint $x_1 \leq 7$ (see Section 1.1) is 0 and remains 0 as long as the perturbation of the capacity is not less than $-2\tfrac{1}{2}$, e.g., as long as the right hand side of $x_1 \leq 7 + \delta$ is not less than $4\tfrac{1}{2}$. As soon as $\delta < -2\tfrac{1}{2}$, then this constraint becomes binding, and therefore the optimal solution changes; compare also Section 1.1.2.

5.3.3 Perturbation of nonnegativity constraints

Recall that the shadow price of a nonnegativity constraint is defined as the rate at which the objective value increases for sufficiently small feasible perturbations of its right hand side; in the case of binding nonnegativity constraints the optimal solution is assumed to be nondegenerate. Since a positive perturbation of the right hand side value of a nonnegativity constraint generally shrinks the feasible region, the shadow price of any nonnegativity constraint (in a maximizing LO-model) is nonpositive.

> **Theorem 5.3.4.** (*Shadow prices of nonnegativity constraints*)
> For any feasible standard primal LO-model with a nondegenerate optimal solution, the shadow price of any nonnegativity constraint exists and is equal to the negative of the optimal value of the corresponding dual slack variable.

Proof of Theorem 5.3.4. Let $k \in \{1, \ldots, n\}$. Suppose that we perturb the nonnegativity constraint $x_k \geq 0$, i.e., we replace it by $x_k \geq \delta$, for δ small enough. The set of nonnegativity constraints then becomes $\mathbf{x} \geq \delta\mathbf{e}_k$, with $\mathbf{e}_k$ the k'th unit vector in $\mathbb{R}^n$. Substituting $\mathbf{w} = \mathbf{x} - \delta\mathbf{e}_k$, the optimal objective value is transformed into:

$$z^*(\delta) = \delta\mathbf{c}^\mathsf{T}\mathbf{e}_k + \max\left\{\mathbf{c}^\mathsf{T}\mathbf{w} \mid \mathbf{Aw} \leq \mathbf{b} - \delta\mathbf{Ae}_k, \mathbf{w} \geq \mathbf{0}\right\}. \tag{5.10}$$

In this new model, the right hand side values of the constraints are perturbed by the vector $-\delta \mathbf{A}\mathbf{e}_k$. Considering the dual of the LO-model in (5.10), we have that:

$$z^*(\delta) = \delta \mathbf{c}^\mathsf{T}\mathbf{e}_k + \min\left\{(\mathbf{b} - \delta \mathbf{A}\mathbf{e}_k)^\mathsf{T}\mathbf{y} \;\middle|\; \mathbf{A}^\mathsf{T}\mathbf{y} \geq \mathbf{c}, \mathbf{y} \geq \mathbf{0}\right\}$$
$$= \delta \mathbf{c}^\mathsf{T}\mathbf{e}_k + (\mathbf{b} - \delta \mathbf{A}\mathbf{e}_k)^\mathsf{T}\mathbf{y}^* = \delta \mathbf{c}^\mathsf{T}\mathbf{e}_k - \delta \mathbf{e}_k^\mathsf{T}\mathbf{A}^\mathsf{T}\mathbf{y}^* + \mathbf{b}^\mathsf{T}\mathbf{y}^*,$$

with $\mathbf{y}^*$ the optimal solution of the dual model. Let $\mathbf{y}_s^*$ be the corresponding vector of values of the slack variables. From $\mathbf{A}^\mathsf{T}\mathbf{y}^* - \mathbf{y}_s^* = \mathbf{c}$, it follows that (recall that $z^*(0) = \mathbf{c}^\mathsf{T}\mathbf{x}^* = \mathbf{b}^\mathsf{T}\mathbf{y}^*$):

$$z^*(\delta) - z^*(0) = \delta \mathbf{e}_k^\mathsf{T}\mathbf{c} - \delta \mathbf{e}_k^\mathsf{T}(\mathbf{c} + \mathbf{y}_s^*) = \delta \mathbf{e}_k^\mathsf{T}\mathbf{c} - \delta \mathbf{e}_k^\mathsf{T}\mathbf{c} - \delta \mathbf{e}_k^\mathsf{T}\mathbf{y}_s^*$$
$$= -\delta \mathbf{e}_k^\mathsf{T}\mathbf{y}_s^* = -\delta (\mathbf{y}_s^*)_k = -\delta y_{m+k}^*.$$

So, we have that $\frac{z^*(\delta) - z^*(0)}{\delta} = -y_{m+k}^*$. Hence, the shadow price of the constraint $x_k \geq 0$ is equal to the negative of the optimal value of the dual variable y_{m+k}. □

It is left to the reader to formulate a similar theorem for the nonstandard case (the signs of the dual variables need special attention).

Similar to the case of technology constraints, the shadow price and the tolerance interval of a nonnegativity constraint can be read directly from the optimal simplex tableau. It follows from Theorem 4.5.1 that the optimal objective coefficients corresponding to the primal decision variables are precisely the negatives of the shadow prices. Let $\mathbf{B}$ be an optimal basis matrix of the LO-model $\max\{\mathbf{c}^\mathsf{T}\mathbf{x} \mid \mathbf{A}\mathbf{x} \leq \mathbf{b}, \mathbf{x} \geq \mathbf{0}\}$. It is shown in the proof of Theorem 5.3.4 that the perturbed model is equivalent to model (5.10). This means that $\mathbf{B}$ is an optimal basis matrix of the perturbed model if and only if it is an optimal basis matrix of (5.10). The feasibility condition (5.1) and the optimality condition (5.2) for model (5.10) are:

$$\mathbf{B}^{-1}(\mathbf{b} - \delta \mathbf{A}\mathbf{e}_k) \geq \mathbf{0}, \qquad \text{(feasibility)}$$
$$\text{and } \mathbf{c}_{NI}^\mathsf{T} - \mathbf{c}_{BI}^\mathsf{T}\mathbf{B}^{-1}\mathbf{N} \leq \mathbf{0}. \qquad \text{(optimality)}$$

The feasibility is affected by the perturbation; the optimality condition remains unaffected. We again distinguish two cases:

x_k is a basic decision variable, i.e., $k = BI_\beta$ for some $\beta \in \{1, \ldots, m\}$, and $k \in \{1, \ldots, n\}$. Since x_k is a basic variable, $\mathbf{B}^{-1}\mathbf{A}\mathbf{e}_k \, (= (\mathbf{B}^{-1}\mathbf{A})_{\star,k})$ is equal to the unit vector $\mathbf{e}_\beta$. Hence, $\mathbf{B}^{-1}(\mathbf{b} - \delta \mathbf{A}\mathbf{e}_k) \geq \mathbf{0}$ is equivalent to $\delta \leq (\mathbf{B}^{-1}\mathbf{b})_\beta$. Therefore, the tolerance interval of the nonnegativity constraint $x_k \geq 0$ satisfies:

$$\delta_{\min} = -\infty, \text{ and } \delta_{\max} = (\mathbf{B}^{-1}\mathbf{b})_\beta.$$

Note that the value of $(\mathbf{B}^{-1}\mathbf{b})_\beta$ can be found in the column of right hand side values in the optimal simplex tableau corresponding to $\mathbf{B}$.

x_k is a nonbasic decision variable, i.e., $k = NI_\alpha$ for some $\alpha \in \{1, \ldots, m\}$, and $k \in \{1, \ldots, n\}$. Using the fact that $\mathbf{B}^{-1}\mathbf{A}\mathbf{e}_k = (\mathbf{B}^{-1}\mathbf{N})_{\star,\alpha}$, the feasibility condition is equivalent to $\mathbf{B}^{-1}\mathbf{b} \geq \delta(\mathbf{B}^{-1}\mathbf{N})_{\star,\alpha}$. Note that the right hand side of this inequality, $\delta(\mathbf{B}^{-1}\mathbf{N})_{\star,\alpha}$, is δ times the column corresponding to x_k in the optimal simplex tableau.

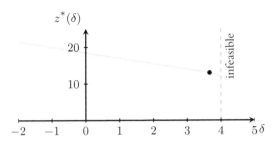

Figure 5.7: Perturbation function for the nonnegativity constraint $x_3 \geq 0$ of the model in Example 5.3.4.

Let $\delta_{\min}$ and $\delta_{\max}$ be the lower and upper bounds, respectively, of the tolerance interval of the right hand side of $x_k \geq 0$. Similar to the formulas in Section 5.3, we find that:

$$\delta_{\min} = \max\left\{ \frac{(\mathbf{B}^{-1}\mathbf{b})_j}{(\mathbf{B}^{-1}\mathbf{N})_{j,\alpha}} \; \middle| \; (\mathbf{B}^{-1}\mathbf{N})_{j,\alpha} < 0, j = 1, \ldots, m \right\}, \text{ and}$$

$$\delta_{\max} = \min\left\{ \frac{(\mathbf{B}^{-1}\mathbf{b})_j}{(\mathbf{B}^{-1}\mathbf{N})_{j,\alpha}} \; \middle| \; (\mathbf{B}^{-1}\mathbf{N})_{j,\alpha} > 0, j = 1, \ldots, m \right\}.$$

The following example illustrates this.

Example 5.3.4. *Consider the following LO-model:*

$$\begin{aligned} \max \quad & 3x_1 + 2x_2 + 3x_3 \\ \text{s.t.} \quad & 2x_1 \qquad + x_3 \leq 5 \qquad &\text{(slack } x_4) \\ & x_1 + x_2 + 2x_3 \leq 8 \qquad &\text{(slack } x_5) \\ & x_1, x_2, x_3 \geq 0. \end{aligned}$$

The optimal simplex tableau reads:

x_1	x_2	x_3	x_4	x_5	$-z$
0	0	$-1\frac{1}{2}$	$-\frac{1}{2}$	-2	$-18\frac{1}{2}$
1	0	$\frac{1}{2}$	$\frac{1}{2}$	0	$2\frac{1}{2}$
0	1	$1\frac{1}{2}$	$-\frac{1}{2}$	1	$5\frac{1}{2}$

The first three entries in the objective coefficient row correspond to the dual slack variables y_3, y_4, and y_5, and the last two to the dual decision variables y_1, y_2. Hence, following Theorem 4.2.4, $\begin{bmatrix} y_1^* & y_2^* & y_3^* & y_4^* & y_5^* \end{bmatrix} = \begin{bmatrix} \frac{1}{2} & 2 & 0 & 0 & 1\frac{1}{2} \end{bmatrix}$ is an optimal dual solution. The shadow prices of the two constraints are $\frac{1}{2}$ and 2, respectively. The shadow prices of the three nonnegativity constraints are 0, 0, and $1\frac{1}{2}$, respectively. The tolerance interval $[\delta_{\min}, \delta_{\max}]$ for the right hand side value of the

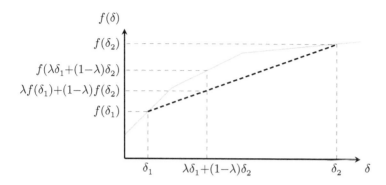

Figure 5.8: Nondecreasing, concave, piecewise linear function.

nonnegative constraint $x_3 \geq 0$ (note that x_3 is a nonbasic variable) satisfies:

$$\delta_{\min} = \max\{\star, \star\} = -\infty, \text{ and } \delta_{\max} = \min\left\{\frac{2\frac{1}{2}}{\frac{1}{2}}, \frac{5\frac{1}{2}}{1\frac{1}{2}}\right\} = \min\{5, 3\tfrac{2}{3}\} = 3\tfrac{2}{3}.$$

Figure 5.7 depicts the perturbation function for the nonnegativity constraint $x_3 \geq 0$. Note that $x_1 \geq 0$, $x_2 \geq 0$, and $x_1 + x_2 + 2x_3 \leq 8$ imply that $x_3 \leq 4$, and hence the model is infeasible for $\delta > 4$.

5.4 Piecewise linearity of perturbation functions

We observed in Section 5.2 and Section 5.3 that the perturbation functions of objective coefficients and of constraints are piecewise linear; see Figure 5.6 and Figure 5.7. An interesting additional feature of perturbation functions of objective coefficients in maximizing LO-models is that they are convex; similarly, perturbation functions of '$\leq$' constraints in maximizing LO-models are concave.

Recall that a function $f(\delta)$ is called *convex* if and only if for each δ_1 and δ_2 with $\delta_1 \leq \delta_2$ it holds that:

$$f(\lambda\delta_1 + (1-\lambda)\delta_2) \leq \lambda f(\delta_1) + (1-\lambda)f(\delta_2),$$

for each λ with $0 \leq \lambda \leq 1$. In other words, the chord between the points $\begin{bmatrix} \delta_1 & f(\delta_1) \end{bmatrix}^\mathsf{T}$ and $\begin{bmatrix} \delta_2 & f(\delta_2) \end{bmatrix}^\mathsf{T}$ 'lies above' the graph of $f(\delta)$. Similarly, a function $f(\delta)$ is called *concave* if and only if for each δ_1 and δ_2 with $\delta_1 \leq \delta_2$ it holds that:

$$f(\lambda\delta_1 + (1-\lambda)\delta_2) \geq \lambda f(\delta_1) + (1-\lambda)f(\delta_2),$$

for each λ with $0 \leq \lambda \leq 1$; see Figure 5.8. It can be easily shown that if the function $f(\delta)$ is convex, then $-f(\delta)$ is concave. For a discussion of 'convexity' and 'concavity', see Appendix D.

We prove these observations in Theorem 5.4.1 and Theorem 5.4.2.

Theorem 5.4.1. (*Piecewise linearity and convexity of objective coefficient perturbation functions*)

Consider perturbing the objective coefficient of the decision variable x_i of an LO-model. The following assertions hold:

(i) The perturbation function is a piecewise linear function; the slope of any of its line segments is equal to the optimal value of x_i in the model in which the objective coefficient is perturbed with a factor δ.

(ii) The perturbation function is convex for maximizing LO-models and concave for minimizing LO-models.

Proof of Theorem 5.4.1. We prove the theorem for a maximizing LO-model; the case of a minimizing LO-model is left to the reader. First note that any LO-model can be transformed into a model of the form:

$$\max\left\{ \mathbf{c}^\mathsf{T}\mathbf{x} \mid \mathbf{A}\mathbf{x} = \mathbf{b}, \mathbf{x} \geq \mathbf{0} \right\},$$

by adding appropriate slack variables; see Section 1.3. Consider the perturbation function $z^*(\delta)$ of the objective coefficient of x_i, i.e.,

$$z^*(\delta) = \max\left\{ \mathbf{c}^\mathsf{T}\mathbf{x} + \delta x_i \mid \mathbf{A}\mathbf{x} = \mathbf{b}, \mathbf{x} \geq \mathbf{0} \right\}.$$

We will first show that the function $z^*(\delta)$ is convex, and that the set of all perturbation factors δ for which the LO-model has a finite-valued optimal solution is a connected interval of $\mathbb{R}$.

Assume that the model has finite optimal solutions for $\delta = \delta_1$ and $\delta = \delta_2$, with $\delta_1 < \delta_2$. Take any λ with $0 \leq \lambda \leq 1$, and let $\delta = \lambda\delta_1 + (1 - \lambda)\delta_2$. Clearly, $\mathbf{c}^\mathsf{T}\mathbf{x} + (\lambda\delta_1 + (1 - \lambda)\delta_2)x_i = \lambda(\mathbf{c}^\mathsf{T}\mathbf{x} + \delta_1 x_i) + (1 - \lambda)(\mathbf{c}^\mathsf{T}\mathbf{x} + \delta_2 x_i)$. Hence, for any feasible $\mathbf{x}$ it holds that $\mathbf{c}^\mathsf{T}\mathbf{x} + (\lambda\delta_1 + (1 - \lambda)\delta_2)x_i \leq \lambda z^*(\delta_1) + (1 - \lambda)z^*(\delta_2)$, and so:

$$z^*(\delta) = z^*(\lambda\delta_1 + (1 - \lambda)\delta_2) \leq \lambda z^*(\delta_1) + (1 - \lambda)z^*(\delta_2).$$

This proves that $z^*(\delta)$ is a convex function for $\delta_1 \leq \delta \leq \delta_2$. Moreover, since both $z^*(\delta_1)$ and $z^*(\delta_2)$ are finite, $z^*(\delta)$ is finite as well. Hence, the set of all values of δ for which the model has a finite optimal solution is in fact a connected interval of $\mathbb{R}$. Let I be this interval. According to Theorem 2.2.4 and Theorem 2.1.5, the model also has an optimal feasible basic solution for each $\delta \in I$.

We will now prove that, if for both δ_1 and δ_2 the matrix $\mathbf{B}$ is an optimal basis matrix, then $\mathbf{B}$ is optimal basis matrix for all $\delta \in [\delta_1, \delta_2]$. Let $\mathbf{x}^*$ with $\mathbf{x}^*_{BI} = \mathbf{B}^{-1}\mathbf{b}$ be the optimal feasible basic solution for both δ_1 and δ_2. Take any λ with $0 \leq \lambda \leq 1$, and let $\delta = \lambda\delta_1 + (1 - \lambda)\delta_2$. Then, $z^*(\delta) \geq \mathbf{c}^\mathsf{T}\mathbf{x}^* + (\lambda\delta_1 + (1-\lambda)\delta_2)x_i^* = \lambda(\mathbf{c}^\mathsf{T}\mathbf{x}^* + \delta_1 x_i^*) + (1 - \lambda)(\mathbf{c}^\mathsf{T}\mathbf{x}^* + \delta_2 x_i^*) = \lambda z^*(\delta_1) + (1 - \lambda)z^*(\delta_2)$. On the other hand, we know already that $z^*(\delta) \leq \lambda z^*(\delta_1) + (1 - \lambda)z^*(\delta_2)$. Hence, $z^*(\delta) = \mathbf{c}^\mathsf{T}\mathbf{x}^* + (\lambda\delta_1 + (1 - \lambda)\delta_2)x_i^*$. This shows that $\mathbf{x}^*$ is an optimal solution for each $\delta \in [\delta_1, \delta_2]$.

A consequence of the above discussions is that for any $\delta \in I$, it holds that, if $\mathbf{B}$ is an optimal basis matrix for this δ, then there is a connected subinterval in I for which $\mathbf{B}$ is optimal. Since there are only finitely many basis matrices in $\mathbf{A}$, the interval I can be partitioned into a finite number of subintervals on which a fixed basis matrix is optimal. For any of these subintervals, there is a feasible basic solution that is optimal for each δ of that subinterval. Note that it may happen that a feasible basic solution is optimal for two (adjacent) subintervals.

Let $\mathbf{B}$ be any basis matrix such that $\mathbf{x}^*$ is a feasible basic solution with $\mathbf{x}^*_{BI} = \mathbf{B}^{-1}\mathbf{b}$. Let $[\delta_1, \delta_2]$ be an interval on which $\mathbf{B}$ is optimal. Then, $z^*(\delta) = \mathbf{c}^\mathsf{T}\mathbf{x}^* + \delta x_i^*$, which is a linear function of δ on $[\delta_1, \delta_2]$. Hence, $z^*(\delta)$ is a piecewise linear function on the interval I.

The slope of the function $z^*(\delta) = \mathbf{c}^\mathsf{T}\mathbf{x} + \delta x_i^*$ is x_i^*, which is in fact the optimal value of the variable x_i of which the objective coefficient is perturbed with a factor δ.   □

The following theorem states a similar fact about right hand side value perturbations.

Theorem 5.4.2. (*Piecewise linearity and concavity of right hand side value perturbation functions*)
For any constraint (including equality, nonnegativity, and nonpositivity constraints) of an LO-model, the following assertions hold.

(i)   The perturbation function is a piecewise linear function; the slope of any of its line segments is equal to the shadow price of the corresponding constraint of which the right hand side is perturbed with a factor corresponding to an interior point of that line segment.

(ii)   The perturbation function is concave for maximizing LO-models, and convex for minimizing LO-models.

Proof of Theorem 5.4.2. We prove the theorem for a maximizing LO-model; the case of a minimizing LO-model is left to the reader. Let $z^*(\delta)$ be the optimal objective value of an LO-model with the perturbation factor δ in the right hand side of constraint j. According to Theorem 4.2.4 (see also Theorem 5.7.1), $z^*(\delta)$ is also the optimal objective value of the corresponding dual LO-model. This dual model has δ in the j'th objective coefficient. Since the dual model is a minimizing LO-model, it follows from Theorem 5.4.1 that $z^*(\delta)$ is in fact a concave piecewise linear function.

Let $z^*(\delta) = \max\{\mathbf{c}^\mathsf{T}\mathbf{x} \mid \mathbf{A}\mathbf{x} = \mathbf{b} + \delta\mathbf{e}_j\}$, with $\mathbf{e}_j$ the j'th unit vector in $\mathbb{R}^m$. Let $\mathbf{x}^*$ be an optimal feasible basic solution of the perturbed model with optimal objective value $z^*(\delta)$, and let $\mathbf{B}$ be the corresponding basis matrix. Let $[\delta_1, \delta_2]$ be an interval on which $\mathbf{B}$ is optimal. Then for each $\varepsilon \in [\delta_1, \delta_2]$, it holds that $z^*(\delta) = \mathbf{c}^\mathsf{T}\mathbf{x}^* = (\mathbf{b} + \delta\mathbf{e}_j)^\mathsf{T}\mathbf{y}^* = \mathbf{b}^\mathsf{T}\mathbf{y}^* + \delta y_j^*$ with

$y^* = (\mathbf{B}^{-1})^\mathsf{T}\mathbf{c}_{BI}$ (see Theorem 4.2.4 and Theorem 5.7.1). Therefore the slope of $z^*(\delta)$ on $[\delta_1, \delta_2]$ is equal to y_j^*, which is in fact the optimal value of the corresponding dual variable. $\square$

Note that the perturbation function for an objective coefficient corresponding to a non-negative variable x_i is nondecreasing, since the slope of this function on each subinterval is equal to the optimal value x_i^* of x_i, which is nonnegative. The perturbation function for a '$\leq$' constraint in a maximizing model is nondecreasing, because the slope of this function is equal to the optimal value of the complementary dual variable, which is nonnegative. The perturbation function for an objective coefficient in a maximizing (minimizing) LO-model is convex (concave). See also Exercise 5.8.20.

5.5 Perturbation of the technology matrix

In this section we consider the situation where the entries of the technology matrix of an LO-model are inaccurate. For instance, if in Model Dovetail the production of boxes of long matches needs more than three cubic meters of wood per box, then the optimal solution of the model may change. However, the way in which the optimal solution changes is much more difficult to compute in general. The easiest way of dealing with perturbations of technology matrix entries is repeating the calculations for each separate change, even though this calculation procedure is in general very time consuming.

Perturbation functions associated with entries of the technology matrix are usually not piece-wise linear functions; see e.g., Figure 5.9, and Section 5.4. Figure 5.9 shows the perturbation function for the coefficient of x_1 in the constraint $3x_1 + x_2 \leq 18$ of Model Dovetail. To calculate this perturbation function, we have used a computer package, and calculated the optimal solutions for several values of this matrix entry. The constraint was replaced by $(3 + \delta)x_1 + x_2 \leq 18$, and $z^*(\delta)$ was calculated for several values of δ. For $\delta = 0$ we have the original model, and so $z^*(0) = 22\frac{1}{2}$. When δ increases from 0 to infinity, the optimal solution moves first along the line $x_1 + x_2 = 9$ from $\begin{bmatrix} 4\frac{1}{2} & 4\frac{1}{2} \end{bmatrix}^\mathsf{T}$ with $z^*(0) = 22\frac{1}{2}$ to $\begin{bmatrix} 3 & 6 \end{bmatrix}^\mathsf{T}$ with $z^*(1) = 21$, and then along the line $x_2 = 6$ towards $\begin{bmatrix} 0 & 6 \end{bmatrix}^\mathsf{T}$ with $z^*(\delta) \to 12$ as $\delta \to \infty$. For decreasing values of δ, the optimal objective value first increases along the line $x_1 + x_2 = 9$ to $\begin{bmatrix} 9 & 0 \end{bmatrix}^\mathsf{T}$ with $z^*(-1) = 27$, and then remains constant, because for $\delta < -1$ the constraint is redundant. The tolerance interval of this matrix entry satisfies $-1 \leq \delta \leq 1$.

There are, however, situations in which the calculations can be done by hand relatively easy. Of course, in case a constraint is nonbinding, the calculations are very easy: since a small change of one of its entries will not influence the optimal solution, the perturbation function is 'horizontal' around $\delta = 0$. A similar observation holds for columns of the technology matrix for which the corresponding dual constraint is nonbinding. (Why?)

A more interesting situation is the case when only one given row of the technology matrix is perturbed. The analysis of such a case is useful for example in product-mix problems when the labor or the materials requirements change due to a change in technology. The following example elaborates on this case.

Example 5.5.1.   *Suppose that the company Dovetail decides to use less wood for the boxes. So the entries in the second row of the technology matrix are perturbed by, say, t_1 and t_2, respectively. The perturbed model of Model Dovetail becomes:*

$$
\begin{array}{llrl}
\max & 3x_1 + & 2x_2 & \\
\text{s.t.} & x_1 + & x_2 \leq 9 & \qquad (\text{slack } x_3) \\
& (3+t_1)x_1 + (1+t_2)x_2 \leq 18 & & \qquad (\text{slack } x_4) \\
& x_1 \qquad\qquad \leq 7 & & \qquad (\text{slack } x_5) \\
& x_2 \leq 6 & & \qquad (\text{slack } x_6) \\
& x_1, x_2 \geq 0. &
\end{array}
$$

Let x_3, x_4, x_5, x_6 be the slack variables of the respective constraints. We assume that the optimal basis matrix remains the same, and so x_1, x_2, x_5, x_6 remain the optimal basic variables under the perturbations t_1, t_2. After straightforward calculations it follows that:

$$
\begin{aligned}
x_1 &= \tfrac{1}{s}(9 - 9t_2 + (1+t_2)x_3 - x_4), \\
x_2 &= \tfrac{1}{s}(9 + 9t_1 - (3+t_1)x_3 + x_4), \\
x_5 &= \tfrac{1}{s}(5 + 7t_1 + 2t_2 - (1+t_2)x_3 + x_4), \\
x_6 &= \tfrac{1}{s}(3 - 3t_1 - 6t_2 + (3+t_1)x_3 - x_4), \text{ and} \\
z &= \tfrac{1}{s}(45 + 18t_1 - 27t_2 - (3 + 2t_1 - 3t_2)x_3 - x_4),
\end{aligned}
$$

with $s = 2 + t_1 - t_2$. The optimality requires that:

$$
2 + t_1 - t_2 > 0, \text{ and } 3 + 2t_1 - 3t_2 \geq 0.
$$

At the optimal solution, it holds that $x_3 = x_4 = 0$ and x_1, x_2, x_5, $x_6 \geq 0$. Hence,

$$
\begin{aligned}
9 - 9t_2 &\geq 0, \\
9 + 9t_1 &\geq 0, \\
5 + 7t_1 + 2t_2 &\geq 0, \text{ and} \\
3 - 3t_1 - 6t_2 &\geq 0.
\end{aligned}
$$

Since the matrix coefficients need to remain nonnegative, we have that:

$$
1 + t_2 \geq 0, \text{ and } 3 + t_1 \geq 0.
$$

In Figure 5.10 the tolerance region, determined by the above five inequalities, is drawn (the shaded region in this figure). The tolerance region, which is determined by the four vertices $\begin{bmatrix} 0 & 0 \end{bmatrix}^\mathsf{T}$, $\begin{bmatrix} -\tfrac{5}{7} & 0 \end{bmatrix}^\mathsf{T}$, $\begin{bmatrix} -\tfrac{3}{7} & -1 \end{bmatrix}^\mathsf{T}$, and $\begin{bmatrix} 0 & -1 \end{bmatrix}^\mathsf{T}$, denotes the feasible values in the case $t_1 \leq 0$ and $t_2 \leq 0$ (i.e., when less wood is used).

There are two interesting special cases:

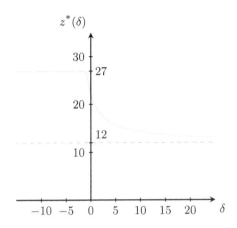

Figure 5.9: Perturbation function for a technology matrix entry.

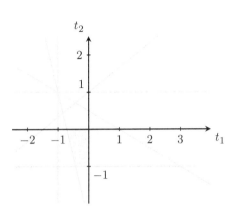

Figure 5.10: Tolerance region.

Suppose that $t_1 = t_2 = t$. Substituting $t_1 = t_2 = t$ in the above expressions for x_1, x_2, x_5, x_6, and z, we find that $x_1 = 4\frac{1}{2} - 4\frac{1}{2}t$, $x_2 = 4\frac{1}{2} + 4\frac{1}{2}t$, $x_5 = 2\frac{1}{2} + 4\frac{1}{2}t$, $x_6 = 1\frac{1}{2} - 4\frac{1}{2}t$, and $z = 22\frac{1}{2} - 4\frac{1}{2}t$, and so the feasible values for t satisfy $-\frac{5}{9} \leq t \leq 0$, and the corresponding objective values z^* satisfy $22\frac{1}{2} \leq z^* \leq 25$.

Suppose that we want to change only one coefficient at a time. For $t_1 = 0$ we find that $x_1 = (9 - 9t_2)/(2 - t_2)$, $x_2 = 9/(2 - t_2)$, $x_5 = (5 + 2t_2)/(2 - t_2)$, $x_6 = (3 - 6t_2)/(2 - t_2)$, and $z = (45 - 27t_2)/(2 - t_2)$. Hence, the feasible values for t_2 satisfy $-1 \leq t_2 \leq 0$, and the corresponding range for the objective values z is $22\frac{1}{2} \leq z \leq 24$. For $t_2 = 0$, we find that $x_1 = 9/(2 + t_1)$, $x_2 = (9 + 9t_1)/(2 + t_1)$, $x_5 = (5 + 7t_1)/(2 + t_1)$, $x_6 = (3 - 3t_1)/(2 + t_1)$, and $z = (45 + 18t_1)/(2 + t_1)$, so that the ranges are $-\frac{5}{7} \leq t_1 \leq 0$, and $22\frac{1}{2} \leq z \leq 25$.

Note that we have not developed a general theory for determining tolerance intervals of technology matrix entries. So we need to use computer packages and calculate a (sometimes long) sequence of optimal solutions.

5.6 Sensitivity analysis for the degenerate case

Recall that the primal LO-model with slack variables can be formulated as follows (see Section 4.2):

$$\max\left\{ \begin{bmatrix} \mathbf{c}^\mathsf{T} & \mathbf{0}^\mathsf{T} \end{bmatrix} \begin{bmatrix} \mathbf{x} \\ \mathbf{x}_s \end{bmatrix} \;\middle|\; \begin{bmatrix} \mathbf{A} & \mathbf{I}_m \end{bmatrix} \begin{bmatrix} \mathbf{x} \\ \mathbf{x}_s \end{bmatrix} = \mathbf{b}, \; \mathbf{x}, \mathbf{x}_s \geq \mathbf{0} \right\},$$

where $\mathbf{c}$ ($\in \mathbb{R}^n$) is the vector of objective coefficients, $\mathbf{x}$ ($\in \mathbb{R}^n$) is the vector of decision variables, $\mathbf{x}_s$ ($\in \mathbb{R}^m$) is the vector of slack variables, $\mathbf{A}$ is the (m, n) technology matrix, $\mathbf{I}_m$ is the (m, m) identity matrix, and $\mathbf{b}$ ($\in \mathbb{R}^m$) is the vector of the right hand side values

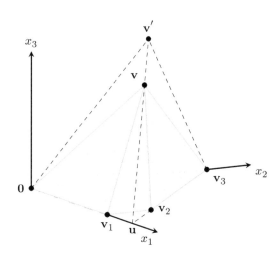

Figure 5.11: Degenerate optimal solution.

of the technology constraints. The corresponding dual model is:

$$\min\left\{ \begin{bmatrix} \mathbf{b}^\top & \mathbf{0}^\top \end{bmatrix} \begin{bmatrix} \mathbf{y} \\ \mathbf{y}_s \end{bmatrix} \,\middle|\, \begin{bmatrix} \mathbf{A}^\top & -\mathbf{I}_n \end{bmatrix} \begin{bmatrix} \mathbf{y} \\ \mathbf{y}_s \end{bmatrix} = \mathbf{c}, \ \mathbf{y}, \mathbf{y}_s \geq \mathbf{0} \right\},$$

where $\mathbf{y}$ ($\in \mathbb{R}^m$) is the vector of dual decision variables, and $\mathbf{y}_s$ ($\in \mathbb{R}^n$) is the vector of dual slack variables. (The sizes of the $\mathbf{0}$'s in the models above are omitted.)

We assume in this section that the feasible region of the primal model is nonempty, and that the model has an optimal solution. Recall from Section 4.3 that, for any basis matrix $\mathbf{B}$ in $\begin{bmatrix} \mathbf{A} & \mathbf{I}_m \end{bmatrix}$ corresponding to an optimal solution, the entries of the vector $\mathbf{x}_{BI}$ ($\in \mathbb{R}^m$) of primal basic variables and the entries of the vector $\mathbf{x}_{NI}$ ($\in \mathbb{R}^n$) of primal nonbasic variables satisfy:

$$\mathbf{x}_{BI} = \mathbf{B}^{-1}\mathbf{b} \geq \mathbf{0}, \text{ and } \mathbf{x}_{NI} = \mathbf{0}.$$

Similarly, for the complementary dual basis matrix $\overline{\mathbf{B}}$ of $\begin{bmatrix} \mathbf{A}^\top & -\mathbf{I}_n \end{bmatrix}$ (corresponding to an optimal solution of the dual model), the entries of the vector $\mathbf{y}_{BI^c}$ ($\in \mathbb{R}^n$) of dual basic variables and the entries of the vector $\mathbf{y}_{NI^c}$ ($\in \mathbb{R}^m$) of dual nonbasic variables satisfy:

$$\mathbf{y}_{BI^c} = \overline{\mathbf{B}}^{-1}\mathbf{c} \geq \mathbf{0}, \text{ and } \mathbf{y}_{NI^c} = \mathbf{0}.$$

Here, BI^c and NI^c are the index sets corresponding to $\overline{\mathbf{B}}$ and $\overline{\mathbf{N}}$, where $\overline{\mathbf{N}}$ is the complement of $\overline{\mathbf{B}}$ in $\begin{bmatrix} \mathbf{A}^\top & -\mathbf{I}_n \end{bmatrix}$ (see, e.g., Section 2.2.3 and Theorem 4.5.1).

5.6.1 Duality between multiple and degenerate optimal solutions

There is an interesting relationship between degeneracy in the primal model and multiplicity in the dual model, and vice versa. This relationship is formulated in Theorem 5.6.1. In the statement of the theorem, 'degeneracy' and 'multiplicity' always refer to vertices of standard LO-models (with inequality constraints). Recall, however, that 'degeneracy' is defined for vertices and feasible basic solutions. Note that an LO-model may have both degenerate and nondegenerate optimal solutions; see Exercise 5.8.8.

Theorem 5.6.1. (*Duality relationships between degeneracy and multiplicity*)
For any pair of standard primal and dual LO-models, where both have optimal solutions, the following implications hold:

	Primal model		**Dual model**
(a)	Multiple optimal solutions	$\Rightarrow$	All solutions optimal degenerate
(b)	Unique and nondegenerate optimal solution	$\Rightarrow$	Unique and nondegenerate optimal solution
(c)	Multiple optimal solutions, at least one nondegenerate	$\Rightarrow$	Unique and degenerate optimal solution
(d)	Unique and degenerate optimal solution	$\Rightarrow$	Multiple optimal solutions

Proof of Theorem 5.6.1. (a) It needs to be shown that each optimal dual basic solution is degenerate. To that end, take any optimal dual basic solution $\begin{bmatrix} \mathbf{y}^*_{BI^c} \\ \mathbf{y}_{NI^c} \end{bmatrix}$ with basis matrix $\overline{\mathbf{B}}$. Let $\mathbf{B}$ be its complementary dual basis matrix, which is in fact a primal basis matrix. Theorem 4.5.2 implies that $\mathbf{y}^*_{BI^c} = (\mathbf{B}^{-1}\mathbf{N})^{\mathsf{T}}\mathbf{c}_{BI} - \mathbf{c}_{NI} \geq \mathbf{0}$. The entries corresponding to $\mathbf{N}$ in the objective coefficients row of the simplex tableau are the entries of the vector $\mathbf{c}^{\mathsf{T}}_{NI} - \mathbf{c}^{\mathsf{T}}_{BI}\mathbf{B}^{-1}\mathbf{N}$, which is $\leq \mathbf{0}$. If all these entries are nonzero, then the optimal primal solution is unique, because no primal nonbasic variable can be increased without decreasing the value of the objective function. (Recall that it is assumed that the primal model is a 'maximizing' model.) Therefore, since there are multiple optimal primal solutions, at least one of these entries must be zero. Hence, at least one of the optimal dual basic variables is zero as well. This proves the degeneracy of all optimal dual basic solutions.

(b) The logical reversion of (a) is: if there exists at least one nondegenerate optimal dual basic solution, then the optimal primal solution is unique. The uniqueness in (b) follows now by interchanging the terms 'primal' and 'dual' in this logical reversion. So we only need to show the nondegeneracy of the optimal dual solution. Suppose, to the contrary, that the optimal dual solution is degenerate. Let $\begin{bmatrix} \mathbf{x}^*_{BI} \\ \mathbf{x}^*_{NI} \end{bmatrix}$ be the unique nondegenerate optimal primal basic solution with respect to the matrix $\mathbf{B}$ in $\begin{bmatrix} \mathbf{A} & \mathbf{I}_m \end{bmatrix}$. Let $\overline{\mathbf{B}}$ be the dual complement of $\mathbf{B}$, and let $\begin{bmatrix} \mathbf{y}^*_{BI^c} \\ \mathbf{y}^*_{NI^c} \end{bmatrix}$ be the corresponding optimal dual basic solution. Since the latter is degenerate, there exists an index $k \in BI^c$ such that $y^*_k = 0$. Let x_{NI_α} be the complementary dual variable of y_k (see Section 4.3). Let $\begin{bmatrix} \mathbf{A} & \mathbf{I}_m \end{bmatrix} \equiv \begin{bmatrix} \mathbf{B} & \mathbf{N} \end{bmatrix}$, and let $\mathbf{e}_\alpha$ be the α'th unit vector in $\mathbb{R}^n$. We will show that the vector $\begin{bmatrix} \hat{\mathbf{x}}_{BI} \\ \hat{\mathbf{x}}_{NI} \end{bmatrix}$ with $\hat{\mathbf{x}}_{BI} = \mathbf{x}^*_{BI} - \delta\mathbf{B}^{-1}\mathbf{N}\mathbf{e}_\alpha$ and $\hat{\mathbf{x}}_{NI} = \delta\mathbf{e}_\alpha$ satisfies $\mathbf{A}\hat{\mathbf{x}} \leq \mathbf{b}$, $\hat{\mathbf{x}} \geq \mathbf{0}$ for small enough $\delta > 0$. Since $\mathbf{x}^*_{BI} > \mathbf{0}$ (because of the nondegeneracy), it follows that

$\hat{\mathbf{x}}_{BI} > \mathbf{0}$ for small enough $\delta > 0$. Moreover,

$$\begin{bmatrix} \mathbf{B} & \mathbf{N} \end{bmatrix} \begin{bmatrix} \hat{\mathbf{x}}_{BI} \\ \hat{\mathbf{x}}_{NI} \end{bmatrix} = \mathbf{B}\hat{\mathbf{x}}_{BI} + \mathbf{N}\hat{\mathbf{x}}_{NI} = \mathbf{B}\mathbf{x}^*_{BI} - \delta \mathbf{N}\mathbf{e}_\alpha + \delta \mathbf{N}\mathbf{e}_\alpha = \mathbf{B}\mathbf{x}^*_{BI} = \mathbf{b}.$$

Hence, $\begin{bmatrix} \hat{\mathbf{x}}_{NI} \\ \hat{\mathbf{x}}_{BI} \end{bmatrix}$ is feasible. Because $\mathbf{y}^*_{NI^c} = \mathbf{0}$, $\hat{x}_{NI_\alpha} \times y^*_k = 0$, and $\hat{x}_i = 0$ for each $i \in NI \setminus \{NI_\alpha\}$, the complementary slackness relations

$$(\hat{\mathbf{x}}_{BI})^\mathsf{T} \mathbf{y}^*_{NI^c} = 0 \text{ and } (\hat{\mathbf{x}}_{NI})^\mathsf{T} \mathbf{y}^*_{BI^c} = 0,$$

hold. Therefore, according to Theorem 4.3.1, it follows that $\begin{bmatrix} \hat{\mathbf{x}}_{BI} \\ \hat{\mathbf{x}}_{NI} \end{bmatrix}$ is an optimal primal solution. So, there are multiple optimal primal solutions, contradicting the assumption that the primal model has a unique solution. So, the optimal dual solution is in fact nondegenerate.

(c) The degeneracy follows from (a), and the uniqueness from the logical reversion of (a) with 'primal' and 'dual' interchanged.

(d) Let $\begin{bmatrix} \mathbf{x}^*_{BI} \\ \mathbf{x}^*_{NI} \end{bmatrix}$ be an optimal primal basic solution and let $\begin{bmatrix} \mathbf{y}^*_{BI^c} \\ \mathbf{y}^*_{NI^c} \end{bmatrix}$ be the corresponding optimal dual basic solution. Since the optimal primal basic solution is degenerate, it follows that $x^*_i = 0$ for some $i \in BI$. Let y_j be the complementary dual variable of x_i. Since x_i is a basic primal variable, y_j is a nonbasic dual variable, and therefore we have that $y^*_j = 0$. On the other hand, Theorem 4.3.3 implies that there exists a pair of optimal (not necessarily basic) primal and dual solutions $\mathbf{x}$ and $\mathbf{y}$ for which either $x_i > 0$ or $y_j > 0$. Since the optimal primal solution is unique and $x^*_i = 0$, it follows that there is an optimal dual solution $\mathbf{y}^*$ for which $y^*_j > 0$. Hence, the dual model has multiple optimal solutions. $\qquad\square$

In case (a) in Theorem 5.6.1, there is either a unique optimal dual solution or there are multiple optimal dual solutions; in case (d), the optimal dual solution is either degenerate or nondegenerate. This can be illustrated by means of the following two examples. We first give an example concerning (d).

Example 5.6.1.   *Consider the following primal LO-model and its dual model:*

$$\begin{array}{lll} \max & x_1 - x_2 & \\ \text{s.t.} & x_1 - x_2 + x_3 \leq 1 & \\ & x_1 \qquad\qquad \leq 1 & \\ & x_1, x_2, x_3 \geq 0, & \end{array} \qquad \begin{array}{lll} \min & y_1 + y_2 & \\ \text{s.t.} & y_1 + y_2 \geq & 1 \\ & -y_1 \qquad \geq & -1 \\ & y_1 \qquad \geq & 0 \\ & y_1, y_2 \geq 0. & \end{array}$$

The primal model has the unique degenerate optimal solution $\begin{bmatrix} 1 & 0 & 0 & 0 & 0 \end{bmatrix}^\mathsf{T}$ (the last two zeros of this vector are the values of the slack variables); see Figure 5.12. Using the graphical solution method (see Section 1.1.2), one can easily verify that the dual model has multiple optimal solutions, namely all points of the line segment with end points $\begin{bmatrix} 1 & 0 \end{bmatrix}^\mathsf{T}$ and $\begin{bmatrix} 0 & 1 \end{bmatrix}^\mathsf{T}$. The end vertex $\begin{bmatrix} 1 & 0 \end{bmatrix}^\mathsf{T}$ is degenerate, while the vertex $\begin{bmatrix} 0 & 1 \end{bmatrix}^\mathsf{T}$ is nondegenerate.

The following example concerns (a) of Theorem 5.6.1.

Figure 5.12: Degenerate vertex $\begin{bmatrix} 1 & 0 & 0 \end{bmatrix}^\mathsf{T}$.

Example 5.6.2. *Consider the model with the same feasible region as in Example 5.6.1, but with objective* $\max x_1$. *From Figure 5.12, it follows that this model has multiple optimal solutions: all points in the 'front' face of the feasible region as drawn in Figure 5.12 are optimal. Mathematically, this set can be described as the cone with apex* $\begin{bmatrix} 1 & 0 & 0 \end{bmatrix}^\mathsf{T}$ *and extreme rays the halflines through* $\begin{bmatrix} 1 & 1 & 0 \end{bmatrix}^\mathsf{T}$ *and* $\begin{bmatrix} 1 & 1 & 1 \end{bmatrix}^\mathsf{T}$. *(See Appendix D for the definitions of the concepts in this sentence.) Note that the optimal solution corresponding to the vertex* $\begin{bmatrix} 1 & 0 & 0 \end{bmatrix}^\mathsf{T}$ *is degenerate. The dual of this model reads:*

$$
\begin{aligned}
\min \quad & y_1 + y_2 \\
\text{s.t.} \quad & y_1 + y_2 \geq 1 \\
& -y_1 \phantom{{}+y_2} \geq 0 \\
& y_1 \phantom{{}+y_2} \geq 0 \\
& y_1, y_2 \geq 0.
\end{aligned}
$$

Obviously, this model has a unique optimal solution, namely $\begin{bmatrix} 0 & 1 & 0 & 0 & 0 \end{bmatrix}^\mathsf{T}$ *which is degenerate (the last three zeros refer to the values of the slack variables).*

5.6.2 Left and right shadow prices

In Section 5.3, shadow prices of constraints were discussed in the case of nondegenerate optimal primal solutions. In this section we will consider the degenerate case.

Let $\delta \in \mathbb{R}$. Recall that a perturbation of a constraint by a factor δ means changing its right hand side b into $b + \delta$. Note that a perturbation of the nonnegativity constraint $x_i \geq 0$ by a factor δ yields $x_i \geq \delta$. If $\delta > 0$, then the perturbation is called *positive*, and if $\delta < 0$ then it is called *negative*. We start with an example to gain some intuition.

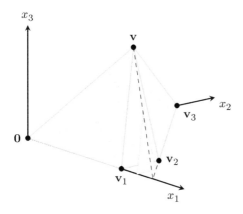

Figure 5.13: Degenerate optimal vertex **v** and a positive perturbation of the redundant constraint $\mathbf{vv_1v_2}$.

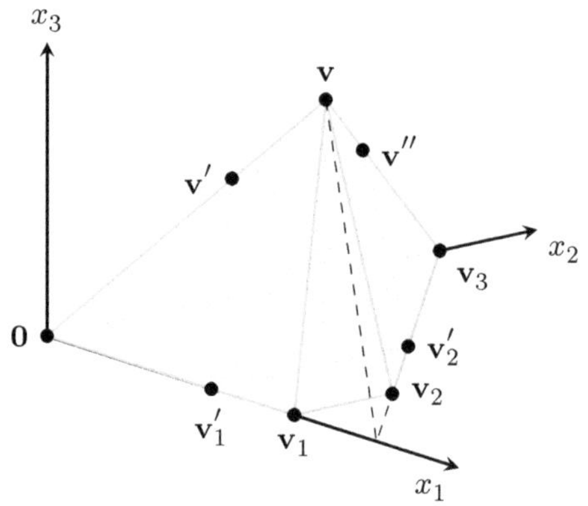

Figure 5.14: Degenerate optimal vertex **v** and a negative perturbation of $\mathbf{vv_1v_2}$.

Example 5.6.3. *Consider again the situation of Example 5.6.5. The LO-model corresponding to Figure 5.11 in Section 5.6.1 is:*

$$\begin{array}{lrcl} \max & x_3 & & \\ \text{s.t.} & 3x_1 + 2x_3 & \leq & 9 \\ & -3x_1 + x_3 & \leq & 0 \\ & -3x_2 + 2x_3 & \leq & 0 \\ & 3x_1 + 3x_2 + x_3 & \leq & 12 \\ & x_1, x_2, x_3 & \geq & 0. \end{array}$$

The feasible region of this model is depicted in Figure 5.13 and in Figure 5.14, namely the region $\mathbf{0vv_1v_2v_3}$. *The optimal degenerate vertex is* **v**. *In Figure 5.13, the constraint* $3x_1 + 2x_3 \leq 9$ *is perturbed by a positive factor, and in Figure 5.14 by a negative factor.*

In the case of Figure 5.13, the constraint corresponding to $\mathbf{vv_1v_2}$ *has moved 'out of' the feasible region. This change enlarges the feasible region, while* **v** *remains the optimal vertex. Note that, once the constraint has moved by a distance* $\varepsilon > 0$, *the optimal vertex* **v** *becomes nondegenerate.*

On the other hand, when moving the constraint through $\mathbf{vv_1v_2}$ *'into' the feasible region, the feasible region shrinks, and the original optimal vertex* **v** *is cut off. So, the objective value decreases; see Figure 5.14.*

In Figure 5.15 the perturbation function for the constraint $3x_1 + 2x_3 \leq 9$ *is depicted. It shows the optimal objective value* z^* *as a function of* δ, *when in the above model the first constraint is replaced by* $3x_1 + 2x_3 \leq 9 + \delta$. *For* $\delta < -9$ *the model is infeasible. Note that* $z^*(\delta) = 3 + \frac{1}{3}\delta$ *if* $\delta \leq 0$, *and* $z^*(\delta) = 3$ *if* $\delta \geq 0$. *Remarkable for this perturbation function is that its graph shows a 'kink' for* $\delta = 0$. *Such kinks can only occur when the corresponding vertex is degenerate; see Theorem 5.6.2. Compare this, for example, to Figure 5.6 and Figure 5.7. The reader may check that, at each kink point in these figures, the corresponding optimal solution is degenerate.*

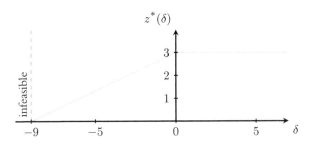

Figure 5.15: Perturbation function for '$\mathbf{v}_1\mathbf{v}_2\mathbf{v}$' in Figure 5.13 and Figure 5.14.

Recall that, assuming that the optimal solutions are nondegenerate, the shadow price of a constraint in a standard LO-model is precisely the slope of the perturbation function at $\delta = 0$. We have seen that shadow prices correspond to the optimal solution of the dual model. The fact that the optimal dual solution is unique is guaranteed by the fact that the optimal primal solution is assumed to be nondegenerate; see Theorem 5.6.1. As soon as the optimal primal solution is degenerate, then (again see Theorem 5.6.1) there may be multiple optimal dual solutions, and so shadow prices cannot, in general, be directly related to optimal dual solutions.

Based on these considerations, we introduce left and right shadow prices. The *right shadow price* of a constraint is the rate at which the objective value increases by positive perturbations of the right hand side of that constraint. Similarly, the *left shadow price* of a constraint is the rate at which the objective value decreases by negative perturbations of its right hand side. Formally, we have that:

$$\text{left shadow price of the constraint} = \lim_{\delta\uparrow 0} \frac{z^*(\delta) - z^*(0)}{\delta}, \text{ and}$$
$$\text{right shadow price of the constraint} = \lim_{\delta\downarrow 0} \frac{z^*(\delta) - z^*(0)}{\delta},$$

where $z^*(\delta)$ is the optimal objective value of the model with the constraint perturbed by a factor δ. In other words, the left shadow price (right shadow price) is the slope of the line segment of $z^*(\delta)$ that lies 'left' ('right') of $\delta = 0$. Compare in this respect the definition (5.5) of the (regular) shadow price of a constraint.

For example, the left shadow price of the constraint $3x_1 + 2x_2 \leq 9$ of the model in Example 5.6.3 is $\frac{1}{3}$ (see Figure 5.15). On the other hand, the right shadow price is zero. This example also shows that a binding constraint may have a zero left or right shadow price. In fact, it may even happen that all binding constraints have zero right shadow prices; it is left to the reader to construct such an example.

In Theorem 5.6.2, we show that the left and right shadow prices can be determined by solving the dual model. In fact, the expression $\max\{y_j \mid \mathbf{y} \text{ is an optimal dual solution}\}$ in the theorem may be computed by determining the optimal objective value z^* of the dual model, and then solving the dual model with the additional constraint $\mathbf{b}^\mathsf{T}\mathbf{y} = z^*$.

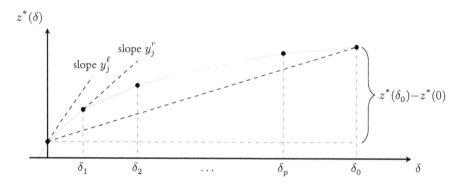

Figure 5.16: Left and right shadow prices.

The theorem applies to the standard primal LO-model. It is left to the reader to make the usual adaptations for nonstandard LO-models, and to formulate similar assertions for nonnegativity and nonpositivity constraints; see Exercise 5.8.21.

Theorem 5.6.2. (*Shadow prices in the degenerate case*)
Consider a standard primal LO-model. Let $j \in \{1, \ldots, m\}$. Assume that the model does not become unbounded or infeasible when the right hand side of constraint j is perturbed by a small amount. The left shadow price y_j^l and the right shadow price y_j^r of constraint j satisfy:

$$y_j^l = \max\{y_j^* \mid \mathbf{y}^* \text{ is an optimal dual solution}\}, \text{ and}$$
$$y_j^r = \min\{y_j^* \mid \mathbf{y}^* \text{ is an optimal dual solution}\},$$

where y_j^* is the j'th entry of $\mathbf{y}^*$.

Proof of Theorem 5.6.2. Take any $j \in \{1, \ldots, m\}$. We only prove the expression in the theorem for y_j^r, because the proof for y_j^l is similar. Perturbing the right hand side value of constraint j is equivalent to perturbing the objective coefficient of y_j in the dual model. So, we have that:

$$z^*(\delta) = \min\left\{\mathbf{b}^\mathsf{T}\mathbf{y} + \delta y_j \mid \mathbf{A}^\mathsf{T}\mathbf{y} \geq \mathbf{c}, \mathbf{y} \geq \mathbf{0}\right\}. \tag{5.11}$$

Recall from Section 5.2 that every optimal feasible solution of the dual model (corresponding to $\delta = 0$) has a tolerance interval with respect to the perturbation of the objective coefficient of y_j. Let $\mathbf{y}^*$ be an optimal dual solution with tolerance interval $[0, \delta_0]$ such that $\delta_0 > 0$. Such a solution exists, because if every optimal dual solution has a tolerance interval $[0, 0]$, then the dual model cannot have an optimal solution for any sufficiently small $\delta > 0$, contrary to the assumptions of the theorem.

Variable	Optimal value	Dual variable (slack var.)	Objective coefficient	Min.	Max.
x_1	9	0	2	1.5	3
x_2	0	1	1	$-\infty$	2
x_3	3	0	3	2	4
x_4	0	3	1	$-\infty$	4
Constraint	**Slack**	**Dual variable (decision var.)**	**Right hand side value**	**Min.**	**Max.**
1	0	1	12	12	15
2	0	0	24	24	∞
3	0	1	15	12	15
4	32	0	17	-15	∞

The optimal objective value is 27.

Table 5.3: Example LO-package output for model (5.12).

Recall from Section 5.2 that for any value of δ in the tolerance interval $[0, \delta_0]$, the optimal objective value of the perturbed dual model (5.11) is equal to $\mathbf{b}^\mathsf{T}\mathbf{y}^* + \delta y_j^*$, i.e., we have that $z^*(\delta) = \mathbf{b}^\mathsf{T}\mathbf{y}^* + \delta y_j^*$ for $\delta \in [0, \delta_0]$. Hence, the right shadow price y_j^r of constraint j is equal to y_j^*.

It remains to show that $y_j^* = \min\{y_j \mid \mathbf{y} \text{ is an optimal dual solution}\}$. Suppose for a contradiction that this is not the case, i.e., suppose that there exists a dual optimal solution $\tilde{\mathbf{y}}^*$ such that $\tilde{y}_j^* < y_j^*$. Since both $\mathbf{y}^*$ and $\tilde{\mathbf{y}}^*$ are optimal dual solutions, their corresponding objective values are equal. Moreover, the objective values corresponding to these dual solutions in the perturbed model (5.11) with $\delta \in (0, \delta_0]$ satisfy:

$$(\mathbf{b} + \delta \mathbf{e}_j)^\mathsf{T}\tilde{\mathbf{y}}^* = \mathbf{b}^\mathsf{T}\tilde{\mathbf{y}}^* + \delta \tilde{y}_j^* < \mathbf{b}^\mathsf{T}\mathbf{y}^* + \delta y_j^* = (\mathbf{b} + \delta \mathbf{e}_j)^\mathsf{T}\mathbf{y}^*.$$

Therefore, $\tilde{\mathbf{y}}^*$ has a smaller corresponding objective value than $\mathbf{y}^*$. It follows that $\mathbf{y}^*$ cannot be an optimal dual solution, a contradiction. $\qquad\square$

The following example illustrates Theorem 5.6.2.

Example 5.6.4. *Consider the following LO-model:*

$$\begin{aligned}
\max \quad & 2x_1 + x_2 + 3x_3 + x_4 & (5.12)\\
\text{s.t.} \quad & x_1 + x_2 + x_3 + x_4 \leq 12 & (1)\\
& x_1 - 2x_2 + 5x_3 - x_4 \leq 24 & (2)\\
& x_1 + x_2 + 2x_3 + 3x_4 \leq 15 & (3)\\
& -2x_1 + 3x_2 + x_3 + x_4 \leq 17 & (4)\\
& x_1, x_2, x_3, x_4 \geq 0.
\end{aligned}$$

Table 5.3 shows a possible output of an LO-package after solving this model. In the third column of Table 5.3 we have written 'slack variable' and 'decision variable' in parentheses. Note that, in the case of a nondegenerate optimal solution, the shadow prices of the technology constraints are the values of the corresponding dual decision variables, and the shadow prices of the nonnegativity constraints are the negatives of the values of the corresponding dual slack variables. The last two columns contain the upper and lower bounds of the respective tolerance intervals.

In Figure 5.17 the perturbation function for the first constraint is drawn. This perturbation function was determined using an LO-package by calculating the optimal solutions for a number of right hand side values of constraint (1), and using the bounds of the tolerance intervals which are available as output of the package.

Based on the fact that there are two decision variables and three slack variables with optimal value zero, it follows that one basic variable must have value zero, and so this optimal solution is degenerate. Moreover, this solution is unique, because otherwise the dual solution would have been degenerate (see Theorem 5.6.1(d)). Since the optimal values of x_1, x_3, and x_8 are nonzero, the optimal solution corresponds to the feasible basic solutions with $BI = \{1, 3, 8, i\}$, where $i = 2, 4, 5, 6, 7$. As it turns out, only the feasible basic solutions with $BI = \{1, 3, 6, 8\}$ and $BI = \{1, 3, 7, 8\}$ are optimal (i.e., have nonpositive objective coefficients). The corresponding simplex tableaus are:

$BI = \{1, 3, 6, 8\}.$

x_1	x_2	x_3	x_4	x_5	x_6	x_7	x_8	$-z$
0	-1	0	-3	-1	0	-1	0	-27
1	1	0	-1	2	0	-1	0	9
0	0	1	2	-1	0	1	0	9
0	-3	0	-10	3	1	-4	0	0
0	5	0	-3	5	0	-3	1	32

The corresponding optimal dual solution reads: $y_1^* = 1$, $y_2^* = 0$, $y_3^* = 1$, and $y_4^* = 0$.

$BI = \{1, 3, 7, 8\}.$

x_1	x_2	x_3	x_4	x_5	x_6	x_7	x_8	$-z$
0	$-\frac{1}{4}$	0	$-\frac{1}{2}$	$-1\frac{3}{4}$	$-\frac{1}{4}$	0	0	-27
1	$1\frac{3}{4}$	0	$1\frac{1}{2}$	$1\frac{1}{4}$	$-\frac{1}{4}$	0	0	9
0	$-\frac{3}{4}$	1	$-\frac{1}{2}$	$-\frac{1}{4}$	$\frac{1}{4}$	0	0	9
0	$\frac{3}{4}$	0	$2\frac{1}{2}$	$-\frac{3}{4}$	$-\frac{1}{4}$	1	0	0
0	$7\frac{1}{4}$	0	$4\frac{1}{2}$	$2\frac{3}{4}$	$-\frac{3}{4}$	0	1	32

The corresponding optimal dual solution reads: $y_1^* = 1\frac{3}{4}$, $y_2^* = -\frac{1}{4}$, $y_3^* = 0$, and $y_4^* = 0$.

Note that the solution shown in Table 5.3 is the feasible basic solution corresponding to $BI = \{1, 3, 6, 8\}$. Which feasible basic solution is found by a computer package depends on the implementation details of the package: it could well be that a different LO-package returns the other feasible basic solution, i.e., the one corresponding to $BI = \{1, 3, 7, 8\}$.

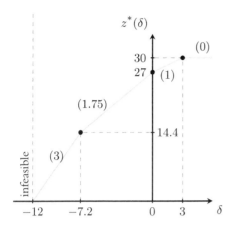

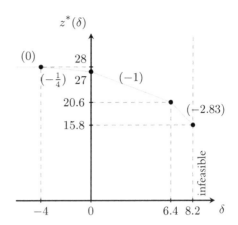

Figure 5.17: Perturbation function for constraint (3) of model (5.12). The numbers in parentheses are the slopes.

Figure 5.18: Perturbation function for nonnegativity constraint $x_2 \geq 0$ of model (5.12). The numbers in parentheses are the slopes.

Using Theorem 5.6.2, it now follows that:

$$y_1^l = \text{the left shadow price of constraint (1)} = \max\left\{1, 1\tfrac{3}{4}\right\} = 1\tfrac{3}{4}, \text{ and}$$
$$y_1^r = \text{the right shadow price of constraint (1)} = \min\left\{1, 1\tfrac{3}{4}\right\} = 1.$$

This can fact can be seen in Figure 5.17, because the line segment left of $\delta = 0$ has slope $1\tfrac{3}{4}$, and the line segment right of $\delta = 0$ has slope 1.

5.6.3 Degeneracy, binding constraints, and redundancy

Recall that only binding constraints may have nonzero shadow prices; the shadow prices of nonbinding constraints are always equal to zero (see Theorem 5.3.1 and Theorem 5.3.2). Consider a standard LO-model with $n \geq 1$ decision variables. While in the nondegenerate case there are precisely n binding constraints at an optimal primal vertex, in the case of degeneracy there are more than n binding constraints.

Example 5.6.5. *In Figure 5.11, a three-dimensional example of degeneracy is depicted. In the figure, we have that $\mathbf{v}_1 = \begin{bmatrix} 3 & 0 & 0 \end{bmatrix}^\mathsf{T}$, $\mathbf{v}_2 = \begin{bmatrix} 3 & 1 & 0 \end{bmatrix}^\mathsf{T}$, $\mathbf{v}_3 = \begin{bmatrix} 0 & 4 & 0 \end{bmatrix}^\mathsf{T}$, and $\mathbf{v} = \begin{bmatrix} 1 & 2 & 3 \end{bmatrix}^\mathsf{T}$. The feasible region consists of a pyramid with base $\mathbf{0}\mathbf{u}\mathbf{v}_3$ and top $\mathbf{v}$, where the region formed by the points $\mathbf{u}$, $\mathbf{v}_1$, $\mathbf{v}_2$, and $\mathbf{v}$ has been 'chopped off'. Let the objective be $\max x_3$; see Section 5.6.2 for the LO-model corresponding to this figure. The point $\mathbf{v}$ in the figure is the optimal vertex. At this optimal vertex, there are four binding constraints. Removing the hyperplane through $\mathbf{0}\mathbf{v}_3\mathbf{v}$ changes the feasible region into $\mathbf{0}\mathbf{v}_1\mathbf{v}_2\mathbf{v}_3\mathbf{v}'$ with $\mathbf{v}' = \begin{bmatrix} 0 & 2\tfrac{2}{3} & 4 \end{bmatrix}^\mathsf{T}$ in the plane $\mathbf{0}x_2x_3$. The line through $\mathbf{v}\mathbf{v}'$ is the intersection of the planes through $\mathbf{0}\mathbf{v}_1\mathbf{v}$ and $\mathbf{v}_2\mathbf{v}_3\mathbf{v}$, and intersects the x_1-axis in $\mathbf{u}$. The optimal vertex of the new region is $\mathbf{v}' = \begin{bmatrix} 0 & 2\tfrac{2}{3} & 4 \end{bmatrix}^\mathsf{T}$. The optimal objective value changes from $z^* = 3$ to $z^* = 4$.*

Recall that a constraint is *redundant with respect to the feasible region* if removing the constraint from the model does not change the feasible region. Removing a nonredundant (with respect to the feasible region) constraint from the model enlarges the feasible region. If the optimal objective value remains unchanged, then more (extreme) points of the feasible region may become optimal. If the optimal objective value changes, a new set of points becomes optimal, or the model becomes unbounded.

A constraint is called *redundant with respect to the optimal point* $\mathbf{v}$, if $\mathbf{v}$ is also an optimal point of the feasible region after removing the constraint from the model. Note that if a constraint is redundant with respect to the feasible region, then it is also redundant with respect to any optimal point. The converse, however, is not true: it may happen that a constraint is redundant with respect to an optimal point $\mathbf{v}$, while it is not redundant with respect to the feasible region. Note that, in the latter case, removing the constraint may also change the set of optimal points. However, the new set of optimal points includes $\mathbf{v}$.

Example 5.6.6. *Consider again the feasible region of Example 5.6.5. The constraint corresponding to the face $\mathbf{v}\mathbf{v}_1\mathbf{v}_2$ is redundant with respect to $\mathbf{v}$, because its removal does not change the optimality of this vertex. The constraint is, however, not redundant with respect to the feasible region, because removing it changes the feasible region into a pyramid with base $\mathbf{0}\mathbf{u}\mathbf{v}_3$ and top $\mathbf{v}$. In fact, none of the constraints are redundant with respect to the feasible region.*

From this example, we may also conclude that, in $\mathbb{R}^n$, if there are more than n binding constraints at an optimal vertex (i.e., if the optimal vertex is degenerate), then this does not mean that all of these n constraints are redundant with respect to the optimal vertex. However, in Theorem 5.6.3, we show that in the case of a degenerate optimal feasible basic solution, corresponding to vertex $\mathbf{v}$, say, there is at least one binding constraint that is redundant with respect to $\mathbf{v}$. Let $k \geq 0$ be an integer. A vertex $\mathbf{v}$ of the feasible region of an LO-model is called k-*degenerate* if the number of binding constraints at $\mathbf{v}$ is $n + k$. For $k = 0$, $\mathbf{v}$ is nondegenerate, and so '0-degenerate' is another way of saying 'nondegenerate'. Similarly, an optimal feasible basic solution is called k-degenerate if and only if the number of zero basic variables is k. Note that the vertex $\mathbf{v}$ of the feasible region $\mathbf{0}\mathbf{v}\mathbf{v}_1\mathbf{v}_2\mathbf{v}_3$ in Figure 5.14 is 1-degenerate, and that the vertex $\mathbf{0}$ is 2-degenerate. There is an interesting duality relationship in this respect: the dimension of the dual optimal face is equal to degeneracy degree of the primal optimal face; see Tijssen and Sierksma (1998).

Theorem 5.6.3. (*Degeneracy and redundancy*)
Consider a standard primal LO-model. Let $k \geq 0$. If the model has a k-degenerate optimal vertex $\mathbf{v}$, then at least k binding constraints at $\mathbf{v}$ are redundant with respect to the optimal solution. The technology constraints among them have zero right shadow prices, and the nonnegativity constraints among them have zero left shadow prices.

Proof of Theorem 5.6.3. Let $k \geq 0$, let $\begin{bmatrix} \mathbf{x}_{BI}^* \\ \mathbf{x}_{NI}^* \end{bmatrix}$ be a k-degenerate optimal basic solution, and let $\mathbf{v}$ be the corresponding optimal vertex. Then, $\mathbf{x}_{NI}^* = \mathbf{0}$, and $\mathbf{x}_{BI}^*$ contains k zero entries. Let $S \subseteq BI$ with $|S| = k$ be the set of indices of the k zero entries, i.e., $\mathbf{x}_S = \mathbf{0}$. The binding constraints at $\mathbf{v}$ can be partitioned into two subsets: the n binding constraints corresponding to NI, namely $\mathbf{x}_{NI} \geq \mathbf{0}$, and the k binding constraints corresponding to S ($\subseteq BI$), namely $\mathbf{x}_S \geq \mathbf{0}$. The objective value of any point $\begin{bmatrix} \hat{\mathbf{x}}_{BI} \\ \hat{\mathbf{x}}_{NI} \end{bmatrix}$ is (see Section 3.2):

$$\mathbf{c}_{BI}^\mathsf{T}\mathbf{B}^{-1}\mathbf{b} + (\mathbf{c}_{NI}^\mathsf{T} - \mathbf{c}_{BI}^\mathsf{T}\mathbf{B}^{-1}\mathbf{N})\hat{\mathbf{x}}_{NI}.$$

Since $\begin{bmatrix} \mathbf{x}_{BI}^* \\ \mathbf{x}_{NI}^* \end{bmatrix}$ is an optimal basic solution, we have that $\mathbf{c}_{NI}^\mathsf{T} - \mathbf{c}_{BI}^\mathsf{T}\mathbf{B}^{-1}\mathbf{N} \leq \mathbf{0}$. It follows that the objective value of any point $\begin{bmatrix} \hat{\mathbf{x}}_{BI} \\ \hat{\mathbf{x}}_{NI} \end{bmatrix}$ that satisfies $\hat{\mathbf{x}}_{NI} \geq \mathbf{0}$ is at most the objective value of $\mathbf{v}$ (which corresponds to $\mathbf{x}_{NI} = \mathbf{0}$). Therefore, if we remove the k hyperplanes defined by $\mathbf{x}_S \geq \mathbf{0}$ (while retaining the ones defined by $\mathbf{x}_{NI} = \mathbf{0}$), the current optimal vertex $\mathbf{v}$ remains optimal, and the optimal objective value does not change. Therefore, the k hyperplanes corresponding to BI are redundant with respect to the optimal solution. It also follows that the technology constraints among these k constraints have zero right shadow prices, and the nonnegativity constraints have zero left shadow prices. Note that, since removing these k constraints may enlarge the feasible region, they are not necessarily redundant with respect to the feasible region (see the beginning of this section). $\qquad \square$

Perturbation functions of nonnegativity constraints can be determined in the same way as those of constraints. As an example, we have calculated the perturbation function for the binding nonnegativity constraint $x_2 \geq 0$ of the model in Example 5.6.4; see Figure 5.18. The left shadow price of $x_2 \geq 0$ is $\frac{1}{4}$, and its right shadow price is 1. This is consistent with the optimal values of the dual variable y_2 in the two optimal dual feasible basic solutions shown in the simplex tableau above (1 and $\frac{1}{4}$, respectively). It is left to the reader to draw the perturbation functions of the remaining constraints. This example shows that, for maximizing models, perturbation functions of nonnegativity constraints are in fact piecewise linear, concave, and nonincreasing; see Theorem 5.4.2.

We close this section with a characterization of general LO-models models with multiple optimal solutions in terms of left and right shadow prices. Theorem 5.6.4 shows that the fact that a model has multiple optimal solutions can be recognized by the existence of a binding constraint for which the perturbation function is 'horizontal' in a neighborhood of $\delta = 0$.

Theorem 5.6.4. (*Multiple optimal solutions*)
Consider an LO-model that has an optimal vertex. The model has multiple optimal solutions if and only if for each optimal feasible basic solution there exists a binding constraint with zero left and right shadow prices.

Proof of Theorem 5.6.4. First, assume that the model has multiple optimal solutions. Let $\mathbf{x}^*$ be any optimal vertex with corresponding optimal objective value z^*. Since the model has multiple optimal solutions, there exists a different optimal solution, $\tilde{\mathbf{x}}^*$, say, that is not a vertex. The set of binding constraints at $\mathbf{x}^*$ is different from the set of binding constraints at $\tilde{\mathbf{x}}^*$. Hence, there exists a constraint j that is binding at $\mathbf{x}^*$ but not binding at $\tilde{\mathbf{x}}^*$. We will prove the statement for the case where constraint j is a technology constraint; the case of nonnegativity constraints is similar and is left for the reader. Because constraint j is not binding at $\tilde{\mathbf{x}}^*$, perturbing it by a sufficiently small factor $\delta < 0$ does not decrease the optimal objective value (because $\tilde{\mathbf{x}}^*$ is still feasible and has the objective value z^*). Because the perturbation shrinks the feasible region, it also does not increase the optimal objective value. This means that constraint j has zero left shadow price, i.e., $y_j^l = 0$. It follows now from Theorem 5.6.2 that $0 \leq y_j^r \leq y_j^l = 0$, and hence $y_j^r = 0$.

The reverse can be shown as follows. Let $\mathbf{x}^*$ be any optimal feasible basic solution, and assume that there is a constraint j with $y_j^r = y_j^l = 0$ that is binding at $\mathbf{x}^*$. Let $z^*(\delta)$ be the optimal objective value of the model with the constraint i perturbed by the factor δ. Since the right shadow price of constraint i is zero, it follows that $z^*(\delta) = z^*(0)$ for small enough $\delta > 0$. Let $\tilde{\mathbf{x}}^*$ be an optimal feasible basic solution corresponding to the perturbation factor δ. Since the unperturbed constraint i is binding at $\mathbf{x}^*$ but not at $\tilde{\mathbf{x}}^*$, it follows that $\mathbf{x}^* \neq \tilde{\mathbf{x}}^*$. Because $\mathbf{x}^*$ and $\tilde{\mathbf{x}}^*$ have the same corresponding objective value $z^*(0)$, it follows that there are multiple optimal solutions. $\square$

Note that the requirement that the model has an optimal vertex is necessary. For example, the nonstandard LO-model $\max\{5x_1 \mid x_1 \leq 1, x_2 \text{ free}\}$ has multiple optimal solutions because any point $\begin{bmatrix} 1 & x_2 \end{bmatrix}^\mathsf{T}$ with $x_2 \in \mathbb{R}$ is optimal, but the left and right shadow price of the only constraint $x_1 \leq 1$ is 5.

The binding constraints with zero shadow prices in the statement of Theorem 5.6.4 have to be binding at the optimal vertices of the feasible region. Consider, for example, the model $\min\{x_2 \mid x_1 + x_2 \leq 1, \ x_1, x_2 \geq 0\}$. The set $\{\begin{bmatrix} x_1 & x_2 \end{bmatrix}^\mathsf{T} \mid 0 \leq x_1 \leq 1, x_2 = 0\}$ is optimal. In the case of the optimal solution $x_1 = \frac{1}{2}$, $x_2 = 0$, the only binding constraint is $x_2 \geq 0$, which has nonzero right shadow price.

5.7 Shadow prices and redundancy of equality constraints

In Section 5.3 we studied perturbations of right hand side values of LO-models for inequality constraints. In the case of a maximizing model, the shadow prices of '$\leq$' inequalities are always nonnegative, and so increments of right hand side values always result in nondecreasing optimal objective values. However, in the case of an equality constraint, it may happen that an increase of a right hand side results in a decrease of the optimal objective value. This phenomenon is called the *more-for-less paradox*; see also Section 1.6.1. It will be explained by means of the following example.

Example 5.7.1. *Consider the LO-model:*

$$\begin{array}{ll} \max & x_1 + x_2 \\ \text{s.t.} & -x_1 + x_2 = 1 \\ & x_2 \leq 2 \\ & x_1, x_2 \geq 0. \end{array}$$

One can easily check (for instance by using the graphical solution method) that the optimal solution reads $x_1^* = 1$, $x_2^* = 2$, and $z^* = 3$. When decreasing the (positive) right hand side of $-x_1 + x_2 = 1$ by $\frac{1}{2}$, the perturbed model becomes:

$$\begin{array}{ll} \max & x_1 + x_2 \\ \text{s.t.} & -x_1 + x_2 = \frac{1}{2} \\ & x_2 \leq 2 \\ & x_1, x_2 \geq 0. \end{array}$$

The optimal solution of this model reads $x_1^* = 1\frac{1}{2}$, $x_2^* = 2$, and $z^* = 3\frac{1}{2}$. Hence, the optimal objective value increases when the right hand side of $-x_1 + x_2 = 1$ decreases; i.e., there is _more_ 'revenue' for _less_ 'capacity'. Recall that increasing the right hand side of the second constraint results in an increase of the optimal objective value.

This 'paradox' is due to the fact that dual variables corresponding to primal equality constraints may be negative. This will be explained in Theorem 5.7.1. The theorem concerns nondegenerate optimal solutions for LO-models with equality constraints.

Degeneracy in the case of LO-models with equality constraints means that the corresponding feasible basic solution contains at least one zero entry, with the basis matrix defined as in Section 1.3. Degeneracy in this case has the same geometric interpretation as in the case of models with only inequality constraints: the number of hyperplanes corresponding to binding equalities or inequalities is at least $n + 1$, with n the number of decision variables.

Note that rewriting the above model in standard form (i.e., with inequality constraints only) leads to a degenerate optimal solution, because equality constraints are split into two linearly dependent inequality constraints.

Theorem 5.7.1. (*Shadow prices for models with equality constraints*)
The shadow price of an equality constraint, in an LO-model with a nondegenerate optimal solution, is equal to the optimal value of the corresponding free dual variable.

Proof of Theorem 5.7.1. Consider the following LO-model with one equality constraint (the other equality constraints may be considered as hidden in $\mathbf{Ax} \leq \mathbf{b}$):

$$\max\left\{\mathbf{c}^\mathsf{T}\mathbf{x} \,\middle|\, \mathbf{Ax} \leq \mathbf{b}, \mathbf{a}^\mathsf{T}\mathbf{x} = b', \mathbf{x} \geq \mathbf{0}\right\},$$

where $\mathbf{A}$ is an (m,n)-matrix, $\mathbf{c} \in \mathbb{R}^n$, $\mathbf{b} \in \mathbb{R}^m$, $\mathbf{a} \in \mathbb{R}^n$, and $b' \geq 0$. The dual of this model is (see Section 4.2.2):

$$\min\left\{\mathbf{b}^\mathsf{T}\mathbf{y} + b'y' \,\middle|\, \mathbf{A}^\mathsf{T}\mathbf{y} + \mathbf{a}y' \geq \mathbf{c}, \mathbf{y} \geq \mathbf{0}, y' \in \mathbb{R}\right\}.$$

Note that the dual variable y', corresponding to the equality constraint, is free. Now let b' be perturbed by a small amount δ ($\in \mathbb{R}$). The perturbed model becomes:

$$\max\left\{\mathbf{c}^\mathsf{T}\mathbf{x} \,\middle|\, \mathbf{Ax} \leq \mathbf{b}, \mathbf{a}^\mathsf{T}\mathbf{x} = b' + \delta, \mathbf{x} \geq \mathbf{0}\right\}.$$

This is equivalent to:

$$\max\left\{\mathbf{c}^\mathsf{T}\mathbf{x} \,\middle|\, \mathbf{Ax} \leq \mathbf{b}, \mathbf{a}^\mathsf{T}\mathbf{x} \leq b' + \delta, -\mathbf{a}^\mathsf{T}\mathbf{x} \leq -b' - \delta, \mathbf{x} \geq \mathbf{0}\right\}.$$

So the change Δz^* of $z = \mathbf{c}^\mathsf{T}\mathbf{x}^*$ satisfies (see the proof of Theorem 5.3.3):

$$\Delta z^* = \delta y_1^* - \delta y_2^* = \delta y'^*,$$

where y_1^*, y_2^*, y'^* are the optimal dual values corresponding to $\mathbf{a}^\mathsf{T}\mathbf{x} \leq b'$, $-\mathbf{a}^\mathsf{T}\mathbf{x} \leq -b'$, and $\mathbf{a}^\mathsf{T}\mathbf{x} = b'$, respectively. Hence y'^* is in fact the shadow price of the equality constraint $\mathbf{a}^\mathsf{T}\mathbf{x} = b'$.

$\square$

Note that the values of the optimal dual variables associated with an equality constraint are not present in optimal simplex tableaus. They can be calculated from $\mathbf{y} = (\mathbf{B}^{-1})^\mathsf{T}\mathbf{c}_{BI}$.

We close this section with a remark on redundancy. Suppose that we have a maximizing LO-model, and $\mathbf{a}^\mathsf{T}\mathbf{x} = b$ is one of the restrictions. It can be seen from the sign of its shadow price whether either $\mathbf{a}^\mathsf{T}\mathbf{x} \leq b$ or $\mathbf{a}^\mathsf{T}\mathbf{x} \geq b$ is redundant with respect to a given optimal solution. Namely, if the shadow price of $\mathbf{a}^\mathsf{T}\mathbf{x} = b$ is nonnegative, then $\mathbf{a}^\mathsf{T}\mathbf{x} = b$ can be replaced in the model by $\mathbf{a}^\mathsf{T}\mathbf{x} \leq b$ while the given optimal solution remains optimal, i.e., $\mathbf{a}^\mathsf{T}\mathbf{x} \geq b$ is redundant with respect to the given optimal solution. Similarly, if $\mathbf{a}^\mathsf{T}\mathbf{x} = b$ has a nonpositive shadow price then $\mathbf{a}^\mathsf{T}\mathbf{x} \leq b$ is redundant. We will explain this by means of the following example.

Example 5.7.2. *Consider Model Dovetail. Suppose the supplier of the shipment boxes for the long matches offers these boxes against a special price, when precisely $500{,}000$ of these boxes are purchased per year. The number of boxes for long matches produced per year is x_2 ($\times 100{,}000$). In the original Model Dovetail (see Section 1.1.1) the constraint $x_2 \leq 6$ has to be replaced by the constraint $x_2 = 5$. One can easily check that the optimal solution of the new model is: $x_1^* = 4$, $x_2^* = 5$, with $z^* = 22$. Moreover, the shadow price of the constraint $x_2 = 5$ is negative (the corresponding dual variable has the optimal value -1). This means that if the constraint $x_2 = 5$ is replaced by the constraints $x_2 \leq 5$ and $x_2 \geq 5$, then both constraints are binding, while $x_2 \leq 5$ has shadow price zero and is redundant. The optimal solution can be improved when less than $500{,}000$ 'long' boxes have to be used. (We know this already, since we have solved the model for $x_2 \leq 6$; the optimal solution satisfies $x_1^* = x_2^* = 4\frac{1}{2}$, $z^* = 22\frac{1}{2}$.)*

Hence, it can be seen from the sign of the shadow price of an equality constraint which corresponding inequality is redundant with respect to the optimal solution.

5.8 Exercises

Exercise 5.8.1.   Consider the LO-model:

$$
\begin{aligned}
\max \quad & (1+\delta)x_1 + x_2 \\
\text{s.t.} \quad & x_2 \leq 5 \\
& 2x_1 + x_2 \leq 8 \\
& 3x_1 + x_2 \leq 10 \\
& x_1, x_2 \geq 0.
\end{aligned}
$$

Construct a table, similar to Table 5.1, for $\delta = -2, -1, 0, 1, 2, 3$. Draw the graph of the perturbation function $z^*(\delta) = (1+\delta)x_1^* + x_2^*$.

Exercise 5.8.2.   Construct an example for which a nonnegativity constraint is binding, and investigate perturbations of its right hand side.

Exercise 5.8.3.   Give a proof of Theorem 5.3.2 (see Section 5.3.3): the shadow price of a nonbinding nonnegativity constraint is zero.

Exercise 5.8.4.

(a) Consider the second constraint of Model Dovetail in Section 1.1, which has shadow price $\frac{1}{2}$. Determine the tolerance interval for a perturbation of the right hand side value of this constraint.

(b) Give an example of an LO-model with a binding nonnegativity constraint; determine the tolerance interval and the perturbation function for this binding nonnegativity constraint.

Exercise 5.8.5.   Consider the following LO-model:

$$
\begin{aligned}
\max \quad & x_1 + x_2 \\
\text{s.t.} \quad & 2x_1 + x_2 \leq 9 \\
& x_1 + 3x_2 \leq 12 \\
& x_1, x_2 \geq 0.
\end{aligned}
$$

(a) Determine the optimal solution and rewrite the model so that the objective function is expressed in terms of the nonbasic variables of the optimal solution.

(b) Determine the tolerance interval of $2x_1 + x_2 \leq 9$. Use this interval to calculate its shadow price. Check the answer by dualizing the model, and calculating the optimal dual solution.

Exercise 5.8.6. Consider the (primal) model

$$\min\{5x_2 \mid 2x_2 \leq 0, \ -x_1 - 4x_2 \leq 3, \ x_1, x_2 \geq 0\}.$$

(a) Check that the primal model has multiple optimal solutions and one of these solutions is degenerate.

(b) Check (directly, i.e., without using Theorem 5.6.1) that the dual of this model has multiple optimal solutions, and contains a degenerate optimal solution.

(c) Which case of Theorem 5.6.1 describes the situation in this exercise?

Exercise 5.8.7. Use Theorem 5.6.1 to determine whether the following models have a unique optimal solution or multiple optimal solutions, and whether these solutions are degenerate or nondegenerate. (Hint: solve the dual models.)

(a) min $-4x_1 + 5x_2 + 8x_3$
   s.t. $\frac{1}{2}x_1 - 2x_2 + \ x_3 \geq 1$
          $-x_1 + \ x_2 + \ x_3 \geq 1$
          $x_1, x_2, x_3 \geq 0.$

(b) min $11x_1 + 3x_2 + 10x_3$
   s.t. $x_1 + \frac{1}{3}x_2 + \ 2x_3 \geq 0$
          $2x_1 + \ x_2 + \ x_3 \geq 1$
          $x_1, x_2, x_3 \geq 0.$

(c) min $6x_1 + 3x_2 + 12x_3$
   s.t. $x_1 + \ x_2 + \ 3x_3 \geq 2$
          $x_1 \qquad + \ x_3 \geq 2$
          $x_1, x_2, x_3 \geq 0.$

Exercise 5.8.8. Construct an LO-model that has both a degenerate optimal solution and a nondegenerate optimal solution.

Exercise 5.8.9. Consider the perturbation function corresponding to a perturbation of an objective coefficient of a maximizing LO-model. For each of the following statements, determine whether the statement is true or false. If the statement is true, give a proof; if it is false, give a counterexample.

(a) The perturbation function is nondecreasing.

(b) Any point of the perturbation function that is not a kink point corresponds to an LO-model with a unique optimal solution.

(c) Any kink point of the perturbation function corresponds to an LO-model with a degenerate optimal solution.

Exercise 5.8.10. Consider the following model:

$$
\begin{aligned}
\max \quad & 3x_1 + 4x_2 - 2x_3 + x_4 - x_5 \\
\text{s.t.} \quad & 4x_1 + 5x_2 - x_3 - x_4 - x_5 \le 30 \\
& x_1 + x_2 + x_3 + x_4 + x_5 \le 15 \\
& x_1 \qquad\quad + 3x_3 + 3x_4 \qquad\quad \le 27 \\
& \qquad\quad x_2 \qquad\quad + 2x_4 \qquad\quad \le 20 \\
& x_1 \qquad\quad + 3x_3 \qquad\qquad\quad \ge 20 \\
& x_1, x_2, x_3, x_4, x_5 \ge 0.
\end{aligned}
$$

(a) Determine optimal primal and dual solutions. Draw and analyze the perturbation function for the objective coefficient of x_4.

(b) Draw and analyze the perturbation function for the second constraint. How can it be seen that the optimal solution is degenerate? Determine its left and right shadow prices.

(c) The constraint $-2x_2 - 4x_4 \ge -10$ is added to the model. Solve the new model. Use the tolerance interval to show that the new constraint is not redundant.

Exercise 5.8.11. The management of the publishing company Book & Co wants to design its production schedule for the next quarter, where three books A, B, and C can be published. Book A can only be published as a paperback, books B and C can also be published in a more fancy form with a hard cover. We label book A by the number 1, book B as paperback by 2 and as hard cover by 4, book C as paperback by 3 and as hard cover by 5. The profits (in \$ units) per book are listed in the second column of the table.

Book	Profit (in \$)	Paper	Processing time	Labor time	Covers (paperback)	Covers (hardcover)
1	10	9	50	80	1	0
2	15	12	50	80	2	0
3	10	15	100	120	1	0
4	30	15	150	160	0	2
5	25	18	200	160	0	1

The objective is the maximization of the total profit. There are the following constraints. The inventory of paper is limited to 15,000 sheets of paper. In the table, the number of sheets needed for the different types of books is listed in the third column. Book & Co has only one machine, and all books are processed on this machine.

The machine processing times for the various types of books are given in the fourth column of the table. During the next quarter, the machine capacity (the total time that the machine is available), is 80,000 time units. The number of employees for finishing the books is limited. During the next quarter, there are 120,000 time units available for finishing the books. The finishing time per book is given in the fifth column of the table.

Finally, each type of book needs a certain number of covers, while the total amount of covers for paperbacks and hard covers is limited to 1,800 and 400 covers, respectively. The amounts of covers required per book are given in the last two columns of the table.

Since the sale of hard cover books is expected to go more slowly than the sale of paperbacks, Book & Co produces less hard cover books than paperbacks in the same time period. The target rule is that the number of paperbacks for each different book must be at least five times the number of hard cover books.

(a) Write this problem as an LO-model, and calculate the maximum amount of profit and the optimal number of books to be produced.

Solve the following problems by using the sensitivity analysis techniques discussed in this chapter.

(b) Book & Co can purchase ten paperback covers from another company for the price of $100. Is this advantageous for Book & Co? From which result can you draw this conclusion?

(c) The government wants book A to be published and is therefore prepared to provide subsidy. What should be the minimum amount of subsidy per book such that it is profitable for Book & Co to publish it.

(d) Consider the effects of fluctuating prices and fluctuating profits on this decision concerning the production. Is this decision very sensitive to a change in profits?

(e) Consider the effects of changing the right hand side values as follows. Vary the number of paperback covers from 400 to 2,000; draw the graph of the profits as a function of the varying number of paperback covers. Are there kinks in this graph? Relate this perturbation function to shadow prices.

Exercise 5.8.12.   Analyze the dual of the LO-model for the company Book & Co from Exercise 5.8.11. What is the optimal value? Perform sensitivity analysis for this model by perturbing successively the right hand side values and the objective coefficients. What do shadow prices mean in this dual model? Determine the basis matrices for the optimal primal and dual solutions and show that they are complementary dual (in the sense of Section 2.2.3).

Exercise 5.8.13.   Consider the following LO-model:

$$\begin{aligned}
\max \ & 5x_1 + 3x_2 + x_3 + 4x_4 \\
\text{s.t.} \ \ & x_1 - 2x_2 + 2x_3 + 3x_4 \le 10 \\
& 2x_1 + 2x_2 + 2x_3 - x_4 \le 6 \\
& 3x_1 + x_2 - x_3 + x_4 \le 10 \\
& \quad\ \ -x_2 + 2x_3 + 2x_4 \le 7 \\
& x_1, x_2, x_3, x_4 \ge 0.
\end{aligned}$$

(a) Determine the optimal solution and perform sensitivity analysis by perturbing the right hand side values. Which constraints are binding?

(b) Determine the dual model, and its optimal solution. Check the optimality of the objective value. Determine the complementary slackness relations for this model.

(c) The vertex with $x_1 = 0$, $x_2 = 4$, $x_3 = 1$, and $x_4 = 4$ is a point of the feasible region. Check this. What is the objective value in this vertex? Compare this value with the optimal value.

Exercise 5.8.14. Consider the LO-model in Section 5.6.2.

(a) Solve this model, and analyze the results.

(b) Show that the perturbation functions $z_i(\delta)$ of all four constraints have a kink for $\delta = 0$.

(c) Draw and analyze the perturbation function for the constraint $-3x_1 + x_3 \leq 0$.

(d) Draw and analyze the perturbation function for the nonnegativity constraint $x_1 \geq 0$.

Exercise 5.8.15. A farm with 25 acres can cultivate the following crops: potatoes, sugar beets, oats, winter wheat, and peas. The farmer wants to determine the optimal cultivating schedule, and should take into account the following factors. The expected yields and the expected costs per acre are given in the table below.

Crop	Expected yield ($ per acre)	Expected costs ($ per acre)
Potatoes	5,586	3,435
Oats	2,824	1,296
Winter wheat	3,308	1,813
Sugar beets	5,311	2,485
Peas	3,760	1,610

Other facts to be taken into account are the crop-rotation requirements: to control the development of diseases and other plagues, the crops should not grow too often on the same piece of land. In the county in which this farm is located, the following rules of thumb are used:

Potatoes: at most once every two years;

Sugar beets: at most once every four years;

Oats and winter wheat: individually, at most once every two years; together, at most three times every four years;

Peas: at most once every six years.

The objective is to make a schedule such that the crop-rotation requirements are satisfied and the profit is as high as possible. The crop-rotation requirements can be modeled in different ways. A simple way is the following. If a certain crop is allowed to be cultivated only once per two years on a given piece of land, then this is the same as to require that the crop is cultivated each year on one half of the land, and the next year on the other half.

(a) Write this problem as an LO-model, and solve it. If a planning period of twelve years is chosen, then there is a second solution with the same profit during these twelve years. Determine this solution. Suppose that the average annual interest rate is 6.5% during the twelve years. Which solution is the most preferable one?

A serious disease for potatoes increasingly infects the soil in the county. To control this process, the government considers compensating farmers that satisfy stricter crop-rotation requirements. The following is proposed. Every farmer that cultivates potatoes once in five years (or even less) receives a subsidy that depends on the size of the farm. The amount of the subsidy is $160 per acre. For our farmer, a reward of $4,000 is possible.

(b) Do you think the farmer will adapt his plan if the above government proposal is carried out? How strict can the government make the crop-rotation requirements to ensure that it is profitable for the farmer to adapt his original plans?

Exercise 5.8.16.   Consider the following LO-model:

$$
\begin{array}{lrl}
\max & 620x_1 + 520x_2 + 720x_3 + 620x_4 - 576 & \\
\text{s.t.} & 2x_1 + \ 2x_2 & \leq \ 600 \\
& 24x_3 + \ 24x_4 & \leq \ 300 \\
& 30x_1 + \ 35x_2 + 120x_3 + 125x_4 & \leq 3375 \\
& 135x_1 \qquad\quad + 135x_3 & \leq 2250 \\
& 30x_1 + \ 30x_2 + \ 30x_3 + \ 30x_4 & \leq 2250 \\
& x_1, x_2, x_3, x_4 \geq 0. &
\end{array}
$$

(a) Solve this LO-model, and analyze the results. Why are there multiple optimal solutions? Determine a second optimal solution. Show that the tolerance interval of $x_3 \geq 0$ is $[-49.07, 9.26]$, and that the model is infeasible for $x_3 > 13.5$.

(b) Determine the tolerance interval of the objective coefficient of x_4. What is the slope of the perturbation function on the tolerance interval?

(c) Determine the tolerance interval of $x_1 \geq 0$, and the slope of the perturbation function on this interval.

(d) Draw and analyze the perturbation function for the constraint $2x_1 + 2x_2 \leq 600$.

Exercise 5.8.17.   Consider the diet problem as formulated in Section 5.7. Perform sensitivity analysis to the model corresponding to this problem by perturbing successively the right hand side values and the objective coefficients. Draw the graphs of the corresponding perturbation functions. Analyze the results in terms of the original diet problem.

Exercise 5.8.18. Consider the following LO-model:

$$\begin{aligned}
\max \quad & 12x_1 + 20x_2 + 18x_3 + 40x_4 \\
\text{s.t.} \quad & 4x_1 + 9x_2 + 7x_3 + 10x_4 \leq 6000 \\
& x_1 + x_2 + 3x_3 + 40x_4 \leq 4000 \\
& x_1, x_2, x_3, x_4 \geq 0.
\end{aligned}$$

(a) Conduct sensitivity analysis by perturbing successively the right hand side values and the objective coefficients; draw the graphs of the corresponding perturbation functions.

(b) Suppose the entries of the first column of the technology matrix are perturbed with the same amount. Determine the tolerance interval. Do the same for the individual coefficients of this column. Draw and analyze the graphs of the corresponding perturbation functions.

Exercise 5.8.19. Consider the following LO-model:

$$\begin{aligned}
\min \quad & -3x_1 + x_2 + x_3 && \text{(P)} \\
\text{s.t.} \quad & x_1 - 2x_2 + x_3 \leq 11 \\
& -4x_1 + x_2 + 2x_3 \geq 3 \\
& 2x_1 - x_3 = -1 \\
& x_1, x_2, x_3 \geq 0.
\end{aligned}$$

(a) Use the graphical solution method to solve model (P).

(b) Show that the feasible region of (P) is unbounded.

(c) Show that (P), extended with the slack variables x_4 and x_5 for the first and second constraint, respectively, has a nondegenerate and unique optimal solution.

(d) Determine the dual of (P), and the signs of the shadow prices of the three constraints.

(e) Use an optimal primal basis matrix to determine an optimal dual solution of (P).

(f) Replace the third constraint of (P) by two inequalities; call the new model (PS). Show that the optimal solution of (PS) is degenerate and unique. Determine all optimal basis matrices of (PS), and the corresponding complementary dual basis matrices.

(g) Show that the dual of model (PS) has multiple optimal solutions.

(h) Draw the simplex adjacency graph of model (PS).

(i) Use the shadow price of the constraint $2x_1 - x_3 = -1$ to determine which corresponding inequality is redundant with respect to the optimal solution.

Exercise 5.8.20. Show that perturbation functions of objective coefficients in maximizing (minimizing) LO-models are convex (concave).

Exercise 5.8.21. Formulate assertions for nonstandard LO-models similar to ones in Theorem 5.6.2.

Exercise 5.8.22.   Let (P) be an LO-model, and (D) its dual. Answer the following questions; explain your answers.

(a) If (P) is nondegenerate, can it be guaranteed that (D) is also nondegenerate?

(b) Is it possible that (P) and (D) are both degenerate?

(c) Is it possible that (P) has a unique optimal solution with finite objective value, but (D) is infeasible?

(d) Is it possible that (P) is unbounded, while (D) has multiple optimal solutions?

Exercise 5.8.23.   A company manufactures four types of TV-sets: TV1, TV2, TV3, and TV4. Department D_1 produces intermediate products which department D_2 assembles to end products. Department D_3 is responsible for packaging the end products. The available number of working hours for the three departments is 4,320, 2,880, and 240 hours per time unit, respectively. Each TV1 set requires in the three departments 12, 9, and $\frac{1}{3}$ hours per time unit, respectively. For each TV2 set these figures are: 15, 12, and 1; for TV3: 20, 12, and 1; and for TV4: 24, 18, and 2. The sales prices are $350 for TV1, $400 for TV2, $600 for TV3, and $900 for TV4. It is assumed that everything will be sold, and that there are no inventories. The objective is to maximize the yield.

(a) Formulate this problem as an LO-model, and determine an optimal solution.

(b) Calculate and interpret the shadow prices.

(c) Suppose that one more employee can be hired. To which department is this new employee assigned?

(d) Which variations in the sales prices of the four products do not influence the optimal production schedule?

(e) What happens if all sales prices increase by 12%?

(f) Formulate the dual model. Formulate the complementary slackness relations, and give an economic interpretation.

Exercise 5.8.24.   A company wants to produce 300 kilograms of cattle fodder with the following requirements. First, it needs to consist of at least 75% so-called digestible nutrients. Second, it needs to contain at least 0.65 grams of phosphorus per kilogram. Third, it needs to consist of at least 15% proteins. (The percentages are weight percentages.) Five types of raw material are available for the production, labeled $1, \dots, 5$; see the table below.

	Raw material				
	1	2	3	4	5
Digestible nutrients (weight percentage)	80	60	80	80	65
Phosphorus (grams per kg)	0.32	0.30	0.72	0.60	1.30
Proteins (weight percentage)	12	9	16	37	14
Price ($ per kg)	2.30	2.18	3.50	3.70	2.22

The problem is to produce a mixture with minimum costs.

(a) Formulate an LO-model for this problem, and determine an optimal solution.

(b) Which constraints are redundant?

(c) Raw material 4 turns out to be relatively expensive. With what amount should its price be decreased in order to make it economically attractive to include it in the mixture? What about raw material 2?

(d) Recall that the fodder needs to contain at least 75% digestible nutrients. If this 75% is slightly increased, what are the minimum costs? Answer similar questions for the proteins and the phosphorus.

Exercise 5.8.25. Consider the following LO-model:

$$\begin{aligned}
\min \quad & 4x_1 - 4x_2 \\
\text{s.t.} \quad & 2x_1 - x_2 \geq 6 \\
& x_1 - x_2 \geq 2 \\
& x_1 - 2x_2 \geq 0 \\
& x_1, x_2 \geq 0.
\end{aligned}$$

Draw the perturbation function for the objective coefficient of x_1. Explain why tolerance intervals need not be equal to the interval for which the rate of change of z with respect to x_1 is constant (i.e., the interval with constant slope of the perturbation function around $\delta = 0$).

Exercise 5.8.26. Consider the LO-model:

$$\begin{aligned}
\min \quad & 4x_3 \\
\text{s.t.} \quad & x_1 \qquad\quad + 2x_3 \leq 7 \\
& -x_1 \qquad\; + x_3 \leq -7 \\
& x_1 - x_2 + 3x_3 \leq 7 \\
& x_1, x_2, x_3 \geq 0.
\end{aligned}$$

Draw the simplex adjacency graph corresponding to all optimal feasible basic solutions. For each feasible basic solution, determine the tolerance interval corresponding to a perturbation of the objective coefficient of x_1. Note that this coefficient is 0. Which optimal feasible basic solution corresponds to a tolerance interval consisting of only one point? Draw the perturbation function.

CHAPTER 6

Large-scale linear optimization

Overview

Although Dantzig's simplex algorithm has proven to be very successful in solving LO-models of practical problems, its theoretical worst-case behavior is nevertheless very bad. In 1972, VICTOR KLEE (1925–2007) and GEORGE J. MINTY (1929–1986) constructed examples for which the simplex algorithm requires an exorbitant amount of computer running time; see Chapter 9. The reason that the simplex algorithm works fast in practice is that LO-models arising from most practical problems are usually so-called 'average' problems and are fortunately not the rare 'worst-case' problems. Besides the theoretical question whether or not there exists a fast algorithm that solves large-scale linear optimization problems, also the practical need for such an efficient algorithm has inspired researchers to try to answer this question.

The first such efficient algorithm (the so-called *ellipsoid algorithm*) was published in 1979 by LEONID G. KHACHIYAN (1952–2005). Although this algorithm is theoretically efficient in the sense that it has polynomial running time (see Chapter 9), its performance in practice is much worse than the simplex algorithm. In 1984, NARENDRA KARMARKAR (born 1957) presented a new algorithm which – as he claimed – could solve large-scale LO-models as much as a hundred times faster than the simplex algorithm. Karmarkar's algorithm is a so-called *interior point algorithm*. The basic idea of this algorithm is to move towards an optimal solution through points in the (relative) interior of the feasible region of the LO-model. Since Karmarkar's publication, a number of variants of this algorithm have been constructed.

Recently, interior point algorithms have received a significant amount of attention because of their applications in non-linear optimization. In particular, interior point algorithms are fruitful for solving so-called *semidefinite optimization* models, which generalize linear optimization models in the sense that the variables in semidefinite optimization models are positive semidefinite matrices. This is a generalization because any nonnegative number can be viewed as a positive semidefinite $(1,1)$-matrix.

In this chapter[1] we will discuss a particular interior point algorithm that is called the *interior path algorithm*, developed by, among others, CEES ROOS (born 1941) and JEAN-PHILIPPE VIAL. Roughly speaking, the interior path algorithm starts at an initial (relative) interior point of the feasible region and uses the so-called interior path as a guide that leads this initial interior point to an optimal point. The number of required iteration steps increases slowly in the number of variables of the model. Experiments with specific versions of the interior path algorithm have shown that, even for very large models, we often do not need more than 60 iterations. In spite of the fact that the time per iteration is rather long, the algorithm is more or less insensitive to the size of the model; this phenomenon makes the interior path algorithm useful for large-scale models.

6.1 The interior path

Instead of following the boundary of the feasible region to an optimal point as is the case when applying the simplex algorithm (see Chapter 3), interior path algorithms start at an interior point of the feasible region and follow an interior path to an optimal point somewhere on the boundary of the feasible region. Interior point algorithms rely strongly on optimization methods from nonlinear constrained optimization. We therefore refer to Appendix E for the relevant background in nonlinear optimization.

6.1.1 The Karush-Kuhn-Tucker conditions for LO-models

Consider the following primal/dual pair of LO-models:

$$\text{Primal model:} \quad \min\{\mathbf{c}^\mathsf{T}\mathbf{x} \mid \mathbf{A}\mathbf{x} = \mathbf{b}, \mathbf{x} \geq \mathbf{0}\} \tag{P}$$
$$\text{Dual model:} \quad \max\{\mathbf{b}^\mathsf{T}\mathbf{y} \mid \mathbf{A}^\mathsf{T}\mathbf{y} \leq \mathbf{c}\} \tag{D}$$

where $\mathbf{x} \in \mathbb{R}^n$, $\mathbf{y} \in \mathbb{R}^m$, $\mathbf{c} \in \mathbb{R}^n$, $\mathbf{b} \in \mathbb{R}^m$, and $\mathbf{A} \in \mathbb{R}^{m \times n}$. The primal and the dual models are both alternative formulations of the standard LO-model; see Section 1.3, and Exercise 4.8.6. We will assume that $\text{rank}(\mathbf{A}) = m$, and that $m \leq n$. Moreover, it is assumed that the feasible regions $F_P = \{\mathbf{x} \in \mathbb{R}^n \mid \mathbf{A}\mathbf{x} = \mathbf{b}, \mathbf{x} \geq \mathbf{0}\}$ and $F_D = \{\mathbf{y} \in \mathbb{R}^m \mid \mathbf{A}^\mathsf{T}\mathbf{y} \leq \mathbf{c}\}$ contain *interior points*; this means that

$$F_P^+ = \{\mathbf{x} \in \mathbb{R}^n \mid \mathbf{A}\mathbf{x} = \mathbf{b}, \mathbf{x} > \mathbf{0}\} \neq \emptyset, \text{ and}$$
$$F_D^+ = \{\mathbf{y} \in \mathbb{R}^m \mid \mathbf{A}^\mathsf{T}\mathbf{y} < \mathbf{c}\} \neq \emptyset.$$

The expression $\mathbf{a} > \mathbf{b}$ means that all entries of $\mathbf{a} - \mathbf{b}$ are strictly positive. In Section 6.4.2, we will see that it is possible to transform any LO-model into a larger equivalent model, whose feasible regions contain interior points.

---

[1]Parts of this chapter are based on lecture notes by Cees Roos of Delft University of Technology.

Suppose that $\hat{\mathbf{x}}$ and $\hat{\mathbf{y}}$ are feasible solutions of (P) and (D), respectively. By the complementary slackness relations (see Section 4.3), it follows that both $\hat{\mathbf{x}}$ and $\hat{\mathbf{y}}$ are optimal for their respective models if and only if $\hat{\mathbf{x}}^\mathsf{T}\hat{\mathbf{w}} = 0$, where $\hat{\mathbf{w}} = \mathbf{c} - \mathbf{A}^\mathsf{T}\hat{\mathbf{y}}$. Note that $\hat{\mathbf{x}} \geq \mathbf{0}$ and $\hat{\mathbf{w}} \geq \mathbf{0}$, because $\hat{\mathbf{x}}$ and $\hat{\mathbf{y}}$ are feasible. This follows directly from Theorem 4.3.1 by rewriting (P) as $-\max\left\{(-\mathbf{c})^\mathsf{T}\mathbf{x} \,\middle|\, \begin{bmatrix} \mathbf{A} \\ -\mathbf{A} \end{bmatrix}\mathbf{x} \leq \begin{bmatrix} \mathbf{b} \\ -\mathbf{b} \end{bmatrix}, \mathbf{x} \geq \mathbf{0}\right\}$. Hence, $\hat{\mathbf{x}}$ and $\hat{\mathbf{y}}$ are optimal solutions of (P) and (D), respectively, if and only if $\hat{\mathbf{x}}, \hat{\mathbf{y}}$, and $\hat{\mathbf{w}}$ form a solution of the following system of equations:

$$\begin{aligned} &\mathbf{A}\mathbf{x} = \mathbf{b}, \ \mathbf{x} \geq \mathbf{0}, &&\text{(primal feasibility)} \\ &\mathbf{A}^\mathsf{T}\mathbf{y} + \mathbf{w} = \mathbf{c}, \ \mathbf{w} \geq \mathbf{0}, &&\text{(dual feasibility)} \quad\text{(PD)}\\ &\mathbf{X}\mathbf{w} = \mathbf{0}, &&\text{(complementary slackness)} \end{aligned}$$

where $\mathbf{X}$ denotes the diagonal (n, n)-matrix with the entries of the vector $\mathbf{x}$ on its main diagonal. The conditions (PD) are called the *Karush-Kuhn-Tucker conditions*; see also Appendix E.4. The system of equations $\mathbf{X}\mathbf{w} = \mathbf{0}$ consists of the complementary slackness relations.

6.1.2 The interior path and the logarithmic barrier function

Let $\mathbf{1} = \begin{bmatrix} 1 & \ldots & 1 \end{bmatrix}^\mathsf{T} \in \mathbb{R}^n$, and let $\mu \geq 0$. Consider the following system, called the *relaxed Karush-Kuhn-Tucker conditions*:

$$\begin{aligned} &\mathbf{A}\mathbf{x} = \mathbf{b}, \mathbf{x} \geq \mathbf{0}, &&\text{(primal feasibility)} \\ &\mathbf{A}^\mathsf{T}\mathbf{y} + \mathbf{w} = \mathbf{c}, \mathbf{w} \geq \mathbf{0}, &&\text{(dual feasibility)} \quad\text{(PD}\mu)\\ &\mathbf{X}\mathbf{w} = \mu\mathbf{1}. &&\text{(relaxed complementary slackness)} \end{aligned}$$

Note that (PD0) = (PD). The equation $\mathbf{X}\mathbf{w} = \mu\mathbf{1}$ can be seen as a relaxation of $\mathbf{X}\mathbf{w} = \mathbf{0}$.

Example 6.1.1. *Consider the model* $\min\{x_1 + x_2 \mid x_1 + x_2 = 1, x_1 \geq 0, x_2 \geq 0\}$. *The dual of this model is* $\max\{y \mid y + w_1 = 1, y + w_2 = 1, w_1 \geq 0, w_2 \geq 0\}$. *It is easy to check that* $\left\{\begin{bmatrix} x_1 & x_2 \end{bmatrix}^\mathsf{T} \,\middle|\, 0 \leq x_1 \leq 1, x_2 = 1 - x_1\right\}$ *is the set of optimal solutions of the primal model. Moreover, the unique optimal dual solution satisfies* $y = 1$, $w_1 = w_2 = 0$. *The relaxed Karush-Kuhn-Tucker conditions are:*

$$\begin{bmatrix} 1 & 1 \end{bmatrix}\begin{bmatrix} x_1 \\ x_2 \end{bmatrix} = 1, x_1, x_2 \geq 0,$$

$$\begin{bmatrix} 1 \\ 1 \end{bmatrix}y + \begin{bmatrix} w_1 \\ w_2 \end{bmatrix} = \begin{bmatrix} 1 \\ 1 \end{bmatrix}, w_1, w_2 \geq 0,$$

$$\begin{bmatrix} x_1 & 0 \\ 0 & x_2 \end{bmatrix}\begin{bmatrix} w_1 \\ w_2 \end{bmatrix} = \mu\begin{bmatrix} 1 \\ 1 \end{bmatrix}.$$

It is left to the reader to check that, for every $\mu > 0$, *the unique solution of the above system is:* $x_1 = x_2 = \frac{1}{2}$, $y = 1 - 2\mu$, $w_1 = w_2 = 2\mu$. *Moreover, for* $\mu = 0$, *the solution set is given by*

$$\left\{\begin{bmatrix} x_1 & x_2 & y & w_1 & w_2 \end{bmatrix}^\mathsf{T} \,\middle|\, 0 \leq x_1 \leq 1, x_2 = 1 - x_1, y = 1, w_1 = w_2 = 0\right\}.$$

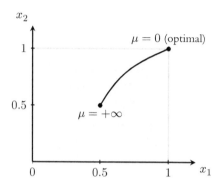

Figure 6.1: Interior path for model (A).

It is no coincidence that the system in Example 6.1.1 has a unique solution: this is actually true in general. The following theorem formalizes this. We postpone proving the theorem until the end of this section.

Theorem 6.1.1. (*Interior path theorem*)
Suppose that the feasible region of (P) is bounded. Then (PDμ) has a unique solution for every fixed $\mu > 0$.

Theorem 6.1.1 also holds without the boundedness condition. However, without the boundedness condition, the proof is more complicated. On the other hand, in practical situations, the boundedness condition is not a real restriction, since all decision variables are usually bounded.

For every $\mu > 0$, we will denote the unique solution found in Theorem 6.1.1 by $\mathbf{x}(\mu)$, $\mathbf{y}(\mu)$, $\mathbf{w}(\mu)$. The sets $\{\mathbf{x}(\mu) \mid \mu > 0\}$, and $\{\mathbf{y}(\mu) \mid \mu > 0\}$ are called the *interior paths* of (P) and (D), respectively. For the sake of brevity, we will sometimes call $\mathbf{x}(\mu)$ and $\mathbf{y}(\mu)$ the interior paths. The parameter μ is called the *interior path*.

In general, interior paths cannot be calculated explicitly. The following simple example presents a case in which calculating the interior path analytically is possible.

Example 6.1.2.   *Consider the following simple LO-model.*

$$\begin{aligned} \max \ & x_1 + 2x_2 \\ \text{s.t.} \ \ & 0 \le x_1 \le 1 \\ & 0 \le x_2 \le 1. \end{aligned} \qquad \text{(A)}$$

The dual of (A), with the slack variables y_3 and y_4, reads:

$$\begin{aligned} \min \ & y_1 + y_2 \\ \text{s.t.} \ \ & y_1 - y_3 = 1 \\ & y_2 - y_4 = 2 \\ & y_1, y_2, y_3, y_4 \ge 0. \end{aligned} \qquad \text{(B)}$$

In the context of this chapter, we consider (B) as the primal model and (A) as its dual. The system (PDμ) *becomes:*

$$\begin{bmatrix} 1 & 0 & -1 & 0 \\ 0 & 1 & 0 & -1 \end{bmatrix} \begin{bmatrix} y_1 \\ y_2 \\ y_3 \\ y_4 \end{bmatrix} = \begin{bmatrix} 1 \\ 2 \end{bmatrix}, y_1, y_2, y_3, y_4 \geq 0$$

$$\begin{bmatrix} 1 & 0 \\ 0 & 1 \\ -1 & 0 \\ 0 & -1 \end{bmatrix} \begin{bmatrix} x_1 \\ x_2 \end{bmatrix} + \begin{bmatrix} w_1 \\ w_2 \\ w_3 \\ w_4 \end{bmatrix} = \begin{bmatrix} 1 \\ 1 \\ 0 \\ 0 \end{bmatrix}, w_1, w_2, w_3, w_4 \geq 0$$

$$y_1 w_1 = \mu, \; y_2 w_2 = \mu, \; y_3 w_3 = \mu, \; y_4 w_4 = \mu.$$

We will solve for x_1 and x_2 as functions of μ. Eliminating the y_i's and the w_i's results in the equations $\frac{1}{1-x_1} - \frac{1}{x_1} = \frac{1}{\mu}$ *and* $\frac{1}{1-x_2} - \frac{1}{x_2} = \frac{2}{\mu}$. *Recall that $\mu > 0$, and so $x_1 > 0$ and $x_2 > 0$. Solving the two equations yields:*

$$x_1 = x_1(\mu) = -\mu + \tfrac{1}{2} + \tfrac{1}{2}\sqrt{1 + 4\mu^2}, \; and$$

$$x_2 = x_2(\mu) = -\tfrac{1}{2}\mu + \tfrac{1}{2} + \tfrac{1}{2}\sqrt{1 + \mu^2}.$$

One can easily check that $\lim_{\mu \downarrow 0} x_1(\mu) = 1 = \lim_{\mu \downarrow 0} x_2(\mu)$. *The expression $\mu \downarrow 0$ means $\mu \to 0$ with $\mu > 0$. Figure 6.1 shows the graph of this interior path.*

The proof of Theorem 6.1.1 uses a so-called *logarithmic barrier function*. The logarithmic barrier function for the primal model is defined as:

$$B_P(\mathbf{x}; \mu) = \frac{1}{\mu}\mathbf{c}^{\mathsf{T}}\mathbf{x} - \sum_{i=1}^{n} \ln(x_i),$$

with domain $F_P^+ = \{\mathbf{x} \in \mathbb{R}^n \mid \mathbf{A}\mathbf{x} = \mathbf{b}, \mathbf{x} > \mathbf{0}\}$. Figure 6.2 depicts the graph of the function $B(x_1, x_2) = -\ln(x_1) - \ln(x_2)$, corresponding to the last term in the definition of $B_P(\mathbf{x}; \mu)$ (for $n = 2$). Since $\ln(x_i)$ approaches $-\infty$ when $x_i \to 0$, it follows that, for fixed $\mu > 0$, $B_P(\mathbf{x}; \mu) \to \infty$ when $x_i \to 0$ $(i = 1, \ldots, n)$. Intuitively, the barrier function is the objective function plus a term that rapidly tends to infinity as we get closer to the boundary of the feasible region. The logarithmic barrier function for the dual model is defined as

$$B_D(\mathbf{y}; \mu) = -\frac{1}{\mu}\mathbf{b}^{\mathsf{T}}\mathbf{y} - \sum_{j=1}^{n} \ln(w_j),$$

with domain $F_D^+ = \{\mathbf{y} \in \mathbb{R}^n \mid \mathbf{A}^{\mathsf{T}}\mathbf{y} + \mathbf{w} = \mathbf{c}, \mathbf{w} > \mathbf{0}\}$. Clearly, for fixed $\mu > 0$, $B_D(\mathbf{y}; \mu) \to \infty$ when $w_j \to 0$ $(j = 1, \ldots, m)$. Therefore, both barrier functions take infinite values at the boundaries of the feasible regions.

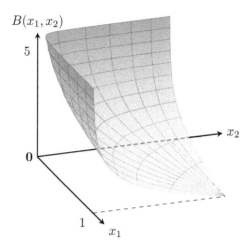

Figure 6.2: Logarithmic barrier function $B(x_1, x_2) = -\ln(x_1) - \ln(x_2)$, drawn for $0 < x_1 \le 1$ and $0 < x_2 \le 1$. The value of $B(x_1, x_2)$ diverges to infinity as $x_1 \downarrow 0$ and as $x_2 \downarrow 0$.

Example 6.1.3.   *Consider the LO-model* $\min\{x_1 + 2x_2 \mid x_1 + x_2 = 1, x_1, x_2 \ge 0\}$. *The logarithmic barrier function for this model is* $B_P(x_1, x_2; \mu) = \frac{1}{\mu}(x_1 + 2x_2) - \ln(x_1 x_2)$. *For* $\mu = 1$, *the minimum of* $B_P(x_1, x_2; 1)$ *is attained at* $x_1 = 0.6$, $x_2 = 0.4$. *In Exercise 6.5.6, the reader is asked to draw the graph of* $B_P(x_1, x_2; \mu)$ *for some values of* $\mu > 0$.

The proof of Theorem 6.1.1 uses an interesting connection between the logarithmic barrier functions $B_P(\cdot\,; \mu)$ and $B_D(\cdot\,; \mu)$, and the system of equations (PDμ): as we will see, the functions $B_P(\cdot\,; \mu)$ and $B_D(\cdot\,; \mu)$ each have a unique minimum that, together, correspond to the unique solution of (PDμ). One of the key observations to show this connections is the fact that the two logarithmic barrier functions are strictly convex (see Appendix D). We show this fact in Theorem 6.1.2.

Theorem 6.1.2. (*Strict convexity of logarithmic barrier functions*)
For every fixed $\mu > 0$, both $B_P(\cdot\,; \mu)$ and $B_D(\cdot\,; \mu)$ are strictly convex functions on F_P^+ and F_D^+, respectively.

Proof of Theorem 6.1.2. We will show that B_P is strictly convex on F_P^+. Take any $\mu > 0$. We need to show that for any $\mathbf{x}$ and $\mathbf{x}'$ in F_P^+ and any λ with $0 < \lambda < 1$, it holds that:

$$B_P(\lambda \mathbf{x} + (1 - \lambda)\mathbf{x}'; \mu) < \lambda B_P(\mathbf{x}; \mu) + (1 - \lambda)B_P(\mathbf{x}'; \mu).$$

We use the fact that $\ln(x_i)$ is strictly concave, i.e., that $\ln(\lambda x_i' + (1 - \lambda)x_i'') > \lambda \ln(x_i') + (1 - \lambda)\ln(x_i'')$. It now follows that:

$$B_P(\lambda \mathbf{x} + (1 - \lambda)\mathbf{x}'; \mu) = \tfrac{1}{\mu}\mathbf{c}^{\mathsf{T}}(\lambda \mathbf{x} + (1 - \lambda)\mathbf{x}') - \sum_{i=1}^{n} \ln(\lambda x_i' + (1 - \lambda)x_i'')$$

$$\begin{aligned}
&= \lambda\left(\tfrac{1}{\mu}\mathbf{c}^\mathsf{T}\mathbf{x}\right) + (1-\lambda)\left(\tfrac{1}{\mu}\mathbf{c}^\mathsf{T}\mathbf{x}'\right) - \sum_i \ln(\lambda x_i' + (1-\lambda)x_i'') \\
&< \lambda\left(\tfrac{1}{\mu}\mathbf{c}^\mathsf{T}\mathbf{x}\right) + (1-\lambda)\left(\tfrac{1}{\mu}\mathbf{c}^\mathsf{T}\mathbf{x}'\right) - \lambda\sum_i \ln(x_i') - (1-\lambda)\sum_i \ln(x_i'') \\
&= \lambda B_P(\mathbf{x};\mu) + (1-\lambda)B_P(\mathbf{x}';\mu).
\end{aligned}$$

The proof of the strict convexity of $B_D(\,\cdot\,;\mu)$ on F_D^+ is similar and is left to the reader. $\qquad\square$

We can now prove Theorem 6.1.1. The main idea of the proof is that, for any $\mu > 0$, a minimizer of $B_P(\,\cdot\,;\mu)$ together with a minimizer of $B_D(\,\cdot\,;\mu)$ form a solution of (PDμ).

Proof of Theorem 6.1.1. Take any $\mu > 0$. Since the feasible region F_P is bounded and closed, and since $B_P(\,\cdot\,;\mu)$ is infinite on the boundary of F_P, it follows that the strictly convex function $B_P(\,\cdot\,;\mu)$ has a unique minimum on F_D^+; see Appendix D. In order to determine this unique minimum, which we call $\mathbf{x}(\mu)$, we use the Lagrange function (see Section E.3):

$$L_\lambda(\mathbf{x},\mu) = B_P(\mathbf{x};\mu) - \lambda_1\left(\sum_{i=1}^n a_{1i}x_i - b_1\right) - \ldots - \lambda_m\left(\sum_{i=1}^n a_{mi}x_i - b_m\right),$$

with $\mathbf{A} = \{a_{ij}\}$, and $\mathbf{b} = [b_1 \ \ldots \ b_m]^\mathsf{T}$. Differentiating $L_\lambda(\mathbf{x},\mu)$ with respect to x_i for each $i = 1,\ldots,n$ yields:

$$\begin{aligned}
\frac{\partial}{\partial x_i}L_\lambda(\mathbf{x},\mu) &= \frac{\partial}{\partial x_i}B_P(\mathbf{x};\mu) - (\lambda_1 a_{1i} + \ldots + \lambda_m a_{mi}) \\
&= \frac{1}{\mu}c_i - \frac{1}{x_i} - (\lambda_1 a_{1i} + \ldots + \lambda_m a_{mi}).
\end{aligned}$$

Hence, the gradient vector of $L_\lambda(\mathbf{x},\mu)$ satisfies:

$$\nabla L_\lambda(\mathbf{x},\mu) = \left[\frac{\partial}{\partial x_1}L_\lambda(\mathbf{x},\mu) \ \ldots \ \frac{\partial}{\partial x_n}L_\lambda(\mathbf{x},\mu)\right]^\mathsf{T} = \frac{1}{\mu}\mathbf{c} - \mathbf{X}^{-1}\mathbf{1} - \mathbf{A}^\mathsf{T}\boldsymbol{\lambda},$$

with $\boldsymbol{\lambda} = [\lambda_1 \ \ldots \ \lambda_m]^\mathsf{T}$, and $\mathbf{X}$ and $\mathbf{1}$ as defined before. From the theory of the Lagrange multiplier method (see Section E.3), it follows that a necessary condition for $\mathbf{x}(\mu)$ being a minimizer for $B_P(\mathbf{x};\mu)$ is that there exists a vector $\boldsymbol{\lambda} = \boldsymbol{\lambda}(\mu)$ such that $\nabla L_\lambda(\mathbf{x}(\mu),\mu) = \mathbf{0}$, $\mathbf{A}\mathbf{x}(\mu) = \mathbf{b}$, and $\mathbf{x}(\mu) \geq \mathbf{0}$. Since for $\mathbf{x} > \mathbf{0}$ there is only one stationary point, it follows that the following system has the unique solution $\mathbf{x} = \mathbf{x}(\mu)$:

$$\frac{1}{\mu}\mathbf{c} - \mathbf{X}^{-1}\mathbf{1} = \mathbf{A}^\mathsf{T}\boldsymbol{\lambda}, \qquad \mathbf{A}\mathbf{x} = \mathbf{b}, \qquad \mathbf{x} > \mathbf{0}. \tag{6.1}$$

Premultiplying the first equation of (6.1) by $\mathbf{A}$, it follows, since $\mathbf{A}\mathbf{A}^\mathsf{T}$ is nonsingular (see Theorem B.4.1 in Appendix B), that $\boldsymbol{\lambda} = (\mathbf{A}\mathbf{A}^\mathsf{T})^{-1}(\frac{1}{\mu}\mathbf{A}\mathbf{c} - \mathbf{A}\mathbf{X}^{-1}\mathbf{1})$, and so $\boldsymbol{\lambda}$ is also uniquely determined; this value of $\boldsymbol{\lambda}$ is called $\boldsymbol{\lambda}(\mu)$. Defining $\mathbf{y}(\mu) = \mu\boldsymbol{\lambda}(\mu)$ and $\mathbf{w}(\mu) = \mu\mathbf{X}^{-1}\mathbf{1}$, it follows that (6.1) is now equivalent to the system:

$$\begin{aligned}
&\mathbf{A}\mathbf{x} = \mathbf{b}, \mathbf{x} > \mathbf{0} \\
&\mathbf{A}^\mathsf{T}\mathbf{y} + \mathbf{w} = \mathbf{c}, \mathbf{w} > \mathbf{0} \\
&\mathbf{X}\mathbf{w} = \mu\mathbf{1}.
\end{aligned}$$

Because for $\mu > 0$, (PDμ) has only solutions that satisfy $\mathbf{x} > \mathbf{0}$ and $\mathbf{w} > \mathbf{0}$ (this follows from $\mathbf{Xw} = \mu\mathbf{1} > \mathbf{0}$), (PD$\mu$) also has the unique solution $\mathbf{x} = \mathbf{x}(\mu) > \mathbf{0}$, $\mathbf{y} = \mathbf{y}(\mu)$, and $\mathbf{w} = \mathbf{w}(\mu) > \mathbf{0}$.   □

6.1.3   Monotonicity and duality

The interior path plays a key role in the interior path algorithm. Theorem 6.1.3 expresses the fact that the objective values that correspond to points on the interior path of the primal (dual) model decrease (increase) monotonically, and converge to the optimal objective value for $\mu \to 0$. Recall that the optimal objective values of the primal and the dual model are the same; see Theorem 4.2.4. In Theorem 6.1.3, we use the term *duality gap*. The interior path algorithm we describe in this chapter is a special type of a so-called *primal-dual algorithm*. Such an algorithm simultaneously keeps track of a primal feasible solution $\mathbf{x}$ and a dual feasible solution $\mathbf{y}$. The duality gap is the difference between the objective value corresponding to current $\mathbf{x}$ (in the primal model) and the objective value corresponding to current $\mathbf{y}$ (in the dual model). Thus, if the duality gap equals zero, then $\mathbf{x}$ is an optimal primal solution, and $\mathbf{y}$ is an optimal dual solution.

Theorem 6.1.3. (*Monotonicity of objective values; optimality on the interior path*)
The objective values on the interior path of the primal (dual) model decrease (increase) monotonically to the optimal objective value of the model. Moreover, for every $\mu > 0$, the duality gap satisfies $\mathbf{c}^\mathsf{T}\mathbf{x}(\mu) - \mathbf{b}^\mathsf{T}\mathbf{y}(\mu) = n\mu$.

Proof of Theorem 6.1.3. We first prove the monotonicity of $\mathbf{c}^\mathsf{T}\mathbf{x}(\mu)$ as a function of μ. Take any μ and μ' with $0 < \mu < \mu'$. We will show that $\mathbf{c}^\mathsf{T}\mathbf{x}(\mu) < \mathbf{c}^\mathsf{T}\mathbf{x}(\mu')$. Since $\mathbf{x}(\mu)$ minimizes $B_P(\mathbf{x}; \mu)$, and $\mathbf{x}(\mu')$ minimizes $B_P(\mathbf{x}; \mu')$, it follows that:

$$\frac{1}{\mu}\mathbf{c}^\mathsf{T}\mathbf{x}(\mu) - \sum_i \ln(x_i(\mu)) < \frac{1}{\mu}\mathbf{c}^\mathsf{T}(\mathbf{x}(\mu')) - \sum_i \ln(x_i(\mu')),$$

$$\text{and } \frac{1}{\mu'}\mathbf{c}^\mathsf{T}\mathbf{x}(\mu') - \sum_i \ln(x_i(\mu')) < \frac{1}{\mu'}\mathbf{c}^\mathsf{T}\mathbf{x}(\mu) - \sum_i \ln(x_i(\mu)).$$

Adding up these inequalities gives:

$$\frac{1}{\mu}\mathbf{c}^\mathsf{T}\mathbf{x}(\mu) + \frac{1}{\mu'}\mathbf{c}^\mathsf{T}\mathbf{x}(\mu') < \frac{1}{\mu}\mathbf{c}^\mathsf{T}\mathbf{x}(\mu') + \frac{1}{\mu'}\mathbf{c}^\mathsf{T}\mathbf{x}(\mu),$$

and so $(\frac{1}{\mu} - \frac{1}{\mu'})(\mathbf{c}^\mathsf{T}\mathbf{x}(\mu) - \mathbf{c}^\mathsf{T}\mathbf{x}(\mu')) < 0$. Since $0 < \mu < \mu'$, it follows that $\frac{1}{\mu} - \frac{1}{\mu'} > 0$, and therefore $\mathbf{c}^\mathsf{T}\mathbf{x}(\mu) < \mathbf{c}^\mathsf{T}\mathbf{x}(\mu')$, as required. The monotonicity of $\mathbf{b}^\mathsf{T}\mathbf{y}(\mu)$ can be shown similarly by using $B_D(\mathbf{y}; \mu)$.

We finally show that $\mathbf{c}^\mathsf{T}\mathbf{x}(\mu) - \mathbf{b}^\mathsf{T}\mathbf{y}(\mu) = n\mu$. The proof is straightforward, namely:

$$\mathbf{c}^\mathsf{T}\mathbf{x}(\mu) - \mathbf{b}^\mathsf{T}\mathbf{y}(\mu) = \left(\mathbf{A}^\mathsf{T}\mathbf{y}(\mu) + \mathbf{w}(\mu)\right)^\mathsf{T}\mathbf{x}(\mu) - \mathbf{b}^\mathsf{T}\mathbf{y}(\mu)$$

$$= \left(\mathbf{y}(\mu)^\mathsf{T}\mathbf{A} + \mathbf{w}(\mu)^\mathsf{T}\right)\mathbf{x}(\mu) - \mathbf{b}^\mathsf{T}\mathbf{y}(\mu)$$

$$= \mathbf{y}(\mu)^\mathsf{T}\mathbf{b} + \mathbf{w}(\mu)^\mathsf{T}\mathbf{x}(\mu) - \mathbf{b}^\mathsf{T}\mathbf{y}(\mu)$$

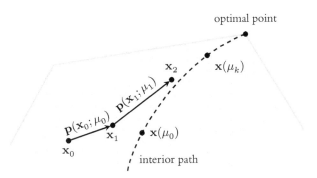

Figure 6.3: The interior path as a guiding hand.

$$= \mathbf{w}(\mu)^{\mathsf{T}} \mathbf{x}(\mu) = \sum_{i=1}^{n} x_i(\mu) w_i(\mu) = n\mu,$$

which proves the theorem. □

In Section 6.4.1, we will see how the duality gap for points on the interior path is used to terminate the calculations of the interior path algorithm with a certain prescribed accuracy for the optimal objective value.

6.2 Formulation of the interior path algorithm

6.2.1 The interior path as a guiding hand

The basic idea of the interior path algorithm is the following. Suppose that $\mathbf{x}_k$ is the current solution (corresponding to an interior point of the feasible region), and that μ_k is the current value of the interior path parameter. Recall that μ_k corresponds to the point $\mathbf{x}(\mu_k)$ on the interior path. In the next step, the algorithm determines an interior point $\mathbf{x}_{k+1}$ that is 'closer to' $\mathbf{x}(\mu_k)$ than $\mathbf{x}_k$. By the expression 'closer to', we mean that the algorithm determines a vector $\mathbf{p}(\mathbf{x}_k; \mu_k)$ which points more or less in the direction of $\mathbf{x}(\mu_k)$. The next interior point $\mathbf{x}_{k+1}$ is then given by:

$$\mathbf{x}_{k+1} = \mathbf{x}_k + \mathbf{p}(\mathbf{x}_k; \mu_k).$$

Notice that $\mathbf{x}_{k+1}$ is in general not equal to $\mathbf{x}(\mu_k)$, so $\mathbf{x}_{k+1}$ need not lie on the interior path. However, it will be 'close to' the interior path. Next, the algorithm decreases the current value of the parameter μ_k with a given updating factor θ that satisfies $0 < \theta < 1$; the new interior path parameter μ_{k+1} is defined by:

$$\mu_{k+1} = (1 - \theta)\mu_k,$$

and corresponds to the interior path point $\mathbf{x}(\mu_{k+1})$. The procedure is repeated for the pair $(\mathbf{x}_{k+1}; \mu_{k+1})$, until a pair $(\mathbf{x}^*, \mu^*)$ is reached for which:

$$n\mu^* \leq e^{-t},$$

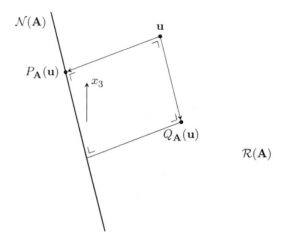

Figure 6.4: Projections on null space and row space.

where t is a prescribed *accuracy parameter*. Since this implies that μ^* is close to 0 and $n\mu^*$ is approximately (recall that $\mathbf{x}^*$ need not be on the interior path) equal to the duality gap $\mathbf{c}^\mathsf{T}\mathbf{x}^* - \mathbf{b}^\mathsf{T}\mathbf{y}^*$ (see Theorem 6.1.3 and Theorem 6.4.1), it follows that $\mathbf{x}^*$ and $\mathbf{y}^*$ are approximate optimal solutions of (P) and (D), respectively; see Theorem 4.2.3. Therefore, the interior points generated by the algorithm, converge to an optimal point. Informally speaking, the interior path serves as a 'guiding hand' that leads an initial interior point to the vicinity of the optimal solution. Figure 6.3 schematically shows the working of the interior path algorithm.

There remain a number of questions to be answered:

How to determine an initial interior point $\mathbf{x}_0$ and an initial interior path parameter μ_0? This will be discussed in Section 6.4.2.

How to determine the search direction $\mathbf{p}(\mathbf{x};\mu)$, and what should be the length of this vector? This is the topic of the next section.

What is a good choice for the updating factor θ? This will also be discussed in the next section.

Last but not least, we should prove that the algorithm is efficient, i.e., it runs in a polynomial amount of time. We postpone this discussion to Section 9.3.

6.2.2   Projections on null space and row space

In this section we develop some fundamental mathematical tools that play an important part in the theory of the interior path algorithm. These tools rely heavily on linear algebra; see Appendix B. Let $\mathbf{A}$ be an (m,n)-matrix with $m \leq n$, and $\mathrm{rank}(\mathbf{A}) = m$. Define:

$$\mathcal{N}(\mathbf{A}) = \{\mathbf{x} \in \mathbb{R}^n \mid \mathbf{A}\mathbf{x} = \mathbf{0}\}, \text{ and } \mathcal{R}(\mathbf{A}) = \{\mathbf{A}^\mathsf{T}\mathbf{y} \in \mathbb{R}^n \mid \mathbf{y} \in \mathbb{R}^m\}.$$

The set $\mathcal{N}(\mathbf{A})$ is called the *null space* of $\mathbf{A}$, and it consists of all vectors $\mathbf{x}$ such that $\mathbf{Ax} = \mathbf{0}$. Clearly, $\mathbf{0} \in \mathcal{N}(\mathbf{A})$. The set $\mathcal{R}(\mathbf{A})$ is called the *row space* of $\mathbf{A}$, and it consists of all vectors $\mathbf{x}$ that can be written as a linear combination of the rows of $\mathbf{A}$.

Take any $\mathbf{u} \in \mathbb{R}^n$. Let $P_{\mathbf{A}}(\mathbf{u})$ be the point in $\mathcal{N}(\mathbf{A})$ that is closest to $\mathbf{y}$. The point $P_{\mathbf{A}}(\mathbf{u})$ is called the *projection* of $\mathbf{u}$ onto $\mathcal{N}(\mathbf{A})$. Similarly, let $Q_{\mathbf{A}}(\mathbf{u})$ be the point in $\mathcal{R}(\mathbf{A})$ that is closest to $\mathbf{u}$. The point $Q_{\mathbf{A}}(\mathbf{u})$ is called the projection of $\mathbf{u}$ onto $\mathcal{R}(\mathbf{A})$. (See Figure 6.4.) By definition, $P_{\mathbf{A}}(\mathbf{u})$ is the optimal solution of the following minimization model (the $\frac{1}{2}$ and the square are added to make the calculations easier):

$$
\begin{aligned}
\min \ & \tfrac{1}{2}\|\mathbf{u} - \mathbf{x}\|^2 \\
\text{s.t. } & \mathbf{Ax} = \mathbf{0} \\
& \mathbf{x} \in \mathbb{R}^n.
\end{aligned} \tag{6.2}
$$

Using the vector $\boldsymbol{\lambda} = \begin{bmatrix} \lambda_1 & \dots & \lambda_m \end{bmatrix}^{\mathsf{T}}$ of Lagrange multipliers, it follows that the optimal solution of (6.2) satisfies the following gradient vector equations:

$$
\nabla \left(\tfrac{1}{2}\|\mathbf{u} - \mathbf{x}\|^2 + \boldsymbol{\lambda}^{\mathsf{T}}(\mathbf{Ax}) \right) = \mathbf{0},
$$

where the gradient is taken with respect to $\mathbf{x}$. One can easily check that this expression is equivalent to $-(\mathbf{u} - \mathbf{x}) + \mathbf{A}^{\mathsf{T}}\boldsymbol{\lambda} = \mathbf{0}$ and, hence, to $\mathbf{A}^{\mathsf{T}}\boldsymbol{\lambda} = (\mathbf{u} - \mathbf{x})$. Premultiplying both sides by $\mathbf{A}$ gives $\mathbf{AA}^{\mathsf{T}}\boldsymbol{\lambda} = \mathbf{A}(\mathbf{u} - \mathbf{x}) = \mathbf{Au}$, where the last equality follows from the fact that $\mathbf{Ax} = \mathbf{0}$. Using the fact that $\mathbf{AA}^{\mathsf{T}}$ is nonsingular (see Theorem B.4.1 in Appendix B), it follows that $\boldsymbol{\lambda} = (\mathbf{AA}^{\mathsf{T}})^{-1}\mathbf{Au}$. Substituting this expression into $\mathbf{A}^{\mathsf{T}}\boldsymbol{\lambda} = (\mathbf{u} - \mathbf{x})$ yields $\mathbf{A}^{\mathsf{T}}(\mathbf{AA}^{\mathsf{T}})^{-1}\mathbf{Au} = \mathbf{u} - \mathbf{x}$, and, hence:

$$
P_{\mathbf{A}}(\mathbf{u}) \ (= \text{the optimal } \mathbf{x}) = (\mathbf{I}_n - \mathbf{A}^{\mathsf{T}}(\mathbf{AA}^{\mathsf{T}})^{-1}\mathbf{A})\mathbf{u}.
$$

Note that $P_{\mathbf{A}}(\mathbf{u})$ is the product of a matrix that only depends on $\mathbf{A}$ and the vector $\mathbf{u}$. Hence, we may write $P_{\mathbf{A}}(\mathbf{u}) = \mathbf{P_A}\mathbf{u}$, where

$$
\mathbf{P_A} = \mathbf{I}_n - \mathbf{A}^{\mathsf{T}}(\mathbf{AA}^{\mathsf{T}})^{-1}\mathbf{A}.
$$

The projection $Q_{\mathbf{A}}(\mathbf{u})$ of $\mathbf{u}$ onto the row space of $\mathbf{A}$ can be determined as follows. By definition, $Q_{\mathbf{A}}(\mathbf{u}) = \mathbf{A}^{\mathsf{T}}\mathbf{y}$, where $\mathbf{y}$ is the optimal solution of the model (the square is again added to simplify the calculations)

$$
\begin{aligned}
\min \ & \tfrac{1}{2}\|\mathbf{u} - \mathbf{A}^{\mathsf{T}}\mathbf{y}\|^2 \\
\text{s.t. } & \mathbf{y} \in \mathbb{R}^m.
\end{aligned} \tag{6.3}
$$

After straightforward calculations, it follows that in this case the gradient expression is equivalent to $(\mathbf{AA}^{\mathsf{T}})\mathbf{y} = \mathbf{Au}$. Therefore, we have that $\mathbf{y} = (\mathbf{AA}^{\mathsf{T}})^{-1}\mathbf{Au}$, so that:

$$
Q_{\mathbf{A}}(\mathbf{u}) = \mathbf{Q_A}\mathbf{u}, \text{ with } \mathbf{Q_A} = \mathbf{A}^{\mathsf{T}}(\mathbf{AA}^{\mathsf{T}})^{-1}\mathbf{A}.
$$

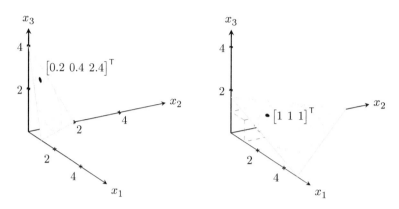

Figure 6.5: Dikin's affine scaling procedure.

Clearly, $\|\mathbf{u} - Q_{\mathbf{A}}(\mathbf{u})\| = \|\mathbf{u} - \mathbf{A}^\mathsf{T}(\mathbf{A}\mathbf{A}^\mathsf{T})^{-1}\mathbf{A}\mathbf{u}\| = \|P_{\mathbf{A}}(\mathbf{u})\|$. Using (6.3), it follows that:

$$\|P_{\mathbf{A}}(\mathbf{u})\| = \min_{\mathbf{y}} \|\mathbf{u} - \mathbf{A}^\mathsf{T}\mathbf{y}\|.$$

6.2.3 Dikin's affine scaling procedure

In order to avoid numerical problems when the current interior point is very close to the boundary of the feasible region, and to prove the computational efficiency of the algorithm (see Section 6.5), we will discuss the so-called *affine scaling method* introduced in 1967 by ILYA I. DIKIN (1936–2008).

Let $F_P = \{\mathbf{x} \in \mathbb{R}^n \mid \mathbf{A}\mathbf{x} = \mathbf{b}, \mathbf{x} \geq \mathbf{0}\}$ denote the feasible region. Let $\mathbf{x}_k$ be the current interior point of F_p, and let μ_k be the current interior path parameter. Since it may happen that $\mathbf{x}_k$ is very close to the boundary of F_P (especially in later iterations because, in general, the optimal solution of an LO-model lies on the boundary), and since the direction in which the new point is searched will be only approximately in the direction of $\mathbf{x}(\mu_k)$ (recall that $\mathbf{x}(\mu_k)$ is not explicitly known), even a small step in the chosen search direction may bring the new point outside F_P. We will explicitly determine the search direction in Section 6.2.4.

Moreover, in order to assure that, at each iteration, a sufficiently big step in the search direction can be made, the current model will be transformed in such a way that the all-ones vector $\mathbf{1}$, together with the unit hyperball around it, is in the interior of the transformed feasible region. The vector $\mathbf{1}$ becomes the starting point of the next iteration. We will show that it is then possible to, starting at $\mathbf{1}$, make a step of length one in any direction inside the transformed feasible region.

Suppose that the current model is:

$$\min\{\mathbf{c}^\mathsf{T}\mathbf{x} \mid \mathbf{A}\mathbf{x} = \mathbf{b}, \mathbf{x} \geq \mathbf{0}\}.$$

Now, replace the vector $\mathbf{x}$ of decision variables by $\mathbf{X}_k \mathbf{x}$, where $\mathbf{X}_k$ denotes the diagonal (n, n)-matrix with $\mathbf{x}_k$ on its main diagonal. Thus, the transformation is

$$\mathbf{x} := \mathbf{X}_k \mathbf{x}.$$

Applying this transformation can be thought of as measuring the value of each decision variable x_i in different units. The transformed model becomes:

$$\min\{(\mathbf{X}_k \mathbf{c})^\mathsf{T} \mathbf{x} \mid (\mathbf{A}\mathbf{X}_k)\mathbf{x} = \mathbf{b}, \mathbf{x} \geq \mathbf{0}\}.$$

Notice that $\mathbf{1}$ is interior point of $\bar{F}_P^+ = \{\mathbf{x} \in \mathbb{R}^n \mid \mathbf{A}\mathbf{X}_k \mathbf{x} = \mathbf{b}, \mathbf{x} > \mathbf{0}\}$. Moreover, the open hyperball

$$B_1(\mathbf{1}) = \{\mathbf{x} \in \mathbb{R}^n \mid \mathbf{A}\mathbf{X}_k \mathbf{x} = \mathbf{b}, \|\mathbf{x} - \mathbf{1}\| < 1\}$$

with radius 1 and center $\mathbf{1}$ is contained in the affine space $\{\mathbf{x} \in \mathbb{R}^n \mid \mathbf{A}\mathbf{X}_k \mathbf{x} = \mathbf{b}\}$. Furthermore, $B_1(\mathbf{1})$ consists of interior points of $\bar{F}_P$. In other words, $B_1(\mathbf{1}) \subset \bar{F}_P^+$. To prove this, take any $\mathbf{x} = \begin{bmatrix} x_1 & \dots & x_n \end{bmatrix}^\mathsf{T} \in B_1(\mathbf{1})$. Hence, $(x_1 - 1)^2 + \dots + (x_n - 1)^2 < 1$. If, say $x_1 \leq 0$, then $(x_1 - 1)^2 \geq 1$, which is not possible. Hence, $B_1(\mathbf{1}) \subset \bar{F}_P^+$.

Example 6.2.1. *Consider the situation as depicted in Figure 6.5, where*

$$F_P = \left\{ \begin{bmatrix} x_1 & x_2 \end{bmatrix}^\mathsf{T} \ \middle| \ x_1 + x_2 = 4, x_1, x_2 \geq 0 \right\}.$$

Take $\mathbf{x}_k = \begin{bmatrix} 0.3 & 3.7 \end{bmatrix}^\mathsf{T}$. After scaling with

$$\mathbf{X}_k = \begin{bmatrix} 0.3 & 0 \\ 0 & 3.7 \end{bmatrix},$$

we find that $\bar{F}_P = \left\{ \begin{bmatrix} x_1 & x_2 \end{bmatrix}^\mathsf{T} \ \middle| \ 0.3x_1 + 3.7x_2 = 4, x_1, x_2 \geq 0 \right\}$. The boundary points of $\bar{F}_P$ are $\begin{bmatrix} 13.3 & 0 \end{bmatrix}^\mathsf{T}$ and $\begin{bmatrix} 0 & 1.1 \end{bmatrix}^\mathsf{T}$. Note that inside $\bar{F}_P$ there is enough room to make a step of length 1 starting at $\mathbf{1} = \begin{bmatrix} 1 & 1 \end{bmatrix}^\mathsf{T}$ in any direction inside $\bar{F}_P$.

6.2.4 Determining the search direction

Let $\mathbf{x}_k$ be the current interior point, and let μ_k be the current value of the interior path parameter. We start with the scaled model:

$$\min\{(\mathbf{X}_k \mathbf{c})^\mathsf{T} \mathbf{x} \mid \mathbf{A}\mathbf{X}_k \mathbf{x} = \mathbf{b}, \mathbf{x} \geq \mathbf{0}\}. \tag{$\bar{P}$}$$

As argued in the previous subsection, the point $\mathbf{1}$ is the current interior point of $(\bar{P})$, and $\mathbf{X}_k^{-1}\mathbf{x}(\mu_k)$ is the current point on the interior path in $\bar{F}_P$. The logarithmic barrier function for $(\bar{P})$ is

$$B_{\bar{P}}(\mathbf{x}; \mu_k) = \frac{1}{\mu_k}(\mathbf{X}_k \mathbf{c})^\mathsf{T} \mathbf{x} - \sum_{i=1}^{n} \ln(x_i),$$

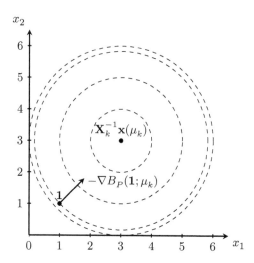

Figure 6.6: Pointing to the interior path.

and the corresponding gradient at $\mathbf{1}$ satisfies:

$$\nabla B_{\bar{P}}(\mathbf{1}; \mu_k) = \frac{1}{\mu_k}(\mathbf{X}_k \mathbf{c}) - \mathbf{1}.$$

Because of the fact that $B_{\bar{P}}(\cdot; \mu)$ is strictly convex (see Theorem 6.1.2), it follows that the vector $-\nabla B_{\bar{P}}(\mathbf{1}; \mu_k) = \mathbf{1} - (\mathbf{X}_k \mathbf{c}/\mu_k)$ points, more or less, in the direction of the point $\mathbf{X}_k^{-1}\mathbf{x}(\mu_k)$ on the interior path; see Figure 6.7.

Example 6.2.2.   *Take* $\mathbf{x}(\mu_k) = \begin{bmatrix} 3 & 3 \end{bmatrix}^\mathsf{T}$, *and* $B_{\bar{P}}(x_1, x_2; \mu_k) = \frac{1}{4}(x_1-3)^2 + \frac{1}{4}(x_2-3)^2$. *The dotted lines in Figure 6.6 correspond to the circles* $\frac{1}{4}(x_1-3)^2 + \frac{1}{4}(x_2-3)^2 = R$ *for different values of* R $(R > 0)$. *The vector* $-B_{\bar{P}}(1, 1; \mu_k) = \begin{bmatrix} 1 & 1 \end{bmatrix}^\mathsf{T}$ *points, starting from* $\mathbf{1}$, *in the direction of* $\begin{bmatrix} 3 & 3 \end{bmatrix}^\mathsf{T}$. *If, however, the level lines of the logarithmic barrier function are not circles, then the negative of the gradient does usually not point precisely to the center* $\mathbf{X}_k^{-1}\mathbf{x}(\mu_k)$. *The only thing that can be said in that case is that the angle between the vectors* $-\nabla B_{\bar{P}}(\mathbf{1}; \mu_k)$ *and* $\mathbf{X}_k^{-1}\mathbf{x}(\mu_k) - \mathbf{1}$ *is acute, i.e., it is less than* $90°$ *(see Figure 6.7).*

Theorem 6.2.1. *(Search direction)*
The angle between the vectors $\mathbf{X}_k^{-1}\mathbf{x}(\mu_k) - \mathbf{1}$ and $-\nabla B_{\bar{P}}(\mathbf{1}; \mu_k)$ is sharp.

Proof of Theorem 6.2.1. Consider an arbitrary differentiable strictly convex function f on a nonempty convex set K. Assume that $\mathbf{x}_0$ is the minimum of f on K, with $\mathbf{x}_0$ not on the boundary of K. We first show that for any $\mathbf{x}, \mathbf{y} \in K$, with $x_i \neq y_i$ for each i, it holds that:

$$f(\mathbf{y}) \geq f(\mathbf{x}) + (\nabla f(\mathbf{x}))^\mathsf{T}(\mathbf{y} - \mathbf{x}).$$

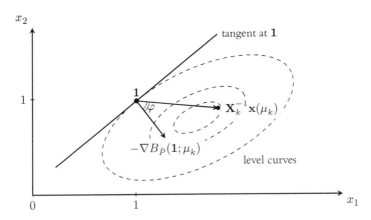

Figure 6.7: The vector $-\nabla B_{\bar{P}}(\mathbf{1}; \mu_k)$ points 'more or less' in the direction of $\mathbf{X}_k^{-1}\mathbf{x}(\mu_k)$.

For all α with $0 < \alpha \leq 1$ and all $i = 1, \ldots, n$, it follows that:

$$\frac{f(x_i + \alpha(y_i - x_i)) - f(x_i)}{\alpha(y_i - x_i)} = \frac{f((1-\alpha)x_i + \alpha y_i) - f(x_i)}{\alpha(y_i - x_i)}$$

$$\leq \frac{(1-\alpha)f(x_i) + \alpha f(y_i) - f(x_i)}{\alpha(y_i - x_i)} \qquad \text{(since } f \text{ is convex)}$$

$$= \frac{f(y_i) - f(x_i)}{y_i - x_i}.$$

Since f is assumed to be differentiable on K, it follows that:

$$\frac{\partial}{\partial x_i} f(\mathbf{x}) = \lim_{\alpha \downarrow 0} \frac{f(x_i + \alpha(y_i - x_i)) - f(x_i)}{\alpha(y_i - x_i)} \leq \frac{f(y_i) - f(x_i)}{y_i - x_i}.$$

Hence, $f(\mathbf{y}) \geq f(\mathbf{x}) + (\nabla f(\mathbf{x}))^{\mathsf{T}}(\mathbf{y} - \mathbf{x})$. Substituting $f(\mathbf{x}) = B_{\bar{P}}(\mathbf{x}; \mu_k)$, $\mathbf{y} = \mathbf{X}_k^{-1}\mathbf{x}(\mu_k)$, and $\mathbf{x} = \mathbf{1}$ in the above expression, we obtain

$$(-\nabla B_{\bar{P}}(\mathbf{1}; \mu_k))^{\mathsf{T}}(\mathbf{X}_k^{-1}\mathbf{x}(\mu_k) - \mathbf{1}) \geq -B_{\bar{P}}(\mathbf{X}_k^{-1}\mathbf{x}(\mu_k); \mu_k) + B_{\bar{P}}(\mathbf{1}; \mu_k) > \mathbf{0}$$

(strict convexity). Using the cosine-rule for vectors, we find that

$$\cos(\phi) = \frac{(-\nabla B_{\bar{P}}(\mathbf{1}; \mu_k))^{\mathsf{T}}(\mathbf{X}_k^{-1}\mathbf{x}(\mu_k) - \mathbf{1})}{\|\nabla B_{\bar{P}}(\mathbf{1}; \mu_k)\| \, \|\mathbf{X}_k^{-1}\mathbf{x}(\mu_k) - \mathbf{1}\|} > 0,$$

where ϕ is the angle between $-\nabla B_{\bar{P}}(\mathbf{1}; \mu_k)$ and $\mathbf{X}_k^{-1}\mathbf{x}(\mu_k) - \mathbf{1}$. Hence, $0 \leq \phi \leq \frac{1}{2}\pi$, and so, in fact, the angle is sharp. $\qquad \square$

Let $\mathbf{p}(\mathbf{1}; \mu_k)$ be the search direction for the next interior point $\mathbf{u}$ in $F_{\bar{P}}$ when starting at $\mathbf{1}$; i.e., $\mathbf{u} = \mathbf{1} + \mathbf{p}(\mathbf{1}; \mu_k)$. In the next section, we will determine the length of $\mathbf{p}(\mathbf{1}; \mu_k)$ such that $\mathbf{u}$ is actually an interior point of $F_{\bar{P}}$. Since $\mathbf{u} \in F_{\bar{P}}$, it follows that $\mathbf{AX}_k(\mathbf{u} - \mathbf{1}) = \mathbf{AX}_k\mathbf{p}(\mathbf{1}; \mu_k) = \mathbf{0}$, and so $\mathbf{p}(\mathbf{1}; \mu_k)$ has to be in the null space $\mathcal{N}(\mathbf{AX}_k)$; see Section 6.2.2. This can be accomplished by taking the projection of $-\nabla B_{\bar{P}}(\mathbf{1}; \mu_k)$ onto the null space of $\mathbf{AX}_k$. So, the search direction in the scaled model (P) becomes:

$$\mathbf{p}(\mathbf{1}; \mu_k) = \mathbf{P}_{AX_k}(-\nabla B_{\bar{P}}(\mathbf{1}; \mu_k)) = \mathbf{P}_{AX_k}(\mathbf{1} - \mathbf{X}_k\mathbf{c}/\mu_k).$$

The search direction $\mathbf{p}(1; \mu_k)$ corresponds to the search direction $\mathbf{X}_k \mathbf{p}(1; \mu_k)$ for the original model, where $\mathbf{x}_k$ is the current interior point, and μ_k is current value of the interior path parameter. Define:

$$\mathbf{p}(\mathbf{x}_k; \mu_k) = \mathbf{X}_k \mathbf{p}(1; \mu_k).$$

Note that $\mathbf{p}(\mathbf{x}_k; \mu_k)$ is in the null space $\mathcal{N}(\mathbf{A})$. It now follows that the search direction for the unscaled model is:

$$\mathbf{p}(\mathbf{x}_k; \mu_k) = \mathbf{X}_k \mathbf{P}_{AX_k} (1 - \mathbf{X}_k \mathbf{c}/\mu_k),$$

and the next point $\mathbf{x}_{k+1}$ is chosen to be $\mathbf{x}_{k+1} = \mathbf{x}_k + \mathbf{p}(\mathbf{x}_k; \mu_k)$; see Figure 6.3.

Example 6.2.3. *Consider again the data used for Figure 6.1 in Section 6.1. In order to apply Dikin's affine scaling procedure, we first need the (P)-formulation of model (A), namely:*

$$-\min - x_1 - 2x_2$$
$$\text{s.t. } x_1 + x_3 = 1$$
$$x_2 + x_4 = 1$$
$$x_1, x_2, x_3, x_4 \geq 0.$$

Take $\mathbf{x}_k = \begin{bmatrix} 0.4 & 0.8 & 0.6 & 0.2 \end{bmatrix}^\mathsf{T}$, and $\mu_k = 2$. After scaling with respect to $\mathbf{x}_k$, we find the equivalent formulation

$$-\min - 0.4x_1 - 1.62x_2$$
$$\text{s.t. } 0.4x_1 + 0.6x_3 = 1$$
$$0.8x_2 + 0.2x_4 = 1$$
$$x_1, x_2, x_3, x_4 \geq 0,$$

and after straightforward calculations we have that

$$\mathbf{P}_{AX_k} = \begin{bmatrix} 0.69 & 0 & -0.46 & 0 \\ 0 & 0.06 & 0 & -0.24 \\ -0.46 & 0 & 0.31 & 0 \\ 0 & -0.24 & 0 & 0.94 \end{bmatrix},$$

and therefore

$$\mathbf{p}(\mathbf{x}_k; \mu_k) = \mathbf{X}_k \mathbf{P}_{AX_k} \left(1 - \frac{1}{\mu_k} \mathbf{X}_k \mathbf{c} \right) = \begin{bmatrix} 0.15 & -0.10 & -0.14 & 0.10 \end{bmatrix}^\mathsf{T}.$$

The projection of $\mathbf{p}(\mathbf{x}_k; \mu_k)$ on the $X_1 O X_2$-plane is the vector $\begin{bmatrix} 0.15 & -0.10 \end{bmatrix}^\mathsf{T}$. Hence, the new point $\mathbf{x}_{k+1}$ satisfies:

$$\mathbf{x}_{k+1} = \mathbf{x}_k + \mathbf{p}(\mathbf{x}_k; \mu_k) = \begin{bmatrix} 0.4 & 0.8 \end{bmatrix}^\mathsf{T} + \begin{bmatrix} 0.15 & -0.10 \end{bmatrix}^\mathsf{T} = \begin{bmatrix} 0.55 & 0.70 \end{bmatrix}^\mathsf{T};$$

see Figure 6.1.

6.3 Convergence to the interior path; maintaining feasibility

6.3.1 The convergence measure

Let $\mathbf{x}_k$ be the current interior point of the feasible region $F_P = \{\mathbf{x} \in \mathbb{R}^n \mid \mathbf{Ax} = \mathbf{b}\}$, and let μ_k be the current value of the interior path parameter. In Section 6.2.4, we defined the search direction $\mathbf{p}(\mathbf{x}_k; \mu_k)$ and, therefore, the next interior point $\mathbf{x}_{k+1} = \mathbf{x}_k + \mathbf{p}(\mathbf{x}_k; \mu_k)$.

The question to be answered in this section is: how to choose the length of $\mathbf{p}(\mathbf{x}_k; \mu_k)$ such that $\mathbf{x}_{k+1}$ is again an interior point of F_P? This question will be answered by choosing the length of $\mathbf{p}(\mathbf{1}; \mu_k)$ in the scaled model such that $\mathbf{u} = \mathbf{1} + \mathbf{p}(\mathbf{1}; \mu_k)$ is an interior point of the scaled feasible region $F_{\bar{P}}$. It can be easily checked that the length of $\mathbf{p}(\mathbf{1}; \mu_k)$ has to be < 1. Recall that $\mathbf{p}(\mathbf{1}; \mu_k) = \mathbf{P}_{AX_k}(\mathbf{1} - \mathbf{X}_k \mathbf{c}/\mu_k)$.

For any interior point $\mathbf{x}$ of F_P and any value $\mu > 0$ of the interior path parameter, define:

$$\delta(\mathbf{x}; \mu) = \left\| \mathbf{P}_{AX} \left(\frac{\mathbf{Xc}}{\mu} - \mathbf{1} \right) \right\|.$$

The entity $\delta(\mathbf{x}; \mu)$ is called the *convergence measure* of the interior path algorithm. It plays a key role in proving both the correctness of the algorithm (meaning that at each iteration of the algorithm the feasibility of the generated points is preserved, and that these points converge to an optimal solution), and the polynomial running time of the algorithm.

Theorem 6.3.1. (*Convergence measure theorem*)
For any interior point $\mathbf{x}$ of F_P, and any interior path parameter μ, it holds that:

$$\delta(\mathbf{x}, \mu) = \min_{\mathbf{y}, \mathbf{w}} \left\{ \left\| \frac{\mathbf{Xw}}{\mu} - \mathbf{1} \right\| \,\middle|\, \mathbf{A}^{\mathsf{T}}\mathbf{y} + \mathbf{w} = \mathbf{c} \right\}.$$

Moreover, for any interior point $\mathbf{x}$ of F_P, it holds that:

$$\delta(\mathbf{x}, \mu) = 0 \text{ if and only if } \mathbf{x} = \mathbf{x}(\mu).$$

Proof of Theorem 6.3.1. For any interior point $\mathbf{x}$ in F_P and any interior path parameter μ, it holds that:

$$\min_{\mathbf{y}, \mathbf{w}} \left\{ \|\mathbf{Xw}/\mu - \mathbf{1}\| \,\middle|\, |\mathbf{A}^{\mathsf{T}}\mathbf{y} + \mathbf{w} = \mathbf{c} \right\}$$
$$= \min_{\mathbf{y}} \|(\mathbf{X}/\mu)(\mathbf{c} - \mathbf{A}^{\mathsf{T}}\mathbf{y}) - \mathbf{1}\|$$
$$= \min_{\mathbf{y}} \|(\mathbf{Xc}/\mu - \mathbf{1}) - (\mathbf{AX})^{\mathsf{T}}(\mathbf{y}/\mu)\| \qquad \text{(see Section 6.2.2)}$$
$$= \|\mathbf{P}_{AX}(\mathbf{Xc}/\mu - \mathbf{1})\| = \delta(\mathbf{x}, \mu).$$

This proves the first part. For the second part, take any $\mathbf{x} > \mathbf{0}$ with $\mathbf{Ax} = \mathbf{b}$. Suppose first that $\mathbf{x}$ is on the interior path, i.e., $\mathbf{x} = \mathbf{x}(\mu)$ for some $\mu > 0$. Let $\mathbf{y}(\mu)$ and $\mathbf{w}(\mu)$ satisfy (PDμ). Then $\mathbf{Xw}(\mu) = \mu\mathbf{1}$ implies that $\delta(\mathbf{x}, \mu) = 0$, as required. Next, assume that $\delta(\mathbf{x}, \mu) = 0$. Let $\mathbf{y}(\mathbf{x}, \mu)$ and $\mathbf{w}(\mathbf{x}, \mu)$ be minimizing values of $\mathbf{y}$ and $\mathbf{w}$ in the above formula for $\delta(\mathbf{x}, \mu)$; i.e.,

$\delta(\mathbf{x}, \mu) = \|\mathbf{X}\mathbf{w}(\mathbf{x}, \mu)\mu^{-1} - \mathbf{1}\|$, and $\mathbf{A}^\top \mathbf{y}(\mathbf{x}, \mu) + \mathbf{w}(\mathbf{x}, \mu) = \mathbf{c}$. Hence, $\mathbf{X}\mathbf{w}(\mathbf{x}, \mu) = \mathbf{1}\mu$, or $\mathbf{w}(\mathbf{x}, \mu) = \mu \mathbf{X}^{-1}\mathbf{1} = \mu\mathbf{x} > \mathbf{0}$. Therefore, $\mathbf{x}$, $\mathbf{y}(\mathbf{x}, \mu)$, and $\mathbf{w}(\mathbf{x}, \mu)$ satisfy the relaxed Karush-Kuhn-Tucker conditions (PDμ). Hence, $\mathbf{x} = \mathbf{x}(\mu)$. This proves the theorem. □

For $\mathbf{w}(\mathbf{x}, \mu)$ and $\mathbf{y}(\mathbf{x}, \mu)$ being minimizing values satisfying $\mathbf{w}(\mathbf{x}, \mu) = \mathbf{c} - \mathbf{A}^\top \mathbf{y}(\mathbf{x}, \mu)$ in the formula for $\delta(\mathbf{x}, \mu)$ in Theorem 6.3.1, we define:

$$\mathbf{s}(\mathbf{x}, \mu) = \mathbf{X}\mathbf{w}(\mathbf{x}, \mu)\mu^{-1}.$$

It then follows that $\mathbf{p}(\mathbf{x}, \mu) = -\mathbf{X}\mathbf{P}_{AX}(\mathbf{X}\mathbf{c}/\mu - \mathbf{1}) = -\mathbf{X}(\mathbf{X}\mathbf{w}(\mathbf{x}, \mu)/\mu - \mathbf{1}) = \mathbf{X}(\mathbf{1} - \mathbf{s}(\mathbf{x}, \mu))$, and $\delta(\mathbf{x}, \mu) = \|\mathbf{1} - \mathbf{s}(\mathbf{x}, \mu)\|$.

Theorem 6.3.2. (*Convergence to the interior path*)
Let $\mathbf{x}_0$ be an interior point of F_P, and $\mu > 0$, such that $\delta(\mathbf{x}_0, \mu) < 1$. For each $k = 0, 1, \ldots$, define $\mathbf{x}_{k+1} = \mathbf{x}_k + \mathbf{p}(\mathbf{x}_k, \mu)$. Then the sequence $\mathbf{x}_0, \mathbf{x}_1, \ldots, \mathbf{x}_k, \ldots$ consists of interior points of F_P and the sequence converges to $\mathbf{x}(\mu)$.

Proof of Theorem 6.3.2. We first show that $\mathbf{x}_{k+1} = \mathbf{x}_k + \mathbf{p}(\mathbf{x}_k, \mu)$ is an interior point of F_P, provided that $\mathbf{x}_k$ is an interior point of F_P. Since $\mathbf{p}(\mathbf{x}_k, \mu) = \mathbf{X}_k \mathbf{P}_{AX_k}(\mathbf{1} - \mathbf{X}_k \mathbf{c}/\mu)$, and $\mathbf{P}_{AX_k}(\mathbf{1} - \mathbf{X}_k \mathbf{c}/\mu)$ is in $\mathcal{N}(\mathbf{A}\mathbf{X}_k)$, it follows directly that $\mathbf{A}\mathbf{x}_{k+1} = \mathbf{b}$. So, we only need to show that $\mathbf{x}_{k+1} > \mathbf{0}$. Clearly, $\mathbf{X}_k^{-1}\mathbf{p}(\mathbf{x}_k, \mu) = \mathbf{P}_{AX_k}(\mathbf{1} - \mathbf{X}_k \mathbf{c}/\mu)$. Since $\|\mathbf{P}_{AX_k}(\mathbf{X}_k \mathbf{c}/\mu) - \mathbf{1}\| < 1$, and $B_1(\mathbf{1}) \subset F_{\bar{P}}^+$ (see Section 6.2.4), it follows that $\mathbf{u} = \mathbf{1} + \mathbf{p}(\mathbf{1}, \mu)$ is an interior point of $F_{\bar{P}}$. Hence, $\mathbf{u} > \mathbf{0}$. Moreover, since $\mathbf{x}_k > \mathbf{0}$, it follows that $\mathbf{x}_{k+1} = \mathbf{X}\mathbf{u} > \mathbf{0}$. Therefore, $\mathbf{x}_{k+1}$ is in fact an interior point of F_P.

We now show that $\delta(\mathbf{x}_k, \mu) < 1$ for each $k = 1, 2, 3, \ldots$, provided that $\delta(\mathbf{x}_0, \mu) < 1$. To that end, we show that:

$$\delta(\mathbf{x}_{k+1}, \mu) \leq (\delta(\mathbf{x}_k, \mu))^2.$$

For brevity, we will write $\mathbf{s}_k = \mathbf{s}(\mathbf{x}_k, \mu)$; $\mathbf{S}_k$ denotes the diagonal matrix associated with $\mathbf{s}_k$. Then, it follows that

$$\delta(\mathbf{x}_{k+1}, \mu) = \left\|\frac{1}{\mu}\mathbf{X}_{k+1}\mathbf{w}(\mathbf{x}_{k+1}, \mu) - \mathbf{1}\right\|$$
$$\leq \left\|\frac{1}{\mu}\mathbf{X}_{k+1}\mathbf{w}(\mathbf{x}_k, \mu) - \mathbf{1}\right\| = \|\mathbf{X}_{k+1}\mathbf{X}_k^{-1}\mathbf{s}_k - \mathbf{1}\|.$$

Since $\mathbf{x}_{k+1} = \mathbf{x}_k + \mathbf{p}(\mathbf{x}_k, \mu) = \mathbf{X}_k\mathbf{1} + \mathbf{X}_k(\mathbf{1} - \mathbf{s}_k) = 2\mathbf{X}_k\mathbf{1} - \mathbf{X}_k\mathbf{s}_k$, we find that

$$\mathbf{X}_{k+1}\mathbf{X}_k^{-1}\mathbf{s}_k - \mathbf{1} = (2\mathbf{X}_k - \mathbf{X}_k\mathbf{S}_k^{-1})\mathbf{X}_k\mathbf{s}_k = 2\mathbf{s}_k - \mathbf{S}_k\mathbf{s}_k - \mathbf{1} = (\mathbf{I}_n - \mathbf{S}_k)(\mathbf{s}_k - \mathbf{1}).$$

Let $\mathbf{s}_k = \begin{bmatrix} s_{k_1} & \cdots & s_{k_n} \end{bmatrix}^\top$. Then,

$$\delta(\mathbf{x}_{k+1}, \mu) \leq \mathbf{I}_n - \mathbf{S}_k\mathbf{s}_k - \mathbf{1} = \left[(1 - s_{k_1})^2 \cdots (1 - s_{k_n})^2\right]^\top$$
$$\leq \|\mathbf{1} - \mathbf{s}_k\|^2 = (\delta(\mathbf{x}_k, \mu))^2.$$

For any $k = 1, 2, \ldots$, it follows that:

$$0 \leq \delta(\mathbf{x}_k, \mu) \leq (\delta(\mathbf{x}_{k-1}, \mu))^2 \leq \cdots \leq (\delta(\mathbf{x}_0, \mu))^{2k}.$$

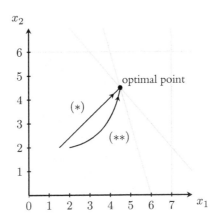

Figure 6.8: Approximations of interior paths. The path denoted (∗) refers to Model Dovetail, and path (∗∗) to the same model without the constraint $x_1 \leq 7$.

Hence, $\lim_{k \to \infty} (\delta(\mathbf{x}_k, \mu))^k = 0$. Since $\delta(\mathbf{x}, \mu) = 0$ if and only if $\mathbf{x} = \mathbf{x}(\mu)$ (see Theorem 6.3.1), it follows that the sequence $\{\mathbf{x}_k\}$ converges to $\mathbf{x}(\mu)$ for $k \to \infty$. $\square$

Theorem 6.3.2 can be used to computationally approximate the interior path of an LO-model. We can do this by choosing various initial interior points $\mathbf{x}_0$ and various values of μ, and approximately calculating $\mathbf{x}(\mu)$. For example, Figure 6.8 shows two interior paths for Model Dovetail (see Section 1.1). The path denoted (∗) refers to Model Dovetail, and path (∗∗) to the same model without the constraint $x_1 \leq 7$. It is left to the reader to repeat some of the calculations for different choices of $\mathbf{x}_0$ and μ. In Exercise 6.5.16, the reader is asked to draw the interior paths for different objective vectors, and for different right hand side vectors.

6.3.2 Maintaining feasibility; the interior path algorithm

In each iteration, the interior path algorithm lowers the current value of the interior path parameter μ_k by a factor θ ($0 < \theta < 1$), such that the new interior path parameter becomes $\mu_{k+1} = (1 - \theta)\mu_k$. A question is: how to choose θ in order to achieve that $\delta(\mathbf{x}_{k+1}, \mu_{k+1}) < 1$, because then it is assured that $\mathbf{x}_{k+1}$ is again an interior point of F_P. The following theorem gives the answer.

Theorem 6.3.3. (*Maintaining feasibility*)
Let $\mathbf{x}_0$ and μ_0 be the initial interior point and the initial value of the interior path parameter, respectively, and suppose that $\delta(\mathbf{x}_0, \mu_0) \leq \frac{1}{2}$. If $\theta = \frac{1}{6\sqrt{n}}$, then $\delta(\mathbf{x}_k, \mu_k) \leq \frac{1}{2}$ for all $k \geq 1$.

Proof of Theorem 6.3.3. Let $\mathbf{x}_k$, μ_k be the current pair, and $\mathbf{x}_{k+1}$, μ_{k+1} the next pair defined as before, with $\theta = \frac{1}{6\sqrt{n}}$. Assuming that $\delta(\mathbf{x}_k, \mu_k) \leq \frac{1}{2}$, we will show that it also holds that $\delta(\mathbf{x}_{k+1}, \mu_{k+1}) \leq \frac{1}{2}$. The proof uses similar ideas as the proof of Theorem 6.3.2. One can easily check that:

$$
\begin{aligned}
\delta(\mathbf{x}_{k+1}, \mu_{k+1}) &= \left\| \frac{1}{\mu_{k+1}} \mathbf{X}_{k+1} \mathbf{w}(\mathbf{x}_{k+1}, \mu_{k+1}) - \mathbf{1} \right\| \leq \left\| \frac{1}{\mu_{k+1}} \mathbf{X}_{k+1} \mathbf{w}(\mathbf{x}_k, \mu_k) - \mathbf{1} \right\| \\
&= \left\| \frac{1}{1-\theta} \mathbf{X}_{k+1} \mathbf{X}_k^{-1} \mathbf{s}_k - \mathbf{1} \right\| = \frac{1}{1-\theta} \left\| \mathbf{X}_{k+1} \mathbf{X}_k^{-1} \mathbf{s}_k - \mathbf{1} - \theta \mathbf{1} \right\| \\
&\leq \frac{1}{1-\theta} \left(\left\| \mathbf{X}_{k+1} \mathbf{X}_k^{-1} \mathbf{s}_k - \mathbf{1} \right\| + \theta \|\mathbf{1}\| \right) \\
&\leq \frac{1}{1-\theta} \left(\| (\mathbf{I}_n - \mathbf{S}_k)(\mathbf{s}_k - \mathbf{1}) \| + \theta \|\mathbf{1}\| \right) \\
&= \frac{1}{1-\theta} \left(\delta(\mathbf{x}_k, \mu_k)^2 + \theta \sqrt{n} \right) \leq \frac{6}{5} \left(\frac{1}{4} + \frac{1}{6} \right) = \frac{1}{2}.
\end{aligned}
$$

Hence, the theorem holds. $\qquad \square$

Theorem 6.3.3 shows that if we choose the initial pair $\mathbf{x}_0$, μ_0 such that $\delta(\mathbf{x}_0, \mu_0) \leq \frac{1}{2}$ and $\theta = \frac{1}{6\sqrt{n}}$, then the value of $\delta(\mathbf{x}_k, \mu_k)$ remains $\leq \frac{1}{2}$ at each iteration step, and so at the end of each iteration step the algorithm is actually back in its initial situation. The algorithm proceeds as long as $n\mu_k > e^{-t}$, where t is the accuracy parameter (see Section 6.2); this will be explained in the next section.

We can now summarize the above discussions by formulating the *interior path algorithm*.

Algorithm 6.3.1. (*Interior path algorithm*)

Input: Values for the parameters in the model $\min\{\mathbf{c}^\mathsf{T}\mathbf{x} \mid \mathbf{A}\mathbf{x} = \mathbf{b}, \mathbf{x} \geq \mathbf{0}\}$, and for the accuracy parameter t.

Output: An optimal solution $\mathbf{x}^*$, or one of the following messages:
(a) the model has no optimal solution, or
(b) the model is unbounded.

Step 1:   *Initialization.* Determine an initial interior point $\mathbf{x}_0$ and an initial value μ for the interior path parameter (see Section 6.4.2). Set $\theta := \frac{1}{6\sqrt{n}}$ and $k := 0$.

Step 2:   *Improvement.* Let $\mathbf{x}_k$ be the current interior point and let μ_k be the current value of the interior path parameter. If $n\mu_k \leq e^{-t}$, then go to step 3; else, set $\mu_{k+1} := (1-\theta)\mu_k$, $\mathbf{x}_{k+1} := \mathbf{x}_k + \mathbf{X}_k \mathbf{P}_{AX_k} (\mathbf{1} - \frac{1}{\mu_k} \mathbf{X}_k \mathbf{c})$, set $k := k+1$, and repeat step 2.

Step 3:   *Optimality.* Stop, the interior point $\mathbf{x}_k$ is close enough to an optimal solution. Set $\mathbf{x}^* := \mathbf{x}_k$.

The expression for the dual slack variables $\mathbf{w}_k$ follows from $\mathbf{X}_k \mathbf{w}_k = \mu \mathbf{1}$, and the expression for the dual decision variables $\mathbf{y}_k$ will be explained in Section 6.4.1.

6.4 Termination and initialization

We have seen how the interior path algorithm iteratively improves the current interior point. In the current section, we discuss two crucial remaining issues: when the interior path algorithm may be terminated, and how to find an initial interior point.

6.4.1 Termination and the duality gap

In Theorem 6.1.3 we saw that the duality gap $\mathbf{c}^\mathsf{T}\mathbf{x}(\mu) - \mathbf{b}^\mathsf{T}\mathbf{y}(\mu)$ for points $\mathbf{x}(\mu)$ and $\mathbf{y}(\mu)$ on the interior path is equal to $n\mu$. If we were able to calculate $\mathbf{x}(\mu)$ explicitly, this would have been the end of the story: just calculate this limit (compare Theorem 6.1.3). However, as we saw before, usually the interior path cannot be determined explicitly. The interior path algorithm generates interior points that get closer and closer to the interior path and, as we will see in Theorem 6.4.1, also closer and closer to an optimal solution. An important by-product is that $\mathbf{y} = \mathbf{y}(\mathbf{x}, \mu)$, defined as the minimizing value in the expression for $\delta(\mathbf{x}, \mu)$ in Theorem 6.3.1, is dual feasible for any current pair $\mathbf{x}_k$, μ_k. So, at each iteration step a corresponding dual solution is known.

Theorem 6.4.1. (*Convergence of the interior path algorithm*)
Let $\mathbf{x}_k$ be the current interior point and μ_k the current interior path parameter. If $\delta(\mathbf{x}_0, \mu_0) \leq 1$, then $\mathbf{y}_k = \mathbf{y}(\mathbf{x}_k, \mu_k)$ is dual feasible, and
$$|(\mathbf{c}^\mathsf{T}\mathbf{x}_k - \mathbf{b}^\mathsf{T}\mathbf{y}_k) - n\mu_k| \leq \mu_k \sqrt{n}\, \delta(\mathbf{x}_k, \mu_k).$$

Proof of Theorem 6.4.1. From Theorem 6.3.3, it follows that $\delta(\mathbf{x}_k, \mu_k) < 1$ for each $k = 1, 2, \ldots$. Hence, $\delta(\mathbf{x}_k, \mu_k) = \|\mathbf{s}(\mathbf{x}_k, \mu_k) - \mathbf{1}\| < 1$. Define $\mathbf{s}_k = \mathbf{s}(\mathbf{x}_k, \mu_k)$. It then follows that $\mathbf{s}_k \geq \mathbf{0}$, because if, to the contrary, $\mathbf{s}_k \leq \mathbf{0}$, then $\|\mathbf{s}_k - \mathbf{1}\| = \|-\mathbf{s}_k + \mathbf{1}\| \geq \|\mathbf{1}\| = \sqrt{n} > 1$, which is clearly false. Since $\mathbf{x}_k > \mathbf{0}$ and $\mu > 0$, it follows that $\mathbf{c} - \mathbf{A}^\mathsf{T}\mathbf{y}_k = \mathbf{w}_k = \mathbf{X}_k^{-1}\mathbf{s}_k\mu_k \geq \mathbf{0}$ and, therefore, $\mathbf{y}_k$ is dual feasible. In order to show the inequality of the theorem, we apply the Cauchy-Schwarz inequality ($\|\mathbf{a}\|\,\|\mathbf{b}\| \geq |\mathbf{a}^\mathsf{T}\mathbf{b}|$, see Appendix B):
$$\sqrt{n}\delta(\mathbf{x}_k, \mu_k) = \|\mathbf{1}\| \left\|\frac{1}{\mu_k}\mathbf{X}_k\mathbf{w}_k - \mathbf{1}\right\| \geq \left|\mathbf{1}^\mathsf{T}\left(\frac{1}{\mu_k}\mathbf{X}_k\mathbf{w}_k - \mathbf{1}\right)\right| = \left|\frac{\mathbf{x}_k^\mathsf{T}\mathbf{w}_k}{\mu_k} - n\right|.$$
Since $\mu_k > 0$, it follows that $|\mathbf{x}_k^\mathsf{T}\mathbf{w}_k - n\mu_k| \leq \mu_k\sqrt{n}\delta(\mathbf{x}_k, \mu_k)$. Recall that $\mathbf{c}^\mathsf{T}\mathbf{x}_k - \mathbf{b}^\mathsf{T}\mathbf{y}_k = (\mathbf{A}^\mathsf{T}\mathbf{y}_k + \mathbf{w}_k)^\mathsf{T}\mathbf{x}_k - \mathbf{b}^\mathsf{T}\mathbf{y}_k = \mathbf{w}_k^\mathsf{T}\mathbf{x}_k = \mathbf{x}_k^\mathsf{T}\mathbf{w}_k$. This proves the theorem. $\square$

The values of $\mathbf{y}_k$ can explicitly be calculated at each iteration k, by solving $\mathbf{c} - \mathbf{A}^\mathsf{T}\mathbf{y}_k = \mathbf{w}_k$. Namely,
$$\mathbf{y}_k = (\mathbf{A}\mathbf{A}^\mathsf{T})^{-1}\mathbf{A}(\mathbf{c} - \mathbf{w}_k).$$
Note that $\mathbf{A}^\mathsf{T}\mathbf{y}_k = \mathbf{A}^\mathsf{T}(\mathbf{A}\mathbf{A}^\mathsf{T})^{-1}\mathbf{A}(\mathbf{c} - \mathbf{w}_k) = \mathbf{Q}_\mathbf{A}(\mathbf{c} - \mathbf{w}_k)$; see Section 6.2.2. Hence, $\mathbf{A}^\mathsf{T}\mathbf{y}_k$ is the projection of $\mathbf{c} - \mathbf{w}_k$ onto the row space $\mathcal{R}(\mathbf{A})$ of $\mathbf{A}$.

Let $(\mathbf{x}_0, \mu_0)$ be an initial pair for the interior path algorithm such that $\delta(\mathbf{x}_0, \mu_0) \leq \frac{1}{2}$, and $\theta = \frac{1}{6\sqrt{n}}$. Moreover, let

$$(\mathbf{x}_0, \mu_0), (\mathbf{x}_1, \mu_1), \ldots, (\mathbf{x}_k, \mu_k), \ldots$$

be a sequence of pairs of interior points and values of the interior path parameter generated in successive steps of the interior path algorithm. Since $\lim\limits_{k \to \infty} \mu_k = \lim\limits_{k \to \infty} (1 - \theta)^k \mu_0 = 0$, Theorem 6.4.1 implies that:

$$0 \leq \lim_{k \to \infty} |\mathbf{c}^\mathsf{T}\mathbf{x}_k - \mathbf{b}^\mathsf{T}\mathbf{y}_k - n\mu_k| \leq \lim_{k \to \infty} \tfrac{1}{2}\mu_k \sqrt{n} = 0.$$

Hence,

$$\lim_{k \to \infty} |\mathbf{c}^\mathsf{T}\mathbf{x}_k - \mathbf{b}^\mathsf{T}\mathbf{y}_k - n\mu_k| = 0.$$

This implies that, for large enough values of k, it holds that:

$$\mathbf{c}^\mathsf{T}\mathbf{x}_k - \mathbf{b}^\mathsf{T}\mathbf{y}_k \approx n\mu_k$$

('$\approx$' means 'approximately equal to'), and so $n\mu_k$ is a good measure for the desired rate of optimality. If t is the desired accuracy parameter, then the algorithm stops as soon as:

$$n\mu_k \leq e^{-t},$$

i.e., as soon as the duality gap is approximately equal to e^{-t}.

6.4.2 Initialization

The interior path algorithm starts with an initial interior point $\mathbf{x}_0$ and an initial value μ_0 for the interior path parameter which satisfies $\delta(\mathbf{x}_0, \mu_0) \leq \frac{1}{2}$. It is not immediately clear how $\mathbf{x}_0$ and μ_0 should be chosen. In this section we will present an auxiliary model, based on the LO-model (P) that needs to be solved. For this auxiliary model, an initial solution is readily available. Moreover, it has the property that, from any optimal solution of the auxiliary model, an optimal solution of (P) can be derived.

The idea behind the initialization procedure is similar to the big-M procedure (see Section 3.6.1) that may be used to initialize the simplex algorithm (see Chapter 3). Similar to the big-M procedure, the LO-model at hand is augmented with an artificial variable with a large objective coefficient. The value of the objective coefficient is taken to be large enough to guarantee that, at any optimal solution, the value of the artificial variable will be zero. A complication compared to the big-M procedure, however, is the fact that such an artificial variable needs to be introduced in both the primal and the dual models. So, an artificial variable with large objective coefficient M_P is introduced in the primal model and an artificial variable with large objective coefficient M_D is introduced in the dual model. Since the objective coefficients of the primal model appear in the right hand side values of the dual model, this means that M_D will appear in the right hand side values of the primal model.

It is not immediately clear how to construct suitable primal and dual models together with constants M_P and M_D. We will show in the remainder of this section that the following auxiliary primal and dual models may be used. Let $\alpha > 0$. Define $M_P = \alpha^2$ and $M_D = \alpha^2(n+1) - \alpha\mathbf{c}^\mathsf{T}\mathbf{1}$. Consider the following pair of primal and dual LO-models:

$$
\begin{aligned}
\min \ & \mathbf{c}^\mathsf{T}\mathbf{x} + M_P x_{n+1} \\
\text{s.t.} \ & \mathbf{A}\mathbf{x} + (\mathbf{b} - \alpha\mathbf{A}\mathbf{1})x_{n+1} = \mathbf{b} \\
& (\alpha\mathbf{1} - \mathbf{c})^\mathsf{T}\mathbf{x} + \alpha x_{n+2} = M_D \\
& \mathbf{x} \geq \mathbf{0}, x_{n+1} \geq 0, x_{n+2} \geq 0,
\end{aligned}
\tag{P$'$}
$$

and

$$
\begin{aligned}
\max \ & \mathbf{b}^\mathsf{T}\mathbf{y} + M_D y_{m+1} \\
\text{s.t.} \ & \mathbf{A}^\mathsf{T}\mathbf{y} + (\alpha\mathbf{1} - \mathbf{c})y_{m+1} + \mathbf{w} = \mathbf{c} \\
& (\mathbf{b} - \alpha\mathbf{A}\mathbf{1})^\mathsf{T}\mathbf{y} + w_{n+1} = M_P \\
& \alpha y_{m+1} + w_{n+2} = 0 \\
& \mathbf{y}, y_{m+1} \text{ free}, \\
& \mathbf{w} \geq \mathbf{0}, w_{n+1} \geq 0, w_{n+2} \geq 0.
\end{aligned}
\tag{D$'$}
$$

We first show that an initial primal solution $\mathbf{x}_0 \ (\in \mathbb{R}^{n+2})$ of (P$'$) that lies on the central path is readily available. The following theorem explicitly gives such an initial primal solution along with an initial dual solution.

Theorem 6.4.2.
Let $\alpha > 0$. Define $\mathbf{x}_0 \in \mathbb{R}^{n+2}$, $\mathbf{y}_0 \in \mathbb{R}^{m+1}$, and $\mathbf{w}_0 \in \mathbb{R}^{n+2}$ by:

$$
\mathbf{x}_0 = \begin{bmatrix} \hat{\mathbf{x}} \\ \hat{x}_{n+1} \\ \hat{x}_{n+2} \end{bmatrix} = \begin{bmatrix} \alpha\mathbf{1} \\ 1 \\ \alpha \end{bmatrix},
$$

$$
\mathbf{y}_0 = \begin{bmatrix} \hat{\mathbf{y}} \\ \hat{y}_{m+1} \end{bmatrix} = \begin{bmatrix} \mathbf{0} \\ -1 \end{bmatrix}, \text{ and } \mathbf{w}_0 = \begin{bmatrix} \hat{\mathbf{w}} \\ \hat{w}_{n+1} \\ \hat{w}_{n+2} \end{bmatrix} = \begin{bmatrix} \alpha\mathbf{1} \\ \alpha^2 \\ \alpha \end{bmatrix}.
$$

Let $\mu_0 = \alpha^2$. Then, $\mathbf{x}_0$ is a feasible solution of (P$'$), $\begin{bmatrix} \mathbf{y}_0 \\ \mathbf{w}_0 \end{bmatrix}$ is a feasible solution of (D$'$), and $\delta(\mathbf{x}_0; \mu_0) = 0$.

Proof. The point $\mathbf{x}_0$ is a feasible solution of (P$'$) because:

$$
\mathbf{A}\hat{\mathbf{x}} + (\mathbf{b} - \alpha\mathbf{A}\mathbf{1})\hat{x}_{n+1} = \alpha\mathbf{A}\mathbf{1} + (\mathbf{b} - \alpha\mathbf{A}\mathbf{1}) = \mathbf{b}, \text{ and}
$$
$$
(\alpha\mathbf{1} - \mathbf{c})^\mathsf{T}\hat{\mathbf{x}} + \alpha\hat{x}_{n+2} = \alpha^2\mathbf{1}^\mathsf{T}\mathbf{1} - \alpha\mathbf{c}^\mathsf{T}\mathbf{1} + \alpha^2 = \alpha^2(n+1) - \alpha\mathbf{c}^\mathsf{T}\mathbf{1} = M_D,
$$

where we have used the fact that $\mathbf{1}^\mathsf{T}\mathbf{1} = n$. The point $\begin{bmatrix} \mathbf{y}_0 \\ \mathbf{w}_0 \end{bmatrix}$ is a feasible solution of (D$'$) because:

$$
\mathbf{A}^\mathsf{T}\hat{\mathbf{y}} + (\alpha\mathbf{1} - \mathbf{c})\hat{y}_{m+1} + \hat{\mathbf{w}} = -\alpha\mathbf{1} + \mathbf{c} + \alpha\mathbf{1} = \mathbf{c},
$$
$$
(\mathbf{b} - \alpha\mathbf{A}\mathbf{1})^\mathsf{T}\hat{\mathbf{y}} + \hat{w}_{n+1} = \alpha^2 = M_P, \text{ and}
$$

$$\alpha \hat{y}_{m+1} + \hat{w}_{n+2} = -\alpha + \alpha = 0.$$

Finally, consider the complementary slackness relations:

$$\hat{x}_i \hat{w}_i = \alpha \times \alpha = \mu_0, \qquad \text{for } i = 1, \ldots, n,$$
$$\hat{x}_{n+1} \hat{w}_{n+1} = \alpha \times \alpha = \mu_0, \text{ and} \qquad (6.4)$$
$$\hat{x}_{n+2} \hat{w}_{n+2} = 1 \times \alpha^2 = \mu_0.$$

Hence, $\mathbf{X}_0 \mathbf{w}_0 = \mu_0 \mathbf{1}$. It follows from Theorem 6.3.1 that:

$$\delta(\mathbf{x}_0, \mu_0) = \min_{\mathbf{y}, \mathbf{w}} \left\{ \left\| \frac{\mathbf{X}_0 \mathbf{w}}{\mu_0} - \mathbf{1} \right\| \mid \mathbf{A}^\mathsf{T} \mathbf{y} + \mathbf{w} = \mathbf{c} \right\} \leq \left\| \frac{\mathbf{X}_0 \mathbf{w}_0}{\mu_0} - \mathbf{1} \right\| = 0.$$

Since $\delta(\mathbf{x}_0, \mu_0) \geq 0$, we have that $\delta(\mathbf{x}_0, \mu_0) = 0$, as required. $\qquad \square$

So, we have constructed auxiliary LO-models for which it is straightforward to find an initial pair of primal and dual solutions that lie on the central path. What remains to show is that, once an optimal solution of (P′) has been found, and the artificial primal variable x_{n+1} and dual variable y_{m+1} have optimal value zero, then this optimal solution can be turned into an optimal solution of the original LO-model.

Theorem 6.4.3.
Consider the primal model (P): $\min\{\mathbf{c}^\mathsf{T} \mathbf{x} \mid \mathbf{A}\mathbf{x} = \mathbf{b}, \mathbf{x} \geq \mathbf{0}\}$, and the corresponding dual model (D): $\max\{\mathbf{b}^\mathsf{T} \mathbf{y} \mid \mathbf{A}^\mathsf{T} \mathbf{y} + \mathbf{w} = \mathbf{c}, \mathbf{w} \geq \mathbf{0}\}$. Let

$$\mathbf{x}_0^* = \begin{bmatrix} \mathbf{x}^* \\ x_{n+1}^* \\ x_{n+2}^* \end{bmatrix}, \text{ and } \mathbf{y}_0^* = \begin{bmatrix} \mathbf{y}^* \\ y_{m+1}^* \end{bmatrix}, \mathbf{w}_0^* = \begin{bmatrix} \mathbf{w}^* \\ w_{n+1}^* \\ w_{n+2}^* \end{bmatrix}$$

be optimal solutions of (P′) and (D′), respectively. If $x_{n+1}^* = 0$ and $y_{m+1}^* = 0$, then $\mathbf{x}^*$ and $\mathbf{y}^*$ are optimal solutions of (P) and (D), respectively.

Proof. Because $x_{n+1}^* = 0$, we have that $\mathbf{A}\mathbf{x}^* = \mathbf{b}, \mathbf{x}^* \geq \mathbf{0}$, and hence $\mathbf{x}^*$ is a feasible solution of the original primal model (P). Similarly, because $y_{m+1}^* = 0$, we have that $\mathbf{A}^\mathsf{T} \mathbf{y}^* + \mathbf{w}^* = \mathbf{c}$ and $\mathbf{w}^* \geq \mathbf{0}$, and hence $\begin{bmatrix} \mathbf{y}^* \\ \mathbf{w}^* \end{bmatrix}$ is a feasible solution of the original dual model (D). Because $\mathbf{x}_0^*, \mathbf{y}_0^*$, and $\mathbf{w}_0^*$ are optimal for (P′) and (D′), respectively, it follows from the complementary slackness relations (6.4) that $x_i^* w_i^* = 0$ for $i = 1, \ldots, n$. Hence, it follows from the complementary slackness relations for (P) and (D) that $\mathbf{x}^*$ and $\mathbf{y}^*$ are optimal solutions of (P) and (D), respectively. $\qquad \square$

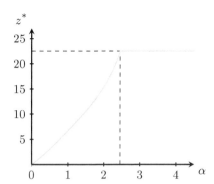

Figure 6.9: The (negative of the) optimal objective value of the auxiliary LO-model for Model Dovetail.

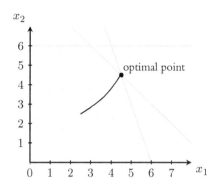

Figure 6.10: Path taken by the interior path algorithm for the auxiliary LO-model for Model Dovetail with $\alpha = 2.5$.

Example 6.4.1. *The auxiliary model for Model Dovetail is:*

$$
\begin{array}{llllllll}
\min & -3x_1 - & 2x_2 & & & + & M_P x_7 & \\
\text{s.t.} & x_1 + & x_2 + & x_3 & & + & (9{-}3\alpha)x_7 & = & 9 \\
& 3x_1 + & x_2 & + x_4 & & + & (18{-}5\alpha)x_7 & = & 18 \\
& x_1 & & + x_5 & & + & (7{-}2\alpha)x_7 & = & 7 \\
& & x_2 & & + x_6 + & (6{-}2\alpha)x_7 & = & 6 \\
& (\alpha{+}3)x_1 + (\alpha{+}2)x_2 + \alpha x_3 + \alpha x_4 + \alpha x_5 + \alpha x_6 & & & & + \alpha x_8 = M_D \\
& x_1, x_2, x_3, x_4, x_5, x_6, x_7, x_8 \geq 0.
\end{array}
$$

Here, $M_P = \alpha^2$ and $M_D = 3\alpha^2 + 5\alpha$. As an initial interior point, we may use $x_1 = x_2 = \ldots = x_6 = x_8 = \alpha$ and $x_7 = 1$. Choosing, e.g., $\alpha = 3$ yields the optimal primal solution:

$$
x_1^* = 4\tfrac{1}{2}, \ x_2^* = 4\tfrac{1}{2}, \ x_3^* = x_4^* = 0, \ x_5^* = 2\tfrac{1}{2}, \ x_6^* = 1\tfrac{1}{2}, \ x_7^* = 0, \ x_8^* = 3\tfrac{1}{2},
$$

and the optimal dual solution:

$$
y_1^* = -1\tfrac{1}{2}, \ y_2^* = -\tfrac{1}{2}, \ y_3^* = y_4^* = y_5^* = 0.
$$

Since $x_3^* = 0$ and $y_5^* = 0$, it follows from Theorem 6.4.3 that these solutions correspond to optimal solutions of Model Dovetail and its dual model. Figure 6.9 illustrates how the optimal objective value of the auxiliary model depends on α. It can be checked that, for $\alpha \geq 2.45$, the optimal solution of the auxiliary model coincides with the optimal solution of Model Dovetail. Figure 6.9 shows the path taken by the interior path algorithm when applied to the auxiliary problem with $\alpha = 2.5$.

The following theorem shows that it is possible to take α sufficiently large so as to guarantee that any pair of optimal solutions of the auxiliary LO-models satisfies $x_{n+1}^* = 0$ and $y_{m+1}^* = 0$.

Theorem 6.4.4.

Consider the primal model (P): $\min\{\mathbf{c}^{\mathsf{T}}\mathbf{x} \mid \mathbf{A}\mathbf{x} = \mathbf{b}, \mathbf{x} \geq \mathbf{0}\}$ and its dual model (D): $\max\{\mathbf{b}^{\mathsf{T}}\mathbf{y} \mid \mathbf{A}^{\mathsf{T}}\mathbf{y} + \mathbf{w} = \mathbf{c}, \mathbf{w} \geq \mathbf{0}\}$. Assume that both models have optimal vertices, and let z^* be their optimal objective value. For large enough values of $\alpha > 0$,

(i)   the models (P$'$) and (D$'$) have optimal objective value $z^* < \alpha$, and

(ii)   any pair of optimal primal and dual solutions

$$
\begin{bmatrix} \mathbf{x}^* \\ x^*_{n+1} \\ x^*_{n+2} \end{bmatrix}, \text{ and } \begin{bmatrix} \mathbf{y}^* \\ y^*_{m+1} \end{bmatrix}, \begin{bmatrix} \mathbf{w}^* \\ w^*_{n+1} \\ w^*_{n+2} \end{bmatrix}
$$

of (P$'$) and (D$'$), respectively, satisfies $x^*_{n+1} = 0$ and $y^*_{m+1} = 0$.

Proof. (i) Let α be such that:

$$
\alpha > 1 + \max\left\{ |\mathbf{c}^{\mathsf{T}}\mathbf{1}|, \ \max_{\hat{\mathbf{x}}} |\mathbf{1}^{\mathsf{T}}\hat{\mathbf{x}}|, \ \max_{\hat{\mathbf{x}}} |\mathbf{c}^{\mathsf{T}}\hat{\mathbf{x}}|, \ \max_{\hat{\mathbf{y}}} |\mathbf{b}^{\mathsf{T}}\hat{\mathbf{y}}|, \ \max_{\hat{\mathbf{y}}} |(\mathbf{A}\mathbf{1})^{\mathsf{T}}\hat{\mathbf{y}}| \right\}, \tag{6.5}
$$

where the maxima are taken over the vertices $\hat{\mathbf{x}}$ ($\in \mathbb{R}^n$) of the feasible region of (P), and over the vertices $\hat{\mathbf{y}}$ ($\in \mathbb{R}^m$) of the feasible region of (D).

We will first show that, for every vertex $\hat{\mathbf{x}}$ ($\in \mathbb{R}^n$) of the feasible region of (P) and every vertex $\hat{\mathbf{y}}$ ($\in \mathbb{R}^m$) of the feasible region of (D), it holds that:

$$
M_P > (\alpha\mathbf{1} - \mathbf{c})^{\mathsf{T}}\hat{\mathbf{x}}, \text{ and } M_D > (\mathbf{b} - \alpha\mathbf{A}\mathbf{1})^{\mathsf{T}}\hat{\mathbf{y}}. \tag{6.6}
$$

To show this, note that, by the definition of α, we have that:

$$
(\alpha\mathbf{1} - \mathbf{c})^{\mathsf{T}}\hat{\mathbf{x}} \leq \alpha|\mathbf{1}^{\mathsf{T}}\hat{\mathbf{x}}| + |\mathbf{c}^{\mathsf{T}}\hat{\mathbf{x}}| < \alpha(\alpha - 1) + \alpha - 1 = \alpha^2 - 1 < \alpha^2 = M_P.
$$

Similarly, we have that:

$$
(\mathbf{b} - \alpha\mathbf{A}\mathbf{1})^{\mathsf{T}}\hat{\mathbf{y}} \leq |\mathbf{b}^{\mathsf{T}}\hat{\mathbf{y}}| + \alpha|(\mathbf{A}\mathbf{1})^{\mathsf{T}}\hat{\mathbf{y}}| < \alpha + \alpha^2 < \alpha^2 + \alpha^2(n - 1)
$$
$$
= \alpha^2(n + 1) - \alpha^2 < \alpha^2(n + 1) - \alpha\mathbf{c}^{\mathsf{T}}\mathbf{1} = M_D.
$$

This proves (6.6).

Now, let $\mathbf{x}^*$ be an optimal vertex of (P), and let $\begin{bmatrix} \mathbf{y}^* \\ \mathbf{w}^* \end{bmatrix}$ be an optimal vertex of (D). Define:

$$
\hat{\mathbf{x}}^* = \begin{bmatrix} \mathbf{x}^* \\ M_P - (\alpha\mathbf{1} - \mathbf{c})^{\mathsf{T}}\mathbf{x}^* \\ 0 \end{bmatrix}, \text{ and } \hat{\mathbf{y}}^* = \begin{bmatrix} \mathbf{y}^* \\ 0 \end{bmatrix}, \hat{\mathbf{w}}^* = \begin{bmatrix} \mathbf{w}^* \\ 0 \\ M_D - (\mathbf{b} - \alpha\mathbf{A}\mathbf{1})^{\mathsf{T}}\mathbf{y}^* \end{bmatrix}.
$$

Note that using (6.6), we have that $M_P - (\alpha\mathbf{1} - \mathbf{c})^{\mathsf{T}}\mathbf{x}^* > 0$ and $M_D - (\mathbf{b} - \alpha\mathbf{A}\mathbf{1})^{\mathsf{T}}\mathbf{y}^* > 0$. It is easy to check feasibility, and optimality follows from the complementary slackness relations. Hence, $\hat{\mathbf{x}}^*$ and $\begin{bmatrix} \hat{\mathbf{y}}^* \\ \hat{\mathbf{w}}^* \end{bmatrix}$ are optimal solutions of (P$'$) and (D$'$), respectively. Moreover, it can be seen that $z^* < \alpha$.

(ii) It suffices to show that $w_{n+1}^* > 0$ and $x_{n+2}^* > 0$. To see that this is sufficient, suppose that $w_{n+1}^* > 0$ and $x_{n+2}^* > 0$. The complementary slackness relation $x_{n+1}^* w_{n+1}^* = 0$ immediately implies that $x_{n+1}^* = 0$. Similarly, the complementary slackness relation $x_{n+2}^* w_{n+2}^* = 0$ implies that $w_{n+2}^* = 0$ and, consequently, the dual constraint $\alpha y_{m+1} + w_{n+2} = 0$ implies that $y_{m+1}^* = 0$.

We first prove that $w_{n+1}^* > 0$. Suppose for a contradiction that $w_{n+1}^* = 0$. In that case, the dual constraint $(\mathbf{b} - \alpha\mathbf{A1})^\mathsf{T}\mathbf{y} + w_{n+1} = M_P$ implies that $(\mathbf{b} - \alpha\mathbf{A1})^\mathsf{T}\mathbf{y}^* = M_P$, and therefore:

$$\mathbf{b}^\mathsf{T}\mathbf{y}^* = M_P + \alpha\mathbf{A1}^\mathsf{T}\mathbf{y}^* = \alpha^2 + \alpha\mathbf{A1}^\mathsf{T}\mathbf{y}^* \geq \alpha^2 - \alpha|\mathbf{A1}^\mathsf{T}\mathbf{y}^*| \geq \alpha^2 - \alpha(\alpha - 1) = \alpha.$$

Hence, the objective value of the dual auxiliary model satisfies $z^* = \mathbf{b}^\mathsf{T}\mathbf{y}^* + M_D y_{m+1}^* \geq \alpha$, a contradiction. So, we have shown that $w_{n+1}^* = 0$.

Next, we prove that $x_{n+2}^* > 0$. Suppose for a contradiction that $x_{n+2}^* = 0$. In that case, the primal constraint $(\alpha\mathbf{1} - \mathbf{c})^\mathsf{T}\mathbf{x} + \alpha x_{n+2} = M_D$ implies that $(\alpha\mathbf{1} - \mathbf{c})^\mathsf{T}\mathbf{x}^* = M_D$, and therefore:

$$\mathbf{c}^\mathsf{T}\mathbf{x}^* = M_D - \alpha\mathbf{1}^\mathsf{T}\mathbf{x}^* = \alpha^2(n+1) - \alpha\mathbf{c}^\mathsf{T}\mathbf{1} - \alpha\mathbf{1}^\mathsf{T}\mathbf{x}^*$$
$$\geq \alpha^2(n+1) - \alpha|\mathbf{c}^\mathsf{T}\mathbf{1}| - \alpha|\mathbf{1}^\mathsf{T}\mathbf{x}^*| \geq \alpha^2(n+1) - \alpha^2 - \alpha^2$$
$$\geq \alpha^2(n-1) \geq \alpha^2.$$

Hence, the objective value of the dual auxiliary model satisfies $z^* = \mathbf{c}^\mathsf{T}\mathbf{x}^* + M_P x_{n+1}^* \geq \alpha^2$, a contradiction. So, we have shown that $x_{n+2}^* = 0$. This proves the theorem. $\square$

Note that we have shown that the initialization procedure is theoretically guaranteed to work for any value of α that is larger than the right hand side of (6.5). In practice, however, much lower values of α suffice.

6.5 Exercises

Exercise 6.5.1. There are several ways to formulate an LO-model in either the form (P) or in the form (D); see Section 6.1.1. Consider for instance the LO-model:

$$\begin{aligned}
\min \quad & x_1 - 2x_2 \\
\text{s.t.} \quad & x_1 \leq 1 \\
& -x_1 + x_2 \leq 1 \\
& x_1, x_2 \geq 0.
\end{aligned}$$

We give two possible formulations. The first possibility is:

(D)-formulation:

$$-\max \quad -x_1 + 2x_2$$
$$\text{s.t.} \quad x_1 \qquad\quad \le 1$$
$$-x_1 \qquad\quad \le 0$$
$$-x_1 + \quad x_2 \le 1$$
$$-x_2 \le 0$$
$$x_1, x_2 \text{ free,}$$

(P)-formulation:

$$-\min \quad y_1 \qquad\quad + y_3$$
$$\text{s.t.} \quad y_1 - y_2 - y_3 \qquad = -1$$
$$y_3 - y_4 = \quad 2$$
$$y_1, y_2, y_3, y_4 \ge 0.$$

The second possibility is:

(P)-formulation:

$$\min \quad x_1 - 2x_2$$
$$\text{s.t.} \quad x_1 \qquad + x_3 \qquad\quad = 1$$
$$-x_1 + \quad x_2 \qquad + x_4 = 1$$
$$x_1, x_2, x_3, x_4 \ge 0,$$

(D)-formulation:

$$\max \quad y_1 + y_2$$
$$\text{s.t.} \quad y_1 - y_2 \le \quad 1$$
$$y_2 \le -2$$
$$y_1 \qquad\quad \le \quad 0$$
$$y_2 \le \quad 0$$
$$y_1, y_2 \text{ free.}$$

The number of variables of the above (P), (D) combinations can be determined as follows.

For the first (P), (D) combination: the (D)-formulation has two decision variables and four slack variables; the (P)-formulation has four decision variables. Hence, the number of variables is $(4 + 6 =)$ 10.

For the second (P), (D) combination: the (P)-formulation has four decision variables; the (D)-formulation has two decision plus four slack variables. So, there are $(4 + 6 =)$ 10 variables.

Determine (P)- and (D)-formulations with the smallest total number of variables (decision variables plus slack variables) for the following LO-models:

(a) $\max \quad x_1 + x_2$
$$\text{s.t.} \quad x_2 = 0$$
$$0 \le x_1 \le 1.$$

(c) $\max \quad 3x_1 + 2x_2$
$$\text{s.t.} \quad x_1 + \quad x_2 \le \quad 9$$
$$3x_1 + \quad x_2 \le 18$$
$$x_1 \qquad\quad \le \quad 7$$
$$x_2 \le \quad 6$$
$$x_1, x_2 \ge 0.$$

(b) $\min \quad x_3$
$$\text{s.t.} \quad x_1 - x_2 \qquad = 0$$
$$x_1 \qquad - x_3 = 0$$
$$x_1 \ge 1, \ x_2 \ge 0, \ x_3 \ge 0.$$

(d) $\min \quad x_1 + x_2$
$$\text{s.t.} \quad x_1 \qquad\quad \ge 1$$
$$x_2 \ge 1$$
$$x_1, x_2 \text{ free.}$$

(e) min x_1
s.t. $x_1 - 2x_2 \geq 0$
$x_2 \geq 1$
x_1, x_2 free.

(g) max x_1
s.t. $x_1 \leq 1$
$x_2 \leq 1$
$x_1, x_2 \geq 0.$

(f) max $x_1 + x_2$
s.t. $x_1 \geq -1$
$x_1 \leq 1$
$x_2 \leq 1$
x_1, x_2 free.

Exercise 6.5.2. Determine the following entities for both a (P)- and a (D)-formulation with the smallest number of variables of the models formulated in Exercise 6.5.1.

(a) The sets of interior points F_P^+ and F_D^+ (as defined in Section 6.1); if the sets are nonempty, determine an interior point.

(b) The Karush-Kuhn-Tucker conditions.

(c) The logarithmic barrier functions.

Exercise 6.5.3. Determine the interior paths for a (P)- and a (D)-formulation with the smallest number of variables of the models of Exercise 6.5.1(b),(d)–(g); draw the interior paths for the cases (e), (f), and (g).

Exercise 6.5.4. Calculate, for each of the models of Exercise 6.5.1(b),(d)–(g), the duality gap as a function of μ (see Theorem 6.1.3), and draw the functions $\mathbf{c}^T\mathbf{x}(\mu)$ and $\mathbf{b}^T\mathbf{y}(\mu)$ in one figure. Use the interior paths to obtain optimal solutions.

Exercise 6.5.5. Let $\mathbf{x}(\mu)$ denote the interior path of the primal model (P), and suppose that the primal feasible region F_P be bounded. Then $\mathbf{x}_P = \lim_{\mu \to \infty} \mathbf{x}(\mu)$ is called the *analytic center* of F_P. The analytic center of the dual feasible region is defined similarly.

(a) Show that $\mathbf{x}_P \in F_P$.

(b) Show that $\mathbf{x}_P = \min\left\{ -\sum_{i=1}^n \ln(x_i) \,\middle|\, \mathbf{x} = [x_1 \ \ldots \ x_n]^T \in F_P \right\}$.

(c) Calculate, if they exist, the analytic centers of the feasible regions of the models of Exercise 6.5.1(b)–(g).

Exercise 6.5.6. Draw the graph of $B_P(x_1, x_2, \mu) = \frac{1}{\mu}(x_1 + 2x_2) - \ln(x_1 x_2)$ for several values of the parameter $\mu > 0$; compare Figure 6.2.

Exercise 6.5.7.

(a) Calculate $\mathbf{P}_A$ (see Section 6.2.2), where $\mathbf{A}$ is the matrix corresponding to a (P)-formulation with the smallest number of variables of the models of Exercise 6.5.1.

(b) Draw the sets $\mathcal{N}(\mathbf{A})$ and $\mathcal{R}(\mathbf{A})$ for the case $\mathbf{A} = \begin{bmatrix} 1 & -1 \end{bmatrix}$. Calculate, and draw in one figure, the points $\begin{bmatrix} 0 & 1 \end{bmatrix}^\mathsf{T}$ and $\mathbf{P}_A(\begin{bmatrix} 0 & 1 \end{bmatrix}^\mathsf{T})$.

Exercise 6.5.8.   Apply Dikin's affine scaling procedure with respect to an interior point $\hat{\mathbf{x}}$ to the following LO-models; show that the unit ball (circle in $\mathbb{R}^2$, line segment in $\mathbb{R}$) with center $\mathbf{1}$ and radius 1 is completely contained in the transformed feasible region (compare the example in Section 6.2.3).

(a)   max $x_1 + 2x_2$
      s.t. $2x_1 + 2x_2 + x_3 = 2$
             $x_1, x_2, x_3 \geq 0$.
      Take $\hat{\mathbf{x}} = \begin{bmatrix} 0.1 & 0.1 & 1.6 \end{bmatrix}^\mathsf{T}$.

(b)   min $3x_1 - 2x_2$
      s.t. $-x_1 + 2x_2 = 4$
             $x_1, x_2 \geq 0$.
      Take $\hat{\mathbf{x}} = \begin{bmatrix} 0.02 & 2.01 \end{bmatrix}^\mathsf{T}$.

Exercise 6.5.9.   Consider Model Dovetail as formulated in Exercise 6.5.1(c). Choose a feasible point $\hat{\mathbf{x}}$ close to the boundary of its feasible region. Apply Dikin's affine scaling procedure with respect to $\hat{\mathbf{x}}$ on a (P)-formulation of Model Dovetail. Draw the transformed feasible region.

Exercise 6.5.10.   Consider the function $f(x_1, x_2) = (x_1 - 1)^2 + 3x_2^2 + 1$.

(a) Show that f is strictly convex, and determine its minimum.

(b) Draw in one figure the level curves $(x_1 - 1)^2 + 3x_2^2 + 1 = \alpha$ for $\alpha = 1, 4, 8, 12$, and 21.

(c) Calculate the gradient $\nabla f(-3, -\frac{2}{3}\sqrt{3})$, and draw this vector in the figure made under (b).

(d) Calculate the angle between $-\nabla f(-3, -\frac{2}{3}\sqrt{3})$ and $\begin{bmatrix} 1 & 0 \end{bmatrix}^\mathsf{T} - \begin{bmatrix} -3 & -\frac{2}{3}\sqrt{3} \end{bmatrix}^\mathsf{T}$.

Exercise 6.5.11.   Consider model (A) in Section 6.1.2; see also Figure 6.1. Calculate the tangent in the point $\begin{bmatrix} 0.4 & 0.8 \end{bmatrix}^\mathsf{T}$ of the level curve of $B_P(\begin{bmatrix} x_1 & x_2 & 1-x_1 & 1-x_2 \end{bmatrix}^\mathsf{T}, 2)$ through the point $\begin{bmatrix} 0.4 & 0.8 \end{bmatrix}^\mathsf{T}$. Draw this tangent in Figure 6.1. Explain why 'starting at $\begin{bmatrix} 0.4 & 0.8 \end{bmatrix}^\mathsf{T}$ and moving in the direction of the vector $\begin{bmatrix} 0.15 & -0.10 \end{bmatrix}^\mathsf{T}$' is moving in the direction of the interior path. Answer the same questions for the points $\begin{bmatrix} 0.1 & 0.2 \end{bmatrix}^\mathsf{T}$ and $\begin{bmatrix} 0.5 & 0.4 \end{bmatrix}^\mathsf{T}$.

Exercise 6.5.12. Consider model (A) in Section 6.1.2; see also Figure 6.1. Determine the search direction when $\mathbf{x}_0 = \begin{bmatrix} 0.55 & 0.70 \end{bmatrix}^\mathsf{T}$ is taken as initial interior point and $\mu_0 = 1$. Draw the new point $\mathbf{x}_1$ in the same figure.

Exercise 6.5.13. Consider model (A) in Section 6.1.2.

(a) Use $\delta(\mathbf{x}, \mu)$ to show that the point $\mathbf{x}_1$ found in Exercise 6.5.12 is again an interior point.

(b) If $\mathbf{x}_1(\mu)$ is not interior, find a value μ' of μ for which $\mathbf{x}_1(\mu')$ is an interior point.

Exercise 6.5.14. Consider Model Dovetail; see Exercise 6.5.1(c).

(a) Extend this model by adding the redundant constraint $3x_1 + x_2 \leq 18$. Draw the interior path of the new model and compare it with the interior path of the original model (see Figure 6.8).

(b) Extend Model Dovetail by adding the constraints $x_1 + x_2 \leq 9$ and $3x_1 + x_2 \leq 18$. Again, draw the interior path and compare it with the interior paths drawn under (a).

Exercise 6.5.15. Consider Model Dovetail extended with $x_1 + x_2 \leq 9$ and $3x_1 + x_2 \leq 18$.

(a) Calculate the interior paths of the dual of Model Dovetail and the dual of the extended model.

(b) Calculate the duality gap $\mathbf{c}^\mathsf{T}\mathbf{x}(\mu) - \mathbf{b}^\mathsf{T}\mathbf{y}(\mu)$ for several values of μ for both Model Dovetail and the extension formulated under (a).

Exercise 6.5.16. Draw the interior paths in case the right hand side values and objective coefficients in Model Dovetail are perturbed. Give special attention to the cases where the objective function is parallel to a constraint.

CHAPTER 7
Integer linear optimization

Overview

An *integer linear optimization model* (notation: *ILO-model*) is a linear optimization model in which the values of the decision variables are restricted to be integers. The reason for investigating integer optimization models is that many practical problems require integer-valued solutions. Solving practical integer optimization models is usually very complicated, and solution techniques usually require excessive amounts of (computer) calculations. We give several examples of ILO-models, and we introduce *binary variables*, which can be used to model certain nonlinear constraints and objective functions, and logical if-then constraints.

The most widely used solution technique for solving ILO-models is the so-called *branch-and-bound algorithm*: a solution procedure that creates a 'tree' of (noninteger) LO-models whose solutions 'grow' to an optimal integer solution of the initial model. The branch-and-bound algorithm is applied to a knapsack problem, a machine scheduling problem (traveling salesman problem), and a decentralization problem. We also discuss the *cutting plane algorithm* discovered by RALPH E. GOMORY (born 1929) for solving *mixed integer linear optimization models* (MILO-models), in which some of the decision variables are constrained to be integer-valued and the others are allowed to take arbitrary (nonnegative real) values.

7.1 Introduction

In this section we introduce a prototype example which will be used to explain and illustrate the new concepts and techniques. We then introduce the standard forms of ILO- and MILO-models, and the corresponding terminology. A round-off procedure is introduced for complicated situations in which a quick solution is demanded (see also Chapter 17). We will see that round-off procedures may lead to suboptimal or even infeasible outputs, and so more sophisticated techniques are needed.

7.1.1 The company Dairy Corp

The dairy factory Dairy Corp produces mainly milk and cheese. Part of the transportation of the milk and cheese to grocery stores is done by the company itself, and the remaining part is contracted out. The present vehicle fleet of Dairy Corp is obsolete and will be replaced. Dairy Corp can choose from two types of new vehicles. These vehicles transport cheese and milk in large crates: the cheese is transported in crates that contain 100 kgs of cheese, and the milk is transported in crates that contain 200 liters of milk. A type A vehicle can transport cheese up to 100 crates, but it does not have the appropriate refrigeration facilities to transport milk. A type B vehicle can transport both cheese and milk. It can simultaneously transport up to 50 crates of cheese and up to 20 crates of milk.

Purchasing a type A vehicle yields a cost reduction of $\$1,000$ per month compared to contracting out the distribution; for a type B vehicle, the cost reduction is $\$700$ per month. Clearly, Dairy Corp wants to maximize the total cost reduction. If x_1 is the number of vehicles of type A and x_2 of type B, then the objective becomes:

$$\max 1000x_1 + 700x_2.$$

Of course, the demand for cheese and milk fluctuates over time. The management of Dairy Corp has decided that the total vehicle capacity to be purchased should not exceed the minimum daily demand, which is $2,425$ crates of cheese and 510 crates of milk. Therefore the constraints are:

$$\begin{aligned} 100x_1 + 50x_2 &\leq 2425 \\ 20x_2 &\leq\ \ 510. \end{aligned}$$

Since it does not make sense to buy parts of vehicles, the variables x_1 and x_2 have to be integer-valued. So the model becomes:

Model Dairy Corp.

$$\begin{aligned} \max\ &1000x_1 + 700x_2 \\ \text{s.t.}\quad &100x_1 +\ \ 50x_2 \leq 2425 \\ &\qquad\qquad 20x_2 \leq\ \ 510 \\ &x_1, x_2 \geq 0,\ \text{and integer}. \end{aligned}$$

The variables of Model Dairy Corp are restricted to be integer-valued. For this reason, the model is an example of an *ILO-model*. In Figure 7.1 the feasible region of Model Dairy Corp is depicted. Note that the feasible region consists of points with integer coordinate values, called *integer points*. The problem is to find a feasible integer point that maximizes the objective function $1000x_1 + 700x_2$; compare Section 1.2.3. This model can be solved by means of the branch-and-bound algorithm, to be formulated and discussed in Section 7.2.

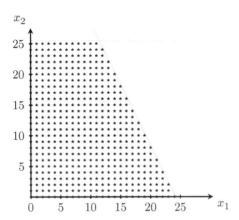

Figure 7.1: Feasible integer points of Model Dairy Corp.

7.1.2 ILO-models

In general, the standard form of an *ILO-model* is:

$$\max\{\mathbf{c}^\mathsf{T}\mathbf{x} \mid \mathbf{A}\mathbf{x} \leq \mathbf{b}, \mathbf{x} \geq \mathbf{0}, \mathbf{x} \text{ integer}\},$$

where $\mathbf{x}, \mathbf{c} \in \mathbb{R}^n$, $\mathbf{b} \in \mathbb{R}^m$, and $\mathbf{A}$ is an (m, n)-matrix. The expression '$\mathbf{x}$ integer' means that the entries of the vector $\mathbf{x}$ are required to be integers; these constraints are also called the *integrality constraints*. An *integer point* is a point whose corresponding vector has integer-valued entries.

The LO-model that arises from an ILO-model by removing the integrality constraints is called the *LO-relaxation* of that ILO-model. In general, a *relaxation* of a mathematical optimization model with constraints is a model that arises from the original model by deleting one or more constraints of that model. Since the LO-relaxation (usually) has a larger feasible region than the corresponding ILO-model, we have that:

$$\max\{\mathbf{c}^\mathsf{T}\mathbf{x} \mid \mathbf{A}\mathbf{x} \leq \mathbf{b}, \mathbf{x} \geq \mathbf{0}\} \geq \max\{\mathbf{c}^\mathsf{T}\mathbf{x} \mid \mathbf{A}\mathbf{x} \leq \mathbf{b}, \mathbf{x} \geq \mathbf{0}, \mathbf{x} \text{ integer}\}.$$

7.1.3 MILO-models

The general form of a maximizing MILO-model can be written as follows:

$$\begin{aligned} \max \ & \mathbf{c}_1^\mathsf{T}\mathbf{x}_1 + \mathbf{c}_2^\mathsf{T}\mathbf{x}_2 \\ \text{s.t.} \ \ & \mathbf{A}_1\mathbf{x}_1 + \mathbf{A}_2\mathbf{x}_2 \leq \mathbf{b} \\ & \mathbf{x}_1 \geq \mathbf{0} \\ & \mathbf{x}_2 \geq \mathbf{0}, \text{ and integer,} \end{aligned}$$

with $\mathbf{A}_1$ an (m, n_1)-matrix and $\mathbf{A}_2$ an (m, n_2)-matrix. The concept of the LO-relaxation of an ILO-model also applies to MILO-models.

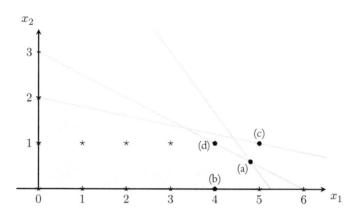

Figure 7.2: Different solutions found by rounding the optimal LO-relaxation solution. Point (a) is the optimal LO-relaxation solution, point (b) is the rounded-down solution, point (c) is the rounded-up solution, and point (d) is the optimal ILO-solution.

7.1.4 A round-off procedure

In order to solve a given ILO-model, a natural first approach is to ignore the integrality constraints and use the simplex algorithm developed in Chapter 3 (or the interior point algorithm from Chapter 6) to solve the LO-relaxation. Recall, however, that the vertices of the feasible region of an LO-model are in general not integer points. Hence, applying the simplex algorithm to the LO-relaxation does usually not generate an optimal (integer) solution of the original ILO-model. Sometimes, rounding off this solution to a feasible integer point may lead to a satisfactory solution of the ILO-model (see also Chapter 17). Rounding an optimal solution of the LO-relaxation corresponding to an ILO-model may lead to a solution that is 'far away' from an optimal solution of the original ILO-model. The following example illustrates this phenomenon.

Example 7.1.1.   *Consider the following ILO-model:*

$$\max 2x_1 + 3x_2$$
$$\text{s.t.} \quad x_1 + 5x_2 \leq 10$$
$$2x_1 + 4x_2 \leq 12$$
$$4x_1 + 3x_2 \leq 21$$
$$x_1, x_2 \geq 0, \text{ and integer.}$$

The feasible region of this model and its LO-relaxation are shown in Figure 7.2. The shaded area is the feasible region of the LO-relaxation and the stars represent the feasible points of the ILO-model.

(a) The optimal solution of the LO-relaxation is: $x_1 = 4\frac{4}{5}$, $x_2 = \frac{3}{5}$ with optimal objective value $z = 11\frac{2}{5}$.

(b) Rounding down $\begin{bmatrix} 4\frac{4}{5} & \frac{3}{5} \end{bmatrix}^\mathsf{T}$ to the closest feasible integer solution leads to $x_1 = 4$, $x_2 = 0$ with objective value $z = 8$.

(c) Rounding up $\begin{bmatrix} 4\frac{4}{5} & \frac{3}{5} \end{bmatrix}^\mathsf{T}$ to the closest integer solution leads to $x_1 = 5$, $x_2 = 1$, which is infeasible because $2 \times 5 + 4 \times 1 = 14 \not\leq 12$. Rounding to the closest integer solution leads to the same infeasible point.

(d) The optimal integer solution is: $x_1 = 4$, $x_2 = 1$ with optimal objective value $z = 11$.

Hence, rounding may lead to a large deviation of the optimal solution. If, for instance, in the above example the production is in batches of $1,000$ units, then the difference between $8,000$ and $11,000$ may be far too large.

Notice that, in the example, the optimal solution $\begin{bmatrix} 4 & 1 \end{bmatrix}^\mathsf{T}$ can be obtained from the optimal solution of the LO-relaxation by rounding down the value found for x_1 and rounding up the value found for x_2. This suggests that an optimal solution of an ILO-model can be found by solving the corresponding LO-relaxation and appropriately choosing for each variable whether to round its value up or down. Unfortunately this, in general, does not work either. The reader is presented with an example in Exercise 7.5.2. The conclusion is that more specific techniques are needed for solving ILO-models.

7.2 The branch-and-bound algorithm

The *branch-and-bound algorithm* is in practice the most widely used method for solving ILO-models and MILO-models. Basically, the algorithm solves a model by breaking up its feasible region into successively smaller regions (*branching*), calculating bounds on the objective value over each corresponding submodel (*bounding*), and using them to discard some of the submodels from further consideration. The bounds are obtained by replacing the current submodel by an easier model (*relaxation*), such that the solution of the latter yields a bound for the former. The procedure ends when each submodel has either produced an infeasible solution, or has been shown to contain no better solution than the one already at hand. The best solution found during the procedure is an optimal solution.

7.2.1 The branch-and-bound tree

We start by explaining the *branch-and-bound algorithm* for maximizing ILO-models. In the first step of the branch-and-bound algorithm, the LO-relaxation (see Section 7.1.2) is solved. The optimal objective value of the LO-relaxation is an upper bound for the optimal objective value of the original ILO-model. If the solution of the LO-relaxation is integer-valued, then we are done, because the solution of the LO-relaxation is an optimal solution of the original ILO-model as well. If not, then the feasible region of the LO-relaxation is partitioned into two subregions giving rise to two new LO-models. This procedure is called *branching*. We illustrate it by means of Model Dairy Corp.

The LO-relaxation of Model Dairy Corp, which we will call model M1, has the unique optimal solution $x_1 = 11\frac{1}{2}$ and $x_2 = 25\frac{1}{2}$, with corresponding optimal objective value $z = 29,350$. Note that neither x_1 nor x_2 has an integer value. In the first iteration step of the branch-and-bound algorithm, one of the non-integer-valued decision variables, say

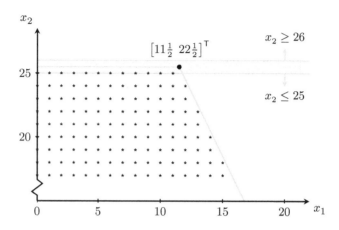

Figure 7.3: Region of Figure 7.1 that is excluded by requiring that $x_2 \leq 25$ or $x_2 \geq 26$.

x_2, is chosen to branch on. This means the following. Since there are no integers between 25 and 26, any optimal integer solution must satisfy either $x_2 \leq 25$, or $x_2 \geq 26$. The feasible region is accordingly split into two parts by adding (a) $x_2 \leq 25$, and (b) $x_2 \geq 26$, respectively, to the current model, resulting in submodels M2 and M3, respectively. So the region determined by $25 < x_2 < 26$, and hence also the solution $x_1 = 11\frac{1}{2}$, $x_2 = 25\frac{1}{2}$, is excluded from the original feasible region. This excluded (shaded) region is shown in Figure 7.3.

In the next step of the branch-and-bound algorithm, the following two LO-models are solved:

$$
\begin{array}{ll}
\text{(M2)} \quad \max & 1000x_1 + 700x_2 \\
\text{s.t.} & 100x_1 + 50x_2 \leq 2425 \\
& 20x_2 \leq 510 \\
& x_2 \leq 25 \\
& x_1, x_2 \geq 0.
\end{array}
\qquad
\begin{array}{ll}
\text{(M3)} \quad \max & 1000x_1 + 700x_2 \\
\text{s.t.} & 100x_1 + 50x_2 \leq 2425 \\
& 20x_2 \leq 510 \\
& x_2 \geq 26 \\
& x_1, x_2 \geq 0.
\end{array}
$$

The optimal solution of LO-model M2 is: $x_1 = 11\frac{3}{4}$, $x_2 = 25$, $z = 29{,}250$. Model M3 is infeasible, and so $z = -\infty$. (recall that, by definition, $\max(\emptyset) = -\infty$). The procedure is now repeated for model M2, resulting in models M4 and M5. In Figure 7.4, the process is schematically represented as a tree. Note that, as we go deeper into the tree, the feasible regions become smaller and smaller, and hence the corresponding optimal objective values z decrease (or, to be precise, do not increase). When solving model M4, we arrive at an integer solution, namely $x_1 = 11$, $x_2 = 25$ with $z = 28{,}500$. Since this solution is a feasible solution to the original ILO-model, we call this a *candidate solution*. Notice that the objective value corresponding to this integer solution is a lower bound for the optimal objective value of the original ILO-model, i.e., $z^* \geq 28{,}500$, where z^* is the optimal objective value of the original ILO-model. Since branching on this model will only lead to solutions with optimal objective values of at most 28,500, there is no reason to continue

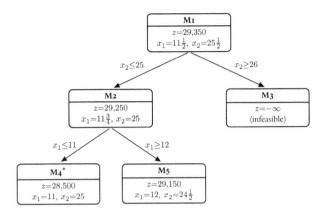

Figure 7.4: The first iterations of the branch-and-bound algorithm for Model Dairy Corp. The models marked with a star have integer-optimal solutions.

branching on M4; therefore, the process is not continued here. However, model M5 (with $x_1 = 12$, $x_2 = 24\frac{1}{2}$) may give rise – by further branching – to a value of z that is larger than the current largest one (namely, $z = 28,500$). Therefore, we continue by branching on model M5.

In Figure 7.5, the full *branch-and-bound tree* of Model Dairy Corp is shown. In model M8, we find a new candidate integer solution, namely $x_1 = 12$ and $x_2 = 24$, with $z = 28,800$, which is better than the integer solution found in model M4. This solution gives the lower bound 28,800, i.e., $z^* \geq 28,800$. Notice that the optimal solution found for model M9 is not integer yet. So we might consider branching on it. However, branching on model M9 will only lead to models that are either infeasible, or have optimal objective values at most 28,750. So, it does not make sense to continue branching on model M9. Since there are no models left to branch on, the branch-and-bound algorithm stops here. The optimal solution of Model Dairy Corp is therefore the one found in model M8, namely $z^* = 28,800$, $x_1^* = 12$, and $x_2^* = 24$.

The branch-and-bound algorithm for ILO-models can easily be modified for MILO-models. Namely, branching should only be carried out on variables that are required to be integer-valued. Also for a solution of a submodel to be a candidate solution, it need only assign integer values to those variables that are required to be integer-valued.

Example 7.2.1. *Consider the following MILO-model:*

$$\begin{aligned} \max \quad & x_1 + 2x_2 \\ \text{s.t.} \quad & x_1 + x_2 \leq 3 \\ & 2x_1 + 5x_2 \leq 8 \\ & x_1, x_2 \geq 0, x_2 \text{ integer.} \end{aligned}$$

We ask for an optimal solution for which x_2 is the only integer-valued variable. First, the LO-relaxation of this model is solved: $x_1 = 2\frac{1}{3}$, $x_2 = \frac{2}{3}$, $z = 3\frac{2}{3}$. Since x_2 is the only variable that

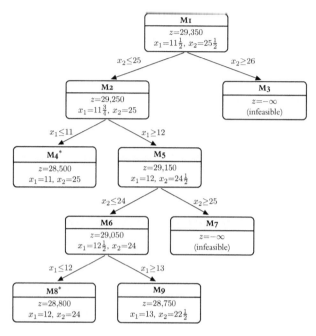

Figure 7.5: Branch-and-bound tree for Model Dairy Corp. The models marked with a star have integer optimal solutions.

has to be an integer, we are forced to branch on x_2. This yields a submodel with $x_2 \leq 0$ (that is, $x_2 = 0$), and $x_2 \geq 1$. Next, we choose to solve the model on the '$x_2 \leq 0$'-branch; the optimal solution is: $x_1 = 3$, $x_2 = 0$, $z = 3$. The LO-relaxation of the submodel on the '$x_2 \geq 1$'-branch has the optimal solution: $x_1 = 1\frac{1}{2}$, $x_2 = 1$, $z = 3\frac{1}{2}$. The submodel with solution $z = 3$ can be excluded from consideration, because the other submodel has already a larger z-value, namely $z = 3\frac{1}{2}$. Therefore, the optimal solution is $x_1^* = 1\frac{1}{2}$, $x_2^* = 1$, $z^* = 3\frac{1}{2}$. It is left to the reader to draw the corresponding branch-and-bound tree.

7.2.2 The general form of the branch-and-bound algorithm

In this section, we formulate the branch-and-bound algorithm for a maximizing optimization model. The algorithm can actually be stated in a more general form than the algorithm we used to solve Model Dairy Corp. To do so, we distinguish for each iteration of the branch-and-bound algorithm a number of distinct phases which we describe below. We take the solution procedure of Model Dairy Corp as reference; see Section 7.1.1.

The algorithm starts by constructing a relaxation of the ILO-model or MILO-model to be solved. This relaxed model serves as the first node of the branch-and-bound tree. As the algorithm progresses, new nodes are added to this tree. These new nodes are always added 'under' existing nodes, i.e., whenever a new node Mj is added, an arc from an existing node Mk to Mj is added as well (so, we have that $j > k$). The new node Mj is then called a *subnode* of Mk. The algorithm also keeps track of upper and lower bounds on the optimal objective value z^* of the original (M)ILO–model.

In the case of Model Dairy Corp, the relaxation concerned removing the restriction that the variables should be integer-valued. In general, the relaxation should be chosen such that the new model is easier to solve than the original problem.

We distinguish the following phases of a branch-and-bound iteration.

Node selection phase. Select a node Mk ($k \geq 1$) of the branch-and-bound tree that has not been selected so far. In the case of Model Dairy Corp, we chose to always select the most recently created unselected node. In general there are a number of different *node selection rules*. We mention three possibilities; see also Exercise 7.5.5:

Backtracking. This is a procedure where branching is performed on the most recently created unselected node of the branch-and-bound tree. Backtracking is a so-called *LIFO search procedure* (last-in-first-out search procedure). Note that the above solution procedure for Model Dairy Corp uses backtracking.

A *FIFO search procedure* (first-in-first-out search procedure) proceeds the branching process on the unselected node that was created earliest.

Jumptracking. Here, branching is performed on the branch with the current best (finite) objective value. Jumptracking is a form of a so-called *best-first search procedure*.

Bounding phase. The optimal objective value z_j of the submodel corresponding to subnode Mj of Mk is at most the optimal objective z_k of the submodel corresponding to node Mk. This means that, if we select node Mj in the node selection phase, and the current lower bound is greater than z_k, then the submodel corresponding to node Mj can be excluded from further consideration, because we have already found a solution that is better than the best that we can obtain from this submodel. In that case, we return to the node selection phase, and select a different node (if possible). In the case of Model Dairy Corp, we used LO-relaxations to determine these upper bounds.

Solution phase. The submodel corresponding to node Mk (which was selected in the node selection phase) is solved. In Model Dairy Corp, this entails solving the LO-model using a general purpose linear optimization algorithm like the simplex algorithm or an interior point algorithm. Sometimes, however, there is a simpler algorithm to solve the submodel (see, for instance, Section 7.2.3). If the solution satisfies the constraints of the original model (e.g., the integrality constraints in the case of Model Dairy Corp), then we have found a candidate solution for the original model, along with a lower bound for the optimal value z^* of the original (M)ILO-model. In that case, we return to the node selection phase. Otherwise, we continue with the branching phase.

Branching phase. Based on the solution found in the solution phase, the feasible region of the submodel corresponding to node Mk is partitioned into (usually) mutually exclusive subsets, giving rise to two or more new submodels. Each of these submodels is represented by a new node that is added to the branch-and-bound tree and that is connected to node Mk by an arc that points in the direction of the new node. In Model Dairy

Corp, we branched by choosing a variable (that is required to be integer-valued) whose optimal value is noninteger, and adding constraints that excluded this noninteger value. If there is more than one choice for this variable, then it is not clear beforehand which variable should be chosen to branch on. Some computer codes choose to branch on the noninteger variable with the lowest index. Often a better strategy is to choose the variable with the greatest 'economic importance', e.g., by taking into account the objective coefficient. Sometimes, there are other ways of branching; see for instance Section 7.2.4. Note also that there is no reason to create exactly two new branches in the branching phase. Section 7.2.4 gives an example where at some point three new branches are created.

Determining the branch-and-bound tree in the case of (M)ILO-models uses these phases. We may therefore formulate the branch-and-bound algorithm for any maximizing optimization model to which the various phases can be applied. The following assumptions are made. First, it is assumed that an algorithm is available for calculating optimal solutions of the submodels. In the case of (M)ILO-models this can be the simplex algorithm, applied to an LO-model of which the feasible region is a subset of the feasible region of the LO-relaxation of the original model. Moreover, it is assumed that a branching rule is available. The process is repeated until a feasible solution is determined whose objective value is no smaller than the optimal objective value of any submodel.

The general form of the branch-and-bound algorithm for maximizing models can now be formulated as follows. We use the set NS to hold the nodes that have not been selected in the node selection phase; the variable z_L denotes the current lower bound on the optimal solution of the original model. The function $f \colon F \to \mathbb{R}$, where F is the feasible region, is the objective function. The feasible region of the LO-relaxation of the original model is denoted by F^R. Optimal solutions of submodels are denoted by $(z, \mathbf{x})$, where $\mathbf{x}$ is the optimal solution of the submodel, and z is the corresponding objective value.

Algorithm 7.2.1. (*Branch-and-bound algorithm for ILO-models*)

Input: Values for the parameters of the model $\max\{f(\mathbf{x}) \mid \mathbf{x} \in F\}$, where F is the feasible region.

Output: Either

      (a) the message: the model has no optimal solution; or

      (b) an optimal solution of the model.

Step 0: *Initialization.* Define $F_0 = F^R$, $NS = \{0\}$, and $z_L = -\infty$.

Step 1: *Optimality test and stopping rule.* If $NS \neq \emptyset$, then go to Step 2. Stop the procedure when $NS = \emptyset$; the current best solution is optimal. If there is no current best solution, i.e., $z_L = -\infty$, then the original model has no optimal solution.

Step 2: *Node selection phase.* Select a node $k \in NS$, and set $NS := NS \setminus \{k\}$.

Step 3: *Bounding phase.* Let z_U be the optimal objective value of the node above node k (or $+\infty$ if $k = 0$).

   (a) If $z_U \leq z_L$, then the model corresponding to node k cannot lead to a solution that is better than the current best solution; go to Step 1.

   (b) If $z_U > z_L$, then let $\max\{f(\mathbf{x}) \mid \mathbf{x} \in F_k\}$, with $F_k \subset F^R$, be the submodel associated with node k; call this submodel S_k, and go to Step 4.

Step 4: *Solution phase.* Determine, if it exists, an optimal solution $(z_k, \mathbf{x}_k)$ of S_k (note that z_k is an upper bound for $\max\{f(\mathbf{x}) \mid \mathbf{x} \in F_k \cap F\}$). There are several possible outcomes:

   (a) If S_k has no optimal solution, then go to Step 5.

   (b) $z_k \leq z_L$ (i.e., z_k is not better than the current best solution). Submodel S_k can be excluded from further consideration, because branching may only lead to nonoptimal solutions or alternative optimal solutions. Go to Step 5.

   (c) $z_k > z_L$ and $\mathbf{x}_k \in F$ (i.e., z_k is better than the current best solution and the solution that was found is a solution of the original model). Set $z_L := z_k$, and go to Step 5.

   (d) $z_k > z_L$ and $\mathbf{x}_k \in F$ (i.e., z_k is better than the current best solution, but the solution that was found is not a solution of the original model). This node needs further branching; go to Step 5.

Step 5: *Branching phase.* Partition F_k into two or more new subsets, say $F_{k_1}, \ldots, F_{k_s}$. Set $NS := NS \cup \{k_1, \ldots, k_s\}$. Go to Step 1.

In the case of a minimizing model, the above described branch–and–bound algorithm can easily be adapted: e.g., z_L has to be replaced by the 'current upper bound' z_U, $-\infty$ by ∞, and the inequality signs reversed.

Regardless of how branch–and–bound algorithms are implemented, their drawback is that often they require excessive computing time and storage. In the next section we will apply the branch–and–bound algorithm to a problem with a specific structure through which the algorithm works very effectively, whereas in Chapter 9 we give an example for which the algorithm needs an enormous amount of computing time.

7.2.3 The knapsack problem

The name *knapsack problem* derives from the following problem setting. We are given a number of objects that have to be packed into a knapsack. Each object has a given value and a given size. The knapsack also has a given size, so in general we cannot pack all objects into it. We want to pack the objects in such a way that we carry with us the most valuable combination of objects, subject to the constraint that the total size of the objects does not

Object	Value (×$1,000)	Volume (liters)
1	5	5
2	3	4
3	6	7
4	6	6
5	2	2

Table 7.1: Knapsack problem inputs.

exceed the size of the knapsack. So we need to decide which objects are to be packed in the knapsack and which ones are not. Knapsack problems can usually effectively be solved by means of the branch-and-bound algorithm.

Example 7.2.2.   *Consider a knapsack of fifteen liters and suppose that we are given five objects, whose values and volumes are listed in Table 7.1. One can easily check that the best selection of objects is an optimal solution to the ILO-model:*

$$\max 5x_1 + 3x_2 + 6x_3 + 6x_4 + 2x_5$$
$$\text{s.t.} \quad 5x_1 + 4x_2 + 7x_3 + 6x_4 + 2x_5 \le 15$$
$$x_1, x_2, x_3, x_4, x_5 \in \{0, 1\},$$

where object i ($i = 1, \ldots, 5$) is selected if and only if $x_i^* = 1$ in the optimal solution $\mathbf{x}^* = \begin{bmatrix} x_1^* & \ldots & x_5^* \end{bmatrix}^\mathsf{T}$.

The general form of a knapsack problem can be written as follows.

Model 7.2.1. *(ILO-model Knapsack problem)*
$$\max c_1 x_1 + \ldots + c_n x_n$$
$$\text{s.t.} \quad a_1 x_1 + \ldots + a_n x_n \le b$$
$$x_1, \ldots, x_n \in \{0, 1\},$$

with n the number of objects, c_i (≥ 0) the value of choosing object i, b (≥ 0) the amount of an available resource (e.g., the volume of the knapsack), and a_i (≥ 0) the amount of the available resource used by object i (e.g., the volume of object i); $i = 1, \ldots, n$.

For knapsack problems in which the decision variables are not required to be in $\{0, 1\}$, see for instance Exercise 7.5.9 and Section 16.2.

This type of knapsack problems can be solved by the branch-and-bound algorithm. In fact, because of the special structure of the model, the branching phase and the solution phase in the branch-and-bound algorithm are considerably simplified. The branching phase is simplified because each variable must be equal to either 0 or 1, which means that branching on the value of x_i ($i = 1, \ldots, n$) yields exactly two branches, namely an '$x_i = 0$'-branch and an '$x_i = 1$'-branch. More importantly, the solution phase can be simplified because the

LO-relaxation of a knapsack problem can be solved by inspection. To see this, observe that c_i/a_i may be interpreted as the value of object i per size unit. Thus, the most promising objects have the largest values of c_i/a_i. In order to solve the LO-relaxation of a knapsack problem, we compute the ratios c_i/a_i and order the objects in nonincreasing order of c_i/a_i; the largest has the best ranking and the smallest the worst. The algorithm continues by first packing the best ranked object in the knapsack, then the second best, and so on, until the best remaining object will overfill the knapsack. The knapsack is finally filled with as much as possible of this last object. In order to make this more precise we show the following theorem.

Theorem 7.2.1.

The LO-relaxation of Model 7.2.1 has the optimal solution:

$$x_1^* = \ldots = x_r^* = 1, \ x_{r+1}^* = \frac{1}{a_{r+1}}(b - a_1 - \ldots - a_r), \ x_{r+2}^* = \ldots = x_n^* = 0,$$

with r such that $a_1 + \ldots + a_r \le b$ and $a_1 + \ldots + a_{r+1} > b$, under the assumption that $c_1/a_1 \ge \ldots \ge c_n/a_n$.

(Informally, the interpretation of r is that the first r objects may leave some space in the knapsack but the $(r+1)$'th object overfills it.)

Proof. Clearly, $0 \le \frac{1}{a_{r+1}}(b - a_1 - \ldots - a_r) < 1$. One can easily check that $\mathbf{x}^*$, as defined in the statement of the theorem, is feasible. We will use Theorem 4.2.3 to prove Theorem 7.2.1. In order to use this theorem, we need a dual feasible solution, say $\begin{bmatrix} y_1^* & \ldots & y_{n+1}^* \end{bmatrix}^{\mathsf{T}}$, for which $c_1 x_1^* + \ldots + c_n x_n^* = b y_1^* + y_2^* + \ldots + y_{n+1}^*$. The dual model reads:

$$\begin{aligned}
\min \quad & b y_1 + y_2 + \ldots + y_{n+1} \\
\text{s.t.} \quad & a_1 y_1 + y_2 \qquad\qquad\qquad \ge c_1 \\
& \qquad \vdots \qquad\quad \ddots \\
& a_n y_1 \qquad\qquad + y_{n+1} \ge c_n \\
& y_1, \ldots, y_{n+1} \ge 0.
\end{aligned}$$

It is left to the reader to check that the choice

$$y_1^* = \frac{c_{r+1}}{a_{r+1}}, \quad y_k^* = c_{k-1} - a_{k-1}\left(\frac{c_{r+1}}{a_{r+1}}\right) \qquad\qquad \text{for } k = 2, \ldots, r+1, \text{ and}$$

$$y_k^* = 0 \qquad\qquad\qquad\qquad\qquad\qquad\qquad\qquad \text{for } k = r+2, \ldots, n+1.$$

is dual feasible, and that the objective values for the points $\mathbf{x}^* = \begin{bmatrix} x_1^* & \ldots & x_n^* \end{bmatrix}^{\mathsf{T}}$ and $\mathbf{y}^* = \begin{bmatrix} y_1^* & \ldots & y_{n+1}^* \end{bmatrix}^{\mathsf{T}}$ are the same. Hence, $\mathbf{x}^*$ is an optimal solution of the LO-relaxation of the knapsack problem. $\qquad\square$

Example 7.2.3. *We will solve the LO-relaxation of the knapsack problem from Example 7.2.2. We start by computing the ratios c_i/a_i and assigning rankings from best to worst (see Table 7.2). The LO-relaxation is now solved as follows. There are three objects with the highest ranking 1; choose object*

Object	Relative value c_i/a_i	Ranking (1=best, 5=worst)
1	1	1
2	$3/4 = 0.75$	5
3	$6/7 \approx 0.86$	4
4	1	1
5	1	1

Table 7.2: Ratio ranking.

1, and so $x_1 = 1$. Then $(b - a_1x_1 = 15 - 5 =)$ 10 liters remain. Next we include the 'second best' object (either object 4, or object 5) in the knapsack; choose $x_4 = 1$. Now $(b - a_1x_1 - a_4x_4 =)$ 4 liters remain. Then we choose object 5 $(x_5 = 1)$, so 2 liters remain. The best remaining object is object 3. We fill the knapsack with as much as possible of object 3. Since only 2 liters of the knapsack remain, we take $x_3 = \frac{2}{7}$, which means that we put only a fraction $\frac{2}{7}$ of object 3 in the knapsack. Now the knapsack is full. Thus a feasible solution of the LO-relaxation is:

$$x_1 = 1, \; x_2 = 0, \; x_3 = \tfrac{2}{7}, \; x_4 = 1, \; x_5 = 1, \; and \; z = 14\tfrac{5}{7}.$$

To show that this solution is actually optimal, we have to calculate y as defined in the proof of Theorem 7.2.1. Clearly, $y_1 = c_3/a_3 = \frac{6}{7}$, $y_2 = c_1 - a_1(c_3/a_3) = \frac{5}{7}$, $y_5 = c_4 - a_4(c_3/a_3) = \frac{6}{7}$, $y_6 = c_5 - a_5(c_3/a_3) = \frac{2}{7}$ (because $c_1/a_1 = c_4/a_4 = c_5/a_5 \geq c_3/a_3 \geq c_2/a_2$), and $y_3 = y_4 = 0$. Note that $z = by_1 + y_2 + y_3 + y_4 + y_5 + y_6 = 14\frac{5}{7}$, and so the feasible solution above is in fact an optimal solution of the LO-relaxation.

We have shown how to solve the LO-relaxation of a knapsack problem. What remains is to deal with the variables that have been branched on. When solving a submodel in an '$x_i = 0$'-branch, we simply ignore object i. On the other hand, when solving a submodel in an '$x_i = 1$'-branch, object i is forced to be in the knapsack; we solve the LO-relaxation of a knapsack problem for the remaining capacity $b - a_i$.

Example 7.2.4. *Consider again the knapsack problem of Example 7.2.2. We illustrate the above by solving a submodel in which we have that $x_1 = x_3 = 1$ and $x_4 = 0$. Since $x_4 = 0$, object 4 is ignored in the analysis. Because $x_1 = x_3 = 1$, objects 1 and 3 are already packed in the knapsack. What remains to be solved is the LO-relaxation of a knapsack problem with knapsack capacity $b - a_1 - a_3 = 15 - 5 - 7 = 3$, where objects 2 and 5 can still be packed into the knapsack. We have that $c_2/a_2 = \frac{3}{4}$ and $c_5/a_5 = 1$, and therefore object 5 has ranking 1 and object 2 has ranking 2. Now it follows that an optimal solution is $x_5 = 1$ and $x_2 = \frac{1}{4}$, along with $x_1 = 1$, $x_3 = 1$, and $x_4 = 0$.*

The process is repeated until an optimal solution is reached in the branch-and-bound tree. A possible(!) tree is presented in Figure 7.6 (the variables with a zero value are not mentioned in the blocks of the tree). Note that if we use the backtracking node selection rule, then we would successively solve submodels M1, M2, M4, M5, and M3, and we could already stop after solving submodel M3. The

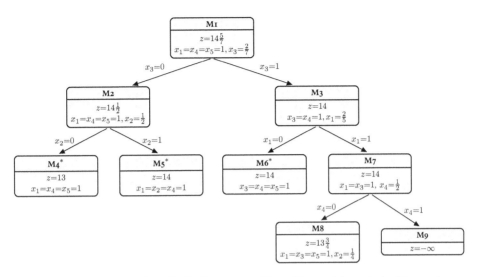

Figure 7.6: Branch-and-bound tree for the knapsack problem. The models marked with a star have integer optimal solutions.

branch-and-bound algorithm has found two optimal solutions:

$$x_1^* = 1, \ x_2^* = 1, \ x_3^* = 0, \ x_4^* = 1, \ x_5^* = 0, \ \text{with } z^* = 14,$$
$$\text{and } x_1^* = 0, \ x_2^* = 0, \ x_3^* = 1, \ x_4^* = 1, \ x_5^* = 1, \ \text{with } z^* = 14.$$

Note that no unit of the knapsack remains unused in these optimal solutions. The reader may check that these are the only optimal solutions.

7.2.4 A machine scheduling problem

This section deals with the following problem. One machine has to process a certain number of jobs, say n (≥ 2). The jobs are labeled $1, \ldots, n$. The processing time of job i is p_i (minutes). After each job the machine has to be set up in order to process the next job. The set up time for each job depends both on the job itself and the job that is processed before it. Let the setup time from job i to job j be c_{ij} (minutes). After processing all jobs, the machine has to be reset to its initial state, which we define to be job 0. The setup time from the initial state to job i is c_{0i} (minutes), and from job j to the initial state c_{j0} (minutes). The problem is: in which order should the jobs be scheduled so as to minimize the total processing time of all jobs? Since all jobs have to be processed on one machine, the total processing time equals the sum of the processing times of the jobs plus the sum of the set up times. Since the sum of the processing times of the jobs is independent of the order in which they are processed, the total processing time only depends on the used setup times. So instead of minimizing the total processing time, it suffices to minimize the sum of the used setup times. The machine scheduling problem in this section uses a branch-and-bound algorithm, where submodels are partitioned into two or more new submodels.

From job i	To job j						
	0	1	2	3	4	5	6
0	–	1	1	5	4	3	2
1	1	–	2	5	4	3	2
2	1	5	–	4	2	5	4
3	5	4	6	–	6	2	5
4	5	2	6	3	–	5	4
5	5	3	5	1	5	–	3
6	6	5	4	6	6	5	–

Table 7.3: Setup and reset times.

Example 7.2.5.   *Suppose that we have the five jobs labeled* 1, 2, 3, 4, *and* 5, *each requiring a processing time of two minutes. Hence,* $n = 5$ *and* $p_i = 2$ *for* $i = 1, \ldots, 5$. *The setup times are given in Table 7.3. One possible schedule is to process the jobs in the order* $3 \to 5 \to 2 \to 4 \to 1$. *The schedule is presented in Figure 7.7. In this figure, time is drawn from left to right. The left edge corresponds to the beginning of the schedule, and the right edge to the end of the schedule. The numbers are minute numbers; for example, the number* 1 *corresponds to the first minute of the schedule. In the first five minutes, the machine is being set up to start processing job* 3. *The next two minutes are spent on processing job* 3. *The two minutes following this are spent on setting up the machine to start processing job* 5, *and so on. It should be clear that the corresponding total processing time is* 10 *and the total setup time is* 17. *This type of chart is called a* Gantt chart; *see also Section 8.2.6.*

We introduce the following variables. For each $i, j = 0, \ldots, n$, define the decision variable δ_{ij}, with the following interpretation:

$$\delta_{ij} = \begin{cases} 1 & \text{if job } j \text{ is processed immediately after job } i \\ 0 & \text{otherwise.} \end{cases}$$

Since δ_{ij} can only take the values 0 and 1, δ_{ij} is called a *binary variable*, or a $\{0, 1\}$-*variable*. The objective of the problem can now be formulated as:

$$\min \sum_{i=0}^{n} \sum_{j=0}^{n} \delta_{ij} c_{ij}.$$

The constraints can also be expressed in terms of the δ_{ij}'s. Since each job has to be processed precisely once, we have that:

$$\sum_{i=0}^{n} \delta_{ij} = 1 \qquad \text{for } j = 0, \ldots, n, \tag{7.1}$$

and because after each job exactly one other job has to be processed, we also have that:

$$\sum_{j=0}^{n} \delta_{ij} = 1 \qquad \text{for } i = 0, \ldots, n. \tag{7.2}$$

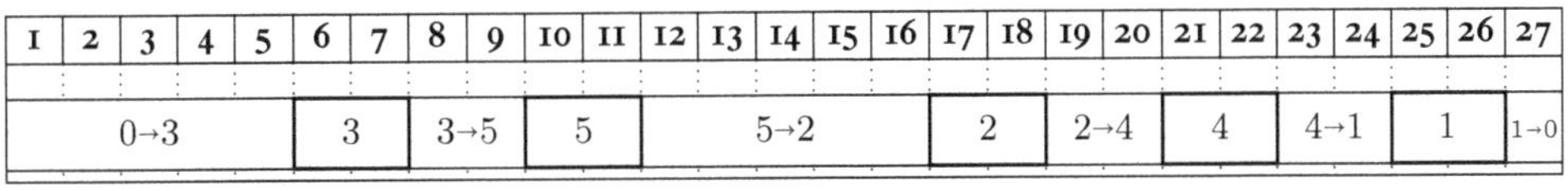

I	2	3	4	5	6	7	8	9	10	II	12	13	14	15	16	17	18	19	20	21	22	23	24	25	26	27
		0→3				3		3→5		5			5→2				2		2→4		4		4→1		1	1→0

Figure 7.7: Gantt chart for Example 7.2.5. The horizontal axis denotes time; the blocks denoted by '$i{\to}j$' are the setup times; the blocks denoted by 'i' are the processing times ($i, j = 0, \ldots, 5$).

Example 7.2.6. *Consider again the machine scheduling problem of Example 7.2.5. The solution depicted in Figure 7.7 can be represented by taking $\delta_{03} = \delta_{35} = \delta_{52} = \delta_{24} = \delta_{41} = \delta_{10} = 1$ and all other δ_{ij}'s equal to zero. It is left to the reader to check that this solution satisfies constraints (7.1) and (7.2). This solution can also be represented as a directed cycle in a graph; see Figure 7.8(a). The nodes of this graph are the jobs $0, \ldots, 5$, and an arc from job i to job j ($i \neq j$) means that job j is processed directly after job i.*

The question is whether the model

$$\min \sum_{i=0}^{n} \sum_{j=0}^{n} \delta_{ij} c_{ij}$$

$$\text{s.t.} \quad \sum_{j=0}^{n} \delta_{ij} = 1 \qquad \text{for } i = 0, \ldots, n$$

$$\sum_{i=0}^{n} \delta_{ij} = 1 \qquad \text{for } j = 0, \ldots, n$$

$$\delta_{ij} \in \{0, 1\} \qquad \text{for } i, j = 0, \ldots, n,$$

gives a complete description of the problem of determining meaningful optimal schedules. The following example illustrates that this is not the case.

Example 7.2.7. *Consider once more the machine scheduling problem of Example 7.2.5. Take $n = 5$. Then an optimal solution to the above model could be $\delta_{03} = \delta_{31} = \delta_{10} = 1$, $\delta_{52} = \delta_{24} = \delta_{45} = 1$, and all other δ_{ij}'s equal to zero. As one can easily check, this solution corresponds to two subtours (see Figure 7.8(b)), and so this solution does not correspond to a feasible schedule.*

A feasible schedule should consist of only one cycle. Therefore, the model needs constraints that exclude subtours from optimal solutions. Such a set of constraints is:

$$\sum_{i,j \in S} \delta_{ij} \leq |S| - 1 \qquad \text{for } S \subset \{0, \ldots, n\}, S \neq \emptyset;$$

$|S|$ denotes the number of elements of the set S, and '$S \subset \{0, \ldots, n\}$' means that S is a subset of $\{0, \ldots, n\}$ with $S \neq \{0, \ldots, n\}$. These constraints are called *subtour-elimination constraints*. In Exercise 7.5.19 the reader is asked to show that the above subtour-elimination constraints really exclude subtours. One may now ask the question whether including the subtour-elimination constraints is sufficient, or whether even more constraints need to be added. Exercise 7.5.19 asks to show that the subtour-elimination constraints are in fact sufficient, i.e., once the subtour-elimination constraints have been included, every feasible

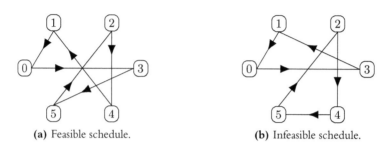

(a) Feasible schedule.   (b) Infeasible schedule.

Figure 7.8: Solutions to the machine scheduling problem. The arrows represent the order in which the jobs are processed on the machine.

solution of the ILO-model corresponds to a feasible schedule, and vice versa. So, the ILO-model for the machine-scheduling problem can now be formulated as follows.

Model 7.2.2. (*ILO-model Machine scheduling problem*)

$$\min \sum_{i=0}^{n} \sum_{j=0}^{n} \delta_{ij} c_{ij}$$

$$\text{s.t.} \quad \sum_{j=0}^{n} \delta_{ij} = 1 \qquad \text{for } i = 0, \ldots, n$$

$$\sum_{i=0}^{n} \delta_{ij} = 1 \qquad \text{for } j = 0, \ldots, n$$

$$\sum_{i,j \in S} \delta_{ij} \leq |S| - 1 \qquad \text{for } S \subset \{0, \ldots, n\}, S \neq \emptyset$$

$$\delta_{ij} \in \{0, 1\} \qquad \text{for } i, j = 0, \ldots, n.$$

(Note that '$\delta \in \{0, 1\}$' is equivalent to '$0 \leq \delta \leq 1$, δ integer', so this model is in fact an ILO-model.)

We will solve this problem by means of a branch-and-bound algorithm. In contrast to Model Dairy Corp, in which the submodels represented by the nodes of the branch-and-bound tree are LO-models, the submodels are ILO-models themselves. However, they are 'easier' than the original machine scheduling problem because we delete the subtour-elimination constraints.

The general idea of the branch-and-bound algorithm as applied to the machine scheduling problem is as follows. First, solve the relaxation of the model in which the subtour-elimination constraints are removed. If the optimal solution to this relaxation is a feasible schedule, then stop: it must be an optimal schedule. If the solution to the relaxation is not a feasible schedule, then we have obtained a lower bound solution. This means that no feasible schedule, including the optimal ones, can have a value smaller than this lower bound. Moreover, at least one of the arcs, which cause the infeasibility, cannot be included in optimal schedules. When solving a model in which one of these arcs is forced out of the schedule

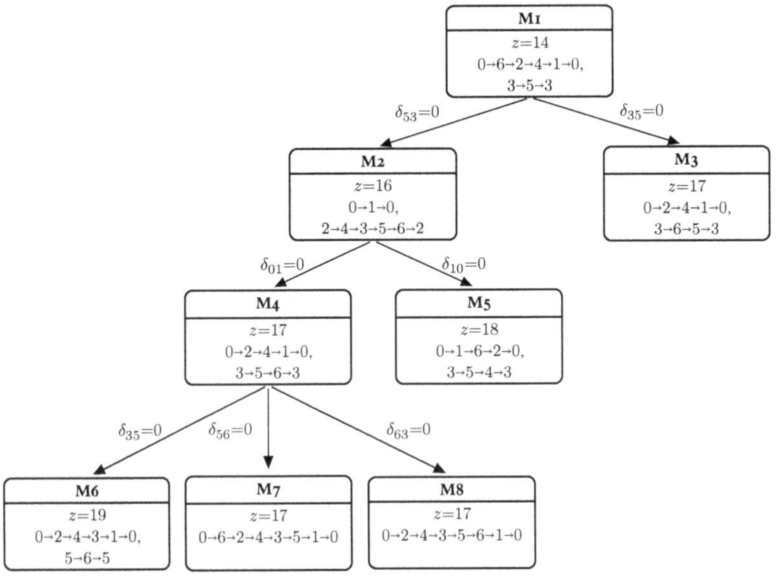

Figure 7.9: Branch-and-bound tree for the machine scheduling problem.

(by giving the weight on this arc a very large value), the value of an optimal solution to the relaxation with this restriction cannot be any better than the previous optimal solution. Hence, the value of the lower bound cannot decrease. If the value of the new lower bound is at least as large as the value of any known feasible schedule, then we need not consider restricting any more arcs, since this can only increase the value of the optimal solution. This process of restricting arcs is repeated until a better feasible schedule is found, or no further restrictions can be better.

Example 7.2.8. *Consider again the machine scheduling problem of Example 7.2.5. We will apply the above described branch-and-bound process to the data of Table 7.3. The branch-and-bound tree is depicted in Figure 7.9. The solution procedure is as follows:*

Iteration 1. *The original model without the subtour-elimination constraints is solved, resulting in a lower bound z for the optimal objective value. The calculations are carried out with a computer package. The solution that is found has $z = 14$ and contains two subtours, namely $0 \to 6 \to 2 \to 4 \to 1 \to 0$ and $3 \to 5 \to 3$. Since optimal solutions do not contain either of these subtours, we may choose an arbitrary one of them to branch on, meaning that all branches of the tree below this branch do not contain the subtour that is chosen for branching. We choose the subtour $3 \to 5 \to 3$, corresponding to $\delta_{35} = 1$ and $\delta_{53} = 1$. In order to exclude this subtour, we branch by adding the constraints $\delta_{35} = 0$ and $\delta_{53} = 0$, respectively, in the next two submodels.*

Iteration 2. *The new submodels correspond to the nodes M2 and M3 in Figure 7.9. Model M2 has optimal objective value $z = 16$ and there are again two subtours, namely $0 \to 1 \to 0$ and $2 \to 4 \to 3 \to 5 \to 6 \to 2$. Model M3 has as optimal objective value $z = 17$ with the subtours $0 \to 2 \to 4 \to 1 \to 0$ and $3 \to 6 \to 5 \to 3$. Since model M2 has a larger optimal objective value than model*

M_3, we select M_2 for further branching. The branches in the tree correspond to $\delta_{01} = 0$ and $\delta_{10} = 0$, respectively.

Iteration 3. Model M_4 has optimal objective value $z = 17$ and subtours $0{\to}2{\to}4{\to}1{\to}0$ and $3{\to}5{\to}6{\to}3$. Model M_5 has as optimal objective value $z = 18$ and subtours $0{\to}1{\to}6{\to}2{\to}0$ and $3{\to}5{\to}4{\to}3$. Note that at this stage of the calculation procedure the submodels M_3, M_4, and M_5 have not been selected yet. We select M_4 for further branching. The cycle $3{\to}5{\to}6{\to}3$ is used to exclude the solution of M_4. This gives three new branches.

Iteration 4. Model M_6 has a larger optimal objective value $(z = 19)$ than model M_7 $(z = 17)$ and M_8 $(z = 17)$, while models M_7 and M_8 both give a feasible solution. Hence, M_6 is excluded from further consideration. Similarly, M_5 is excluded. Model M_3 may give rise to a feasible solution with objective value $z = 17$, but since models M_7 and M_8 already have feasible solutions with the same objective value $z = 17$, M_3 is excluded from further consideration as well.

The conclusion is that M_7 provides an optimal solution of the original model; the solution corresponds to the schedule $0{\to}6{\to}2{\to}4{\to}3{\to}5{\to}1{\to}0$, and the total reset and setup time is 17 time units. Note that M_8 gives an alternative optimal solution. It is left to the reader to try whether or not M_3 also gives rise to alternative optimal solutions.

7.2.5   The traveling salesman problem; the quick and dirty method

The ILO-model for the machine scheduling problem in Section 7.2.4 can also be viewed as a model for the famous *traveling salesman problem*, which can be described as follows. Given a number of cities along with distances between them, the objective is to determine a shortest route through the cities, such that each city is visited precisely once, and the route starts and ends in the same city. Let the cities be labeled 1 through n (with $n \geq 1$), and let c_{ij} be the distance between city i and city j.

Traveling salesman problems (TSPs), such as the machine scheduling problem in Section 7.2.4, can usually not be solved to optimality within reasonable time limits. Part of the reason of this phenomenon is the fact that when the number n of 'cities' increases, the number of constraints in the ILO-formulation of the problem increases by an exorbitant amount. In particular, the number of constraints that avoid subtours may become very large. Note that the number of subtour-elimination constraints is $2^n - 2$, because the number of subsets of $\{1, \ldots, n\}$ is 2^n, minus the empty set and the set $\{1, \ldots, n\}$ itself. In Example 7.2.5, we have $n = 7$ (six jobs plus the dummy job 0), and so there are $2^7 - 2 = 126$ subtour-elimination constraints. Note that the number of subtour-elimination constraints grows exponentially in n; for instance, if $n = 40$, then $2^{40} - 2 = 1,099,511,627,774 \approx 10^{12}$. The computational problems arising from *exponentiality* are discussed in more detail in Chapter 9.

Although the formulation requires a large number of subtour-elimination constraints, it turns out that many of them are nonbinding at any given optimal solution. This suggests a so-called *quick and dirty method* (Q&D method), which can be formulated as follows.

Step 1. Solve the ILO-model for the traveling salesman problem without the subtour-elimination constraints. To be precise, solve the following model:

$$\min \sum_{i=1}^{n} \sum_{j=1}^{n} \delta_{ij} c_{ij}$$

$$\text{s.t.} \quad \sum_{j=1}^{n} \delta_{ij} = 1 \qquad \text{for } i = 1, \ldots, n$$

$$\sum_{i=1}^{n} \delta_{ij} = 1 \qquad \text{for } j = 1, \ldots, n$$

$$\delta_{ij} \in \{0, 1\} \qquad \text{for } i, j = 1, \ldots, n.$$

Step 2. The current solution may contain subtours. If so, we first remove all possible subtours of length two by adding the following constraints:

$$\delta_{ij} + \delta_{ji} \leq 1 \qquad \text{for } 1 \leq i < j \leq n. \tag{7.3}$$

After adding these constraints to the model, the resulting model is solved.

Step 3. If, after Step 2, the solution still contains subtours, then we eliminate them one by one. For instance, if $n = 8$, then the subtour $1 \to 2 \to 3 \to 4 \to 1$ can be eliminated by the requirement that from at least one of the cities 1, 2, 3, 4 a city different from 1, 2, 3, 4 (i.e., one of the cities 5, 6, 7, 8) has to be visited. This is achieved by the following constraint:

$$\sum_{i=1}^{4} \sum_{j=5}^{8} \delta_{ij} \geq 1. \tag{7.4}$$

In general, let $X \subset \{1, \ldots, n\}$ be the subset of cities visited by some subtour in the current optimal solution. Then, we add the following constraint:

$$\sum_{i \in X} \sum_{j \in \{1, \ldots, n\} \setminus X} \delta_{ij} \geq 1. \tag{7.5}$$

Step 3 is repeated until the solution consists of only one tour, which is then an optimal solution of the problem.

When prohibiting certain subtours, some new subtours may appear. So when the Q&D method is applied, it is possible that a lot of constraints have to be added. Also notice that if the number of cities is large, then the subtour-elimination constraints as described above may contain a lot of terms. Indeed, the number of terms in (7.5) is $|X| \times (n - |X|)$. So, if at some point during the Q&D method, we encounter a subtour that contains, e.g., $n/2$ cities, then the number of terms in (7.5) is $n^2/4$.

7.2.6 The branch-and-bound algorithm as a heuristic

The fact that any ILO- or MILO-model can be solved using the branch-and-bound algorithm does not mean that this algorithm is always the most appropriate algorithm. In fact, the number of iterations might be enormous; see Section 9.4. For many practical problems, an optimal solution cannot be found within reasonable time limits. This fact has led to the development of algorithms that do not guarantee optimality, but (hopefully) good feasible solutions within (hopefully) reasonable time limits. Such algorithms are called *heuristics*. Although heuristics generally do not guarantee anything regarding the quality of the solution it finds or how much time it will take to find a solution, they do work well in many practical situations.

Branch-and-bound algorithms can be used as heuristics, when one stops the calculation process before it is finished (provided that a feasible solution has been found when the calculation is stopped). The current best feasible solution (if there is one) is then used as the solution of the problem. There are two commonly used criteria to stop the branch-and-bound algorithm before it finishes. The first one simply sets a time limit so that the algorithm terminates after the given amount of time. If a feasible solution has been found, then the current best solution is used. Notice that, because the branch-and-bound algorithm keeps track of upper bounds for the (unknown) optimal objective value z^* of the original (M)ILO-model, we are able to assess the quality of the current best solution. Let z_U be the current best known upper bound for the optimal objective value, and let z_L be the objective value corresponding to the best known feasible solution. Then, clearly, $z_L \leq z^* \leq z_U$. Hence, assuming that $z_L > 0$, the objective value of the current best feasible solution lies within

$$\frac{z_U - z_L}{z_U} \times 100\%$$

of the optimal objective value. To see this, notice that

$$\left(1 - \frac{z_U - z_L}{z_U}\right) z^* = \frac{z_L}{z_U} z^* \leq \frac{z_L}{z_U} z_U = z_L.$$

Example 7.2.9.   *Suppose that we had set a time limit for finding a solution for the example knapsack problem in Section 7.2.3, and suppose that the branch-and-bound algorithm was terminated after solving models M1, M2, M3, and M4; see Figure 7.6. Then the current best solution is the one found in M4 with $z_L = 13$. The best known upper bound at that point is the one given by M2, i.e., $z_U = 14\frac{1}{2}$ (recall that, since we have not solved M5 yet, its optimal objective value is not known). Thus, we know at this point that the objective value of the current best solution lies within $100 \times (14\frac{1}{2} - 13)/14\frac{1}{2} \approx 10.4\%$ of the optimal objective value. Notice that we have not used the value of z^*, the actual optimal objective value, to establish this percentage. The solution with $z = 13$ actually lies within $100 \times (14 - 13)/14 \approx 7.2\%$ of the optimal solution.*

The discussion about the quality of the current best solution motivates the second stopping criterion: sometimes we are satisfied (or we are forced to be satisfied!) with a solution that lies within a certain fraction α of the optimal solution. In that case, we can stop the branch-and-bound algorithm as soon as it finds a feasible solution with objective value z_L and an upper bound z_U satisfying $(z_U - z_L)/z_U \leq \alpha$.

In practice, it often happens that the branch-and-bound algorithm finds a feasible solution that is actually very close to an optimal solution (or, even better, it is an optimal solution). However, the algorithm cannot conclude optimality without a good upper bound on the optimal objective value, and as a result it sometimes spends an excessive amount of time branching nodes without finding improved solutions. So the branch-and-bound algorithm often spends much more time on proving that a solution is optimal than on finding this solution. One way to improve the upper bounds is by using the so-called cutting plane algorithm which will be introduced in Section 7.4.

7.3 Linearizing logical forms with binary variables

In this section it is shown how *binary variables*, also called $\{0, 1\}$-*variables*, can be used to linearize a variety of complex logical forms, such as 'either-or' constraints, 'if-then' constraints, and piecewise linear functions.

7.3.1 The binary variables theorem

Recall that Dairy Corp (see Section 7.1.1) needs to decide how to transport the cheese and milk it produces to grocery stores, either by buying vehicles, or by contracting out some of their transportation. The management of Dairy Corp realizes that they have overlooked a third option of transport. As it turns out, there is a container train running from a convenient location near the production factory to the city that the factory serves. The company can rent containers on this train, each of which can hold up to 200 ($\times 1,000$ kgs) of cheese and 100 ($\times 1,000$ liters) of milk. Renting the container saves the company $\$5,000$ per container, but in order to rent containers, the railroad company charges a $\$2,000$ fixed cost. So, if a total of x_3 storage units are used, the total cost reduction for transportation by train, denoted by $T(x_3)$, is given by:

$$T(x_3) = \begin{cases} 0 & \text{if } x_3 = 0 \\ 5000x_3 - 2000 & \text{if } x_3 > 0. \end{cases}$$

Thus, the problem that Dairy Corp needs to solve is the following:

$$\max 1000x_1 + 700x_2 + T(x_3)$$
$$\text{s.t.} \quad 100x_1 + \ 50x_2 + 200x_3 \leq 2425$$
$$20x_2 + 100x_3 \leq \ 510$$
$$x_1, x_2, x_3 \geq 0, \text{ and integer.}$$

However, because the function $T(x_3)$ is nonlinear, this model is not an ILO-model. So the problem is: how do we include such a nonlinear function in an ILO-model? To solve this problem, we introduce a *binary variable* δ, which will have the following interpretation:

$$\delta = \begin{cases} 0 & \text{if } x_3 = 0 \\ 1 & \text{if } x_3 > 0. \end{cases}$$

The function $T(x_3)$ becomes $T(\delta, x_3) = 5000x_3 - 2000\delta$. The nonlinearity is now hidden in the fact that δ can only take the values 0 and 1. This turns the problem into:

$$\begin{aligned}
\max\ & 1000x_1 + 700x_2 + 5000x_3 - 2000\delta \\
\text{s.t.}\ \ & 100x_1 + 50x_2 + 200x_3 && \leq 2425 \\
& 20x_2 + 100x_3 && \leq 510 \\
& x_1, x_2, x_3 \geq 0, \text{integer, and } \delta \in \{0, 1\}.
\end{aligned}$$

However, this is not enough: in the current formulation, it is still feasible to have $\delta = 0$ and $x_3 \geq 1$ at the same time, which clearly contradicts the intended interpretation of δ. So we need to rule out such solutions by adding constraints in terms of δ and x_3. In order to determine the correct constraints, first note that the intended interpretation of δ can be rewritten as:

$$x_3 \geq 1 \Rightarrow \delta = 1, \text{ and } x_3 = 0 \Rightarrow \delta = 0.$$

The symbol '$\Rightarrow$' means 'if-then' or 'implies'; see Section 7.3.2. Note that $[x_3 \geq 1 \Rightarrow \delta = 1]$ is equivalent to $[\delta = 0 \Rightarrow x_3 \leq 0]$, and that $[x_3 = 0 \Rightarrow \delta = 0]$ is equivalent to $[\delta = 1 \Rightarrow x_3 \geq 1]$. The following theorem offers a useful tool for linearizing expressions of this kind.

Theorem 7.3.1. (*Binary variables theorem*)
Let $D \subseteq \mathbb{R}^n$ ($n \geq 1$) and $f : D \to \mathbb{R}$, and let M be a nonzero real number such that $M \geq \max\{f(\mathbf{x}) \mid \mathbf{x} \in D\}$. Then, for each $\delta \in \{0, 1\}$ and each $\mathbf{x} \in D$ the following assertions are equivalent:

(i) $\delta = 0 \Rightarrow f(\mathbf{x}) \leq 0$;

(ii) $f(\mathbf{x}) - M\delta \leq 0$.

Proof of Theorem 7.3.1. (i) $\Rightarrow$ (ii): We need to show that $f(\mathbf{x}) \leq M\delta$ for each $\mathbf{x} \in D$ and each $\delta \in \{0, 1\}$, given that the implication $[\delta = 0 \Rightarrow f(\mathbf{x}) \leq 0]$ holds for each $\mathbf{x} \in D$ and each $\delta \in \{0, 1\}$. Take any $\mathbf{x} \in D$. For $\delta = 1$, we have to show that $f(\mathbf{x}) \leq M$. This follows from the definition of M. For $\delta = 0$, we have to show that $f(\mathbf{x}) \leq 0$. But this follows from the given implication.

(ii) $\Rightarrow$ (i): We now have to show that the implication $[\delta = 0 \Rightarrow f(\mathbf{x}) \leq 0]$ holds for each $\mathbf{x} \in D$ and each $\delta \in \{0, 1\}$, given that $f(\mathbf{x}) \leq M\delta$ for each $\mathbf{x} \in D$ and each $\delta \in \{0, 1\}$. Again take any $\mathbf{x} \in D$. If $\delta = 0$, then $f(\mathbf{x}) \leq 0$ follows from $f(\mathbf{x}) \leq M\delta$ with $\delta = 0$. For $\delta = 1$, the implication becomes $[1 = 0 \Rightarrow f(\mathbf{x}) \leq 0]$, which is true because $1 = 0$ is false; see Table 7.5. $\qquad\square$

We can now linearize $[\delta = 0 \Rightarrow x_3 \leq 0]$ as follows. In practical situations, the values of decision variables are always bounded. In our example, the second constraint of the model for Dairy Corp implies that x_3 can be at most 5. So we may take $M_1 = 5$. Replacing

$f(x_3)$ by x_3 in Theorem 7.3.1, we obtain that $[\delta = 0 \Rightarrow x_3 \leq 0]$ is equivalent to:

$$x_3 - M_1 \delta \leq 0.$$

Linearizing $[\delta = 1 \Rightarrow 1 - x_3 \leq 0]$ can be done as follows. Observe that the expression is equivalent to $[1 - \delta = 0 \Rightarrow x_3 \geq 1]$. Now, by Theorem 7.3.1 with $f(x_3) = 1 - x_3$, we find that $[\delta = 0 \Rightarrow x_3 \leq 0]$ is equivalent to:

$$(1 - x_3) - M_2(1 - \delta) \leq 0, \text{ or, equivalently, } x_3 - M_2 \delta \geq 1 - M_2.$$

Notice that $f(x_3) \leq 1$ because $x_3 \geq 0$, so we may take $M_2 = 1$. Summarizing, the equivalence $[\delta = 0 \Leftrightarrow x_3 = 0]$ can be written in terms of linear constraints by means of the linear inequalities $x_3 - M_1 \delta \leq 0$ and $x_3 - M_2 \delta \geq 1 - M_2$, with $M_1 = 5$ and $M_2 = 1$. So, the correct model for Dairy Corp is:

$$
\begin{array}{llllll}
\max & 1000x_1 + 700x_2 + 5000x_3 - 2000\delta & & \\
\text{s.t.} & 100x_1 + 50x_2 + 200x_3 & & \leq 2425 \\
& 20x_2 + 100x_3 & & \leq 510 \\
& x_3 - 5\delta & \leq & 0 \\
& x_3 - \delta & \geq & 0 \\
\end{array}
$$
$$x_1, x_2, x_3 \geq 0, \text{integer, and } \delta \in \{0, 1\}.$$

Recall that '$\delta \in \{0, 1\}$' is equivalent to '$0 \leq \delta \leq 1$, δ integer', so this model is in fact an ILO-model. The optimal solution of this model is $x_1^* = 14$, $x_2^* = 0$, $x_3^* = 5$, and $\delta^* = 1$, with objective value $z^* = 37{,}000$.

When we linearized the expression $[x_3 = 0 \Rightarrow \delta = 0]$, we used the fact that x_3 is an integer variable, because we rewrote it as $[\delta = 1 \Rightarrow x_3 \geq 1]$. The situation when x_3 is not an integer variable needs some more attention. Since the nonnegativity constraint $x_3 \geq 0$ is explicitly included in the model, $[\delta = 1 \Rightarrow x_3 \neq 0]$ is equivalent to $[\delta = 1 \Rightarrow x_3 > 0]$. In order to write this in the form of implication (i) of Theorem 7.3.1, we have to formulate $x_3 > 0$ in the form $f(x_3) \leq 0$. This can only be done by introducing a small positive real number m, and using $x_3 \geq m$ instead of $x_3 > 0$. Mathematically, the expressions '$x_3 > 0$' and '$x_3 \geq m$ for small $m > 0$' are not equivalent, but in practical situations they are, when taking m small enough. Now the implication $[\delta = 1 \Rightarrow x_3 > 0]$ is equivalent to $[1 - \delta = 0 \Rightarrow m - x_3 \leq 0]$. Using Theorem 7.3.1 with $\delta := 1 - \delta$ and $M := m \geq \max\{m - x_3 \mid m \leq x_3 \leq M\}$, it follows that the implication is equivalent to $(m - x_3) - m(1 - \delta) \leq 0$, which can be written as:

$$x_3 - m\delta \geq 0.$$

7.3.2 Introduction to the theory of logical variables

In this section we will give a short introduction to the theory of logical variables. A *simple proposition* is a logical variable which can take on only the values 'true' and 'false'. *Compound propositions* can be formed by modifying a simple proposition by the word 'not', or by con-

Name of the logical connective	Symbol	Interpretation	Constraint
Negation	$\neg P_1$	not P_1	$\delta_1 = 0$
Conjunction	$P_1 \wedge P_2$	both P_1 and P_2	$\delta_1 + \delta_2 = 2$
Inclusive disjunction	$P_1 \vee P_2$	either P_1 or P_2	$\delta_1 + \delta_2 \geq 1$
Exclusive disjunction	$P_1 \veebar P_2$	exactly one of P_1 or P_2	$\delta_1 + \delta_2 = 1$
Implication	$P_1 \Rightarrow P_2$	if P_1 then P_2	$\delta_1 - \delta_2 \leq 0$
Equivalence	$P_1 \Leftrightarrow P_2$	P_1 if and only if P_2	$\delta_1 - \delta_2 = 0$

Table 7.4: Propositional connectives and constraints.

necting propositions with *logical connectives* such as 'and', 'or', 'if-then', and 'if and only if'. Table 7.4 lists a number of logical connectives with their names and interpretations; P_1 and P_2 are propositions. The meaning of the 'Constraint' column will be explained in Section 7.3.3.

Example. For $i = 1, 2$, let $P_i = $ 'item i is produced'. Then, $\neg P_1 = $ 'item 1 is not produced', $(P_1 \wedge P_2) = $ 'both item 1 and item 2 are produced', $(P_1 \vee P_2) = $ 'either item 1 or item 2 is produced, or both', $(P_1 \veebar P_2) = $ 'either item 1 or item 2 is produced, but not both', $(P_1 \Rightarrow P_2) = $ 'if item 1 is produced then item 2 is produced', $(P_1 \Leftrightarrow P_2) = $ 'item 1 is produced if and only if item 2 is produced'.

In Table 7.4, the connective 'disjunction' is used in two meanings. The term *inclusive disjunction* means that at least one proposition in the disjunction is true, allowing for the possibility that both are true, and the term *exclusive disjunction* means that exactly one of the propositions in the disjunction is true. The exclusive disjunction can be written as:

$$(P_1 \veebar P_2) = (\neg P_1 \wedge P_2) \vee (P_1 \wedge \neg P_2). \tag{7.6}$$

In words, the expression on the right hand side states that either (1) P_1 is false and P_2 is true, or (2) P_1 is true and P_2 is false, or (3) both. Clearly, (1) and (2) cannot be true at the same time.

The proofs of relationships of this kind lie beyond the scope of this book. However, since the theory needed for such proofs is so elegant and easy, we will give a short introduction. Such proofs can be given by using so-called *truth tables*. The truth tables for the connectives introduced above are called the *connective truth tables*; they are listed in Table 7.5. Table 7.5 means the following. P_1 and P_2 denote propositions, '1' means 'true' and '0' means 'false'. Now consider, for instance, the implication $P_1 \Rightarrow P_2$. If P_1 is 'true' and P_2 is 'true' then $P_1 \Rightarrow P_2$ is 'true', if P_1 is 'true' and P_2 is 'false' then $P_1 \Rightarrow P_2$ is 'false', if P_1 is 'false' then $P_1 \Rightarrow P_2$ is 'true', regardless of whether P_2 is 'true' or 'false'. In order to prove that two compound propositions are equal (also called *logically equivalent*), we have to show that the corresponding truth tables are the same.

We will give a proof for (7.6). To that end, we have to determine the truth table of $(\neg P_1 \wedge P_2) \vee (P_1 \wedge \neg P_2)$; see Table 7.6. This truth table is determined as follows. The first two columns of Table 7.6 represent all 'true' and 'false' combinations for P_1 and P_2. The fifth

P_1	P_2	$\neg P_1$	$P_1 \wedge P_2$	$P_1 \vee P_2$	$P_1 \veebar P_2$	$P_1 \Rightarrow P_2$	$P_1 \Leftrightarrow P_2$
1	1	0	1	1	0	1	1
1	0	0	0	1	1	0	0
0	1	1	0	1	1	1	0
0	0	1	0	0	0	1	1

Table 7.5: Connective truth table.

P_1	P_2	$(\neg P_1$	$\wedge$	$P_2)$	$\vee$	$(P_1$	$\wedge$	$\neg P_2)$
1	1	0	0	1	0	1	0	0
1	0	0	0	0	1	1	1	1
0	1	1	1	1	1	0	0	0
0	0	1	0	0	0	0	0	1

Table 7.6: Truth table of the compound proposition $(\neg P_1 \wedge P_2) \vee (P_1 \wedge \neg P_2)$.

column, labeled 'P_2', follows immediately; it is just a copy of the second column. The seventh column, labeled 'P_1', follows similarly. The columns corresponding to $\neg P_1$ and $\neg P_2$ then follow by using the connective truth table for '$\neg$'. The fourth column, whose column label is '$\wedge$', corresponds to the formula $\neg P_1 \wedge P_2$; it is obtained from the third and the fifth column by using the connective truth table for '$\wedge$'. The seventh column, also labeled '$\wedge$', is determined similarly. Finally, the sixth column, labeled '$\vee$', is the main column. It follows from the fourth and the eighth column by applying the connective truth table for '$\vee$'. This sixth column is the actual truth table for this compound proposition. It is equal to column corresponding to $P_1 \veebar P_2$ in Table 7.6, as required.

7.3.3 Logical forms represented by binary variables

In this section we will give a number of different examples of logical forms that can be reformulated as a linear expression. First, we will give the transformations of the logical forms of Table 7.4 into linear constraints. Let P_1 and P_2 be propositions, i.e., logical variables that can take on only the values 'true' and 'false'. Let $i = 1, 2$. With each value of P_i we associate a binary decision variable in the following way:

$$\delta_i = \begin{cases} 0 & \text{if and only if } P_i \text{ is true} \\ 1 & \text{if and only if } P_i \text{ is false.} \end{cases}$$

It is left to the reader to check the formulas in the 'Constraint' column of Table 7.4. Note that $P_1 \wedge P_2$ is also equivalent to $\delta_1 = \delta_2 = 1$.

Let k and n be integers with $1 \le k \le n$. Let $P_1, \ldots, P_n$ be propositions. Generalizations of the basic propositional connectives and its corresponding linear constraints are (the symbol

'≡' means 'is equivalent to'):

$$
\begin{aligned}
P_1 \vee \ldots \vee P_n &\equiv \delta_1 + \ldots + \delta_n \geq 1; \\
P_1 \veebar \ldots \veebar P_n &\equiv \delta_1 + \ldots + \delta_n = 1; \\
P_1 \wedge \ldots \wedge P_k \Rightarrow P_{k+1} \vee \ldots \vee P_n &\equiv (1-\delta_1) + \ldots + (1-\delta_k) + \delta_{k+1} + \ldots + \delta_n \geq 1; \\
\text{at least } k \text{ out of } n \text{ are 'true'} &\equiv \delta_1 + \ldots + \delta_n \geq k; \\
\text{exactly } k \text{ out of } n \text{ are 'true'} &\equiv \delta_1 + \ldots + \delta_n = k.
\end{aligned}
$$

Example 7.3.1.   *Consider the following implication:*

If a company decides to produce a certain item, then at least 10 units of it need to be produced. The production capacity is 25 units.

Let x be the number of units being produced, i.e., $x \geq 0$. The logical expression in the above question can be formulated as: 'exactly one of $x = 0$ and $x \geq 10$ is true'. Using the connective '$\veebar$', this is equivalent to $[x = 0 \veebar x \geq 10]$.

Before linearizing this logical form, we consider a more general formulation. Let $f : D \to \mathbb{R}$ and $g : D \to \mathbb{R}$ be two linear functions with $D \subseteq \mathbb{R}$. Let $M_1 = \max\{f(\mathbf{x}) \mid \mathbf{x} \in D\}$, $M_2 = \max\{g(\mathbf{x}) \mid \mathbf{x} \in D\}$, and let m_1 and m_2 be the respective minima. We will formulate the logical form

$$
f(\mathbf{x}) \leq 0 \veebar g(\mathbf{x}) \leq 0
$$

in terms of linear constraints. To that end, we introduce the binary variables δ_1 and δ_2 with the following interpretation:

$$
\begin{aligned}
\delta_1 = 1 &\Leftrightarrow f(\mathbf{x}) \leq 0 \\
\delta_2 = 1 &\Leftrightarrow g(\mathbf{x}) \leq 0.
\end{aligned}
$$

Clearly, $[f(\mathbf{x}) \leq 0 \veebar g(\mathbf{x}) \leq 0]$ is equivalent to $\delta_1 + \delta_2 = 1$. The implication $[\delta_1 = 1 \Rightarrow f(\mathbf{x}) \leq 0]$ is, according to Theorem 7.3.1, equivalent to $f(\mathbf{x}) + M_1\delta_1 \leq M_1$. Moreover, $[\delta_1 = 1 \Leftarrow f(\mathbf{x}) \leq 0] \equiv [\delta_1 = 0 \Rightarrow f(\mathbf{x}) > 0] \equiv [\delta_1 = 0 \Rightarrow f(\mathbf{x}) \geq \varepsilon]$ for small enough $\varepsilon > 0$. According to Theorem 7.3.1, this is equivalent to $f(\mathbf{x}) + (\varepsilon - m_1)\delta_1 \geq \varepsilon$. Similar relations hold for $g(\mathbf{x})$. Since $\delta_1 + \delta_2 = 1$, we can replace δ_1 by δ, and δ_2 by $1 - \delta$. Hence, $[f(\mathbf{x}) \leq 0 \veebar g(\mathbf{x}) \leq 0]$ is equivalent to the following set of constraints:

$$
\begin{aligned}
f(\mathbf{x}) + M_1\delta &\leq M_1 \\
g(\mathbf{x}) - M_2\delta &\leq 0 \\
f(\mathbf{x}) + (\varepsilon - m_1)\delta &\geq \varepsilon \\
g(\mathbf{x}) + (m_2 - \varepsilon)\delta &\geq m_2.
\end{aligned}
\tag{7.7}
$$

We now apply the above inequalities to the expression $[x = 0 \veebar x \geq 10]$. Let $\mathbf{x} = [x]$, and take $f(x) = x$ and $g(x) = 10 - x$. Note that $x = 0$ is equivalent to $x \leq 0$. Since $0 \leq x \leq 25$, we have $M_1 = 25$, $M_2 = 10$. Moreover, $m_1 = \min\{f(x) \mid x \in D\} = \min\{x \mid x \in D\} = 0$, and $m_2 = \min\{10 - x \mid x \in D\} = 10 - \max\{x \mid x \in D\} = 10 - 25 = -15$.

Substituting these values into (7.7), it follows that the expression $[x = 0 \veebar x \geq 10]$ *is equivalent to:*

$$
\begin{aligned}
x + 25\delta &\leq 25 \\
x + 10\delta &\geq 10 \\
x + \varepsilon\delta &\geq \varepsilon \\
x + (15 + \varepsilon)\delta &\leq 25.
\end{aligned}
$$

One can easily check that $\delta = 0$ *corresponds to* $10 \leq x \leq 25$, *and* $\delta = 1$ *to* $x = 0$. *So the last two inequalities are actually redundant. It is left to the reader to check that the above formulation is equivalent to* $[x > 0 \iff x \geq 10]$.

Example 7.3.2. *Consider the following implication:*

If item A *and/or item* B *are produced, then at least one of the items* C, D, *or* E *have to be produced.*

In order to linearize this implication, we introduce the proposition X_i, *meaning that 'item* i *is produced'* $(i = A, B, C, D, E)$. *The implication can now be written as:*

$$
(X_A \vee X_B) \Rightarrow (X_C \vee X_D \vee X_E).
$$

For $i \in \{A, B, C, D, E\}$, *let* δ_i *be defined by:*

$$
\delta_i = 1 \text{ if and only if item } i \text{ is produced.}
$$

The compound proposition $X_A \vee X_B$ *can then be represented as* $\delta_A + \delta_B \geq 1$, *and* $X_C \vee X_D \vee X_E$ *as* $\delta_C + \delta_D + \delta_E \geq 1$; *see Table 7.4. In order to linearize the above implication, we introduce a dummy proposition* X *which satisfies:*

$$
X_A \vee X_B \Rightarrow X \Rightarrow X_C \vee X_D \vee X_E,
$$

and a corresponding binary variable δ *with the interpretation:*

$$
\delta = 1 \text{ if and only if } X \text{ is true.}
$$

(Explain why we need a dummy variable.) Hence, we need to model the implications:

$$
\delta_A + \delta_B \geq 1 \Rightarrow \delta = 1 \Rightarrow \delta_C + \delta_D + \delta_E \geq 1.
$$

Using Theorem 7.3.1 one can easily check that these implications are equivalent to:

$$
\delta_A + \delta_B - 2\delta \leq 0, \text{ and } \delta - \delta_C - \delta_D - \delta_E \leq 0.
$$

Example 7.3.3. *Consider the following expression:*

The binary variables δ_1, δ_2, *and* δ_3 *satisfy* $\delta_1 \delta_2 = \delta_3$.

It may happen that an expression of the form $\delta_1 \delta_2$ appears somewhere in the model, for instance in the objective function (see Section 7.3.4). Note that the value of $\delta_1 \delta_2$ is also binary. Clearly, $\delta_1 \delta_2 = \delta_3$ is equivalent to $[\delta_3 = 1 \Leftrightarrow \delta_1 = 1$ and $\delta_2 = 1]$, which is equivalent to $[\delta_3 = 1 \Leftrightarrow \delta_1 + \delta_2 = 2]$, and can be split into:

$$1 - \delta_3 = 0 \Rightarrow 2 - \delta_1 - \delta_2 \leq 0, \ \text{together with } \delta_1 + \delta_2 = 2 \Rightarrow \delta_3 = 1.$$

Since $\max(2 - \delta_1 - \delta_2) = 2 - \min(\delta_1 + \delta_2) = 2$, it follows from Theorem 7.3.1 that the first one is equivalent to $(2 - \delta_1 - \delta_2) - 2(1 - \delta_3) \leq 0$, or to:

$$\delta_1 + \delta_2 - 2\delta_3 \geq 0.$$

The second implication $[\delta_1 + \delta_2 = 2 \Rightarrow \delta_3 = 1]$ is equivalent to:

$$\delta_1 + \delta_2 - \delta_3 \leq 1.$$

These two inequalities linearize the equation $\delta_1 \delta_2 = \delta_3$.

Example 7.3.4.   *Consider the following equivalence:*

The inequality $\sum_i a_i x_i \leq b$ holds if and only if the condition X holds.

Assume that $m \leq \sum_i a_i x_i - b \leq M$. Define the binary variable δ with the interpretation that $\delta = 1$ if and only if X holds. So we have to linearize:

$$\delta = 1 \Leftrightarrow \sum_i a_i x_i \leq b.$$

We first linearize $[\delta = 1 \Rightarrow \sum_i a_i x_i \leq b]$. Applying Theorem 7.3.1, we find that the implication is equivalent to $(\sum_i a_i x_i - b) - M(1 - \delta) \leq 0$, which can be rewritten as:

$$\sum_i a_i x_i + M\delta \leq M + b. \tag{7.8}$$

Next we consider $[\sum_i a_i x_i \leq b \Rightarrow \delta = 1]$, which is equivalent to $[\delta = 0 \Rightarrow \sum_i a_i x_i > b]$. Here we run into the same difficulty as with the expression $x > 0$; see Section 7.3.1. We replace $\sum_i a_i x_i - b > 0$ by $\sum_i a_i x_i - b \geq \varepsilon$, with ε a small positive number such that for numbers smaller than ε the inequality does not hold. In case the a_i's and the x_i's are integers, we can take $\varepsilon = 1$. The implication can now be written as $[\delta = 0 \Rightarrow -\sum_i a_i x_i + b + \varepsilon \leq 0]$. Applying Theorem 7.3.1, we find that $(-\sum_i a_i x_i + b + \varepsilon) - (\varepsilon - m)\delta \leq 0$, which can be rewritten as:

$$\sum_i a_i x_i - m\delta \geq b + \varepsilon(1 - \delta). \tag{7.9}$$

The inequalities (7.8) and (7.9) linearize the expression of this example.

Example 7.3.5.   *Recall that the graph of a piecewise linear function is composed of a number of line segments connected by kink points. We will illustrate the linearization of a piecewise linear function by*

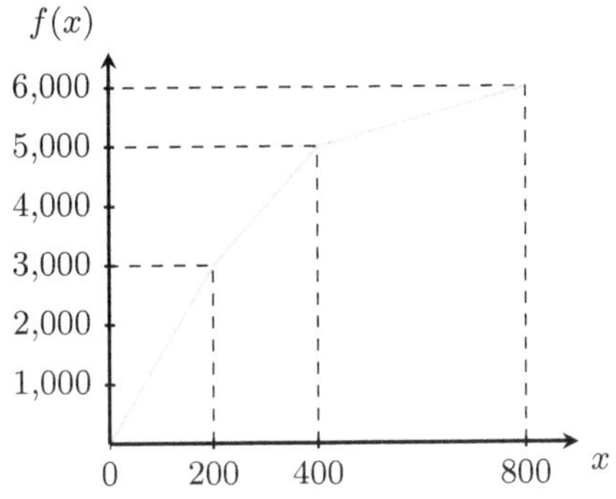

Figure 7.10: Piecewise linear function.

means of the following example. Suppose that at most 800 pounds of a commodity may be purchased. The cost of the commodity is $\$15$ per pound for the first 200 pounds, $\$10$ for the next 200 pounds, and $\$2.5$ for the remaining 400 pounds. Let x be the amount of the commodity to be purchased (in pounds), and $f(x)$ the total costs (in dollars) when x pounds are purchased. The function f can be represented as a piecewise linear function in the following way:

$$f(x) = \begin{cases} 15x & \text{if } 0 \leq x \leq 200 \\ 1000 + 10x & \text{if } 200 < x \leq 400 \\ 4000 + 2.5x & \text{if } 400 < x \leq 800. \end{cases}$$

The graph of this function is given in Figure 7.10. Note that the kink points occur at $x_1 = 200$, and $x_2 = 400$. Let x_0 and x_3 be the start and end points, respectively, of the definition interval of the function; i.e., $x_0 = 0$ and $x_3 = 800$. For any x with $0 \leq x \leq 800$, the following equation holds.

$$x = \lambda_0 x_0 + \lambda_1 x_1 + \lambda_2 x_2 + \lambda_3 x_3,$$

where $\lambda_0, \lambda_1, \lambda_2, \lambda_3 \geq 0$, $\lambda_0 + \lambda_1 + \lambda_2 + \lambda_3 = 1$, and at most two adjacent λ_i's are positive (λ_i and λ_j are adjacent if, either $j = i + 1$, or $j = i - 1$). It can also be easily checked that

$$f(x) = \lambda_0 f(x_0) + \lambda_1 f(x_1) + \lambda_2 f(x_2) + \lambda_3 f(x_3),$$

for each x with $x_0 \leq x \leq x_3$. The statement that at most two adjacent λ_i's are positive can be linearized by applying Theorem 7.3.1. To that end we introduce binary variables δ_1, δ_2, and δ_3, one for each segment, with the following interpretation (with $i = 1, 2, 3$):

$$\delta_i = 1 \Rightarrow \lambda_k = 0 \text{ for each } k \neq i - 1, i.$$

Moreover, we introduce the constraint:

$$\delta_1 + \delta_2 + \delta_3 = 1.$$

This definition means that if, for instance, $\delta_2 = 1$ then $\delta_1 = \delta_3 = 0$ and $\lambda_0 = \lambda_3 = 0$, and so $x = \lambda_1 x_1 + \lambda_2 x_2$ and $f(x) = \lambda_1 f(x_1) + \lambda_1 f(x_2)$ with $\lambda_1, \lambda_2 \geq 0$ and $\lambda_1 + \lambda_2 = 1$.

	A	B	C
Large DC	1,000	1,200	1,500
Small DC	400	700	800

	A	B	C
A	-	1,100	1,500
B	1,100	-	1,600
C	1,500	1,600	-

(a) The cost reductions.   (b) The transportation costs.

Table 7.7: Data for the decentralization problem.

Since $[\delta_1 = 1 \Rightarrow \lambda_2 = 0 \wedge \lambda_3 = 0]$ is equivalent to $[1 - \delta_1 = 0 \Rightarrow \lambda_2 + \lambda_3 \leq 0]$, Theorem 7.3.1 implies that this is equivalent to:

$$\lambda_2 + \lambda_3 + 2\delta_1 \leq 2. \tag{7.10}$$

Similarly, we find that:

$$\lambda_0 + \lambda_3 + 2\delta_2 \leq 2, \tag{7.11}$$
$$\lambda_0 + \lambda_1 + 2\delta_3 \leq 2. \tag{7.12}$$

For instance, if $\delta_2 = 1$, and so $\delta_1 = \delta_3 = 0$, then (7.10) becomes the constraint $\lambda_2 + \lambda_3 \leq 2$, (7.11) implies that $\lambda_0 = \lambda_3 = 0$, and (7.12) becomes the constraint $\lambda_0 + \lambda_1 \leq 2$. Hence, $\delta_2 = 1$ refers in fact to the interval $x_1 \leq x \leq x_2$.

It is left to the reader to generalize these formulas for piecewise linear functions with an arbitrary finite number of kink points (see Exercise 7.5.14).

7.3.4   A decentralization problem

The decentralization problem to be discussed in this section uses the linearization of Example 7.3.3 in Section 7.3.3. The problem can be stated as follows. A supermarket chain wants to decentralize its distribution of provisions. The management is considering opening two distribution centers, a large center (DC 1) and a small center (DC 2) in two different cities. The cities A, B, and C are candidate locations for a distribution center. Table 7.7(a) lists the cost reduction compared to the current situation (e.g., due to subsidies from local governments) for each of these cities. There is traffic between the distribution centers because of common services (e.g., they share a common repair department). The transportation costs due to common services between the three cities are listed in Table 7.7(b).

Taking into account the cost reductions and the transportation costs, the management of the supermarket chain wants to find the best possible locations of both distribution centers. This decision problem will be solved by means of an ILO-model with binary variables. To that end we introduce the following symbols. For each $i \in \{1, 2\}$, and $j, l \in \{1, 2, 3\}$, define (where $A = 1$, $B = 2$, $C = 3$):

B_{ij} = the cost reduction when DC i is in city j;

K_{jl} = the transportation costs between j and l;

$$\delta_{ij} = \begin{cases} 1 & \text{if DC } i \text{ is in city } j \\ 0 & \text{otherwise.} \end{cases}$$

The objective function consists of two parts. The first part concerns the cost reductions with respect to the present-day situation. The total cost reduction is:

$$\sum_{i=1}^{2} \sum_{j=1}^{3} B_{ij} \delta_{ij}.$$

Each term in this summation is B_{ij} if $\delta_{ij} = 1$, and zero otherwise. The second part concerns the transportation costs. These are:

$$\frac{1}{2} \sum_{i=1}^{2} \sum_{k=1}^{2} \sum_{j=1}^{3} \sum_{l=1}^{3} \delta_{ij} \delta_{kl} K_{jl}.$$

Each term in this summation is K_{jl} if $\delta_{ij} = 1$ and $\delta_{kl} = 1$, and zero otherwise. Note that if $\delta_{ij} \delta_{kl} = 1$, then also $\delta_{i'j} \delta_{k'l}$ for $i' = k$ and $k' = i$, and hence each K_{jl} appears in the sum with coefficient either 0 or 2. To counter this double-counting, we multiply the sum by $\frac{1}{2}$. The objective becomes:

$$\max \left(\sum_{i=1}^{2} \sum_{j=1}^{3} \delta_{ij} B_{ij} - \frac{1}{2} \sum_{i=1}^{2} \sum_{k=1}^{2} \sum_{j=1}^{3} \sum_{l=1}^{3} \delta_{ij} \delta_{kl} K_{jl} \right)$$

There are two sets of constraints. First, each DC has to be assigned to precisely one city. Hence,

$$\sum_{j=1}^{3} \delta_{ij} = 1 \qquad \text{for } i = 1, 2.$$

Secondly, no two DC's are to be assigned to the same city. Hence,

$$\sum_{i=1}^{2} \delta_{ij} \leq 1 \qquad \text{for } j = 1, 2, 3.$$

In the objective function the quadratic term $\delta_{ij} \delta_{kl}$ appears. In Section 7.3.3 (Example 7.3.3), it is described how such a term can be linearized by introducing binary variables. Namely, define binary variables ν_{ijkl} (for $i, k = 1, 2$ and $j, l = 1, 2, 3$) with the interpretation:

$$\nu_{ijkl} = \begin{cases} 1 & \text{if } \delta_{ij} = 1 \text{ and } \delta_{kj} = 1 \\ 0 & \text{otherwise.} \end{cases}$$

This can be rewritten as $[\nu_{ijkl} = 1 \Leftrightarrow \delta_{ij} = 1 \wedge \delta_{kl} = 1]$, which is equivalent to:

$$\delta_{ij} + \delta_{kl} - 2\nu_{ijkl} \geq 0 \text{ and } \delta_{ij} + \delta_{kl} - \nu_{ijkl} \leq 1.$$

The ILO-model of the decentralization problem becomes:

Model 7.3.1. (*ILO-model Decentralization problem*)

$$\max \left(\sum_{i=1}^{2} \sum_{j=1}^{3} \delta_{ij} B_{ij} - \frac{1}{2} \sum_{i=1}^{2} \sum_{k=1}^{2} \sum_{j=1}^{3} \sum_{l=1}^{3} \delta_{ij} \delta_{kl} K_{jl} \right)$$

$$\text{s.t.} \quad \sum_{j=1}^{3} \delta_{ij} = 1 \quad \text{for } i = 1, 2$$

$$\sum_{i=1}^{2} \delta_{ij} \leq 1 \quad \text{for } j = 1, 2, 3$$

$$\delta_{ij} + \delta_{kl} - 2\nu_{ijkl} \geq 0 \quad \text{for } i, k = 1, 2 \text{ and } j, l = 1, 2, 3$$

$$\delta_{ij} + \delta_{kl} - \nu_{ijkl} \leq 1 \quad \text{for } i, k = 1, 2 \text{ and } j, l = 1, 2, 3$$

$$\delta_{ij}, \delta_{kl}, \nu_{ijkl} \in \{0, 1\} \quad \text{for } i, k = 1, 2 \text{ and } j, l = 1, 2, 3.$$

This ILO-model can be solved by means of a computer package. It turns out that there are two optimal solutions:

	Cost reduction
DC 1 in C	1,500
DC 2 in B	700
B – C	−1,600
Net cost reduction	600

	Cost reduction
DC 1 in A	1,000
DC 2 in B	700
A – B	−1,100
Net cost reduction	600

The management of the supermarket chain can choose either of the two optimal solutions, and take − if desired − the one that satisfies as much as possible any criterion that has not been included in the model.

It is left to the reader to formulate the general form of the decentralization problem for m DCs and n locations ($m, n \geq 1$).

7.4 Gomory's cutting-plane algorithm

This section is devoted to a description of the *cutting plane algorithm* for solving ILO-models and MILO-models. The algorithm was developed in 1963 by RALPH E. GOMORY (born 1929). In Section 7.4.1, we will illustrate the algorithm by means of a simple example, and then give a formal description of the cutting algorithm for ILO-models. In Section 7.4.2, we extend the algorithm to the setting of MILO-models.

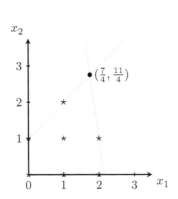

Figure 7.11: Feasible region consisting of integer points. The stars are the integer feasible points, the shaded area is the feasible region of the LO-relaxation, and the dot is the optimal solution of the LO-relaxation.

7.4.1 The cutting-plane algorithm

Consider the following ILO-model:

$$
\begin{aligned}
\max \quad & x_1 + x_2 \\
\text{s.t.} \quad & 7x_1 + x_2 \le 15 \\
& -x_1 + x_2 \le 1 \\
& x_1, x_2 \ge 0, \text{ and integer.}
\end{aligned} \tag{P}
$$

The feasible region of this model consists of the integer points depicted in Figure 7.11.

Gomory's algorithm starts by solving the LO-relaxation of the original ILO-model. Let x_3 and x_4 be the slack variables. The optimal simplex tableau corresponds to the following model. It is left to the reader to carry out the calculations leading to the final simplex tableau.

$$
\begin{aligned}
\max \quad & -\tfrac{1}{4}x_3 - \tfrac{3}{4}x_4 + 4\tfrac{1}{2} \\
\text{s.t.} \quad x_1 \quad & + \tfrac{1}{8}x_3 - \tfrac{1}{8}x_4 = 1\tfrac{3}{4} \\
& x_2 + \tfrac{1}{8}x_3 + \tfrac{7}{8}x_4 = 2\tfrac{3}{4} \\
& x_1, x_2, x_3, x_4 \ge 0.
\end{aligned} \tag{7.13}
$$

So an optimal solution of the LO-relaxation of model (P) is $x_1^0 = 1\tfrac{3}{4}$, $x_2^0 = 2\tfrac{3}{4}$, $x_3^0 = x_4^0 = 0$, and $z^0 = 4\tfrac{1}{2}$. Note that this solution does not correspond to an integer point of the feasible region.

Gomory's algorithm now derives a constraint that is satisfied by all feasible integer solutions, but that is not satisfied by the optimal solution of the LO-relaxation. In other words, the halfspace corresponding to this new constraint contains all feasible integer points, but not the optimal vertex of the LO-relaxation. So adding the constraint 'cuts off' the optimal solution of the LO-relaxation. This new constraint is therefore called a *cut constraint*. The cut constraint is added to the current LO-model (the LO-relaxation of the original ILO-model), and the new model is solved. The process is then repeated until an integer solution is found.

Such a cut constraint is calculated as follows:

Step 1. Choose one of the noninteger-valued variables, say x_1. Notice that x_1 is a basic variable (because its value is noninteger, and hence nonzero). We select the constraint in model (7.13) in which x_1 appears (there is a unique choice for this constraint because every basic variable appears in exactly one constraint):

$$x_1 + \tfrac{1}{8}x_3 - \tfrac{1}{8}x_4 = 1\tfrac{3}{4}. \tag{7.14}$$

The variables (other than x_1) that appear in this equation are all nonbasic. Note that this constraint is binding for every feasible solution, including the solution of the LO-relaxation, $x_1^0 = 1\tfrac{1}{4}$, $x_2^0 = 2\tfrac{3}{4}$, $x_3^0 = x_4^0 = 0$. We call this constraint the *source constraint*.

Step 2. Rewrite (7.14) by separating the integer and fractional parts of the coefficients of the nonbasic variables. That is, we write each coefficient, p, say, as $p = \lfloor p \rfloor + \varepsilon$ with $\varepsilon \in [0, 1)$. Doing this gives the following equivalent formulation of (7.14):

$$x_1 + (0 + \tfrac{1}{8})x_3 + (-1 + \tfrac{7}{8})x_4 = 1\tfrac{3}{4}.$$

Separating the terms of this expression into integer and fractional parts, we obtain:

$$x_1 + 0x_3 + (-1)x_4 = 1\tfrac{3}{4} - \tfrac{1}{8}x_3 - \tfrac{7}{8}x_4. \tag{7.15}$$

Because this constraint is equivalent to (7.14), it holds for *all* feasible points of the LO-relaxation. So, up to this point, we have merely derived a constraint that is satisfied by all feasible points of the LO-relaxation. But now we are going to use the fact that the variables may only take on integer values. First notice that, because x_3 and x_4 are nonnegative variables, the right hand side of (7.15) is at most $1\tfrac{3}{4}$. That is,

$$1\tfrac{3}{4} - \tfrac{1}{8}x_3 - \tfrac{7}{8}x_4 \leq 1\tfrac{3}{4}. \tag{7.16}$$

On the other hand, because x_1, x_3, and x_4 are only allowed to take on integer values, the left hand side of (7.15) has to be integer-valued, which implies that the right hand side has to be integer-valued as well. Because there are no integers in the interval $(1, 1\tfrac{3}{4}]$, this means that, in fact, the right hand side has to be at most 1. This gives the first cut constraint:

$$1\tfrac{3}{4} - \tfrac{1}{8}x_3 - \tfrac{7}{8}x_4 \leq 1. \tag{7.17}$$

This constraint is satisfied by every integer feasible point, but not by the optimal solution of the LO-relaxation (observe that $1\tfrac{3}{4} - \tfrac{1}{8}x_3^0 = 1\tfrac{3}{4} \not\leq 1$).

Geometrically, using the equations $x_3 = 15 - 7x_1 - x_2$ and $x_4 = 1 + x_1 - x_2$, we find that cut constraint (7.17) is equivalent to $x_2 \leq 2$. Similarly, (7.15) is equivalent to $x_2 \leq 2\tfrac{3}{4}$. We have drawn these two constraints in Figure 7.12(a). Note that the cut constraint 'cuts off' a triangular area from the feasible region which includes the solution

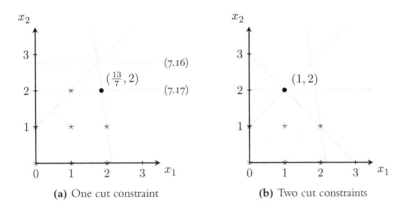

Figure 7.12: Adding cut constraints. The points marked by a dot represent the optimal solutions of the corresponding LO-relaxations.

$\begin{bmatrix} 1\frac{3}{4} & 2\frac{3}{4} \end{bmatrix}^{\mathsf{T}}$. However, since all integer feasible points satisfy the cut constraint $x_2 \le 2$, no integer points are cut off.

The procedure to derive a cut constraint from its source constraint is described formally in the next section. We will first finish the example.

Step 3. Introducing the slack variable x_5 for this cut constraint, the new LO-model becomes:

$$\begin{array}{llll} \max & -\frac{1}{4}x_3 - \frac{3}{4}x_4 & +\, 4\frac{1}{2} \\ \text{s.t.} \quad x_1 & +\frac{1}{8}x_3 - \frac{1}{8}x_4 & = 1\frac{3}{4} \\ & x_2 + \frac{1}{8}x_3 + \frac{7}{8}x_4 & = 2\frac{3}{4} \\ & -\frac{1}{8}x_3 - \frac{7}{8}x_4 + x_5 & = -\frac{3}{4} \\ & x_1, x_2, x_3, x_4, x_5 \ge 0. \end{array}$$

The basic solution with $BI = \{1, 2, 5\}$, i.e., $x_1 = 1\frac{3}{4}$, $x_2 = 2\frac{3}{4}$, $x_5 = -\frac{3}{4}$, is not feasible. On the other hand, all objective coefficients are nonpositive, and so we may apply the simplex algorithm to the dual model (see Section 4.6). The solution is $x_1 = 1\frac{6}{7}$, $x_2 = 2$, $x_4 = \frac{6}{7}$, $x_3 = x_5 = 0$, and the objective value is $z = 3\frac{6}{7}$. The optimal simplex tableau corresponds to the model:

$$\begin{array}{llll} \max & -\frac{1}{7}x_3 & -\frac{6}{7}x_4 & +\, 3\frac{3}{7} \\ \text{s.t.} \quad x_1 & +\frac{1}{7}x_3 & -\frac{1}{7}x_5 = 1\frac{6}{7} \\ & x_2 & +\ x_5 = 2 \\ & \frac{1}{7}x_3 & +\ x_4 - 1\frac{1}{7}x_5 = \frac{6}{7} \\ & x_1, x_2, x_3, x_4, x_5 \ge 0. \end{array}$$

The optimal solution of this model is still noninteger. The source constraint will now be $x_1 + \frac{1}{7}x_3 - \frac{1}{7}x_5 = 1\frac{6}{7}$. The corresponding cut constraint in terms of the nonbasic variables x_3 and x_5 is $-\frac{1}{7}x_3 - \frac{6}{7}x_5 \le -\frac{6}{7}$. One can easily check that this cut constraint is equivalent to $x_1 + x_2 \leq 3$; see Figure 7.12(b). After introducing the slack variable x_6, the equation $-\frac{1}{7}x_3 - \frac{6}{7}x_5 + x_6 = -\frac{6}{7}$ is added to the model.

Step 4. Applying again the simplex algorithm to the dual model, one can easily check that the optimal solution of the new model (i.e., the 'optimal model' of Step 2 augmented with the second cut constraint) has the optimal solution $x_1 = 1$, $x_2 = 2$, $x_3 = 6$, $x_4 = x_5 = x_6 = 0$. This solution is integer-valued, and so $x_1 = 1, x_2 = 2$ is in fact an optimal solution of the original model.

We now give a formal description of Gomory's cutting plane algorithm. Consider the standard form ILO-model:

$$\max\{\mathbf{c}^\mathsf{T}\mathbf{x} \mid \mathbf{A}\mathbf{x} = \mathbf{b}, \mathbf{x} \geq \mathbf{0}, \text{ and } \mathbf{x} \text{ integer}\}.$$

Let BI be the index set of the (current) basic variables and NI the index set of (current) non-basic variables (see Section 2.1.1) corresponding to an optimal solution of the LO-relaxation. Let x_k be a variable that has a noninteger value at the current optimal feasible basic solution. Consider the unique constraint in the optimal simplex tableau in which x_k appears:

$$x_k = b_k - \sum_{j \in NI} a_{kj} x_j, \tag{7.18}$$

where a_{kj} is the (k, j)'th entry of $\mathbf{A}$. This is the (current) *source constraint*. Notice that b_k is necessarily noninteger because $x_k = b_k$ (recall that, for all $j \in NI$, the nonbasic variable x_j has value zero). Define:

$$a_{kj} = \lfloor a_{kj} \rfloor + \varepsilon_{kj} \qquad \text{for } j \in NI,$$

where $\lfloor p \rfloor$ is the largest integer such that $\lfloor p \rfloor \leq p$. Clearly, $0 \leq \varepsilon_{kj} < 1$. By using (7.4.1) and rearranging, the source constraint (7.18) can equivalently be written as:

$$x_k + \sum_{j \in NI} \lfloor a_{kj} \rfloor x_j = b_k - \sum_{j \in NI} \varepsilon_{kj} x_j. \tag{7.19}$$

Since $x_j \geq 0$ and $\varepsilon_{kj} \geq 0$ for all $k \in BI$ and $j \in NI$, it follows that every feasible solution satisfies $\sum_{j \in NI} \varepsilon_{kj} x_j \geq 0$, and so every feasible solution satisfies:

$$b_k - \sum_{j \in NI} \varepsilon_{kj} x_j \leq b_k.$$

On the other hand, because x_k, x_j, and $\lfloor a_{kj} \rfloor$ are all integers, the left hand side of (7.19) has to be integer-valued. Therefore, $b_k - \sum_{j \in NI} \varepsilon_{kj} x_j$ is integer-valued as well. Hence,

$$b_k - \sum_{j \in NI} \varepsilon_{kj} x_j \leq \lfloor b_k \rfloor,$$

which is the so-called *cut constraint* that is added to the model.

In conclusion, the cutting plane algorithm consists of two phases:

Phase 1. The LO-phase. The current LO-relaxation is solved. If the optimal solution $\mathbf{x}^0$ is integer-valued, then $\mathbf{x}^0$ is an optimal solution of the original ILO-model.

Phase 2. The cutting plane phase. A cut constraint is derived that does not satisfy $\mathbf{x}^0$, but satisfies the original feasible region. This cut constraint is added to the current model and solved by means of the simplex algorithm applied to the dual model (see Section 4.6). If the resulting optimal solution is integer-valued, then the algorithm stops. Otherwise, Phase 2 is repeated. If, at some iteration, the simplex algorithm indicates that no feasible solution exists, then the original ILO-model has no solution.

In practice, solving ILO-models using the cutting plane algorithm alone needs a lot of computer time and memory. Also, it is not immediately clear from the above that the algorithm will ever terminate. It turns out that the algorithm can be adapted in such a way that it will always terminate and find an optimal solution (provided that it exists). This analysis lies beyond the scope of this book; the interested reader is referred to, e.g., Wolsey (1998).

After a number of iterations one can stop the algorithm (for instance, if not much is gained anymore in the objective value by adding new cut constraints), and use the current objective value as an upper bound. However, unlike in the case of the branch-and-bound algorithm, intermediate stopping does not yield a feasible integer solution, and so intermediate stopping can mean having invested a lot of computer time with no useful result. To overcome this situation, algorithms have been developed that produce suboptimal solutions that are both integer and feasible in the original constraints.

The cutting plane algorithm can also be used in combination with the branch-and-bound algorithm. In particular, cut constraints can be added during the iterations of the branch-and-bound algorithm to yield better upper bounds on the optimal objective value than the branch-and-bound algorithm alone can provide.

7.4.2 The cutting-plane algorithm for MILO-models

In Section 7.2.1 we already encountered a mixed integer linear optimization model (MILO-model), which is an LO-model where some of the variables are integers and others are not. We will discuss the mixed-integer version of Gomory's algorithm in this section.

Let x_k be an integer variable in a MILO-model. Consider the LO-relaxation of the MILO-model and an optimal feasible basic solution to it. Suppose that x_k is not integer-valued at this optimal solution. Hence, x_k is nonzero and therefore a basic variable. As in Section 7.4.1, consider the unique constraint in the optimal simplex tableau in which x_k appears:

$$x_k = b_k - \sum_{j \in NI} a_{kj} x_j,$$

with NI the index set of nonbasic variables. Clearly, b_k is noninteger. We will write $b_k = \lfloor b_k \rfloor + \beta$, where $0 < \beta < 1$. Hence,

$$x_k - \lfloor b_k \rfloor = \beta - \sum_{j \in NI} a_j x_j. \qquad \text{(source constraint)}$$

In a MILO-model, some of the nonbasic variables may not be restricted to integer values, and so we cannot simply apply the cutting plane algorithm from Section 7.4.1. However, a new cut constraint can be obtained as follows.

Because x_k is an integer variable, every integer feasible point satisfies either $x_k \leq \lfloor b_k \rfloor$, or $x_k \geq \lfloor b_k \rfloor + 1$. Using the source constraint, this condition is equivalent to:

$$\text{either } \sum_{j \in NI} a_{kj} x_j \geq \beta, \text{ or } \sum_{j \in NI} a_{kj} x_j \leq \beta - 1. \qquad (7.20)$$

In its current form, this constraint states that at least one out of the two constraints should be satisfied, but not necessarily both. Such a constraint is called a *disjunctive constraint* (see also Table 7.4). In particular, it is not linear, and, therefore, it cannot be used as a cut constraint. In what follows, we will use the disjunctive constraint (7.20) to construct a linear constraint that can be used as a cut constraint.

We first divide both sides of the first inequality in (7.20) by β, and both sides of the second one by $1 - \beta$. This gives a disjunctive constraint that is equivalent to (7.20) (recall that both $\beta > 0$ and $1 - \beta > 0$):

$$\text{either } \sum_{j \in NI} \frac{a_{kj}}{\beta} x_j \geq 1, \text{ or } -\sum_{j \in NI} \frac{a_{kj}}{1 - \beta} x_j \geq 1. \qquad (7.21)$$

Since the variables x_j $(j \in NI)$ have nonnegative values, (7.21) implies that:

$$\sum_{j \in NI} \max\left\{ \frac{a_{kj}}{\beta}, -\frac{a_{kj}}{1 - \beta} \right\} x_j \geq \max\left\{ \sum_{j \in NI} \left(\frac{a_{kj}}{\beta} \right) x_j, \sum_{j \in NI} \left(-\frac{a_{kj}}{1 - \beta} \right) x_j \right\} \geq 1,$$

where the last inequality follows from the fact that at least one of the sums $\sum_{j \in NI} \frac{a_{kj}}{\beta} x_j$ and $-\sum_{j \in NI} \frac{a_{kj}}{1-\beta} x_j$ is greater than or equal to 1. Therefore, the disjunctive constraint (7.21) implies the linear constraint:

$$\sum_{j \in NI} \max\left\{ \frac{a_{kj}}{\beta}, -\frac{a_{kj}}{1 - \beta} \right\} x_j \geq 1. \qquad (7.22)$$

Determining the maximum of a_{kj}/β and $-a_{kj}/(1 - \beta)$ is straightforward, because for each $j \in NI$, the values of these expressions have opposite signs, which depend only on the sign of a_{kj}. To be precise, define:

$$J^+ = \{ j \in NI \mid a_{kj} \geq 0 \} \text{ and } J^- = \{ j \in NI \mid a_{kj} < 0 \}.$$

Note that $NI = J^+ \cup J^-$. Then we have that:

$$\max\left\{\frac{a_{kj}}{\beta}, -\frac{a_{kj}}{1-\beta}\right\} = \begin{cases} \dfrac{a_{kj}}{\beta} & \text{if } j \in J^+ \\[2mm] -\dfrac{a_{kj}}{1-\beta} & \text{if } j \in J^-. \end{cases}$$

Therefore, (7.22) reduces to:

$$\sum_{j \in J^+} \frac{a_{kj}}{\beta} x_j - \sum_{j \in J^-} \frac{a_{kj}}{1-\beta} x_j \geq 1.$$

This is called a *mixed cut constraint*. It represents a necessary condition for x_k to be integer-valued. Since $x_j = 0$ for each $j \in NI$, it follows that the slack variable of this mixed cut constraint has a negative value (namely, -1) in the current optimal primal solution. So, the dual simplex algorithm (see Section 4.7) can be applied to efficiently reoptimize the model with the added cut constraint.

The following example illustrates the procedure.

Example 7.4.1. *Suppose that x_1 is the only integer variable in the example of Section 7.4. The constraint corresponding to the basic variable x_1 is:*

$$x_1 + \tfrac{1}{8}x_3 - \tfrac{1}{8}x_4 = 1\tfrac{3}{4}.$$

Then, $J^+ = \{3\}$, $J^- = \{4\}$, $b_1 = 1\tfrac{3}{4}$, and $\beta = \tfrac{3}{4}$. Hence, the mixed cut constraint is:

$$\frac{\tfrac{1}{8}}{\tfrac{3}{4}}x_3 - \frac{-\tfrac{1}{8}}{1-\tfrac{3}{4}}x_4 \geq 1,$$

which is equivalent to

$$\tfrac{1}{6}x_3 + \tfrac{1}{2}x_4 \geq 1.$$

Since $x_3 = 15 - 7x_1 - x_2$, and $x_4 = 1 + x_1 - x_2$, it follows that the mixed cut constraint in terms of the decision variables reads $x_1 + x_2 \leq 3$.

7.5 Exercises

Exercise 7.5.1. A company wants to produce two commodities, X and Y. Commodity X will be produced on type A machines, each with a capacity of six units per hour. Commodity Y will be produced on type B machines, each with a capacity of four units per hour. The costs for the production of X and Y are partly fixed (machine costs), namely $30 per hour for a type A machine and $40 per hour for a type B machine. The variable (raw material) costs are $40 per unit of X and also $40 per unit of Y. The returns are $37 per unit for commodity X and $55 per unit for commodity Y. The maximum number of machines that can be deployed in the factory is ten. The company wants to maximize the production such that the costs do not exceed the returns.

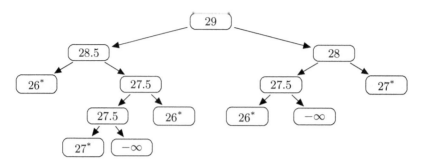

Figure 7.13: Branch-and-bound tree of Exercise 7.5.4.

(a)   Write an ILO-model for this problem.

(b)   Determine an optimal solution of the LO-relaxation of the ILO-model.

(c)   Solve the MILO-model where the production quantities need not be integer-valued.

(d)   Show that rounding down the optimal solution of the MILO-model leads to an optimal solution of the ILO-model.

Exercise 7.5.2.   Consider the following model:

$$
\begin{aligned}
\min \quad & x_1 \\
\text{s.t.} \quad & 15x_1 - 20x_2 \le 4 \\
& 21x_1 - 20x_2 \ge 10 \\
& x_1, x_2 \ge 0, \text{ and integer.}
\end{aligned}
$$

Show that rounding off the optimal solution of the LO-relaxation of this model to a closest integer solution does not lead to an optimal integer solution.

Exercise 7.5.3.   Draw the branch-and-bound tree that corresponds to a branch-and-bound solution of the ILO-model:

$$
\begin{aligned}
\max \quad & 2x_1 + x_2 \\
\text{s.t.} \quad & x_1 + 2x_2 \le 8 \\
& x_1 - x_2 \le \tfrac{1}{2} \\
& x_1, x_2 \ge 0, \text{ and integer.}
\end{aligned}
$$

Exercise 7.5.4.   Consider the branch-and-bound tree in Figure 7.13. The numbers in the nodes are the various values of the objective function. The optimal objective values marked with a star indicate that the corresponding solution is integer-valued.

(a)   Show that further branching is not needed and that therefore the branch-and-bound algorithm stops.

(b)   Which of the two optimal solutions will be found first in the case of 'backtracking'?

(c) Which optimal solution is found in the case of 'jumptracking'?

(d) Does one of the two methods, 'backtracking' or 'jumptracking', need less branches?

Exercise 7.5.5. Consider the following model:

$$
\begin{array}{rl}
\max & 3x_1 + 3x_2 + 13x_3 \\
\text{s.t.} & -3x_1 + 6x_2 + 7x_3 \le 8 \\
& 6x_1 - 3x_2 + 7x_3 \le 8 \\
& x_1, x_2, x_3 \ge 0, \text{ and integer.}
\end{array}
$$

(a) Solve the problem using the jumptracking version of the branch-and-bound algorithm (in case of ties, always branch on the variable with the lowest index that is not an integer).

(b) When the backtracking version is applied to this problem, there is freedom in first choosing either the subproblem with $x_i \le \star$, or the subproblem with $x_i \ge \star + 1$. Analyze the difference between these two versions of backtracking. Compare the results with the solution found under (a).

Exercise 7.5.6. Verify the complementary slackness relations for the pair of primal and dual solutions given in the proof of Theorem 7.2.1.

Exercise 7.5.7. In this exercise, ILO-models have to be solved by using the branch-and-bound algorithm. Draw the branch-and-bound trees and use a computer package to solve the subproblems. Compare your solutions by directly using an ILO-package.

(a)
$$
\begin{array}{rl}
\max & 23x_1 + 14x_2 + 17x_3 \\
\text{s.t.} & -x_1 + x_2 + 2x_3 \le 7 \\
& 2x_1 + 2x_2 - x_3 \le 9 \\
& 3x_1 + 2x_2 + 2x_3 \le 13 \\
& x_1, x_2, x_3 \ge 0, \text{ and integer.}
\end{array}
$$

(c)
$$
\begin{array}{rl}
\max & 19x_1 + 27x_2 + 23x_3 \\
\text{s.t.} & 4x_1 - 3x_2 + x_3 \le 7 \\
& 2x_1 + x_2 - 2x_3 \le 11 \\
& -3x_1 + 4x_2 + 2x_3 \le 5 \\
& x_1, x_2, x_3 \ge 0, \text{ and integer.}
\end{array}
$$

(d)
$$
\begin{array}{rl}
\max & 3x_1 + 3x_2 + 13x_3 \\
\text{s.t.} & -3x_1 + 6x_2 + 7x_3 \le 8 \\
& 6x_1 - 3x_2 + 7x_3 \le 8 \\
& 0 \le x_1, x_2, x_3 \le 5, \text{ and integer.}
\end{array}
$$

(b)
$$
\begin{array}{rl}
\max & 26x_1 + 17x_2 + 13x_3 \\
\text{s.t.} & 3x_1 + x_2 + 4x_3 \le 9 \\
& x_1 + 2x_2 - 3x_3 \le 6 \\
& 4x_1 - x_2 + x_3 \le 17 \\
& x_1, x_2, x_3 \ge 0, \text{ and integer.}
\end{array}
$$

(e)
$$
\begin{array}{rl}
\max & 34x_1 + 29x_2 + 2x_3 \\
\text{s.t.} & 7x_1 + 5x_2 - x_3 \le 16 \\
& -x_1 + 3x_2 + x_3 \le 10 \\
& -x_2 + 2x_3 \le 3 \\
& x_1, x_2, x_3 \ge 0, \text{ and integer.}
\end{array}
$$

(f)   max $135x_1 + 31x_2 + 142x_3$
    s.t. $\quad x_1 + 2x_2 - x_3 \le 13$
    $\qquad 2x_1 - x_2 + 3x_3 \le 5$
    $\qquad -x_1 + 2x_3 \le 6$
    $\qquad x_1, x_2, x_3 \ge 0,$ and integer.

(g)   max $39x_1 + 40x_2 + 22x_3$
    s.t. $\quad 4x_1 + 5x_2 + x_3 \le 16$
    $\qquad 3x_1 + 2x_2 + 4x_3 \le 9$
    $\qquad 2x_1 + 3x_2 - 2x_3 \le 4$
    $\qquad x_1, x_2, x_3 \ge 0,$ and integer.

Exercise 7.5.8.   Use the branch-and-bound algorithm to solve the following MILO-models.

(a)   max $3x_1 + x_2$
    s.t. $\quad 5x_1 + 2x_2 \le 10$
    $\qquad 4x_1 + x_2 \le 7$
    $\qquad x_1, x_2 \ge 0, x_2$ integer.

(b)   min $3x_1 + x_2$
    s.t. $\quad x_1 + 5x_2 \ge 8$
    $\qquad x_1 + 2x_2 \ge 4$
    $\qquad x_1, x_2 \ge 0, x_1$ integer.

Exercise 7.5.9.   A vessel has to be loaded with batches of N items ($N \ge 1$). Each unit of item i has a weight w_i and a value v_i ($i = 1, \ldots, N$). The maximum cargo weight is W. It is required to determine the most valuable cargo load without exceeding the maximum weight W.

(a)   Formulate this problem as a knapsack problem and determine a corresponding ILO-model.

(b)   Determine, by inspection, the optimal solution for the case where $N = 3$, $W = 5$, $w_1 = 2$, $w_2 = 3$, $w_3 = 1$, $v_1 = 65$, $v_2 = 80$, and $v_3 = 30$.

(c)   Carry out sensitivity analysis for the value of v_3.

Exercise 7.5.10.   A parcel delivery company wants to maximize its daily total revenue. For the delivery, the company has one car with a volume of eleven (volume units). On the present day, the following packages have to be delivered: package 1 with volume two, package 2 with volume three, package 3 with volume four, package 4 with volume six, and package 5 with volume eight. The revenues of the packages are $10, $14, $31, $48, and $60, respectively.

(a)   Formulate this problem in terms of an ILO-model. What type of problem is this? Solve it by means of the branch-and-bound algorithm, and draw the branch-and-bound tree.

(b)   Formulate the dual model of the LO-relaxation of the ILO-model, together with the corresponding complementary slackness relations. Use these relations and a result obtained in (a) to determine an optimal solution of this dual model.

(c)   Determine the optimal integer solution of the dual of (b), and show that the complementary slackness relations do not hold for the primal and dual integer optimal solutions.

Exercise 7.5.11. Let P, Q, and R be simple propositions. Prove the following equivalencies:

(a) $P \Rightarrow Q \equiv \neg P \vee Q$

(b) $\neg(P \vee Q) \equiv \neg P \wedge \neg Q$ *(De Morgan's first law[1])*

(c) $\neg(P \wedge Q) \equiv \neg P \vee \neg Q$ *(De Morgan's second law)*

(d) $P \Rightarrow Q \wedge R \equiv (P \Rightarrow Q) \wedge (P \Rightarrow R)$

(e) $(P \wedge Q) \Rightarrow R \equiv (P \Rightarrow R) \vee (Q \Rightarrow R)$

(f) $(P \vee Q) \Rightarrow R \equiv (P \Rightarrow R) \wedge (Q \Rightarrow R)$

(g) $P \wedge (Q \vee R) \equiv (P \wedge Q) \vee (P \wedge R)$ *(first distributive law)*

(h) $P \vee (Q \wedge R) \equiv (P \vee Q) \wedge (P \vee R)$ *(second distributive law)*

Exercise 7.5.12. Let $1 \leq k < n$, and let P, Q, R, and $P_1, \ldots, P_n$ be simple propositions. Write the following expressions as linear constraints.

(a) $\neg(P \vee Q)$.

(b) $\neg(P \wedge Q)$.

(c) $P \Rightarrow \neg Q$.

(d) $P \Rightarrow (Q \wedge R)$.

(e) $P \Rightarrow (Q \vee R)$.

(f) $(P \wedge Q) \Rightarrow R$.

(g) $(P \vee Q) \Rightarrow R$.

(h) $P \wedge (Q \vee R)$.

(i) $P \vee (Q \wedge R)$.

(j) At most k of $P_1, \ldots, P_n$ are true.

(k) $P_n \Leftrightarrow (P_1 \vee \ldots \vee P_k)$.

(l) $P_n \Leftrightarrow (P_1 \wedge \ldots \wedge P_k)$.

Exercise 7.5.13. The functions f and g are defined by: $f(x) = 10 + 2x$ if $0 \leq x \leq 5$, $f(x) = 15 + x$ if $x \geq 5$, and $g(y) = 8 + y$ if $0 \leq y \leq 2$, $g(y) = 4 + 3y$ if $y \geq 2$. Formulate the following maximization model as a MILO-model.

$$\max \ f(x) + g(y)$$
$$\text{s.t.} \quad x, y \geq 0$$

At least two of the following three constraints hold:
$$x + y \leq 9$$
$$x + 3y \leq 12$$
$$2x + y \leq 16.$$

Exercise 7.5.14. In this exercise, we consider models that include arbitrary piecewise linear functions and investigate two methods for rewriting such models as MILO-models. Let $a, b \in \mathbb{R}$ with $a < b$. Let $g \colon [a, b] \to \mathbb{R}$ be a (continuous) piecewise linear function defined on the interval $[a, b]$, with $n \ (\geq 1)$ kink points $\alpha_1, \ldots, \alpha_n$. Write $\alpha_0 = a$ and $\alpha_{n+1} = b$. Consider a (nonlinear) optimization model that includes $g(x)$. First, introduce

---

[1]Named after the British mathematician and logician AUGUSTUS DE MORGAN (1806–1871).

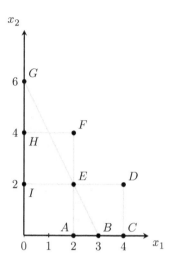

Figure 7.14: Nonconvex region of Exercise 7.5.15.

$n + 2$ variables $\lambda_0, \ldots, \lambda_{n+1}$, satisfying $\sum_{i=0}^{n+1} \lambda_i = 1$ and $\lambda_i \geq 0$ for $i = 0, \ldots, n + 1$, and introduce $n + 1$ binary variables $\delta_1, \ldots, \delta_{n+1}$, satisfying $\sum_{i=1}^{n+1} \delta_i = 1$.

(a) Show that the original model may be solved by adding the above variables, replacing every occurrence of $g(x)$ by $\sum_{i=0}^{n+1} \lambda_i g(\alpha_i)$, every occurrence of x by $\sum_{i=0}^{n+1} \lambda_i \alpha_i$, and adding the following set of linear constraints:

$$\begin{aligned} \lambda_0 &\leq \delta_1 \\ \lambda_i &\leq \delta_i + \delta_{i+1} \qquad \text{for } i = 1, \ldots, n \\ \lambda_{n+1} &\leq \delta_{n+1}. \end{aligned}$$

(b) Show that, instead of the set of linear constraints under (a), the following smaller set of constraints may be used:

$$\lambda_{i-1} + \lambda_i \geq \delta_i \qquad \text{for } i = 1, \ldots, n + 1.$$

(c) Derive the $n + 1$ constraints in part (b) from the $n + 2$ constraints in part (a).

Exercise 7.5.15.   Describe the nonconvex shaded region of Figure 7.14 in terms of linear constraints.

Exercise 7.5.16.   Which of the following statements are true, and which are not? Explain the answers.

(a) If an ILO-model has an optimal solution, then its LO-relaxation has an optimal solution.

(b) If the LO-relaxation of an ILO-model has an optimal solution, then the ILO-model itself has an optimal solution.

(c) If the feasible region of the LO-relaxation is unbounded, then the feasible region of the corresponding ILO-model is unbounded. *Remark:* Recall that the feasible region of an ILO-model is a set of integer points. The feasible region of an ILO-model is bounded (unbounded) if the convex hull (see Appendix D) of the feasible region (i.e., the convex hull of the integer points) is bounded (unbounded).

(d) If an ILO-model has multiple optimal solutions, then the corresponding LO-relaxation has multiple optimal solutions.

Exercise 7.5.17. Consider the following model:

$$\max 2x_1 + 4x_2 + 3\delta$$
$$\text{s.t.} \quad x_1 + 2x_2 \leq 4$$
$$[\delta = 1 \Leftrightarrow x_1 - x_2 \geq -1]$$
$$x_1, x_2 \geq 0, \text{and } \delta \in \{0, 1\}.$$

Write this model as an MILO-model.

Exercise 7.5.18. Model the following expressions as constraints of a mixed ILO-model.

(a) $3\delta_1(1 - \delta_2) \leq 2$, $\delta_1, \delta_2 \in \{0, 1\}$.

(b) For each $i, j, k \in \{1, \ldots, n\}$ (with $n \geq 3$), let $\delta_i, \delta_j, \delta_k \in \{0, 1\}$, and

$$x_{ijk} = \begin{cases} 1 & \text{if } \delta_i = 0, \delta_j = 1, \text{and } \delta_k = 1, \\ 0 & \text{otherwise.} \end{cases}$$

Exercise 7.5.19. The machine scheduling problem of Section 7.2.4 can be interpreted as a traveling salesman problem (TSP).

(a) What is the meaning of the variables, the objective function and the constraints, when considering this machine scheduling problem as a TSP?

(b) Show that the subtour-elimination constraints exclude subtours (including loops; see Appendix C) from optimal solutions.

(c) Show that every feasible solution of Model 7.2.2 (which includes the subtour-elimination constraints) corresponds to a traveling salesman tour, and vice versa.

Exercise 7.5.20.

(a) Explain why the constraints in (7.3) exclude all subtours of length 2.

(b) Explain why constraint (7.4) eliminates the subtour $1 \to 2 \to 3 \to 4 \to 1$. Does it also eliminate the subtour $1 \to 3 \to 2 \to 4 \to 1$?

(c) Explain why constraint (7.5) eliminates any subtour that visits exactly the cities in X.

Exercise 7.5.21. The company Whattodo has $100,000 available for investments in six projects. If the company decides to invest in a project at all, it has to invest a multiple of a given minimum amount of money (in $). These minimum amounts along with the (expected) rate of return for each project are given in Table 7.8. For example, if the company decides to invest in project 1, it may invest $10,000, $20,000, $30,000, ..., or $100,000 in the project. The *rate of return* is the amount of money gained as a percentage of the investment. For example, if the company decides to invest $20,000 in project 1, then it receives $20,000 × 1.13 = $26,600 when the project finishes.

Project	1	2	3	4	5	6
Minimum investment (in $)	10,000	8,000	20,000	31,000	26,000	32,000
Rate of return	13%	12%	15%	18%	16%	18%

Table 7.8: Data for company Whattodo.

(a) Formulate an ILO-model that can be used to find an optimal investment strategy. Determine such an optimal strategy.

(b) This problem can be seen as a special case of a problem described in this chapter. Which problem is that?

Type			Weekly amount of pigment						Price (€ per kg)
1	3,275	80	1,988	4,152	1,332	2,150	1,395	869	2.03
2	1,154	1,091	2,855	706	556	2,768	2,735	1,459	1.11
3	50	623	1,388	375	313	756	313	0	1.88
4	4,265	1,669	1,864	1,785	2,027	3,065	1,625	0	2.52
5	1,401	275	275	1,650	1,650	1,948	1,113	1,938	2.16
6	550	0	0	0	1,100	550	0	1,100	2.54

Table 7.9: Weekly use of different types of pigment and corresponding prices per kg.

Exercise 7.5.22. A paint factory in Rotterdam needs pigments for its products and places an order with a supplier in Germany. The delivery time is four weeks. The transportation of the pigments is done by trucks with a capacity of 15,000 kgs. The company always orders a quantity of 15,000 kgs or multiples of it. The trucks can transport several types of pigments. The question is when and how much should be ordered, so that the inventory costs are as low as possible and the amount of pigment in stock is always sufficient for the daily production of paint. For a planning period of a given number of weeks, the production scheme of paint is fixed, and the amount of pigment necessary for the production of paint is known for each week of the planning period. In Table 7.9, the weekly use of the different types of pigment during a planning period of eight weeks is given, together with the prices per kg. We assume that there are six different types of pigment. In order to formulate this problem as a mathematical model, the pigments are labeled from 1 through n, and the weekly periods from 1 through T. Hence, $n = 6$ and $T = 8$. For $i = 1, \ldots, n$ and $t = 1, \ldots, T$, define the following variables:

d_{it} = the amount of pigment i used during week t,
s_{it} = the inventory of pigment i at the end of week t,
x_{it} = the amount of pigment i delivered at the beginning of week t.

Furthermore, we assume that the orders will be placed at the beginning of the week and that the pigments are delivered four weeks later at the beginning of the week. The requirement that enough pigment is delivered to meet the weekly demands can be expressed as follows:

$$s_{i0} + \sum_{k=1}^{t} x_{ik} \geq \sum_{k=1}^{t} d_{ik} \qquad \text{for } i = 1, \ldots, n, \text{ and } t = 1, \ldots, T,$$

where s_{i0} is the inventory of pigment i at the beginning of the planning period. It is assumed that the initial inventory is zero for all pigments. All trucks have the same capacity w (i.e., $w = 15{,}000$). For $t = 1, \ldots, T$, we define:

p_t = the number of trucks that deliver pigment at the beginning of week t.

In order to use the full capacity of the trucks, the following equation must hold:

$$\sum_{i=1}^{n} x_{it} = p_t w \qquad \text{for } t = 1, \ldots, T.$$

The inventory costs depend on the amount of inventory and on the length of the time period that the pigment is in stock. We assume that the weekly cost of holding one kilogram of pigment in the inventory is a fixed percentage, $100\beta\%$, say, of the price of the pigment. It is assumed that $\beta = 0.08$. The average inventory level of pigment i during week t is (approximately) $(s_{i,t-1} + s_{it})/2$. Thus, the average inventory cost for pigment i during week t is:

$$\beta c_i (s_{i,t-1} + s_{it})/2,$$

where c_i is the price per kilogram of pigment i; see Table 7.9. The inventory of pigment i at the end of week t is given by:

$$s_{it} = s_{i0} + \sum_{k=1}^{t} x_{ik} - \sum_{k=1}^{t} d_{ik} \qquad \text{for } i = 1, \ldots, n, \text{ and } t = 1, \ldots, T.$$

The total inventory costs during the planning period can be written as:

$$\beta \sum_{i=1}^{n} \sum_{t=1}^{T} c_i (s_{i,t-1} + s_{it})/2.$$

The costs of transportation are taken to be constant and hence do not play a part in determining the delivery strategy. Also the administration costs of placing orders are not taken into account. So the objective is to minimize the total inventory costs.

(a) Write the above problem as an ILO-model, and solve it.

The management of the company does not want to have too much inventory of pigment at the end of the planning period of eight weeks. This constraint can be built into the model by increasing the value of the objective function by 1,000 if the final inventory of pigment i is larger than 3,000 kgs. For example, if the final inventory of pigment 4 is larger than 3,000 kgs, then the value of the objective function is increased by 1,000.

(b) Solve this problem by adding the term $1000 \sum_i \delta_i$ to the objective function, where δ_i is equal to 1 if the final inventory s_{i8} of pigment i is larger than 3,000 kgs, and 0 otherwise.

(c) Consider the models formulated under (a) and (b). For which of these models is the value (in €) of the final inventory smallest? Is it possible to decrease the final inventory?

Exercise 7.5.23. A *lot sizing problem* is a production planning problem for which one has to determine a schedule of production runs when the product's demand varies over time. Let $d_1, \ldots, d_n$ be the demands over an n-period planning horizon. At the beginning of each period i, a setup cost a_i is incurred if a positive amount is produced, as well as a unit production cost c_i. Any amount in excess of the demand for period i is held in inventory until period $i+1$, incurring a per-unit cost of h_i; $i = 1, \ldots, n$ (period $n+1$ is understood as the first period of the next planning horizon).

(a) Formulate this problem as an MILO-model.

(b) Assuming that c_i is independent of i (i.e., $c_1 = \ldots = c_n$), show that any optimal solution has the property (optimality property) that the production in period i is the sum of the demands in some set of successive future periods $i, i + 1, \ldots, i + k$.

(c) Show that if c_i depends on i, then the optimality property does not necessarily hold.

Exercise 7.5.24. Consider the following ILO-model.

$$
\begin{aligned}
\max \quad & 77x_1 + 6x_2 + 3x_3 + 6x_4 + 33x_5 + 13x_6 + 110x_7 + 21x_8 + 47x_9 \\
\text{s.t.} \quad & 77x_1 + 76x_2 + 22x_3 + 42x_4 + 21x_5 + 760x_6 + 818x_7 + 62x_8 + 785x_9 \leq 1500 \\
& 67x_1 + 27x_2 + 794x_3 + 53x_4 + 234x_5 + 32x_6 + 792x_7 + 97x_8 + 435x_9 \leq 1500 \\
& x_1, \ldots, x_9 \in \{0, 1\}.
\end{aligned}
$$

(a) Solve the LO-relaxation of this model, and determine a feasible integer solution by rounding off the solution of the LO-relaxation.

(b) Show that $z^* = 176$, $x_1^* = x_3^* = x_6^* = x_9^* = 0$, $x_2^* = x_4^* = x_5^* = x_7^* = x_8^* = 1$ is an optimal solution of the model, by using the branch-and-bound algorithm. Draw the corresponding branch-and-bound tree.

(c) Draw the perturbation functions for the right hand side of the first constraint.

(d) Draw the perturbation function for the objective coefficient 77 of x_1.

Project	Employee						
	A	B	C	D	E	F	G
1	★	★					
2	★	★					
3		★	★				
4			★	★			
5			★	★			
6				★			
7		★			★		
8		★				★	★
9		★				★	★
10				★	★		
11			★				
12			★	★	★		
13	★		★				
14					★	★	

Table 7.10: Project data for Exercise 7.5.27.

Exercise 7.5.25. Consider the following MILO-model.

$$\max \quad -x_1 - 9x_2$$
$$\text{s.t.} \quad -x_1 + x_2 + x_3 = \tfrac{1}{2}$$
$$x_1, x_2 \geq 0, \, x_3 \in \{0, 1\}.$$

(a) Determine an optimal solution by means of the branch-and-bound algorithm, and draw the corresponding branch-and-bound tree. Is the optimal solution unique?

(b) Draw the perturbation function for the right hand side of the '=' constraint. Why is this a piecewise linear function? Explain the kink points.

Exercise 7.5.26. This question is about the quick and dirty (Q&D) method for the traveling salesman problem. (See Section 7.2.5.)

(a) Explain why the constraints in (7.3) exclude all subtours of length 2.

(b) Explain why constraint (7.4) eliminates the subtour 3→4→5→6→3.

(c) Show that the constraint:

$$\delta_{24} + \delta_{25} + \delta_{26} + \delta_{42} + \delta_{45} + \delta_{46} + \delta_{52} + \delta_{54} + \delta_5 + \delta_{62} + \delta_{64} + \delta_{65} \leq 3$$

eliminates the subtour 2→4→6→5→2 from the optimal solution.

Exercise 7.5.27. Mr. T is the supervisor of a number of projects. From time to time he reserves a full day to discuss all his projects with the employees involved. Some employees are involved in more than one project. Mr. T wants to schedule the project discussions in such a way, that the number of employees entering and leaving his office is as small as possible.

There are fourteen projects (labeled 1 to 14) and seven employees (labeled A to G). The stars in Table 7.10 show which employees are involved in which projects. This problem can be formulated as a traveling salesman problem in which each 'distance' c_{ij} denotes the number of employees entering and leaving Mr. T's office if project j is discussed immediately after project i.

(a) Formulate the above problem as a traveling salesman problem. (Hint: introduce a dummy project 0.)

(b) Use both the Q&D method from Section 7.2.5 and the branch-and-bound algorithm to solve the meeting scheduling problem.

Exercise 7.5.28.   Every year, a local youth association organizes a bridge drive for 28 pairs in 14 locations in downtown Groningen, the Netherlands. In each location, a table for two pairs is reserved. One session consists of three games played in one location. After three games, all pairs move to another location and meet a different opponent. At each table, one pair plays in the North-South direction (NS-pair) and the other pair in the East-West direction (EW-pair).

The locations are labeled 1 through 14. After each session, the NS-pairs move from location i to location $i+1$, and the EW-pairs from location i to location $i-1$. During the whole bridge drive, the NS-pairs always play in the NS direction and the EW-pairs in the EW direction. The problem is to determine an order in which the locations should be visited such that the total distance that each pair has to walk from location to location is minimized.

This problem can be formulated as a traveling salesman problem. The distances between the 14 locations, in units of 50 meters, are given in Table 7.11.

Use both the Q&D method and the branch-and-bound algorithm to find an optimal solution of this problem.

Exercise 7.5.29.   The organizer of a cocktail party has an inventory of the following drinks: 1.2 liters of whiskey, 1.5 liters of vodka, 1.6 liters of white vermouth, 1.8 liters of red vermouth, 0.6 liters of brandy, and 0.5 liters of coffee liqueur. He wants to make five different cocktails: Chauncy (2/3 whiskey, 1/3 red vermouth), Black Russian (3/4 vodka, 1/4 coffee liqueur), Sweet Italian (1/4 brandy, 1/2 red vermouth, 1/4 white vermouth), Molotov Cocktail (2/3 vodka, 1/3 white vermouth), and Whiskey-on-the-Rocks (1/1 whiskey). The numbers in parentheses are the amounts of liquor required to make the cocktails. All cocktails are made in glasses of 0.133 liters. The organizer wants to mix the different types of drinks such that the number of cocktails that can be served is as large as possible. He expects that Molotov Cocktails will be very popular, and so twice as many Molotov Cocktails as Chauncies have to be mixed.

(a) Formulate this problem as an ILO-model.

| | | | | | | | To | | | | | | |
From	1	2	3	4	5	6	7	8	9	10	11	12	13	14
1	–	4	10	10	11	10	9	7	8	11	12	12	12	11
2		–	6	6	7	6	5	6	7	9	11	11	11	13
3			–	1	1	2	2	4	11	10	12	14	15	19
4				–	1	2	2	5	12	11	13	15	16	20
5					–	1	2	5	12	11	14	16	16	20
6						–	1	4	11	10	13	15	15	19
7							–	4	11	10	12	14	15	19
8								–	7	6	9	11	11	15
9									–	2	4	4	4	8
10										–	2	4	5	9
11											–	2	3	6
12												–	1	4
13													–	4
14														–

Table 7.11: The distances between the 14 locations (in units of 50 meters) for Exercise 7.5.28.

The organizer himself likes to drink vodka. He decides to decrease the number of served cocktails by five, if at least 0.25 liters of vodka are left over.

(b) How would you use binary variables to add this decision to the model formulated under (a)? Solve the new model.

Exercise 7.5.30. Solve the following ILO-model by using Gomory's cutting plane algorithm.

$$\begin{aligned}
\max\ & 2x_1 + x_2 \\
\text{s.t.}\ & 3x_1 + x_2 \leq 21 \\
& 7x_1 + 4x_2 \leq 56 \\
& x_1 + x_2 \leq 12 \\
& x_1, x_2 \geq 0, \text{ and integer.}
\end{aligned}$$

Exercise 7.5.31. Consider the following knapsack problem:

$$\begin{aligned}
\max\ & 10x_1 + 24x_2 + 10x_3 + 2x_4 \\
\text{s.t.}\ & 2x_1 + 4x_2 + 3x_3 + x_4 \leq 23 \\
& x_1, x_2, x_3, x_4 \geq 0 \text{ and integer.}
\end{aligned}$$

(a) Solve this model by using Gomory's cutting plane algorithm.

(b) Transform the model into a $\{0,1\}$ knapsack problem (see Section 7.2.3). Use the branch-and-bound algorithm to solve this $\{0,1\}$ knapsack problem.

Chapter 8
Linear network models

Overview

A *network LO-model* is a model of which the technology matrix is based on a network (i.e., a graph). In this chapter we consider network models that can be formulated as ILO-models, and use methods from linear optimization to solve them. Many network models, such as the transportation problem, the assignment problem, and the minimum cost flow problem, can be formulated as an ILO-model with a technology matrix that has a special property, namely the so-called 'total unimodularity' property. It is shown in this chapter that, for these types of problems, any optimal solution produced by the simplex algorithm (see Chapter 3) when applied to the LO-relaxation is integer-valued. Moreover, we derive an important theorem in combinatorial optimization, namely the so-called max-flow min-cut theorem. We also show how project scheduling problems can be solved by employing dual models of network models.

Finally, we introduce the network simplex algorithm, which is a modification of the usual simplex algorithm that is specifically designed for network models, including the transshipment problem. The network simplex algorithm is in practice up to 300 times faster than the usual simplex algorithm. We will see that, like the usual simplex algorithm, the network simplex algorithm may cycle, and we will show how to remedy cycling behavior.

8.1 LO-models with integer solutions; total unimodularity

We have seen so far that, in general, LO-models have noninteger-valued solutions, in the sense that one or more entries of the optimal solution vector has a noninteger value (see, for example, Model Dovetail). If we insist on finding an integer-valued solution, then we should employ the more computationally intensive algorithms from integer optimization (see Chapter 7).

In this chapter, we will show that some LO-models have a special structure which guarantees that they have an integer-valued optimal solution. Besides the fact that the simplex algorithm applied to these problems generates integer-valued solutions, the usual sensitivity analysis (see Chapter 5) can also be applied for these problems.

8.1.1 Total unimodularity and integer vertices

An *integer vector* (*integer matrix*) is a vector (matrix) consisting of integers. We call a vertex an *integer vertex* if the vector corresponding to that vertex consists of integer-valued entries. In this section we will formulate a broad class of LO-models that have the special property that every vertex of the corresponding feasible region is an integer vertex. Since any optimal solution that is found by the simplex algorithm is a feasible basic solution (corresponding to a vertex), this means that the simplex algorithm, when applied to such an LO-model, always finds an integer-valued optimal solution (provided that the model is feasible or not unbounded).

To describe this class of models, we introduce the concepts of 'unimodularity' and 'total unimodularity'. A matrix $\mathbf{A}$ of rank r (≥ 1) is called *unimodular* if all entries of $\mathbf{A}$ are integers, and the determinant of every (r, r)-submatrix of $\mathbf{A}$ is either 0, $+1$, or -1. The following theorem shows that if, in the LO-model $\max\{\mathbf{c}^\mathsf{T}\mathbf{x} \mid \mathbf{A}\mathbf{x} = \mathbf{b}, \mathbf{x} \geq \mathbf{0}\}$, the matrix $\mathbf{A}$ is unimodular and the vector $\mathbf{b}$ is an integer vector, then every vertex of the feasible region is integer-valued.

> **Theorem 8.1.1.** (*Unimodularity and integer vertices*)
> Let $m, n \geq 1$ and let $\mathbf{A}$ be an integer (m, n)-matrix of full row rank m. The matrix $\mathbf{A}$ is unimodular if and only if, for every integer m-vector $\mathbf{b}$, every vertex of the feasible region $\{\mathbf{x} \in \mathbb{R}^n \mid \mathbf{A}\mathbf{x} = \mathbf{b}, \mathbf{x} \geq \mathbf{0}\}$ is integer-valued.

Proof of Theorem 8.1.1. Suppose that $\mathbf{A}$ is unimodular and $\mathbf{b}$ is an integer vector. Consider any vertex of the region $\{\mathbf{x} \in \mathbb{R}^n \mid \mathbf{A}\mathbf{x} = \mathbf{b}, \mathbf{x} \geq \mathbf{0}\}$. Since $\mathbf{A}$ has full row rank, this vertex corresponds to a feasible basic solution with $\mathbf{x}_{BI} = \mathbf{B}^{-1}\mathbf{b} \geq \mathbf{0}$ and $\mathbf{x}_{NI} = \mathbf{0}$ for some nonsingular basis matrix $\mathbf{B}$ in $\mathbf{A}$; see Section 3.8. We will show that $\mathbf{x}_{BI}$ is an integer vector. By Cramer's rule (see Appendix B), we have that, for each $i \in BI$:

$$x_i = \frac{\det(\mathbf{A}(i; \mathbf{b}))}{\det(\mathbf{B})},$$

where $\mathbf{A}(i; \mathbf{b})$ is the matrix constructed from $\mathbf{A}$ by replacing the i'th column by $\mathbf{b}$. Since $\mathbf{A}$ is unimodular, it follows that the nonsingular matrix $\mathbf{B}$ satisfies $\det(\mathbf{B}) = \pm 1$. Moreover, all entries of $\mathbf{A}(i; \mathbf{b})$ are integers, and hence $\det(\mathbf{A}(i; \mathbf{b}))$ is an integer. It follows that x_i is integer-valued for all $i \in BI$. So, $\mathbf{x}_{BI}$ is an integer vector.

To show the converse, suppose that, for every integer vector $\mathbf{b}$ ($\in \mathbb{R}^m$), all vertices of the feasible region $\{\mathbf{x} \in \mathbb{R}^n \mid \mathbf{A}\mathbf{x} = \mathbf{b}, \mathbf{x} \geq \mathbf{0}\}$ are integer vertices. Let $\mathbf{B}$ be any basis matrix in $\mathbf{A}$. To show that $\mathbf{A}$ is unimodular, we will show that $\det(\mathbf{B}) = \pm 1$. We claim that a sufficient condition for $\det(\mathbf{B}) = \pm 1$ is:

$$\mathbf{B}^{-1}\mathbf{t} \text{ is an integer vector for every integer vector } \mathbf{t} \ (\in \mathbb{R}^m). \tag{$*$}$$

The proof of "$(*) \implies \det(\mathbf{B}) = \pm 1$" is as follows. Suppose that $(*)$ holds. Choosing successively the unit vectors $\mathbf{e}_1, \ldots, \mathbf{e}_m$ for $\mathbf{t}$, it follows from $(*)$ that $\mathbf{B}^{-1}$ is an integer matrix. From $\mathbf{B}\mathbf{B}^{-1} = \mathbf{I}_n$, it follows that $\det(\mathbf{B}) \times \det(\mathbf{B}^{-1}) = 1$. Since $\mathbf{B}$ and $\mathbf{B}^{-1}$ are both integer matrices, both $\det(\mathbf{B})$ and $\det(\mathbf{B}^{-1})$ are integers, and so we have that in fact $\det(\mathbf{B}) = \pm 1$.

So it remains to show that $(*)$ holds. Take any integer vector $\mathbf{t} \in \mathbb{R}^m$. Then there obviously exists an integer vector $\mathbf{s}$ such that $\mathbf{B}^{-1}\mathbf{t} + \mathbf{s} \geq \mathbf{0}$. Define $\mathbf{b}' = \mathbf{t} + \mathbf{B}\mathbf{s}$, and define the feasible basic solution $\begin{bmatrix} \tilde{\mathbf{u}} \\ \tilde{\mathbf{u}}_s \end{bmatrix} \equiv \begin{bmatrix} \tilde{\mathbf{u}}_{BI} \\ \tilde{\mathbf{u}}_{NI} \end{bmatrix}$ of $\{\mathbf{x} \in \mathbb{R}^n \mid \mathbf{A}\mathbf{x} = \mathbf{b}', \mathbf{x} \geq \mathbf{0}\}$ by $\tilde{\mathbf{u}}_{BI} = \mathbf{B}^{-1}\mathbf{b}' = \mathbf{B}^{-1}\mathbf{t} + \mathbf{s}$, $\tilde{\mathbf{u}}_{NI} = \mathbf{0}$. Because $\mathbf{b}'$ is an integer vector, all vertices of $\{\mathbf{x} \in \mathbb{R}^n \mid \mathbf{A}\mathbf{x} = \mathbf{b}', \mathbf{x} \geq \mathbf{0}\}$ are integer vertices. Hence, it follows that $\begin{bmatrix} \tilde{\mathbf{u}}_{BI} \\ \tilde{\mathbf{u}}_{NI} \end{bmatrix}$ is an integer vector as well. Therefore, $\mathbf{B}^{-1}\mathbf{t} + \mathbf{s} = \tilde{\mathbf{u}}_{BI}$ is an integer vector, as required.

This proves $(*)$ and, hence, $\det(\mathbf{B}) = \pm 1$. Hence, $\mathbf{A}$ is unimodular. $\qquad\square$

Note that Theorem 8.1.1 applies to models with equality constraints. The analogous concept for models with inequality constraints is 'total unimodularity'. A matrix $\mathbf{A}$ is called *totally unimodular* if the determinant of every square submatrix of $\mathbf{A}$ is either 0, $+1$, or -1. Any totally unimodular matrix consists therefore of 0 and ± 1 entries only (because the entries are determinants of $(1, 1)$-submatrices). Note that total unimodularity implies unimodularity, and that any submatrix of a totally unimodular matrix is itself totally unimodular. Also note that a matrix $\mathbf{A}$ is totally unimodular if and only if $\mathbf{A}^\mathsf{T}$ is totally unimodular.

The following theorem deals with LO-models in standard form, i.e., with models for which the feasible region is defined by inequality constraints.

Theorem 8.1.2. (*Total unimodularity and integer vertices*)
Let $m, n \geq 1$ and let $\mathbf{A}$ be an integer (m, n)-matrix. Then, $\mathbf{A}$ is totally unimodular if and only if, for every integer m-vector $\mathbf{b}$, every vertex of the feasible region $\{\mathbf{x} \mid \mathbf{A}\mathbf{x} \leq \mathbf{b}, \mathbf{x} \geq \mathbf{0}\}$ is integer-valued.

Proof of Theorem 8.1.2. Let $\mathbf{A}$ be any (m, n) integer matrix. We first show that

$$\mathbf{A} \text{ is totally unimodular if and only if } \begin{bmatrix} \mathbf{A} & \mathbf{I}_m \end{bmatrix} \text{ is unimodular.} \tag{$*$}$$

First, assume that $\begin{bmatrix} \mathbf{A} & \mathbf{I}_m \end{bmatrix}$ is unimodular. Take any (k, k)-submatrix $\mathbf{C}$ of $\mathbf{A}$ ($k \geq 1$). We will show that $\det(\mathbf{C}) \in \{0, 1, -1\}$. If $\mathbf{C}$ is singular, then $\det(\mathbf{C}) = 0$. So, we may assume that $\mathbf{C}$ is nonsingular, because otherwise there is nothing left to prove. If $k = m$, then $\det(\mathbf{C}) = \pm 1$, because $\mathbf{C}$ is an (m, m)-submatrix of $\begin{bmatrix} \mathbf{A} & \mathbf{I}_m \end{bmatrix}$. So we may also assume that $k < m$. Let I be the set of row indices of $\mathbf{C}$ in $\mathbf{A}$, and J the set of column indices of $\mathbf{C}$ in $\mathbf{A}$. So, $\mathbf{C} = \mathbf{A}_{I,J}$.

Let $\bar{I} = \{1, \ldots, m\} \setminus I$. Then, $\begin{bmatrix} \mathbf{A} & \mathbf{I}_m \end{bmatrix}$ has a submatrix $\mathbf{C}'$ that satisfies:

$$\mathbf{C}' \equiv \begin{bmatrix} \mathbf{C} & \mathbf{0} \\ \mathbf{A}_{\bar{I}, J} & \mathbf{I}_{m-|I|} \end{bmatrix},$$

where '$\equiv$' means, as usual, equality up to an appropriate permutation of rows and columns, $\mathbf{I}_{m-|I|}$ is an identity matrix, and $\mathbf{0}$ is the zero matrix with $|I|$ rows and $m - |I|$ columns. Note that we have that:

$$\det(\mathbf{C}') = \det(\mathbf{C}) \underbrace{\det(\mathbf{I}_{m-|I|})}_{=1} = \det(\mathbf{C}),$$

(see Appendix B.7). Because $\det(\mathbf{C}) \neq 0$, it follows that $\mathbf{C}'$ has full row rank. Hence, because $\mathbf{C}'$ is an (m, m)-submatrix of $\begin{bmatrix} \mathbf{A} & \mathbf{I}_m \end{bmatrix}$, we have that $\det(\mathbf{C}') = \pm 1$. Hence, $\det(\mathbf{C}) = \pm 1$, as required. Therefore, $\mathbf{A}$ is totally unimodular. The converse follows by similar reasoning and is left as an exercise (Exercise 8.4.3). Hence, (*) holds.

It is left to the reader to check that for any integer vector $\mathbf{b} \in \mathbb{R}^m$, it holds that the vertices of $\{\mathbf{x} \mid \mathbf{A}\mathbf{x} \leq \mathbf{b}, \mathbf{x} \geq \mathbf{0}\}$ are integer if and only if the vertices of

$$\left\{ \begin{bmatrix} \mathbf{x} \\ \mathbf{x}_s \end{bmatrix} \;\middle|\; \begin{bmatrix} \mathbf{A} & \mathbf{I}_m \end{bmatrix} \begin{bmatrix} \mathbf{x} \\ \mathbf{x}_s \end{bmatrix} = \mathbf{b}, \begin{bmatrix} \mathbf{x} \\ \mathbf{x}_s \end{bmatrix} \geq \mathbf{0} \right\}$$

are integer. The theorem now follows directly from Theorem 8.1.1 by taking $\begin{bmatrix} \mathbf{A} & \mathbf{I}_m \end{bmatrix}$ instead of $\mathbf{A}$. $\qquad\square$

8.1.2   Total unimodularity and ILO-models

Besides the mathematical beauty of LO-models that have feasible regions with only integer vertices, total unimodularity has important implications for a special class of ILO-models. Namely, suppose that we want to solve the standard ILO-model:

$$\max\{\mathbf{c}^\mathsf{T}\mathbf{x} \mid \mathbf{A}\mathbf{x} \leq \mathbf{b}, \mathbf{x} \geq \mathbf{0}, \mathbf{x} \text{ integer}\}. \tag{8.1}$$

In general, in order to solve such an ILO-model, we have to use the techniques from Chapter 7. We have seen in that chapter that the general techniques for solving ILO-models can be computationally very time consuming.

Suppose, however, that we know that the matrix $\mathbf{A}$ is totally unimodular and that $\mathbf{b}$ is an integer vector. Then, solving (8.1) can be done in a more efficient manner. Recall that the LO-relaxation (see Section 7.1.2) of (8.1) is found by removing the restriction that $\mathbf{x}$ should be an integer vector. Consider what happens when we apply the simplex algorithm to the LO-relaxation $\max\{\mathbf{c}^\mathsf{T}\mathbf{x} \mid \mathbf{A}\mathbf{x} \leq \mathbf{b}, \mathbf{x} \geq \mathbf{0}\}$. If the algorithm finds an optimal solution, then Theorem 8.1.2 guarantees that this optimal solution corresponds to an integer vertex of the feasible region, because the technology matrix $\mathbf{A}$ of this model is totally unimodular and $\mathbf{b}$ is integer-valued. Clearly, any optimal solution of the LO-relaxation is also a feasible solution of the original ILO-model. Since the optimal objective value of the LO-relaxation is an upper bound for the optimal objective value of the original ILO-model, this means that the solution that we find by solving the LO-relaxation is actually an optimal solution of the original ILO-model. So the computer running time required for determining an optimal solution is not affected by the fact that the variables are required to be integer-valued.

The following theorem makes this precise. In the statement of the theorem, 'two-phase simplex algorithm' means the simplex algorithm together with the two-phase procedure; see Section 3.6.2.

> **Theorem 8.1.3.**
> Consider the standard ILO-model (8.1). If $\mathbf{A}$ is totally unimodular and $\mathbf{b}$ is an integer vector, then applying the two-phase simplex algorithm to the LO-relaxation of (8.1) either produces an optimal solution of the ILO-model, or establishes that the model is infeasible or unbounded.

Proof. Let (P) and (RP) denote the ILO-model and its LO-relaxation, respectively. Applying the two-phase simplex algorithm to (RP) results in one of the following three outcomes: either the algorithm establishes that (RP) is infeasible, or it finds an optimal vertex $\mathbf{x}^*$ of (RP), or it establishes that (RP) is unbounded.

First, suppose that the algorithm establishes that (RP) is infeasible. Since (RP) is the LO-relaxation of (P), it immediately follows that (P) is infeasible as well.

Next, suppose that the algorithm finds an optimal vertex $\mathbf{x}^*$ of (RP). Since $\mathbf{A}$ is totally unimodular and $\mathbf{b}$ is an integer vector, it follows from Theorem 8.1.2 that $\mathbf{x}^*$ is in fact an integer vertex. Let z_P^* and z_{RP}^* be the optimal objective values of (P) and (RP), respectively. Since $z_P^* \leq z_{RP}^*$, $\mathbf{x}^*$ is a feasible point of (P), and $\mathbf{x}^*$ has objective value z_{RP}^*, it follows that $\mathbf{x}^*$ is in fact an optimal solution of (P).

Finally, suppose that the algorithm establishes that (RP) is unbounded. Let $\mathbf{B}$ be the basis matrix at the last iteration of the algorithm, and let $\alpha \in \{1, \ldots, n\}$ be the index found in step 2 in the last iteration of the simplex algorithm (Algorithm 3.3.1). We have that $(\mathbf{c}_{NI}^\mathsf{T} - \mathbf{c}_{BI}^\mathsf{T}\mathbf{B}^{-1}\mathbf{N})_\alpha > 0$ and $(\mathbf{B}^{-1}\mathbf{N})_{\star,\alpha} \leq \mathbf{0}$. Theorem 3.5.2 implies that the set L' defined by:

$$L' = \left\{ \mathbf{x} \in \mathbb{R}^n \, \middle| \, \begin{bmatrix} \mathbf{x} \\ \mathbf{x}_s \end{bmatrix} \equiv \begin{bmatrix} \mathbf{x}_{BI} \\ \mathbf{x}_{NI} \end{bmatrix} = \begin{bmatrix} \mathbf{B}^{-1}\mathbf{b} \\ \mathbf{0} \end{bmatrix} + \lambda \begin{bmatrix} -(\mathbf{B}^{-1}\mathbf{N})_{\star,\alpha} \\ \mathbf{e}_\alpha \end{bmatrix}, \lambda \geq 0, \lambda \text{ integer} \right\}$$

is contained in the feasible region of (RP). Since $\mathbf{A}$ is totally unimodular, $\mathbf{B}^{-1}\mathbf{t}$ is integer-valued for any integer vector $\mathbf{t} \in \mathbb{R}^m$; see the proof of Theorem 8.1.1. It follows that both $\mathbf{B}^{-1}\mathbf{b}$ and $(\mathbf{B}^{-1}\mathbf{N})_{\star,\alpha} = \mathbf{B}^{-1}(\mathbf{N}_{\star,\alpha})$ are integer-valued. Hence, all points in L' are integer-valued, and therefore L' is also contained in the feasible region of (P). Because the objective function takes on arbitrarily large values on L', it follows that (P) is unbounded. $\square$

Note that the interior path algorithm described in Chapter 6 does not necessarily terminate at a vertex. Hence, Theorem 8.1.3 does not hold if the simplex algorithm is replaced by the interior path algorithm.

8.1.3 Total unimodularity and incidence matrices

In this section we will study total unimodularity of incidence matrices associated with undirected and directed graphs; see Appendix C. Before we do this, we present a sufficient condition for total unimodularity that is easy to check. This sufficient condition will be used later on when we investigate the relationship between total unimodularity and incidence matrices.

8.1.3.1 A sufficient condition for total unimodularity

Determining whether or not a given matrix is totally unimodular may be difficult. In fact, the fastest known algorithm that determines whether a given (m, n)-matrix is totally unimodular is due to KLAUS TRUEMPER (born 1942); see Truemper (1990). His (very complicated) algorithm takes $O\big((m + n)^3\big)$ computation time; see also Chapter 9 for a discussion of computational complexity in general. The following theorem gives a sufficient condition that is easy to check. We also show by means of an example that this condition is not a necessary condition.

> **Theorem 8.1.4.** (*Sufficient condition for total unimodularity*)
> Let $m, n \geq 1$ and let $\mathbf{A}$ be an (m, n)-matrix. Suppose that all of the following conditions hold:
>
> (i)   each entry of $\mathbf{A}$ is either 0, 1, or -1, and
>
> (ii) each column of $\mathbf{A}$ contains not more than two nonzero entries, and
>
> (iii) the rows of $\mathbf{A}$ can be 'partitioned' into two (possibly empty) subsets such that:
>
> > (a) if a column contains two entries with the same signs, then the corresponding rows belong to different subsets, and
> >
> > (b) if a column contains two entries with opposite signs, then the corresponding rows belong to the same subset.
>
> Then, $\mathbf{A}$ is totally unimodular.

(Note that the word 'partitioned' is in quotes because, usually, the parts of a partition are required to be nonempty.)

> **Proof of Theorem 8.1.4.** We will show that, if $\mathbf{A}$ is a matrix that satisfies conditions (i), (ii), (iii)a, and (iii)b, then every (k, k)-submatrix $\mathbf{C}_k$ of $\mathbf{A}$ has determinant 0, ± 1 ($k \geq 1$). The proof is by induction on k. For $k = 1$, we have that every $(1, 1)$-submatrix $\mathbf{C}_1$ consists of a single entry of $\mathbf{A}$, and hence $\det(\mathbf{C}_1) \in \{0, \pm 1\}$ for every such $\mathbf{C}_1$. So, the statement is true for $k = 1$.
>
> Now assume that $\det(\mathbf{C}_k) \in \{0, \pm 1\}$ for all (k, k)-submatrices $\mathbf{C}_k$ of $\mathbf{A}$. Take any $(k+1, k+1)$-submatrix $\mathbf{C}_{k+1}$ of $\mathbf{A}$. If $\mathbf{C}_{k+1}$ contains a column with precisely one nonzero entry, say a_{ij},

then $\det(\mathbf{C}_{k+1})$ can be developed with respect to this column, i.e.,

$$\det(\mathbf{C}_{k+1}) = (-1)^{i+j} a_{ij} \det(\mathbf{C}_k^{i,j}),$$

where $\mathbf{C}_k^{i,j}$ is the matrix obtained from $\mathbf{C}_{k+1}$ by deleting its i'th row and j'th column, and so $\mathbf{C}_k^{i,j}$ is a (k, k)-submatrix; see Appendix B. It follows from the induction hypothesis that $\det(\mathbf{C}_k^{i,j}) \in \{0, \pm 1\}$. Since $a_{ij} \in \{0, \pm 1\}$, we have that $\det(\mathbf{C}_{k+1}) \in \{0, \pm 1\}$.

So we may assume that each column of $\mathbf{C}_{k+1}$ has precisely two nonzero entries. We will prove that this implies that $\det(\mathbf{C}_{k+1}) = 0$. According to condition (iii), the rows of $\mathbf{C}_{k+1}$ can be partitioned into two subsets, corresponding to two submatrices, say $\mathbf{C}'$ and $\mathbf{C}''$, such that each column of $\mathbf{C}_{k+1}$ either has two equal nonzero entries with one entry in $\mathbf{C}'$ and the other one in $\mathbf{C}''$, or has two different nonzero entries which are either both in $\mathbf{C}'$ or both in $\mathbf{C}''$. Thus, for each column $(\mathbf{C}_{k+1})_{\star,j}$, we have the following four possibilities:

	#1	#2	#3	#4
$C'_{\star,j} \equiv$	$\begin{bmatrix} 1 & -1 & 0 & \dots & 0 \end{bmatrix}^\mathsf{T}$	$\begin{bmatrix} 0 & 0 & 0 & \dots & 0 \end{bmatrix}^\mathsf{T}$	$\begin{bmatrix} 1 & 0 & \dots & 0 \end{bmatrix}^\mathsf{T}$	$\begin{bmatrix} -1 & 0 & \dots & 0 \end{bmatrix}^\mathsf{T}$
$C''_{\star,j} \equiv$	$\begin{bmatrix} 0 & 0 & 0 & \dots & 0 \end{bmatrix}^\mathsf{T}$	$\begin{bmatrix} 1 & -1 & 0 & \dots & 0 \end{bmatrix}^\mathsf{T}$	$\begin{bmatrix} 1 & 0 & \dots & 0 \end{bmatrix}^\mathsf{T}$	$\begin{bmatrix} -1 & 0 & \dots & 0 \end{bmatrix}^\mathsf{T}$

Now notice that, for each of these four possibilities, the sum of the entries of $\mathbf{C}'_{\star,j}$ is equal to the sum of the entries of $\mathbf{C}''_{\star,j}$. This means that the sum of the rows of $\mathbf{C}'$ is equal to the sum of the rows in $\mathbf{C}''$ and, hence, the rows of $\mathbf{C}_{k+1}$ are linearly dependent. Therefore, $\det(\mathbf{C}_{k+1}) = 0$, as required. $\qquad\qquad\square$

This theorem provides a useful way to check whether a given matrix (with entries $0, \pm 1$) is totally unimodular. We will use the conditions in the next few sections. The following example shows, however, that the conditions of Theorem 8.1.4 are sufficient but not necessary.

Example 8.1.1. *The following matrix is totally unimodular, but does not satisfy the conditions of Theorem 8.1.4:*

$$\mathbf{A} = \begin{bmatrix} 1 & 0 & 0 & 1 & 0 & 0 \\ 1 & 1 & 0 & 0 & 1 & 0 \\ 1 & 0 & 1 & 0 & 0 & 1 \end{bmatrix}.$$

In Exercise 8.4.4, the reader is asked to show that all vertices of $\{\mathbf{x} \in \mathbb{R}^6 \mid \mathbf{Ax} \le \mathbf{b}, \mathbf{x} \ge \mathbf{0}\}$ are integer if $\mathbf{b}$ is an integer vector.

8.1.3.2 Undirected graphs and their node-edge incidence matrices

Let $G = (\mathcal{V}, \mathcal{E})$ be an undirected graph, where $\mathcal{V}$ is the node set and $\mathcal{E}$ is the edge set. The edge set $\mathcal{E}$ consists of unordered pairs $\{i, j\}$ with $i, j \in \mathcal{V}$. The *node-edge incidence matrix* $\mathbf{M}_G$ associated with G is the $\{0, 1\}$-matrix in which the rows correspond to the nodes and the columns to the edges; moreover, letting k denote the index of the column of $\mathbf{M}_G$ corresponding to $\{i, j\} \in \mathcal{E}$, we have that $(\mathbf{M}_G)_{i,k} = 1$ and $(\mathbf{M}_G)_{j,k} = 1$, and all other entries in column k are 0.

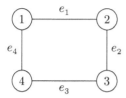
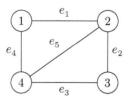

Figure 8.1: Graph G_1 with a totally unimodular node-edge incidence matrix.

Figure 8.2: Graph G_2 with a node-edge incidence matrix that is not totally unimodular.

Example 8.1.2.   *Consider the graph G_1 drawn in Figure 8.1. The node-edge incidence matrix $\mathbf{M}_{G_1}$ associated with this graph is:*

$$
\mathbf{M}_{G_1} = \begin{array}{c} \\ 1 \\ 2 \\ 3 \\ 4 \end{array}
\begin{array}{cccc}
e_1 & e_2 & e_3 & e_4 \\
\left[\begin{array}{cccc}
1 & 0 & 0 & 1 \\
1 & 1 & 0 & 0 \\
0 & 1 & 1 & 0 \\
0 & 0 & 1 & 1
\end{array}\right]
\end{array}.
$$

It is left to the reader to check that the matrix $\mathbf{M}_G$ is totally unimodular.

Before stating and proving Theorem 8.1.6, which gives necessary and sufficient conditions for a node-edge incidence matrix being totally unimodular, the following example shows that not all undirected graphs have a node-edge incidence matrix that is totally unimodular.

Example 8.1.3.   *Consider the graph G_2 drawn in Figure 8.2. The node-edge incidence matrix $\mathbf{M}_{G_2}$ associated with this graph is:*

$$
\mathbf{M}_{G_2} = \begin{array}{c} \\ 1 \\ 2 \\ 3 \\ 4 \end{array}
\begin{array}{ccccc}
e_1 & e_2 & e_3 & e_4 & e_5 \\
\left[\begin{array}{ccccc}
1 & 0 & 0 & 1 & 0 \\
1 & 1 & 0 & 0 & 1 \\
0 & 1 & 1 & 0 & 0 \\
0 & 0 & 1 & 1 & 1
\end{array}\right]
\end{array}.
$$

The matrix $\mathbf{M}_{G_2}$ is not totally unimodular, because, for instance, the submatrix corresponding to the columns with labels e_2, e_3, e_4, e_5 has determinant -2. In fact, the submatrix corresponding to columns with labels e_2, e_3, e_5, and the rows with labels 2, 3, 4 has determinant 2. Note that the edges e_2, e_3, e_5, together with the nodes 2, 3, 4 form the cycle 2-e_2-3-e_3-4-e_5-2 of length three in the graph in Figure 8.2.

The fact that G_2 in Figure 8.2 contains a cycle of length three is the reason why the node-edge incidence of G_2 is not totally unimodular. In general, as soon as an undirected graph contains a cycle of odd length ≥ 3, the node-edge incidence matrix is not totally unimodular. The following theorem shows that the determinant of the node-edge incidence of any odd cycle is equal to ± 2. A *cycle graph* is a graph consisting of a single cycle.

Theorem 8.1.5. (*Determinant of the node-edge incidence matrix of a cycle graph*)
Let $k \geq 3$ and let $\mathbf{M}_k$ be the node-edge incidence matrix of a cycle graph of length k. Then,

$$|\det(\mathbf{M}_k)| = \begin{cases} 0 & \text{if } k \text{ is even} \\ 2 & \text{if } k \text{ is odd.} \end{cases}$$

Proof. Let $k \geq 3$ and let G_k be a cycle graph of length k. The node-edge incidence matrix of G_k can be written as:

$$\mathbf{M}_k \equiv \begin{bmatrix} 1 & 0 & \cdots & \cdots & 0 & 1 \\ 1 & 1 & 0 & \cdots & 0 & 0 \\ 0 & \ddots & \ddots & \ddots & & \vdots \\ \vdots & \ddots & \ddots & \ddots & 0 & \vdots \\ \vdots & & \ddots & \ddots & 1 & 0 \\ 0 & \cdots & \cdots & 0 & 1 & 1 \end{bmatrix}.$$

Moreover, for $n \geq 1$, define:

$$\mathbf{D}_n = \begin{bmatrix} 1 & 1 & 0 & \cdots & 0 \\ 0 & \ddots & \ddots & \ddots & \vdots \\ \vdots & \ddots & \ddots & \ddots & 0 \\ \vdots & & \ddots & \ddots & 1 \\ 0 & \cdots & \cdots & 0 & 1 \end{bmatrix}.$$

Note that $\det(\mathbf{D}_n) = 1$ for all $n \geq 1$. Next, by developing the determinant of $\mathbf{M}_k$ along its first row, we have that $|\det(\mathbf{M}_k)| = |1 \times \det(\mathbf{D}_{k-1}) + (-1)^{k-1}\det(\mathbf{D}_{k-1})| = 1 + (-1)^{k-1}$. Thus, $\det(\mathbf{M}_k) = 0$ if k is even, and $|\det(\mathbf{M}_k)| = 2$ if k is odd. $\qquad\square$

Theorem 8.1.5 shows that the node-edge incidence matrix of an odd cycle graph is not totally unimodular. Hence, an undirected graph that contains an odd cycle corresponds to a node-edge incidence matrix that is not totally unimodular, because the node-edge incidence matrix corresponding to the odd cycle has a determinant that is not equal to 0, ± 1. Therefore, if we have a graph with a totally unimodular node-edge incidence matrix, then it does not contain any odd cycles, i.e., it must be *bipartite* (see Appendix C). The following theorem makes this precise, and shows that the converse also holds.

Theorem 8.1.6. (*Total unimodularity and undirected graphs*)
The node-edge incidence matrix associated with an undirected graph is totally unimodular if and only if the graph is bipartite.

Proof of Theorem 8.1.6. Let $G = (\mathcal{V}, \mathcal{E})$ be any undirected graph. First assume that G is a bipartite graph. Then $\mathcal{V}$ can be partitioned into two disjoint subsets $\mathcal{V}_1$ and $\mathcal{V}_2$ with $\mathcal{V}_1 \cup \mathcal{V}_2 = \mathcal{V}$, and every edge in $\mathcal{E}$ has one end in $\mathcal{V}_1$ and the other end in $\mathcal{V}_2$. It can be easily checked that the node-edge incidence matrix $\mathbf{M}_G$ associated with G is a $\{0, 1\}$-matrix with precisely two 1's per column, namely one in the rows corresponding to $\mathcal{V}_1$ and one in the rows corresponding to $\mathcal{V}_2$. Applying Theorem 8.1.4 with row partition $\mathcal{V}_1$ and $\mathcal{V}_2$, it follows that $\mathbf{M}_G$ is totally unimodular.

Now, assume that G it not bipartite. Then, because a graph is bipartite if and only if it does not contain an odd cycle, G must contain a cycle v_1-e_1-v_2-e_2-$\ldots$-v_k-e_k-v_1 with k odd. It follows from Theorem 8.1.5 that the node-edge submatrix associated with this cycle has determinant equal to ± 2. This contradicts the definition of total unimodularity of $\mathbf{M}_G$. Hence, $\mathbf{M}_G$ is not totally unimodular. □

8.1.3.3 Directed graphs and their node-arc incidence matrices

Let $G = (\mathcal{V}, \mathcal{A})$ be a *directed graph* (also called *digraph*), where $\mathcal{V}$ is the node set and $\mathcal{A}$ is the arc set of G. The arc set is a collection of ordered pairs (i, j), where $i, j \in \mathcal{V}$. Each element $(i, j) \in \mathcal{A}$ represents an arc from its *tail node i* to its *head node j*. An element $(i, j) \in \mathcal{A}$ is called a *loop* if $i = j$. We assume that G is *loopless*, i.e., it does not have any loops. The *node-arc incidence matrix* associated with G is a $\{0, \pm 1\}$-matrix $\mathbf{M}_G$ in which the rows correspond to the nodes and the columns to the arcs; moreover, letting k be the index of the column of $\mathbf{M}_G$ corresponding to $(i, j) \in \mathcal{A}$, we have that $(\mathbf{M}_G)_{i,k} = 1$, $(\mathbf{M}_G)_{j,k} = -1$, and all other entries in column k are 0. Note that we require the graph G to be loopless, because otherwise the definition of the node-arc incidence matrix does not make sense.

The following theorem deals with node-arc incidence matrices of loopless digraphs.

Theorem 8.1.7. (*Total unimodularity and digraphs*)
The node-arc incidence matrix of any loopless digraph is totally unimodular.

Proof of Theorem 8.1.7. Let $G = (\mathcal{V}, \mathcal{A})$ be a digraph without loops, and let $\mathbf{M}_G$ be its node-arc incidence matrix. Since each column of $\mathbf{M}_G$ corresponds to some arc $(i, j) \in \mathcal{A}$, we have that each column of $\mathbf{M}_G$ contains exactly one 1 (namely, the entry corresponding to i) and one -1 (namely, the entry corresponding to j), and all other entries are 0. Hence, if we choose the (trivial) partition $\{\mathcal{V}, \emptyset\}$ of $\mathcal{V}$, all conditions of Theorem 8.1.4 are satisfied. Therefore, Theorem 8.1.4 implies that $\mathbf{M}_G$ is totally unimodular. □

Applications of Theorem 8.1.6 and Theorem 8.1.7 are given in the next section. The following example illustrates Theorem 8.1.7.

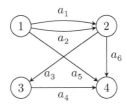

Figure 8.3: Digraph whose node-arc incidence matrix is totally unimodular.

Example 8.1.4. *Consider the loopless digraph G of Figure 8.3. The node-arc incidence matrix $\mathbf{M}_G$ associated with G reads:*

$$
\mathbf{M}_G =
\begin{matrix}
\phantom{{}-}1 \\ 2 \\ 3 \\ 4
\end{matrix}
\begin{array}{c}
\begin{matrix} a_1 & a_2 & a_3 & a_4 & a_5 & a_6 \end{matrix} \\
\left[
\begin{matrix}
1 & 1 & 0 & 0 & 1 & 0 \\
-1 & -1 & 1 & 0 & 0 & 1 \\
0 & 0 & -1 & 1 & 0 & 0 \\
0 & 0 & 0 & -1 & -1 & -1
\end{matrix}
\right]
\end{array}.
$$

According to Theorem 8.1.7, $\mathbf{M}_G$ is totally unimodular. Note that this also follows directly from Theorem 8.1.4.

Node-arc incidence matrices of loopless digraphs do not have full row rank, since the sum of the rows is the all-zero vector (why is this true?). However, if the digraph G is connected, then deleting any row of $\mathbf{M}_G$ leads to a matrix with full row rank, as the following theorem shows.

> **Theorem 8.1.8.** (*The rank of node-arc incidence matrices*)
> Let $\mathbf{M}_G$ be the node-arc incidence matrix associated with the connected loopless digraph $G = (\mathcal{V}, \mathcal{A})$ with m (≥ 2) nodes. Then, deleting any row of $\mathbf{M}_G$ leads to a matrix of full row rank, i.e., $\mathrm{rank}(\mathbf{M}_G) = m - 1$.

> **Proof of Theorem 8.1.8.** Let $m \geq 2$. We will first prove the special case that if T_m is a directed tree with m nodes, then deleting any row from $\mathbf{M}_{T_m}$ results in an invertible matrix. Note that $\mathbf{M}_{T_m}$ is an $(m, m-1)$-matrix, and therefore deleting one row results in an $(m-1, m-1)$-matrix. The proof is by induction on m. The statement is obviously true for $m = 2$. So suppose that the statement holds for directed trees on $m - 1$ nodes. Let T_m be a directed tree with m nodes and let r be the index of the row that is deleted. Since T_m is a tree, it has at least two leaf nodes (a *leaf node* is a node with exactly one arc incident to it). So, T has a leaf node v that is different from r. Let a be the index of the unique arc incident with v. Row v consists of zeros, except for $(\mathbf{M}_T)_{v,a} = \pm 1$. Note that $\mathbf{M}_{T_{m-1}}$ can be obtained from $\mathbf{M}_{T_m}$ by deleting row v

and column a. Thus, we may write:

$$\mathbf{M}_{T_m} \equiv \begin{bmatrix} \mathbf{M}_{T_{m-1}} & \mathbf{u} \\ \mathbf{0}^{\mathsf{T}} & \pm 1 \end{bmatrix}, \tag{8.2}$$

where $\mathbf{u}$ satisfies $\begin{bmatrix} \mathbf{u} \\ \pm 1 \end{bmatrix} \equiv (\mathbf{M}_{T_m})_{\star,a}$. Construct T_{m-1} from T_m by deleting v and a, and let $\mathbf{M}_{T_{m-1}}$ be its node-arc incidence matrix. Let $\mathbf{M}'_{T_m}$ be obtained from $\mathbf{M}_{T_m}$ by deleting row r. Since $r \neq v$, it follows from (8.2) that:

$$\mathbf{M}'_{T_m} \equiv \begin{bmatrix} \mathbf{M}'_{T_{m-1}} & \mathbf{u}' \\ \mathbf{0}^{\mathsf{T}} & \pm 1 \end{bmatrix}, \tag{8.3}$$

where $\mathbf{M}'_{T_{m-1}}$ and $\mathbf{u}'$ are constructed from $\mathbf{M}_{T_{m-1}}$ and $\mathbf{u}$, respectively, by deleting row r. By the induction hypothesis, $\mathbf{M}'_{T_{m-1}}$ is invertible. Therefore, it follows from (8.3) that $\det(\mathbf{M}'_{T_m}) = (\pm 1) \times \det(\mathbf{M}'_{T_{m-1}}) \neq 0$, which proves that $\mathbf{M}'_{T_m}$ is invertible, as required.

Now, consider the general case. Let $G = (\mathcal{V}, \mathcal{A})$ be a connected digraph with node set $\mathcal{V}$ and arc set $\mathcal{A}$, and let $m = |\mathcal{V}|$. Furthermore, let $\mathbf{M}_G$ be the node-arc incidence matrix associated with G. We need to show that, after removing an arbitrary row from $\mathbf{M}_G$, the remaining submatrix has full row rank. Because G is connected, G contains a spanning tree T (see Appendix C). Let $\mathbf{M}_T$ denote the node-arc incidence matrix corresponding to T. Clearly, $\mathbf{M}_T$ is an $(m, m-1)$-submatrix of $\mathbf{M}_G$. Let r be the index of the row that is deleted, and let $\mathbf{M}'_G$ and $\mathbf{M}'_T$ be obtained from $\mathbf{M}$ and $\mathbf{M}_T$, respectively, by deleting row r. Then, $\mathbf{M}'_T$ is an $(m-1, m-1)$-submatrix of $\mathbf{M}'_G$. It follows from the result for trees (above) that $\mathbf{M}'_T$ has rank $m-1$, and hence $\mathbf{M}'_G$ has rank $m-1$, as required. □

Example 8.1.5. *Consider the node-arc incidence matrix $\mathbf{M}_G$ of the graph G in Figure 8.3. Suppose that we take $r = 4$, i.e., the fourth row of $\mathbf{M}_G$ will be deleted. Let $\mathbf{M}'_G$ be the matrix obtained from $\mathbf{M}_G$ by deleting the fourth row. In order to show that $\mathbf{M}'_G$ is invertible, the second part of the proof shows that it suffices to consider an arbitrary spanning tree in G. So, we may use the tree T with arc set $\{a_2, a_3, a_5\}$. The node-arc incidence matrix $\mathbf{M}_T$ of T is:*

$$\mathbf{M}_T = \begin{bmatrix} \overset{a_2}{1} & \overset{a_3}{0} & \overset{a_5}{1} \\ -1 & 1 & 0 \\ 0 & -1 & 0 \\ 0 & 0 & -1 \end{bmatrix}.$$

To show that deleting the fourth row from $\mathbf{M}_T$ results in a matrix of rank 3, the proof of Theorem 8.1.8 proceeds by choosing a leaf node v of T that is not vertex $(r =)$ 4, and the unique arc a incident with v. Since T has only two leaf nodes, we have to take $v = 3$, and hence $a = a_3$. Deleting the fourth row yields:

$$\mathbf{M}'_T = \begin{bmatrix} \overset{a_2}{1} & \overset{a_3}{0} & \overset{a_5}{1} \\ -1 & 1 & 0 \\ 0 & -1 & 0 \end{bmatrix}, \text{ and } \det(\mathbf{M}'_T) = -1 \times \det \begin{bmatrix} 1 & 1 \\ -1 & 0 \end{bmatrix} = -1 \neq 0.$$

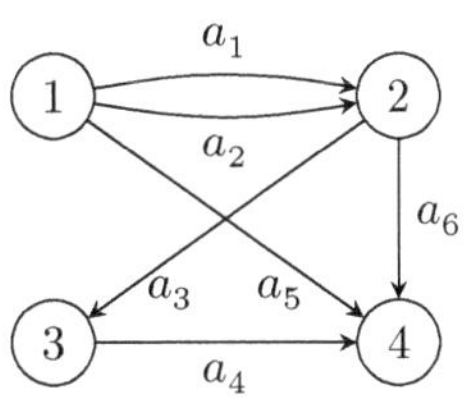

Figure 8.3: Digraph whose node-arc incidence matrix is totally unimodular.

Example 8.1.4. *Consider the loopless digraph G of Figure 8.3. The node-arc incidence matrix $\mathbf{M}_G$ associated with G reads:*

$$
\mathbf{M}_G = \begin{array}{c} \\ 1 \\ 2 \\ 3 \\ 4 \end{array}
\begin{bmatrix}
\overset{a_1}{1} & \overset{a_2}{1} & \overset{a_3}{0} & \overset{a_4}{0} & \overset{a_5}{1} & \overset{a_6}{0} \\
-1 & -1 & 1 & 0 & 0 & 1 \\
0 & 0 & -1 & 1 & 0 & 0 \\
0 & 0 & 0 & -1 & -1 & -1
\end{bmatrix}.
$$

According to Theorem 8.1.7, $\mathbf{M}_G$ is totally unimodular. Note that this also follows directly from Theorem 8.1.4.

Node-arc incidence matrices of loopless digraphs do not have full row rank, since the sum of the rows is the all-zero vector (why is this true?). However, if the digraph G is connected, then deleting any row of $\mathbf{M}_G$ leads to a matrix with full row rank, as the following theorem shows.

Theorem 8.1.8. (*The rank of node-arc incidence matrices*)
Let $\mathbf{M}_G$ be the node-arc incidence matrix associated with the connected loopless digraph $G = (\mathcal{V}, \mathcal{A})$ with m (≥ 2) nodes. Then, deleting any row of $\mathbf{M}_G$ leads to a matrix of full row rank, i.e., $\mathrm{rank}(\mathbf{M}_G) = m - 1$.

Proof of Theorem 8.1.8. Let $m \geq 2$. We will first prove the special case that if T_m is a directed tree with m nodes, then deleting any row from $\mathbf{M}_{T_m}$ results in an invertible matrix. Note that $\mathbf{M}_{T_m}$ is an $(m, m-1)$-matrix, and therefore deleting one row results in an $(m-1, m-1)$-matrix. The proof is by induction on m. The statement is obviously true for $m = 2$. So suppose that the statement holds for directed trees on $m - 1$ nodes. Let T_m be a directed tree with m nodes and let r be the index of the row that is deleted. Since T_m is a tree, it has at least two leaf nodes (a *leaf node* is a node with exactly one arc incident to it). So, T has a leaf node v that is different from r. Let a be the index of the unique arc incident with v. Row v consists of zeros, except for $(\mathbf{M}_T)_{v,a} = \pm 1$. Note that $\mathbf{M}_{T_{m-1}}$ can be obtained from $\mathbf{M}_{T_m}$ by deleting row v

and column a. Thus, we may write:

$$\mathbf{M}_{T_m} \equiv \begin{bmatrix} \mathbf{M}_{T_{m-1}} & \mathbf{u} \\ \mathbf{0}^{\mathsf{T}} & \pm 1 \end{bmatrix}, \tag{8.2}$$

where $\mathbf{u}$ satisfies $\begin{bmatrix} \mathbf{u} \\ \pm 1 \end{bmatrix} \equiv (\mathbf{M}_{T_m})_{*,a}$. Construct T_{m-1} from T_m by deleting v and a, and let $\mathbf{M}_{T_{m-1}}$ be its node-arc incidence matrix. Let $\mathbf{M}'_{T_m}$ be obtained from $\mathbf{M}_{T_m}$ by deleting row r. Since $r \neq v$, it follows from (8.2) that:

$$\mathbf{M}'_{T_m} \equiv \begin{bmatrix} \mathbf{M}'_{T_{m-1}} & \mathbf{u}' \\ \mathbf{0}^{\mathsf{T}} & \pm 1 \end{bmatrix}, \tag{8.3}$$

where $\mathbf{M}'_{T_{m-1}}$ and $\mathbf{u}'$ are constructed from $\mathbf{M}_{T_{m-1}}$ and $\mathbf{u}$, respectively, by deleting row r. By the induction hypothesis, $\mathbf{M}'_{T_{m-1}}$ is invertible. Therefore, it follows from (8.3) that $\det(\mathbf{M}'_{T_m}) = (\pm 1) \times \det(\mathbf{M}'_{T_{m-1}}) \neq 0$, which proves that $\mathbf{M}'_{T_m}$ is invertible, as required.

Now, consider the general case. Let $G = (\mathcal{V}, \mathcal{A})$ be a connected digraph with node set $\mathcal{V}$ and arc set $\mathcal{A}$, and let $m = |\mathcal{V}|$. Furthermore, let $\mathbf{M}_G$ be the node-arc incidence matrix associated with G. We need to show that, after removing an arbitrary row from $\mathbf{M}_G$, the remaining submatrix has full row rank. Because G is connected, G contains a spanning tree T (see Appendix C). Let $\mathbf{M}_T$ denote the node-arc incidence matrix corresponding to T. Clearly, $\mathbf{M}_T$ is an $(m, m-1)$-submatrix of $\mathbf{M}_G$. Let r be the index of the row that is deleted, and let $\mathbf{M}'_G$ and $\mathbf{M}'_T$ be obtained from $\mathbf{M}$ and $\mathbf{M}_T$, respectively, by deleting row r. Then, $\mathbf{M}'_T$ is an $(m-1, m-1)$-submatrix of $\mathbf{M}'_G$. It follows from the result for trees (above) that $\mathbf{M}'_T$ has rank $m-1$, and hence $\mathbf{M}'_G$ has rank $m-1$, as required. $\qquad\square$

Example 8.1.5. *Consider the node-arc incidence matrix $\mathbf{M}_G$ of the graph G in Figure 8.3. Suppose that we take $r = 4$, i.e., the fourth row of $\mathbf{M}_G$ will be deleted. Let $\mathbf{M}'_G$ be the matrix obtained from $\mathbf{M}_G$ by deleting the fourth row. In order to show that $\mathbf{M}'_G$ is invertible, the second part of the proof shows that it suffices to consider an arbitrary spanning tree in G. So, we may use the tree T with arc set $\{a_2, a_3, a_5\}$. The node-arc incidence matrix $\mathbf{M}_T$ of T is:*

$$\mathbf{M}_T = \begin{bmatrix} \overset{a_2}{1} & \overset{a_3}{0} & \overset{a_5}{1} \\ -1 & 1 & 0 \\ 0 & -1 & 0 \\ 0 & 0 & -1 \end{bmatrix}.$$

To show that deleting the fourth row from $\mathbf{M}_T$ results in a matrix of rank 3, the proof of Theorem 8.1.8 proceeds by choosing a leaf node v of T that is not vertex $(r =) 4$, and the unique arc a incident with v. Since T has only two leaf nodes, we have to take $v = 3$, and hence $a = a_3$. Deleting the fourth row yields:

$$\mathbf{M}'_T = \begin{bmatrix} \overset{a_2}{1} & \overset{a_3}{0} & \overset{a_5}{1} \\ -1 & 1 & 0 \\ 0 & -1 & 0 \end{bmatrix}, \text{ and } \det(\mathbf{M}'_T) = -1 \times \det \begin{bmatrix} 1 & 1 \\ -1 & 0 \end{bmatrix} = -1 \neq 0.$$

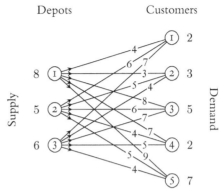

	Customer				
	1	2	3	4	5
Depot 1	4	3	8	7	9
Depot 2	6	5	6	4	5
Depot 3	7	4	7	5	4

(a) Transportation costs (×$100).   (b) Digraph of the transportation problem.

Figure 8.4: Input for the transportation problem.

Therefore $\mathbf{M}_T'$ *is indeed invertible. It is left to the reader to check that deleting any other row of* $\mathbf{M}_G$ *results in an invertible matrix as well.*

It is also left to the reader to check that the node-arc incidence matrices of the transportation problem (see Section 8.2.1), the minimum cost flow problem (see Section 8.2.4), and the transshipment problem (see Section 8.3.1) have full row rank when an arbitrary row is deleted from the corresponding incidence matrix.

8.2 ILO-models with totally unimodular matrices

In this section we will discuss five problems that can be formulated as ILO-models and whose LO-relaxations are LO-models with integer optimal solutions. These problems are: the transportation problem, the assignment problem, the minimum cost flow problem, the maximum flow problem, and a project scheduling problem with precedence constraints.

8.2.1 The transportation problem

A truck rental company has three depots, labeled 1, 2, 3, at which a total of 8, 5, and 6 vehicles, respectively, are parked. Furthermore, there are five customers, labeled $1, \ldots, 5$, demanding 2, 3, 5, 2, and 7 vehicles, respectively. The transportation costs (per vehicle) from the depots to the customers are listed in Table 8.4(a). The problem is to transport the vehicles from the depots to the customers, making sure that the demand of the customers is satisfied, and in such a way that the total transportation costs are minimized. Figure 8.4(b) illustrates the situation by means of a graph.

For $i = 1, 2, 3$ and $j = 1, 2, 3, 4, 5$, let c_{ij} be the transportation cost from depot i to customer j, and let x_{ij} be the number of vehicles that are to be transported from depot i to customer j; the cost of transporting x_{ij} vehicles from i to j is therefore $c_{ij}x_{ij}$. Moreover, let a_i be the number of vehicles in depot i (the supply). Hence, $a_1 = 8$, $a_2 = 5$, $a_3 = 6$.

Let b_j be the number of vehicles ordered by customer j (the demand). Hence, $b_1 = 2$, $b_2 = 3, b_3 = 5, b_4 = 2, b_5 = 7$. This problem can formulated as an ILO-model as follows:

$$\min 4x_{11} + 3x_{12} + 8x_{13} + 7x_{14} + 9x_{15} + 6x_{21} + 5x_{22} + 6x_{23} +$$
$$4x_{24} + 5x_{25} + 7x_{31} + 4x_{32} + 7x_{33} + 5x_{34} + 4x_{35}$$

$$\begin{aligned}
\text{s.t.} \quad & x_{11} + x_{12} + x_{13} + x_{14} + x_{15} \leq 8 \\
& x_{21} + x_{22} + x_{23} + x_{24} + x_{25} \leq 5 \\
& x_{31} + x_{32} + x_{33} + x_{34} + x_{35} \leq 6 \\
& x_{11} + x_{21} + x_{31} \geq 2 \qquad\qquad\qquad\qquad (8.4) \\
& x_{12} + x_{22} + x_{32} \geq 3 \\
& x_{13} + x_{23} + x_{33} \geq 5 \\
& x_{14} + x_{24} + x_{34} \geq 2 \\
& x_{15} + x_{25} + x_{35} \geq 7 \\
& x_{11}, \ldots, x_{35} \geq 0, \text{ and integer.}
\end{aligned}$$

The first three constraints express the fact that each depot can only deliver as many vehicles as it has available, and the other five constraints express the fact that the customers should receive at least the number of vehicles that they demand. The node-arc incidence matrix $\mathbf{T}$ of the transportation problem corresponding to the digraph of Figure 8.4(b) is:

$(1,1)$	$(1,2)$	$(1,3)$	$(1,4)$	$(1,5)$	$(2,1)$	$(2,2)$	$(2,3)$	$(2,4)$	$(2,5)$	$(3,1)$	$(3,2)$	$(3,3)$	$(3,4)$	$(3,5)$	
1	1	1	1	1	0	0	0	0	0	0	0	0	0	0	Depot 1
0	0	0	0	0	1	1	1	1	1	0	0	0	0	0	Depot 2
0	0	0	0	0	0	0	0	0	0	1	1	1	1	1	Depot 3
−1	0	0	0	0	−1	0	0	0	0	−1	0	0	0	0	Customer 1
0	−1	0	0	0	0	−1	0	0	0	0	−1	0	0	0	Customer 2
0	0	−1	0	0	0	0	−1	0	0	0	0	−1	0	0	Customer 3
0	0	0	−1	0	0	0	0	−1	0	0	0	0	−1	0	Customer 4
0	0	0	0	−1	0	0	0	0	−1	0	0	0	0	−1	Customer 5

The matrix $\mathbf{T}$ is almost, but not exactly, equal to the technology matrix of the ILO-model (8.4). By multiplying both sides of each of the last five constraints by -1 and flipping the inequality signs from '$\geq$' to '$\leq$', we obtain an equivalent (standard) ILO-model that has technology matrix $\mathbf{T}$. Since $\mathbf{T}$ is totally unimodular according to Theorem 8.1.7, this implies that (8.4) is equivalent to an ILO-model with a totally unimodular technology matrix. Since the right hand side values are integers, it follows from Theorem 8.1.3 that (8.4) can be solved by applying the simplex algorithm to its LO-relaxation. However, it is not immediately obvious that the LO-relaxation is feasible. Finding an initial feasible basic solution for the simplex algorithm can in general be done using the big-M or the two-phase procedure, but in this case it can also be done directly. In Exercise 8.4.6, the reader is asked to find a feasible solution that shows that this model is feasible.

Using a linear optimization computer package, one may verify that the minimum transportation costs are 9,000; the optimal nonzero values of the x_{ij}'s are attached to the arcs

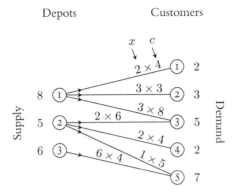

Figure 8.5: Optimal transportation schedule.

in Figure 8.5. In Section 8.3, this problem will be solved more efficiently by means of the so-called network simplex algorithm.

We now formulate the general form of the *transportation problem*. Let m (≥ 1) be the number of depots and n (≥ 1) the number of customers to be served. The cost of transporting goods from depot i to customer j is c_{ij}, the supply of depot i is a_i, and the demand of customer j is b_j ($i = 1, \ldots, m$ and $j = 1, \ldots, n$). The model can now be formulated as follows:

Model 8.2.1. (*Transportation problem*)

$$\min \sum_{i=1}^{m} \sum_{j=1}^{n} c_{ij} x_{ij}$$

$$\text{s.t.} \sum_{j=1}^{n} x_{ij} \leq a_i \qquad \text{for } i = 1, \ldots, m$$

$$\sum_{i=1}^{m} x_{ij} \geq b_j \qquad \text{for } j = 1, \ldots, n$$

$$x_{ij} \geq 0, \text{ and integer} \qquad \text{for } i = 1, \ldots, m \text{ and } j = 1, \ldots, n.$$

Since the right hand side values a_i and b_j are integers for all i and j, it follows from Theorem 8.1.1 and Theorem 8.1.7 that the above ILO-model of the transportation problem has an integer optimal solution. (Why is the model not unbounded?) In Exercise 8.4.8, the reader is asked to find a method that finds a feasible solution and an initial feasible basic solution for the LO-relaxation.

8.2.2 The balanced transportation problem

Consider again the transportation problem of Section 8.2.1. If we want to make sure that the demand of all customers is satisfied, then the total supply in the various depots should be at least as large as the total demand. Suppose now that the total supply and demand are actually exactly equal, i.e.,

$$\sum_{i=1}^{m} a_i = \sum_{j=1}^{n} b_j. \tag{8.5}$$

This equation is called a *supply-demand balance equation*. If this equation holds, then it should be clear that no customer can receive more than the corresponding demand, and every depot has to exhaust its supply. This means that, under these circumstances, the inequalities in the technology constraints of Model 8.2.1 may without loss of generality be replaced by equality constraints. The resulting model is called the *balanced transportation problem*.

Model 8.2.2. (*Balanced transportation problem*)

$$\min \sum_{i=1}^{m} \sum_{j=1}^{n} c_{ij} x_{ij}$$

$$\text{s.t.} \sum_{j=1}^{n} x_{ij} = a_i \qquad \text{for } i = 1, \ldots, m$$

$$\sum_{i=1}^{m} x_{ij} = b_j \qquad \text{for } j = 1, \ldots, n$$

$$x_{ij} \geq 0, \text{ and integer} \qquad \text{for } i = 1, \ldots, m \text{ and } j = 1, \ldots, n.$$

The reader is asked in Exercise 8.4.7 to show that Model 8.2.2 has an optimal solution if and only if the supply-demand balance equation (8.5) holds. Because the technology matrix of Model 8.2.2 is again the node-arc incidence matrix $\mathbf{T}$, it is totally unimodular. We can therefore use the simplex algorithm for models with equality constraints (see Section 3.8) to solve the model. Note that the model has $m + n$ constraints. The simplex algorithm requires an initial basis matrix, i.e., an invertible $(m+n, m+n)$-submatrix of $\mathbf{T}$. However, according to Theorem 8.1.8, $\mathbf{T}$ does not have full row rank. Hence, no such invertible submatrix exists. So, in order to be able to apply the simplex algorithm, we should drop at least one of the constraints. Following Theorem 8.1.8, we may drop any one of the constraints in order to obtain a new technology matrix that has full row rank. This results in an LO-model that can be solved using the simplex algorithm.

The important advantage of the formulation of Model 8.2.2 compared to Model 8.2.1 is the fact that the formulation of Model 8.2.2 is more compact. In fact, the LO-relaxation of Model 8.2.2 has nm variables and $n + m - 1$ constraints, whereas the LO-relaxation of Model 8.2.1 has $nm + n + m$ variables and $n + m$ constraints.

Note that each of the $m+n$ constraints of Model 8.2.2 is automatically satisfied once the remaining constraints are satisfied. In other words, each individual constraint is redundant with respect to the feasible region. Suppose, for example, that we drop the very last constraint, i.e., the constraint with right hand side b_n. Assume that we have a solution $\mathbf{x} = \{x_{ij}\}$ that satisfies the first $m + n - 1$ constraints. We have that:

$$\sum_{i=1}^{m}\sum_{j=1}^{n} x_{ij} = \sum_{j=1}^{n}\sum_{i=1}^{m} x_{ij} = \sum_{j=1}^{n-1}\sum_{i=1}^{m} x_{ij} + \sum_{i=1}^{m} x_{in}.$$

Since the first $m + n - 1$ constraints of Model 8.2.2 are satisfied, this implies that:

$$\sum_{i=1}^{m} a_i = \sum_{j=1}^{n-1} b_j + \sum_{i=1}^{m} x_{in}.$$

Using (8.5), this in turn implies that:

$$\sum_{i=1}^{m} x_{in} = \sum_{i=1}^{m} a_i - \sum_{j=1}^{n-1} b_j = \left(\sum_{i=1}^{m} a_i - \sum_{j=1}^{n} b_j\right) + b_n = b_n,$$

so the last constraint of Model 8.2.2 is indeed satisfied.

8.2.3 The assignment problem

The *assignment problem* can be formulated as follows. Let $n \geq 1$. There are n jobs to be carried out by n persons. Each person has to do precisely one of the n jobs, and each job has to be done by one person only. If job i is done by person j, then the associated cost is c_{ij} $(i, j = 1, \ldots, n)$. The objective is to determine an optimal schedule, meaning that all jobs are carried out by one person, each person has precisely one job to do, and the total cost is as small as possible. In order to formulate this problem as an ILO-model, we introduce the binary variables x_{ij} for $i, j = 1, \ldots, n$, with the following meaning:

$$x_{ij} = \begin{cases} 1 & \text{if job } i \text{ is performed by person } j \\ 0 & \text{otherwise.} \end{cases}$$

The assignment problem can now be modeled as follows:

Model 8.2.3. (*Assignment problem*)

$$\min \sum_{i=1}^{n}\sum_{j=1}^{n} c_{ij} x_{ij}$$

$$\text{s.t.} \sum_{j=1}^{n} x_{ij} = 1 \qquad\qquad \text{for } i = 1, \ldots, n \qquad\qquad (8.6)$$

$$\sum_{i=1}^{n} x_{ij} = 1 \qquad\qquad \text{for } j = 1, \ldots, n \qquad\qquad (8.7)$$

$$x_{ij} \in \{0, 1\} \qquad\qquad \text{for } i, j = 1, \ldots, n. \qquad\qquad (8.8)$$

Note that the assignment problem is a special case of the transportation problem; see Section 8.2.1. The constraints of (8.6) express the fact that each person has precisely one job to do, while the constraints of (8.7) express the fact that each job has to be done by precisely one person. Because of (8.6) and (8.7), the expression (8.8) can be replaced by '$x_{ij} \geq 0$ and integer for $i, j = 1, \ldots, n$'. (Why this is true?) One can easily verify that the technology matrix corresponding to (8.6) and (8.7) is precisely the same as the one of the transportation problem. Therefore, Theorem 8.1.1 and Theorem 8.1.6 can be applied, and so the assignment problem has an integer optimal solution. (Why is this model feasible and bounded?) Also note that the ILO-model of the assignment problem is the same as the one for the traveling salesman problem without the subtour-elimination constraints; see Section 7.2.4.

8.2.4   The minimum cost flow problem

The *minimum cost flow problem* is to find a flow of minimum cost from a source node to a sink node in a network. One can think of the problem of sending quantities of commodities from a depot, through a network, to a destination while minimizing the total costs. The network can, for example, represent a sewage system, where the arcs represent water pipes, and the nodes represent connection points. Or the network can represent a computer network, where the arcs represent optical fiber cables and the nodes represent hubs.

Consider for instance the network in Figure 8.6. There are seven nodes: the source node 1, the sink node 7, and the intermediate nodes labeled 2, 3, 4, 5, and 6. Precisely 11 units of flow are to be sent into the network at node 1, along the arcs of the network, towards node 7. At node 7, 11 units of flow are to be received. We say that there is an *external flow* of 11 into node 1 and an external flow of 11 out of node 7. For each arc (i, j), define the decision variable $x_{(i,j)}$ that denotes the number of units of flow that are sent along arc (i, j). Define the *flow into node i* as the total amount of flow that enters node i. For example, the flow into node 4 is $x_{(2,4)} + x_{(3,4)}$. Similarly, the *flow out of node i* is the total amount of flow that leaves node i. It is assumed to no flow is lost, meaning that, for each $i = 1, \ldots, 7$, the flow into node i and the flow out of node i are equal. For instance, for node 4, we require $x_{(2,4)} + x_{(3,4)} = x_{(4,7)}$ or, equivalently, $x_{(2,4)} + x_{(3,4)} - x_{(4,7)} = 0$. Moreover, associated with each arc (i, j) is a pair $c_{(i,j)}, k_{(i,j)}$. Here, $c_{(i,j)}$ denotes the cost per unit of flow, and $k_{(i,j)}$ denotes the capacity of arc (i, j). Sending $x_{(i,j)}$ units of flow along arc (i, j) costs $x_{(i,j)} c_{(i,j)}$. At most $k_{(i,j)}$ units of flow can be sent along arc (i, j).

In general, let $G = (\mathcal{V}, \mathcal{A})$ be a loopless directed graph that has node set $\mathcal{V}$ and arc set $\mathcal{A}$. Let $m = |\mathcal{V}|$ and $n = |\mathcal{A}|$ ($m \geq 2$, $n \geq 1$). We will assume without loss of generality that node 1 is the source node and node m is the sink node of the network. We assume that there are no arcs entering the source node 1, and there are no arcs leaving the sink node m. An arc is denoted as an ordered pair (i, j) with tail node i and head node j. Let $\mathbf{M}_G$ be the node-arc incidence matrix associated with graph G. Let $\mathbf{k}$ be the vector of the arc capacities, and $\mathbf{c}$ the vector of the costs per unit of flow on the arcs. Let $\mathbf{b} = \begin{bmatrix} b_1 & \ldots & b_m \end{bmatrix}^{\mathsf{T}}$ be the external flow vector, i.e., the entries of $\mathbf{b}$ are the differences between the external

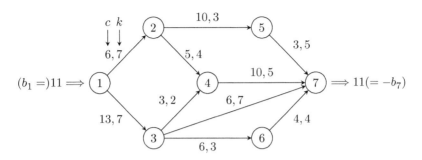

Figure 8.6: Network with capacities and costs on the arcs.

flow out of the nodes and the external flow into the nodes. This implies that the value of b_m is negative. For example, in Figure 8.6, we have that $b_7 = -11$. It is assumed that there is no loss of flow on the various nodes, so that $b_2 = \ldots = b_{m-1} = 0$. This implies that the following *supply-demand balance equation* holds:

$$\sum_{i=1}^{m} b_i = 0.$$

In particular, because there is one source node and one sink node, it follows that $b_1 = -b_m$. We will write $b_1 = v_0$ (the flow that leaves the source node 1), and $b_m = -v_0$ (the flow that enters the sink node m). Let $x_{(i,j)}$ be the flow (number of units of flow) through the arc (i,j). The vector $\mathbf{x}$ ($\in \mathbb{R}^n$) with entries $x_{(i,j)}$, $(i,j) \in \mathcal{A}$, is called the *flow vector*. The minimum cost flow problem can now be formulated as follows.

Model 8.2.4. (*Minimum cost flow problem*)

$$\min \quad \sum_{(i,j)\in\mathcal{A}} c_{(i,j)} x_{(i,j)}$$

$$\text{s.t.} \quad \sum_{(1,j)\in\mathcal{A}} x_{(1,j)} = v_0$$

$$-\sum_{(i,m)\in\mathcal{A}} x_{(i,m)} = -v_0$$

$$-\sum_{(i,j)\in\mathcal{A}} x_{(i,j)} + \sum_{(j,i)\in\mathcal{A}} x_{(j,i)} = 0 \qquad \text{for } i = 2, \ldots, m-1$$

$$0 \le x_{(i,j)} \le k_{(i,j)} \qquad \text{for } (i,j) \in \mathcal{A}.$$

The reason why we have written a '$-$' sign in the second constraint is that we want to have the technology matrix, excluding the capacity constraints, to be equal to the corresponding node-arc incidence matrix. The objective function $\mathbf{c}^{\mathsf{T}}\mathbf{x} = \sum_{(i,j)\in\mathcal{A}} c_{(i,j)} x_{(i,j)}$ is the total cost of the flow vector $\mathbf{x}$. The first and the second constraints guarantee that both the total flow out of the source node and the total flow into the sink node are v_0. The third set of

						Arcs					
b	i	(1,2)	(1,3)	(2,5)	(2,4)	(3,4)	(3,7)	(3,6)	(4,7)	(5,7)	(6,7)
11	1	1	1	0	0	0	0	0	0	0	0
0	2	-1	0	1	1	0	0	0	0	0	0
0	3	0	-1	0	0	1	1	1	0	0	0
0	4	0	0	0	-1	-1	0	0	1	0	0 $= \mathbf{M}_G$
0	5	0	0	-1	0	0	0	0	0	1	0
0	6	0	0	0	0	0	0	-1	0	0	1
-11	7	0	0	0	0	0	-1	0	-1	-1	-1
		7	7	3	4	2	7	3	5	5	4 $= \mathbf{k}^\mathsf{T}$
		6	13	10	5	3	6	6	10	3	4 $= \mathbf{c}^\mathsf{T}$

Table 8.1: Input data corresponding to the network of Figure 8.6.

constraints guarantees that there is no loss of flow on the intermediate nodes. To understand these constraints, notice that, for $i = 1, \ldots, m$,

$$\sum_{(i,j)\in\mathcal{A}} x_{(i,j)} = \text{total flow into node } i, \quad \text{and} \quad \sum_{(j,i)\in\mathcal{A}} x_{(j,i)} = \text{total flow out of node } i.$$

The last constraints of Model 8.2.4 are the nonnegativity constraints and the capacity constraints.

Note that if both $v_0 = 1$ and the constraints $x_{(i,j)} \leq k_{(i,j)}$ are removed from the model, then the minimum cost flow problem reduces to the problem of determining a *shortest path* from the source node to the sink node; see Appendix C.

Consider again the network of Figure 8.6. The values of the input data vectors $\mathbf{b}$, $\mathbf{c}$, $\mathbf{k}$, and the matrix $\mathbf{M}_G$ of the minimum cost flow problem of Figure 8.6 are listed in Table 8.1. The corresponding ILO-model now reads as follows:

$$
\begin{aligned}
\min \ & 6x_{(1,2)} + 13x_{(1,3)} + 10x_{(2,5)} + 5x_{(2,4)} + 3x_{(3,4)} \\
& + 6x_{(3,7)} + 6x_{(3,6)} + 10x_{(4,7)} + 3x_{(5,7)} + 4x_{(6,7)} \\
\text{s.t.} \quad & x_{(1,2)} + x_{(1,3)} & = \ & 11 \\
& -x_{(1,2)} + x_{(2,5)} + x_{(2,4)} & = \ & 0 \\
& -x_{(1,3)} + x_{(3,4)} + x_{(3,6)} + x_{(3,7)} & = \ & 0 \\
& -x_{(2,4)} - x_{(3,4)} + x_{(4,7)} & = \ & 0 \\
& -x_{(2,5)} + x_{(5,7)} & = \ & 0 \\
& -x_{(3,6)} + x_{(6,7)} & = \ & 0 \\
& -x_{(3,7)} - x_{(4,7)} - x_{(5,7)} - x_{(6,7)} & = \ & -11 \\
& 0 \leq x_{(1,2)} \leq 7, 0 \leq x_{(1,3)} \leq 7, 0 \leq x_{(2,5)} \leq 3, 0 \leq x_{(2,4)} \leq 4, \\
& 0 \leq x_{(3,4)} \leq 2, 0 \leq x_{(3,7)} \leq 7, 0 \leq x_{(3,6)} \leq 3, 0 \leq x_{(4,7)} \leq 5, \\
& 0 \leq x_{(5,7)} \leq 5, 0 \leq x_{(6,7)} \leq 4.
\end{aligned}
$$

According to Theorem 8.1.7, the matrix $\mathbf{M}_G$ that corresponds to the first seven constraints is totally unimodular, because it is precisely the node-arc incidence matrix of the network of Figure 8.6. The capacity constraints give rise to the identity matrix $\mathbf{I}_{|\mathcal{A}|}$ ($|\mathcal{A}| = 10$ in the above example), and so the technology matrix is $\begin{bmatrix} \mathbf{M}_G \\ \mathbf{I}_{|\mathcal{A}|} \end{bmatrix}$. If we are able to prove that $\begin{bmatrix} \mathbf{M}_G & \mathbf{0} \\ \mathbf{I}_{10} & \mathbf{I}_{10} \end{bmatrix}$ is totally unimodular, then Theorem 8.1.8 and Theorem 8.1.1 can be used to show that there is an integer optimal solution. (Note that the model contains both equality and inequality constraints, and the 'second' $\mathbf{I}_{10}$ in the latter matrix refers to the slack variables of the capacity constraints. Also note that the model is bounded.) That is exactly what the following theorem states.

Theorem 8.2.1.
If $\mathbf{M}$ is a totally unimodular (p, q)-matrix $(p, q \geq 1)$, then so is $\begin{bmatrix} \mathbf{M} & \mathbf{0} \\ \mathbf{I}_q & \mathbf{I}_q \end{bmatrix}$, where $\mathbf{0}$ is the all-zero (p, q)-matrix.

Proof. Take any square submatrix $\mathbf{B}$ of $\begin{bmatrix} \mathbf{M} & \mathbf{0} \\ \mathbf{I}_q & \mathbf{I}_q \end{bmatrix}$. We may write $\mathbf{B} = \begin{bmatrix} \mathbf{B}_1 & \mathbf{0} \\ \mathbf{B}_2 & \mathbf{B}_3 \end{bmatrix}$, where $\mathbf{B}_1$ is a submatrix of $\mathbf{M}$, and $\mathbf{B}_2$ and $\mathbf{B}_3$ are submatrices of $\mathbf{I}_q$. If $\mathbf{B}_1$ is an empty matrix, then $\mathbf{B} = \mathbf{B}_3$, and hence $\det \mathbf{B} = \det \mathbf{B}_3 \in \{0, \pm 1\}$. So we may assume that $\mathbf{B}_1$ is not an empty matrix. If $\mathbf{B}_3$ contains a column of zeroes, then $\mathbf{B}$ contains a column of zeroes, and hence $\det \mathbf{B} = 0$. So, we may assume that $\mathbf{B}_3$ is an identity matrix, say $\mathbf{B}_3 = \mathbf{I}_r$, with $r \geq 0$. It follows that $\mathbf{B}_1$ is a square submatrix of $\mathbf{M}$. Hence, $\mathbf{B} = \begin{bmatrix} \mathbf{B}_1 & \mathbf{0} \\ \mathbf{B}_3 & \mathbf{I}_r \end{bmatrix} \sim \begin{bmatrix} \mathbf{B}_1 & \mathbf{0} \\ \mathbf{0} & \mathbf{I}_r \end{bmatrix}$, so that $\det \mathbf{B} = \pm \det \mathbf{B}_1 \in \{0, \pm 1\}$, because $\mathbf{B}$ is a square submatrix of $\mathbf{M}$, and $\mathbf{M}$ is totally unimodular. $\square$

So, it follows from Theorem 8.2.1 that the technology matrix of the minimum cost flow problem is totally unimodular.

8.2.5 The maximum flow problem; the max-flow min-cut theorem

In this section we describe two network problems. The first network problem is the so-called 'maximum flow problem', which is a variation of the minimum cost flow problem (see Section 8.2.4). In the maximum flow problem, the objective is not to send a given amount of flow from the source node to the sink node at minimum total costs, but to send as much flow as possible from the source node to the sink node, taking into account the capacities of the arcs. The second network problem that we consider is the 'minimum cut problem'. Informally speaking, this problem asks for the minimum cost way to disconnect the source node and the sink node of the network by deleting arcs. The optimal objective value of this problem can be seen as a measure of the reliability of the network.

We will explore the connection between these two problem. We will see that, as is the case with all network problems in this chapter, both problems can be written as LO-models. More interestingly, we will see that the two resulting LO-models are each other's dual models. This fact is expressed in the so-called 'max-flow min-cut theorem'.

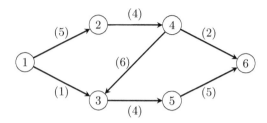

Figure 8.7: Network with capacities. The numbers in parentheses are the capacities $k_{(i,j)}$.

8.2.5.1 **The maximum flow problem**

Consider a network $G = (\mathcal{V}, \mathcal{A})$, where $\mathcal{V} = \{1, \ldots, m\}$ is the set of nodes and $\mathcal{A}$ is the set of arcs. The network has a single source node (with only outgoing arcs) and a single sink node (with only incoming arcs). All other nodes are called intermediate nodes. Without loss of generality, we assume that node 1 is the source node and node n is the sink node. Define the following decision variables:

$$x_{(i,j)} = \text{the amount of flow from node } i \text{ to node } j \text{ along arc } (i,j), \text{ for } (i,j) \in \mathcal{A};$$
$$z = \text{the net flow out of node } 1.$$

For every arc (i,j), we have a *capacity* $k_{(i,j)}$ (> 0). The flow variables should satisfy

$$0 \leq x_{(i,j)} \leq k_{(i,j)} \qquad \text{for } (i,j) \in \mathcal{A}.$$

We assume that flow is preserved at the intermediate nodes, meaning that, at each intermediate node k (i.e., $k \in \{2, \ldots, m-1\}$), the total flow $\sum_{(i,k)\in\mathcal{A}} x_{(i,k)}$ into node k equals the total flow $\sum_{(k,j)\in\mathcal{A}} x_{(k,j)}$ out of node k. The objective of the maximum flow problem is to find the maximum amount of flow that can be sent from the source node 1 to the sink node m. Since flow is preserved at the intermediate nodes, it should be intuitively clear that the amount of flow that is sent through the network from node 1 to node k is equal to the total flow z out of the source node, and also to the total flow into the sink node. The reader is asked to prove this fact in Exercise 8.4.17.

The *maximum flow problem* can now be formulated as the following LO-model:

Model 8.2.5. (*Maximum flow problem*)

$$\max z$$
$$\text{s.t.} \sum_{(i,j)\in\mathcal{A}} x_{(i,j)} - \sum_{(j,i)\in\mathcal{A}} x_{(j,i)} - \alpha_i z = 0 \qquad \text{for } i = 1, \ldots, m \qquad (8.9)$$
$$x_{(i,j)} \leq k_{(i,j)} \qquad \text{for } (i,j) \in \mathcal{A} \qquad (8.10)$$
$$x_{(i,j)} \geq 0 \qquad \text{for } (i,j) \in \mathcal{A},$$

where $\alpha_1 = 1$, $\alpha_m = -1$, and $\alpha_i = 0$ for $i = 2, \ldots, m-1$.

The first set of constraints (8.9) has the following meaning. For each $i = 1, \ldots, m$, the first summation in the left hand side is the total flow out of node i, and the second summation is the total flow into node i. Hence, the left hand side is the net flow out of node i. So, for the intermediate nodes $2, \ldots, m-1$, these constraints are exactly the flow preservation constraints. For $i = 1$, the constraint states that the net flow out of the source node should be equal to z. Similarly, for $i = m$, the constraint states that the net flow into the sink node should be equal to z. The second set of constraints (8.10) contains the capacity constraints and the nonnegativity constraints.

Example 8.2.1. *Consider the network in Figure 8.7. The network has six nodes: the source node 1, the sink node 6, and the intermediate nodes 2, 3, 4, and 5. The objective is to send z units of flow from node 1 to node 6, while maximizing z. Analogously to the minimum cost flow problem, every arc (i, j) has a capacity $k_{(i,j)}$. However, there are no costs associated with the arcs. The LO-model corresponding to this network is:*

$$
\begin{aligned}
\max \quad & z \\
\text{s.t.} \quad & x_{(1,2)} + x_{(1,3)} && - z = 0 \\
& -x_{(1,2)} + x_{(2,4)} && = 0 \\
& -x_{(1,3)} + x_{(3,5)} - x_{(4,3)} && = 0 \\
& -x_{(2,4)} + x_{(4,3)} + x_{(4,6)} && = 0 \\
& -x_{(3,5)} + x_{(5,6)} && = 0 \\
& -x_{(4,6)} - x_{(5,6)} + z = 0 \\
& x_{(1,2)} && \leq 5 \\
& x_{(1,3)} && \leq 1 \\
& x_{(2,4)} && \leq 4 \\
& x_{(3,5)} && \leq 4 \\
& x_{(4,3)} && \leq 6 \\
& x_{(4,6)} && \leq 2 \\
& x_{(5,6)} && \leq 5 \\
& x_{(1,2)}, x_{(1,3)}, x_{(2,4)}, x_{(3,5)}, x_{(4,3)}, x_{(4,6)}, x_{(5,6)}, z \geq 0.
\end{aligned}
$$

Figure 8.8(a) shows an example of a feasible solution for the network in Figure 8.7. Notice that the capacities are respected and no flow is lost. Figure 8.8(b) shows a feasible solution with objective value 5. We will see in the next section that, in fact, this feasible solution is optimal.

8.2.5.2 The minimum cut problem

In Figure 8.8(b), we have depicted a feasible solution with objective value 5 for the maximum flow problem depicted in Figure 8.7. In fact, this feasible solution is optimal. But how do we know that this feasible solution is indeed optimal? There is an elegant way to determine this. To that end, we introduce the notion of a 'cut'. A *cut* in a network $G = (\mathcal{V}, \mathcal{A})$ is a subset C

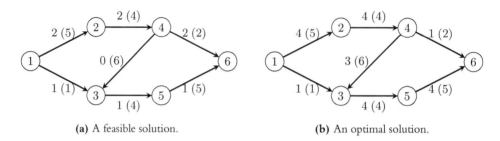

(a) A feasible solution.　　　　　　　　(b) An optimal solution.

Figure 8.8: A feasible and an optimal solution of the maximum flow problem of Figure 8.7. The numbers on the arcs are the flow values and the numbers in parentheses are the capacities.

of $\mathcal{A}$, such that after deleting the arcs of C from the network G, there are no directed paths (see Appendix C) from the source node 1 to the sink node m (with $m = |\mathcal{V}|$). For example, Figure 8.9 depicts two cuts in the network of Figure 8.7. The sets $C_1 = \{(1,2),(1,3)\}$ (Figure 8.9(a)) and $C_2 = \{(2,4),(5,6)\}$ (Figure 8.9(b)) are cuts. The arcs in the cuts are drawn with solid lines. The dotted lines show how the sets C_1 and C_2 'cut' the paths from the source node to the sink node. Note that the definition of a cut states that after removing the arcs in the cut, there are no <u>directed</u> paths from node 1 to node m left. Hence, in Figure 8.9(b), the arc $(4,3)$ is not part of the cut C_2 (although adding $(4,3)$ results in another cut). On the other hand, for instance, $\{(1,3),(3,5),(5,6)\}$ is not a cut, because after deleting the arcs in that set there is still a path from node 1 to node 6 in the graph, namely the path $1{\to}2{\to}4{\to}6$.

Cuts have the following useful property. If C is a cut, then every unit of flow from the source node to the sink node has to go through at least one of the arcs in C. This fact is also illustrated by the cuts in Figure 8.9: every unit of flow from node 1 to node m has to cross the dotted line, and hence has to go through one of the arcs in the cut. This property implies that if C is a cut, then we can use the capacities of the arcs of C to find an upper bound on the maximum amount of flow that can be sent from node 1 to node m. To be precise, define the *capacity* $k(C)$ of any cut C as:

$$k(C) = \sum_{(i,j) \in C} k_{(i,j)}.$$

Then, the total amount of flow from node 1 to node m can be at most $k(C)$. Therefore, the optimal objective value z^* of Model 8.2.5 is at most $k(C)$. So, we have that:

$$z^* \leq k(C) \text{ for every cut } C, \tag{8.11}$$

or, equivalently, that:
$$z^* \leq \min\{k(C) \mid C \text{ is a cut of } G\}.$$

Therefore, every cut implies an upper bound on the optimal value of the maximum flow problem.

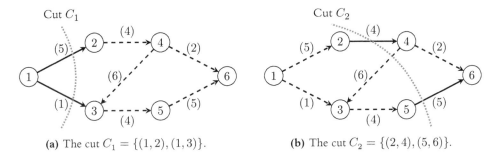

(a) The cut $C_1 = \{(1,2),(1,3)\}$.   **(b)** The cut $C_2 = \{(2,4),(5,6)\}$.

Figure 8.9: Cuts in a network.

Example 8.2.2. *Consider again Figure 8.8(b), and take $C^* = \{(1,3),(2,4)\}$. The reader should check that C^* is a cut. The capacity of C^* is $(1 + 4 =)\ 5$. Hence, we know from (8.11) that no feasible solution has an objective value larger than 5. Since the objective value of the feasible solution in Figure 8.8(b) is exactly 5, this implies that this feasible solution is in fact optimal.*

The above discussion begs the question: can we find the best upper bound on z^*? In other words, can we find a cut that has the minimum capacity? This is the so-called *minimum cut problem*. It asks the question: which cut C minimizes $k(C)$? In other words, it asks to determine a cut $C^* \in \operatorname{argmin}\{k(C) \mid C \text{ is a cut of } G\}$. We will show that the minimum cut problem corresponds to the dual model of the maximum flow problem.

We first derive the dual model of Model 8.2.5 as follows. Recall that this primal model has two sets of constraints: the flow preservation constraints (8.9), and the 'capacity constraints' (8.10). Since there is a dual decision variable for each constraint of the primal problem, we introduce the following two sets of dual variables:

$$y_i \quad = \text{ the dual variable for (8.9) corresponding to node } i\ (\in \mathcal{V});$$
$$w_{(i,j)} \quad = \text{ the dual variable for (8.10) corresponding to arc } (i,j)\ (\in \mathcal{A}).$$

Note that the constraints in (8.9) are '$=$' constraints, and hence the dual variable y_i (for each $i \in \mathcal{V}$) is unrestricted in sign. The constraints in (8.10) are '$\leq$' constraints, and so the value of $w_{(i,j)}$ should be nonnegative for all $(i,j) \in \mathcal{A}$. The right hand side values of the constraints in (8.9) are all 0, and the right hand side values of the constraints in (8.10) are the $k_{(i,j)}$'s. Since the right hand side values of the primal model become the objective coefficients of the dual model, this means that the objective of the dual model reads:

$$\min \sum_{(i,j)\in\mathcal{A}} k_{(i,j)} w_{(i,j)}. \tag{8.12}$$

In order to find the constraints of the dual model, we need to transpose the technology matrix of the primal model. This primal technology matrix $\mathbf{A}$ reads:

$$\mathbf{A} = \begin{bmatrix} \mathbf{T} & \mathbf{e}_m {-} \mathbf{e}_1 \\ \mathbf{I}_n & \mathbf{0} \end{bmatrix},$$

where $m = |\mathcal{V}|$, $n = |\mathcal{A}|$, $\mathbf{T}$ is the node-arc incidence matrix of G, and $\mathbf{e}_i$ is the usual i'th unit vector in $\mathbb{R}^m$ $(i \in \mathcal{V})$. The columns of $\mathbf{A}$ correspond to the primal decision variables $x_{(i,j)}$ with $(i,j) \in \mathcal{A}$, except for the last column, which corresponds to z. The rows of $\mathbf{A}$ correspond to the primal constraints.

Recall that there is a constraint in the dual model for every decision variable in the primal model, and that the coefficients of any dual constraint are given by the corresponding column of the primal technology matrix. Consider the decision variable $x_{(i,j)}$ with $(i,j) \in \mathcal{A}$. The column in $\mathbf{A}$ corresponding to $x_{(i,j)}$ contains three nonzero entries: the entry corresponding to the flow preservation constraint for node i (which is 1), the entry corresponding to the flow preservation constraint for node j (which is -1), and the entry corresponding to the capacity constraint $x_{(i,j)} \leq k_{(i,j)}$ (which is 1). Thus, the left hand side of the dual constraint corresponding to the primal decision variable $x_{(i,j)}$ is $y_i - y_j + w_{(i,j)}$. The primal decision variable $x_{(i,j)}$ is nonnegative, so the constraint is a '$\geq$' constraint. The right hand side is the primal objective coefficient of $x_{(i,j)}$, which is 0. Hence, the dual constraint corresponding to the primal decision variable $x_{(i,j)}$ is:

$$y_i - y_j + w_{(i,j)} \geq 0 \qquad \text{for } (i,j) \in \mathcal{A}. \tag{8.13}$$

The remaining primal decision variable is z. Consider the column of $\mathbf{A}$ corresponding to z. It contains a -1 for the entry corresponding to node 1, and a $+1$ for the entry corresponding to node m. The primal objective coefficient of z is 1, and hence the corresponding dual constraint is:

$$y_m - y_1 \geq 1. \tag{8.14}$$

Combining (8.12), (8.13), and (8.14), we find the following dual model:

Model 8.2.6. (*Dual model of Model 8.2.5*)

$$\min \sum_{(i,j) \in \mathcal{A}} k_{(i,j)} w_{(i,j)} \tag{8.12}$$

$$\text{s.t. } y_i - y_j + w_{(i,j)} \geq 0 \qquad \text{for } (i,j) \in \mathcal{A} \tag{8.13}$$

$$y_m - y_1 \geq 1 \tag{8.14}$$

$$z_{(i,j)} \geq 0 \qquad \text{for } (i,j) \in \mathcal{A}.$$

The following theorem relates the optimal solutions of Model 8.2.6 to the minimum cut(s) in the graph G. The theorem states that Model 8.2.6 solves the minimum cut problem.

Theorem 8.2.2.
Any integer-valued optimal solution of Model 8.2.6 corresponds to a minimum cut in G, and vice versa.

Proof. Consider any integer-valued optimal solution $\begin{bmatrix} \mathbf{y}^* \\ \mathbf{w}^* \end{bmatrix}$ of Model 8.2.6, with corresponding optimal objective value z^*. Let N_1 be the set of nodes i such that $y_i^* \leq y_1^*$, and let N_2 be the set of nodes i such that $y_i^* \geq y_1^* + 1$. Note that since $\begin{bmatrix} \mathbf{y}^* \\ \mathbf{w}^* \end{bmatrix}$ is integer-valued, we have that $N_1 \cup N_2 = \{1, \ldots, m\}$. Clearly, we also have that $1 \in N_1$ and, since $y_m^* - y_1^* \geq 1$, we have that $m \in N_2$. Define $C = \{(i, j) \in \mathcal{A} \mid i \in N_1, j \in N_2\}$. Because $1 \in N_1$ and $m \in N_2$, any path $v_1 \to v_2 \to \ldots \to v_k$ with $v_1 = 1$ and $v_k = m$, must contain an arc $(v_p, v_{p+1}) \in C$ (with $p \in \{1, \ldots, k-1\}$). Therefore, removing the arcs of C from G removes all directed paths from node 1 to node m, and hence C is a cut.

To show that C is a minimum cut, we first look at the capacity of C. Observe that it follows from the above definition of C that $y_j^* - y_i^* \geq (y_1^* + 1) - y_1^* \geq 1$ for $(i, j) \in C$. Moreover, (8.13) implies that $w_{(i,j)}^* \geq y_j^* - y_i^*$ for all $(i, j) \in \mathcal{A}$. Hence, $w_{(i,j)}^* \geq 1$ for $(i, j) \in C$. Using this observation, we find that:

$$
\begin{aligned}
z^* = \sum_{(i,j) \in \mathcal{A}} k_{(i,j)} w_{(i,j)}^* &\geq \sum_{(i,j) \in C} k_{(i,j)} w_{(i,j)}^* \qquad \text{(because } k_{(i,j)}, w_{(i,j)}^* \geq 0 \text{ for } (i,j) \in \mathcal{A}) \\
&\geq \sum_{(i,j) \in C} k_{(i,j)} = k(C). \qquad \text{(because } w_{(i,j)}^* \geq 1 \text{ for } (i,j) \in C)
\end{aligned}
$$

Now suppose for a contradiction that there exists some other cut C' with $k(C') < k(C)$. Let N_1' be the set of nodes of G that are reachable from node 1 by using a directed path without using any arc in C, and let N_2' be the set of all other nodes. Define $y_i' = 0$ for $i \in N_1'$, $y_i' = 1$ for $i \in N_2'$, and $w_{(i,j)}' = \max\{0, y_j' - y_i'\}$ for $(i,j) \in \mathcal{A}$. We have that $y_1' = 0$ and, because C' is a cut, we have that $m \in N_2'$ and hence $y_m' = 1$. Thus, $\begin{bmatrix} \mathbf{y}' \\ \mathbf{w}' \end{bmatrix}$ satisfies the constraints of Model 8.2.6. Moreover,

$$
y_j' - y_i' = \begin{cases} 1 & \text{if } i \in N_1', j \in N_2' \\ 0 & \text{if either } i, j \in N_1', \text{ or } i, j \in N_2' \\ -1 & \text{if } i \in N_2', j \in N_1'. \end{cases}
$$

Hence,

$$
w_{(i,j)}' = \max\{0, y_j' - y_i'\} = \begin{cases} 1 & \text{if } (i,j) \in C' \\ 0 & \text{otherwise.} \end{cases}
$$

This means that the objective value z' corresponding to $\begin{bmatrix} \mathbf{y}' \\ \mathbf{w}' \end{bmatrix}$ satisfies:

$$
z' = \sum_{(i,j) \in \mathcal{A}} k_{(i,j)} \max\{0, y_j' - y_i'\} = \sum_{(i,j) \in C'} k_{(i,j)} = k(C') < k(C) \leq z^*,
$$

contrary to the fact that z^* is the optimal objective value of Model 8.2.6. So, C is indeed a minimum cut. $\qquad \square$

It now follows from the duality theorem, Theorem 4.2.4, that Model 8.2.5 and Model 8.2.6 have the same objective value. So, we have proved the following important theorem, the so-called *max-flow min-cut theorem*, which is a special case of Theorem 4.2.4:

Project i	Revenue r_i	Dependencies	Machine j	Cost c_j
1	100	Requires machines 1 and 2	1	200
2	200	Requires machine 2	2	100
3	150	Requires machine 3	3	50

Table 8.2: Projects and machines.

Theorem 8.2.3. (*Max-flow min-cut theorem*)
Let G be a network with arc weights, arc capacities, and a source and a sink node. The maximum amount of flow that can be sent from the source node to the sink node is equal to the minimum cut capacity of that network.

The following example illustrates the maximum flow problem and Theorem 8.2.3.

Example 8.2.3. (Project selection problem) *In the project selection problem, there are m projects, labeled $1, \dots, m$, and n machines, labeled $1, \dots, n$. Each project i yields revenue r_i and each machine j costs c_j to purchase. Every project requires the usage of a number of machines that can be shared among several projects. The problem is to determine which projects should be selected and which machines should be purchased in order to execute the selected projects. The projects and machines should be selected so that the total profit is maximized. Table 8.2 lists some example data for the project selection problem.*

Let $P \subseteq \{1, \dots, m\}$ be the set of projects that are <u>not</u> selected, and let $Q \subseteq \{1, \dots, n\}$ be the set of machines that should be purchased. Then, the project selection problem can be formulated as follows:

$$z^* = \max \sum_{i=1}^{m} r_i - \sum_{i \in P} r_i - \sum_{j \in Q} c_j.$$

Note that the term $\sum_{i=1}^{m} r_i$ is a constant, and hence it is irrelevant for determining an optimal solution. Also, recall that instead of maximizing the objective function, we may equivalently minimize the negative of that function. So, the above problem can be formulated as a minimizing model as follows:

$$\min \sum_{i \in P} r_i + \sum_{j \in Q} c_j. \tag{8.15}$$

The above minimizing model can now be formulated as a minimum cut problem as follows. Construct a network G with nodes s, $u_1, \dots, u_m$, $v_1, \dots, v_n$, and t. The nodes $u_1, \dots, u_m$ represent the projects, and the nodes $v_1, \dots, v_n$ represent the machines. Let $i = 1, \dots, m$ and $j = 1, \dots, n$. The source node s is connected to the 'project node' u_i by an arc with capacity r_i. The sink t is connected to the 'machine node' v_j by an arc with capacity c_j $(j = 1, \dots, n)$. The arc (u_i, v_j) with infinite capacity is added if project i requires machine j.

Since the arc from u_i to v_j has infinite capacity, this arc does not appear in a minimum cut (otherwise the capacity of that cut would be infinite, whereas taking all arcs incident with s constitute already a cut with finite capacity). Hence, a minimum cut consists of a set of arcs incident with s and arcs incident

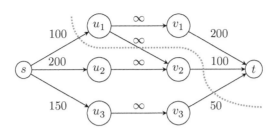

Figure 8.10: The project selection problem as a maximum flow model. The dotted curve represents the unique minmum cut.

with t. The arcs in the cut that are incident with s represent the projects in P, and the arcs in the cut incident with t represent the machines in Q. The capacity of the cut is $\sum_{i \in P} r_i + \sum_{j \in Q} c_j$, which is the objective function of (8.15). By Theorem 8.2.3, one can solve this minimum cut problem as a maximum flow problem.

Figure 8.10 shows a network representation of the project selection problem corresponding to the data in Table 8.2. The minimum cut $C^* = \{(s, u_1), (v_2, t), (v_3, t)\}$ is represented by the dotted curve in Figure 8.10. Note that the arc (u_1, v_2) crosses the dotted line, but it crosses from the side containing t to the side containing s. So, its capacity (∞) is not included in the capacity of the cut. Hence, the corresponding capacity $k(C^*)$ is 250. The corresponding set P consists of the indices i such that $(s, u_i) \in C^*$, and the set Q consists of the indices j such that $(v_j, t) \in C^*$. So, $P = \{1\}$ and $Q = \{2, 3\}$. The total revenue of the projects is 450. Therefore, the corresponding maximum profit z^* is $(450 - 250 =)$ 200, which is attained by selecting projects 2 and 3, and purchasing machines 2 and 3.

8.2.6 Project scheduling; critical paths in networks

A *project* is a combination of interrelated *activities* that need to be carried out. Each activity requires a certain amount of time, called the *execution time* of the activity. In addition, there are *precedence constraints*, meaning that some activities cannot start until others have been completed. An activity does not have to be started immediately after the necessary preceding activities are completed. We want to construct a *schedule* that describes the order in which the activities need to be carried out, while respecting the precedence constraints. The objective is to determine a schedule such that the total amount of time required to execute this schedule, the so-called *total completion time*, is minimized.

A project can be represented by a so-called *project network*, which is a network with the following properties:

(i) The nodes refer to *goals* (or *milestones*).

(ii) Each activity of the project is represented by precisely one arc in the network.

(iii) The network does not contain a directed cycle.

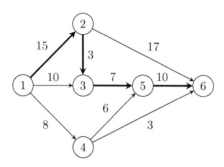

Figure 8.11: Project network. The thick arcs form a critical path.

A goal is realized when all activities corresponding to its incoming arcs have been completed. Moreover, the activities corresponding to a goal's outgoing arcs can start only when the goal has been realized. Clearly, the goals should be chosen in such a way that the precedence constraints of the activities are represented in the network.

Example 8.2.4. *Consider the network of Figure 8.11. There are six nodes, corresponding to the goals 1, 2, 3, 4, 5, and 6. The numbers attached to the arcs refer to the execution times, i.e., the number of time units (days) required to perform the activities corresponding to the arcs. For instance, the activity of arc $(4, 5)$ requires six days. The network depicts the precedence relationships between the activities. For instance, activity $(3, 5)$ cannot start until both the activities $(1, 3)$ and $(2, 3)$ are completed.*

As the following example illustrates, it is not always immediately clear how to construct a project network for a given project,

Example 8.2.5. *Consider the following project with four activities, A, B, C, and D. Activity C can be started when activity A is completed, and activity D can be started when both activities A and B are completed. How should the arcs corresponding to the four activities be arranged? Because of the precedence constraints, it seems that the arcs corresponding to A and C should share a node (representing a goal) in the network, and similarly for A and D, and for B and D. There is only one arrangement of the arcs that respects these requirements, and this arrangement is depicted in the network of Figure 8.12(a). This network, however, does not correctly model the precedence constraints of the project, because the network indicates that activity C can only be started when B is completed, which is not a requirement of the project. To model the project's precedence constraints correctly, a dummy arc needs to be added. This dummy arc corresponds to a dummy activity with zero execution time. The resulting network is depicted in Figure 8.12(b). The reader should verify that the precedence constraints are correctly reflected by this project network.*

The arcs cannot form a directed cycle, because otherwise some activities can never be carried out. For example, if the network has three nodes, labeled 1, 2, and 3, and the arcs are $(1, 2)$, $(2, 3)$, $(3, 1)$, then goal 3 needs to be realized before goal 1, goal 2 before goal 3, and goal 1 before 3; hence, goal 1 must be realized before itself, which is clearly impossible. Thus,

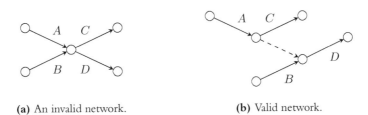

(a) An invalid network.   **(b)** Valid network.

Figure 8.12: Invalid and valid project networks for Example 8.2.5.

the network cannot contain a directed cycle. Such a network is also called an *acyclic directed graph*; see Appendix C.

Project networks contain *initial goals* and *end goals*. An initial goal is a node without incoming arcs, and so an initial goal is realized without doing any activities. An end goal is a node with only incoming arcs. It follows from the fact that project networks do not contain directed cycles that any such network has at least one initial goal and at least one end goal. If there is more than one initial goal, then we can introduce a dummy initial goal connected to the initial goals by arcs with zero execution time. So we may assume without loss of generality that there is exactly one initial goal. Similarly, we may assume that there is exactly one end goal. The objective is then to determine a schedule such that the time between realizing the initial goal and realizing the end goal, called the *total completion time*, is minimized.

Example 8.2.6. *An optimal solution for the project network of Figure 8.11 can be found by inspection. Assume that the time units are days. Realize goal 1 on day 0. Then goal 2 can be realized on day 15, and not earlier than that. Goal 3 can be realized on day $(15 + 3 =)$ 18. Note that, although activity $(1, 3)$ takes ten days, goal 3 cannot be realized on day 10, because activities $(1, 2)$ and $(2, 3)$ need to be completed first. Goal 4 can be realized on day 8. In the case of goal 5, there are two conditions. Its realization cannot take place before day $(18 + 7 =)$ 25 (because goal 3 cannot be realized earlier than on day 18), and also not before day $(8 + 6 =)$ 14 (compare goal 4). So, goal 5 can be realized on day 25. In the case of goal 6, three conditions need to be fulfilled, and these lead to a realization on day $(25 + 10 =)$ 35. Hence, the optimal solution is 35 days, and all calculated realization times are as early as possible.*

Figure 8.13 shows this solution in the form of a so-called Gantt chart[1]. This chart can be read as follows. Time runs from left to right. The days 1 through 35 are shown in the top-most row. The bottom row shows when the several goals of the project are realized. The three rows in between show when the various activities are carried out. So, the project starts at the left of the chart by performing activities $(1, 2)$, $(1, 3)$, and $(1, 4)$. Since goal 1 is the initial goal, it is realized immediately. This fact is illustrated by the flag labeled 1 in the bottom row. After eight days, activity $(1, 4)$ is completed. This means that, at that time, goal 4 is realized. At that time, activities $(1, 2)$ and $(1, 3)$ continue to be carried out, and activity $(4, 5)$ is started. The end goal, goal 6, is realized on day 35.

---

[1]Named after the American mechanical engineer and management consultant HENRY L. GANTT (1861–1919).

It is not necessary to complete all activities as early as possible. For example, for the project schedule in Figure 8.13, it should be clear that activities $(1, 4)$ and $(4, 5)$ may both be delayed by a few days without affecting the other activities and the total completion time of the project. Hence, goal 4 may also be delayed by a few days without affecting the total completion time.

In general, given an optimal solution for the project network, the *earliest starting time* of an activity is the earliest time at which that activity may be started without violating the precedence constraints, and the *latest starting time* of an activity is the latest time at which that activity may be started without increasing the total completion time of the project. The *earliest completion time* of an activity is the earliest start time plus the execution time of that activity, and similarly for the *latest completion time*. The difference between the earliest and the latest starting time of an activity is called the *slack* of that activity.

Example 8.2.7.   *The earliest starting time of activity $(4, 5)$ in Figure 8.13 is the beginning of day 9, because $(4, 5)$ can only be started when activity $(1, 4)$ is finished, and activity $(1, 4)$ requires eight days. In order to not delay the project, activity $(4, 5)$ needs to be completed before the end of day 25, because otherwise the start of activity $(5, 6)$ would have to be delayed, which in turn would increase the total completion time of the project. This means that the latest starting time for activity $(4, 5)$ is the beginning of day 20. Hence, the slack of activity $(5, 6)$ is $(20 - 9 =) 11$. On the other hand, activity $(2, 3)$ has zero slack. This activity cannot start earlier, because of activity $(1, 2)$ that precedes it, and it cannot start later because that would delay activity $(3, 5)$, hence activity $(5, 6)$, and hence the whole project. So, in order to realize the end goal as early as possible, it is crucial that activity $(2, 3)$ starts on time.*

So, an important question is: which activities cannot be postponed without affecting the total completion time? These are exactly the activities with zero slack. For the example above, one can easily check that activities $(1, 2)$, $(2, 3)$, $(3, 5)$, and $(5, 6)$ have zero slack, while the activities $(1, 3)$, $(1, 4)$, $(2, 6)$, $(4, 5)$, and $(4, 6)$ have nonzero slack.

A sequence of goals, from the initial goal to the end goal, consisting of goals that cannot be postponed without increasing the total completion time is called a *critical path* of the activity network. For instance, in the project network of Figure 8.11, $1 \rightarrow 2 \rightarrow 3 \rightarrow 5 \rightarrow 6$ is a critical path. Note that critical paths are in general not unique. Activities on a critical path have zero slack, and – conversely – an activity with zero slack is on a critical path (Exercise 8.4.29). It should be clear that determining the minimum total completion time is equivalent to determining the total completion time of a critical path.

In order to formulate this problem as an LO-model, let 1 be the initial goal and n the end goal; the set of nodes is $\mathcal{V} = \{1, \ldots, n\}$, and the set of arcs is $\mathcal{A}$. The time needed to carry out the activity corresponding to the arc (i, j) is denoted by $c_{(i,j)}$, and it is assumed that $c_{(i,j)} \geq 0$ $(i, j \in \mathcal{V})$. Let t_i be the realization time of goal i $(i \in \mathcal{V})$. The LO-model can

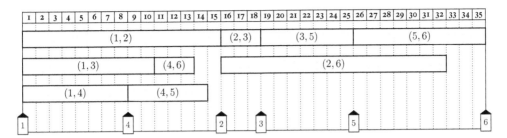

Figure 8.13: Gantt chart for the solution of Figure 8.11.

be formulated as follows:

$$\min\ t_n - t_1$$
$$\text{s.t.}\ \ t_j - t_i \geq c_{(i,j)}\ \text{for}\ (i,j) \in \mathcal{A}. \tag{P}$$

Note that model (P) does not contain nonnegativity constraints, and we do not require that $t_1 = 0$. This is because the model contains only differences $t_j - t_i$ with $(i,j) \in \mathcal{A}$, and so an optimal solution with $t_1 \neq 0$ can always be changed into one with $t_1 = 0$ by subtracting the optimal value of t_1 from the optimal value of t_i for all $i = 2, \ldots, n$.

Example 8.2.8. *In the case of the data of Figure 8.11, the LO-model becomes:*

$$\min\ t_6 - t_1$$
$$\begin{array}{llll}
\text{s.t.} & t_2 - t_1 \geq 15 & t_3 - t_1 \geq 10 & t_4 - t_1 \geq 8 \\
& t_3 - t_2 \geq 3 & t_6 - t_2 \geq 17 & t_5 - t_3 \geq 7 \\
& t_5 - t_4 \geq 6 & t_6 - t_4 \geq 3 & t_6 - t_5 \geq 10.
\end{array} \tag{P'}$$

Using a computer package, we found the following optimal solution of (P'):

$$t_1^* = 0,\ t_2^* = 15,\ t_3^* = 18,\ t_4^* = 8,\ t_5^* = 25,\ t_6^* = 35.$$

The corresponding optimal objective value is $z^* = 35$. Compare this with the Gantt chart of Figure 8.13.

The optimal solution of model (P) contains the realization times of the several goals. However, it does not list which arcs are on a critical path. These arcs may be determined by solving the dual of model (P) as follows. Note that we may increase the execution time $c_{(i,j)}$ of any activity (i,j) with nonzero slack by a small amount without affecting the total completion time of the project, and hence without affecting the optimal objective value of model (P). On the other hand, increasing the execution time $c_{(i,j)}$ of an activity (i,j) that has zero slack leads to an increase of the total completion time, and hence to an increase of the optimal objective value of model (P). This means that we can determine whether an arc $(i,j) \in \mathcal{A}$ is on a critical path of the project by determining whether the (nonnegative) shadow price of the constraint $t_j - t_i \geq c_{(i,j)}$ is strictly positive or zero. Recall from Section 5.3.2 that the shadow price of a constraint is the optimal value of the corresponding dual value (assuming nondegeneracy). Therefore, arc $(i,j) \in \mathcal{A}$ is on a critical path if and only if the dual variable corresponding to the constraint $t_j - t_i \geq c_{(i,j)}$ has an optimal value that is strictly greater than zero.

To construct the dual of (P), we introduce for each arc (i,j) a dual decision variable $y_{(i,j)}$. The dual model then reads:

$$
\begin{aligned}
\max \quad & \sum_{(i,j)\in\mathcal{A}} c_{(i,j)} y_{(i,j)} \\
\text{s.t.} \quad & -\sum_{(1,j)\in\mathcal{A}} y_{(1,j)} = -1 \\
& \sum_{(i,k)\in\mathcal{A}} y_{(i,k)} - \sum_{(k,j)\in\mathcal{A}} y_{(k,j)} = 0 \quad \text{for } k = 2,\ldots,n-1 \\
& \sum_{(i,n)\in\mathcal{A}} y_{(i,n)} = 1 \\
& y_{(i,j)} \geq 0 \qquad\qquad\qquad\qquad\qquad \text{for } (i,j) \in \mathcal{A}.
\end{aligned}
\tag{D}
$$

So the dual model of the original LO-model is the model of a minimum cost flow problem without capacity constraints, in which one unit of flow is sent through the network against maximum costs; see Section 8.2.4. This is nothing else than a model for determining a *longest path* from the source node to the sink node in the network.

Example 8.2.9. *The dual of model (P') is:*

$$
\max 15y_{(1,2)} + 10y_{(1,3)} + 8y_{(1,4)} + 3y_{(2,3)} + 17y_{(2,6)} + \\
7y_{(3,5)} + 6y_{(4,5)} + 3y_{(4,6)} + 10y_{(5,6)}
$$

$$
\text{s.t.} \quad
\begin{bmatrix}
-1 & -1 & -1 & 0 & 0 & 0 & 0 & 0 & 0 \\
1 & 0 & 0 & -1 & -1 & 0 & 0 & 0 & 0 \\
0 & 1 & 0 & 1 & -1 & 0 & 0 & 0 & 0 \\
0 & 0 & 1 & 0 & 0 & 0 & -1 & -1 & 0 \\
0 & 0 & 0 & 0 & 0 & 1 & 1 & 0 & -1 \\
0 & 0 & 0 & 0 & 1 & 0 & 0 & 1 & 1
\end{bmatrix}
\begin{bmatrix}
y_{(1,2)} \\ y_{(1,3)} \\ y_{(1,4)} \\ y_{(2,3)} \\ y_{(2,6)} \\ y_{(3,5)} \\ y_{(4,5)} \\ y_{(4,6)} \\ y_{(5,6)}
\end{bmatrix}
=
\begin{bmatrix}
-1 \\ 0 \\ 0 \\ 0 \\ 0 \\ 1
\end{bmatrix}
\tag{D'}
$$

$$
y_{(1,2)}, y_{(1,3)}, y_{(1,4)}, y_{(2,3)}, y_{(2,6)}, y_{(3,5)}, y_{(4,5)}, y_{(4,6)}, y_{(5,6)} \geq 0.
$$

The optimal solution of (D') turns out to be:

$$
y^*_{(1,2)} = y^*_{(2,3)} = y^*_{(3,5)} = y^*_{(5,6)} = 1, \text{ and}
$$
$$
y^*_{(1,3)} = y^*_{(1,4)} = y^*_{(2,6)} = y^*_{(4,5)} = y^*_{(4,6)} = 0.
$$

From this optimal solution, it follows that arcs $(1,2)$, $(2,3)$, $(3,5)$, and $(5,6)$ are on a critical path. Since they form a path, there is a unique critical path, namely $1 \rightarrow 2 \rightarrow 3 \rightarrow 5 \rightarrow 6$.

The reader may have noticed that both the primal and the dual model have integer-valued optimal solutions. The fact that there is an integer-valued optimal dual solution follows from Theorem 8.1.1. Moreover, since the technology matrix of (P') is the transpose of the technology matrix of (D') and the right hand side of (P') is integer-valued, it follows from Theorem 8.1.2 that (P') also has an integer-valued optimal solution. So, an optimal solution can be determined by means of the usual simplex algorithm. However, because of the network structure of the model, the network simplex algorithm, to be introduced in Section 8.3, is a more efficient method to solve this model. So the dual formulation of the project scheduling problem has two advantages: the dual model can be solved more efficiently than the primal model, and it allows us to directly determine the critical paths of the project.

The project scheduling problem described in this section is an example of the problems that are typically studied in the field of *scheduling*, which deals with problems of assigning groups of tasks to resources or machines under a wide range of constraints. For example, the number of tasks that can be carried out simultaneously may be restricted, there may be setup times involved (see also the machine scheduling problem of Section 7.2.4), or it may be possible to interrupt a task and resume it at a later time. Also, the objective may be varied. For example, in the problem described in the current section, the objective is to complete the project as quickly as possible, but alternatively there may be a deadline, and as many tasks as possible need to be completed before the deadline. There exists a large literature on scheduling; the interested reader is referred to Pinedo (2012).

8.3 The network simplex algorithm

In this chapter we consider LO-models of which the technology matrices are node-arc incidence matrices of digraphs, usually called *network matrices*. The corresponding LO-models are called *network LO-models*. Although these models can be solved directly by means of the usual simplex algorithm, it is far more efficient to use the so-called *network simplex algorithm*, which explicitly uses the network structure of the technology matrix. Empirical tests have shown that the algorithm is up to 300 times faster than the usual simplex algorithm when using it to solve large-scale network LO-models. The details of the network simplex algorithm will be demonstrated for so-called transshipment problems. An example of a transshipment problem is the so-called catering service problem, which is discussed in Chapter 18.

8.3.1 The transshipment problem

The *transshipment problem* is the problem of sending a flow through a network against minimum costs. It is a generalization of the minimum cost flow problem discussed in Section 8.2.4. The network consists of a number of supply nodes, called *source nodes*, a number of demand nodes, called *sink nodes*, and a number of *intermediate nodes* (or *transshipment nodes*) with no supply and demand. There is a shipping cost associated with each arc of the network. In terms of linear optimization, the transshipment problem can be formulated as follows. Let $G = (\mathcal{V}, \mathcal{A})$ be a network with $|\mathcal{V}| = m$ and $|\mathcal{A}| = n$. Throughout, m and n are positive integers. For $(i,j) \in \mathcal{A}$, let c_{ij} be the unit shipping cost along arc (i,j), and, for $i \in \mathcal{V}$, let b_i be the supply at i if i is a source node, and the negative of the demand at i if i is a sink node. We assume that they satisfy the supply-demand balance equation $\sum_{i\in\mathcal{V}} b_i = 0$. The LO-model of the transshipment problem then reads:

Model 8.3.1. (*Transshipment problem*)

$$\min \quad \sum_{(i,j)\in\mathcal{A}} c_{(i,j)} x_{(i,j)}$$

$$\text{s.t.} \quad \sum_{(j,i)\in\mathcal{A}} x_{(j,i)} - \sum_{(i,j)\in\mathcal{A}} x_{(i,j)} = b_i \qquad \text{for } i = 1,\ldots,m \qquad (8.16)$$

$$x_{(i,j)} \geq 0 \qquad \qquad \text{for } (i,j) \in \mathcal{A}.$$

If $x_{(i,j)}$ is interpreted as the flow on the arc (i,j), then any feasible solution satisfies the following condition: for each node i, the total flow out of node i minus the total flow into node i is equal to b_i. These conditions, which are expressed in (8.16), are called the *flow-preservation conditions*.

The model can be written in a compact way using matrix notation as follows. Let $\mathbf{A}$ be the node-arc incidence matrix associated with G, and let $\mathbf{x} \in \mathbb{R}^n$ be the vector of the flow variables. Moreover, let $\mathbf{c}$ ($\in \mathbb{R}^n$) be the *flow cost vector* consisting of the entries $c_{(i,j)}$ for each arc $(i,j) \in \mathcal{A}$, and let $\mathbf{b}$ ($\in \mathbb{R}^m$) be the supply-demand vector consisting of the entries b_i. Then, Model 8.3.1 can be compactly written as:

$$\begin{aligned} \min \quad & \mathbf{c}^{\mathsf{T}}\mathbf{x} \\ \text{s.t.} \quad & \mathbf{A}\mathbf{x} = \mathbf{b} \\ & \mathbf{x} \geq \mathbf{0}. \end{aligned} \qquad (\text{P})$$

Recall that we assume that $\sum_{i\in\mathcal{V}} b_i = 0$. Without this assumption, the system $\mathbf{A}\mathbf{x} = \mathbf{b}$ may not have a solution at all, because the rows of $\mathbf{A}$ sum to the all-zero vector; see Appendix B, and Theorem 8.1.8.

The rows of $\mathbf{A}$ correspond to the nodes of G, and the columns to the arcs of G. Each column of $\mathbf{A}$ contains two nonzero entries. Let $i,j \in \mathcal{V}$ and $(i,j) \in \mathcal{A}$. The column corresponding to the arc (i,j) contains a $+1$ in the row corresponding to node i, and a -1 in the row corresponding to node j. In this section, $\mathbf{A}_{(i,j)}$ refers to the column of $\mathbf{A}$ corresponding to the arc $(i,j) \in \mathcal{A}$. Hence, the columns of $\mathbf{A}$ satisfy:

$$\mathbf{A}_{(i,j)} = \mathbf{e}_i - \mathbf{e}_j$$

with $\mathbf{e}_i$ the i'th unit vector in $\mathbb{R}^m$ ($i = 1,\ldots,m$).

Example 8.3.1.   *Figure 8.14 shows the network of a transshipment problem. This network has six nodes and nine arcs. The values of the supply-demand vector* $\mathbf{b}$ *are written next to the corresponding nodes, and the values of the flow costs next to the corresponding arcs. So the nodes 1 and 3 are supply*

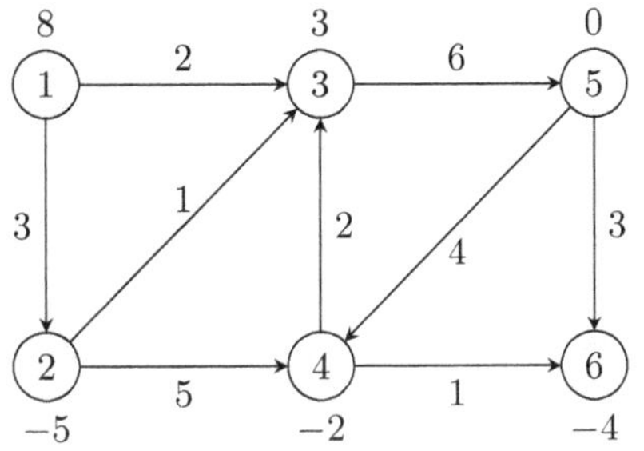

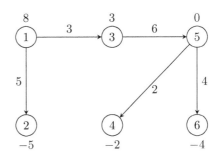

Figure 8.14: Network of a transshipment problem.

Figure 8.15: Feasible tree solution.

nodes; 2, 4, and 6 are demand nodes; and 5 is a transshipment node. It is left to the reader to determine the node-arc incidence matrix $\mathbf{A}$ of the transshipment problem of Figure 8.14. Note that the sum of the rows of $\mathbf{A}$ is the all-zero vector.

8.3.2 Network basis matrices and feasible tree solutions

In the remainder of this section, we formulate the network simplex algorithm, and show its relationship with the usual simplex algorithm. Instead of basic solutions, the network simplex algorithm uses so-called *tree solutions*. In the case of the network simplex algorithm, the technology matrix $\mathbf{A}$ (i.e., the node-incidence matrix of the graph G) is called the *network matrix*. As with the usual simplex algorithm, the network simplex algorithm starts with an initial solution, called an *initial feasible tree solution*. In Section 8.3.7, we will explain how to determine such an initial feasible tree solution. Until then, it is assumed that an initial feasible tree solution is available.

The basic idea of the network simplex algorithm is made precise in the following theorem. It formulates the relationship between spanning trees in a network and so-called *network basis matrices*, which are $(m, m - 1)$-submatrices of rank $m - 1$ of the network matrix. Since each column of the network matrix corresponds to an arc of the graph G, any $(m, m - 1)$-submatrix of $\mathbf{A}$ corresponds to a choice of $m - 1$ arcs of G. Before we prove the theorem, we illustrate the structure of the network basis matrix by means of an example.

Example 8.3.2. *Consider again Figure 8.15. The matrix $\mathbf{B}$ is a $(6, 5)$-submatrix of the node-arc incidence matrix $\mathbf{A}$ of the directed graph of Figure 8.14. In particular, $\mathbf{B}$ consists of the columns of $\mathbf{A}$ associated with the arcs of the tree depicted in Figure 8.15. Thus, we have that:*

$$
\mathbf{B} = \begin{array}{c} 1 \\ 2 \\ 3 \\ 4 \\ 5 \\ 6 \end{array}
\begin{array}{c} {\scriptstyle (1,2)\ (1,3)\ (3,5)\ (5,4)\ (5,6)} \\ \left[\begin{array}{ccccc}
1 & 1 & 0 & 0 & 0 \\
-1 & 0 & 0 & 0 & 0 \\
0 & -1 & 1 & 0 & 0 \\
0 & 0 & 0 & -1 & 0 \\
0 & 0 & -1 & 1 & 1 \\
0 & 0 & 0 & 0 & -1
\end{array} \right] \end{array}.
$$

Note that $\mathbf{B}$ has rank 5, and removing any row from $\mathbf{B}$ results in an invertible $(5,5)$-submatrix of $\mathbf{B}$.

Theorem 8.3.1. (*Network basis matrices and spanning trees*)
Let $G = (\mathcal{V}, \mathcal{A})$ be a network, and $\mathbf{A}$ the corresponding network matrix. Any spanning tree in G corresponds to a network basis matrix in $\mathbf{A}$, and vice versa.

Proof of Theorem 8.3.1. Let T be any spanning tree in the network $G = (\mathcal{V}, \mathcal{A})$ with network matrix $\mathbf{A}$. We first show that the node-arc incidence matrix $\mathbf{B}$ associated with T is a network basis matrix in $\mathbf{A}$. Clearly, $\mathbf{B}$ is an $(m, m-1)$-submatrix of $\mathbf{A}$. Since $\mathbf{B}$ is the node-arc incidence matrix of a connected digraph (a tree is connected; see Appendix C), it follows from Theorem 8.1.8 that the rank of $\mathbf{B}$ is $m - 1$. Hence, $\mathbf{B}$ is a network basis matrix of $\mathbf{A}$.

We now show that each network basis matrix $\mathbf{B}$ in $\mathbf{A}$ corresponds to a spanning tree in G. Let $\mathbf{B}$ be a network basis matrix, and let $G_{\mathbf{B}}$ be the subgraph of G whose node-arc incidence matrix is $\mathbf{B}$. Since $\mathbf{B}$ is an $(m, m-1)$-matrix of full column rank $m - 1$, $G_{\mathbf{B}}$ has m nodes and $m - 1$ arcs. Hence, $G_{\mathbf{B}}$ is a spanning subgraph of G. If $G_{\mathbf{B}}$ has no cycles, then it follows that $G_{\mathbf{B}}$ is a tree (see Appendix C) and we are done. Therefore, we need to show that $G_{\mathbf{B}}$ contains no cycles.

Assume for a contradiction that $G_{\mathbf{B}}$ does contain some cycle C. Let $v_1, \ldots, v_p$ be the nodes of C in consecutive order, and define $v_{p+1} = v_1$. Let $i \in \{1, \ldots, p\}$. Because C is a cycle, we have that either $(v_i, v_{i+1}) \in C$, or $(v_{i+1}, v_i) \in C$. Define:

$$\mathbf{w}(v_i, v_{i+1}) = \begin{cases} \mathbf{A}_{(v_i, v_{i+1})} & \text{if } (v_i, v_{i+1}) \in C \\ -\mathbf{A}_{(v_{i+1}, v_i)} & \text{if } (v_{i+1}, v_i) \in C. \end{cases}$$

Note that $\mathbf{w}(v_i, v_{i+1})$ is a column of $\mathbf{B}$, multiplied by 1 or -1. Recall that $\mathbf{A}_{(i,j)} = \mathbf{e}_i - \mathbf{e}_j$, and observe that, for any two consecutive nodes v_i, v_{i+1}, we have two possibilities:

(a) $(v_i, v_{i+1}) \in C$. Then, $\mathbf{w}(v_i, v_{i+1}) = \mathbf{A}_{(v_i, v_{i+1})} = \mathbf{e}_{v_i} - \mathbf{e}_{v_{i+1}}$.

(b) $(v_{i+1}, v_i) \in C$. Then, $\mathbf{w}(v_i, v_{i+1}) = -\mathbf{A}_{(v_i, v_{i+1})} = -(\mathbf{e}_{v_{i+1}} - \mathbf{e}_{v_i}) = \mathbf{e}_{v_i} - \mathbf{e}_{v_{i+1}}$.

Thus, in either case, we have that $\mathbf{w}(v_i, v_{i+1}) = \mathbf{e}_{v_i} - \mathbf{e}_{v_{i+1}}$ and hence:

$$\sum_{i=1}^{p} \mathbf{w}(v_i, v_{i+1}) = \sum_{i=1}^{p} \mathbf{e}_{v_i} - \mathbf{e}_{v_{i+1}} = \sum_{i=1}^{p} \mathbf{e}_{v_i} - \sum_{i=1}^{p} \mathbf{e}_{v_{i+1}} = \mathbf{0}.$$

This means that the columns of $\mathbf{B}$ corresponding to the arcs of C are linearly dependent, contradicting the fact that $\mathbf{B}$ has full column rank $m - 1$. $\square$

In the following theorem, BI is, as usual, the index set of the columns of $\mathbf{B}$, and NI is the index set of the remaining columns of $\mathbf{A}$. The theorem states that for any network basis matrix of the network matrix $\mathbf{A}$, the set of m equalities $\mathbf{B}\mathbf{x}_{BI} = \mathbf{b}$ has a (not necessarily nonnegative) solution, provided that the sum of all entries of $\mathbf{b}$ is zero. If $\mathbf{N}$ consists of the columns of $\mathbf{A}$ that are not in $\mathbf{B}$, then we can write $\mathbf{A} \equiv [\mathbf{B} \ \mathbf{N}]$. Note that $\mathcal{A} = BI \cup NI$. The vector $\mathbf{x} \equiv \begin{bmatrix} \mathbf{x}_{BI} \\ \mathbf{x}_{NI} \end{bmatrix}$, satisfying $\mathbf{B}\mathbf{x}_{BI} = \mathbf{b}$, with $\mathbf{x}_{BI} \geq \mathbf{0}$ and $\mathbf{x}_{NI} = \mathbf{0}$, is called a *tree solution* of $\mathbf{A}\mathbf{x} = \mathbf{b}$ corresponding to $\mathbf{B}$. The entries of the vector $\mathbf{x}_{BI}$ are called *basic variables*, and the entries of $\mathbf{x}_{NI}$ are called *nonbasic variables*. The matrix $\mathbf{B}$ is the node-arc incidence matrix of a spanning tree in the network G (according to Theorem 8.3.1). Since $\mathbf{B}$ is not a square matrix, we cannot simply take the inverse of $\mathbf{B}$ in order to calculate the value of the entries of $\mathbf{x}_{BI}$, as was done in the case of basis matrices in standard LO-models; see Section 2.2.2. We will show that the fact that the entries of $\mathbf{b}$ sum to zero implies that the system $\mathbf{B}\mathbf{x}_{BI} = \mathbf{b}$ (without the nonnegativity constraints!) has a unique solution; see Appendix B.

Theorem 8.3.2. (*Existence of tree solutions*)
Let $\mathbf{B}$ be any $(m, m-1)$ network basis matrix, and let $\mathbf{b} \in \mathbb{R}^m$ be such that $\sum_{i=1}^{m} b_i = 0$. The system $\mathbf{B}\mathbf{x} = \mathbf{b}$ has a unique solution.

Proof of Theorem 8.3.2. Let $G = (\mathcal{V}, \mathcal{A})$ be a network with m nodes, and let $\mathbf{A}$ be the associated network matrix. Moreover, let $\mathbf{B}$ be a network basis matrix, and let $T_{\mathbf{B}}$ be the corresponding spanning tree in G. Let $v \in \mathcal{V}$, and construct $\mathbf{B}'$ from $\mathbf{B}$ by deleting the row corresponding to v. So, $\mathbf{B}'$ is an $(m - 1, m - 1)$-matrix. Similarly, construct $\mathbf{b}'$ from $\mathbf{b}$ by deleting the entry corresponding to v. Let b_v be that entry. We have that:

$$\mathbf{B} \equiv \begin{bmatrix} \mathbf{B}' \\ \mathbf{B}_{v,\star} \end{bmatrix}, \text{ and } \mathbf{b} \equiv \begin{bmatrix} \mathbf{b}' \\ b_v \end{bmatrix}.$$

(Recall $\mathbf{B}_{v,\star}$ denotes the v'th row of $\mathbf{B}$.) Since $\mathbf{B}$ is the node-arc incidence matrix of a connected digraph (namely, the spanning tree $T_{\mathbf{B}}$), it follows from Theorem 8.1.8 that $\mathbf{B}'$ is nonsingular and has rank $m - 1$. Let $\hat{\mathbf{x}} = (\mathbf{B}')^{-1}\mathbf{b}'$. We need to show that $\hat{\mathbf{x}}$ satisfies $\mathbf{B}\hat{\mathbf{x}} = \mathbf{b}$. Obviously, it satisfies all equations of $\mathbf{B}\hat{\mathbf{x}} = \mathbf{b}$ that do not correspond to node v. So it remains to show that $\mathbf{B}_{v,\star}\hat{\mathbf{x}} = b_v$. To see that this holds, recall that the rows of $\mathbf{B}$ add up to the all-zero vector, and that the entries of $\mathbf{b}$ add up to zero. Hence, $\mathbf{B}_{v,\star} = -\sum_{i \neq v} \mathbf{B}_{i,\star}$, and $b_v = -\sum_{i \neq v} b_i$. The equation $\mathbf{B}_{v,\star}\hat{\mathbf{x}} = b_v$ holds because

$$\mathbf{B}_{v,\star}\hat{\mathbf{x}} = \left(-\sum_{i \neq v} \mathbf{B}_{i,\star}\right)\hat{\mathbf{x}} = -\sum_{i \neq v} \mathbf{B}_{i,\star}\hat{\mathbf{x}} = -\sum_{i \neq v} b_i = b_v,$$

as required. $\qquad\square$

There is an elegant way to calculate a solution of the system $\mathbf{B}\mathbf{x} = \mathbf{b}$ by using the tree structure of the matrix $\mathbf{B}$. Recall that the separate equations of the system $\mathbf{A}\mathbf{x} = \mathbf{b}$ represent the flow-preservation conditions for the network. Since we want to solve the system $\mathbf{B}\mathbf{x} = \mathbf{b}$, this means that we want to find a flow vector $\mathbf{x}$ such that the flow-preservation conditions hold, and all nonbasic variables (i.e., the variables corresponding to arcs that are not in the tree $T_{\mathbf{B}}$) have value zero. This means that we can only use the arcs in the tree $T_{\mathbf{B}}$ to satisfy the flow-preservation conditions.

Consider the tree of Figure 8.15. It is a spanning tree in the network of Figure 8.14. We can find a value for $x_{(i,j)}$ with $(i,j) \in BI$ as follows. The demand at node 2 is 5. The flow-preservation condition corresponding to node 2 states that this demand has to be satisfied by the flow vector $\mathbf{x}$. Since $(1,2)$ is the only arc incident with node 2 that is allowed to ship a positive number of units of flow, arc $(1,2)$ has to be used to satisfy the demand at node 2. This means that we are forced to choose $x_{(1,2)} = 5$. Similarly, we are forced to choose $x_{(5,4)} = 2$ and $x_{(5,6)} = 4$. This process can now be iterated. We have that $x_{(1,2)} = 5$, the total supply at node 1 is 8, and all supply at node 1 should be used. There are two arcs incident with node 1 in the tree $T_\mathbf{B}$, namely $(1,2)$ and $(1,3)$. Since we already know the value of $x_{(1,2)}$, this means that arc $(1,3)$ has to be used to satisfy the flow-preservation condition for node 1. That is, we are forced to choose $x_{(1,3)} = 3$. This leaves only one arc with an undetermined flow value, namely arc $(3,5)$. Because $b_3 = 3$ and $x_{(1,3)} = 3$, we have again no choice but to set $x_{(3,5)} = 6$ in order to satisfy the flow-preservation condition for node 3. Thus, we have determined a solution of the system $\mathbf{Bx} = \mathbf{b}$. Note that while constructing this solution, for each $(i,j) \in \mathcal{A}$, $x_{(i,j)}$ was forced to have a particular value. So, in fact, the solution that we find is unique.

8.3.3   Node potential vector; reduced costs; test for optimality

A crucial concept in the development of the network simplex algorithm is the concept of 'node potential vector'. A node potential vector associates real numbers with the nodes of the graph G which are used to easily determine whether a current network basis matrix is optimal and, if not, to find a new network basis matrix with a smaller objective value. Node potential vectors are related to the dual model of Model 8.3.1.

Formally, let $\mathbf{B}$ be a network basis matrix and let $T_\mathbf{B}$ be the corresponding spanning tree. Let $\mathbf{y} = \begin{bmatrix} y_1 & \cdots & y_m \end{bmatrix}^\mathsf{T}$. The vector $\mathbf{y}$ is called a *node potential vector* with respect to $\mathbf{B}$ if

$$c_{(i,j)} - y_i + y_j = 0 \quad \text{for } (i,j) \in T_\mathbf{B}. \tag{8.17}$$

Note that the system of equations (8.17) is equivalent to $\mathbf{B}^\mathsf{T}\mathbf{y} = \mathbf{c}_{BI}$. The *reduced cost* of arc $(i,j) \in \mathcal{A}$ (with respect to the node potential vector $\mathbf{y}$) is denoted and defined as:

$$\bar{c}_{(i,j)} = c_{(i,j)} - y_i + y_j. \tag{8.18}$$

According to (8.17), the reduced cost of any arc (i,j) in the tree $T_\mathbf{B}$ is zero, whereas the reduced cost of any other arc may be positive, negative, or zero.

The node potential vector and the reduced costs of the arcs can be calculated in a simple manner. First note that, if $\mathbf{y}$ is a node potential vector, then we may add the same number to each of its entries, and obtain a new node potential vector. Therefore, there are in general infinitely many node potential vectors for any given network basis matrix. However, setting one of the entries of the node potential vector (e.g., y_1) equal to zero results in a unique solution. Now, expression (8.18) together with (8.17) enable us to calculate $\bar{c}_{(i,j)}$ for all $(i,j) \in \mathcal{A}$, and y_i for all $i \in \mathcal{N}$. This process is best illustrated with an example.

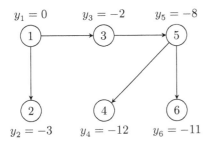

(a) The entries of the node potential vector.

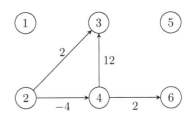

(b) Reduced costs $\bar{c}_{(i,j)}$.

Figure 8.16: Calculating the node potential vector and the reduced costs.

Example 8.3.3. *Consider the network basis matrix* $\mathbf{B}$ *associated with the tree in Figure 8.15. The entries of the node potential vector are written next to the corresponding nodes in Figure 8.16. The values of the reduced costs* $\bar{c}_{(i,j)}$ *are written next to the arcs. The calculations are as follows (see Figure 8.16(a)). Take* $y_1 = 0$. *Using* (8.17) *applied to arc* $(1, 2)$, *yields that* $y_2 = y_1 - c_{(1,2)} = 0 - 3 = -3$. *Hence,* $y_2 = -3$. *Doing this for all arcs in the tree* $T_{\mathbf{B}}$, *we find that:*

$$
\begin{aligned}
y_2 &= y_1 - c_{(1,2)} = & 0 - 3 &= & -3, \\
y_3 &= y_1 - c_{(1,3)} = & 0 - 2 &= & -2, \\
y_5 &= y_3 - c_{(3,5)} = & -2 - 6 &= & -8, \\
y_4 &= y_5 - c_{(5,4)} = & -8 - 4 &= -12, \\
y_6 &= y_5 - c_{(5,6)} = & -8 - 3 &= -11.
\end{aligned}
$$

For each (i, j) *that is not in the tree* $T_{\mathbf{B}}$, *the value of* $c_{(i,j)}$ *is calculated as follows (see Figure 8.16(b)):*

$$
\begin{aligned}
\bar{c}_{(2,3)} &= c_{(2,3)} - y_2 + y_3 = 1 + 3 - 2 = & 2, \\
\bar{c}_{(2,4)} &= c_{(2,4)} - y_2 + y_4 = 5 + 3 - 12 = & -4, \\
\bar{c}_{(4,3)} &= c_{(4,3)} - y_4 + y_3 = 2 + 12 - 2 = & 12, \\
\bar{c}_{(4,6)} &= c_{(4,6)} - y_4 + y_6 = 1 + 12 - 11 = & 2.
\end{aligned}
$$

As we will show below, if all arcs have nonnegative reduced cost, then the current tree solution is optimal.

Theorem 8.3.3. (*Optimality criterion*)
Let $\mathbf{B}$ be a feasible network basis matrix with corresponding tree $T_{\mathbf{B}}$. Let $\hat{\mathbf{y}} = \begin{bmatrix} \hat{y}_1 & \dots & \hat{y}_m \end{bmatrix}^{\mathsf{T}}$ be a node potential vector, and let $\bar{c}_{(i,j)} = c_{(i,j)} - \hat{y}_i + \hat{y}_j$ for $(i, j) \in \mathcal{A}$. If

$$
\begin{aligned}
\bar{c}_{(i,j)} &= 0 \quad \text{for } (i, j) \in T_{\mathbf{B}}, \text{ and} \\
\bar{c}_{(i,j)} &\geq 0 \quad \text{for } (i, j) \in \mathcal{A} \setminus T_{\mathbf{B}},
\end{aligned}
\tag{8.19}
$$

then $\mathbf{B}$ corresponds to an optimal feasible solution.

Proof. Let $\hat{\mathbf{x}}$ ($\in \mathbb{R}^n$) be the tree solution corresponding to $\mathbf{B}$. We will use the complementary slackness relations to show that $\hat{\mathbf{x}}$ is an optimal solution of Model 8.3.1. The dual model of Model 8.3.1 is:

$$\max \sum_{i \in \mathcal{V}} b_i y_i$$
$$\text{s.t.} \quad y_i - y_j \leq c_{(i,j)} \text{ for } (i,j) \in \mathcal{A} \qquad (8.20)$$
$$y_i \text{ free} \qquad \text{for } i \in \mathcal{V}.$$

Note that $\hat{y}_i - \hat{y}_j = c_{(i,j)} - \bar{c}_{(i,j)} \leq c_{(i,j)}$ for $(i,j) \in \mathcal{A}$. Hence, the node potential vector $\hat{\mathbf{y}}$ is a feasible solution of this dual model. The complementary slackness relations state that, for each arc $(i,j) \in \mathcal{A}$, we should have that either $\hat{x}_{(i,j)} = 0$, or $\hat{y}_i - \hat{y}_j = c_{(i,j)}$. Note that if $(i,j) \in \mathcal{A} \setminus T_{\mathbf{B}}$, then $x_{(i,j)} = 0$. Moreover, for each arc $(i,j) \in T_{\mathbf{B}}$, we have that $\hat{y}_i - \hat{y}_j = c_{(i,j)}$. Hence, the complementary slackness relations hold, and Theorem 4.3.1 implies that $\hat{\mathbf{x}}$ is an optimal solution of Model 8.3.1, and $\hat{\mathbf{y}}$ is an optimal solution of (8.20). $\square$

If (8.19) is not satisfied, then this means that there is an arc (i,j) that is not in the tree and that satisfies $\bar{c}_{(i,j)} = c_{(i,j)} - y_i + y_j < 0$. This is, for instance, the case in the example above, where $\bar{c}_{(2,4)} < 0$, and so $\mathbf{B}$ does not correspond to an optimal feasible tree solution. In the next section it will be explained how to proceed from here.

8.3.4   Determining an improved feasible tree solution

We saw in Section 8.3.3 that if $\bar{c}_{(k,l)} \geq 0$ for all arcs (k,l) that are not in the tree $T_{\mathbf{B}}$, then the current feasible tree solution is optimal. If the current feasible tree solution is not optimal, then there is an arc (k,l) not in the tree $T_{\mathbf{B}}$ for which $\bar{c}_{(k,l)} < 0$. In case $\bar{c}_{(i,j)} < 0$ for more than one arc (i,j), we may choose any of them. Such an arc is called an *entering arc*. In this section we will use the following pivot rule: choose an arc corresponding to the reduced cost which is most negative (see Section 3.3); if there is still a tie, then we will choose, among the arcs with the smallest reduced cost, the lexicographically smallest arc.

Including arc (k,l) in the current feasible tree solution results in a unique cycle, say $C_{(k,l)}$ (see Appendix C). The nonbasic variable $x_{(k,l)}$, corresponding to the entering arc (k,l), is given a nonnegative value, say t. We add t to the current values of the basic variables that correspond to the arcs of $C_{(k,l)}$ pointing into the direction of (k,l); t is subtracted from the values of the basic variables that correspond to arcs of $C_{(k,l)}$ pointing into the direction opposite to (k,l); see Figure 8.17(a). The question is: what will be an appropriate value of t?

It can be seen from Figure 8.17(a) that the net flow into any node (and, hence, the net flow out of any node) remains the same; see also the remark at the end of Section 8.3.2. This means that changing the flow on each arc in the way described above maintains feasibility as long as the new values of $x_{(i,j)}$ on the cycle remain nonnegative. Let $C_{(k,l)}^-$ represent the

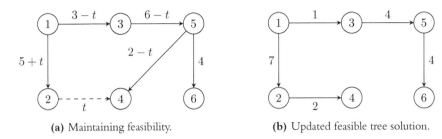

(a) Maintaining feasibility.   (b) Updated feasible tree solution.

Figure 8.17: Determining an improved feasible tree solution.

set of arcs of $C_{(k,l)}$ that point into the direction opposite to the direction of (k, l). Define:

$$
\Delta = \begin{cases} \infty & \text{if } C^-_{(k,l)} = \emptyset \\ \min\left\{ x_{(i,j)} \mid (i,j) \in C^-_{(k,l)} \right\} & \text{otherwise.} \end{cases}
$$

Note that this is a special case of the minimum-ratio test of the simplex algorithm (see Section 3.3). The reader is asked to show this in Exercise 8.4.12. An arc $(u, v) \in C_{(k,l)}$ for which $x_{(u,v)} - \Delta = 0$ is a candidate for leaving the tree $T_{\mathbf{B}}$ in favor of (k, l), creating a new spanning tree in G. The objective value corresponding to the new feasible tree solution is at most the current objective value, since $\bar{c}_{(k,l)} < 0$. The new value of $x_{(k,l)}$ satisfies $x_{(k,l)} = \Delta \geq 0$.

If $\Delta = 0$, then the entering arc (k, l) maintains a zero flow, and so the objective value does not improve. The case $\Delta = 0$ happens if the current tree $T_{\mathbf{B}}$ contains an arc for which the current flow is zero; such a tree is called a *degenerate spanning tree*. Clearly, the corresponding feasible tree solution is then degenerate, i.e., at least one of the basic variables has a zero value.

If all arcs in the cycle $C_{(k,l)}$ point into the direction of (k, l), then the value of t can be chosen arbitrarily large, i.e., $\Delta = \infty$. This means that the objective value can be made arbitrarily large, i.e., the model is unbounded; see Section 3.7. On the other hand, if $\mathbf{c} \geq \mathbf{0}$, then for every feasible tree solution $\mathbf{x}$, it holds that $\mathbf{c}^\mathsf{T} \mathbf{x} \geq 0$. Therefore the model is never unbounded if $\mathbf{c} \geq \mathbf{0}$.

Example 8.3.4. *In the case of the cycle $C_{(2,4)}$ of Figure 8.17(a), the entering arc is $(5, 4)$, because*

$$
\Delta = \max\{t \mid 2 - t \geq 0, 6 - t \geq 0, 3 - t \geq 0\} = 2;
$$

the maximum is attained for arc $(5, 4)$. The new tree, with the updated feasible tree solution, is depicted in Figure 8.17(b).

8.3.5 The network simplex algorithm

The network simplex algorithm can now be formulated as follows:

Algorithm 8.3.1. (*Network simplex algorithm*)

Input: Values for the entries of the (m, n) network matrix $\mathbf{A}$ associated with a network with m nodes and n arcs, the cost vector $\mathbf{c}$ ($\in \mathbb{R}^n$), and the supply-demand vector $\mathbf{b}$ ($\in \mathbb{R}^m$) with the sum of the entries equal to 0, and an initial feasible tree solution for the LO-model $\min\{\mathbf{c}^\mathsf{T}\mathbf{x} \mid \mathbf{A}\mathbf{x} = \mathbf{b}, \mathbf{x} \geq \mathbf{0}\}$.

Output: Either

the message: the model is unbounded; or

an optimal solution of the model.

Step 1: *Calculation of the node potential vector.*

Let $\mathbf{B}$ be the current network basis matrix. Determine values of the node potentials $y_1, \ldots, y_m$ such that $y_1 = 0$ and $y_i - y_j = c_{(i,j)}$ for each arc (i, j) in the tree $T_\mathbf{B}$; $i, j \in \{1, \ldots, m\}$; see Section 8.3.3.

Step 2: *Selection of an entering arc.*

Calculate $\bar{c}_{(i,j)} = c_{(i,j)} - y_i + y_j$ for each arc (i, j) not in $T_\mathbf{B}$. Select an arc (k, l) with $\bar{c}_{(k,l)} < 0$. If such an arc does not exist, then stop: the current feasible tree solution is optimal.

Step 3: *Selection of a leaving arc.*

Augment the current tree with the arc (k, l). Let $C_{(k,l)}$ be the resulting cycle, and let $C_{(k,l)}^-$ be the set of arcs of $C_{(k,l)}$ that point in the direction opposite to (k, l). Define:

$$
\Delta = \begin{cases} \infty & \text{if } C_{(k,l)}^- = \emptyset \\ \min\left\{ x_{(i,j)} \mid (i,j) \in C_{(k,l)}^- \right\} & \text{otherwise.} \end{cases}
$$

If $\Delta = \infty$, then stop: the model is unbounded. Otherwise, let $(u, v) \in C_{(k,l)}^-$ with $\Delta = x_{(u,v)}$. Delete (u, v) from the current tree $T_\mathbf{B}$ and add (k, l); i.e., set $T_\mathbf{B} := (T_\mathbf{B} \cup \{(k, l)\}) \setminus \{(u, v)\}$. Set

$x_{(k,l)} := \Delta$;

$x_{(i,j)} := x_{(i,j)} + \Delta$ for $(i, j) \in C_{(k,l)}^+$ (the set of arcs of $C_{(k,l)}$ that have the same direction as (k, l));

$x_{(i,j)} := x_{(i,j)} - \Delta$ for $(i, j) \in C_{(k,l)}^-$.

Return to Step 1.

Example 8.3.5.   *We will apply Algorithm 8.3.1 to the example of Figure 8.14. The procedure starts with the spanning tree drawn in Figure 8.15. Figure 8.18 shows the various iterations of the network simplex algorithm that lead to an optimal solution. Column 1 of Figure 8.18 contains the (primal) feasible tree solutions. The supply and demand values in the vector $\mathbf{b}$ are written next to the*

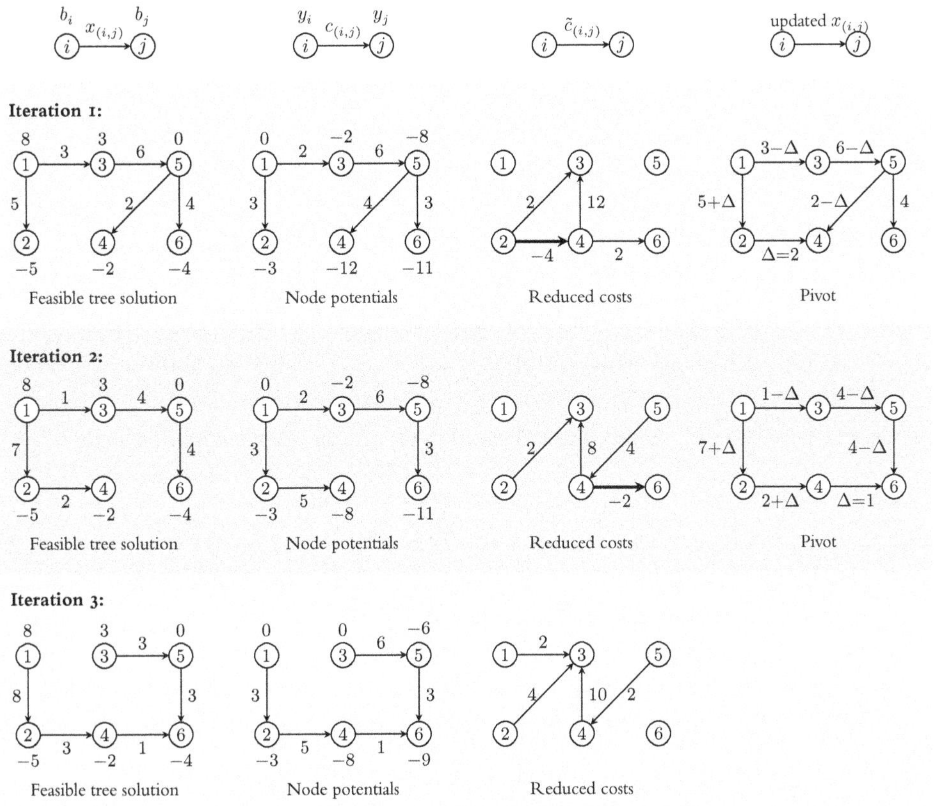

Iteration 1:

Iteration 2:

Iteration 3:

Figure 8.18: Three iterations of the network simplex algorithm, and an optimal solution. The arc above each column explains the meaning of the node and arc labels in that column.

nodes, and the flow values $x_{(i,j)}$ are written next to the arcs. The arcs corresponding to the nonbasic variables, which have flow value zero, are omitted. Column 2 contains the current node potential vector. The values of the y_i's are written next to the nodes. The numbers next to the arcs are the values of the flow cost $c_{(i,j)}$. The node potential vector is calculated by the formula $c_{(i,j)} = y_i - y_j$ (see Step 1). Column 3 contains the values of the current reduced cost $\bar{c}_{(i,j)}$, which are calculated by the formula $\bar{c}_{(i,j)} = c_{(i,j)} - y_i + y_j$. An arc with the most negative value of $\bar{c}_{(i,j)}$ enters the current tree, resulting in a unique cycle (see Step 2). Column 4 shows this unique cycle. The values of $x_{(i,j)}$ are updated by subtracting and adding Δ. An arc (i,j) with $x_{(i,j)} = \Delta$ leaves the current tree (see Step 3).

We will briefly describe the different iterations for the data of Figure 8.14.

Iteration 1. There is only one negative value of $\bar{c}_{(i,j)}$, namely $\bar{c}_{(2,4)} = -4$. Hence, $(k,l) = (2,4)$ is the entering arc. The leaving arc is $(u,v) = (5,4)$. The values of $x_{(i,j)}$ are updated and used as input for Iteration 2.

Iteration 2. One can easily check that $(k,l) = (4,6)$, and $(u,v) = (1,3)$.

Iteration 3. *Now all values of $\bar{c}_{(i,j)}$ are nonnegative, and so an optimal solution has been reached.* *The optimal solution satisfies:*

$$x^*_{(1,2)} = 8, \; x^*_{(2,4)} = 3, \; x^*_{(3,5)} = 3, \; x^*_{(4,6)} = 1, \; x^*_{(5,6)} = 3,$$
$$\text{and } x^*_{(1,3)} = x^*_{(2,3)} = x^*_{(5,4)} = 0.$$

The optimal objective value is $z^* = 67$.

8.3.6 Relationship between tree solutions and feasible basic solutions

There is a strong relationship between the network simplex algorithm and the regular simplex algorithm that is described in Chapter 3. In fact, the network simplex algorithm is a special case of the regular simplex algorithm. This section describes why this is the case.

Let $\mathbf{B}$ be a network basis matrix corresponding to the feasible tree solution $\begin{bmatrix} \mathbf{x}_{BI} \\ \mathbf{x}_{NI} \end{bmatrix}$ satisfying $\mathbf{B}\mathbf{x}_{BI} = \mathbf{b}, \mathbf{x}_{BI} \geq \mathbf{0}$, and $\mathbf{x}_{NI} = \mathbf{0}$. Let $\mathbf{A} \equiv \begin{bmatrix} \mathbf{B} & \mathbf{N} \end{bmatrix}$. Recall that $\mathbf{B}$ is an $(m, m-1)$-matrix. Hence, $\mathbf{N}$ is an $(m, n-m+1)$-matrix. Since $\mathbf{B}$ is associated with a connected digraph (namely, the spanning tree $T_{\mathbf{B}}$), deleting any row from $\mathbf{A}$ leads to a nonsingular $(m-1, m-1)$-submatrix $\tilde{\mathbf{B}}$ of $\mathbf{B}$; see Theorem 8.1.8. We delete row m (the last row) from $\mathbf{A}$. The resulting matrices are denoted by $\tilde{\mathbf{A}}$, $\tilde{\mathbf{B}}$, and $\tilde{\mathbf{N}}$, and the right hand side vector by $\tilde{\mathbf{b}}$. Note that this amounts to deleting one of the constraints. Thus, we have the following two models:

$$\begin{array}{ll} \min \; \mathbf{c}^\mathsf{T}\mathbf{x} \\ \text{s.t.} \; \mathbf{A}\mathbf{x} = \mathbf{b} \quad \text{(P)} \\ \quad\quad \mathbf{x} \geq \mathbf{0}. \end{array} \qquad\qquad \begin{array}{ll} \min \; \mathbf{c}^\mathsf{T}\mathbf{x} \\ \text{s.t.} \; \tilde{\mathbf{A}}\mathbf{x} = \tilde{\mathbf{b}} \quad \text{(}\tilde{\text{P}}\text{)} \\ \quad\quad \mathbf{x} \geq \mathbf{0}. \end{array}$$

Model (P) is the original transshipment model, and model $(\tilde{\text{P}})$ is the same model, but with the first constraint removed. The following theorem shows that the two models are equivalent, in the sense that there is a one-to-one correspondence between the feasible tree solutions of (P) and the feasible basic solutions of $(\tilde{\text{P}})$, and the corresponding objective values coincide.

> **Theorem 8.3.4.**
> Every feasible tree solution of (P) corresponds to a feasible basic solution of $(\tilde{\text{P}})$, and vice versa.

Proof. Let $\mathbf{B}$ be a network basis matrix corresponding to the feasible tree solution $\begin{bmatrix} \mathbf{x}_{BI} \\ \mathbf{x}_{NI} \end{bmatrix}$ of (P). Construct $\tilde{\mathbf{B}}$ from $\mathbf{B}$ by removing row m; let $\mathbf{a}^\mathsf{T}$ be this row. Construct $\tilde{\mathbf{b}}$ analogously. Clearly,

$$\mathbf{B}\mathbf{x}_{BI} = \begin{bmatrix} \tilde{\mathbf{B}}\mathbf{x}_{BI} \\ \mathbf{a}^\mathsf{T}\mathbf{x}_{BI} \end{bmatrix} = \begin{bmatrix} \tilde{\mathbf{b}} \\ b_m \end{bmatrix},$$

and hence we have that $\tilde{\mathbf{B}}\mathbf{x}_{BI} = \tilde{\mathbf{b}}$. Therefore, $\begin{bmatrix} \mathbf{x}_{BI} \\ \mathbf{x}_{NI} \end{bmatrix}$ is a feasible basic solution of model $(\tilde{P})$. Moreover, since the objective vectors of (P) and $(\tilde{P})$ are the same, the objective value of the feasible tree solution of (P) equals the objective value of the feasible basic solution of $(\tilde{P})$. The converse is left to the reader. □

Since we consider minimizing models in this section, feasible basic solutions are optimal if and only if the corresponding objective coefficients are nonnegative. Applying this to $(\tilde{P})$, this means that the feasible basic solution $\begin{bmatrix} \mathbf{x}_{BI} \\ \mathbf{x}_{NI} \end{bmatrix}$ with corresponding basis matrix $\tilde{\mathbf{B}}$ is optimal if and only if:

$$c_{NI_\alpha} - \mathbf{c}_{BI}^\mathsf{T} \tilde{\mathbf{B}}^{-1} \tilde{\mathbf{N}}_{\star,\alpha} \geq 0 \qquad \text{for } \alpha = 1, \ldots, n - m + 1. \tag{8.21}$$

In order to determine optimality using the network simplex algorithm, we use the system (8.19) in Theorem 8.3.3, which states that as soon as we have found a feasible tree solution such that, for all $(i, j) \in \mathcal{A}$, the reduced cost $\bar{c}_{(i,j)}$ has a nonnegative value, we have a found an optimal tree solution. Theorem 8.3.5 shows that optimality criterion (8.19) that is used in the network simplex algorithm is in fact equivalent to (8.21) that is used in the (regular) simplex algorithm. This means that the network simplex algorithm can be viewed a special case of the regular simplex algorithm.

Theorem 8.3.5.
A feasible tree solution of (P) together with a node potential vector satisfies (8.19) if and only if the corresponding feasible basic solution of $(\tilde{P})$ satisfies conditions (8.21).

Proof. We prove the 'only if' direction. The opposite direction is left to the reader. Consider any optimal feasible tree solution $\begin{bmatrix} \mathbf{x}_{BI} \\ \mathbf{x}_{NI} \end{bmatrix}$ with $\mathbf{x}_{NI} = \mathbf{0}$. Let $\mathbf{B}$ be the corresponding network basis matrix, and let $\mathbf{y}$ $(\in \mathbb{R}^m)$ be a node potential vector with respect to $\mathbf{B}$ satisfying (8.19). Construct $\tilde{\mathbf{B}}$ from $\mathbf{B}$ by removing row m; let $\mathbf{a}^\mathsf{T}$ be this row. Similarly, construct $\tilde{\mathbf{N}}$ from $\mathbf{N}$ by removing row m; let $\mathbf{u}^\mathsf{T}$ be this row. Let BI be the set of column indices corresponding to the arcs of $T_{\mathbf{B}}$, and let NI be the set of column indices corresponding to the arcs in $\mathcal{A} \setminus T_{\mathbf{B}}$. Note that $|NI| = n - m + 1$. We will show that (8.21) holds.

Let $\alpha \in \{1, \ldots, n - m + 1\}$. We may assume without loss of generality that $y_m = 0$. Recall that, since $\mathbf{y}$ is a node potential vector, we have that $\mathbf{y}$ is a solution of the system $\mathbf{B}^\mathsf{T}\mathbf{y} = \mathbf{c}_{BI}$; see Section 8.3.3. Since $y_m = 0$, we may write $\mathbf{y} = \begin{bmatrix} \tilde{\mathbf{y}} \\ 0 \end{bmatrix}$, and hence we have that:

$$\mathbf{c}_{BI}^\mathsf{T} = \mathbf{y}^\mathsf{T}\mathbf{B} = \begin{bmatrix} \tilde{\mathbf{y}}^\mathsf{T} & 0 \end{bmatrix} \begin{bmatrix} \tilde{\mathbf{B}} \\ \mathbf{a}^\mathsf{T} \end{bmatrix} = \tilde{\mathbf{y}}^\mathsf{T}\tilde{\mathbf{B}}.$$

Postmultiplying both sides of this equation by $\tilde{\mathbf{B}}^{-1}$ gives $\tilde{\mathbf{y}} = \mathbf{c}_{BI}^\mathsf{T}\tilde{\mathbf{B}}^{-1}$. Recall that NI_α is the index of arc (i, j) in $\mathcal{A} \setminus T_{\mathbf{B}}$. Hence, substituting $\mathbf{c}_{BI}^\mathsf{T}\tilde{\mathbf{B}}^{-1} = \tilde{\mathbf{y}}^\mathsf{T}$, we find that (8.21) is equivalent to:

$$
c_{NI_\alpha} - \mathbf{c}_{BI}^{\mathsf{T}} \tilde{\mathbf{B}}^{-1} \tilde{\mathbf{N}}_{\star,\alpha} = c_{NI_\alpha} - \tilde{\mathbf{y}}^{\mathsf{T}} \tilde{\mathbf{N}}_{\star,\alpha} = c_{NI_\alpha} - \begin{bmatrix} \tilde{\mathbf{y}}^{\mathsf{T}} & 0 \end{bmatrix} \begin{bmatrix} \tilde{\mathbf{N}}_{\star,\alpha} \\ \mathbf{u}^{\mathsf{T}} \end{bmatrix}
$$

$$
= c_{NI_\alpha} - \mathbf{y}^{\mathsf{T}} \mathbf{N}_{\star,\alpha} = c_{NI_\alpha} - \mathbf{y}^{\mathsf{T}} (\mathbf{e}_j - \mathbf{e}_i)
$$

$$
= c_{(i,j)} + \tilde{y}_i - \tilde{y}_j \geq 0,
$$

where we have used the fact that $\mathbf{N}_{\star,\alpha} = \mathbf{e}_i - \mathbf{e}_j$. This proves the theorem.   □

Note also that it follows from the proof of Theorem 8.3.5 that the value of the reduced cost $\bar{c}_{(i,j)}$ of arc (i,j) is exactly the current objective coefficient of the variable $x_{(i,j)}$.

8.3.7 Initialization; the network big-M and two-phase procedures

In Section 3.6, the big-M and the two-phase procedures are used to determine an initial feasible basic solution for the simplex algorithm. When using the big-M procedure, the original model is augmented with a number of artificial (or dummy) variables, in such a way that an initial solution for the modified model can be easily constructed, and an optimal solution for the modified model provides an optimal solution for the original model. In the two-phase procedure, a similar augmented model is solved to optimality in Phase 1, and the resulting optimal solution is used as an initial solution for the original model in Phase 2. The same idea can be used in the case of the network simplex algorithm. We will illustrate the so-called *network big-M procedure* by means of a simple example.

Example 8.3.6.   *Consider the transshipment network depicted in Figure 8.19. Constructing an initial feasible tree solution for this network can certainly be done using trial and error, but such an approach tends to become hard for models with larger networks. To find an initial solution efficiently, a dummy transshipment node 0 is introduced through which initially all units are shipped. For each node i with $b_i > 0$, a dummy arc $(i, 0)$ is introduced pointing from i to 0. Similarly, for each node j with $b_j < 0$, a dummy arc $(0, j)$ is introduced pointing from 0 to j. Finally, for each node i with $b_i = 0$, a dummy arc either $(i, 0)$ or $(0, i)$ is introduced. In Figure 8.19(b) we have drawn the various dummy arcs for the network of Figure 8.19(a). We have chosen to add the arc $(2, 0)$ for node 2 which satisfies $b_2 = 0$.*

The flow costs on all dummy arcs are taken to be some sufficiently large number M. This means that these arcs are so highly 'penalized' that the flow on these arcs should be zero for any optimal solution. If an optimal solution is found that has a positive flow value on a dummy arc, then the original problem is infeasible. Compare the working of the big-M procedure in Section 3.6 in this respect.

As an initial feasible tree solution we take the flow values on the dummy arcs equal to the absolute values of the corresponding b_i's, and the remaining flows are taken to be zero. Note that, if one of the b_i's has value zero, then the initial solution is degenerate. In Figure 8.19(c) the initial spanning tree and the corresponding flow values are depicted. Since the total supply and demand are (assumed to be) balanced, the net flow into node 0 is exactly zero.

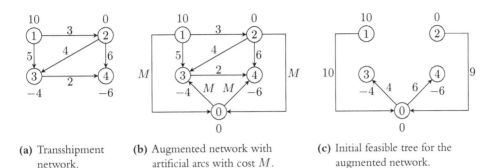

(a) Transshipment network.

(b) Augmented network with artificial arcs with cost M.

(c) Initial feasible tree for the augmented network.

Figure 8.19: Initial feasible tree for the network big-M procedure.

It is left to the reader to carry out the different iterations of the network simplex algorithm applied to the network of Figure 8.19(b) with the initial feasible tree solution of Figure 8.19(c); see Exercise 8.4.13. Analogously to the network big-M procedure, the *network two-phase procedure* can be formulated. The reader is asked to describe this procedure in Exercise 8.4.14, and to use it to find an optimal feasible tree solution for the network of Figure 8.19.

8.3.8 Termination, degeneracy, and cycling

At each iteration, the network simplex algorithm moves from one feasible tree solution to another. If the algorithm makes a cycle, i.e., it encounters a feasible tree solution that has already been encountered before, then the algorithm does not terminate (compare Section 3.5.5). However, if cycling does not occur, then trees are used only once, and since there are only finitely many of them (see Exercise 8.4.16), the algorithm terminates in a finite number of iterations. In Section 3.5.5, we introduced the so-called perturbation procedure, which prevents the ordinary simplex algorithm from cycling. In the case of the network simplex algorithm, we will discuss a simpler cycling-avoiding procedure. This procedure uses so-called strongly feasible spanning trees. But first we give an example of a network in which cycling of the network simplex algorithm occurs. Note that cycling depends on the pivot rule that is being used. We therefore use a specific pivot rule in the example.

Example 8.3.7. *Consider the network of Figure 8.20. It consists of three nodes, labeled 1, 2, and 3, and ten arcs, labeled $a, \ldots, j$. The arc costs (written next to the arcs in Figure 8.20) are:*

$$c_a = 1,\ c_b = -1,\ c_c = 3,\ c_d = 1,\ c_e = -1,$$
$$c_f = 3,\ c_g = 3,\ c_h = -3,\ c_i = 3,\ c_j = -3.$$

The supply-demand vector is $\begin{bmatrix} 0 & 0 & 0 \end{bmatrix}^\top$. The problem is to determine, by means of the network simplex algorithm, a flow through the network satisfying the supplies and demands against minimum cost. We will show that the network simplex algorithm might make a cycle when the following pivot rule is used for the selection of the entering arc: arbitrarily select an arc (k, l) for which the value of $\bar{c}_{(k,l)}$ is negative (see Step 2 in the network simplex algorithm).

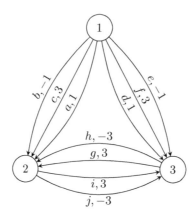

Figure 8.20: Cunningham-Klincewicz network that may cause cycling of the network simplex algorithm.

This example is due to WILLIAM H. CUNNINGHAM (*born 1947*) *and* JOHN G. KLINCEWICZ (*born 1954*). *In Table 8.3, we have listed the feasible trees that occur at each iteration, in case the entering and leaving arcs are the ones listed in the third and fifth column, respectively. In this example, the selection rule for the entering arc is not relevant; the leaving arcs are selected arbitrarily. The fourth column of Table 8.4(a) contains the cycle that occurs when the feasible tree is augmented with the entering arc. It is left to the reader to carry out the various calculations. After ten iterations, we arrive at the initial feasible tree, consisting of the arcs* a *and* i.

It is also left to the reader to determine the network simplex adjacency graph *of the Cunningham-Klincewicz example. This graph consists of* $(3 \times 3 \times 4 =)$ 36 *nodes, namely all possible spanning trees, and there is an arc between two nodes* u *and* v, *if* v *can be derived from* u *by means of an iteration step of the network simplex algorithm. Compare in this respect the definition of the simplex adjacency graph; see Section 3.4.*

Now that we have seen that cycling may occur, we will show how it can be avoided. As said, strongly feasible spanning trees (to be defined below) play a key role.

In Step 1 of the network simplex algorithm, we have chosen $y_1 = 0$ as the start of the calculations of the remaining entries of the node potential vector. This choice is arbitrary, since choosing any value of any entry y_i leads to a node potential vector; see Section 8.3.3. Crucial in the cycle-avoiding procedure is the fact that one node is kept fixed during the whole iteration process, and that for this node (called the *root node* of the network) the value of the corresponding entry of the node potential vector is also kept fixed during the process. Node 1 is taken as the root node.

From Theorem C.3.1 in Appendix C, we know that there exists a unique path between any pair of nodes in the tree, where the directions of the arcs are not taken into account (so in such a path the arcs may have opposite directions). We call the arc (i, j) *directed away from the root* in the spanning tree $T_\mathbf{B}$, if node i is encountered before node j when traversing the unique path from the root. For any network basis matrix $\mathbf{B}$, a feasible tree $T_\mathbf{B}$ is called a

Iteration	Feasible tree	Entering arc	Cycle	Leaving arc
1	a, i	j	3-2-3	i
2	a, j	f	1-3-2-1	a
3	f, j	b	1-2-3-1	j
4	f, b	g	2-3-1-2	f
5	g, b	d	1-3-2-1	b
6	g, d	h	2-3-2	g
7	h, d	c	1-2-3-1	d
8	h, c	e	1-3-2-1	h
9	e, c	i	3-2-1-3	c
10	e, i	a	1-2-3-1	e
11	a, i	j	3-2-3	i

Table 8.3: Network simplex cycle.

strongly feasible spanning tree if every arc in $T_{\mathbf{B}}$ that has zero flow is directed away from the root in $T_{\mathbf{B}}$. The following theorem holds.

Theorem 8.3.6. (*Anti-cycling theorem*)
If at each iteration a strongly feasible spanning tree is maintained, then the network simplex algorithm does not cycle.

Proof of Theorem 8.3.6. It can be easily checked that cycling can only occur in a sequence of degenerate steps; i.e., the corresponding feasible tree solutions contain at least one arc with a zero flow. After a degenerate iteration, the deletion of the last entered arc, say (k,l), splits the current strongly feasible spanning tree $T_{\mathbf{B}}$ into two disjoint subtrees. Let T_r denote the subtree containing the root, and let T_o denote the other subtree. Note that $T_{\mathbf{B}} \setminus \{(k,l)\} = T_r \cup T_o$, and that T_r and T_o are disjoint trees, both containing one end node of the arc (k,l).

Since $T_{\mathbf{B}}$ is a strongly feasible spanning tree, both the root node 1 and node k are in T_r, and $l \in T_o$. The fact that $1 \in T_r$ implies that the entries of the node potential vector corresponding to the nodes in T_r did not change in the last iteration. For the 'new' value of y_l, denoted by $y_{l,\text{new}}$, the following holds (the variables $y_{k,\text{new}}, y_{k,\text{old}}$, and $y_{l,\text{old}}$ have obvious meanings):

$$y_{l,\text{new}} = y_{k,\text{new}} - c_{(k,l)} = y_{k,\text{old}} - c_{(k,l)} = y_{l,\text{old}} - \bar{c}_{(k,l)},$$

with $\bar{c}_{(k,l)} = c_{(k,l)} - y_{k,\text{old}} + y_{l,\text{old}}$. For each node $j \in T_o$, the 'new' value of y_j can be determined by subtracting $\bar{c}_{(k,l)}$ from its 'old' value. Hence,

$$y_{j,\text{new}} = \begin{cases} y_{j,\text{old}} & \text{if } j \in T_r \\ y_{j,\text{old}} - \bar{c}_{(k,l)} & \text{if } j \in T_o. \end{cases}$$

Since $\bar{c}_{(k,l)} < 0$ (because (k,l) is an entering arc), it follows that:

$$\sum_{j \in \mathcal{V}} y_{j,\text{new}} > \sum_{j \in \mathcal{V}} y_{j,\text{old}}.$$

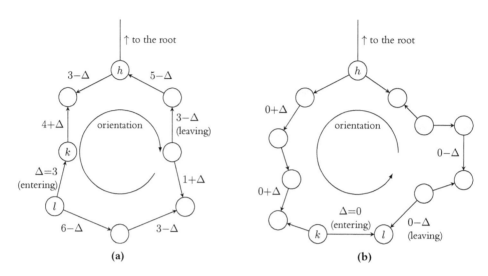

Figure 8.21: Leaving arc rule. The orientation of the cycle is in the direction of the entering arc (k, l).

Recall that, in each iteration, we choose $y_1 = 0$, and so for any spanning tree, the values of the y_j's are fixed. Hence, the 'new' spanning tree is different from the 'old' one. In fact, we have proved that, in each iteration, the sum of the entries of the node potential vector strictly increases. Therefore, the algorithm never encounters the same spanning tree twice and, hence, cycling does not occur. □

The remaining question is how to find and maintain a strongly feasible spanning tree. Determining an initial strongly feasible spanning tree poses no problem when enough artificial arcs are added to the network. This can be done by following the approach of the big-M procedure (see Section 3.6.1). Once a strongly feasible spanning tree solution has been obtained, it can be maintained by selecting the leaving arc according to the following *leaving arc rule*.

Leaving arc rule: Let (k, l) be the entering arc of the current strongly feasible spanning tree $T_\mathbf{B}$, and let $C_{(k,l)}$ be the unique cycle in $T_\mathbf{B} \cup \{(k, l)\}$. There are two cases to be considered:

(a) The iteration is nondegenerate. In order to maintain a strongly feasible spanning tree, the first zero-flow arc in $C_{(k,l)}$, when traversing $C_{(k,l)}$ in the direction of (k, l) and starting in the node closest to the root, has to be the leaving arc. At least one arc obtains a zero flow after adding and subtracting Δ (> 0), as in Step 3 of the network simplex algorithm. For example, consider Figure 8.21(a), where $\Delta = 3$ and three arcs obtain zero flow after subtracting Δ. Note that, in fact, all arcs that have been assigned zero flow are directed away from the root.

(b) The iteration is degenerate, and at least one arc has zero flow in the direction opposite to (k, l). The leaving arc should be the first one that is encountered when traversing $C_{(k,l)}$ in the direction of (k, l) and starting in (k, l). In this case $\Delta = 0$. The zero-flow arcs are directed away from the root. For example, in Figure 8.21(b), there are four zero-flow arcs of which two are in the direction opposite to (k, l).

In both cases (a) and (b), the new spanning tree has all its zero-flow arcs pointing away from the root, and so it is in fact a strongly feasible spanning tree.

Example 8.3.8. *We apply the anti-cycling procedure described above to the example of Figure 8.20. The following iterations may occur. In all iterations, we take $y_1 = 0$.*

Iteration 1. Start with the strongly feasible spanning tree $\{a, d\}$. Then $y_2 = -1$ and $y_3 = -1$. Since $\bar{c}_b = -2$, arc a leaves and b enters the tree.

Iteration 2. The new strongly (!) feasible spanning tree is $\{b, d\}$. Then $y_2 = 1$ and $y_3 = -1$. Since $\bar{c}_h = -5$, arc d leaves and h enters the tree.

Iteration 3. The new strongly feasible spanning tree is $\{b, h\}$. Then $y_2 = 1$ and $y_3 = 4$. Take j with $\bar{c}_j = -6$ as the entering arc. Extending the tree $\{b, h\}$ with j, leads to the cycle consisting of j and h. This cycle does not contain arcs that point opposite to j, and so the flow on this cycle can increase without bound, while decreasing the cost without bound. Hence, as could have been expected, this problem has no bounded solution.

Note that in fact a strongly feasible tree has been maintained, and that cycling did not occur.

8.4 Exercises

Exercise 8.4.1. Determine whether or not the following matrices are totally unimodular.

(a) $\begin{bmatrix} 1 & 0 & -1 \\ 1 & 1 & 0 \\ 1 & 0 & -1 \end{bmatrix}$

(b) $\begin{bmatrix} 1 & -1 & 0 & -1 & 0 \\ -1 & 0 & 0 & 0 & 1 \\ 0 & 0 & 1 & 1 & 0 \\ 0 & 0 & -1 & 0 & 0 \end{bmatrix}$

(c) $\begin{bmatrix} 0 & 1 & 0 & 0 & 1 \\ -1 & 1 & 0 & 0 & 0 \\ 0 & 0 & 1 & -1 & 0 \\ 0 & 0 & -1 & -1 & 1 \\ 1 & 0 & 0 & 0 & 0 \end{bmatrix}$

(d) $\begin{bmatrix} 1 & 0 & -1 & 0 & 0 \\ 1 & -1 & 0 & -1 & 0 \\ 0 & 1 & -1 & 0 & -1 \\ 0 & 0 & 0 & 1 & -1 \end{bmatrix}$

(e) $\begin{bmatrix} 1 & 1 & 0 \\ 0 & 1 & 0 \\ 1 & 1 & -1 \end{bmatrix}$

(f) $\begin{bmatrix} -1 & 1 & 1 & 1 \\ 0 & -1 & -1 & -1 \\ -1 & 0 & 0 & -1 \end{bmatrix}$

Exercise 8.4.2.

(a) Give an example of a full row rank unimodular matrix that is not totally unimodular.

(b) Give an example of a full row rank totally unimodular matrix with at least one positive and at least one negative entry in each column, that does not satisfy the conditions of Theorem 8.1.4.

Exercise 8.4.3.   Let $\mathbf{A}$ be an (m, n)-matrix. Show that if $\mathbf{A}$ is totally unimodular, then $\begin{bmatrix} \mathbf{A} & \mathbf{I}_m \end{bmatrix}$ is unimodular.

Exercise 8.4.4.   Show, by calculating all feasible basic solutions, that the region

$$\left\{ \mathbf{x} \in \mathbb{R}^6 \; \middle| \; \begin{bmatrix} 1 & 0 & 0 & 1 & 0 & 0 \\ 1 & 1 & 0 & 0 & 1 & 0 \\ 1 & 0 & 1 & 0 & 0 & 1 \end{bmatrix} \mathbf{x} \le \mathbf{b}, \mathbf{x} \ge \mathbf{0} \right\}$$

has only integer vertices when $\mathbf{b}$ is an integer vector (see Section 8.1.1).

Exercise 8.4.5.   Consider the transportation problem as formulated in Section 4.3.3 for $m = 3$ and $n = 4$. Consider the problem with the following data:

$$\mathbf{C} = \begin{bmatrix} 8 & 6 & 10 & 9 \\ 9 & 12 & 13 & 7 \\ 14 & 9 & 16 & 5 \end{bmatrix}, \mathbf{a} = \begin{bmatrix} 35 & 50 & 40 \end{bmatrix}^\mathsf{T}, \text{ and } \mathbf{b} = \begin{bmatrix} 45 & 20 & 30 & 30 \end{bmatrix}^\mathsf{T},$$

with $\mathbf{C}$ the cost matrix, $\mathbf{a}$ the supply vector, and $\mathbf{b}$ the demand vector. Find an optimal feasible solution by using the complementary slackness relations, given that an optimal dual solution is $\{y_i\} = \begin{bmatrix} 2 & 5 & 5 \end{bmatrix}^\mathsf{T}$ with i the index of the supply nodes, and $\{y_j\} = \begin{bmatrix} 4 & 4 & 8 & 0 \end{bmatrix}^\mathsf{T}$ with j the index of the demand nodes.

Exercise 8.4.6.   Consider the transportation problem in Figure 8.4 of Section 8.2.1 and the corresponding ILO-model (8.4).

(a) Find a feasible solution of the ILO-model (and hence of its LO-relaxation).

(b) Find an initial feasible basic solution for the LO-relaxation that can start the simplex algorithm.

Exercise 8.4.7.   Show that Model 8.2.2 has an optimal solution if and only if the supply-demand balance equation (8.5) holds.

Exercise 8.4.8.   This exercise is a continuation of Exercise 8.4.6. Consider the (general) transportation problem of Section 8.2.1 and the corresponding ILO-model, Model 8.2.1.

(a) Show how to construct a feasible solution for Model 8.2.1.

(b) Show how to construct an initial feasible basic solution for the LO-relaxation of Model 8.2.1.

Exercise 8.4.9. How can a transportation problem, as described in Section 8.2.1, be solved if the total demand is not equal to the total supply?

Exercise 8.4.10. Show that for every feasible flow in the network of a transshipment problem, there exists a spanning tree corresponding to a flow with at most the same cost.

Exercise 8.4.11. Give two examples of feasible flows that contain a cycle, and show that the corresponding node-arc incidence matrices have linearly dependent columns; namely in the following two cases.

(a) The network has two nodes and two arcs.

(b) The network has three nodes, four arcs, and two cycles.

Exercise 8.4.12. Show that step 3 of the network simplex algorithm, Algorithm 8.3.1, when applied to (P) is equivalent to the minimum-ratio test when the simplex algorithm is applied to $(\tilde{P})$.

Exercise 8.4.13. Use the big-M procedure to calculate an optimal solution of the transshipment problem corresponding to the data of the network of Figure 8.19.

Exercise 8.4.14. Describe how the two-phase procedure can be applied to the transshipment problem.

Exercise 8.4.15. Consider a digraph $G = (V, A)$. Let s and t be different nodes in V, and let c_{ij} be defined as the distance from node i to node j. Show that the problem of finding a shortest path from s to t can be formulated as a minimum cost flow problem.

Exercise 8.4.16. In Section 8.3.8, it is claimed that the number of spanning trees in any given network is finite.

(a) Consider again the network in Figure 8.20. Calculate the total number of spanning trees in this network.

(b) Let G be any network with m nodes and n edges. Show that the number of spanning trees in G is finite.

Exercise 8.4.17. Consider the network G with node set $\{1, \ldots, n\}$, where 1 is the source node, and n is the sink node. Assume that flow is preserved at all nodes $2, \ldots, n-1$. Show that the net flow out of node 1 equals the net flow into node n.

Exercise 8.4.18. Consider the supply-demand problem with the following data:

	Supply	Demand
Node A	12	0
Node B	0	0
Node C	0	3
Node D	0	9

There are only deliveries from A to B (notation AB), AC, BC, BD, and CD; the costs per unit flow are 3, 5, 4, 6, 2, respectively. The problem is to satisfy the demand against minimum costs.

(a) Formulate this problem as a transshipment problem, and determine the node-arc incidence matrix.

(b) Determine an initial feasible tree solution by applying the network big-M procedure.

(c) Determine an optimal solution by means of the network simplex algorithm. Why is this solution optimal?

Exercise 8.4.19. Consider a transshipment problem with the following data. There are four nodes A, B, C, and D. The supply vector is $\begin{bmatrix} b_A & b_B \end{bmatrix} = \begin{bmatrix} 8 & 6 \end{bmatrix}$ and the demand vector is $\begin{bmatrix} b_C & b_D \end{bmatrix} = \begin{bmatrix} 3 & 11 \end{bmatrix}$; the cost vector is $\begin{bmatrix} c_{(A,B)} & c_{(A,C)} & c_{(B,C)} & c_{(B,D)} & c_{(C,D)} \end{bmatrix} = \begin{bmatrix} 2 & 6 & 1 & 5 & 2 \end{bmatrix}$.

(a) Give an LO-formulation of this problem.

(b) Show that the technology matrix is totally unimodular.

(c) Solve the problem by means of the network simplex algorithm. Why is the solution optimal?

(d) Determine the tolerance interval of $c_{(A,C)}$.

(e) Determine an optimal solution for $c_{(B,D)} = 2$.

(f) Suppose that the arc AD is included in the network with $c_{(A,D)} = 4$; determine an optimal solution of this new problem.

Exercise 8.4.20. Consider the following network. The nodes are A, B, C, D. The cost vector is $\begin{bmatrix} c_{(A,B)} & c_{(A,C)} & c_{(B,D)} & c_{(C,B)} & c_{(C,D)} & c_{(D,A)} \end{bmatrix} = \begin{bmatrix} 1 & 3 & 3 & 7 & 4 & 1 \end{bmatrix}$. The supply vector is $\begin{bmatrix} b_A & b_C \end{bmatrix} = \begin{bmatrix} 2 & 2 \end{bmatrix}$, and the demand vector is $\begin{bmatrix} b_B & b_D \end{bmatrix} = \begin{bmatrix} 1 & 3 \end{bmatrix}$. The objective is to determine a flow in this network that satisfies the supply and demand against minimum total cost.

(a) Transform this problem into a minimum cost flow problem with one supply node and one demand node.

(b) Transform this problem into a transportation problem.

(c) Determine an optimal solution. Why is it optimal?

Exercise 8.4.21. A company produces a commodity under two different circumstances, namely, in regular working time and in overtime. The cost of producing one unit in regular working time is \$8. The cost per unit produced in overtime is \$11. The production capacities (in the case of 'regular' and 'overtime') for three consecutive months are given in the table below (left). The demand values are given in the last row of this table. The price of keeping one unit of the product in stock is \$2 per month. The total costs of production and inventory (per unit) are summarized in the table below (right).

	Month		
	I	2	3
Regular time	100	200	150
Overtime	40	80	60
Demand	90	200	270

		Month		
		I	2	3
Month 1	Regular time	8	10	12
	Overtime	11	13	15
Month 2	Regular time		8	10
	Overtime		11	13
Month 3	Regular time			8
	Overtime			11

(a) Formulate this problem as an ILO-model, and determine an optimal production schedule.

(b) This problem can be considered as a special case of a problem described in this chapter. Which problem is that?

Exercise 8.4.22. A salesperson has to visit five companies in five different cities. She decides to travel from her home town to the five cities, one after the other, and then returning home. The distances between the six cities (including the salesperson's home town) are given in the following matrix:

	Home	I	2	3	4	5
Home	–	12	24	30	40	16
I	12	–	35	16	30	25
2	24	35	–	28	20	12
3	30	16	28	–	12	40
4	40	30	20	12	–	32
5	16	25	12	40	32	–

(a) Write an ILO-model that can be used to find a shortest route along the six cities with the same begin and end city; determine such a shortest route.

(b) This problem can be seen as a special case of a problem described in this chapter. Which problem is that?

(c) Suppose that the salesperson wants to start her trip in her home town and finish her trip in town 4, while visiting all other cities. Formulate an ILO-model of this problem and determine and optimal solution.

Exercise 8.4.23.   The distances between four cities are given in the following table:

	Cities			
	1	2	3	4
City 1	0	6	3	9
City 2	6	0	2	3
City 3	3	2	0	6
City 4	9	3	6	0

(a) Write an ILO-model that can be used to determine a shortest path from city 1 to 4; determine a shortest path from 1 to 4.

(b) Why are all feasible basic solutions integer-valued?

(c) Formulate this problem as a minimum cost flow problem. Show that the technology matrix is totally unimodular. Solve the problem.

Exercise 8.4.24.   In the harbor of Rotterdam, five ships have to be unloaded by means of five cranes. The ships are labeled $(i =) 1, \ldots, 5$, and the cranes $(j =) 1, \ldots, 5$. Each ship is unloaded by one crane, and each crane unloads one ship. The cargo of ship i is w_i volume units. Each crane j can handle c_j volume units of cargo per hour. The problem is to assign ships to cranes such that the total amount of unloading time is minimized.

(a) What type of problem is this?

(b) Determine the coefficients in the objective function, and formulate the problem as an ILO-model.

In the table below, values for w_i and c_j are given.

i, j	1	2	3	4	5
w_i	5	14	6	7	10
c_j	4	2	8	1	5

(c) In order to determine a solution for this data set, one may expect that the ship with the largest amount of cargo should be unloaded by the crane with the highest capacity. Check this idea by determining an optimal solution.

(d) Another idea could be: minimize the maximum amount of time needed for a crane to unload a ship. Formulate this new objective in mathematical terms, and determine an optimal solution of the model with this new objective.

Exercise 8.4.25. During the summer season, a travel agency organizes daily sightseeing tours. For each day, there are m (≥ 1) trips planned during the morning hours, and m trips during the afternoon hours. There are also exactly m buses (with bus driver) available. So, each bus has to be used twice per day: once for a morning trip, and once for an afternoon trip.

As an example, take $m = 5$, and consider the following time table:

	Morning trips			Afternoon trips	
	Departure time	Return time		Departure time	Return time
1	8:30AM	12:00PM	1	1:00PM	5:00PM
2	8:45AM	12:45PM	2	1:00PM	5:30PM
3	8:50AM	12:30PM	3	1:15PM	5:00PM
4	9:00AM	1:00PM	4	1:35PM	5:00PM
5	9:05AM	1:05PM	5	12:55PM	4:50PM

The bus drivers normally work eight hours per day. Sometimes, however, working more than that is unavoidable and one or more bus drivers need to work overtime. (Overtime is defined as the number of hours worked beyond the regular hours on a day.) Since overtime hours are more expensive than regular hours, we want to keep the amount of overtime to a minimum.

The problem is to find combinations of morning and afternoon trips for the bus drivers, such that the total overtime is minimized.

(a) This problem can be seen as a special case of a problem described in this chapter. Which problem is that?

(b) Formulate this problem as an ILO-model. Explain what the constraints mean in terms of the original problem. Also pay attention to the fact that several combinations of trips are not allowed.

Now assume that the bus drivers must have a 30-minute break between two trips.

(c) How can this constraint be incorporated into the model formulated under (b)?

(d) Solve the problem formulated in (c).

(e) Perform sensitivity analysis on the input parameters.

Exercise 8.4.26. Draw perturbation graphs of the values of the completion times for the activities corresponding to the arcs $(2,3)$, $(4,5)$, and $(5,6)$ of the project network in Section 8.2.6.

Exercise 8.4.27. Consider the project scheduling problem described in Section 8.2.6. Prove or disprove the following statements.

(a) Increasing the execution time of an activity (i, j) on a critical path always increases the total completion time of the project.

(b) Decreasing the execution time of an activity (i, j) on a critical path always decreases the total completion time of the project.

Exercise 8.4.28.   Consider a project network. Show that the fact that the network has no directed cycles implies that there is at least one initial and at least one end goal.

Exercise 8.4.29.   Consider the project scheduling problem described in Section 8.2.6.

(a) Show that any activity with zero slack is on a critical path.

(b) Show that every critical path of a project has the same total completion time.

Exercise 8.4.30.   The publishing company Book & Co has recently signed a contract with an author to publish and market a new textbook on linear optimization. The management wants to know the earliest possible completion date for the project. The relevant data are given in the table below. A total of eight activities, labeled A1, ..., A8, have to be completed. The descriptions of these activities are given in the second column of this table. The third column lists the estimated numbers of weeks needed to complete the activities. Some activities can only be started when others have been finished. The last column of the table lists the immediate predecessors of each activity.

Label	Description	Time estimate	Immediate predecessors
A1	Author preparation of manuscript	25	None
A2	Copy edit the manuscript	3	A1
A3	Correct the page proofs	10	A2
A4	Obtain all copyrights	15	A1
A5	Design marketing materials	9	A1
A6	Produce marketing materials	5	A5, A4
A7	Produce the final book	10	A3, A4
A8	Organize the shipping	2	A6, A7

(a) Draw a project network of this problem.

(b) Compute the total completion time of the project, together with the earliest starting and earliest completion times of all activities.

(c) Compute the latest starting and latest completion times of all activities.

(d) What is a critical path for this project? Is it unique?

(e) Draw the perturbation graph of the activities A2, A4, and A8.

Exercise 8.4.31.   The company PHP produces medical instruments, and has decided to open a new plant. The management has identified eleven major project activities, to be completed before the actual production can start. The management has also specified the activities (the immediate predecessors) that must be completed before a given activity can begin. For each of the eleven activities, the execution time has been estimated. In the table below, the results are listed.

Activity	Description	Time estimate	Immediate predecessors
A	Select staff	13	None
B	Select site	26	None
C	Prepare final construction plans and layout	11	B
D	Select equipment	11	A
E	Bring utilities to site	39	B
F	Interview applicants and fill positions	11	A
G	Purchase equipment	36	C
H	Construct the building	41	D
I	Develop information system	16	A
J	Install equipment	5	E, G, H
K	Train staff	8	F, I, J

(a) Draw a project network of this problem.

(b) Give an LO-model of this problem, and solve it with a computer package.

(c) Give the dual of the LO-model found under (b). Show that its solution corresponds to a longest path in the project network.

(d) Determine the shadow prices of all activities. Why are shadow prices of activities that are not on any critical path equal to zero?

(e) Determine the earliest starting and completion times, and the latest starting and completion times for all activities.

(f) Determine the tolerance intervals for all activities.

(g) Draw perturbation functions of an activity on a critical path, and of an activity that is not on any critical path.

Exercise 8.4.32. There are three cities, labeled A, B, and C. The families in these three cities are partitioned into three categories: "no children", "one child", and "at least two children". The table below lists census data on these families in the cities A, B, and C in the form of percentages.

City	Number of children			Total
	≥ 2	1	0	
A	3.14	6.8	7.3	17.24
B	9.6	2.4	0.7	12.7
C	3.6	1.2	6.5	11.3
Total	16.34	10.4	14.5	41.24

The entries of the table are percentages of the population in different categories. The bold entries show the row and column sums, except for the bottom right entry which gives the total of all entries.

The government wants to publish this table, but it wants to round all percentages in the table (i.e., both the bold and the nonbold entries) to an integer. To make sure that the table does not display any incorrect numbers, the numbers may only be rounded up or down, but not necessarily to the nearest integer. At the same time, however, the resulting table needs to be consistent, in the sense that the rows and columns have to add up to the listed totals.

(a) Show that the following simple approaches do not work: (i) rounding all percentages to the nearest integer, (ii) rounding down all percentages, and (iii) rounding up all percentages.

This problem can be viewed as a *maximum flow problem* on a network $G = (\mathcal{V}, \mathcal{A})$ with $\mathcal{V} = \{1, \ldots, n\}$ and $\mathcal{A}$ the set of arcs of G. In addition to the capacity $k_{(i,j)}$, this problem also has a lower bound $l_{(i,j)}$ on each arc (i, j). The general LO-formulation of this problem reads:

$$\max \sum_{(1,j)\in\mathcal{A}} x_{(s,j)}$$

$$\text{s.t. } \sum_{(j,i)\in\mathcal{A}} x_{(j,i)} = \sum_{(i,j)\in\mathcal{A}} x_{(i,j)} \qquad \text{for } i = 2, \ldots, n - 1$$

$$l_{(i,j)} \le x_{(i,j)} \le k_{(i,j)} \qquad \text{for } (i, j) \in \mathcal{A}.$$

(b) Formulate a maximum flow model with lower bounds that solves the rounding problem. (Hint: construct a graph that has a source and a sink node, one node for every row, and one node for every column.) Why is it true that this model solves the rounding problem?

Exercise 8.4.33.   In open-pit mining, blocks of earth are dug, starting from the surface, to excavate the ore contained in them. During the mining process, the surface of the land is excavated, forming a deeper and deeper pit until the mining operation terminates. Usually, the final shape of this so-called open pit is determined before the mining operation begins. A common approach in designing an optimal pit (i.e., one that maximizes profit) is to divide the entire mining area into 3-dimensional blocks. Using geological information from drill cores, the value of the ore in each block is estimated. Clearly, there is also a cost of excavating each particular block. Thus, to each block in the mine, we can assign a profit value. The objective of designing an optimal pit is then to choose blocks to be excavated while maximizing the total profit. However, there are also constraints on which blocks can be dug out: blocks underlying other blocks can only be excavated after the blocks on top of them have been excavated.

As a special case, consider the two-dimensional pit drawn in Figure 8.22. The values in the figure refer to the value of the ore in each block ($\times\$1,000$). Suppose that excavating one block costs the same for each block, namely $\$2,500$ per block.

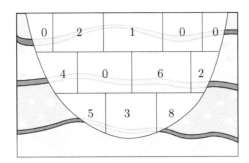

Figure 8.22: Two-dimensional open pit.

This pit design problem can be represented using a directed graph, $G = (\mathcal{V}, \mathcal{A})$. Let n ($= 12$) be the number of blocks. Let $i = 1, \ldots, n$. We create a node for each block in the mining area, and assign a weight b_i representing the profit value of excavating block i. There is a directed arc from node i to node j if block i cannot be excavated before block j, which is on a layer immediately above block i. If there is an arc from i to j, then j is called a *successor* of i. To decide which blocks to excavate in order to maximize profit, we need to find a maximum weight set of nodes in the graph such that all successors of all nodes in the set are also included in the set. Such a set is called a *maximum closure* of G.

(a) Draw a network G for the mining problem described above.

The maximum weight closure problem can be solved by a minimum cut algorithm on a related graph $G' = (\mathcal{V}', \mathcal{A}')$. To construct the graph G', we add a source node s and sink node t, and $\mathcal{A}'$ consists of all arcs in $\mathcal{A}$, as well as an arc from s to each node in $\mathcal{V}$ with positive value ($b_i > 0$), and an arc from each node in $\mathcal{V}$ with negative value ($b_i < 0$) to t. The capacity of each arc in $\mathcal{A}$ is set to ∞. For each arc (s, i), we set the capacity to b_i and for arc (j, t) we set the capacity to $-b_j$.

(b) Draw the network G' for the mining problem described above.

(c) Show that solving the mining problem as described above indeed solves the open pit (maximum closure) problem.

(d) Solve the problem using a computer package.

CHAPTER 9

Computational complexity

Overview

In this chapter we give a brief introduction to the theory of computational complexity. For a more extensive account on this subject we refer the reader to Schrijver (1998). We will successively examine DANTZIG's simplex algorithm from Chapter 3, the interior path algorithm from Chapter 6, and the branch-and-bound algorithm from Chapter 7.

9.1 Introduction to computational complexity

One of the challenges, when using a computer to solve a problem, is to use a solution technique with which the problem can be solved within reasonable time limits. For instance, when a production schedule for the next week is due, it is usually realistic to allow the computer to spend one night of calculations to determine a schedule. However, when online decisions have to be taken, the computer should generate solutions usually within seconds. The important question therefore is to make a realistic estimation of the computer running time needed. Of course, the running time depends on the problem itself, the solution technique used, the way the problem is implemented, and the computer hardware available.

We use the term *algorithm* for a method that generates a solution of a problem in a step-by-step manner. An algorithm is called *iterative* if it contains a sequence of steps that is carried out a number of times; one call of such a sub algorithm is called an *iteration*. An algorithm is said to *solve* a problem if that algorithm can be applied to any input instance of that problem and is guaranteed to produce a solution for that instance in finite time. For example, an input instance of a linear optimization problem is a set of values for the input parameters of that problem, i.e., values for the number of decision variables and the constraints, and values for the entries of the technology matrix, the right hand side vector, and the objective vector. The question asked by linear optimization problems is: 'What is an optimal solution to the problem, or is the problem unbounded, or is it infeasible?' In other words: 'If the problem has a solution, determine a point in the feasible region of the model that optimizes the objective function'. Both the simplex algorithm (see Chapter 3) and the interior path algorithm (see Chapter 6) solve linear optimization problems. Even an algorithm that simply enumerates all (exponentially, but finitely, many) basic solutions and saves the current best one, solves LO models.

In this chapter, we are not interested in the details of a specific implementation of the algorithm used, but rather in its running time, defined as the number of arithmetic operations (such as additions, subtractions, multiplications, divisions, and comparisons) required by the algorithm. For instance, when computing the inner product $\mathbf{a}^\top \mathbf{b}$ of two n-vectors $\mathbf{a}$ and $\mathbf{b}$, we need n multiplications and $n - 1$ additions, resulting in a total of $2n - 1$ arithmetic operations. The parameter n in this example is called the *size* of the problem. In the case of linear optimization, the size is the total number of entries in the technology matrix, the objective coefficient vector, and the right hand side vector. So, the size of a standard LO-model with n variables and m constraints is $nm + n + m$.

A measure for the computer running time of an algorithm required to solve a certain instance of a problem is its *computational complexity*, denoted using the so-called *big-O notation*. Let n ($n \geq 1$) be the size of the problem, and let f and g be functions that map positive numbers to positive numbers. The notation

$$f(n) = O(g(n))$$

means that there exist positive numbers n_0 and α such that $f(n) \leq \alpha g(n)$ for each $n \geq n_0$. If the running time of a certain algorithm applied to a certain instance of a problem of size n needs $f(n)$ arithmetic operations and $f(n) = O(g(n))$, then we say that this algorithm *runs in order $g(n)$* (notation: $O(g(n))$ time. For instance, the above 'algorithm' used to calculate $\mathbf{a}^\top \mathbf{b}$ for the n-vectors $\mathbf{a}$ and $\mathbf{b}$ has a running time of order n, because $2n - 1 = O(n)$. The latter holds because for $f(n) = 2n - 1$ and $g(n) = n$, we may take $n_0 = 1$ and $\alpha = 2$. Note that $3n^3 + 3567n^2 = O(n^3)$, $2766n^8 + 2^n = O(2^n)$, and $n! = O(2^n)$.

The above definition of running time depends on the numerical values of the input data. Instead of trying to estimate the running time for all possible choices of the input (i.e., all possible instances of the problem), it is generally accepted to estimate the running time for a *worst-case* instance of the problem, i.e., for an input data set of size n for which the running time of the algorithm is as large as possible. If this worst-case running time is $f(n)$ and $f(n) = O(g(n))$, then we say that the algorithm has a (*worst-case*) *complexity of order* $g(n)$. The notation is again $O(g(n))$, and $g(n)$ is now called the *complexity* of the algorithm applied to the problem of size n. This worst-case complexity, however, has the obvious disadvantage that 'pathological' instances determine the complexity of the algorithm. In practice, the *average-case complexity* might be more relevant. However, one of the drawbacks of using the average-case approach is that the average running time over all instances is in general very difficult to estimate. For this reason we shall take the worst-case complexity approach when comparing algorithms with respect to their computer running times.

n	Running time							
	1	$\log n$	n	n^2	n^3	n^{10}	2^n	$n!$
1	10^{-13} s	0	10^{-13} s	10^{-13} s	10^{-13} s	10^{-13} s	2×10^{-13} s	10^{-13} s
10	10^{-13} s	2×10^{-13} s	10^{-12} s	10^{-11} s	10^{-10} s	10^{-3} s	10^{-10} s	4×10^{-7} s
100	10^{-13} s	5×10^{-13} s	10^{-11} s	10^{-9} s	10^{-7} s	115 d	4×10^{9} y	10^{137} y
1,000	10^{-13} s	7×10^{-13} s	10^{-10} s	10^{-7} s	10^{-4} s	10^{9} y	10^{280} y	10^{2547} y
10,000	10^{-13} s	9×10^{-13} s	10^{-9} s	10^{-5} s	10^{-1} s	10^{19} y	10^{2989} y	10^{35638} y
100,000	10^{-13} s	10^{-12} s	10^{-8} s	10^{-3} s	1 m	10^{29} y	10^{30082} y	10^{456552} y

Table 9.1: Growth of functions. ('s' = seconds, 'm' = minutes, 'd' = days, 'y' = years.) We assume that the computer processor can perform 10^{13} operations per second. All times are approximate. For comparison, the number of atoms in the universe has been estimated at approximately 10^{80}, and the age of the universe has been estimated at approximately 1.38×10^{9} years.

As a measure of performance of an algorithm applied to a problem of size n, we have chosen the number of arithmetic operations spent on a worst-case instance of that problem, and expressed as a function of n, called the complexity of the algorithm. We now specify a relationship between the complexity and the question of whether the algorithm is 'good' or 'bad'. The distinction between 'good' and 'bad' algorithms is rather difficult to make. A widely accepted definition of a 'good' algorithm is an algorithm of which the worst-case complexity is a polynomial function of the input size of the problem. We call an algorithm *good* (the terms *polynomial* and *efficient* are frequently used instead of 'good'), if it has a complexity function that is a polynomial in the size of the model; otherwise, the algorithm is called 'bad', *exponential* or *inefficient*. See, e.g., Schrijver (1998) for more specific definitions of these concepts. These definitions were introduced by JACK R. EDMONDS (born 1934) in the mid-1960s; see, e.g., Lenstra *et al.* (1991). As an example, consider the algorithm used above for the calculation of the inner product $\mathbf{a}^{\top}\mathbf{b}$. This algorithm is 'good', because the complexity function n is a polynomial in n. In Section 9.2, we will see that the simplex algorithm for linear optimization problems is an exponential algorithm, whereas – as we will prove in Section 9.3 – the interior path algorithm is polynomial. On the other hand, it can be shown (see again Schrijver (1998)) that the simplex algorithm is efficient from the average-case complexity point of view. Since most practical instances of linear optimization problems are not 'pathological', it is precisely this point that explains the success of the simplex algorithm in practice.

The differences between polynomials (e.g., n and n^3) and exponentials (e.g., 2^n and $n!$) may become clear when studying Table 9.1. Of particular interest is the remarkable increase in the values of exponential functions as compared to the polynomial functions. For instance, a high-end computer processor in 2014 was able to perform roughly 10^{13} arithmetic operations per second (also called *floating-point operations per second*, or *flops*). Solving a problem for $n = 75$ by running an algorithm that requires 2^n operations would already require around 120 years of computation. From a practical point of view, however, the difference between an efficient (polynomial) algorithm and an inefficient (exponential) algorithm is not always significant. For example, an algorithm with complexity $O(n^{100})$ can hardly be considered as practically efficient.

9.2 Computational aspects of Dantzig's simplex algorithm

Empirical testing of the simplex algorithm using large problem instances show that the number of iterations increases proportionally with the number of constraints and is relatively insensitive to the number of decision variables. It is this remarkable fact that accounts for the enormous practical success of the simplex algorithm. With the current level of computer technology, models with millions of variables and constraints can be handled successfully.

The question remains whether or not it is possible to solve any LO-model instance within reasonable time limits using the simplex algorithm. The answer unfortunately is negative. In 1972, VICTOR KLEE (1925–2007) and GEORGE J. MINTY (1929–1986) constructed instances of linear optimization problems for which the simplex algorithm needs exponential time. In other words, the number of arithmetic operations needed to solve an instance of the problem is an exponential function of the size of the instance (which, in this case, is the number of decision variables plus the number of constraints). The general form of the Klee-Minty example is formulated below ($n \geq 1$ and integer).

Model 9.2.1. (*The Klee-Minty example*)

$$
\begin{aligned}
\max \quad & x_1 + x_2 + x_3 + \ldots + x_{n-1} + x_n \\
\text{s.t.} \quad & x_1 && \leq 1 \\
& 2x_1 + x_2 && \leq 3 \\
& 2x_1 + 2x_2 + x_3 && \leq 9 \\
& \quad \vdots && \vdots \\
& 2x_1 + 2x_2 + 2x_3 + \ldots + 2x_{n-1} + x_n && \leq 3^{n-1} \\
& x_1, \ldots, x_n \geq 0.
\end{aligned}
$$

The j'th constraint (with $j = 1, \ldots, n$) in this model is: $x_j + 2\sum_{k=1}^{j-1} x_k \leq 3^{j-1}$.

Figure 9.1 shows the feasible region of the Klee-Minty example for $n = 3$. The corresponding simplex adjacency graph (see Section 3.5.1) has eight nodes (namely, one for each vertex of the feasible region), and twelve arcs (depicted in the figure). The general feasible region has 2^n vertices. We will show that the simplex algorithm, when using the following pivot rules, passes through all 2^n vertices before reaching the optimal solution. These pivot rules are:

Rule 1. Choose the currently largest objective coefficient.

Rule 2. In the case of a tie in Rule 1, choose from the currently largest objective coefficients the one with the lowest index, i.e., the leftmost one in the tableau.

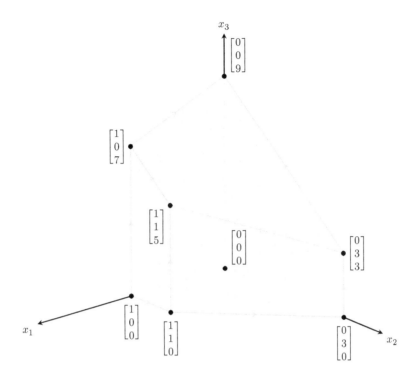

Figure 9.1: The feasible region of the Klee-Minty example for $n = 3$. The arrows show the arcs of the simplex adjacency graph.

Since the feasible region contains no degenerate vertices, no tie-breaking pivot rule is needed for the variables that leave the current set of basic variables.

We calculate the first four simplex tableaus for the case $n = 3$:

Initialization. $BI = \{4, 5, 6\}$.

x_1	x_2	x_3	x_4	x_5	x_6	$-z$
1	1	1	0	0	0	0
①	0	0	1	0	0	1
2	1	0	0	1	0	3
2	2	1	0	0	1	9

Iteration 1. Exchange x_1 and x_4.

x_4	x_2	x_3	x_1	x_5	x_6	$-z$
-1	1	1	0	0	0	-1
1	0	0	1	0	0	1
-2	①	0	0	1	0	1
-2	2	1	0	0	1	7

Iteration 2. Exchange x_2 and x_5.

x_4	x_5	x_3	x_1	x_2	x_6	$-z$
1	-1	1	0	0	0	-2
$\boxed{1}$	0	0	1	0	0	1
-2	1	0	0	1	0	1
2	-1	1	0	0	1	5

Iteration 3. There is a tie. Based on Rule 2, x_4 and x_1 are exchanged.

It is left to the reader to carry out the remaining iterations. It takes $2^3 - 1 = 7$ iterations to reach the optimum $\begin{bmatrix} 0 & 0 & 9 \end{bmatrix}^{\mathsf{T}}$ with optimal objective value $z^* = 9$. Note that the x_3 column remains nonbasic during the first three iterations, so that the simplex algorithm first passes through the vertices in the $x_1 x_2$-plane before jumping to the vertices that are formed by the third constraint. The simplex algorithm successively visits the vertices:

$$\begin{bmatrix} 0 \\ 0 \\ 0 \end{bmatrix}, \begin{bmatrix} 1 \\ 0 \\ 0 \end{bmatrix}, \begin{bmatrix} 1 \\ 1 \\ 0 \end{bmatrix}, \begin{bmatrix} 0 \\ 3 \\ 0 \end{bmatrix}, \begin{bmatrix} 0 \\ 3 \\ 3 \end{bmatrix}, \begin{bmatrix} 1 \\ 1 \\ 5 \end{bmatrix}, \begin{bmatrix} 1 \\ 0 \\ 7 \end{bmatrix}, \begin{bmatrix} 0 \\ 0 \\ 9 \end{bmatrix}.$$

This example, for $n = 3$, hints at the situation we face for arbitrary n: The simplex algorithm first passes through the 2^k vertices formed by the first k constraints, and then passes through the 2^k vertices in the hyperplane formed by the $(k + 1)$'th constraint (for $k = 2, \ldots, n - 1$).

The 'usual' pivot rules force the simplex algorithm to perform $2^n - 1$ iterations, and to traverse all 2^n vertices of the feasible region. The Klee-Minty example therefore shows that the simplex algorithm is not a theoretically efficient method for solving linear optimization models, and we have the following theorem.

Theorem 9.2.1.
The worst-case complexity of the simplex algorithm for solving linear optimization problems is exponential.

At present, for almost all pivot rules that have been formulated for the simplex algorithm, there exist examples (often pathological) in which the algorithm is forced to traverse through an exponential number of vertices. These examples also show that the simplex algorithm is already exponential in the number of decision variables, even if we do not count the number of arithmetic operations and the number of constraints. Hence, the simplex algorithm is theoretically 'bad' (although in most practical situations very effective, as we have mentioned already!). For a detailed account on this subject, we refer the reader to Terlaky and Zhang (1993).

One may ask the question, whether or not there exists a pivot rule (different from Rules 1 and 2 above) for which the simplex algorithm finds an optimal solution of the Klee-Minty example in polynomial time. Actually, when pivoting on the x_n-row in the Klee-Minty example, the optimum is reached in only one pivot step.

Because of such disappointing discoveries, and also because an increasing number of problems (including many large-scale crew scheduling problems) could not be solved by means of the simplex algorithm within reasonable time limits, the scientific hunt for a theoretical, as well as practical, efficient algorithm for linear optimization was opened. Among other algorithms, this hunt has yielded the class of interior point algorithms (see Chapter 6).

9.3 The interior path algorithm has polynomial running time

In 1979, LEONID G. KHACHIYAN (1952–2005) presented a theoretically efficient algorithm for linear optimization, the so-called *ellipsoid algorithm*. Khachiyan's algorithm is only theoretically efficient, and so the development of a practically and theoretically efficient algorithm remained attractive. In 1984, NARENDRA KARMARKAR (born 1957) presented his famous *interior point algorithm*, which is – as he claimed – efficient in theory and in practice. In Chapter 6 we discussed the *interior path* version of Karmarkar's algorithm. Besides this new approach for solving LO-models, there is still interest in trying to design a pivot rule that makes the simplex algorithm theoretically efficient.

In this section we will show that the interior path algorithm is a 'good' *approximation algorithm*, by which we mean that the number of iterations needed to reach an optimal solution with a prescribed accuracy t (meaning that the absolute value of the difference between the approximate objective value and the optimal objective value is at least t) is bounded from above by a polynomial function in the size of the model. Here, we do not count the number of arithmetic operations for the various iteration steps; it can be shown that this number is also a polynomial in the size of the model (see Exercise 9.5.6).

We refer the reader to Chapter 6 for the meaning of the concepts used in the following theorem. In this theorem we also show that the duality gap between the final optimal primal solution and the corresponding dual solution is exponentially small. So, both the primal and dual solution are approximately optimal.

> **Theorem 9.3.1.** (*Polynomiality of the interior path algorithm*)
> Let $\mathbf{x}_0$ be the initial interior point and μ_0 the initial interior path parameter such that $\delta(\mathbf{x}_0, \mu_0) \leq \frac{1}{2}$. Let t be the accuracy parameter, and let the updating factor satisfy $\theta = 1/(6\sqrt{n})$. Then the algorithm stops after at most $6\sqrt{n}(t + \ln(n\mu_0)) = O(\sqrt{n}\ln(n))$ iterations. Let μ^* be the last generated interior path parameter. Then the last generated point $\mathbf{x}^*$ and the corresponding dual point $\mathbf{y}(\mathbf{x}^*, \mu^*)$ are both interior points; moreover the duality gap satisfies:
>
> $$\mathbf{c}^\mathsf{T}\mathbf{x}^* - \mathbf{b}^\mathsf{T}\mathbf{y}(\mathbf{x}^*, \mu^*) \leq \tfrac{3}{2}e^{-t}.$$

Proof of Theorem 9.3.1. Let κ be the number of iterations needed to reach optimality with an accuracy of t, and let μ^* be the interior path parameter at the κ'th iteration. Hence,

$$n\mu^* = n(1 - \theta)^\kappa \mu_0 < e^{-t}.$$

Taking logarithms on both sides in the above inequality, we find that $\ln(n\mu_0) + \kappa \ln(1 - \theta) < -t$. Since $\ln(1 - \theta) < -\theta$, this inequality certainly holds if $\kappa\theta > t + \ln(n\mu_0)$. This last inequality is equivalent to $\kappa > (6\sqrt{n})(t + \ln(n\mu_0))$. Since t and μ_0 are constants, it follows that $(6\sqrt{n})(t + \ln(n) + \ln(\mu_0)) = O(\sqrt{n}\ln(n))$.

Let $\mathbf{x}^*$ be the last generated point. Since $\delta(\mathbf{x}^*, \mu^*) \leq \frac{1}{2}$, it follows from Theorem 6.4.1 that $\mathbf{y}(\mathbf{x}^*, \mu^*)$ is dual feasible and that:

$$
\begin{aligned}
\mathbf{c}^\mathsf{T}\mathbf{x}^* - \mathbf{b}^\mathsf{T}\mathbf{y}(\mathbf{x}^*, \mu^*) &\leq n\mu^* + \mu^*\sqrt{n}\,\delta(\mathbf{x}^*, \mu^*) \\
&\leq n\mu^*\left(1 + n^{-\frac{1}{2}}\delta(\mathbf{x}^*, \mu^*)\right) \\
&\leq e^{-t}\left(1 + \frac{1}{2}n^{-\frac{1}{2}}\right) \\
&\leq \tfrac{3}{2}e^{-t}.
\end{aligned}
$$

This proves the theorem. $\qquad\square$

As an illustration, consider the example of Section 6.4.2. The extended matrix $\mathbf{A}$ has six decision variables; so in Theorem 9.3.1, $n = 6$. Let $\mu_0 = 3$, and $t = 5$. Then, $6\sqrt{n}(t + \ln(n\mu_0)) \approx 113$. So 116 is a very rough upper bound for the number of iterations needed to reach an optimum with an accuracy of 5. An accuracy of 5 means that the duality gap is about $\frac{3}{2}e^{-5} \approx 0.0067$ when the algorithm stops.

9.4 Computational aspects of the branch-and-bound algorithm

In the two previous sections we discussed complexity aspects of general linear optimization problems. In this section we consider integer linear optimization problems.

In the examples of Chapter 7, the number of LO-relaxations solved was very small: nine in the case of Model Dairy Corp (Section 7.2.1), and again nine in the case of the knapsack problem (Section 7.2.3). In general, however, the number of subproblems to be solved may be exorbitantly large. In this section it is shown that the number of subproblems to be solved is in general an exponential function of the number of (decision) variables. To that end, we construct a worst-case example for which the branch-and-bound algorithm needs to solve an exponential number of subproblems.

Consider the following knapsack model:

$$
\begin{aligned}
\max \quad & x_1 + \ldots + x_n \\
\text{s.t.} \quad & 2x_1 + \ldots + 2x_n \leq n \\
& x_i \in \{0, 1\} \text{ for } i = 1, \ldots, n,
\end{aligned}
$$

where n is an odd integer ≥ 3. An optimal solution of this model is easily found; namely, take $\lfloor n/2 \rfloor = (n-1)/2$ decision variables x_i equal to 1, and the remaining $\lfloor n/2 \rfloor + 1 = (n+1)/2$ decision variables x_i equal to 0. The optimal objective value z^* satisfies $z^* = \lfloor n/2 \rfloor$. However, solving this model by means of the branch-and-bound algorithm, takes an exorbitant amount of computer time, since the number of subproblems it solves turns out to be exponential in n.

As an illustration, consider the case $n = 3$. The branch-and-bound tree for this case is depicted in Figure 9.2. Note that both backtracking and jumptracking give rise to the same tree. If a finite solution of an LO-relaxation of this problem is not integer-valued, then exactly one variable has the noninteger value $\frac{1}{2}$. In the next iteration, branching is applied on this noninteger variable, say x_i, giving rise to a '$x_i = 0$'-branch and a '$x_i = 1$'-branch. In our illustration, the algorithm does not terminate before all integer solutions are calculated, leading to eleven solved subproblems.

We will now show that, for arbitrary n (odd, ≥ 3), the number of subproblems solved is at least 2^n. In order to prove this assertion, we show that an integer solution is found on a node of the branch-and-bound tree for which the corresponding subproblem has at least $(n+1)/2$ decision variables with the value zero obtained from previous branchings. This means that the path in the tree from the initial node to this node has at least $(n+1)/2$ zero-variable branches. Assume, to the contrary, that an integer solution was found for which less than $(n+1)/2$ variables x_i are 0 on the previous branches in the tree. Then there are at least $(n+1)/2$ variables with a value either 1 or $\frac{1}{2}$. Hence, the optimal solution of the current subproblem contains precisely $(n-1)/2$ variables with the value 1 and one equal to $\frac{1}{2}$, and is therefore not integer-valued as was assumed. Hence, the subproblems with an integer solution occur on a path on which at least $(n+1)/2$ previous branches correspond to a zero variable. On the other hand, infeasible subproblems occur on a path on which at least $(n+1)/2$ previous branches correspond to variables with the value 1. There is no need to branch subproblems that are either infeasible or have an integer optimal solution. Hence the branch-and-bound tree is 'complete' for the first $(n+1)/2$ 'levels', i.e., contains $2^{(n+1)/2}$ subproblems. So, the total number of subproblems in the branch-and-bound tree of our example is at least $2^{(n+1)/2}$, and this is at least 2^n, which is an exponential function of n.

We have now shown that there exist problems for which the branch-and-bound algorithm demands an exponential number of subproblems to solve. This fact is expressed in the following theorem.

Theorem 9.4.1.
The worst-case complexity of the branch-and-bound algorithm for solving integer linear optimization problems is exponential.

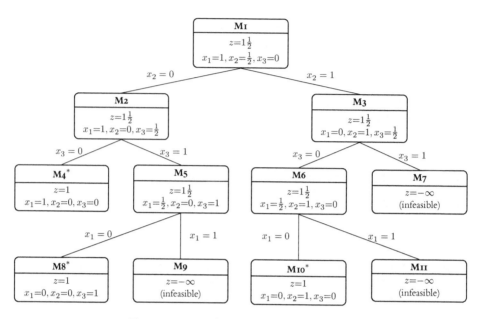

Figure 9.2: Branch-and-bound tree for $n = 3$.

Recall that the size of the branch-and-bound tree depends on the branching rule used (compare, in this respect, the role of pivot rules in Dantzig's simplex algorithm; see Section 9.2).

Branch-and-bound algorithms are almost as frequently applied to integer linear optimization problems as the simplex algorithm (see Section 9.2) to linear optimization problems. But, frequently, users cannot wait until an optimal solution has been reached. In such situations, the calculations can be stopped at some stage and the best solution obtained thus far can be used. This illustrates a positive aspect to branch-and-bound algorithms: we often have a feasible solution at hand at a very early stage of the algorithmic calculations. However, similarly to the simplex algorithm, the worst-case complexity of branch-and-bound algorithms is exponential.

So it could be expected that a hunt for an efficient algorithm for ILO-models is on. However, there are very few hunters working in that direction. This is because the belief that such an algorithm exists has decreased drastically. Why is that so?

In the early 1970s, a class of optimization problems was defined for which efficient algorithms are extremely unlikely. An interesting feature of this class of problems is their strong interrelation: if one finds an efficient algorithm to solve <u>any one</u> problem in this class, one can easily modify it to solve <u>all</u> the problems in this class efficiently. This class is usually referred to as the class of NP-*hard problems*. ILO problems have been one of the earliest members of this class. Other NP-hard problems include the traveling salesman problem (see Section 7.2.4), the machine scheduling problem (see Section 7.2.4), and the knapsack problem (see Section 7.2.3). So far, everyone that has tried to find a 'good' algorithm for, e.g., the traveling salesman problem has failed. The conjecture that no NP-hard problem is efficiently solvable is denoted by

$$P \neq NP,$$

where P refers to the class of problems for which a polynomial algorithm is known. Hence, linear optimization problems are in this class (see Section 9.3). So, what remains in this particular area of research is to prove (or disprove) the $P \neq NP$ conjecture. Until someone disproves this conjecture, algorithms for NP-hard problems are designed that generate good feasible solutions in acceptable computer running times. Such algorithms may be classified into *heuristics* and *approximation algorithms*. A heuristic is an algorithm that in practice finds reasonably good feasible solutions in a reasonable amount of time. Heuristics are often based on intuition and practical experience, and they usually do not have any performance guarantees in terms of computation time and solution quality. An approximation algorithm, on the other hand, is an algorithm that yields feasible solutions of a provable quality within a provable time limit.

For example, a special case of the traveling salesman problem is the so-called *Euclidean traveling salesman problem*. This problem imposes the additional restriction on the input that the distances between cities should obey the *triangle inequality* $d(A, C) \leq d(A, B) + d(B, C)$ for any three cities A, B, and C. This means that the distance required to travel from city A to city C should be at most the distance required to travel from city A, through city B, to city C; see also Section 17.3. Just like the general traveling salesman problem, the Euclidean traveling salesman problem is NP-hard. An often used heuristic for the Euclidean traveling salesman problem is the *nearest neighbor heuristic*, which chooses an arbitrary initial city, iteratively adds a nearest unvisited city to the list of visited cities, and finally returns to the initial city; see also Exercise 17.10.5. Although the nearest neighbor heuristic usually yields reasonably good traveling salesman tours, there are situations in which the heuristic yields very bad (i.e., extremely long) tours. It may even happen that the heuristic does not yield a feasible solution at all. On the other hand, the approximation algorithm discovered by NICOS CHRISTOFIDES for the Euclidean traveling salesman problem yields, in polynomial time, a traveling salesman tour that is guaranteed to be at most 50% longer than the optimal tour. For the general traveling salesman problem, however, it is known that no such approximation algorithm exists.

The popularity of heuristics and approximation algorithms has led to a new branch of research, in which these algorithms are classified and tested, and their performance is compared. It is beyond the scope of this book to elaborate on the design, application, and performance of heuristics and approximation algorithms. The interested reader is referred to, e.g., Williamson and Shmoys (2011).

9.5 Exercises

Exercise 9.5.1. Write the following function in the big-O notation $O(f(n))$ with $f(n)$ a simple polynomial or exponential function (e.g., n^p, q^n, $n!$, or $(\log n)^p$).

(a) $800n + n^4$

(b) $n^3 3^n + 2^{2n}$

(c) $10^{800n} + (n-1)!$

(d) $n^4 \log n$

Exercise 9.5.2.   The distinction in complexity between polynomial and exponential complexity (in other words, between efficiency and inefficiency), is somewhat artificial. For instance, an algorithm with complexity $O(n^{10})$ (with $n \geq 1$) is theoretically efficient, but practically inefficient. Compare for a number of values the functions n^{10} and 1.6^n, and explain why both are practically inefficient. Show that $1.6^n \gg n^{10}$ for $n > 100$.

Exercise 9.5.3.   Let $\mathbf{A}$ and $\mathbf{B}$ be (n, n)-matrices (with $n \geq 1$). A simple (but not the most efficient) algorithm to compute the product $\mathbf{AB}$ (see Appendix B) is to calculate the inner product of row i of $\mathbf{A}$ and column j of $\mathbf{B}$ to obtain entry (i, j) of $\mathbf{AB}$ for each $i = 1, \ldots, n$ and $j = 1, \ldots, n$. Determine the complexity of this algorithm. Is it 'good' or 'bad'?

Exercise 9.5.4.   Let $\mathbf{Ax} = \mathbf{b}$ be a set of n equations with n unknowns (with $n \geq 1$). Gaussian elimination, where one variable at a time is eliminated, may be used to either calculate a solution of this set of equations, or to decide that no such solution exists; see Appendix B. Show that this algorithm has complexity $O(n^3)$.

Exercise 9.5.5.   Let $\mathbf{A}$ be an (n, n)-matrix (with $n \geq 1$). Use Gaussian elimination to transform $\begin{bmatrix} \mathbf{A} & \mathbf{I}_n \end{bmatrix}$ into $\begin{bmatrix} \mathbf{I}_n & \mathbf{A}^{-1} \end{bmatrix}$ to determine the inverse of $\mathbf{A}$, or to show that the inverse does not exist; see Appendix B. Count the number of arithmetic operations as a function of n, and show that the complexity of this algorithm is $O(n^3)$.

Exercise 9.5.6.   Let $\mathbf{A}$ be any (m, n)-matrix with $m \leq n$, and with rank m. Show that $\mathbf{P}_A$ can be calculated with $O(n^3)$ arithmetic operations $(+, -, \times, /)$.

Exercise 9.5.7.   Let $n \geq 1$ be the size of some problem, and let $n^{\log n}$ be the number of calculations (as a function of n) needed by some algorithm to calculate a solution of a worst-case instance of the problem. Show that this algorithm has exponential complexity.

Exercise 9.5.8.   Solve Model Dovetail from Section 1.1.1 by considering all basic solutions. How many iterations are needed in a worst-case situation?

Exercise 9.5.9.   In a worst-case situation, how many steps are needed when all integer points are considered for calculating an optimal solution of Model Dairy Corp in Section 7.1.1?

Exercise 9.5.10. Consider the standard LO-model $\max\{\mathbf{x} \in \mathbb{R}^n \mid \mathbf{A}\mathbf{x} \le \mathbf{b}, \mathbf{x} \ge \mathbf{0}\}$, where $\mathbf{A}$ is an (m, n)-matrix. Assume that $n \ge m$. This LO-model may be solved by considering all (feasible and infeasible) basic solutions. Show that this algorithm has a worst-case complexity with lower bound 2^m.

Part II

Linear optimization practice
Advanced techniques

CHAPTER 10

Designing a reservoir for irrigation

Overview

This case study is based on a research project in Tanzania carried out by CASPAR SCHWEIGMAN (born 1938); see Schweigman (1979). The LO-model used in this section can partly be solved by inspection; the more advanced analysis of the model needs computer calculations.

10.1 The parameters and the input data

The government of a tropical country wants to built a water reservoir in a river in order to irrigate the land. The location for the reservoir is known, as well as the region to be irrigated. During the wet season there is considerably more water in the river than during the dry season. The reservoir will collect the water of the wet season, which can then be used during the dry season. The problem we will study is, determining the size of the reservoir, and the area of the land that can be irrigated with it. The solution to this problem depends, among others, on the quantity of water needed for the irrigation during the wet and dry seasons. Also the quantity of rain and the quantity of water in the river play essential parts. The relevant data have to be obtained from meteorological and hydrological observation stations. In the present case study, the mathematical model is pre-eminently suitable to give a systematic analysis of the problem.

Since the rainfall and the flow through the river vary in time, we partition the planning period into subperiods. The planning period is one year and the subperiods are the twelve months, to be labeled by $t = 1, \dots, T$ (where $T = 12$). We will first introduce the different parameters of the model.

Define:

V = the capacity of the reservoir (in m^3);
x = the area of the land to be irrigated (in m^2).

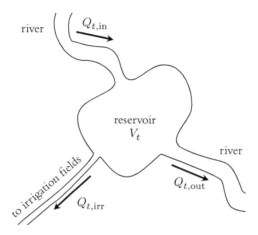

Figure 10.1: Sketch of the reservoir and the river.

The level of water in the reservoir will vary during the year. Define for $t = 1, \ldots, T$:

V_t = the amount of water in the reservoir at the end of month t (in m³);
V_0 = the amount of water in the reservoir at the beginning of the period (in m³).

River water flows into the reservoir and can be made to flow through lock gates. For $t = 1, \ldots, T$, define:

$Q_{t,\text{in}}$ = the amount of water that flows into the reservoir during month t (in m³);
Q_t = the amount of water that leaves the reservoir in month t (in m³);
$Q_{t,\text{irr}}$ = the amount of water that flows to the irrigation fields in month t (in m³);
$Q_{t,\text{out}}$ = the amount of water that flows downstream back into the river (in m³).

In Figure 10.1, this situation is depicted schematically. It follows immediately from the above definitions that:

$$Q_t = Q_{t,\text{irr}} + Q_{t,\text{out}} \qquad \text{for } t = 1, \ldots, T. \tag{10.1}$$

The water in the reservoir consists of rain water and river water. The water leaving the reservoir is water that either leaves it through the locks, or evaporates, or penetrates the soil, or leaves the reservoir via leaks. In the present case study, the contributions of rainfall and evaporation are negligible with respect to the quantity of river water that flows into the reservoir. We therefore assume that at the end of month t the quantity of water in the reservoir is equal to the quantity of water in the reservoir at the end of the previous month $t - 1$, plus the quantity of river water that flows into the reservoir, minus the quantity of water that leaves the reservoir, i.e.:

$$V_t = V_{t-1} + Q_{t,\text{in}} - Q_t \qquad \text{for } t = 1, \ldots, T. \tag{10.2}$$

Month t	α_t (in m^3 water per m^2 land)	$Q_{t,\text{in}}$ ($\times 10^6$ m^3)
January	0.134	41
February	0.146	51
March	0.079	63
April	0.122	99
May	0.274	51
June	0.323	20
July	0.366	14
August	0.427	12
September	0.421	2
October	0.354	14
November	0.140	34
December	0.085	46

Table 10.1: Monthly quantities of irrigation water and inflow water.

The 'inventory equations' (10.2) can be rewritten as:

$$V_t = V_0 + \sum_{k=1}^{t} Q_{k,\text{in}} - \sum_{k=1}^{t} Q_k \qquad \text{for } t = 1, \ldots, T. \tag{10.3}$$

For each $t = 1, \ldots, T$, the parameter $Q_{t,\text{in}}$ is supposed to be known, whereas Q_t and V_t are decision variables. Also V_0 has to be calculated. The out flowing water Q_t is partly used for the irrigation. The quantity of irrigation water depends of course on the area of land that is irrigated. Define for $t = 1, \ldots, T$:

α_t = the quantity of water needed in month t for the irrigation of 1 m^2 of land.

It then follows that:

$$Q_{t,\text{irr}} = \alpha_t x \qquad \text{for } t = 1, \ldots, T. \tag{10.4}$$

The most complicated aspect of building a model for this problem is to find reliable values for the coefficients α_t. These values depend on the assimilation of water by the plants, the growth phase of the plants, the loss of water, the efficiency of the irrigation system, and several other factors. One more reason for this difficulty is that the values need to be known before the actual irrigation takes place. Sometimes it is possible to obtain reasonable estimates from experiments on test fields. In the present case study, we use estimates obtained for growing corn in a region in the western part of Tanzania; see Table 10.1. In Figure 10.2 these data are depicted in graphs. These graphs show a certain negative correlation between $Q_{t,\text{in}}$ and α_t: in April the inflow is maximum while the least irrigation water is needed, and in September the situation is the other way around.

Besides the water that is used for irrigation, there should be sufficient water flowing down the river, because people, cattle, and crops are dependent on it. Determining the minimum quantity M of water that should go down the river is a profound and difficult political

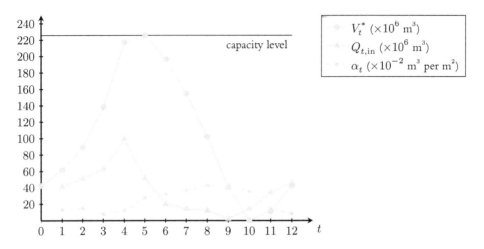

Figure 10.2: Needed irrigation water versus the inflow.

question. One has to take into account a number of aspects, such as the population density, the future plans, and all kinds of conflicting interests. In the present case study, the minimum quantity is fixed at $2.3 \times 10^6 \, \text{m}^3$, and so:

$$Q_{t,\text{out}} \geq M = 2.3 \times 10^6 \qquad \text{for } t = 1, \ldots, T. \tag{10.5}$$

Note that only in September the inflow is less than 2.3×10^6. This means that in September more water flows out of the reservoir than flows into it. The quantity of water in the reservoir cannot exceed its capacity, i.e.:

$$V_t \leq V \qquad \text{for } t = 1, \ldots, T. \tag{10.6}$$

In order to keep the differences between successive planning periods as small as possible, we assume that at the end of the planning period the quantity of water in the reservoir is the same as at the beginning of the next period. Hence,

$$V_0 = V_T. \tag{10.7}$$

This formula implies that, during the whole planning period, the quantity of water that flows into the reservoir is equal to the quantity that leaves it. Moreover, all parameters introduced so far need to be nonnegative, i.e.:

$$x \geq 0, V \geq 0, V_t \geq 0 \qquad \text{for } t = 1, \ldots, T. \tag{10.8}$$

The decision variables Q_t, $Q_{t,\text{irr}}$, and $Q_{t,\text{out}}$ are implicitly nonnegative because of (10.1), (10.4), (10.5), and $x \geq 0$. Since $V_0 = V_T$, V_0 is also nonnegative.

10.2 Maximizing the irrigation area

In the present case study, we do not take into consideration the cost of building the reservoir and the construction of the irrigation system. The revenues of the irrigation are also not taken into account. We will restrict ourselves to the simplified situation of determining the maximum area of the irrigation region, and the corresponding size of the reservoir. This simplified objective can be formulated as in the following model:

$$\max x$$
$$\text{s.t.} \quad (10.1), (10.3)-(10.8).$$

(10.9)

This is an LO-model that can be solved as soon as the values of $Q_{t,\text{in}}$ are available. These values are given in Table 10.1. Before solving (10.9), we will investigate (10.3) in more detail. It follows from (10.3), (10.7), and Table 10.1 that:

$$\sum_{t=1}^{12} Q_t = \sum_{t=1}^{12} Q_{t,\text{in}} = 447 \times 10^6.$$

(10.10)

Using (10.1) and (10.4), it follows from (10.10) that:

$$\sum_{t=1}^{12} Q_t = x \sum_{t=1}^{12} \alpha_t + \sum_{t=1}^{12} Q_{t,out}.$$

Eliminating x yields:

$$x = \frac{\sum_{t=1}^{12} Q_t - \sum_{t=1}^{12} Q_{t,\text{out}}}{\sum_{t=1}^{12} \alpha_t} = \frac{447 \times 10^6 - \sum_{t=1}^{12} Q_{t,\text{out}}}{2.871}$$

(10.11)

$$= 155.7 \times 10^6 - 0.348 \sum_{t=1}^{12} Q_{t,\text{out}}.$$

(10.12)

The LO-model (10.9) can now also be formulated as:

$$\min \sum_{t=1}^{12} Q_{t,\text{out}}$$
$$\text{s.t.} \quad (10.1), (10.3)-(10.8).$$

(10.13)

This result is not quite unexpected: if we want to maximize the area of land to be irrigated, then the amount of water that flows downstream has to be minimized. We also know that $Q_{t,\text{out}} \geq 2.3 \times 10^6$, and so the minimum in (10.13) is never lower than $12 \times (2.3 \times 10^6) = 27.6 \times 10^6$. Based on these considerations, we try the minimum solution:

$$Q_{t,\text{out}} = 2.3 \times 10^6 \qquad \text{for } t = 1, \ldots, T.$$

From (10.11) it follows that $x^* = 146.1 \times 10^6$. This means that $146.1\,\mathrm{km}^2$ is irrigated. We will show that the above choice of $Q_{t,\mathrm{out}}$ yields an optimal solution of the LO-model. To that end we only have to check its feasibility.

The conditions $0 \leq V_t \leq V$ for $t = 1, \ldots, 12$ become, after the substitution:

$0 \leq V_0 \leq V$, and

$$0 \leq V_0 + \sum_{k=1}^{t} Q_{k,\mathrm{in}} - x^* \sum_{k=1}^{t} \alpha_k - (2.3 \times 10^6)t \leq V \qquad \text{for } t = 1, \ldots, T.$$

Using Table 10.1, we find that:

$$\begin{aligned}
0 &\leq V_0 \leq V \\
0 &\leq V_0 + 19.1 \times 10^6 &\leq V \\
0 &\leq V_0 + 46.5 \times 10^6 &\leq V \\
0 &\leq V_0 + 95.7 \times 10^6 &\leq V \\
0 &\leq V_0 + 174.5 \times 10^6 &\leq V \\
0 &\leq V_0 + 183.2 \times 10^6 &\leq V \\
0 &\leq V_0 + 153.7 \times 10^6 &\leq V \\
0 &\leq V_0 + 112.0 \times 10^6 &\leq V \\
0 &\leq V_0 + 59.3 \times 10^6 &\leq V \\
0 &\leq V_0 - 2.5 \times 10^6 &\leq V \\
0 &\leq V_0 - 42.5 \times 10^6 &\leq V \\
0 &\leq V_0 - 31.3 \times 10^6 &\leq V \\
0 &\leq V_0 - 0.1 \times 10^6 &\leq V.
\end{aligned} \qquad (10.14)$$

In order to assure feasibility, we have to determine values of V_0 and V such that all inequalities in (10.14) are satisfied. Actually, there are a lot of possibilities: a large reservoir can certainly contain all the irrigation water. We are interested in the minimum value for V, and so the building costs are minimum. It is immediately clear from (10.14) that the minimum value V^* for V is found for $V_0^* = 42.5 \times 10^6$ (m³), and so

$$V^* = 42.5 \times 10^6 + 183.2 \times 10^6 = 225.7 \times 10^6 \text{ (m}^3\text{)}.$$

The optimal values V_t^* of V_t are:

$$\begin{array}{lll}
V_1^* = 61.7 \times 10^6, & V_5^* = 225.7 \times 10^6, & V_9^* = 40.0 \times 10^6, \\
V_2^* = 89.0 \times 10^6, & V_6^* = 196.3 \times 10^6, & V_{10}^* = 0, \\
V_3^* = 138.2 \times 10^6, & V_7^* = 154.5 \times 10^6, & V_{11}^* = 11.2 \times 10^6, \\
V_4^* = 217.1 \times 10^6, & V_8^* = 101.8 \times 10^6, & V_{12}^* = 42.5 \times 10^6.
\end{array}$$

The graph of $V_1^*, \ldots, V_{12}^*$ is depicted in Figure 10.2. Note that the reservoir is full in May, and empty in October.

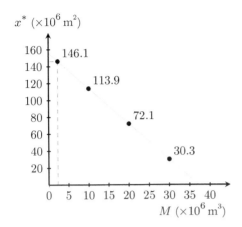

Figure 10.3: Sensitivity with respect to M.

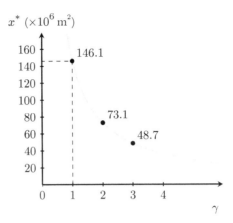

Figure 10.4: Sensitivity with respect to γ.

10.3 Changing the input parameters of the model

In this section we will investigate the relationship between the optimal solution x^* and the optimal capacity V^* of the reservoir on the one hand, and the parameters M, α_t, and $Q_{t,\text{in}}$ on the other hand.

The relationship between x^* and M is linear, namely:

$$x^* = 155.7 \times 10^6 - 4.18M. \tag{10.15}$$

Formula (10.15) follows immediately from (10.5) and (10.11). The graph is depicted in Figure 10.3. If the value of M was taken to be 0, then $x^* = 155.7 \times 10^6$ (m³), so the choice $M = 2.3 \times 10^6$ (see (10.5)) is certainly not a bad one. Note that for $M = 37.25 \times 10^6$, it is not possible to use water for irrigation. In Exercise 10.5.2, the reader is asked to calculate the relationship between M and V^*.

Suppose we change all values of α_t by a certain fraction γ; so instead of α_t we take $\gamma\alpha_t$. Formula (10.4) then becomes: $Q_{t,\text{irr}} = (\gamma\alpha_t)x$. Hence, the relationship between x^* and γ is $x^* = x^*(\gamma) = 146/\gamma$. If, for instance, all values of α_t are 10% higher, then $\gamma = 1.1$ and $x^* = 146/1.1 \times 10^6$, which corresponds to 133×10^6 km². The relationship between γ and x^* is depicted in Figure 10.4.

When we change the values of $Q_{t,\text{in}}$ by a certain fraction β, then formula (10.3) becomes $V_t = V_{t-1} + \beta Q_{t,\text{in}} - Q_t$. Summation over t yields:

$$\beta \sum_{t=1}^{12} Q_{t,\text{in}} = \sum_{t=1}^{12} Q_t = \sum_{t=1}^{12} Q_{t,\text{irr}} + \sum_{t=1}^{12} Q_{t,\text{out}} = x^* \sum_{t=1}^{12} \alpha_t + 27.6 \times 10^6.$$

Hence,

$$x^* = 155.695\beta - 9.6134. \tag{10.16}$$

So the relationship between x^* and β, the rate of the change of the $Q_{t,\text{in}}$'s, is linear. Note that $x^* = 0$ for $\beta = 0.06175$, and so the annual inflow is $0.062 \times 447 \times 10^6 = 27.6 \times 10^6 = M$, as expected. For $\beta < 0.06175$ the model is infeasible. In Exercise 10.5.4 the reader is asked to draw a graph showing the relationship between x^* and β.

We finally consider the case that one of the $Q_{t,\text{in}}$'s is changed. If one of the $Q_{t,\text{in}}$'s is changed into $Q_{t,\text{in}} + \delta$, then the total $\sum_{t=1}^{12} Q_{t,\text{in}}$ changes into $\delta + \sum_{t=1}^{12} Q_{t,\text{in}}$. It now follows from (10.11) that:

$$x^* = 146.1 \times 10^6 + 0.348\delta. \tag{10.17}$$

For instance, if in April the inflow changes from 99×10^6 to 60×10^6, and so the peek in the graph of Figure 10.2 disappears, then $\delta = 39 \times 10^6$, and $x^*(\delta = 39 \times 10^6) = 125.1 \times 10^6$. The change of δ is $39/99$, which is approximately 39.4%. On the other hand, the change of x^* is $(146.1 - 125.1)/146.1$, which is approximately 14.4%. Hence, the area that can be irrigated is certainly dependent on the inflow, but the rate of the change of the area is significantly lower than the rate of the change of the inflow.

The dependency for the capacity V^* is somewhat more complicated. Suppose the inflow in April is changed by δ. The formulas (10.14) then become:

$$
\begin{aligned}
0 &\leq V_0 \leq V \\
0 &\leq V_0 + 19.1 \times 10^6 && - 0.047\delta \leq V \\
0 &\leq V_0 + 46.5 \times 10^6 && - 0.097\delta \leq V \\
0 &\leq V_0 + 95.7 \times 10^6 && - 0.125\delta \leq V \\
0 &\leq V_0 + 174.5 \times 10^6 + \delta - 0.167\delta \leq V \\
0 &\leq V_0 + 183.2 \times 10^6 + \delta - 0.263\delta \leq V \\
0 &\leq V_0 + 153.7 \times 10^6 + \delta - 0.375\delta \leq V \\
0 &\leq V_0 + 112.0 \times 10^6 + \delta - 0.503\delta \leq V \\
0 &\leq V_0 + 59.3 \times 10^6 \ \ + \delta - 0.615\delta \leq V \\
0 &\leq V_0 - 2.5 \times 10^6 \ \ + \delta - 0.798\delta \leq V \\
0 &\leq V_0 - 42.5 \times 10^6 \ \ + \delta - 0.921\delta \leq V \\
0 &\leq V_0 - 31.3 \times 10^6 \ \ + \delta - 0.970\delta \leq V \\
0 &\leq V_0 - 0.1 \times 10^6 \ \ + \delta - 0.999\delta \leq V.
\end{aligned}
$$

For $\delta \leq (42.5 \times 10^6)/(1 - 0.921) = 538.0 \times 10^6$, it follows that $V_0^* = 42.5 \times 10^6 - 0.079\delta$, and so $V^* = V_0^* + 183.2 \times 10^6 + \delta - 0.263\delta = 225.7 \times 10^6 + 0.658\delta$. So, if $\delta = 39 \times 10^6$, then the value of V^* changes by $0.658 \times 39/225.7$ (11.4%). Hence, x^* and V^* are more or less relatively equally dependent on δ.

10.4 GMPL model code

```
1        T;
2        Qin   {t   1..T} >= 0;
3        alpha {t   1..T} >= 0;
4        M;
```

```
 5
 6      Qout {t    1..T} >= 0;
 7      Qirr {t    1..T} >= 0;
 8      Q    {t    1..T} >= 0;
 9      V    {t    0..T} >= 0;
10      W >= 0;
11      x >= 0;
12
13          objective:
14      {t    1..T} Qout[t];
15              defQt {t    1..T}:              # (10.1)
16   Q[t] = Qirr[t] + Qout[t];
17              defVt {t    1..T}:              # (10.3)
18    V[t] = V[0] +    {k    1..t} (Qin[k] - Q[k]);
19              defQirr {t    1..T}:            # (10.4)
20   Qirr[t] = alpha[t] * x;
21              Qout_lb {t    1..T}:            # (10.5)
22   Qout[t] >= M;
23              Vt_ub {t    1..T}:              # (10.6)
24   V[t] <= W;
25              defV0:                          # (10.7)
26   V[0] = V[T];
27
28      ;
29
30      T := 12;
31      M := 2.3;
32      alpha :=
33   1 0.134    2 0.146    3 0.079    4 0.122    5 0.274    6 0.323
34   7 0.366    8 0.427    9 0.421   10 0.354   11 0.140   12 0.085;
35      Qin :=
36   1 41       2 51       3 63       4 99       5 51       6 20
37   7 14       8 12       9 2       10 14      11 34      12 46;
38      ;
```

10.5 Exercises

Exercise 10.5.1. Consider the irrigation problem of this chapter.

(a) Calculate the number of constraints and decision variables of model (10.9).

(b) Calculate the optimal solution of model (10.9), and draw in one figure the optimal values of V_t, Q_t, $Q_{t,\mathrm{irr}}$, and $Q_{t,\mathrm{out}}$.

Exercise 10.5.2. Draw the graph of the relationship between M and V^*. Why is the model (10.9) infeasible for $M > 37.25$? Calculate V^* for $M = 37.25$.

Exercise 10.5.3. Consider model (10.9). If the values of α_t are changed by the same (multiplicative) factor γ, then the value of x^* changes as well; see Section 10.3. Show that the optimal values of V and V_t do not change.

Exercise 10.5.4.   Consider model (10.9). Draw the graph of the relationship between β (see Section 10.3) and x^*, and between β and V^*.

Exercise 10.5.5.   Consider model (10.9). Draw the graph of the relationship between δ (see Section 10.3) and x^*, and between δ and V^*. Show that, for $\delta \geq 537.97 \times 10^6$, the maximum value of V_t^* is attained for $t = 4$ (in April), and $V^* = 174.5 \times 10^6 + 0.83\delta$.

Exercise 10.5.6.   As in Exercise 10.5.2, change the value of $Q_{t,\text{in}}$ for $t \neq 4$; pay extra attention to the cases $t = 9, 10, 11, 12$.

Exercise 10.5.7.   Consider model (10.9). As in Exercise 10.5.5, let δ be the change of $Q_{4,\text{in}}$. Show that $V_{10} = 0$ for $0 \leq \delta \leq 230.68 \times 10^6$, and that $V_{11} = 0$ for $230.68 \times 10^6 \leq \delta \leq 611.81 \times 10^6$. For the other values of t, calculate the range of values of δ (possibly including negative values) for which $V_t = 0$.

CHAPTER II

Classifying documents by language

Overview

In this chapter we will show how linear optimization can be used in *machine learning*. Machine learning is a branch of artificial intelligence that deals with algorithms that identify ('learn') complex relationships in empirical data. These relationships can then be used to make predictions based on new data. Applications of machine learning include spam email detection, face recognition, speech recognition, webpage ranking in internet search engines, natural language processing, medical diagnosis based on patients' symptoms, fraud detection for credit cards, control of robots, and games such as chess and backgammon.

An important drive behind the development of machine learning algorithms has been the commercialization of the internet in the past two decades. Large internet businesses, such as search engine and social network operators, process large amounts of data from around the world. To make sense of these data, a wide range of machine learning techniques are employed. One important application is ad click prediction; see, e.g., McMahan *et al.* (2013).

In this chapter, we will study the problem of automated language detection of text documents, such as newspaper articles and emails. We will develop a technique called a *support vector machine* for this purpose. For an elaborate treatment of support vector machines in machine learning, we refer to, e.g., Cristianini and Shawe-Taylor (2000).

II.1 Machine learning

Machine learning is a branch of applied mathematics and computer science that aims at developing computational methods that use experience in the form of data to make accurate predictions. Usually, the 'experience' comes in the form of *examples*. In general, a machine learning algorithm uses these examples to 'learn' about the relationships between the examples. Broadly speaking, there are a few different but related types of such algorithms. The present chapter is about a so-called *classification algorithm*. A classification algorithm re-

423

quires a set of examples, each of which has a *label*. An example of a classification problem is the problem faced by email providers to classify incoming email messages into spam and non-spam messages. The examples for such a classification problem would be a number of email messages, each labeled as either 'spam' or 'not spam'. The goal of the classification algorithm is to find a way to accurately predict labels of new observations. The labels for the examples are usually provided by persons. This could be someone who explicitly looks at the examples and classifies them as 'spam' or 'not spam', or this could be provided by the users of the email service. For example, when a user clicks on the 'Mark this message as spam' button in an email application, the message at hand is labeled as 'spam', and this information can be used for future predictions.

Other areas of machine learning include: *regression*, where the goal is to predict a number rather than a label (e.g., tax spending based on income, see also Section 1.6.2); *ranking*, where the goal is to learn how to rank objects (for example in internet search engines); *clustering*, which means determining groups of objects that are similar to each other. In the cases of classification and regression, the examples usually have labels attached to them, and these labels are considered to be the 'correct' labels. The goal of the algorithms is then to 'learn' to predict these labels. Such problems are sometimes categorized as *supervised machine learning*. In contrast, in ranking and clustering, the examples usually do not have labels, and they are therefore categorized as *unsupervised machine learning*.

The current chapter is a case study of a (supervised) *classification algorithm*. To make a classification algorithm successful, certain features of the messages are determined that (hopefully) carry predictive knowledge. For example, for spam classification, it can be helpful to count the number of words in the message that relate to pharmaceutical products, or whether or not the email message is addressed to many recipients rather than to one particular recipient. Such features are indicative for the message being classified as spam. Other words, such as the name of the recipient's friends may be indicative for the message not being spam. The features are represented as numbers, and the features of a single example can hence be grouped together as a vector. Usually, it is not one particular feature that determines the label of an example, but it is rather the combination of them. For example, an email message that is sent to many different recipients and that contains five references to pharmaceutical products can probably be classified as spam, whereas a message that has multiple recipients and includes ten of the main recipient's friends should probably be classified as non-spam. A classification algorithm attempts to make sense of the provided features and labels, and uses these to classify new, unlabeled, examples.

Clearly, the design of features is crucial and depends on the problem at hand. Features should be chosen that have predictive power, and hence the design uses prior knowledge about the problem. In many cases it may not be immediately clear how to choose the features.

11.2 Classifying documents using separating hyperplanes

In automated text analysis, a basic problem is to determine the language in which a given text document (e.g., a newspaper article, or an email) is written in. In this chapter, we will show how linear optimization can be used to classify text documents into two languages, English and Dutch.

Suppose that we have a (potentially very large) set D of text documents, some of which are written in English and the others are written in Dutch. For each document d in D, we calculate m so-called *features* denoted by $f_1^d, \ldots, f_m^d$. We will use as features the relative frequency of each letter, i.e., the number of times each letter appears divided by the total number of letters in the document. Of course, one can think of many more features that may be relevant, e.g., the average word length, the relative frequency of words of length $1, 2, 3, \ldots$, or the relative frequency of certain letter combinations. We will restrict ourselves to the relative frequency of individual letters. For simplicity, we will also treat capitalized and small case letters equally. So, in our case we have that $m = 26$. Thus, for each document, we construct an m-dimensional vector containing these features. Such a vector is called a *feature vector*. Since we have $|D|$ documents, we have $|D|$ of these m-dimensional feature vectors. For each document $d \in D$, let $\mathbf{f}^d$ ($\in \mathbb{R}^m$) be the feature vector for d. For any subset $D' \subseteq D$, define $F(D') = \{\mathbf{f}^d \mid d \in D'\}$.

As an example, we have taken 31 English and 39 Dutch newspaper articles from the internet and calculated the letter frequencies. Table 11.1 shows the relative frequencies of the 26 letters for six English and six Dutch newspaper articles. The columns of the table are the twelve corresponding feature vectors $\mathbf{f}^d$ ($d = 1, \ldots, 12$).

Our goal is to construct a function $g \colon \mathbb{R}^m \to \mathbb{R}$, a so-called *classifier* (also called a *support vector machine*), which, for each document $d \in D$, assigns to the feature vector $\mathbf{f}^d$ a real number that will serve as a tool for deciding in which language document d was written. The interpretation of the value $g(\mathbf{f}^d)$ is as follows. For any document $d \in D$, if $g(\mathbf{f}^d) > 0$, then we conclude that the text was written in English; if $g(\mathbf{f}^d) < 0$, then we conclude that the text was written in Dutch.

To construct such a classifier, we assume that for a small subset of the documents, the language is known in advance (for example, the articles have been read and classified by a person). We partition this subset into two subsets, L and V. The subset L is called the *learning set*, and it will be used to construct a classifier. The subset V is called the *validation set*, and it will be used to check that the classifier constructed from the learning set correctly predicts the language of given documents. If the classifier works satisfactorily for the validation set, then it is accepted as a valid classifier, and it may be used to determine the language of the documents that are not in $L \cup V$ (i.e., for the documents for which the language is currently unknown). Let L_1 be the subset of L that are known to be written in English and, similarly, let L_2 be the subset of L that are known to be written in Dutch. Define V_1 and V_2 analogously. In our example of newspaper articles, we will use the data in Table 11.1 as the learning set.

Letter	Articles written in English						Articles written in Dutch					
	1	2	3	4	5	6	7	8	9	10	11	12
A	10.40	9.02	9.48	7.89	8.44	8.49	8.68	9.78	12.27	7.42	8.60	10.22
B	1.61	1.87	1.84	1.58	1.41	1.55	2.03	1.08	0.99	1.82	1.79	2.62
C	2.87	2.95	4.86	2.78	3.85	3.13	0.80	1.37	1.10	1.03	2.15	1.83
D	4.29	3.52	3.16	4.18	3.91	5.04	5.50	6.05	6.13	7.11	5.97	6.46
E	12.20	11.75	12.69	11.24	11.88	11.82	18.59	17.83	17.74	17.85	15.29	18.25
F	2.12	2.31	2.97	2.00	2.55	1.90	0.76	0.89	0.77	1.26	1.19	1.48
G	2.23	1.67	2.17	2.16	1.79	2.58	2.75	2.99	2.96	2.21	2.39	4.10
H	4.45	4.36	4.25	5.56	4.45	5.43	1.69	3.29	2.96	1.66	2.27	2.62
I	9.20	7.61	7.59	7.62	8.33	8.25	5.25	6.89	6.24	8.14	7.29	5.68
J	0.13	0.22	0.14	0.25	0.22	0.40	1.35	1.30	0.88	2.05	1.55	1.57
K	0.75	0.64	0.57	0.91	0.33	0.91	3.90	1.90	1.86	2.13	1.91	2.10
L	4.05	3.28	4.81	4.74	4.31	3.77	4.11	4.44	3.50	3.40	3.82	3.76
M	2.41	3.46	2.55	3.10	2.74	2.58	2.50	2.21	3.40	3.00	1.67	1.75
N	7.03	7.70	6.60	7.02	7.16	7.34	10.63	8.80	10.30	11.45	9.32	11.53
O	5.85	6.82	8.11	6.74	6.76	6.54	6.48	5.51	4.27	4.82	4.78	4.28
P	1.53	2.51	1.79	1.93	2.41	1.43	1.31	1.23	1.42	1.11	2.51	0.52
Q	0.11	0.02	0.14	0.12	0.19	0.08	0.00	0.05	0.00	0.00	0.00	0.09
R	6.44	7.00	6.04	5.82	6.35	6.07	7.75	6.24	5.04	6.24	6.69	6.20
S	7.35	7.68	5.28	7.22	6.35	6.66	3.43	4.66	3.18	5.13	6.57	3.14
T	8.50	8.10	8.92	9.03	8.98	7.61	5.42	6.77	7.12	4.82	6.33	5.94
U	2.25	3.17	2.12	2.87	2.88	3.09	1.74	1.32	1.20	1.58	2.75	1.40
V	0.80	1.01	1.08	0.89	1.22	0.95	3.30	2.56	3.61	4.19	2.75	2.45
W	1.26	1.21	1.51	1.61	1.47	2.62	0.97	1.51	1.53	0.55	1.31	0.87
X	0.05	0.22	0.09	0.05	0.19	0.28	0.04	0.00	0.00	0.00	0.00	0.00
Y	1.72	1.87	1.04	2.56	1.71	1.43	0.08	0.14	0.11	0.00	0.24	0.09
Z	0.40	0.02	0.19	0.14	0.14	0.08	0.93	1.18	1.42	1.03	0.84	1.05

Table 11.1: Relative letter frequencies (in percentages) of several newspaper articles.

We will restrict our attention to *linear classifiers*, i.e., the classifier g is restricted to have the form

$$g(\mathbf{f}) = \sum_{j=1}^{m} w_j f_j + b = \mathbf{w}^{\mathsf{T}} \mathbf{f} + b \quad \text{for } \mathbf{f} = \begin{bmatrix} f_1 & \cdots & f_m \end{bmatrix}^{\mathsf{T}} \in \mathbb{R}^m,$$

where $\mathbf{w}$ ($\in \mathbb{R}^m \setminus \{\mathbf{0}\}$) is called the *weight vector* of the classifier, b ($\in \mathbb{R}$) is the *intercept*, and $\mathbf{f}$ is any feature vector. Note that we exclude the possibility that $\mathbf{w} = \mathbf{0}$, because the corresponding classifier does not take into account any feature, and therefore is not of any use to predict the language of a document. Our goal is to construct a weight vector $\mathbf{w}$ and an intercept b such that:

$$\begin{aligned} d \in L_1 &\implies \mathbf{w}^{\mathsf{T}} \mathbf{f}^d + b > 0, \text{ and} \\ d \in L_2 &\implies \mathbf{w}^{\mathsf{T}} \mathbf{f}^d + b < 0. \end{aligned} \tag{11.1}$$

Linear classifiers have the following geometric interpretation. For any $\mathbf{w} \in \mathbb{R}^m \setminus \{\mathbf{0}\}$ and $b \in \mathbb{R}$, define the hyperplane $H(\mathbf{w}, b) = \{\mathbf{f} \in \mathbb{R}^m \mid \mathbf{w}^{\mathsf{T}} \mathbf{f} + b = 0\}$, and the two (strict) halfspaces $H^+(\mathbf{w}, b) = \{\mathbf{f} \in \mathbb{R}^m \mid \mathbf{w}^{\mathsf{T}} \mathbf{f} + b < 0\}$ and $H^-(\mathbf{w}, b) = \{\mathbf{f} \mid \mathbf{w}^{\mathsf{T}} \mathbf{f} + b > 0\}$ corresponding to $H(\mathbf{w}, b)$. Hence, (11.1) is equivalent to:

$$\begin{aligned} F(L_1) &= \{\mathbf{f}^d \mid d \in L_1\} \subseteq H^+(\mathbf{w}, b), \text{ and} \\ F(L_2) &= \{\mathbf{f}^d \mid d \in L_2\} \subseteq H^-(\mathbf{w}, b). \end{aligned} \tag{11.2}$$

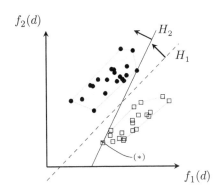

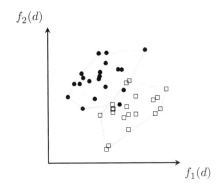

Figure 11.1: Separable learning set with 40 documents. The solid and the dashed lines are separating hyperplanes.

Figure 11.2: Nonseparable learning set with 40 documents. The convex hulls of the learning sets intersect.

So, we want to construct a hyperplane in $\mathbb{R}^m$ such that the feature vectors corresponding to documents in L_1 lie in the halfspace $H^+(\mathbf{w}, b)$, and the vectors corresponding to L_2 in $H^-(\mathbf{w}, b)$.

If there exist a weight vector $\mathbf{w}$ and an intercept b such that the conditions of (11.2) are satisfied, then $F(L_1)$ and $F(L_2)$ are said to be *separable*; they are called *nonseparable* otherwise. The corresponding hyperplane $H(\mathbf{w}, b)$ is called a *separating hyperplane* for $F(L_1)$ and $F(L_2)$, and the function $\mathbf{w}^\top \mathbf{f}^d + b$ is called a *separator* for $F(L_1)$ and $F(L_2)$; see also Appendix D. We make the following observations (see Exercise 11.8.2):

$H^+(-\mathbf{w}, -b) = H^-(\mathbf{w}, b)$ for $\mathbf{w} \in \mathbb{R}^m \setminus \{\mathbf{0}\}, b \in \mathbb{R}$.

$H(\lambda \mathbf{w}, \lambda b) = H(\mathbf{w}, b)$ for $\mathbf{w} \in \mathbb{R}^m \setminus \{\mathbf{0}\}, b \in \mathbb{R}$, and $\lambda \neq 0$.

If $\mathbf{w}$ and b define a separating hyperplane for $F(L_1)$ and $F(L_2)$ such that $F(L_1) \subseteq H^+(\mathbf{w}, b)$ and $F(L_2) \subseteq H^-(\mathbf{w}, b)$, then we also have that $\mathrm{conv}(F(L_1)) \subseteq H^+(\mathbf{w}, b)$ and $\mathrm{conv}(F(L_2)) \subseteq H^-(\mathbf{w}, b)$; therefore, $\mathbf{w}$ and b also define a separating hyperplane for $\mathrm{conv}(F(L_1))$ and $\mathrm{conv}(F(L_2))$.

Note that even for a small learning set L, it is not beforehand clear whether or not $F(L_1)$ and $F(L_2)$ are separable. So the first question that needs to be addressed is: does there exist a separating hyperplane for $F(L_1)$ and $F(L_2)$? Figure 11.1 shows an example of a separable learning set with $(m =) 2$ features. The squares correspond to the feature vectors in $F(L_1)$, and the circles to the feature vectors in $F(L_2)$. Also, the convex hulls of square points and the circle points are shown. The solid and the dashed lines represent two possible hyperplanes. Figure 11.2 shows a learning set which is not separable.

Figure 11.1 illustrates another important fact. Suppose that we discard feature f_2 and only consider feature f_1. Let $F'(L_1)$ ($\subset \mathbb{R}^1$) and $F'(L_2)$ ($\subset \mathbb{R}^1$) be the feature 'vectors' obtained from discarding feature f_2. Then, the vectors in $F'(L_1)$ and $F'(L_2)$ are one-dimensional and can be plotted on a line; see Figure 11.3. (This graph can also be constructed by moving all points in Figure 11.1 straight down onto the horizontal axis.) A hyperplane

Figure 11.3: The learning set of Figure 11.1 after discarding feature f_2.

in one-dimensional Euclidean space is a point. Hence, the one-dimensional learning set is separable if and only if there exists a point P on the line such that all vectors in $F'(L_1)$ are strictly to the left of P, and all vectors in $F'(L_2)$ are strictly on the right of P. From this figure, it is clear that such a point does not exist. Hence, the learning set has become nonseparable.

This fact can also be seen immediately from Figure 11.1 as follows. Discarding feature f_2 is equivalent to requiring that the weight w_2 assigned to f_2 is zero. This, in turn, is equivalent to requiring that the separating hyperplane is 'vertical' in the figure. Clearly, there is no vertical separating hyperplane for the learning set drawn in Figure 11.1. The same holds when discarding feature f_1 and only considering f_2, which is equivalent to requiring that the separating hyperplane is horizontal.

In general, the following observations hold (see Exercise 11.8.1):

If a learning set with a certain set of features is separable, then adding a feature keeps the learning set separable. However, removing a feature may make the learning set nonseparable.

If a learning set with a certain set of features is nonseparable, then removing a feature keeps the learning set nonseparable. However, adding a feature may make the learning set separable.

In the context of the English and Dutch documents, consider the frequencies of the letters A, B, and N. Figure 11.4 shows scatter plots of the letter frequencies of the pairs (A, B), (A, N), and (B, N). It is clear from the plots that if the letter frequencies of only two of the three letters A, B, N are used as features, then the corresponding learning sets are not separable. However, if the letter frequencies of all three letters are taken into account, the learning set is separable. Therefore, a question to consider is: given a learning set, what is a minimal set of features that makes the learning set separable?

11.3 LO-model for finding separating hyperplanes

Constructing a separating hyperplane for the learning set $L_1 \cup L_2$ can be done by designing and solving an LO-model as follows. The decision variables of the LO-model will be the entries $w_1, \ldots, w_m$ of the weight vector $\mathbf{w}$, and the intercept b. Since the value of the classifier should be strictly positive if document d was written in English (i.e., $d \in L_1$), and strictly negative if document d was written in Dutch (i.e., $d \in L_2$), we have the constraints:

$$\begin{aligned}
\mathbf{w}^\mathsf{T}\mathbf{f}^d + b > 0 \qquad &\text{for } d \in L_1, \text{ and} \\
\mathbf{w}^\mathsf{T}\mathbf{f}^d + b < 0 \qquad &\text{for } d \in L_2.
\end{aligned} \tag{11.3}$$

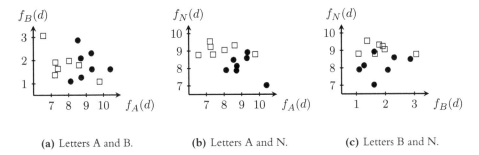

(a) Letters A and B. **(b)** Letters A and N. **(c)** Letters B and N.

Figure 11.4: Scatter plots of relative letter frequencies (in percentages). The squares represent the vectors in $F(L_1)$ and the circles are the vectors in $F(L_2)$. Here, L_1 is the set of English documents, and L_2 is the set of Dutch documents.

Because these inequalities are strict inequalities, they cannot be used in an LO-model. To circumvent this 'limitation', we will show that it suffices to use the following '$\geq$' and '$\leq$' inequalities instead:

$$\begin{aligned} \mathbf{w}^\mathsf{T}\mathbf{f}^d + b &\geq 1 && \text{for } d \in L_1, \text{ and} \\ \mathbf{w}^\mathsf{T}\mathbf{f}^d + b &\leq -1 && \text{for } d \in L_2. \end{aligned}$$

(11.4)

Clearly, the solution set (in terms of $\mathbf{w}$ and b) of (11.4) is in general a strict subset of the solution set of (11.3). However, the sets of hyperplanes defined by (11.3) and (11.4) coincide. To be precise, let $\mathcal{H}_1 = \{H(\mathbf{w}, b) \mid \mathbf{w} \text{ and } b \text{ satisfy (11.3)}\}$, i.e., $\mathcal{H}_1$ is the collection of hyperplanes defined by the solutions of (11.3). Let $\mathcal{H}_2 = \{H(\mathbf{w}, b) \mid \mathbf{w} \text{ and } b \text{ satisfy (11.4)}\}$. We claim that $\mathcal{H}_1 = \mathcal{H}_2$. It is easy to check that $\mathcal{H}_2 \subseteq \mathcal{H}_1$. To see that $\mathcal{H}_1 \subseteq \mathcal{H}_2$, take any $\mathbf{w}$ and b that satisfy (11.3). Then, because L_1 and L_2 are finite sets, there exists $\varepsilon > 0$ such that $\mathbf{w}^\mathsf{T}\mathbf{f}^d + b \geq \varepsilon$ for $d \in L_1$ and $\mathbf{w}^\mathsf{T}\mathbf{f}^d + b \leq -\varepsilon$ for $d \in L_2$. Define $\hat{\mathbf{w}} = \frac{1}{\varepsilon}\mathbf{w}$ and $\hat{b} = \frac{1}{\varepsilon}b$. Then, it is straightforward to check that $\hat{\mathbf{w}}$ and $\hat{b}$ satisfy (11.4) and that $H(\hat{\mathbf{w}}, \hat{b}) = H(\mathbf{w}, b)$, as required.

From now on, we will only consider the inequalities of (11.4). For each $\mathbf{w} \in \mathbb{R}^m \setminus \{\mathbf{0}\}$ and $b \in \mathbb{R}$, define the following halfspaces:

$$\begin{aligned} H^{+1}(\mathbf{w}, b) &= \{\mathbf{f} \in \mathbb{R}^m \mid \mathbf{w}^\mathsf{T}\mathbf{f} + b \geq 1\}, \text{ and} \\ H^{-1}(\mathbf{w}, b) &= \{\mathbf{f} \in \mathbb{R}^m \mid \mathbf{w}^\mathsf{T}\mathbf{f} + b \leq -1\}. \end{aligned}$$

Then, (11.4) is equivalent to:

$$F(L_1) \subseteq H^{+1}(\mathbf{w}, b), \text{ and } F(L_2) \subseteq H^{-1}(\mathbf{w}, b).$$

(11.5)

If the halfspaces $H^{+1}(\mathbf{w}, b)$ and $H^{-1}(\mathbf{w}, b)$ satisfy the conditions of (11.5), then the set $\{\mathbf{f} \in \mathbb{R}^m \mid -1 \leq \mathbf{w}^\mathsf{T}\mathbf{f} + b \leq -1\}$ is called a *separation* for $F(L_1)$ and $F(L_2)$, because it 'separates' $F(L_1)$ from $F(L_2)$. Figure 11.5 illustrates this concept.

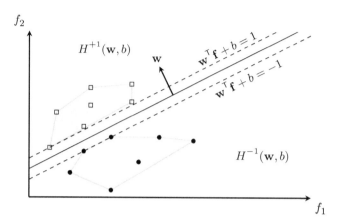

Figure 11.5: Separation for a learning set. The area between the dashed lines is the separation.

It follows from the discussion above that, in order to find a separating hyperplane for $F(L_1)$ and $F(L_2)$, the system of inequalities (11.4) needs to be solved. This can be done by solving the following LO-model:

$$
\begin{aligned}
\min \; & 0 \\
\text{s.t.} \;\; & w_1 f_1^d + \ldots + w_m f_m^d + b \geq 1 \quad \text{for } d \in L_1 \\
& w_1 f_1^d + \ldots + w_m f_m^d + b \leq -1 \quad \text{for } d \in L_2 \\
& w_1, \ldots, w_m, b \text{ free.}
\end{aligned}
\tag{11.6}
$$

In this LO-model, the decision variables are the weights $w_1, \ldots, w_m$ and the intercept b of the classifier. The values of f_i^d with $i \in \{1, \ldots, m\}$ and $d \in L_1 \cup L_2$ are parameters of the model.

Once a classifier (equivalently, a separating hyperplane) for the learning set $L_1 \cup L_2$ has been constructed by solving the LO-model (11.6), this classifier may be used to predict the language of any given document $d \in D$. This prediction is done as follows. Let $w_1^*, \ldots, w_m^*, b^*$ be an optimal solution of model (11.6). This optimal solution defines the classifier value $w_1^* f_1^d + \ldots + w_m^* f_m^d + b^*$ for document d, based on the feature values of that document. If the classifier value is ≥ 1, then the document is classified as an English document; if the value is ≤ -1, then the document is classified as a Dutch document. If the value lies between -1 and 1, then the classifier does not clearly determine the language of the document. In that case, the closer the value lies to 1, the more confident we can be that d is an English document. Similarly, the closer the value lies to -1, the more confident we can be that d is a Dutch document.

Example 11.3.1. *Consider the learning set of Table 11.1, where L_1 is the set of the six newspaper articles written in English, and L_2 is the set of the six newspaper articles written in Dutch. Solving model (11.6) using a computer package (e.g., the online solver for this book) yields the following optimal* *solution:*

$$b^* = 0, \ w_8^* = 0.296, \ w_{15}^* = 0.116, \ w_{17}^* = 1.978, \ w_{21}^* = -0.163, \ w_{26}^* = -2.116.$$

(See Section 11.7 for the GMPL code for this model.) All other decision variables have value zero at this optimal solution. The corresponding classifier is:

$$g(\mathbf{f}) = w_8^* = 0.296 f_8 + 0.116 f_{15} + 1.978 f_{17} - 0.163 f_{21} - 2.116 f_{26}.$$

The weights correspond to the letters H, O, Q, U, and Z, respectively. Thus, the classifier bases its calculations only on the relative frequencies of the letters H, O, Q, U, and Z. Note that the weight w_{17}^* assigned to the letter Q is positive and relatively large compared to the other positive weights. This means that, for any given document $d \in D$, the expression $w_1^* f_1^d + \ldots + w_m^* f_m^d + b$ tends to be more positive if the document contains relatively many occurrences of the letter Q. This means that such a document is more likely to be classified as an English newspaper article. On the other hand, the weight w_{26}^* assigned to the letter Z is negative, and so a document containing relatively many occurrences of the letter Z is likely to be classified as a Dutch newspaper article.

11.4 Validation of a classifier

Recall that the set of documents for which the language is known is partitioned into two parts, namely the learning set L and the validation set V. Before using a classifier to predict the language of any document $d \in D$, it is good practice to *validate* it by comparing its predictions to the expected results for the validation set V. This validation step is done in order to check that the classifier found by the LO-model makes sensible predictions for documents that are not in the learning set. We illustrate this process with an example.

Example 11.4.1. *For the validation of the classifier found in Example 11.3.1, we have a validation set consisting of twenty-five English and thirty-three Dutch newspaper articles, i.e., $|V_1| = 25$ and $|V_2| = 33$. Table 11.2 lists the relevant letter frequencies, the classifier value $w_1^* f_1^d + \ldots + w_m^* f_m^d + b^*$, and the language prediction found for a number of documents d in $L \cup V$. The row 'Predicted language' indicates the language of the article predicted by the classifier. A question mark indicates that the classifier is inconclusive about the language; in that case the sign of the classifier value determines whether the classifier leans towards one of the two languages.*

The documents 1, 2, 7, and 8, which are in the learning set, are correctly predicted. This should come as no surprise, as the constraints of model (11.6) ensure this fact. For the validation set, the results are not as clear-cut. The classifier correctly predicts the language of most newspaper articles in the validation set; these cases have been omitted from Table 11.2 (except article 21). The classifier is inconclusive about articles 30 and 66, but at least the sign of the classifier value is correct, meaning that the classifier leans towards the correct language. However, for articles 57 and 67, even the sign of the classifier is wrong.

The above example illustrates the fact that the validation step may reveal problems with the classifier constructed using model (11.6). One way to improve the classifier is to increase

Letter	English newspaper articles				Dutch newspaper articles				
	1	2	21	30	7	8	57	66	67
H	4.45	4.36	5.14	4.47	1.69	3.29	1.68	3.40	2.33
O	5.85	6.82	8.00	7.35	6.48	5.51	6.15	3.93	5.37
Q	0.11	0.02	0.06	0.00	0.00	0.05	0.00	0.00	0.00
U	2.25	3.17	2.69	2.32	1.74	1.32	2.24	1.17	2.15
Z	0.40	0.02	0.00	0.46	0.93	1.18	0.34	0.85	0.18
Classifier value	1.00	1.56	2.13	0.82	−1.00	−1.00	0.13	−0.53	0.58
Predicted language	Eng	Eng	Eng	Eng?	Dut	Dut	Eng?	Dut?	Eng?

Table 11.2: Validation results for the classifier. The articles 1, 2, 7, and 8 are in the learning set; the articles 21, 30, 57, 66, and 67 are in the validation set. The question marks in the row 'Predicted language' indicate that the classifier is inconclusive about the language.

the learning set. In the example, we used only six documents per language. In real-life applications the learning set is usually taken to be much larger.

In the next sections, we present another way to improve the classification results. Note that the objective function of model (11.6) is the zero function, which means that any feasible solution of the model is optimal. So, the objective is in a sense 'redundant', because it can be replaced by maximizing or minimizing any constant objective function. In fact, in general, the model has multiple optimal solutions. Hence, there are in general multiple separating hyperplanes. Figure 11.1 shows two hyperplanes corresponding to two feasible solutions, namely a dotted line and a solid line. In the next section, we study the 'quality' of the hyperplanes.

II.5 Robustness of separating hyperplanes; separation width

The LO-model (11.6) generally has multiple optimal solutions. In fact, since the objective function is the zero function, any feasible solution (i.e., any choice of separating hyperplane) is optimal. Recall that the goal of constructing a classifier is to use this function to automatically classify the documents that are not in the learning set into English and Dutch documents.

Figure 11.1 shows two hyperplanes, H_1 and H_2, corresponding to two feasible solutions. Suppose that we find as an optimal solution the hyperplane H_1 (drawn as a solid line), and suppose that we encounter a text document $\hat{d}$ ($\in D$) whose feature vector $\mathbf{f}(\hat{d}) = \left[f_1(\hat{d}) \ f_2(\hat{d}) \right]^\mathsf{T}$ is very close to the feature vector marked with ($*$), but just on the other side of H_1. Based on H_1, we would conclude that the new text document is written in Dutch. However, the feature vector $\mathbf{f}(\hat{d})$ is much closer to a vector corresponding to an English document than to a vector corresponding to a Dutch document. So it makes more sense to conclude that document $\hat{d}$ was written in English. Observe also that hyperplane H_2 does not suffer as much from this problem. In other words, hyperplane H_2 is more *robust* with respect to perturbations than hyperplane H_1.

To measure the robustness of a given separating hyperplane, we calculate its so-called separation width. Informally speaking, the separation width is the m-dimensional generalization of the width of the band between the dashed lines in Figure 11.5. For given $\mathbf{w} \in \mathbb{R}^m \setminus \{\mathbf{0}\}$ and $b \in \mathbb{R}$, the *separation width* of the hyperplane $H(\mathbf{w}, b) = \{\mathbf{f} \in \mathbb{R}^m \mid \mathbf{w}^\mathsf{T}\mathbf{f} + b = 0\}$ is defined as the distance between the halfspaces $H^{+1}(\mathbf{w}, b)$ and $H^{-1}(\mathbf{w}, b)$, i.e.,

$$\text{width}(\mathbf{w}, b) = \min\{\|\mathbf{f} - \mathbf{f}'\| \mid \mathbf{f} \in H^{+1}(\mathbf{w}, b), \mathbf{f}' \in H^{-1}(\mathbf{w}, b)\},$$

where $\|\mathbf{f} - \mathbf{f}'\|$ is the Euclidean distance between the vectors $\mathbf{f}$ and $\mathbf{f}'$ ($\in \mathbb{R}^m$). Note that, for any $\mathbf{w} \in \mathbb{R}^m \setminus \{\mathbf{0}\}$ and $b \in \mathbb{R}$, $\text{width}(\mathbf{w}, b)$ is well-defined because the minimum in the right hand side in the above expression is attained. In fact, the following theorem gives an explicit formula for the separation width.

Theorem 11.5.1.
For any $\mathbf{w} \in \mathbb{R}^m \setminus \{\mathbf{0}\}$ and $b \in \mathbb{R}$, it holds that $\text{width}(\mathbf{w}, b) = \frac{2}{\|\mathbf{w}\|}$.

Proof. Take any point $\hat{\mathbf{f}} \in \mathbb{R}^m$ such that $\mathbf{w}^\mathsf{T}\hat{\mathbf{f}} + b = -1$. Note that $\hat{\mathbf{f}} \in H^{-1}(\mathbf{w}, b)$. Define $\hat{\mathbf{f}}' = \hat{\mathbf{f}} + \mathbf{w}^*$, with $\mathbf{w}^* = \frac{2}{\|\mathbf{w}\|^2}\mathbf{w}$. Then, we have that $\|\mathbf{w}^*\| = \frac{2}{\|\mathbf{w}\|}$. It follows that:

$$\mathbf{w}^\mathsf{T}\hat{\mathbf{f}}' + b = \mathbf{w}^\mathsf{T}\left(\hat{\mathbf{f}} + \frac{2}{\|\mathbf{w}\|^2}\mathbf{w}\right) + b = \mathbf{w}^\mathsf{T}\hat{\mathbf{f}} + b + \frac{2\mathbf{w}^\mathsf{T}\mathbf{w}}{\|\mathbf{w}\|^2} = -1 + 2 = 1,$$

where we have used the fact that $\mathbf{w}^\mathsf{T}\mathbf{w} = \|\mathbf{w}\|^2$. Therefore, $\hat{\mathbf{f}}' \in H^{+1}(\mathbf{w}, b)$. So, we have that $\hat{\mathbf{f}} \in H^{-1}(\mathbf{w}, b)$ and $\hat{\mathbf{f}}' \in H^{+1}(\mathbf{w}, b)$. Hence, $\text{width}(\mathbf{w}, b) \leq \|\hat{\mathbf{f}} - \hat{\mathbf{f}}'\| = \|\mathbf{w}^*\| = \frac{2}{\|\mathbf{w}\|}$. To show that $\text{width}(\mathbf{w}, b) \geq \frac{2}{\|\mathbf{w}\|}$, take any $\hat{\mathbf{f}} \in H^{+1}(\mathbf{w}, b)$ and $\hat{\mathbf{f}}' \in H^{-1}(\mathbf{w}, b)$. By the definitions of $H^{+1}(\mathbf{w}, b)$ and $H^{-1}(\mathbf{w}, b)$, we have that:

$$\mathbf{w}^\mathsf{T}\hat{\mathbf{f}} + b \geq 1, \text{ and } \mathbf{w}^\mathsf{T}\hat{\mathbf{f}}' + b \leq -1.$$

Subtracting the second inequality from the first one gives the inequality $\mathbf{w}^\mathsf{T}(\hat{\mathbf{f}}' - \hat{\mathbf{f}}) \geq 2$. The cosine rule (see Appendix B) implies that:

$$\cos\theta = \frac{\mathbf{w}^\mathsf{T}(\hat{\mathbf{f}}' - \hat{\mathbf{f}})}{\|\mathbf{w}\|\,\|\hat{\mathbf{f}}' - \hat{\mathbf{f}}\|} \geq \frac{2}{\|\mathbf{w}\|\,\|\hat{\mathbf{f}}' - \hat{\mathbf{f}}\|},$$

where θ is the angle between the vectors $\mathbf{w}$ and $\hat{\mathbf{f}}' - \hat{\mathbf{f}}$. Since $\cos\theta \leq 1$, we have that:

$$\frac{2}{\|\mathbf{w}\|\,\|\hat{\mathbf{f}}' - \hat{\mathbf{f}}\|} \leq 1.$$

Rearranging, this yields that $\|\hat{\mathbf{f}}' - \hat{\mathbf{f}}\| \geq \frac{2}{\|\mathbf{w}\|}$ for all $\hat{\mathbf{f}} \in H^{+1}(\mathbf{w}, b)$ and all $\hat{\mathbf{f}}' \in H^{-1}(\mathbf{w}, b)$. Hence, also $\min\{\|\mathbf{f} - \mathbf{f}'\| \mid \mathbf{f} \in H^{+1}(\mathbf{w}, b), \mathbf{f}' \in H^{-1}(\mathbf{w}, b)\} \geq \frac{2}{\|\mathbf{w}\|}$, i.e., $\text{width}(\mathbf{w}, b) \geq \frac{2}{\|\mathbf{w}\|}$, as required. $\square$

We conclude that the separation direction is determined by the direction of the vector $\mathbf{w}$ and that, according to Theorem 11.5.1, the separation width is inversely proportional to the length of $\mathbf{w}$. Figure 11.1 depicts two separating hyperplanes. From this figure, we can see that the separation width corresponding to hyperplane H_2 is much smaller than the separation width corresponding to hyperplane H_1.

11.6 Models that maximize the separation width

In order to find a hyperplane that is as robust as possible with respect the separation, the values of the weight vector $\mathbf{w}$ and the intercept b should be chosen so as to maximize the separation width. According to Theorem 11.5.1, the separation width is $\frac{2}{\|\mathbf{w}\|}$. Note that minimizing $\|\mathbf{w}\|$ yields the same optimal values for $\mathbf{w}$ and b as maximizing $\frac{2}{\|\mathbf{w}\|}$. Hence, it suffices to solve the following optimization model in order to find a maximum-width separation for the learning set:

$$
\begin{aligned}
\min \ &\|\mathbf{w}\| \\
\text{s.t.} \ \ &\mathbf{w}^\mathsf{T}\mathbf{f}^d + b \geq \ \ 1 \ \text{for } d \in L_1 \\
&\mathbf{w}^\mathsf{T}\mathbf{f}^d + b \leq -1 \ \text{for } d \in L_2 \\
&b, w_j \ \text{free} \qquad \text{for } j = 1, \ldots, m.
\end{aligned} \tag{11.7}
$$

The objective function $\|\mathbf{w}\| = \sqrt{\sum_{i=1}^m w_i^2}$ is obviously a nonlinear function of the decision variables $w_1, \ldots, w_m$, so that (11.7) is a nonlinear optimization model. Such models may be hard to solve, especially when the number of documents (and, hence, the number of variables) is very large. Therefore, we look for a linear objective function. In general, this will result in a classifier of less quality, i.e., the hyperplane corresponding to the resulting $(\mathbf{w}, b)$ has smaller separation width than the optimal hyperplane corresponding to an optimal solution $(\mathbf{w}^*, b^*)$ of (11.7).

The objective function of the above (nonlinear) optimization model is the Euclidean norm (see Appendix B.1) of the vector $\mathbf{w}$. A generalization of the Euclidean norm is the so-called p-norm. The p-norm of a vector $\mathbf{w} = \begin{bmatrix} w_1 \ \ldots \ w_m \end{bmatrix}^\mathsf{T} \in \mathbb{R}^m$ is denoted and defined as ($p \geq 1$ and integer):

$$
\|\mathbf{w}\|_p = \left(\sum_{i=1}^m |w_i|^p \right)^{1/p}.
$$

Clearly, the Euclidean norm corresponds to the special case $p = 2$, i.e., the Euclidean norm is the 2-norm. Since the 2-norm is a nonlinear function, it cannot be included in an LO-model. Below, however, we will see that two other choices for p lead to LO-models, namely the choices $p = 1$ and $p = \infty$. In the remainder of this section, we consecutively discuss LO-models that minimize the 1-norm and the ∞-norm of the weight vector.

11.6.1 Minimizing the 1-norm of the weight vector

In this section, we consider minimizing the 1-norm of the weight vector. The 1-norm of a vector $\mathbf{w} \in \mathbb{R}^m$ is defined as:

$$\|\mathbf{w}\|_1 = \sum_{i=1}^{m} |w_i|.$$

So, we replace the objective $\min \sqrt{w_1^2 + \ldots + w_m^2}$ of model (11.7) by the objective:

$$\min \sum_{j=1}^{m} |w_j|.$$

The function $\sum_{j=1}^{m} |w_j|$ is still not a linear function, but we already saw in Section 1.5.2 how to deal with absolute values in the context of linear optimization. In order to turn the objective into a linear objective, as in Section 1.5.2, define $w_j = w_j^+ - w_j^-$ for $j = 1, \ldots, m$. Hence, $|w_j| = w_j^+ - w_j^-$, with $w_j^+ \geq 0$ and $w_j^- \geq 0$. This leads to the following LO-model:

$$\min \sum_{j=1}^{m}(w_j^+ + w_j^-)$$

$$\text{s.t.} \quad \sum_{j=1}^{m} w_j^+ f_j^d - w_j^- f_j^d + b \geq 1 \quad \text{for } d \in L_1$$

$$\sum_{j=1}^{m} w_j^+ f_j^d - w_j^- f_j^d + b \leq -1 \quad \text{for } d \in L_2$$

$$w_j^+ \geq 0, w_j^- \geq 0, b \text{ free} \quad \text{for } j = 1, \ldots, m. \tag{11.8}$$

The constraints are still linear, because the values of the f_j^i's are parameters of the model, and hence the left hand sides of the constraints are linear functions of the decision variables b, w_j^+, and w_j^- ($j = 1, \ldots, m$).

11.6.2 Minimizing the ∞-norm of the weight vector

We now consider minimizing the ∞-norm of the weight vector. Mathematically, the ∞-norm of a vector is defined as the limit of its p-norm as p goes to infinity. The following theorem states that this limit is well-defined, and it is in fact equal to the entry with the largest absolute value.

> **Theorem 11.6.1.**
> Let $\mathbf{w} = \begin{bmatrix} w_1 & \ldots & w_m \end{bmatrix}^\mathsf{T}$ be a vector. Then, $\lim_{p \to \infty} \|\mathbf{w}\|_p = \max\{|w_1|, \ldots, |w_m|\}$.

Proof. Define $M = \max\{|w_i| \mid i = 1, \ldots, m\}$, and let p be any positive integer. We have that:

$$\|\mathbf{w}\|_p = \left(\sum_{i=1}^{m} |w_i|^p\right)^{1/p} \geq \left(M^p\right)^{1/p} = M.$$

On the other hand, we have that:

$$\|\mathbf{w}\|_p = \left(\sum_{i=1}^{m} |w_i|^p\right)^{1/p} \leq \left(mM^p\right)^{1/p} = m^{1/p}M.$$

It follows that $M \leq \|\mathbf{w}\|_p \leq m^{1/p}M$. Letting $p \to \infty$ in this expression, we find that $M \leq \lim_{p \to \infty} \|\mathbf{w}\|_p \leq M$, which is equivalent to $\lim_{p \to \infty} \|\mathbf{w}\|_p = M$, as required. $\square$

So, according to Theorem 11.6.1, we should consider the following objective:

$$\min \max\{|w_1|, \ldots, |w_m|\}.$$

The objective function $\max\{|w_1|, \ldots, |w_m|\}$ is clearly not linear. However, it can be incorporated in an LO-model by using the following 'trick'. First, a new decision variable z is introduced, which will represent $\max\{|w_1|, \ldots, |w_m|\}$. The objective is then replaced by: 'min x', and the following constraints are added:

$$|w_j| \leq x \qquad \text{for } j = 1, \ldots, m.$$

Because the value of the variable x is minimized at any optimal solution, we will have that the optimal value x^* will be as small as possible, while satisfying $x^* \geq |w_j^*|$ for $j = 1, \ldots, m$. This means that in fact $x^* = \max\{|w_1^*|, \ldots, |w_m^*|\}$ at any optimal solution. Combining this 'trick' with the treatment of absolute values as in model (11.8), we find the following LO-model:

$$
\begin{aligned}
\min \quad & x \\
\text{s.t.} \quad & \sum_{j=1}^{m} w_j^+ f_j^d - w_j^- f_j^d + b \geq 1 && \text{for } d \in L_1 \\
& \sum_{j=1}^{m} w_j^+ f_j^d - w_j^- f_j^d + b \leq -1 && \text{for } d \in L_2 \\
& w_j^+ + w_j^- \leq x && \text{for } j = 1, \ldots, m. \\
& x \geq 0,\ w_j^+ \geq 0,\ w_j^- \geq 0,\ b \text{ free} && \text{for } j = 1, \ldots, m.
\end{aligned}
\tag{11.9}
$$

The values of the f_j^i's are parameters of the model, and the decision variables are b, x, w_j^+, and w_j^- for $j = 1, \ldots, m$.

11.6.3 Comparing the two models

We have constructed two LO-models that 'approximately' solve the problem of maximizing the separation width. It is interesting to compare the results of the two models. To do so, we have used a computer package to find optimal solutions of models (11.8) and (11.9) for the learning set of Table 11.1. (See Section 11.7 for the GMPL code corresponding to model (11.8).)

Model (11.8), corresponding to minimizing the 1-norm of the weight vector, has the optimal solution (for the learning set of Table 11.1):

$$w_5^* = -0.602, \ w_{15}^* = 0.131, \ b^* = 7.58.$$

All other decision variables have optimal value zero. Therefore, the corresponding linear classifier $g_1(\mathbf{f})$ for this learning set is:

$$g_1(\mathbf{f}) = -0.602 f_5 + 0.131 f_{19} + 7.58.$$

Note that this classifier uses very little information from the feature vector $\mathbf{f}$. Only two features are taken into account, namely the relative frequencies of occurrences of the letters E and O. Because $w_{15}^* > 0$, a newspaper article with relatively many occurrences of the letter O is more likely to be categorized as written in English, whereas the letter E is considered an indication that the article is written in Dutch.

In contrast, consider model (11.9), corresponding to minimizing the ∞-norm of the weight vector. This model has the optimal solution:

$$
\begin{aligned}
w_j^* &= 0.0765 &&\text{for } j = 1, 3, 6, 8, 9, 13, 15, 17, 19, 20, 21, 23, 24, 25, \\
w_j^* &= -0.0765 &&\text{for } j = 2, 4, 5, 10, 11, 12, 14, 16, 18, 22, 26, \\
w_7^* &= 0.0463, \\
b^* &= -0.5530.
\end{aligned}
$$

Let $g_\infty(\mathbf{f})$ be the corresponding linear classifier. As opposed to $g_1(\mathbf{f})$, the classifier $g_\infty(\mathbf{f})$ takes into account all features to make a prediction for the language of a given article. The first set of weights in the solution above corresponds to the letters A, C, F, H, I, M, O, Q, S, T, U, W, X, and Y. Since these weights are all positive, the classifier treats a relatively high frequency of occurrences of any of these letters in a given article as evidence that the article may be written in English. On the other hand, the second set of weights, corresponding to the letters B, D, E, J, K, L, N, P, R, V, and Z, are negative. This means that a relatively high frequency of occurrences of any of these is treated as evidence that the article may be written in Dutch. Note that weight w_7^* corresponds to the letter G.

It is left to the reader to carry out the validation steps (see Section 11.4) for these classifiers, i.e., to verify that these classifiers correctly predict the language of all newspaper articles in the learning set and in the validation set, although both classifiers have values between -1 and 1 for some newspaper articles. (The data are available on the website of this book.) Note that both models assign a negative weight to the letter E, and a positive weight to the letter O. Hence, both classifiers treat frequent occurrences of the letter E as evidence towards the article being written in Dutch, and frequent occurrences of the letter O as evidence towards it being written in English.

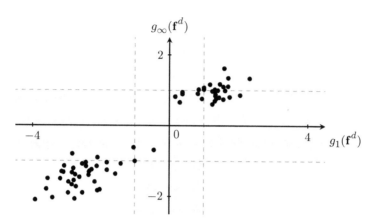

Figure 11.6: Comparison of classifiers based on minimizing the 1-norm of the weight vector, versus minimizing the ∞-norm.

An interesting question is: is one of the two classifiers significantly better than the other? To answer this question, we have calculated the values of the two classifiers for all documents in the learning set and in the validation set. Figure 11.6 shows the results of these calculations. Each point in the figure represents a newspaper article. On the horizontal axis we have plotted the values of $g_1(\mathbf{f}^d)$, and on the vertical axis we have plotted the values of $g_\infty(\mathbf{f}^d)$ $(d \in D)$. From the figure, we see that the 'north west' and 'south east' quadrants contain no points at all. This means that the two classifiers have the same sign for each $d \in D$: whenever $g_1(\mathbf{f}^d)$ is positive, $g_\infty(\mathbf{f}^d)$ is positive, and vice versa. The 'north east' quadrant contains points for which both classifiers are positive, i.e., these are the newspaper articles that are predicted to be in English by both classifiers. Similarly, the 'south west' quadrant contains the newspaper articles that are predicted to be in Dutch. It can be seen from the figure that the values of the two classifiers are more or less linearly related, meaning that they result in roughly the same predictions.

The horizontal gray band in the figure is the area in which the classifier $g_1(\mathbf{f})$ has a value between -1 and 1, i.e., the area in which the classifier does not give a clear-cut prediction. Similarly, the vertical gray band is the area in which the classifier $g_\infty(\mathbf{f})$ does not give a clear-cut prediction. The horizontal gray band contains 25 points, whereas the vertical gray band contains only 9 points. From this, we can conclude that the classifier $g_\infty(\mathbf{f})$ tends to give more clear-cut predictions than $g_1(\mathbf{f})$. So, in that sense, the classifier $g_\infty(\mathbf{f})$ is a better classifier than $g_1(\mathbf{f})$.

11.7 GMPL model code

The following listing gives the GMPL model code for model (11.8).

```
1        N1;                      # number of English documents in learning set
2        N2;                      # number of Dutch documents in learning set
```

```
3
4      DOCUMENTS := 1..(N1+N2);    # set of all documents
5      L1 := 1..N1;                # set of English documents
6      L2 := (N1+1)..(N1+N2);      # set of Dutch documents
7      FEATURES;                   # set of features
8
9        f{FEATURES, DOCUMENTS};   # values of the feature vectors
10
11     wp{FEATURES} >= 0;          # positive part of weights
12     wm{FEATURES} >= 0;          # negative part of weights
13     b;                          # intercept
14
15        obj:                     # objective
16     {j   FEATURES} (wp[j] + wm[j]);
17
18          cons_L1{i   L1}:  # constraints for English documents
19     {j   FEATURES} (wp[j] — wm[j]) * f[j, i] + b >= 1;
20
21          cons_L2{i   L2}:  # constraints for Dutch documents
22     {j   FEATURES} (wp[j] — wm[j]) * f[j, i] + b <= —1;
23
24   ;
25
26     N1 := 6;
27     N2 := 6;
28
29     FEATURES := A B C D E F G H I J K L M N O P Q R S T U V W X Y Z;
30
31     f :
32        1     2     3     4     5     6     7     8     9    10    11    12 :=
33 A 10.40  9.02  9.48  7.89  8.44  8.49  8.68  9.78 12.27  7.42  8.60 10.22
34 B  1.61  1.87  1.84  1.58  1.41  1.55  2.03  1.08  0.99  1.82  1.79  2.62
35 C  2.87  2.95  4.86  2.78  3.85  3.13  0.80  1.37  1.10  1.03  2.15  1.83
36 D  4.29  3.52  3.16  4.18  3.91  5.04  5.50  6.05  6.13  7.11  5.97  6.46
37 E 12.20 11.75 12.69 11.24 11.88 11.82 18.59 17.83 17.74 17.85 15.29 18.25
38 F  2.12  2.31  2.97  2.00  2.55  1.90  0.76  0.89  0.77  1.26  1.19  1.48
39 G  2.23  1.67  2.17  2.16  1.79  2.58  2.75  2.99  2.96  2.21  2.39  4.10
40 H  4.45  4.36  4.25  5.56  4.45  5.43  1.69  3.29  2.96  1.66  2.27  2.62
41 I  9.20  7.61  7.59  7.62  8.33  8.25  5.25  6.89  6.24  8.14  7.29  5.68
42 J  0.13  0.22  0.14  0.25  0.22  0.40  1.35  1.30  0.88  2.05  1.55  1.57
43 K  0.75  0.64  0.57  0.91  0.33  0.91  3.90  1.90  1.86  2.13  1.91  2.10
44 L  4.05  3.28  4.81  4.74  4.31  3.77  4.11  4.44  3.50  3.40  3.82  3.76
45 M  2.41  3.46  2.55  3.10  2.74  2.58  2.50  2.21  3.40  3.00  1.67  1.75
46 N  7.03  7.70  6.60  7.02  7.16  7.34 10.63  8.80 10.30 11.45  9.32 11.53
47 O  5.85  6.82  8.11  6.74  6.76  6.54  6.48  5.51  4.27  4.82  4.78  4.28
48 P  1.53  2.51  1.79  1.93  2.41  1.43  1.31  1.23  1.42  1.11  2.51  0.52
49 Q  0.11  0.02  0.14  0.12  0.19  0.08  0.00  0.05  0.00  0.00  0.00  0.09
50 R  6.44  7.00  6.04  5.82  6.35  6.07  7.75  6.24  5.04  6.24  6.69  6.20
51 S  7.35  7.68  5.28  7.22  6.35  6.66  3.43  4.66  3.18  5.13  6.57  3.14
52 T  8.50  8.10  8.92  9.03  8.98  7.61  5.42  6.77  7.12  4.82  6.33  5.94
53 U  2.25  3.17  2.12  2.87  2.88  3.09  1.74  1.32  1.20  1.58  2.75  1.40
54 V  0.80  1.01  1.08  0.89  1.22  0.95  3.30  2.56  3.61  4.19  2.75  2.45
55 W  1.26  1.21  1.51  1.61  1.47  2.62  0.97  1.51  1.53  0.55  1.31  0.87
56 X  0.05  0.22  0.09  0.05  0.19  0.28  0.04  0.00  0.00  0.00  0.00  0.00
```

```
57 Y  1.72  1.87  1.04  2.56  1.71  1.43  0.08  0.14  0.11  0.00  0.24  0.09
58 Z  0.40  0.02  0.19  0.14  0.14  0.08  0.93  1.18  1.42  1.03  0.84  1.05;
59
60     ;
```

11.8 Exercises

Exercise 11.8.1.

(a) Show that if a learning set with a certain set of features is separable, then adding a feature keeps the learning set separable.

(b) Give an example of a separable learning set with the property that removing a feature makes the learning set nonseparable.

Exercise 11.8.2. Prove the following statements:

(a) $H^+(-\mathbf{w}, -b) = H^-(\mathbf{w}, b)$ for $\mathbf{w} \in \mathbb{R}^m \setminus \{\mathbf{0}\}, b \in \mathbb{R}$.

(b) $H(\lambda\mathbf{w}, \lambda b) = H(\mathbf{w}, b)$ for $\mathbf{w} \in \mathbb{R}^m \setminus \{\mathbf{0}\}, b \in \mathbb{R}$, and $\lambda \neq 0$.

(c) If $\mathbf{w}$ and b define a separating hyperplane for $F(L_1)$ and $F(L_2)$ such that $F(L_1) \subseteq H^+(\mathbf{w}, b)$ and $F(L_2) \subseteq H^-(\mathbf{w}, b)$, then also $\text{conv}(F(L_1)) \subseteq H^+(\mathbf{w}, b)$ and $\text{conv}(F(L_2)) \subseteq H^-(\mathbf{w}, b)$; therefore $\mathbf{w}$ and b also define a separating hyperplane for $\text{conv}(F(L_1))$ and $\text{conv}(F(L_2))$.

Exercise 11.8.3. Let $n \geq 1$ and $i \in \{1, 2\}$. For $S_i \subseteq \mathbb{R}^n$, define:

$$S_i' = \left\{ \begin{bmatrix} x_1 & \cdots & x_{n-1} \end{bmatrix}^\mathsf{T} \,\middle|\, \begin{bmatrix} x_1 & \cdots & x_{n-1} & x_n \end{bmatrix}^\mathsf{T} \in S_i \right\}.$$

(a) Prove that if S_i is convex, then S_i' is convex.

(b) Prove that if S_1' and S_2' are separable, then S_1 and S_2 are separable.

(c) Is it true that if S_1 and S_2 are separable, then S_1' and S_2' are separable? Prove this or give a counterexample.

Exercise 11.8.4. In real-life applications, the learning set is usually not separable. One way to deal with this problem is to allow that some points in the data set lie 'on the wrong side' of the hyperplane. Whenever this happens, however, this should be highly penalized. How can model (11.8) be generalized to take this into account?

CHAPTER 12

Production planning: a single
product case

Overview

Linear optimization is applied on a large scale in the field of production planning. In particular, when there is demand for a broad variety of products or when the demand fluctuates strongly, the design of a good production plan may be very difficult, and linear optimization may be a useful analysis and solution tool. When planning the production, both social and economic factors have to be taken into account. A number of questions may be asked in this respect. Is working in overtime acceptable? Is it profitable to hire employees on a temporary basis? Is idle time acceptable?

In this chapter, we restrict ourselves to the production of a single product. In Chapter 13, we describe the situation of several products.

12.1 Model description

Consider a production company that produces certain nonperishable products. By *nonperishable*, we mean that the product does not spoil when it is kept in stock. The company faces a certain future amount of demand for its products. Taking into account the company's current inventory, we want to decide how much the company should produce in the near future so that the future demand is satisfied.

The *planning period* is the time span for which a production plan has to be designed. The planning period is partitioned into T *periods* which are labeled $1, \ldots, T$. For $t = 1, \ldots, T$, we define:

$$
\begin{aligned}
d_t &= \text{the demand in period } t; \\
x_t &= \text{the production in period } t; \\
s_t &= \text{the inventory of the product at the end of period } t;
\end{aligned}
$$

s_0 = the inventory of the product at the beginning of the planning period, i.e., the initial inventory.

It is assumed that the demand d_t is known for all $t \in \{1, \ldots, T\}$ and, within each period, this demand is uniformly distributed over the time span of the period. Also, the initial inventory s_0 is assumed to be known. Moreover, we assume that the amount of production x_t becomes uniformly available during the planning period. The decision variables are x_t and s_t for $t = 1, \ldots, T$.

The inventory s_t at the end of period t satisfies:

$$s_t = s_{t-1} + x_t - d_t \quad \text{for } t = 1, \ldots, T. \tag{12.1}$$

The equations (12.1) are called the *inventory equations*. Note that when the demand exceeds the production, then the value of s_t in (12.1) is negative; $-s_t$ is then the *shortage* at the end of period t. We will consider a shortage as a negative inventory, and therefore continue to use the term inventory. The demand has to be satisfied immediately, so that shortages are not allowed. Hence, s_t has to satisfy $s_t \geq 0$ for $t = 1, \ldots, T$.

The equations (12.1) can be rewritten as follows:

$$s_t = s_0 + \sum_{k=1}^{t} x_k - \sum_{k=1}^{t} d_k \quad \text{for } t = 1, \ldots, T. \tag{12.2}$$

The demand considered in this chapter is the *expected demand*. So assuming that we have 'good' predictions, we have that in about fifty percent of the periods the realized demand will be higher than the expected demand, and therefore in about fifty percent of the periods there will be a shortage. When a high shortage rate is unacceptable, one can build in a so-called *safety stock*. The safety stock for period $t + 1$ has to be available at the end of period t. Define, for each $t = 1, \ldots, T$:

$$\sigma_t = \text{the safety stock at the end of period } t.$$

Hence the following equations should hold:

$$s_t \geq \sigma_t \quad \text{for } t = 1, \ldots, T. \tag{12.3}$$

One way to determine the value of σ_t is to make it dependent on the expected demand in period $t + 1$. For instance, by means of the following relationships:

$$\sigma_t = \gamma d_{t+1} \quad \text{for } t = 1, \ldots, T, \tag{12.4}$$

where d_{T+1} is the expected demand in the first period after the planning period; i.e., σ_T is the initial inventory for the next planning period. It is assumed that $\sigma_T \geq 0$. The parameter γ may depend on, for instance, the service level (the percentage of the number of periods without shortages), or the variance of the expected demand. Alternative relationships are

$\sigma_t = \gamma_t d_{t+1}$ for $t = 1, \ldots, T$, or $\sigma_t = \gamma \operatorname{Var}(d_{t+1})$ for $t = 1, \ldots, T$. We assume that:

$$0 \leq \gamma \leq 1.$$

Define:

$$a_t = \sum_{k=1}^{t} d_k + \sigma_t - s_0 \qquad \text{for } t = 1, \ldots, T. \tag{12.5}$$

We call a_t the *desired production* during the first t periods. Note that if $a_t < 0$ for certain t, then the initial inventory s_0 is large enough to satisfy the total demand during the first t periods plus the safety inventory σ_t. Although a_t does not agree with our intuition of 'desired production' if a_t is negative, we nevertheless use the term 'desired production' in this case. Note that the values of a_t are known in advance for each period t, because they depend only on the values of the d_t's and σ_t's. If $a_t \geq 0$, then a_t is the minimum amount that has to be produced in the periods $1, \ldots, t$ in order to assure that there is no shortage at the end of period t and the desired safety stock is available. Using (12.5), we may combine (12.2) and (12.3) into:

$$x_1 + x_2 + \ldots + x_t \geq a_t \qquad \text{for } t = 1, \ldots, T; \tag{12.6}$$

i.e., the total production during the first t periods is at least equal to the desired production during the first t periods.

For each $t = 2, \ldots, T$, it follows that:

$$\begin{aligned} a_t - a_{t-1} &= \sum_{k=1}^{t} d_k + \sigma_t - s_0 - \sum_{k=1}^{t-1} d_k - \sigma_{t-1} + s_0 \\ &= d_t + \sigma_t - \sigma_{t-1} \\ &= (1-\gamma)d_t + \gamma d_{t+1} \geq 0. \end{aligned}$$

Hence,

$$a_t \geq a_{t-1} \qquad \text{for } t = 2, \ldots, T;$$

i.e., the sequence $a_1, \ldots, a_T$ is nondecreasing. Throughout, we will assume that $a_T > 0$, because, since a_t is a nondecreasing function of t, $a_T \leq 0$ would mean that the company does not need to produce anything at all during the planning period. Define:

w_t = the number of working hours required for the production in period t.

We assume that, for each period, the number of working hours is proportional to the production, i.e.:

$$w_t = \alpha x_t \qquad \text{for } t = 1, \ldots, T, \tag{12.7}$$

where the parameter $\alpha \ (> 0)$ denotes the number of working hours required to produce one unit of the product. The labor time w_t can be supplied by regular employees during regular working hours, or by means of overtime, and also by means of temporary employees. In the next section, we will determine the number of regular employees required for the production in each period during regular working hours. We do not consider restrictions on the machine capacity. So it is assumed that in each period the production is feasible, provided that the number of regular employees is sufficiently large.

12.2 Regular working hours

In this section we consider the situation where the production during the planning period is done by only regular employees. So, we do not consider overtime. Moreover, we do not take into account holidays, illness, or idle time. Therefore, the so-called *regular production* is constant in each period, and we define:

$$x = \text{the regular production in any period.}$$

In the case of regular production, we have that:

$$x = x_t \qquad \text{for each } t = 1, \dots, T. \tag{12.8}$$

The question is how to realize the production with a minimum number of regular employees. Based on (12.6), (12.7), and (12.8), we may write the following LO-model to answer that question:

$$\begin{aligned}
\min \ & x \\
\text{s.t.} \ & tx \geq a_t \qquad \text{for } t = 1, \dots, T \\
& x \geq 0.
\end{aligned} \tag{PP1}$$

The minimum number of regular employees is given by the optimal value x^* of the decision variable x in (PP1). Using our assumption that $a_T > 0$, it follows directly that this optimal value x^* satisfies:

$$x^* = \max\left\{ \frac{a_t}{t} \mid t = 1, \dots, T \right\} = \frac{a_{t^*}}{t^*}, \tag{12.9}$$

with t^* the value of t for which a_t/t attains its maximum value.

Example 12.2.1. *As an illustration, consider the following example. The relevant data are listed in Table 12.1. The periods are the twelve months in a year, i.e., $T = 12$. The largest number in the column labeled 'a_t/t' of Table 12.1 is 29, so that $x^* = 29$. Note that σ_{12} is the desired initial inventory for the next planning period. In Table 12.1, we assume that $d_{13} = d_1$, so that $\sigma_{12} = 0.2 \times 30 = 6$. This means that the actual total demand is 416 units. The columns labeled 'tx^*' and '$tx^* - a$' follow by taking $x^* = 29$. In Figure 12.1 we have drawn the graphs of a_t and of tx^*. Using formula (12.2), it follows that:*

$$s_t = s_0 + tx^* - \sum_{k=1}^{t} d_k = tx^* - a_t + \sigma_t,$$

and so:

$$tx^* - a_t = s_t - \sigma_t,$$

$s_0 = 140, \gamma = 0.2$							
t	d_t	σ_t	$\sum_{k=1}^{t} d_k$	a_t	a_t/t	tx^*	$tx^* - a_t$
1	30	8	30	-102	-102	29	131
2	40	14	70	-56	-28	58	114
3	70	12	140	12	4	87	75
4	60	14	200	74	18.5	116	42
5	70	8	270	138	27.6	145	7
6	40	4	310	174	29	174	0
7	20	4	330	194	27.7	203	7
8	20	2	350	212	26.5	232	20
9	10	2	360	222	24.7	261	39
10	10	4	370	234	23.4	290	56
11	20	4	390	254	23.1	319	65
12	20	6	410	276	23	348	72

Table 12.1: Input data.

which is the additional inventory on top of the safety stock in month t, or the overproduction *in month t. In month 6, the overproduction is zero, while at the end of the planning period the overproduction is 72 units.*

If the total production Tx^* equals the desired production a_T, then the production exactly matches the total demand for the planning period. However, if $Tx^* > a_T$, then too much is produced. This is usually not a desired situation, particularly because a_T already contains the safety stock demanded at the end of the planning period. If one does not want to produce the surplus $Tx^* - a_T$, then some employees should not take part in the production. In other words, we accept *idle time*. In order to minimize the idle time, we need to solve model (PP2) below.

Define:

x = the number of units that can be produced when all employees are working during the regular working hours; i.e., the regular production.

In order to plan the production process with the least possible number of employees, the following LO-model has to be solved.

$$
\begin{aligned}
\min \ & x \\
\text{s.t.} \ & x_1 + \ldots + x_t \geq a_t && \text{for } t = 1, \ldots, T-1 \\
& x_1 + \ldots + x_T = a_T \\
& 0 \leq x_t \leq x && \text{for } t = 1, \ldots, T.
\end{aligned}
\qquad \text{(PP2)}
$$

Let $\bar{t}$ be such that $\max\{a_t/t \mid t = 1, \ldots, T\} = a_{\bar{t}}/\bar{t}$, and let $\bar{x} = a_{\bar{t}}/\bar{t}$. An optimal solution of (PP2) is easy to find if $a_t - a_{t-1} \leq \bar{x}$ for $t = \bar{t}+1, \ldots, T$. It can be easily verified that these conditions hold for the input data of Table 12.1, where $\bar{t} = 6$; see Exercise 12.7.5. Let x^* be the optimal value of the decision variable x. Since $tx^* \geq x_1 + \ldots + x_t \geq a_t$,

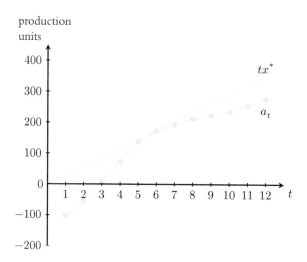

Figure 12.1: Actual versus desired production for model (PP1).

we find that $x^* \geq \bar{x}$. It can be easily checked that an optimal solution in this case is (see Exercise 12.7.5):

$$x^* = \bar{x}, \text{ and } x_t^* = \begin{cases} \bar{x} & \text{if } t \in \{1, \ldots, \bar{t}\} \\ a_t - a_{t-1} & \text{if } \{t = \bar{t}+1, \ldots, T\}. \end{cases}$$

Example 12.2.2. *An optimal solution of model* (PP2), *using the data of Table 12.1, can now readily be calculated:* $x^* = x_1^* = \ldots = x_6^* = 29$, $x_7^* = 20$, $x_8^* = 18$, $x_9^* = 10$, $x_{10}^* = 12$, $x_{11}^* = 20$, $x_{12}^* = 22$. *This means that, in each period, sufficiently many employees are available to produce* 29 *units of the product during regular hours. Since in periods* 7, \ldots, 12, *the production level is lower than that, some employees will have nothing to do during those periods. In Exercise 12.7.2, the reader is asked to check that the given optimal solution is not unique.*

Since α is the number of working hours required for the production of one unit, it follows that the idle time in period t equals $\alpha(x - x_t)$. The total idle time for the whole planning period is therefore $\alpha(x-x_1)+\ldots+\alpha(x-x_T) = \alpha Tx-\alpha(x_1+\ldots+x_T) = \alpha(Tx-a_T)$. Since α, T, and a_T are known parameters, it follows that the objective of (PP2) can be replaced by $\min(\alpha Tx - \alpha a_T)$. So, as soon as the minimum value of x^* in (PP2) is found, we also know the minimum idle time $\alpha Tx^* - \alpha a_T$. The total number of working hours is therefore αx^*.

Example 12.2.3. *Using the input data of Table 12.1, we found that* $x^* = 29$, *and so* $\alpha x^* = 29\alpha$. *Assuming that the number of working hours in one month is* 180 *hours, it follows that the minimum number of regular employees required to take care of the production in the case of model* (PP2) *is* $\alpha x^*/180 = 0.16\alpha$. *In case the company does not want to work with part-time employees,* 0.16α *has to be an integer. When this constraint is taken into consideration, the model becomes a mixed integer LO-model; see Section 12.5.*

12.3 Overtime

A possible way of avoiding idle time is to control the fluctuations in demand by using overtime. This is of course only possible in case the employees are willing to work more than the regular working hours. The management could make overtime attractive by raising the hourly wage during overtime hours. We assume that all employees have the same hourly wage during regular hours, and the hourly wages during overtime are β (> 1) times the hourly wage during the regular working hours. Define:

$$c = \text{the hourly wage per employee } (c > 0).$$

If x is again the production by regular employees during regular working hours, then the total labor costs per period for regular hours is $c\alpha x$. Moreover, the company needs $x_t - x$ hours of overtime during period t, which means additional labor costs of $c\alpha\beta(x_t - x)$. Hence, the total wage cost over the whole planning period is:

$$c\alpha x T + \sum_{t=1}^{T} c\alpha\beta(x_t - x).$$

It is the interest of the management to know how the production can be realized with the total wage costs as low as possible. This problem can be formulated as the following LO-model.

$$
\begin{aligned}
\min \ & c\alpha T x + \sum_{t=1}^{T} c\alpha\beta(x_t - x) \\
\text{s.t.} \ & x_1 + \ldots + x_t \geq a_t \qquad \text{for } t = 1, \ldots, T-1 \qquad \text{(PP3)} \\
& x_1 + \ldots + x_T = a_T \\
& 0 \leq x \leq x_t \qquad \text{for } t = 1, \ldots, T.
\end{aligned}
$$

Note the difference: we have $x_t \leq x$ in (PP2), whereas we have $x_t \geq x$ in (PP3). By using $x_1 + \ldots + x_T = a_T$, the objective function in (PP3) can be rewritten as $c\alpha\beta a_T - c\alpha x T$. This means that, since $\alpha > 0$, $\beta > 1$, and $c > 0$, instead of (PP3), we may equivalently solve the following LO-model.

$$
\begin{aligned}
\max \ & x \\
\text{s.t.} \ & x_1 + \ldots + x_t \geq a_t \qquad \text{for } t = 1, \ldots, T-1 \\
& x_1 + \ldots + x_T = a_T \qquad\qquad\qquad\qquad \text{(PP4)} \\
& 0 \leq x \leq x_t \qquad \text{for } t = 1, \ldots, T.
\end{aligned}
$$

Note that the parameters α, β, and c do not appear in model (PP4), and hence the optimal solution does not depend on their values. We will show that the optimal objective value x^* of (PP4) satisfies:

$$x^* = \min\left\{ \frac{a_T}{T}, \ \min_{t=1,\ldots,T-1} \frac{a_T - a_t}{T - t} \right\}. \qquad (12.10)$$

First observe that for $t = 1, \ldots, T-1$, it holds that $(T-t)x^* \leq x^*_{t+1} + \ldots + x^*_T \leq a_T - a_t$, so that $x^* \leq \min\{(a_T - a_t)/(T-t) \mid t = 1, \ldots, T-1\}$. Moreover, from

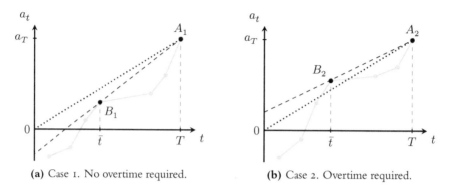

(a) Case 1. No overtime required.   **(b)** Case 2. Overtime required.

Figure 12.2: Two cases: overtime versus no overtime.

$a_T = x_1^* + \ldots + x_T^* \geq Tx$, it follows that $x^* \leq a_T/T$. Therefore,

$$x^* \leq \min\left\{ \frac{a_T}{T},\ \min_{t=1,\ldots,T-1} \frac{a_T - a_t}{T - t} \right\}.$$

Two cases can now be considered:

Case 1: $a_t/t \leq a_T/T$ for each $t \in \{1,\ldots,T-1\}$.

Case 2: $a_t/t > a_T/T$ for some $t \in \{1,\ldots,T-1\}$.

In Figure 12.2, the two situations are schematically depicted. In case 1, the graph of a_t is completely 'below' the line $\mathbf{0}A_1$; see Figure 12.2(a). It can be easily verified that $a_t/t \leq a_T/T$ is equivalent to $a_T/T \leq (a_T - a_t)(T - t)$ (see Exercise 12.7.9), and that:

$$x^* = x_1^* = \ldots = x_T^* = a_T/T$$

is an optimal solution of (PP4) in case 1. Clearly, no overtime is required, the production can be achieved during regular working hours.

In case 2, the average desired production during the first t periods exceeds the average desired total production. Now, there is a point on the graph of a_t that is 'above' the line $\mathbf{0}A_2$; see Figure 12.2(b).

Define $\bar{t}$ as the point for which:

$$\min_{t=1,\ldots,T-1} \frac{a_T - a_t}{T - t} = \frac{a_T - a_{\bar{t}}}{T - \bar{t}}.$$

Since $a_{\bar{t}}/\bar{t} > a_T/T$ is equivalent to $(a_T - a_{\bar{t}})(T - \bar{t}) < a_T/T$, the optimal objective value of (PP4) is:

$$x^* = (a_T - a_{\bar{t}})(T - \bar{t}).$$

For $x_{\bar{t}+1},\ldots,x_T$ we can take as an optimal solution $x_{\bar{t}+1}^* = \ldots = x_T^* = x^*$. For $t = 1,\ldots,\bar{t}$, the situation is a bit more complicated. If $a_t/t \leq a_{\bar{t}}/\bar{t}$ for each $t = 1,\ldots,\bar{t}$, then we can take $x_1^* = \ldots = x_{\bar{t}}^* = a_{\bar{t}}/\bar{t}$, which means that the graph of a_t 'left of' $\bar{t}$ in

Figure 12.2(b) is completely 'below' the line $\mathbf{0}B_2$. So in this particular case, all the overtime work can be done in the first $\bar{t}$ periods. The overproduction during the first $\bar{t}$ periods is then equal to:

$$\frac{a_{\bar{t}}}{\bar{t}} - \frac{a_T - a_{\bar{t}}}{T - \bar{t}}.$$

The case that the graph of a_t is somewhere 'above' $\mathbf{0}B_2$ can be considered similarly as the situation above when taking $T := \bar{t}$, i.e., $\bar{t}$ is taken as the last period instead of T. It is left to the reader to check that the production function is a piecewise linear function; see Exercise 12.7.10.

Example 12.3.1. *In the case of the data of Table 12.1, we are in the situation that overtime can be used to avoid the idle time. A possible optimal solution of model (PP4) is: $x^* = 16$, $x_1^* = 100$, $x_2^* = \ldots = x_{12}^* = 16$ (with $\bar{t} = 8$). Note that in this solution all the overtime is carried out in the first period; the overtime production is $100 - 16 = 84$ units. Taking $\alpha = 8$ and $c = 10$, then the total wage cost is $\$28,800$, which is $\$15,360$ for the regular employees and $\$13,440$ for the overtime hours.*

There are several reasons to be careful when designing a model like model (PP4). For instance, if $a_T = a_{T-1}$ then there is no demand in period T. It then follows from (12.10) that $x^* = 0$, which means that the whole production should be done in overtime. It is left to the reader to argue that this result is not acceptable, and how the model can be adapted such that this situation is avoided; see also Exercise 12.7.11.

12.4 Allowing overtime and idle time

The models in the previous sections take two different approaches to the production planning process. Model (PP2) in Section 12.2 requires that all production is carried out during regular hours, while model (PP4) in Section 12.3 requires that employees are always busy during their regular hours (i.e., they have no idle time).

A more realistic model tries to find a balance between the amount of idle time for employees during regular hours, and the amount of overtime. We take as the criterion for this balance the total labor costs. In that case the problem can again be formulated as an LO-model. For $t = 1, \ldots, T$, define:

x = the number of units to be produced in each period with regular hour labor;
y_t = the number of units to be produced in period t with overtime labor.

Then, the problem of minimizing labor costs can be formulated as follows:

$$\min \; c\alpha T x + \sum_{t=1}^{T} c\alpha\beta y_t$$

$$\begin{aligned}
\text{s.t.} \quad & x_1 + \ldots + x_t \geq a_t && \text{for } t = 1, \ldots, T-1 \\
& x_1 + \ldots + x_T = a_T \\
& x + y_t \geq x_t && \text{for } t = 1, \ldots, T \\
& x_t, y_t \geq 0 && \text{for } t = 1, \ldots, T \\
& x \geq 0.
\end{aligned}$$
(PP5)

Since $\alpha > 0$ and $c > 0$, we may rewrite the objective of this model as:

$$\min \; T x + \sum_{t=1}^{T} \beta y_t.$$

This means that, as with the model in Section 12.3, the values of the parameters α and c are irrelevant. However, the parameter β (i.e., the ratio between the overtime hourly wage and regular time hourly wage) remains in the objective function and therefore β plays a role in the optimal solution.

Example 12.4.1. *Suppose that $\beta = 1.7$. Since α and c are irrelevant, let us take $\alpha = c = 1$. We use a computer package to find an optimal solution of model (PP5), using the data of Table 12.1. The optimal solution that we find satisfies $x^* = 18$, and the values of x_t^* and y_t^* are given in the following table:*

t	1	2	3	4	5	6	7	8	9	10	11	12
x_t^*	84	18	18	18	18	18	20	18	18	18	18	10
y_t^*	66	0	0	0	0	0	2	0	0	0	0	0

The corresponding optimal objective value is 331.6. It can be seen from the table that the overtime production is carried out during periods 1 and 7. Moreover, period 12 is the only period in which idle time occurs.

In model (PP4), the optimal number of products produced during regular hours is 16, two fewer than the optimal solution of (PP5). So, the optimal solution of (PP5) shows that it is cheaper to produce two more products during regular hours (i.e., to hire more regular workers), at the expense of having some idle time in period 12. In fact, it can be checked that the labor costs of producing 16 products during regular hours, rather than 18, are 334.8, which is about 1% more expensive.

Similarly, in (PP2), the optimal number of products produced during regular hours is 29. It can be checked that the labor costs of producing 29 products during regular hours, rather than 18, are 348.0, which is about 5% more expensive.

Note that (PP4) and (PP5) do not specify an upper bound on the amount of overtime. This may not be correct, since employees may be willing to work only a certain amount of

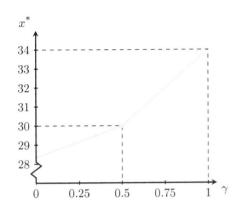

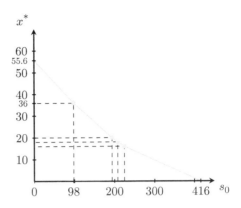

Figure 12.3: Perturbation function for γ in model (PP1).

Figure 12.4: Perturbation function for s_0 in model (PP1).

overtime. For example, they may be willing to spend at most 20% of the regular working hours as overtime. This constraint can be included in (PP4) as:

$$x \leq x_t \leq 1.2x \qquad \text{for } t = 1, \ldots, T.$$

On the other hand, (PP4) may now become infeasible. For instance when using the input data of Table 12.1, then in fact model (PP4) is infeasible. In fact, this model can only be solved when the percentage of overtime is at least 82%. This is a rather high percentage, and so overtime does not seem to be an appropriate solution tool in this particular case. In the case of model (PP5), the maximum fraction of overtime hours can be incorporated by including the following set of constraints:

$$y_t \leq 0.20x \qquad \text{for } t = 1, \ldots, T.$$

12.5 Sensitivity analysis

In this section we will carry out sensitivity analysis for a number of parameters of the models discussed in the previous sections. It is assumed that $d_{13} = 30$, i.e., the demand in the first month of the next planning period is 30. In the case of model (PP1), it is of interest to know how the choice of the value of the parameter γ in (12.4) influences the minimum number of regular employees, i.e., the minimum production level x^*. In Figure 12.3 this relationship is depicted. It is seen that if the value of γ increases, i.e., all safety inventories increase by the same percentage, then the value of x^* increases as well. For $\frac{1}{2} \leq \gamma < 1$, the increase of the value of x^* is sharper than for $0 < \gamma \leq \frac{1}{2}$. In Exercise 12.7.3, the reader is asked to show that the kink point of this perturbation function occurs at $\gamma = \frac{1}{2}$, and to explain why the graph of the perturbation function is steeper for $\frac{1}{2} \leq \gamma < 1$ than for $0 < \gamma \leq \frac{1}{2}$.

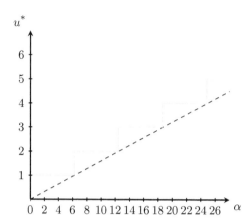

Figure 12.5: The labor input versus the number of regular employees in model (PP2).

The perturbation function for s_0 in model (PP1) is depicted in Figure 12.4. Notice that it follows from (12.9) that the optimal objective value of (PP1) as a function of s_0 satisfies:

$$x^*(s_0) = \max\left\{\frac{\sum_{k=1}^t d_k - \sigma_t + s_0}{t} \,\middle|\, t = 1, \ldots, T\right\}$$

$$= \max\left\{\frac{38-s_0}{1}, \frac{84-s_0}{2}, \frac{152-s_0}{3}, \frac{214-s_0}{4}, \frac{278-s_0}{5}, \frac{314-s_0}{6},\right.$$

$$\left.\frac{334-s_0}{7}, \frac{352-s_0}{8}, \frac{362-s_0}{9}, \frac{374-s_0}{10}, \frac{394-s_0}{11}, \frac{416-s_0}{12}\right\}.$$

The reader is asked in Exercise 12.7.4 to check that:

$$x^*(s_0) = \begin{cases} \frac{278-s_0}{5} & \text{if } 0 \le s_0 \le 98 \\ \frac{314-s_0}{6} & \text{if } 98 < s_0 \le 194 \\ \frac{334-s_0}{7} & \text{if } 194 < s_0 \le 208 \\ \frac{352-s_0}{8} & \text{if } 208 < s_0 \le 224 \\ \frac{416-s_0}{12} & \text{if } s_0 > 224. \end{cases} \qquad (12.11)$$

If $s_0 = 0$, then the total demand plus the initial inventory σ_{12} for the next planning period (416 units) needs to be satisfied completely by the production. It follows from the expression above that $x^*(0) = \frac{278}{5} = 55.6$. If $s_0 = 416$, then the initial inventory is enough, and nothing needs to be produced. As the expression for $x^*(s_0)$ shows, the kink points of the perturbation graph in Figure 12.4 occur at $s_0 = 98$, $s_0 = 194$, $s_0 = 208$, and $s_0 = 224$. For $s_0 = 98$, a_t/t is maximum for $t = 5, 6$; for $s_0 = 194$, a_t/t is maximum for $t = 6, 7$; for $s_0 = 208$, a_t/t is maximum for $t = 7, 8$; for $s_0 = 224$, a_t/t is maximum for $t = 8, 12$.

The perturbation functions of γ and s_0 in the case of model (PP2) are the same as the ones for model (PP1), since the optimal solution x^* for both models is determined by the tangent tx^* to the graph of a_t; see Figure 12.1. The relationship between the minimum

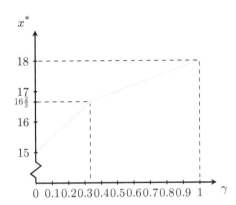

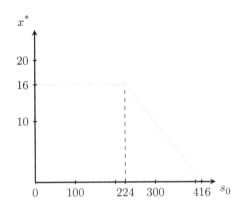

Figure 12.6: Perturbation function for γ in model (PP4).

Figure 12.7: Perturbation function for s_0 in model (PP4).

number $u^* = \alpha x^*/180$ of regular employees and the number α of working hours per unit production is the dotted line determined by $u^* = 0.16\alpha$ (since $x^* = 29$) in Figure 12.5. In case we demand that u^* is an integer, then the graph becomes the 'staircase' in Figure 12.5. Clearly, if the labor input increases, then the number of regular employees must increase in order to reach the demanded production.

In the case of model (PP4) the perturbation function for γ is more or less constant. The graph of Figure 12.6 has a kink at $\gamma = \frac{1}{3}$; see also Exercise 12.7.12.

The perturbation function for s_0 for model (PP4), depicted in Figure 12.7, is constant for $0 \leq s_0 \leq 224$, because for these values of s_0 the right hand side values in model (PP4) all change by the same amount, and so $x^* = 16$ for $0 \leq s_0 \leq 224$.

When the value of s_0 increases to 224, then the line through $\begin{bmatrix} 8 & 212 \end{bmatrix}^{\mathsf{T}}$ and $\begin{bmatrix} 12 & 276 \end{bmatrix}^{\mathsf{T}}$ with slope $x^* = 16$ in Figure 12.1 moves 'downwards' until $\begin{bmatrix} 0 & 0 \end{bmatrix}^{\mathsf{T}}$ is on this line. This happens for $s_0 = a_{12} - 12x^* + 140 = 224$. Increasing the value of s_0 from 224, then x^* is the slope of the line through $\begin{bmatrix} 0 & 0 \end{bmatrix}^{\mathsf{T}}$ and $\begin{bmatrix} 12 & a_{12} \end{bmatrix}$. For $s_0 = 416$, we have that $a_{12} = 0$, and the values of $a_1, \ldots, a_{11}$ are all negative. For $s_0 > 416$, model (PP4) is infeasible.

We conclude the section with the perturbation function for β in model (PP5). As was mentioned, the optimal solution depends on the value of β. Since β is the ratio between regular working hours and overtime hours, only values of β with $\beta \geq 1$ realistically make sense. However, as a model validation step we can also consider values below 1. Figure 12.8 shows how the optimal objective value changes as β changes, with $\beta \in [0.9, 2.5]$. Figure 12.9 shows the corresponding optimal values of x. Clearly, if $\beta < 1$, then overtime hours are cheaper than regular hours, and therefore in the optimal solution the number of regular hours equals zero. If, on the other hand, $\beta > 2$, then it can be seen that overtime hours are too expensive, and hence the optimal solution satisfies $x^* = 29$, i.e., it is optimal to use no overtime at all.

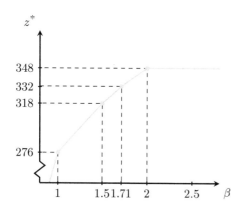

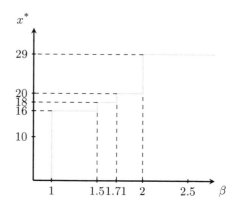

Figure 12.8: Perturbation function for β in model (PP5).

Figure 12.9: x^* for different values of β in model (PP5).

12.6 GMPL model code

12.6.1 Model (PP2)

```
 1       T;
 2       s0;
 3       gamma;
 4       d{1..T};
 5       sigma{t    1..T} := gamma * d[1 + t     T];
 6       a{t    1..T} :=       {k     1..t} d[k] + sigma[t] — s0;
 7
 8    xstar    >= 0;
 9    x{1..T} >= 0;
10
11         z:
12   xstar;
13           acons{t    1..T—1}:
14      {k    1..t} x[k] >= a[t];
15           aconsT:
16      {k    1..T} x[k] = a[T];
17           xcons{t    1..T}:
18   x[t] <= xstar;
19
20    ;
21
22      T     := 12;
23      s0    := 140;
24      gamma := 0.2;
25    :  d    :=
26    1   30      2   40      3   70      4   60      5   70      6   40
27    7   20      8   20      9   10     10   10     11   20     12   20 ;
28    ;
```

12.6.2 Model (PP3)

```
1          T;
2          s0;
3          c;
4          alpha;
5          gamma;
6          d{1..T};
7          sigma{t    1..T} := gamma * d[1 + t     T];
8          a{t    1..T}     :=     {k    1..t} d[k] + sigma[t] - s0;
9
10    xstar    >= 0;
11    x{1..T} >= 0;
12
13         z:
14    c * alpha * T * xstar + 2 * c * alpha *     {t    1..T} (x[t] - xstar);
15              acons{t    1..T-1}:
16      {k    1..t} x[k] >= a[t];
17              aconsT:
18      {k    1..T} x[k] = a[T];
19              xcons{t    1..T}:
20    xstar <= x[t];
21
22    ;
23
24       T     := 12;
25       s0    := 140;
26       alpha := 8;
27       c     := 10;
28       gamma := 0.2;
29      : d    :=
30    1   30    2   40    3   70    4   60    5   70    6   40
31    7   20    8   20    9   10   10   10   11   20   12   20 ;
32    ;
```

12.7 Exercises

Exercise 12.7.1. Determine the dual of model (PP1), as well as its optimal solution. Why is the optimal dual solution nondegenerate and unique?

Exercise 12.7.2. Show that the optimal solution of model (PP2) found in Example 12.2.2 is not the only optimal solution, i.e., the model has multiple optimal solutions.

Exercise 12.7.3. Consider the perturbation function of the parameter γ in model (PP1). The function is depicted Figure 12.3.

(a) Show that the kink point of the perturbation function is located at $\gamma = \frac{1}{2}$.

(b) Show that the perturbation function is steeper for $\frac{1}{2} \leq \gamma < 1$ than for $0 < \gamma \leq \frac{1}{2}$.

Exercise 12.7.4. Show that equation (12.11) holds.

Exercise 12.7.5. Define $\max\{a_t/t \mid t = 1, \ldots, T\} = a_{\bar{t}}/\bar{t} = \bar{x}$, as in Section 12.2.

(a) Show that, for the data in Table 12.1, $a_t - a_{t-1} \leq \bar{x}$ for $t = \bar{t} + 1, \ldots, T$.

(b) Under the conditions of (a), show that an optimal solution of model (PP2) in Section 12.2 is:

$$x^* = \bar{x}, \quad \text{and} \quad x_t^* = \begin{cases} \bar{x} & \text{if } t \in \{1, \ldots, \bar{t}\} \\ a_t - a_{t-1} & \text{if } t \in \{\bar{t} + 1, \ldots, T\}. \end{cases}$$

Verify the optimality of the solution of model (PP2) given in Section 12.2.

Exercise 12.7.6. Show that the shadow price of the constraint $x_1 + \ldots + x_T = a_T$ of model (PP2) in Section 12.2 is nonnegative. (Hint: prove that model (PP2) is equivalent to the model with $x_1 + \ldots + x_T = a_T$ replaced by $x_1 + \ldots + x_T \geq a_T$.)

Exercise 12.7.7. Draw the perturbation function for σ_T in $x_1 + \ldots + x_T = a_T$ of model (PP2) in Section 12.2. Show that $x^* = 29$ for $-16 \leq \sigma_T \leq 78$, and $x^* = 29 + \frac{1}{12}(\sigma_T - 78)$ for $\sigma_T \geq 78$. Why is the model infeasible for $\sigma_T < -16$?

Exercise 12.7.8. Consider model (PP2), extended with slack variables.

(a) Dualize this model.

(b) Why is the optimal dual solution degenerate and unique?

(c) Calculate the optimal dual solution. (Hint: use the complementary slackness relations.)

(d) Explain why $x_1 + \ldots + x_{12} = a_{12}$ has shadow price 0.

(e) Explain why the strong complementary slackness relation does not hold for $x_1 + \ldots + x_{12} = a_{12}$.

Exercise 12.7.9. Show that $a_t/t \leq a_T/T$ is equivalent to $a_T/T \leq (a_T - a_t)(T - t)$; see Section 12.3. Calculate the values of t for which $a_t/t \leq a_T/T$ for the data of Table 12.1; give an economic interpretation of these values of t.

Exercise 12.7.10. Consider Figure 12.2. The case that the graph of a_t is somewhere 'above' the line $\mathbf{0}A_2$ can be treated similarly to the situation described in Section 12.3, when taking $\bar{t}$ instead of T; i.e., $\bar{t}$ is taken as the last period. Draw an example of such a graph, and show that the production function is a piecewise linear function.

Exercise 12.7.11. Consider model (PP4) in Section 12.3. Assume that $a_T = a_{T-1}$.

(a) Show that the safety inventory at the end of the planning period is zero, and that all production can be done in overtime.

(b) How can model (PP4) be adapted to avoid the unacceptable situation of (a)?

Exercise 12.7.12. Consider model (PP4) in Section 12.3. Draw the graphs of a_t for $\gamma = 0$ and $\gamma = 1$. Show that $x^*(\gamma{=}0) = 15$, and that $x^*(\gamma{=}1) = 18$; see Figure 12.6. Why does the perturbation function for Figure 12.6 have a kink at $\gamma = \frac{1}{3}$?

Exercise 12.7.13. Consider model (PP4) in Section 12.3.

(a) Calculate an optimal solution different from the one in Section 12.3.

(b) Show that the shadow prices of the constraints $x_1 + \ldots + x_t \geq a_t$ for $t = 1, \ldots, T{-}1$ are all nonpositive, while the shadow price of $x_1 + \ldots + x_T = a_T$ is nonnegative. (Hint: show that (PP4) is equivalent to the model with $x_1 + \ldots + x_T \leq a_T$ instead of $x_1 + \ldots + x_T = a_T$.)

(c) Use Theorem 5.6.1 to show that the optimal dual solution is degenerate and unique.

(d) Dualize model (PP4), and calculate its optimal solution.

Exercise 12.7.14. Consider model (PP4) in Section 12.3. Show that, for any optimal solution, the production level x_T^* is never carried out in overtime, i.e., it is cheaper to produce during the last period in regular working time than in overtime.

Exercise 12.7.15. The optimal solution of model (PP4) given in Example 12.3.1 has a large amount of overtime production during the first period. Perhaps a more desirable solution is: $x^* = 16$, $x_1^* = \ldots = x_6^* = 29$, $x_7^* = 20$, $x_8^* = 18$, $x_9^* = \ldots = x_{12}^* = 16$. In this solution, the overtime production is distributed more equally over the periods. Modify model (PP4) so that its optimal solution has more equally distributed overtime production. Explain how your model achieves this.

CHAPTER 13

Production of coffee machines

Overview

In this chapter we will describe the situation where more than one product is produced. This case study is based on the M.S. thesis (University of Groningen, The Netherlands): JAN WEIDE, *The Coffee Makers Problem; Production Planning at Philips* (in Dutch). The data used in this section are fictional.

13.1 Problem setting

The multinational company PLS produces coffee machines in 21 different designs which are sold in several countries. The production of coffee machines takes place on five conveyor belts, each conveyor belt having the capability of producing any design. The capacity of each conveyor belt is 2,600 coffee machines per week. At the beginning of the year, the central planning department of PLS makes demand forecasts. These forecasts span the whole year and specify the forecasted weekly demand for each design. In addition to these forecasts, there are initial inventories and backlogs (i.e., accumulated unsatisfied demand from the previous year that needs to be satisfied this year) that the company needs to take into account.

For illustration purposes, we will consider a planning period of fourteen weeks instead of one year. Table 13.1 lists the initial inventories and the forecasted weekly demand for all 21 coffee machine designs. The negative numbers in the s_{i0} column represent backlogs.

The various coffee machine designs are labeled $1, \ldots, m$, and the periods are labeled $t = 1, \ldots, T$. Thus, $m = 21$ and $T = 14$ in our case. For $i = 1, \ldots, m$ and $t = 1, \ldots, T$, we introduce the following parameters and decision variables (see also Section 12.1):

$$
\begin{aligned}
d_{it} &= \text{the demand for design } i \text{ in period } t; \\
x_{it} &= \text{the production of design } i \text{ in period } t; \\
s_{it} &= \text{the inventory of design } i \text{ at the end of period } t;
\end{aligned}
$$

s_{i0} = the initial inventory of design i, i.e., the inventory at the beginning of the planning period.

The decision variables of the model that we will construct in this chapter are x_{it} and s_{it}, whereas the values of d_{it} and s_{i0} are assumed to be known for $i = 1, \ldots, m$ and $t = 1, \ldots, T$.

13.2 An LO-model that minimizes backlogs

We first discuss the situation in which the objective is to minimize the total backlog during the planning period, so that the demand is satisfied 'as much as possible'. The production schedule will be based on this objective. In each period t, the total production is equal to $\sum_{i=1}^{m} x_{it}$, and cannot exceed the available production capacity. Let p be the number of available conveyor belts (i.e., $p = 5$), and let c be the per-period production capacity of one conveyor belt (i.e., $c = 2{,}600$). Then the following inequalities must hold:

$$\sum_{i=1}^{m} x_{it} \leq cp \qquad \text{for } t = 1, \ldots, T. \tag{13.1}$$

Next, we need to deal with inventory. If the inventory of design i at the end of period $t - 1$ is $s_{i,t-1}$, then the inventory at the end of period t will be $s_{i,t-1}$ plus the number x_{it} of coffee machine of design i produced in period t, minus the number of such coffee machines that are sold in period t. Therefore we have the following *inventory equations*:

$$s_{it} = s_{i,t-1} + x_{it} - d_{it} \qquad \text{for } i = 1, \ldots, m, \text{ and } t = 1, \ldots, T,$$

or, equivalently:

$$s_{it} = s_{i0} + \sum_{k=1}^{t} x_{ik} - \sum_{k=1}^{t} d_{ik} \qquad \text{for } i = 1, \ldots, m, \text{ and } t = 1, \ldots, T. \tag{13.2}$$

The value of s_{it} can become positive if too much is produced, and negative if too little is produced (i.e., there is a *backlog*). In order to minimize the total backlog, we introduce the following new variables:

$$s_{it}^{+} = \begin{cases} s_{it} & \text{if } s_{it} \geq 0 \\ 0 & \text{otherwise,} \end{cases} \qquad \text{and} \qquad s_{it}^{-} = \begin{cases} -s_{it} & \text{if } s_{it} \leq 0 \\ 0 & \text{otherwise.} \end{cases}$$

It then follows that:

$$s_{it} = s_{it}^{+} - s_{it}^{-}, \text{ and } s_{it}^{+} \geq 0, \ s_{it}^{-} \geq 0, \text{ for } i = 1, \ldots, m, \text{ and } t = 1, \ldots, T. \tag{13.3}$$

Moreover, we have that either $s_{it}^{+} = 0$ or $s_{it}^{-} = 0$. This can be formulated as:

$$s_{it}^{+} \times s_{it}^{-} = 0 \qquad \text{for } i = 1, \ldots, m, \text{ and } t = 1, \ldots, T. \tag{13.4}$$

If $s_{it}^- > 0$ (hence, $s_{it}^+ = 0$) for some i and t, then s_{it}^- is a backlog. The restrictions stated in (13.4) cannot be included in an LO-model, since they are not linear. Fortunately, we do not need to include constraints (13.4) in the model, because the optimal solution always satisfies (13.4); see Section 1.5.2, where a similar statement is proved.

The objective of minimizing backlogs can be formulated as $\min \sum_{i=1}^m \sum_{t=1}^T s_{it}^-$. However, backlogs of some particular products may be more severe than backlogs of other products. Therefore we introduce the positive weight coefficient c_{it} for s_{it}^-. The LO-model becomes:

$$\min \sum_{i=1}^m \sum_{t=1}^T c_{it}s_{it}^-$$
$$\text{s.t.} \quad (13.1),(13.2),(13.3)$$
$$x_{it} \geq 0 \text{ and integer, for } i = 1,\ldots,m, \text{ and } t = 1,\ldots,T. \tag{13.5}$$

It is not difficult to show that in an optimal solution of (13.5), the restrictions (13.4) are always satisfied; the reader is asked to show this in Exercise 13.7.8.

Table 13.2 and Table 13.3 list the results of computer calculations for the solution of model (13.5), where we have taken $c_{it} = 1$ for all i and t. The first row contains the labels of the fourteen weeks, and the first column the labels of the different designs. The figures in the central part of Table 13.2 are the optimal values of x_{it}; the central part of Table 13.3 contains the optimal values of s_{it}. Note that design 6 is demanded only in week 9, where the total production (1,600 coffee machines) is executed, and there is no inventory or backlog of design 6 during the whole planning period. Design 17 is demanded only in week 1; the production (6,476) in week 1 covers the backlog (2,476) plus the demand (4,000), and there is no inventory or backlog during the planning period.

i	s_{i0}	1	2	3	4	5	6	7	8	9	10	11	12	13	14
1	813	0	0	1500	7400	1300	1200	1000	1000	1000	1000	1000	1300	1200	1300
2	−272	3400	3600	0	9700	1400	700	800	700	700	700	900	1000	900	900
3	−2500	1500	0	5000	0	0	0	0	0	0	0	0	0	0	0
4	0	0	0	0	2900	0	1000	0	400	0	400	0	400	0	0
5	220	800	2600	2400	1200	1300	1200	700	700	700	700	700	400	400	500
6	0	0	0	0	0	0	0	0	0	1600	0	0	0	0	0
7	−800	400	300	300	300	300	200	100	200	200	100	200	200	200	200
8	16	0	1000	2500	3200	600	600	700	600	700	500	600	500	600	300
9	0	0	0	0	2500	0	0	0	1500	0	0	0	0	0	0
10	1028	0	0	0	800	200	200	200	200	200	100	200	200	200	100
11	1333	600	600	900	1700	1000	1200	500	1000	500	1000	500	1200	600	500
12	68	600	0	0	0	0	0	0	0	0	0	0	0	0	0
13	97	1400	0	0	600	300	300	300	500	400	300	200	300	200	100
14	1644	0	1000	0	0	0	300	300	400	300	400	400	300	400	400
15	0	0	0	0	1200	300	100	100	100	100	200	0	100	100	100
16	0	80	0	0	1300	400	400	400	500	400	300	400	300	400	200
17	−2476	4000	0	0	0	0	0	0	0	0	0	0	0	0	0
18	0	0	0	0	0	0	0	0	0	0	2500	0	2500	0	0
19	86	900	0	0	1000	500	200	300	200	200	300	300	300	300	200
20	1640	0	0	0	0	0	1500	0	0	0	0	0	200	0	0
21	0	0	0	0	2250	0	750	0	0	0	0	0	0	0	0

Table 13.1: Initial inventories and demands.

$i \backslash t$	1	2	3	4	5	6	7	8	9	10	11	12	13	14
1	0	0	687	2250	0	0	8650	1000	1000	1000	1000	1300	1200	1300
2	0	7272	0	600	10500	633	0	1567	2200	2000	0	0	0	900
3	3859	0	5141	0	0	0	0	0	0	0	0	0	0	0
4	0	0	0	0	0	3900	0	400	0	400	0	400	0	0
5	0	3180	2400	1200	1300	1200	0	1400	700	700	700	400	400	500
6	0	0	0	0	0	0	0	0	1600	0	0	0	0	0
7	0	1500	300	0	600	200	0	300	500	0	0	400	0	200
8	0	984	2500	3200	600	600	0	1300	700	1600	0	0	600	300
9	0	0	0	0	0	0	2500	1500	0	0	0	0	0	0
10	0	0	0	0	0	172	0	400	200	100	200	200	200	100
11	0	0	1372	0	0	3045	750	1000	500	1000	500	1200	600	500
12	532	0	0	0	0	0	0	0	0	0	0	0	0	0
13	1303	0	600	0	0	600	300	500	400	500	0	300	200	100
14	0	0	0	0	0	0	0	356	300	800	0	300	400	400
15	0	0	0	1200	0	400	100	100	100	200	0	100	100	100
16	80	0	0	1300	0	800	400	500	400	1000	0	0	400	200
17	6476	0	0	0	0	0	0	0	0	0	0	0	0	0
18	0	0	0	0	0	0	0	0	0	2500	0	2500	0	0
19	750	64	0	1000	0	700	300	200	200	1200	0	0	0	200
20	0	0	0	0	0	0	0	0	0	0	0	60	0	0
21	0	0	0	2250	0	750	0	0	0	0	0	0	0	0

Table 13.2: Optimal values of x_{it} for model (13.5).

The computer calculations show that the optimal objective value of model (13.5) is $55{,}322$. In Section 13.4 we will perform sensitivity analysis on the number of conveyor belts used for the production. The result shows that the total backlog is rather sensitive to the number of conveyor belts.

13.3 Old and recent backlogs

In the formulation of model (13.5), we have not taken into account the fact that some backlogs may exist already for a long time and others are more recent. When the production of a certain design of coffee machine is a couple of weeks behind its due date, it does not seem acceptable to proceed with the production of a design with a more recently expired due date. We will discuss how to include the requirement of first satisfying 'old' backlogs in the model. To that end, we define the concepts of 'old' and 'recent' backlogs. Take any $i \in \{1, \ldots, m\}$ and $t \in \{2, \ldots, T\}$. An *old backlog* of design i in period t occurs when there is a backlog $s^-_{i,t-1}$ in period $t-1$ and the production x_{it} in period t is not enough to eliminate this backlog; i.e., if $s^-_{i,t-1} > 0$ and $s^-_{i,t-1} > x_{it}$, then $s^-_{i,t-1} - x_{it}$ is the old backlog of design i in period t; in all other cases the old backlog is defined as zero. A *recent backlog* of design i in period t occurs when there is a backlog s^-_{it} in period t and the difference between this backlog and the old backlog in t is nonnegative; i.e., if $s^-_{it} > 0$ and $s^-_{it} - (s^-_{i,t-1} - x_{it}) > 0$, then the recent backlog in period t is $s^-_{it} - s^-_{i,t-1} + x_{it}$, and in all other cases the recent backlog is defined as zero. Hence, the old backlog is $\max\{0, s^-_{i,t-1} - x_{it}\}$, and the recent backlog is $\max\{0, s^-_{it} - \max\{0, s^-_{i,t-1} - x_{it}\}\}$.

$i \setminus t$	1	2	3	4	5	6	7	8	9	10	11	12	13	14
1	813	813	0	−5150	−6450	−7650	0	0	0	0	0	900	0	0
2	−3672	0	0	−9100	0	−67	−867	0	1500	2800	1900	0	0	0
3	−141	−141	0	0	0	0	0	0	0	0	0	0	0	0
4	0	0	0	−2900	−2900	0	0	0	0	0	0	0	0	0
5	−580	0	0	0	0	0	−700	0	0	0	0	0	0	0
6	0	0	0	0	0	0	0	0	0	0	0	0	0	0
7	−1200	0	0	−300	0	0	−100	0	300	200	0	200	0	0
8	16	0	0	0	0	0	−700	0	0	1100	500	0	0	0
9	0	0	0	−2500	−2500	−2500	0	0	0	0	0	0	0	0
10	1028	1028	1028	228	28	0	−200	0	0	0	0	0	0	0
11	733	133	605	−1095	−2095	−250	0	0	0	0	0	0	0	0
12	0	0	0	0	0	0	0	0	0	0	0	0	0	0
13	0	0	600	0	−300	0	0	0	0	200	0	0	0	0
14	1644	644	644	644	644	344	44	0	0	400	0	0	0	0
15	0	0	0	0	−300	0	0	0	0	0	0	0	0	0
16	0	0	0	0	−400	0	0	0	0	700	300	0	0	0
17	0	0	0	0	0	0	0	0	0	0	0	0	0	0
18	0	0	0	0	0	0	0	0	0	0	0	0	0	0
19	−64	0	0	0	−500	0	0	0	0	900	600	300	0	0
20	1640	1640	1640	1640	1640	140	140	140	140	140	140	0	0	0
21	0	0	0	0	0	0	0	0	0	0	0	0	0	0

Table 13.3: Optimal values of s_{it} for model (13.5).

For $i = 1, \ldots, m$ and $t = 2, \ldots, T$, define:

$$r_{it} = s_{i,t-1} + x_{it},$$
$$r_{it} = r_{it}^+ - r_{it}^-, \tag{13.6}$$
$$r_{it}^+ \geq 0, r_{it}^- \geq 0,$$

with $r_{it}^+ = r_{it}$ if $r_{it} \geq 0$ and $r_{it}^+ = 0$ if $r_{it} < 0$, and $r_{it}^- = -r_{it}$ if $r_{it} \leq 0$ and $r_{it}^- = 0$ if $r_{it} > 0$, for $i = 1, \ldots, m$ and $t = 1, \ldots, T$. So r_{it}^+ is the quantity of design i available for the demand at the end of period t.

Actually, in (13.6) we also need $r_{it}^+ \times r_{it}^- = 0$. By arguments similar to the ones used for not including (13.4) in (13.5), we do not need to include $r_{it}^+ \times r_{it}^- = 0$ in (13.7) either; see the remark below model (13.7).

Clearly, for each i and t it holds that:

$$r_{it}^- \qquad = \text{the old backlog at the end of period } t, \text{ and}$$
$$s_{it}^- - r_{it}^- = \text{the recent backlog at the end of period } t.$$

In order to become more familiar with these concepts, the following discussion may be useful. Let i be any design. Assuming that there is backlog of i at the end of period t, i.e., $s_{it} = -s_{it}^- < 0$, we can distinguish the following cases:

(a) There is no backlog at the beginning of period t. Then, $s_{i,t-1} = s_{i,t-1}^+ \geq 0$, and $s_{i,t-1}^- = 0$. If $r_{it}^- > 0$, then $r_{it}^+ - r_{it}^- = s_{it}^+ - s_{i,t-1}^- + x_{it}$ implies that $-r_{it}^- = s_{i,t-1}^+ + x_{it} < 0$, which is not true. Hence, $r_{it}^- = 0$. Since $s_{it}^+ = 0$ and $r_{it}^- = 0$, we have that $s_{i,t-1}^+ + x_{it} - d_{it}$ and $r_{it}^+ = s_{i,t-1}^+ + x_{it}$, which implies that $r_{it}^+ - d_{it} \leq 0$.

Hence, $d_{it} \geq r_{it}^+$, i.e., the demand for design i in period t is at least the available quantity r_{it}^+ in t (namely, $s_{i,t-1} + x_{it} \geq 0$).

(b) There is backlog both at the beginning of period t, but no old backlog at the end of period t. Then $s_{i,t-1}^- > 0, s_{i,t-1}^+ = 0, s_{it}^- > 0, s_{it}^+ = 0$, and $r_{it}^- = 0, r_{it}^+ \geq 0$. Hence, $r_{it}^+ = -s_{i,t-1}^- + x_{it} \geq 0$, and $s_{it}^- = -r_{it}^+ + d_{it} > 0$, and so the demand at the end of period t is larger than r_{it}^+.

(c) There is backlog at the beginning of period t, as well as an old backlog at the end of period t. Then, $s_{i,t-1}^- > 0, s_{i,t-1}^+ = 0$, and $r_{it}^- > 0, r_{it}^+ = 0$. Hence, $-r_{it}^- = -s_{i,t-1}^- + x_{it} \leq 0$, and so $s_{it}^+ - s_{it}^- = -s_{i,t-1}^- + x_{it} - d_{it} \leq 0$. Since $s_{it}^+ = 0$, we find that the recent backlog is $s_{it}^- - r_{it}^- = d_{it} \geq 0$.

The expression 'old backlogs are more severe than recent backlogs', can mathematically be formulated by introducing the parameters γ_1 and γ_2 with $\gamma_1 > \gamma_2 > 0$ in the following LO-model:

$$\min \ \gamma_1 \sum_{i=1}^{m} \sum_{t=1}^{T} c_{it} r_{it}^- + \gamma_2 \sum_{i=1}^{m} \sum_{t=1}^{T} c_{it}(s_{it}^- - r_{it}^-)$$

$$\text{s.t.} \quad (13.1),(13.2),(13.3),(13.6)$$
$$x_{it} \geq 0 \text{ and integer, for } i = 1,\ldots,m, \text{ and } t = 1,\ldots,T.$$

(13.7)

$i \backslash t$	1	2	3	4	5	6	7	8	9	10	11	12	13	14
1	0	0	687	7400	0	428	2300	1772	3000	0	0	3800	0	0
2	0	7131	141	5600	3200	3000	800	700	2300	0	0	2800	0	0
3	2829	1171	5000	0	0	0	0	0	0	0	0	0	0	0
4	0	0	0	0	2900	1000	0	400	400	0	0	400	0	0
5	580	2600	2400	0	2500	1200	700	700	2500	0	0	0	900	0
6	0	0	0	0	0	0	0	0	1600	0	0	0	0	0
7	1200	300	300	0	600	200	100	200	700	0	0	0	400	0
8	0	984	2500	0	3800	600	700	600	2500	0	0	0	700	0
9	0	0	0	0	0	2500	0	1500	0	0	0	0	0	0
10	0	0	0	0	0	172	200	477	0	423	0	0	300	0
11	0	0	767	0	0	3900	500	1500	0	2700	0	0	1100	0
12	532	0	0	0	0	0	0	0	0	0	0	0	0	0
13	1303	0	0	0	0	0	1500	900	0	1000	0	0	0	100
14	0	0	0	0	0	0	0	656	0	1777	0	0	0	123
15	0	0	0	0	0	0	1700	200	0	500	0	0	0	0
16	80	0	0	0	0	0	2500	900	0	1600	0	0	0	0
17	6476	0	0	0	0	0	0	0	0	0	0	0	0	0
18	0	0	0	0	0	0	0	0	0	5000	0	0	0	0
19	0	814	0	0	0	0	2000	700	0	0	1100	0	0	0
20	0	0	0	0	0	0	0	0	0	0	0	60	0	0
21	0	0	1205	0	0	0	0	1795	0	0	0	0	0	0

Table 13.4: Optimal values of x_{it} for model (13.7).

The objective function in (13.7) can be rewritten as:

$$\sum_i \sum_t (\gamma_1 - \gamma_2) c_{it} r_{it}^- + \sum_i \sum_t \gamma_2 c_{it} s_{it}^-.$$

$i \backslash t$	1	2	3	4	5	6	7	8	9	10	11	12	13	14
1	0	813	1500	7400	0	−872	228	1000	3000	2000	1000	3800	2500	1300
2	0	3459	0	5600	−900	700	800	700	2300	1600	900	2800	1800	900
3	0	0	5000	0	0	0	0	0	0	0	0	0	0	0
4	0	0	0	0	0	1000	0	400	400	400	0	400	0	0
5	0	2600	2400	0	1300	1200	700	700	2500	1800	1100	400	900	500
6	0	0	0	0	0	0	0	0	1600	0	0	0	0	0
7	0	300	300	0	300	200	100	200	700	500	400	200	400	200
8	0	1000	2500	0	600	600	700	600	2500	1800	1300	700	900	300
9	0	0	0	0	−2500	0	0	1500	0	0	0	0	0	0
10	0	1028	1028	1028	288	200	200	477	277	500	400	200	300	100
11	0	733	900	0	−1700	1200	500	1500	500	2700	1700	1200	1100	500
12	0	0	0	0	0	0	0	0	0	0	0	0	0	0
13	0	0	0	0	−600	−900	300	900	400	1000	700	500	200	100
14	0	1644	644	644	644	644	344	700	300	1777	1377	977	677	400
15	0	0	0	0	−1200	−1500	100	200	100	500	300	300	200	100
16	0	0	0	0	−1300	−1700	400	900	400	1600	1300	900	600	200
17	0	0	0	0	0	0	0	0	0	0	0	0	0	0
18	0	0	0	0	0	0	0	0	0	5000	2500	2500	0	0
19	0	0	0	0	−1000	−1500	300	700	500	300	1100	800	500	200
20	0	1640	1640	1640	1640	1640	140	140	140	140	140	200	0	0
21	0	0	1205	1205	−1045	−1045	−1795	0	0	0	0	0	0	0

Table 13.5: Optimal values of r_{it} for model (13.7).

The LO-model (13.7) looks rather complicated. However, it is not too difficult to calculate a feasible solution with a low objective value. This can be done as follows. Since the objective is to avoid backlogs, there is a tendency to use the full capacity. Therefore, we replace $\sum_i x_{it} \leq cp$ by $\sum_i x_{it} = cp$. The rewritten objective of (13.7) suggests that it seems profitable to choose many r_{it}^-'s equal to 0. A solution for which many r_{it}^-'s are 0 can be constructed quite easily. Rearrange all designs with respect to the value of $s_{i,t-1}^-$; start with the design with the largest backlog. For the time being, choose $x_{it} = s_{i,t-1}^-$. If $\sum_i s_{i,t-1}^- \leq cp$, then this choice can be done for all designs. Then consider the designs with the largest demand in period t, and choose, according to this new ordering, $x_{it} = s_{i,t-1}^- + d_{it} - s_{i,t-1}^+$. This production can be realized for all designs as long as $\sum_i (s_{i,t-1}^- + d_{it} - s_{i,t-1}^+) \leq cp$. If $\sum_i (s_{i,t-1}^- + d_{it} - s_{i,t-1}^+) > cp$, then the demand of a number of designs cannot be satisfied in period t, because only $s_{i,t-1}^-$ units are produced. If $\sum_i s_{i,t-1}^- > cp$, then for a number of designs even the backlog $s_{i,t-1}^-$ cannot be replenished.

The reader is asked to formalize the above described procedure, and to apply it to the data of Table 13.1. One should take into account that in a number of periods there are no backlogs but only inventories. Also compare this solution with the optimal solution presented below; see Exercise 13.7.5.

In Table 13.4, Table 13.5, and Table 13.6 we present an optimal solution of model (13.7) in the case $\gamma_1 = \gamma_2 = 1$. In Section 13.4 we present optimal solutions for various values of the γ's. The reason for taking $\gamma_1 = \gamma_2 = 1$, is that we are now able to compare this solution with the optimal solution of model (13.5); especially, the occurrences of old backlogs in model (13.5) can be investigated; see Exercise 13.7.6.

i	1	2	3	4	5	6	7	8	9	10	11	12	13	14
1	813	813	0	0	-1300	-2072	-772	0	2000	1000	0	2500	1300	0
2	-3672	-141	0	-4100	-2300	0	0	0	1600	900	0	1800	900	0
3	-1171	0	0	0	0	0	0	0	0	0	0	0	0	0
4	0	0	0	-2900	0	0	0	0	400	0	0	0	0	0
5	0	0	0	-1200	0	0	0	0	1800	1100	400	0	500	0
6	0	0	0	0	0	0	0	0	0	0	0	0	0	0
7	0	0	0	-300	0	0	0	0	500	400	200	0	200	0
8	16	0	0	-3200	0	0	0	0	1800	1300	700	200	300	0
9	0	0	0	-2500	-2500	0	0	0	0	0	0	0	0	0
10	1028	1028	1028	288	28	0	0	277	77	400	200	0	100	0
11	733	133	0	-1700	-2700	0	0	500	0	1700	1200	0	500	0
12	0	0	0	0	0	0	0	0	0	0	0	0	0	0
13	0	0	0	-600	-900	-1200	0	400	0	700	500	200	0	0
14	1644	644	644	644	644	344	44	300	0	1377	977	677	277	0
15	0	0	0	-1200	-1500	-1600	0	100	0	300	300	200	100	0
16	0	0	0	-1300	-1700	-2100	0	400	0	1300	900	600	200	0
17	0	0	0	0	0	0	0	0	0	0	0	0	0	0
18	0	0	0	0	0	0	0	0	0	2500	2500	0	0	0
19	-814	0	0	-1000	-1500	-1700	0	500	300	0	800	500	200	0
20	1640	1640	1640	1640	1640	140	140	140	140	140	140	0	0	0
21	0	0	1205	-1045	-1045	-1795	-1795	0	0	0	0	0	0	0

Table 13.6: Optimal values of s_{it} for model (13.7).

In Section 13.4 we calculate the influence of different choices of γ_1 and γ_2 on the total optimal backlog.

13.4 Full week productions

The production department of PLS sometimes has reasons to make only full week productions. This may happen, for instance, when less employees are available. It means that if a certain design is being produced on certain conveyor belts, then it is produced on these conveyor belts during full week periods, and that the production will not be interrupted during the week. This additional constraint can be formulated mathematically as:

$$x_{it} = c\delta_{it} \text{ with } \delta_{it} \in \{0, 1, \ldots, p\}, \text{ and } i = 1, \ldots, m, \ t = 1, \ldots, T.$$

For instance, $\delta_{it} = 3$ means that design i in period t is manufactured on three conveyor belts, during full weeks; so the total production is $2,600 \times 3$ units ($2,600$ is the maximum weekly production on any conveyor belt). Condition (13.1) in model (13.7) should be replaced by:

$$\sum_{i=1}^{m} \delta_{it} \leq p \qquad \text{for } t = 1, \ldots, T,$$

$$\delta_{it} \in \{0, 1, \ldots, p\} \qquad \text{for } i = 1, \ldots, m, \text{ and } t = 1, \ldots, T. \tag{13.8}$$

The model has now become a mixed integer linear optimization model; see Chapter 7. The optimal solution, which was obtained by using the commercial solver Cplex, is given in Table 13.7 (for $\gamma_1 = \gamma_2 = 1$, and $c_{it} = 1$ for all i and t). The entries of Table 13.7 refer to the number of conveyor belts used for the production of the design with label

$i \backslash t$	1	2	3	4	5	6	7	8	9	10	11	12	13	14
1	–	–	1	1	1	1	–	1	–	1	–	1	1	–
2	1	2	–	3	1	–	1	–	–	1	–	1	–	1
3	2	–	1	–	–	–	–	1	–	–	–	–	–	–
4	–	–	–	1	–	–	1	–	–	–	–	–	1	–
5	–	1	1	–	1	–	1	–	1	–	–	1	–	–
6	–	–	–	–	–	–	–	–	1	–	–	–	–	–
7	–	1	–	–	–	–	–	–	–	–	1	–	–	–
8	–	–	2	–	–	1	–	1	–	–	1	–	–	–
9	–	–	–	–	1	–	–	1	–	–	–	–	–	–
10	–	–	–	–	–	–	–	–	–	1	–	–	–	–
11	–	–	–	–	1	–	1	–	1	–	–	1	–	1
12	–	–	–	–	–	–	–	–	–	1	–	–	–	–
13	–	1	–	–	–	–	–	–	1	–	–	–	–	1
14	–	–	–	–	–	–	–	–	–	1	–	–	–	1
15	–	–	–	–	–	–	1	–	–	–	–	–	1	–
16	–	–	–	–	–	1	–	–	1	–	–	–	–	1
17	2	–	–	–	–	–	–	1	–	–	–	–	–	–
18	–	–	–	–	–	–	–	–	1	–	1	1	1	–
19	–	–	–	–	–	1	–	–	–	1	–	–	–	–
20	–	–	–	–	–	–	–	–	–	–	–	–	1	–
21	–	–	–	–	–	1	–	–	–	–	1	–	–	–

Table 13.7: Optimal full-week production values δ_{it}.

number in the corresponding row and in the week corresponding to the label number in the corresponding column. For instance, the 3 in row 2 and column 4 means that design 2 is manufactured during the whole week 4 on three conveyor belts. The total backlog for the schedule of Table 13.7 is 84,362. The sum of the entries of each column is 5, and so – as may be expected – the full capacity is used.

13.5 Sensitivity analysis

13.5.1 Changing the number of conveyor belts

We start this section by performing sensitivity analysis to the number of conveyor belts used for the production in model (13.5) in Section 13.2. One may argue that the total backlog 55,322 in the optimal solution of model (13.5) is rather high, and that it makes sense to investigate the total backlog for several values of p (= the number of conveyor belts used). Table 13.8 lists the optimal backlogs for $p = 4, 5, 6, 7$, and 8.

In the current situation, there are five conveyor belts in use. It follows from Table 13.8 that when one belt breaks down, then the total backlog increases drastically (with about 300%). There are no backlogs when eight conveyor belts are in use. This might suggest that there is overcapacity. However, this is not the case. In week 4, the total demand is 36,050, and the maximum possible total production is $8 \times 2,600 = 20,800$. So, a sufficiently large inventory had been built up to cope with the latter situation when demand became larger than production.

13.5.2 Perturbing backlog-weight coefficients

The largest backlog in Table 13.3 is $s_{24}^- = 9{,}100$. One may expect that this backlog has the largest influence on the optimal objective value (the total backlog). In order to investigate this situation, we vary the value of the backlog weight coefficient c_{24}. Table 13.9 lists the results of our calculations. The third column of this table contains the optimal objective values z^* for model (13.5). Since z^* is the optimal total cost of the backlogs, the listed values of z^* are not equal to the total backlog s_{total} (except for the special case $c_{24} = 1$). However, since $c_{it} = 1$ for all $i = 1, \ldots, m$ and $t = 1, \ldots, T$, except the case $i = 2$ and $t = 4$, we have that:

$$z^* = \sum_{i=1}^{m} \sum_{t=1}^{T} c_{it} s_{it}^- = \sum_{i=1}^{m} \sum_{t=1}^{T} s_{it}^- + (c_{24} - 1)s_{24}^-,$$

and, hence, the total backlog s_{total} can be derived from z^* by using the equation:

$$s_{\text{total}} = z^* - (c_{24} - 1)s_{24}^-.$$

It is remarkable that all total backlogs, except for the case $c_{24} = 0$, are equal to 55,322. This is due the fact that the model has many optimal solutions. The fact that $s_{24}^- = 0$ for $c_{24} > 1$ is not unexpected, because c_{24} is then the only c_{it} with a value larger than 1, and the value of each other c_{it} is 1, and so c_{24} has the largest value. For $c_{24} = 0$, the backlog of design 2 in week 4 can take on any value without affecting the optimal solution. However, more of design 2 should be produced in week 5 in order to reduce the backlog of week 4. The values of 'Min.' and 'Max.' in Table 13.9 determine the tolerance interval of c_{24} (see also Section 5.2). For instance, for the values of c_{24} with $0.00 \leq c_{24} \leq 1.00$ (the situation in the second row of Table 13.9), it holds that $s_{24}^- = 9{,}700$; outside this interval the value of s_{24}^- may be different from 9,700.

For $c_{24} \leq 1$, it holds that $s_{24}^- > 0$, so s_{24}^- is a basic variable, which implies that its optimal dual value (shadow price) is 0 (see Theorem 4.3.1). On the other hand, s_{24}^- is nonbasic for $c_{24} > 1$, since its shadow price is positive (see Table 13.9). Recall that the dual value in the case of nondegeneracy is precisely the amount that needs to be subtracted from the corresponding objective coefficient in order to change s_{24}^- into a basic variable.

p	Total backlog
4	189,675
5	55,322
6	16,147
7	702
8	0

Table 13.8: Number of conveyor belts versus total backlog.

c_{24}	s_{24}^-	Obj. value	Total backlog	Shadow price	Min.	Max.
0.00	9,841	45,622	55,463	0.00	0.00	0.00
0.10	9,700	46,592	55,322	0.00	0.00	1.00
0.50	9,700	50,472	55,322	0.00	0.00	1.00
0.90	9,700	54,352	55,322	0.00	0.00	1.00
1.00	9,100	55,322	55,322	0.00	1.00	1.00
1.10	0	55,322	55,322	0.10	1.00	∞
2.00	0	55,322	55,322	1.00	1.00	∞
3.00	0	55,322	55,322	2.00	1.00	∞

Table 13.9: Perturbing c_{24} in model (13.5).

γ_1	γ_2	Obj. value	Number of coffee machines
1.0000	1.0000	55,322	55,322
1.0000	0.5000	32,906	55,322
2.0000	1.0000	65,812	55,322
1.0000	0.0625	13,292	55,322

Table 13.10: Old and new backlogs.

13.5.3 Changing the weights of the 'old' and the 'new' backlogs

Consider model (13.7). The values of the weight parameters γ_1 and γ_2 determine which backlogs are eliminated first. In Table 13.10, we have calculated the objective values for several combinations of values of γ_1 and γ_2.

As in Table 13.9, the total backlog in terms of the number of coffee machines is the same for the values of γ_1 and γ_2 of Table 13.10. Again the reason is that productions can be transferred from one week to the other. This multiplicity of the optimal solution implies that the model is rather insensitive to changes in the values of γ_1 and γ_2.

13.6 GMPL model code

The following listing gives the GMPL model code for model (13.5). We have substituted $s_{it}^+ - s_{it}^-$ for s_{it} for $i = 1, \ldots, m$ and $t = 1, \ldots, T$. So, the GMPL code below describes the following model:

$$\min \sum_{i=1}^{m} \sum_{t=1}^{T} c_{it} s_{it}^-$$

$$\text{s.t.} \quad \sum_{i=1}^{m} x_{it} \leq cp \qquad \text{for } t = 1, \ldots, T$$

$$s_{it}^+ - s_{it}^- = s_{i,0} + x_{it} - d_{it} \qquad \text{for } i = 1, \ldots, m, \text{ and } t = 1$$

$$s_{it}^+ - s_{it}^- = s_{i,t-1} + x_{it} - d_{it} \qquad \text{for } i = 1, \ldots, m, \text{ and } t = 2, \ldots, T$$

$$x_{it} \geq 0, s_{it}^+ \geq 0, s_{it}^- \geq 0 \qquad \text{for } i = 1, \ldots, m, \text{ and } t = 1, \ldots, T.$$

```
 1        m;
 2        nperiods;
 3      I := {1 .. m};
 4      T := {1 .. nperiods};
 5        c;
 6        p;
 7        s0{I};
 8        d{I, T};
 9
10      x{i    I, t    T} >= 0;
11      sp{i    I, t    T} >= 0;
12      sm{i    I, t    T} >= 0;
13
14          total_shortage:
15      {i    I}    {t    T} sm[i,t];
16            beltcap {t    T}:
17      {i    I} x[i,t] <= c * p;
18            inventory {i    I, t    T}:
19      sp[i,t] — sm[i,t] =
20      (    t = 1    s0[i]    sp[i,t—1] — sm[i,t—1])
21      + x[i,t] — d[i,t];
22
23      ;
24
25        m := 21;
26        nperiods := 14;
27        c := 2600;
28        p := 5;
29        s0 :=
30      1     813     2 —272     3 —2500     4     0     5   220     6     0     7 —800     8  16
31      9       0    10  1028    11  1333    12    68    13    97    14  1644    15     0    16   0
32      17 —2476    18     0    19    86    20  1640    21     0;
33        d : 1      2      3      4      5      6      7      8      9     10     11     12     13     14
34   := 1     0      0   1500   7400   1300   1200   1000   1000   1000   1000   1000   1300   1200   1300
35      2  3400   3600      0   9700   1400    700    800    700    700    700    900   1000    900    900
36      3  1500      0   5000      0      0      0      0      0      0      0      0      0      0      0
37      4     0      0      0   2900      0   1000      0    400      0    400      0    400      0      0
38      5   800   2600   2400   1200   1300   1200    700    700    700    700    700    400    400    500
39      6     0      0      0      0      0      0      0      0   1600      0      0      0      0      0
40      7   400    300    300    300    300    200    100    200    200    100    200    200    200    200
41      8     0   1000   2500   3200    600    600    700    600    700    500    600    500    600    300
42      9     0      0      0   2500      0      0      0   1500      0      0      0      0      0      0
43     10     0      0      0    800    200    200    200    200    200    100    200    200    200    100
44     11   600    600    900   1700   1000   1200    500   1000    500   1000    500   1200    600    500
45     12   600      0      0      0      0      0      0      0      0      0      0      0      0      0
46     13  1400      0      0    600    300    300    300    500    400    300    200    300    200    100
47     14     0   1000      0      0      0    300    300    400    300    400    400    300    400    400
48     15     0      0      0   1200    300    100    100    100    100    200      0    100    100    100
49     16    80      0      0   1300    400    400    400    500    400    300    400    300    400    200
50     17  4000      0      0      0      0      0      0      0      0      0      0      0      0      0
51     18     0      0      0      0      0      0      0      0      0   2500      0   2500      0      0
52     19   900      0      0   1000    500    200    300    200    200    300    300    300    300    200
53     20     0      0      0      0      0      0   1500      0      0      0      0    200      0      0
54     21     0      0      0   2250      0    750      0      0      0      0      0      0      0      0;
```

55   ;

Note that the `if ... then ... else` expression in the inventory equations in this listing is evaluated when the solver constructs the technology matrix. See also Section F.3.

13.7 Exercises

Exercise 13.7.1. What are the decision variables of model (13.5)? Calculate the number of decision variables, and the number of constraints.

Exercise 13.7.2. Consider Table 13.2 and Table 13.3.

(a) Explain the large backlogs of design 1 in the weeks 4, 5, and 6.

(b) Explain why design 20 is only produced in week 12.

(c) Explain why there is no inventory and backlog of design 21 during the whole planning period.

(d) Check using Table 13.3 that the optimal objective value of model (13.5) is in fact 55,322.

(e) Construct an alternative optimal solution of model (13.5).

(f) Check whether it is true that the full capacity of the conveyor belts has been used in the optimal solutions of model (13.5).

Exercise 13.7.3. Show that any optimal solution of model (13.5) satisfies $r_{it}^+ \times r_{it}^- = 0$ for all i and t.

Exercise 13.7.4. Calculate the number of decision variables and constraints of model (13.7). Show that if $\gamma_1 = \gamma_2$, then model (13.7) is equivalent to model (13.5). Explain why the optimal solutions of model (13.7) with $\gamma_1 = \gamma_2$ and of model (13.5) need not be the same; compare Table 13.2 and Table 13.3 with Table 13.4 and Table 13.6.

Exercise 13.7.5. Apply the method described in Section 13.3 to find a feasible solution of model (13.7) for the data in Table 13.1.

Exercise 13.7.6. Consider Table 13.5 and Table 13.6 in Section 13.3.

(a) Why is it plausible that old backlogs usually only occur in the first part of the planning period?

(b) Calculate the total backlog. Why is this quantity the same as in the optimal solution of model (13.5)?

Exercise 13.7.7. Consider Section 13.4.

(a) Calculate the number of binary variables of model (13.7), with (13.1) replaced by (13.8).

(b) How can the constraint that 'only certain designs are produced during full week periods' be modeled?

(c) Why is the total backlog in the case of full week productions larger than without this constraint? Recall that the total backlog of 97,114, found in Section 13.4, might be lower.

Exercise 13.7.8.   Show that, in any optimal solution of (13.5), the restrictions (13.4) are always satisfied.

Exercise 13.7.9.   Consider Table 13.9 in Section 13.5.

(a) Explain why model (13.5) is not degenerate for values of c_{24} satisfying $c_{24} > 0$.

(b) Explain why the shadow price of the nonnegativity constraint $s_{24}^{-} \geq 0$ is positive for $c_{24}^{-} > 1$.

Chapter 14

Conflicting objectives: producing versus importing

Overview

This case study describes an example of *multiobjective optimization*. In multiobjective optimization, models are considered that involve multiple objectives subject to both 'hard' and 'soft' constraints. The multiple objectives usually conflict with each other, and so optimizing one objective function is at the expense of the others. Therefore, one cannot expect to achieve optimal values for all objective functions simultaneously, but a 'best' value somewhere in between the individual optimal values. We will use a method called *goal optimization* to determine this 'best' value.

14.1 Problem description and input data

The company Plastics International produces, among other things, high quality plastic tubes in three standard sizes. Tube type T_1 sells for $20 per meter, T_2 for $24 per meter, and T_3 for $18 per meter. To manufacture one meter of tube type T_1 requires 0.55 minutes of processing time, one meter of T_2 needs 0.40 minutes, and one meter of T_3 needs 0.60 minutes. After production, each meter of any type of tube requires one gram of finishing material. The total production costs are estimated to be $4, $6, and $7 per meter of types T_1, T_2, and T_3, respectively.

For the next week, the company has received a large order for 3,000 meters of T_1, 5,000 meters of T_2, and 7,000 meters of T_3. There are 40 hours of machine time available this week, and 6,000 grams of finishing material are in inventory. This implies that the production department will not be able to satisfy these demands, since these demands require a total of $(3{,}000 \times 0.55 + 5{,}000 \times 0.40 + 7{,}000 \times 0.60 \text{ mins.} =)$ 130.83 hours machine time and $((3{,}000 + 5{,}000 + 7{,}000) \times 1 =)$ 15,000 grams of finishing material. The management does not expect this high level of demand to continue, and decides not to

Tube type	Selling price ($/m)	Demand	Production time (min/m)	Finishing material (g/m)	Production cost ($/m)	Purchase cost ($/m)
T_1	20	3,000	0.55	1	4	7
T_2	24	5,000	0.40	1	6	7
T_3	18	7,000	0.60	1	7	9
Amount available:			40 hours	6,000 g		

Table 14.1: Input data for Plastics International.

expand the production facilities. It also does not want to delay the deliveries or to lose this order.

In order to meet the demand, the management considers purchasing tubes from suppliers abroad at a delivered cost of $7 per meter of tube type T_1, $7 per meter of T_2, and $9 per meter of T_3. The imported tubes arrive ready-made, and so no finishing material is needed. The data are summarized in Table 14.1.

There are two objectives:

The first objective is to determine how much of each tube type need be produced and how much purchased from abroad, in order to meet the demands and to maximize the company's profit.

A second objective is due to a request by the government, who wants to reduce the amount of money spent on imports. So in addition to maximizing the total profit, the management of Plastics International also needs to minimize the total costs of the imports.

14.2 Modeling two conflicting objectives; Pareto optimal points

The production-import problem of Section 14.1 can be formulated in terms of linear optimization with two objectives. To that end, we first introduce the following six decision variables:

$$x_i = \text{the production of tube type } T_i \text{ (in meters), for } i = 1, 2, 3;$$
$$t_i = \text{the purchase of tube type } T_i \text{ (in meters), for } i = 1, 2, 3.$$

The two objectives can now be formulated as follows.

Objective 1. The profit, to be maximized, is the sum of the profit from production (selling price minus production cost) plus the profit from products purchased from abroad (selling price minus purchase cost). Hence, this objective is:

$$\max \left(16x_1 + 18x_2 + 11x_3\right) + \left(13t_1 + 17t_2 + 9t_3\right). \tag{O1}$$

	Objective (O1)	Objective (O2)
Profit	$198,500.00	$193,333.33
Cost of import	$86,750.00	$78,000.00
x_1	3,000	0
x_2	0	5,000
x_3	1,250	666.67
t_1	0	3,000
t_2	5,000	0
t_3	5,750	6,333.33
s_1	0	0
s_2	1,750	333.33

Table 14.2: Optimal solutions for the separate objectives.

Objective 2. The cost of imports, to be minimized, is equal to the total cost of importing the tube types T_1, T_2, and T_3. Hence, this objective becomes:

$$\min\ 7t_1 + 7t_2 + 9t_3. \tag{O2}$$

The demand constraints are:

$$x_1 + t_1 = 3,000 \qquad \text{(demand for } T_1) \tag{1}$$
$$x_2 + t_2 = 5,000 \qquad \text{(demand for } T_2) \tag{2}$$
$$x_3 + t_3 = 7,000 \qquad \text{(demand for } T_3). \tag{3}$$

The resource constraints are:

$$0.55x_1 + 0.40x_2 + 0.60x_3 \leq 2,400 \qquad \text{(machine time)} \tag{4}$$
$$x_1 + x_2 + x_3 \leq 6,000 \qquad \text{(finishing material).} \tag{5}$$

The nonnegativity constraints are:

$$x_1, x_2, x_3, t_1, t_2, t_3 \geq 0.$$

The conflict between the two objectives (O1) and (O2) can be illustrated by solving the above model, separately for the two objectives. The optimal solutions are obtained by using a linear optimization computer package. In Table 14.2 we have summarized the results. The variables s_1 and s_2 are the slack variables of the 'machine time' constraint and the 'finishing material' constraint, respectively.

The following conclusions can be drawn from Table 14.2.

To maximize the profit, while neglecting the amount of import, the company should produce 3,000 meters of tube type T_1, produce 1,250 meters of T_3, import 5,000 meters of T_2, and import 5,750 meters of T_3. The profit is $198,500. The total cost of importing these quantities of tubes is equal to $0 \times \$7 + 5,000 \times \$7 + 5,750 \times \$9 = \$86,750$.

To minimize the amount of import, while neglecting the profit, the company should produce 5,000 meters of T_2, produce 666.67 meters of T_3, import 3,000 meters of T_1, and import 6,333.33 meters of T_3. The profit then is $0 \times \$16 + 5{,}000 \times \$18 + 666.67 \times \$11 + 3{,}000 \times \$13 + 0 \times \$17 + 6{,}333.33 \times \$9 = \$193{,}333$.

In both cases, there is no machine idle-time (i.e., $s_1 = 0$). The inventory of finishing material is better used in the second case (namely, $s_2 = 333.33$ in the case of (O2) against $s_2 = 1{,}750$ in the case of (O1)), because in the second case more is produced, and so more finishing material is used, while the imported tubes are ready-made.

If the cost of the imports is neglected, then the profit drops from \$198,500 to \$193,333. If the profit is neglected, then the costs of the imports increase from \$78,000 to \$86,750. Hence, the two objectives are in fact conflicting.

For LO-models with multiple objectives, it might be very useful if one knows all possible values of the different objectives. The present model has only two objectives, and is therefore small enough to determine the feasible region for the two objective functions. We use the following model:

$$\max \ \alpha z + \beta w$$
$$\text{s.t.} \quad \text{Constraints } (1) - (5)$$
$$z = 16x_1 + 18x_2 + 11x_3 + 13t_1 + 17t_2 + 9t_3$$
$$w = 7t_1 + 7t_2 + 9t_3$$
$$x_1, x_2, x_3, t_1, t_2, t_3 \geq 0$$
$$z, w \text{ free.}$$

The choice $\alpha = 1$ and $\beta = 0$ corresponds to objective (O1), and the choice $\alpha = 0$ and $\beta = -1$ corresponds to objective (O2). Since the demand is fixed, the two objective functions are bounded. The feasible region, as depicted in Figure 14.1, consists of the optimal points $[w^* \ z^*]^\mathsf{T}$ that arise when the values of α and β are varied. The vertices of the feasible region are determined by trial and error. It turns out that the vertices are:

Point	α	β	w^*	z^*
A	0	-1	78,000.00	193,333.33
B	3	-2	78,909.09	194,181.82
C	2	-1	84,875.00	197,875.00
D	1	0	86,750.00	198,500.00
E	3	1	98,000.00	196,000.00
F	-1	1	119,000.00	187,000.00
G	-5	-1	84,000.00	192,000.00

Note that points in the interior of the feasible region never correspond to optimal solutions, since the objectives are maximizing the profit and minimizing the import. Also note that the points of interests are the ones where neither of the objectives can be improved without losing something on another objective. They correspond to nonnegative values of α and nonpositive values of β. In Figure 14.1, this is the piecewise linear curve $ABCD$;

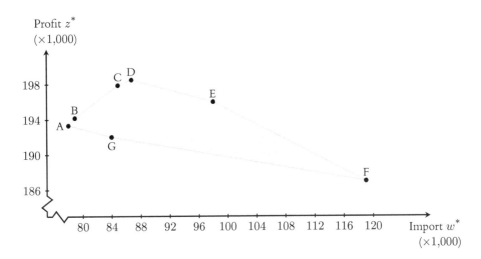

Figure 14.1: The feasible region.

these points are called *Pareto optimal points*[1]. In general, this set of points is very difficult to determine.

In the following section, we will show how Plastics International can deal with the conflicting objectives of maximizing the profits and minimizing the costs of imports, by using the technique of goal optimization.

14.3 Goal optimization for conflicting objectives

The basic idea of *goal optimization* is the formulation of goals and penalties in the following way. A *goal* is a specified numerical target value for an objective to be achieved. A *penalty* is a relative value, representing the dissatisfaction with each unit the optimal objective value is below (above) the goal if the objective function is maximized (minimized). Once the goals and penalties are identified, a solution that minimizes the total penalty associated with all objectives can be determined.

The goals are the values that the management would ideally like to achieve for each objective individually. From Table 14.2, we know that the maximum achievable profit is $198,500. The management of Plastics International may therefore choose to set this value as the target for achieving the highest possible profit. On the other hand, since the minimum achievable costs of imports are $78,000, this value may be chosen as the other goal. The goal of $78,000 can be violated only if doing so results in a significant increase of the profit.

The penalties reflect the relative importance of the unwanted deviations from the goals associated with the objectives. A larger value for a penalty indicates that meeting the cor-

---

[1]Named after the Italian engineer, sociologist, economist, political scientist, and philosopher Vilfredo F.D. Pareto (1848–1923).

responding goal has a higher priority. If a deviation from the goal is not unwanted, then there need not be a penalty. However, if there is an unwanted deviation from the goal, then there should be a penalty; the larger the unwanted deviation, the higher the penalty. The function that describes the relationship between the deviation from the goal and the penalty, is called the *penalty function*. We will assume that penalty functions are linearly dependent on the deviation from the corresponding goals. So the penalty is the slope of the penalty function: the steeper the slope, the higher the penalty.

The management of Plastics International has decided to incur the following penalties:

$$\text{Profit penalty} \ = 2 \text{ for each dollar of profit below } \$198{,}500;$$
$$\text{Import penalty} = 1 \text{ for each dollar of import above } \$78{,}000.$$

This means that the management considers it as two times as important to meet the goal of $198,500 for the profit, as it is to meet the goal of $78,000 for the import. We could have chosen a ratio of $4 : 2$ or $8 : 4$, et cetera; the penalties have no physical meaning other than to indicate the relative importance of meeting the goals. So one of the penalties can always be taken equal to 1.

Now that we have determined values for the goals and for the penalties, we can formulate an LO-model whose solution will provide the best values of the decision variables in terms of minimizing the total penalty of the unwanted deviations from the goals.

In addition to the decision variables defined above, we need to define two new decision variables for each objective: one to represent the amount by which the objective value exceeds the goal, and the other for the amount by which it falls short of that goal. Since there are two objectives, we need four new decision variables:

z^+ = the amount by which the profit is above the goal of $198,500;
z^- = the amount by which the profit is below the goal of $198,500;
w^+ = the amount by which the import cost is above the goal of $78,000;
w^- = the amount by which the import cost is below the goal of $78,000.

Obviously, these decision variables have to be nonnegative. Moreover, we must ensure that at least one of z^+ and z^-, and one of w^+ and w^- is 0. In other words:

$$z^+ \times z^- = 0, \text{ and } w^+ \times w^- = 0.$$

However, similar to the discussion in Section 1.5.2, it can be shown that these (nonlinear) constraints are redundant (with respect to the optimal solution) in the LO-model to be formulated below. This means that any solution that does not satisfy these constraints cannot be optimal. See also Exercise 14.9.1.

The objective is to minimize the total penalty for the unwanted deviations from the goals, which is equal to two times the sum of the penalty for being under the profit goal plus the penalty for being above the import cost goal. Hence, the penalty function to be minimized is $2z^- + 1w^+$. Besides the three demand constraints and the two resource constraints, we need so-called *goal constraints*, one for each goal. To understand how z^+ and z^- represent the amount by which the profit is above or below the goal of $198,500, the following can be said. First recall that the profit is:

$$z = 16x_1 + 18x_2 + 11x_3 + 13t_1 + 17t_2 + 9t_3;$$

see Section 14.2. If the profit z is above the goal, then the value of z^- should be 0 and the value of z^+ should satisfy $z^+ = z - 198,500$. Hence,

$$z - z^+ = 198,500, \text{ and } z^- = 0.$$

On the other hand, if the profit z is below the goal, then the value of z^+ should be 0, and the value of z^- should satisfy $z^- = 198,500 - z$. Hence,

$$z + z^- = 198,500, \text{ and } z^+ = 0.$$

Since $z^+ \times z^- = 0$, the above two cases can be written as a single constraint, namely:

$$z - z^+ + z^- = 198,500.$$

The general form of a goal constraint is:

(value obj. function) $-$ (amount above goal) $+$ (amount below goal) $=$ goal.

Similarly, the goal constraint
$$w = 7t_1 + 7t_2 + 9t_3$$

corresponding to the import cost satisfies:

$$w - w^+ + w^- = 78,000.$$

The LO-model, called model (O3), can now be formulated as follows.

$$
\begin{array}{lllll}
\min\ 2z^- + w^+ & & & & \text{(O3)} \\
\text{s.t. } x_1 + t_1 & & = & 3,000 & \text{(1)} \\
\quad x_2 + t_2 & & = & 5,000 & \text{(2)} \\
\quad x_3 + t_3 & & = & 7,000 & \text{(3)} \\
\quad 0.55x_1 + 0.40x_2 + 0.60x_3 & & \leq & 2,400 & \text{(4)} \\
\quad x_1 + x_2 + x_3 & & \leq & 6,000 & \text{(5)} \\
\quad 16x_1 + 18x_2 + 11x_3 + 13t_1 + 17t_2 + 9t_3 - z^+ + z^- & & = & 198,500 & \text{(6)} \\
\quad 7t_1 + 7t_2 + 9t_3 - w^+ + w^- & & = & 78,000 & \text{(7)} \\
\quad x_1, x_2, x_3, t_1, t_2, t_3, z^+, z^-, w^+, w^- \geq 0.
\end{array}
$$

We have solved Model (O3) by means of a linear optimization package. In Table 14.3 the results of the calculations are listed. From this table it can be concluded that the optimal production-import schedule is the one given in Table 14.4. It also follows from Table 14.3

Variable	Optimal value	Dual value (Shadow price)	Objective coefficient	Minimum	Maximum
x_1	3,000	0	0	$-\infty$	0.6250
x_2	1,875	0	0	-0.4545	0.3333
x_3	0	0.5	0	-0.5	∞
t_1	0	0.6250	0	-0.6250	∞
t_2	3,125	0	0	-0.3333	0.4545
t_3	7,000	0	0	$-\infty$	0.5
z^+	0	2	0	-2	∞
z^-	625	0	2	1.6154	3
w^+	6,875	0	1	0.666767	1.2381
w^-	0	1	0	-1	∞

Constraint	Slack	Dual value (Shadow price)	Right hand side	Minimum	Maximum
1	0	-19.6250	3,000	2,285	3,042
2	0	-27	5,000	4,017	5,036
3	0	-9	7,000	6,236	7,069
4	0	-22.5	2,400	1,650	2,650
5	1,125	0	6,000	4,875	∞
6	0	2	198,500	197,875	∞
7	0	-1	78,000	$-\infty$	84,875

Optimal objective value: 8,125.

Table 14.3: Computer output, optimal solution of (O3).

that both goals are not completely achieved. The fact that $z^- = 625$ indicates that the profit goal of \$198,500 is not met by \$625 (note that $198,500 - 625 = 197,875$), and the fact $w^+ = 6,875$ indicates that the import goal of \$78,000 is exceeded by \$6,875 (note that $78,000 + 6,875 = 84,875$). Hence, this goal optimization model achieves lower import cost at the expense of decreasing profit, when comparing this to the profit-maximization model (O1); see Table 14.2. Similarly, this goal optimization model achieves a larger profit at the expense of increasing import cost, when compared to the import-minimization model (O2); see also Table 14.2.

Since the value of the slack variable of constraint (5) is $1,125$, it follows that $1,125$ grams of the initial inventory of $6,000$ grams is left over. Also notice that this solution contains no machine idle-time.

The dual values in the above solution can be interpreted as shadow prices, because this solution is nondegenerate; see Chapter 5. Increasing any of the demand values leads to a lower penalty, since the corresponding shadow price is negative. This also holds for the (binding!) 'machine time' constraint. The 'finishing material' constraint is nonbinding, which means a zero shadow price. Increasing the profit goal leads to a higher penalty (the situation becomes worse), while increasing the import goal lowers the penalty (the situation improves). See also Section 14.4.

Production:	3,000	meters of tube type T_1
	1,875	meters of tube type T_2
Import:	3,125	meters of tube type T_2
	7,000	meters of tube type T_3
Profit:	\$197,875	
Import cost:	\$84,875	

Table 14.4: Optimal production-import schedule for model (O3).

In Exercise 14.9.2, the reader is asked to explain the signs of the shadow prices of Table 14.3.

14.4 Soft and hard constraints

In the previous sections we have modeled the case of two conflicting objectives by means of penalties on the deviations from the objective goals. In this section we will use penalties to model so-called hard and soft constraints. A constraint is called *hard* if the goal associated with the constraint (usually its right hand side) has to be satisfied, and is called *soft* if some deviation on the associated goal is allowed (for instance, by working in overtime). Consider, for example, the 'machine time' constraint:

$$0.55x_1 + 0.40x_2 + 0.60x_3 \leq 2,400, \qquad (4)$$

and suppose that this constraint is soft. This means that some deviation from the goal $2,400$ is tolerable. In order to indicate the deviation from the right hand side of (4), we introduce nonnegative decision variables m^+ and m^-, such that:

$$0.55x_1 + 0.40x_2 + 0.60x_3 - m^+ + m^- = 2,400, \text{ with } m^+, m^- \geq 0. \qquad (4')$$

Since the machine time constraint is a '$\leq$' constraint with a nonnegative right hand side value, there is no penalty for being below the goal $2,400$, but there should be a penalty for exceeding this goal. Hence, m^+ reflects the unwanted deviation from the goal, and this deviation multiplied by a certain penalty has to be minimized. This penalty is a subjective value, reflecting the relative importance of exceeding the goal when compared to the deviation from the profit and the import goals. The importance of not meeting the machine time goal (2,400 minutes) is compared to that of not meeting the profit goal (\$198,500) and the import goal (\$78,000); see Section 14.3. The question then is, what is a reasonable decrease of the value of $2z^- + w^+$ if during one extra hour is produced. Suppose the management has decided that one extra minute production per week is allowed for every \$20 improvement on the previous penalty function. Hence, the machine time penalty is set to 20, and the new penalty function becomes $2z^- + w^+ + 20m^+$. The corresponding LO-model, called (O4), can now be formulated as follows.

$$\min 2z^- + w^+ + 20m^+ \qquad \text{(O4)}$$
$$\text{s.t. } x_1 + t_1 \qquad\qquad\qquad = \quad 3,000 \qquad \text{(1)}$$

$$x_2 + t_2 \qquad\qquad\qquad\qquad\qquad\qquad\qquad\qquad = \quad 5{,}000 \qquad (2)$$
$$x_3 + t_3 \qquad\qquad\qquad\qquad\qquad\qquad\qquad\qquad = \quad 7{,}000 \qquad (3)$$
$$0.55x_1 + 0.40x_2 + 0.60x_3 - m^+ + m^- \qquad = \quad 2{,}400 \qquad (4')$$
$$x_1 + x_2 + x_3 \qquad\qquad\qquad\qquad\qquad\qquad \leq \quad 6{,}000 \qquad (5)$$
$$16x_1 + 18x_2 + 11x_3 + 13t_1 + 17t_2 + 9t_3 - z^+ + z^- = 198{,}500 \qquad (6)$$
$$7t_1 + 7t_2 + 9t_3 - w^+ + w^- \qquad\qquad\qquad = \quad 78{,}000 \qquad (7)$$
$$x_1, x_2, x_3, t_1, t_2, t_3, z^+, z^-, w^+, w^-, m^+, m^- \geq 0.$$

The optimal solution of model (O4), which has again been calculated using a computer package, is as follows.

Optimal solution model (O4)

$z = 198{,}500, w = 78{,}000$
$x_1 = 2{,}821.43, x_2 = 3{,}035.71, x_3 = 0$
$t_1 = 178.57, t_2 = 1{,}964.29, t_3 = 7{,}000$
$z^+ = 0, z^- = 0, w^+ = 0, w^- = 0, m^+ = 366.07, m^- = 0$

This production-import schedule can only be realized if $(366.07/60 =)$ 6.10 extra hours per week are worked. The solution reflects the situation in which both the production and the import are optimal ((O1) and (O2), respectively), at the cost of expanding the capacity with 6.10 hours.

Suppose that the management of Plastics International considers 6.10 additional hours as unacceptable, and has decided that only four hours extra per week can be worked. This leads to the constraint:

$$m^+ \leq 240. \qquad (8)$$

Adding constraint (8) to model (O4) results in model (O5), and solving it leads to the following optimal solution.

Optimal solution model (O5)

$z = 198{,}475, w = 80{,}675$
$x_1 = 3{,}000, x_2 = 2{,}475, x_3 = 0$
$t_1 = 0, t_2 = 2{,}525, t_3 = 7{,}000$
$z^+ = 0, z^- = 25, w^+ = 2{,}675, w^- = 0, m^+ = 240, m^- = 0$

When comparing the solution of model (O3) with this solution, it appears that both the profit and the import goal are now better satisfied:

In the case of model (O3): $\quad z^- = 625, w^+ = 6{,}875,$ and
in the case of model (O5): $\quad z^- = 25, \quad w^+ = 2{,}675.$

Also the inventory of finishing material is much better used: in the case of (O1) the slack of constraint (5) is 1,125, and in the case of (O5) the slack is 525. Based on these facts, the management may consider extending the man-power capacity by three hours per week. In Exercise 14.9.6 the reader is asked to calculate the influence on the optimal solution of model (O5), when the penalty 20 of m^+ is changed.

14.5 Sensitivity analysis

In this section we investigate the influence of changing the penalties and the goals on the optimal solution of model (O3).

14.5.1 Changing the penalties

In this case study there are only two penalties: a profit penalty and an import penalty. One of the penalties can always be taken equal to 1. In our case the import penalty is set to 1. Define:

$$\alpha = \text{profit penalty.}$$

Recall that z is defined as the total profit, and w as the total import cost. In Figure 14.2 we have drawn, in one figure, the perturbation functions $z^*(\alpha)$ and $w^*(\alpha)$ for $\alpha \geq 0$. In Exercise 14.9.7, the reader is asked to explain why these functions are nondecreasing 'staircase' functions. For $\alpha = 0$ we are in the situation of only one objective, namely objective (O2); see Table 14.2. Clearly, $w^+ = w^- = 0$ for $\alpha = 0$. So, only the import cost is minimized. This situation is maintained for $0 \leq \alpha \leq 1.0714$. For large values of α the import penalty becomes relatively small, and so we are again in the situation of one objective. In this case, the optimal solution satisfies $z^+ = z^- = 0$. The profit has reached its maximum possible value (namely, \$198,500), in conjunction with the highest import cost (namely, \$86,750). In Figure 14.2 one can see that this situation holds for $\alpha \geq 3$.

The tolerance interval of the profit penalty can also be read from Figure 14.2. All values of α for which $1.6154 \leq \alpha \leq 3$ yield the same optimal solution. Hence, this solution is not very sensitive to small changes of the profit penalty 2. Note that the solution is more sensitive to values of the profit penalty around 1.6.

14.5.2 Changing the goals

In this section we parameterize the goals. We leave it to the reader to discuss the practical relevance of these parameterizations; we will mainly focus on the mathematical aspects. Define:

$$PG = \text{profit goal,}$$
$$IG = \text{import goal.}$$

Recall that in Table 14.3, $PG = 198,500$, and $IG = 78,000$.

In Figure 14.3, the 'perturbation functions' $z^+ = z^+(PG)$, $z^- = z^-(PG)$, $w^+ = w^+(PG)$, and $w^- = w^-(PG)$ of model (O3) are depicted. Notice that $w^-(PG) = 0$

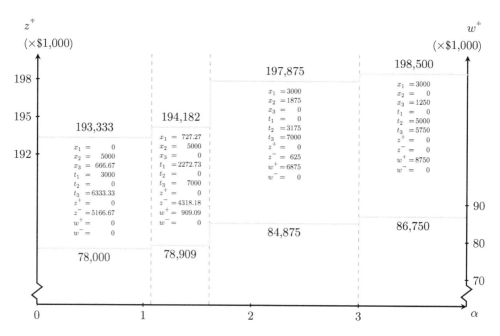

Figure 14.2: Profit and import costs as functions of the profit penalty α.

for all values of PG; see Exercise 14.9.8. The function z^+ decreases for $PG \leq 193,333$, and is 0 for all other values of PG. If both z^+ and z^- are 0, while $PG > 193,333$ then $w^+ \geq w^- - 78,000 + w$, and so, since w^+ is minimized, $w^- = 0$. Because (6) is now equivalent to $z = PG$, it follows that the values of t_1, t_2, t_3 increase when PG increases, because (4) remains binding. Hence, w^+ will increase. Similar arguments can be used to explain the behavior of the other functions in Figure 14.3. Note that a decrease of the original profit goal with at least $(198,500 - 197,875 =)$ 625 units changes the penalty values of z^- and w^+.

Figure 14.4 shows the relationships between the import goal IG and the penalties. In this case, $z^+(IG) = 0$ for all values of IG. Moreover, the penalties are not sensitive to small changes of the original value (78,000) of IG. The analysis of the behavior of the four penalty functions is left to the reader.

14.6 Alternative solution techniques

There are several alternative techniques for solving multiobjective optimization models. We will discuss the following methods: lexicographic goal optimization, and fuzzy linear optimization.

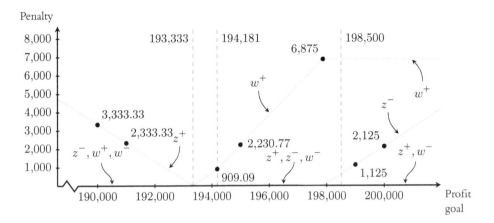

Figure 14.3: Penalties as functions of the profit goal.

14.6.1 Lexicographic goal optimization

In applying lexicographic goal optimization, the objective 'function' consists of at least two parts. The first part contains the sum of the unwanted deviations from the soft goals. For the other part, a number of different options are available. The objective can be formulated by means of the symbol lexmin, called the *lexicographic minimum*, defined as follows. The vector $[u_1 \ldots u_n]$ is called *lexicographically smaller* (larger) than the vector $[v_1 \ldots v_n]$, if and only if there is an index $k \in \{1, \ldots, n\}$ for which $u_k < v_k$ ($u_k > v_k$) and $u_i = v_i$ for each $i \in \{1, \ldots, k-1\}$. A lexicographic minimum of a collection S of vectors is a vector $\mathbf{u}^*$ in S that is lexicographically smaller than all vectors in S different from $\mathbf{u}^*$; notation $\mathbf{u}^* = $ lexmin$\{S\}$. For example, lexmin$\{[1\ 2\ 4], [1\ 2\ 3], [1\ 3\ 6], [1\ 2\ 3]\} = [1\ 2\ 3]$.

The general form of a lexmin model can be formulated as follows.

$$\text{lexmin}\{\mathbf{u} \mid \mathbf{u} \in P\},$$

where $\mathbf{u} \in P$ denotes a set of constraints (including nonnegativity constraints) in the variables $u_1, \ldots, u_n$. Lexmin models for n-vectors can be solved by solving n LO-models, in the following way. Let $1 \leq k \leq n$. Determine, successively:

$$u_1^* = \min\{u_1 \mid \mathbf{u} \in P\},$$
$$u_2^* = \min\{u_2 \mid \mathbf{u} \in P, u_1 = u_1^*\},$$
$$\vdots \qquad \qquad \vdots$$
$$u_k^* = \min\{u_k \mid \mathbf{u} \in P, u_1 = u_1^*, u_2 = u_2^*, \ldots, u_{k-1} = u_{k-1}^*\}.$$

The optimal values of the remaining variables $u_{k+1}, \ldots, u_n$ are determined in the k'th model. The sequence stops after k steps, where k is the smallest index for which the corresponding LO-model has a unique solution.

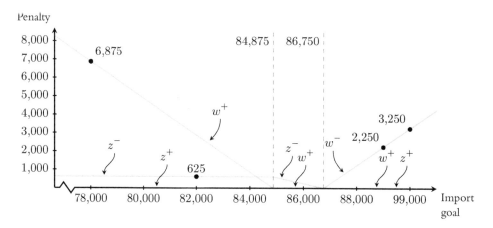

Figure 14.4: Penalties as functions of the import goal.

In lexicographic optimization, u_1 is always the sum of the unwanted deviations from the goals of the soft constraints. If $u_1^* = 0$, then there is a solution for which all the unwanted deviations in the 'soft' constraints are zero. This means that there is a feasible solution if all the 'soft' constraints are considered as 'hard' constraints. For the remaining part, $u_2, \ldots, u_n$, we discuss the following two options.

Option 1. Ranking the objectives.

The decision maker assigns a ranking in order of importance to the different objectives. First, an optimal solution is determined for the most important objective while ignoring the other objectives. In considering the next most important objective, the optimal value of the first objective should not get worse. This can be achieved by adding a constraint to the model that ensures that the first optimal objective value remains the same. The final solution is obtained by repeating the process of optimizing for each successive objective and ensuring that no previous objective value gets worse. Note that the algorithm proceeds to a next (less important) objective only if the current model has multiple optimal solutions.

Option 2. Weighted sum of deviations from objective goals.

In this option there are only two variables u_1 and u_2 in the optimization function; u_1 is the sum of all unwanted deviations from the soft goals, and u_2 is the sum of all unwanted deviations from the objective goals where the weight factors are the corresponding penalties.

As an illustration of option 1, consider the following situation. Suppose that all constraints of model (O4) are soft. The unwanted deviations are a^-, b^-, c^-, m^+, n^+, z^-, and w^+. The management considers deviations from the profit goal as more unwanted than deviations from the import goal. A problem may arise, because in (1'), (2'), and (3') it is allowed that the production exceeds the demand. We leave the discussion of this problem to the reader.

The model obtained by choosing option 1, which we call model (O6), then becomes:

$$\text{lexmin} \left[(a^- + b^- + c^- + m^+ + n^+) \ (z^-) \ (w^+) \right] \qquad \text{(O6)}$$

$$
\begin{aligned}
\text{s.t. } & x_1 + t_1 - a^+ + a^- && = \ 3{,}000 && (1') \\
& x_2 + t_2 - b^+ + b^- && = \ 5{,}000 && (2') \\
& x_3 + t_3 - c^+ + c^- && = \ 7{,}000 && (3') \\
& 0.55x_1 + 0.40x_2 + 0.60x_3 - m^+ + m^- && = \ 2{,}400 && (4') \\
& x_1 + x_2 + x_3 - n^+ + n^- && \leq \ 6{,}000 && (5') \\
& 16x_1 + 18x_2 + 11x_3 + 13t_1 + 17t_2 + 9t_3 - z^+ + z^- && = 198{,}500 && (6) \\
& 7t_1 + 7t_2 + 9t_3 - w^+ + w^- && = 78{,}000 && (7)
\end{aligned}
$$

$$x_1, x_2, x_3, t_1, t_2, t_3, z^+, z^-, w^+, w^- \geq 0$$
$$a^+, a^-, b^+, b^-, c^+, c^-, m^+, m^-, n^+, n^- \geq 0.$$

Since the objective 'function' consists of three factors, the solution procedure can be carried out in at most three steps.

Step 1. The objective is $\min(a^- + b^- + c^- + m^+ + n^+)$. An optimal solution is:

$z = 210{,}000, w = 84{,}000$
$x_1 = 0, x_2 = 6{,}000, x_3 = 0$
$t_1 = 3{,}000, t_2 = 0, t_3 = 7{,}000$
$a^+ = 0, a^- = 0, b^+ = 1{,}000, b^- = 0, c^+ = 0, c^- = 0$
$m^+ = 0, m^- = 0, n^+ = 0, n^- = 0$
$z^+ = 11{,}500, z^- = 0, w^+ = 6{,}000, w^- = 0$

Note that the unwanted deviations are all zero.

Step 2. The objective is $\min z^-$; the constraint $a^- + b^- + c^- + m^+ + n^+ = 0$ is added to the model in Step 1. The optimal solution in Step 1 is also an optimal solution here.

Step 3. The objective is $\min w^+$; the constraint $z^- = 0$ is added to the model in Step 2. The final result is:

Optimal solution model (O6)

$z = 198{,}500, w = 79{,}860$
$x_1 = 0, x_2 = 5{,}310, x_3 = 460$
$t_1 = 3{,}000, t_2 = 0, t_3 = 6{,}540$
$a^+ = 0, a^- = 0, b^+ = 310, b^- = 0, c^+ = 0, c^- = 0$
$m^+ = 0, m^- = 0, n^+ = 0, n^- = 230$
$z^+ = 0, z^- = 0, w^+ = 1{,}860, w^- = 0$

When comparing this solution to the optimal solution of model (O3) we see that both the profit and the import goals are better approximated (compare also the optimal solutions of (O1) and (O2)), mainly at the cost of producing an additional 310 meters of T_2 tubes

$(b^+ = 310)$. We leave to the reader the discussion of the problem that there is no immediate demand for these 310 meters.

The model obtained by choosing option 2, which we call model (O7), reads as follows:

$$\text{lexmin } \left[(a^- + b^- + c^- + m^+ + n^+) \ (2z^- + w^+) \right] \tag{O7}$$

$$\text{s.t. The constraints of model (O6).} \tag{14.1}$$

The optimal solution to model (O7) turns out to be the same as the optimal solution of model (O7).

14.6.2 Fuzzy linear optimization

The basic idea of fuzzy linear optimization is to determine a feasible solution that minimizes the largest weighted deviation from any goal. Assume that all objectives are of the maximization type. The number of objectives is denoted by m. Define:

δ = the decision variable representing a worst deviation level;
z_k = the k'th objective function ($k = 1, \ldots, m$);
U_k = maximum value of z_k, when solving all LO-models separately;
L_k = minimum value of z_k, when solving all LO-models separately;
d_k = the 'fuzzy parameter'.

The expression 'solving all LO-models separately' means solving the LO-models successively for all objectives. In the case of our production-import problem, the two models are (O1) and (O2), where in the latter case the objective is $\max(-7t_1 - 7t_2 - 9t_3)$. Hence, $U_1 = 198,500$, $L_1 = 193,333$, $U_2 = -78,000$, and $L_2 = -86,750$.

The general form of a fuzzy LO-model can then be formulated as follows.

$$\min \delta$$

$$\text{s.t. } \delta \geq \frac{U_k - z_k}{d_k} \qquad \text{for } k = 1, \ldots, m$$

The original constraints

$$\delta \geq 0.$$

The right hand side vector, $\left[(U_1 - z_1)/d_1 \ \ldots \ (U_m - z_m)/d_m \right]$, is called the *fuzzy membership vector function* of the model. We consider two choices for the fuzzy parameter, namely:

$$d_k = U_k - L_k, \text{ and } d_k = 1.$$

First, take $d_k = U_k - L_k$ for each $k = 1, \ldots, m$. The production-import problem in terms of fuzzy linear optimization can then be formulated as follows; we call it model (O8).

$$\min \delta \tag{O8}$$

$$\text{s.t. } \delta \geq \frac{198,500 - 16x_1 - 18x_2 - 11x_3 - 13t_1 - 17t_2 - 9t_3}{5,166.67}$$

$$\delta \geq \frac{-78{,}000 + 7t_1 + 7t_2 + 9t_3}{8{,}750}$$

Constraints $(1) - (5)$

$x_1, x_2, x_3, t_1, t_2, t_3, \delta \geq 0.$

Optimal solution model (O8)

$\delta = 0.4612$
$x_1 = 1{,}918.26, x_2 = 3{,}362.39, x_3 = 0$
$t_1 = 1{,}081.74, t_2 = 1{,}637.61, t_3 = 7{,}000$
$z = 196{,}117.17, w = 82{,}035.43$

The LO-model, called model (O9), in case $d_k = 1$ for all $k = 1, \ldots, m$, reads:

$$\min \delta \qquad\qquad\qquad\qquad\qquad\qquad\qquad\qquad\qquad\qquad \text{(O9)}$$

$$\text{s.t. } \delta \geq 198{,}500 - 16x_1 - 18x_2 - 11x_3 - 13t_1 - 17t_2 - 9t_3$$

$$\delta \geq -78{,}000 + 7t_1 + 7t_2 + 9t_3$$

Constraints $(1) - (5)$

$x_1, x_2, x_3, t_1, t_2, t_3, \delta \geq 0.$

Optimal solution model (O9)

$\delta = 3{,}014.71$
$x_1 = 1{,}529.41, x_2 = 3{,}897.06, x_3 = 0$
$t_1 = 1{,}470.59, t_2 = 1{,}102.94, t_3 = 7{,}000$
$z = 195{,}485.29, w = 81{,}014.71$

14.7 A comparison of the solutions

Figure 14.5 shows the optimal solutions of the various models discussed in the previous sections. The horizontal axis shows the values of net profit z, and the vertical axis the values of the cost of imports w.

The points labeled O1, ..., O9 refer to the optimal solutions of the models (O1), ..., (O9), respectively. The points labeled $\alpha(1)$, $\alpha(2)$, $\alpha(3)$, and $\alpha(4)$ refer to the optimal solutions corresponding to the four intervals of Figure 14.2. Note that O1, O2, and O4 are the three extreme cases. In the case of (O4) and (O5), the production capacity needs to be extended, for instance by means of working in overtime. The optimal solution of model (O6) can only be realized when extra demand is created. Furthermore, the points O8 and O9 belong to the same set of Pareto optimal points. Note that O8, which is the optimal solution of the first fuzzy LO-model in Section 14.6.2, lies more or less in the middle between the points O1 and O2. So, the optimal solution of model (O8) might be considered as the best solution for the production-import problem of Plastics International.

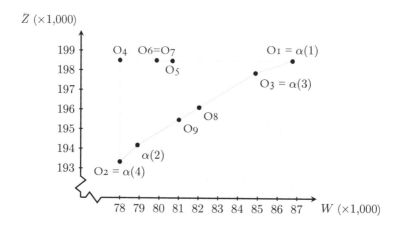

Figure 14.5: Optimal production-import solutions.

14.8 GMPL model code

The following listing gives the GMPL model code for model (O1).

```
1        K;                                      # number of tubes
2        d{k     1..K};                               # demand
3        p{k     1..K};                         # selling price
4        cx{k    1..K};                        # production cost
5        ct{t    1..K};                        # purchasing cost
6        rm{t    1..K};                          # machine time
7        rf{t    1..K};                     # finishing material
8        RM;                          # total available machine time
9        RF;                    # total available finishing material
10
11    x{k     1..K} >= 0;              # amount of production of tube T[k]
12    t{k     1..K} >= 0;          # amount of tube T[k] to be purchased
13    z;                                          # total profit
14    w;                                       # total import costs
15
16         obj:                                     # objective
17   z;
18          demand{k     1..K}:                  # demand constraints
19   x[k] + t[k] = d[k];
20          time_constr:                          # time constraints
21      {k     1..K} rm[k] * x[k] <= RM;
22          material_constr:                  # material constraints
23      {k     1..K} rf[k] * x[k] <= RF;
24          z_def:                             # definition of z
25   z =     {k     1..K} ((p[k] − cx[k]) * x[k] + (p[k] − ct[k]) * t[k]);
26          w_def:                             # definition of w
27   w =     {k     1..K} ct[k] * t[k];
28
29    ;
30
31       K   := 3;
```

```
32          : d   := 1 3000   2 5000   3 7000;
33          : p   := 1 20      2 24     3 18;
34          : cx := 1 4        2 6      3 7;
35          : ct := 1 7        2 7      3 9;
36          : rm := 1 0.55    2 0.40   3 0.60;
37          : rf := 1 1        2 1      3 1;
38          RM := 2400;
39          RF := 6000;
40
41      ;
```

14.9 Exercises

Exercise 14.9.1. Consider model (O3). Determine a feasible solution of model (O3) that does not satisfy the conditions $z^+ \times z^- = 0$ and $w^+ \times w^- = 0$. Show that this solution is not optimal.

Exercise 14.9.2. Explain the signs of the optimal dual values in Table 14.3.

Exercise 14.9.3. What are the effects on the optimal solutions of the models (O1) – (O9), when the quantity of finishing material is restricted to 4,000 grams? (The original quantity is 6,000 grams; see Table 14.1.)

Exercise 14.9.4. Show that the optimal solution of model (O3) is nondegenerate and unique.

Exercise 14.9.5. Consider the production-import problem (O3). Suppose that the constraint $x_1 + t_1 = 3,000$ is soft. Let a^+ and a^- be the deviations from the goal 3,000, such that $x_1 + t_1 - a^+ + a^- = 3,000$. Determine the smallest positive value of the penalty γ in the objective min $2z^- + w^+ + \gamma a^+$ such that the optimal production-import schedule is equal to the optimal schedule when the constraint $x_1 + t_1 = 3,000$ was still hard.

Exercise 14.9.6. Consider model (O5). Determine the smallest positive value of the penalty β of m^+ in min $2z^- + w^+ + \beta m^+$ such that in the optimal solution it holds that $z = 197,875$ and $w = 84,785$ (being the optimal solution of (O3)).

Exercise 14.9.7. Consider Figure 14.2.

(a) Explain why $z^*(\alpha)$ and $w^*(\alpha)$ are nondecreasing staircase functions.

(b) Determine all values of α for which $x_3 \geq 1,000$.

(c) Analyze model (O2) with the extra constraints: $x_1, x_2, x_3 \geq 1,000$.

(d) Explain why $z^+ = 0$ for all values of α.

Exercise 14.9.8. Consider Figure 14.3.

(a) Determine the kink point between $193{,}333.33$ and $197{,}875$.

(b) Determine for each interval the basic and nonbasic variables, and whether or not the corresponding optimal solution is unique or nondegenerate.

(c) Show that $w^- = 0$.

(d) Derive an analytic expression for w^+.

Exercise 14.9.9. Repeat the analysis in Section 14.5.1 for the case that the profit penalty is set to 1, and the import penalty to a parameter λ.

CHAPTER 15

Coalition formation and profit
distribution

Overview

This case study is based on Bröring (1996), and deals with a 'game theoretical' approach to a problem where a group of farmers considers to cooperate as a result of favorable negotiations about profit distribution.

15.1 The farmers cooperation problem

Consider the following situation. Farmers in a certain geographical region produce flour and sugar. In addition, each farmer has a number of cows whose milk is used for producing butter. The farmers can roughly be divided into three different groups. We will refer to the groups as Type 1, Type 2, and Type 3. For simplicity, we assume that farmers that are of the same type work under the same circumstances and have the same annual yields. Table 15.1(a) lists the annual yields of butter, flour, and sugar for the three types of farmers. A complicating factor is that, although only three types of farmers can be distinguished, the number of farmers per type is not known in advance.

The butter, flour, and sugar produced by the farmers are used to produce a variety of products. In this case study, we assume that these products are packets of biscuits. We assume that there are three kinds of biscuits, which we call Bis1, Bis2, and Bis3. Table 15.1(b) lists the amounts of butter, flour, and sugar that are required for the production of one packet of each type of biscuit. It is assumed that these amounts are the only restrictive factors; factors such as labor, electricity and/or gas for the ovens, and packaging material are not taken into consideration. The packets of biscuits are sold to retailers. We assume that packets of biscuits can always be sold to a retailer against a fixed price. The bottom row of Table 15.1(b) contains the corresponding per-packet profits for the three products (in € per packet).

	Type 1	Type 2	Type 3
Butter	2,000	2,000	2,000
Flour	8,000	2,000	4,000
Sugar	2,000	1,000	3,000

(a) The annual production (in ounces).

	Bis1	Bis2	Bis3
Butter	2	0	1
Flour	2	4	3
Sugar	1	0	1
Profit	5	2	7

(b) The amounts needed (in ounces per packet), and the profits (in € per packet).

Table 15.1: Resources required for the production of the biscuits.

Each farmer can individually decide on his biscuit production. For example, a Type 2 farmer can use his butter, flour, and sugar to produce $1,000$ packets of Bis1 biscuits, which results in a $(1,000 \times €5 =) €5,000$ profit. Similarly, a Type 1 farmer has enough butter, flour, and sugar to produce $2,000$ packets of Bis3 biscuits, resulting in a profit of $(2,000 \times €7 =) €14,000$. In fact, the production of $2,000$ packets of Bis3 biscuits leaves $2,000$ ounces of flour unused, which can be used for an additional 500 packets of Bis2 biscuits, resulting in an additional profit of $(500 \times €2 =) €1,000$. So, a Type 1 farmer can make a total profit of $€15,000$.

However, since each farmer produces different amounts of butter, flour, and sugar, it may be more profitable for farmers to combine their yields in order to increase the total production (i.e., the number of packets of biscuits). Consider again the Type 1 and Type 2 farmers. If a Type 1 farmer and a Type 2 farmer decide to produce biscuits individually, the two farmers will have a joint profit of $(€5,000 + 15,000 =) €20,000$. Suppose, however, that they decide to combine their yields. This means that they will have combined yields of $4,000$ ounces of butter, $10,000$ ounces of flour, and $3,000$ ounces of sugar. With the combined yields, they can jointly produce $3,000$ packets of Bis3 biscuits, which gives a joint profit of $(3,000 \times €7 =) €21,000$. In addition, they can produce 250 packets of Bis2 biscuits, yielding another $€500$. So, combining their yields leads to a joint profit of $€21,500$, which is $€1,500$ more than their combined profit if they decide to work separately. Therefore, it is profitable for them to work together.

An important question that arises is: if farmers decide to combine their yields, how do we find a fair distribution of the total earned profit? Clearly, the Type 1 farmer will only agree to combine his yields with the Type 2 farmer if he receives at least as much profit as when he produces biscuits individually. That is, the Type 1 farmer will want to receive at least $€15,000$. Similarly, the Type 2 farmer will want to receive at least $€5,000$. As long as the farmers agree on a profit distribution that satisfies these restrictions, the Type 1 farmer and the Type 2 farmer will agree to cooperate. So, the two farmers could agree to split the profits so that the Type 1 farmer receives, say, $€16,000$, and the Type 2 farmer receives $€5,500$.

Another question is whether or not a farmer should cooperate with just one farmer, or with a number of farmers, or even with all other farmers. A group of farmers that work together is called a *coalition*. We argued above that a farmer will only cooperate with a coalition if his total profit turns out to be at least as large as his profit when he organizes the production of biscuits individually. If this is not the case, the farmer will 'split off' and work by himself. More generally, it could happen that a group of farmers within a coalition (a so-called *subcoalition*) realizes that it can make more profit by splitting off and forming a coalition by itself. In general, a subcoalition will only cooperate with a larger coalition if its joint profit turns out to be at least as large as its joint profit when it splits off and organizes the production of biscuits within the subcoalition.

So, the question is to determine a distribution of the total profit, such that no subcoalition has an incentive to split off. This means that no subcoalition can make each of its members better off by working together. In the following section we will show how this problem can be formulated as a so-called *game theory* problem.

15.2 Game theory; linear production games

Mathematical game theory is a mathematical tool for describing conflict situations, in which a number of parties with (usually) conflicting interests have to take one or more decisions. The parties have a well-specified goal, which is usually based on profit maximization. Moreover, each party's outcome is affected by the other parties' decisions. Such a conflict situation is called a *game*. The parties or individuals are called *players*. It is assumed that the players act rationally, meaning that they are selfish and they try to maximize their individual profits. The following entities are relevant in mathematical game theory.

The set of players.
The number of players need not be equal to the number of individuals; a player may be a group of individuals as well.

The decision possibilities of the players.
Each player has a certain set of possible decision choices. These choices may depend on the player. Also the time point in the game at which a decision is made may vary.

The pay-offs (or profits).
At the end of the game, each player receives a pay-off. The pay-off generally depends on the decisions made by all players.

The information available to each player.
Each player has to make a decision based on the information available to him/her. It is assumed that the players do not know in advance what the other players' decisions will be. (However, by reasoning, it is sometimes possible for a player to rule out one or more of the other players' decision possibilities.) It is also possible that players have private information that is not known to the other players. A player's decision crucially depends on the information that is available to him/her. If all aspects of the game are known to all

players, we say that the players have *complete information*. If, in addition, all players know that all players have complete information, then we speak of *common knowledge*.

The possibility of making binding agreements.

One can distinguish between *cooperative* and *noncooperative games*. In noncooperative games, every player attempts to maximize his/her individual pay-off by competing with his opponents. In cooperative games, however, it is possible to make mutual agreements. A set of players that cooperate is called a *coalition*. Within a coalition, the agreements about the decisions to be made are binding for all players in the coalition.

The current case study deals with a cooperative game in which the players are the three types of farmers. The number of farmers of the individual types is not known. It is assumed that if one farmer of a certain type is in a coalition, then all farmers of that type are in that coalition. Before the actual production of biscuits starts, the farmers decide about their cooperation with other farmers. These decisions depend on the agreed upon distribution of the total profit. Since each farmer offers a given amount of input goods, to be transformed into output products with a certain market value, we are in the situation of a zero-sum game. Furthermore, since the number of farmers per type is not known there is incomplete information.

A linear production game is a cooperative game in which the players have the possibility to form coalitions, and there are no limitations concerning the possibility of communication between the players. Each player has a bundle of resources, which are used for production of output goods. It is assumed that the production process is linear, meaning that the amounts of required input goods are linear functions of the desired production level. This means, for instance, that doubling the amount of input goods results in a doubling of the production level. Furthermore, we assume that the products can be sold at fixed market prices and that the selling process is linear. The owners of the resources can work either individually, or cooperate and possibly increase their total profits.

We define the following symbols and concepts.

n = the number of players.

m = the number of types of input goods.

r = the number of types of output goods.

$\mathbf{a}$ = the *coalition vector*, where the k'th entry a_k ($\in \mathbb{N}$) denotes the number of farmers of Type k ($k = 1, \ldots, n$).

$v(\mathbf{a})$ = the optimal total profit for the coalition vector $\mathbf{a}$. We will write $v([a_1 \ \ldots \ a_n]^{\mathsf{T}}) = v(a_1, \ldots, a_n)$. The value of $v(\mathbf{a})$ is the *coalition value* of the coalition vector $\mathbf{a}$.

$\mathbf{A}$ = the input-output matrix, which is an (m, r)-matrix, where each entry a_{ij} denotes the amount of input good i required for the production of one unit of output good j (in ounces per packet).

$\mathbf{x}$ = the output vector, which is an r-vector, where the j'th entry denotes the amount of output good j to be produced (in numbers of packets).

$$\mathbf{c} \quad = \text{the profit vector, which is an } r\text{-vector, where the } j\text{'th entry denotes the profit}$$
of output good j (in € per packet).

$\mathbf{P} \quad =$ the capacity matrix, which is an (m, n)-matrix, where entry (i, k) is the amount of input good i of the farmers of Type k.

We assume that none of the rows of $\mathbf{A}$ consist of all zeroes, since this would imply that the corresponding input good is not used at all for production, and it can just be eliminated. Similarly, we assume that none of the columns of $\mathbf{A}$ consist of all zeros, since this would imply that the corresponding output good does not use any input goods.

For two nonnegative n-vectors $\mathbf{a} = \begin{bmatrix} a_1 & \dots & a_n \end{bmatrix}^\mathsf{T}$ and $\mathbf{a}' = \begin{bmatrix} a_1' & \dots & a_n' \end{bmatrix}^\mathsf{T}$, we say that $\mathbf{a}'$ is a *subvector* of $\mathbf{a}$ if, for each $k = 1, \dots, n$, either $a_k' = a_k$ or $a_k' = 0$. If $\mathbf{a}'$ is a *subvector* of $\mathbf{a}$, we write $\mathbf{a}' \subseteq \mathbf{a}$. For instance $\begin{bmatrix} 7 & 0 & 5 \end{bmatrix}^\mathsf{T} \subseteq \begin{bmatrix} 7 & 2 & 5 \end{bmatrix}^\mathsf{T}$.

The coalition that corresponds to the vector $\mathbf{a}$ is called the *grand coalition*. Any coalition that corresponds to a nonnegative integer-valued subvector $\mathbf{a}'$ of $\mathbf{a}$ is called a *subcoalition* of the grand coalition. For instance, in the subcoalition corresponding to the vector $\begin{bmatrix} 7 & 0 & 5 \end{bmatrix}^\mathsf{T}$ (which is a subvector of the grand coalition vector $\begin{bmatrix} 7 & 2 & 5 \end{bmatrix}^\mathsf{T}$): only the Type 1 and the Type 3 farmers form a coalition. Moreover, the coalition corresponding to the vector $\mathbf{0}$ is called the *empty (sub)coalition*.

For any coalition vector $\mathbf{a}$, $v(\mathbf{a})$ is the maximum possible total profit if the farmers form the coalition corresponding to $\mathbf{a}$. This value can be calculated by solving the following LO-model:

$$v(\mathbf{a}) = \max \mathbf{c}^\mathsf{T}\mathbf{x}$$
$$\text{s.t. } \mathbf{A}\mathbf{x} \leq \mathbf{P}\mathbf{a} \qquad \qquad \text{(GP)}$$
$$\mathbf{x} \geq \mathbf{0},$$

where $\mathbf{c}^\mathsf{T}\mathbf{x} = c_1 x_1 + \dots + c_r x_r$ is the total profit (to be maximized), and $(\mathbf{A}\mathbf{x})_i = a_{i1}x_1 + \dots + a_{ir}x_r$ the total amount of input good i used for the production. Clearly, the value of $(\mathbf{A}\mathbf{x})_i$ can be at most the total available amount of input good i, which is $(\mathbf{P}\mathbf{a})_i = p_{i1}a_1 + \dots + p_{in}a_n$ ($i = 1, \dots, m$).

It follows from the information in Table 15.1 that, in this case study, we have that:

$$\mathbf{A} = \begin{bmatrix} 2 & 0 & 1 \\ 2 & 4 & 3 \\ 1 & 0 & 1 \end{bmatrix}, \mathbf{c} = \begin{bmatrix} 5 \\ 2 \\ 7 \end{bmatrix}, \text{ and } \mathbf{P} = \begin{bmatrix} 2 & 2 & 2 \\ 8 & 2 & 4 \\ 2 & 1 & 3 \end{bmatrix}.$$

Hence, the LO-model that determines the optimal value of the coalition vector $\mathbf{a} = \begin{bmatrix} a_1 \ a_2 \ a_3 \end{bmatrix}^{\mathsf{T}}$ reads:

$$
\begin{aligned}
v(\mathbf{a}) = \max \quad & 5x_1 + 2x_2 + 7x_3 \\
\text{s.t.} \quad & 2x_1 \quad\quad\quad + \quad x_3 \leq 2a_1 + 2a_2 + 2a_3 && (15.1) \\
& 2x_1 + 4x_2 + 3x_3 \leq 8a_1 + 2a_2 + 4a_3 && (15.2) \\
& x_1 \quad\quad\quad + \quad x_3 \leq 2a_1 + \quad a_2 + 3a_3 && (15.3) \\
& x_1, x_2, x_3 \geq 0.
\end{aligned}
$$

Note that x_1, x_2, and x_3 are expressed in amounts of $1{,}000$ units. Constraints (15.1), (15.2), and (15.3) are the constraints for the input goods butter, flour, and sugar, respectively.

We first consider the special case in which there is exactly one farmer of each type. This means that the grand coalition vector is $\mathbf{1} = \begin{bmatrix} 1 \ 1 \ 1 \end{bmatrix}^{\mathsf{T}}$. Then, the right hand side vector is $\mathbf{P1} = \begin{bmatrix} 6 \ 14 \ 6 \end{bmatrix}^{\mathsf{T}}$. In this special case, the above model can easily be solved for all subcoalitions. There are seven subcoalition vectors $\mathbf{a}' = \begin{bmatrix} a_1 \ a_2 \ a_3 \end{bmatrix}^{\mathsf{T}} \subseteq \begin{bmatrix} 1 \ 1 \ 1 \end{bmatrix}^{\mathsf{T}}$ (including the grand coalition, and excluding the empty coalition). Table 15.2 lists these seven possibilities, together with the corresponding optimal production schedule $\mathbf{x}^*$, and the corresponding value $v(\mathbf{a}')$ of the coalition.

The maximum value of $v(\mathbf{a}')$ in Table 15.2 is attained by the grand coalition; see Exercise 15.6.1. In this case the production schedule consists of $1{,}000$ packets of Bis1, 0 packets of Bis2, and $4{,}000$ packets of Bis3. The corresponding total profit is €$33{,}000$. The last column of Table 15.2 contains, for each subcoalition, the average profit $v_{\text{avg}}(\mathbf{a}')$ ($\times$€$1{,}000$) per farmer, i.e., the profit for each farmer when the total profit is equally divided over the farmers in the corresponding subcoalition.

From Table 15.2, we can deduce that if the farmers decide to divide profits equally, then not all farmers will be happy with the coalition. Indeed, suppose that they decide to do so. In the grand coalition, each farmer receives €$11{,}000$ profit. However, consider the subcoalition corresponding to $\mathbf{a}' = \begin{bmatrix} 1 \ 0 \ 1 \end{bmatrix}^{\mathsf{T}}$, in which the Type 1 farmer cooperates with the Type 3 farmer, and the Type 2 farmer works alone. In that subcoalition, the Type 1 and Type 3 farmers both receive €$14{,}000$. So, it is profitable for the Type 1 and 3 farmers to abandon the grand coalition and start their own subcoalition. But, in fact, this subcoalition will not work either: the Type 1 farmer can decide to work alone and make a €$15{,}000$ profit. This corresponds to the subcoalition $\begin{bmatrix} 1 \ 0 \ 0 \end{bmatrix}^{\mathsf{T}}$. Note that the Type 1 farmer will not cooperate with any other farmer unless his share is at least €$15{,}000$.

The question is: how to divide the maximum possible total profit of €$33{,}000$, in such a way that the farmers are satisfied with the grand coalition? Is this even possible? In general, suppose that it has been decided to cooperate according to the (sub)coalition vector $\mathbf{a}$. The question then is: how to distribute the total profit earned by this coalition, such that the farmers of the coalition are all satisfied with their shares of the total profit?

$\mathbf{a}'$	$\mathbf{x}^*$	$v(\mathbf{a}')$	$v_{\text{avg}}(\mathbf{a}')$
$[\ 1\quad 1\quad 1\ ]$	$[\ 1\quad 0\quad 4\ ]$	33	11
$[\ 1\quad 1\quad 0\ ]$	$[\ 0\quad \frac{1}{4}\quad 3\ ]$	$21\frac{1}{2}$	$10\frac{3}{4}$
$[\ 1\quad 0\quad 1\ ]$	$[\ 0\quad 0\quad 4\ ]$	28	14
$[\ 0\quad 1\quad 1\ ]$	$[\ 1\frac{1}{2}\quad 0\quad 1\ ]$	$14\frac{1}{2}$	$7\frac{1}{4}$
$[\ 1\quad 0\quad 0\ ]$	$[\ 0\quad \frac{1}{2}\quad 2\ ]$	15	15
$[\ 0\quad 1\quad 0\ ]$	$[\ 1\quad 0\quad 0\ ]$	5	5
$[\ 0\quad 0\quad 1\ ]$	$[\ \frac{1}{2}\quad 0\quad 1\ ]$	$9\frac{1}{2}$	$9\frac{1}{2}$

Table 15.2: Optimal production schedules in the case of three farmers.

15.3 How to distribute the total profit among the farmers?

In order to answer the question in the title of this section, we will introduce the concept of 'the core of the game'. Let $\mathbf{a}$ be any (sub)coalition vector. The *core* with respect to $\mathbf{a}$ is a set of n-vectors, where each vector represents a distribution of the total profit among the farmers in such a way that the farmers are willing to cooperate according to the (sub)coalition vector $\mathbf{a}$.

Clearly, a farmer is willing to cooperate if his share of the total profit is at least as much as in the case where he joins any other coalition. Actually, this has to be true not only for a single farmer, but for each subcoalition. Define:

$$u_k = \text{the payout for a farmer of Type } k \ (k = 1, \ldots, n).$$

The vector $\mathbf{u} = \begin{bmatrix} u_1 & \ldots & u_n \end{bmatrix}^{\mathsf{T}}$ is called the *profit distribution vector*. For any (sub)coalition n-vector $\mathbf{a}$, the *core* is denoted and defined by

$$C(\mathbf{a}) = \left\{ \mathbf{u} \in \mathbb{R}^n \ \middle|\ \mathbf{a}^{\mathsf{T}}\mathbf{u} = v(\mathbf{a}), \text{ and } (\mathbf{a}')^{\mathsf{T}}\mathbf{u} \geq v(\mathbf{a}') \text{ for each } \mathbf{a}' \geq \mathbf{0} \text{ with } \mathbf{a}' \subseteq \mathbf{a} \right\}.$$

The set of equalities $\mathbf{a}^{\mathsf{T}}\mathbf{u} = v(\mathbf{a})$ means that the full profit is distributed, while the inequalities in $C(\mathbf{a})$ mean that there is no incentive for any coalition to split off: $(\mathbf{a}')^{\mathsf{T}}\mathbf{u}$ is the total profit received by the members of coalition $\mathbf{a}'$ when all of them are collaborating according to the coalition $\mathbf{a}$, and the payouts are given by $\mathbf{u}$. The right hand side of the inequalities, $v(\mathbf{a}')$, is the total profit that coalition $\mathbf{a}'$ can make if they split off. Instead of $C(\begin{bmatrix} a_1 & \ldots & a_n \end{bmatrix}^{\mathsf{T}})$ we simply write $C(a_1, \ldots, a_n)$. If there is one farmer of each type and there are three types, then the core is the solution set of the following collection of equations

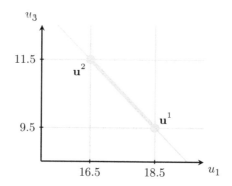

Figure 15.1: Core of the game with $\mathbf{a} = \begin{bmatrix} 1 & 1 & 1 \end{bmatrix}^{\mathsf{T}}$. The figure shows the values of u_1 and u_3. All points in the core satisfy $u_2 = 5$.

and equalities:

$$
\begin{aligned}
u_1 + u_2 + u_3 &= 33 \\
u_1 + u_2 &\geq 21\tfrac{1}{2} \\
u_1 \qquad\; + u_3 &\geq 28 \\
u_2 + u_3 &\geq 14\tfrac{1}{2} \\
u_1 &\geq 15 \\
u_2 &\geq 5 \\
u_3 &\geq 9\tfrac{1}{2}.
\end{aligned}
$$

It is easy to verify that in this simple case the core satisfies:

$$
C(1,1,1) = \left\{ \begin{bmatrix} u_1 & u_2 & u_3 \end{bmatrix}^{\mathsf{T}} \in \mathbb{R}^3 \;\middle|\; \begin{array}{l} 16\tfrac{1}{2} \leq u_1 \leq 18\tfrac{1}{2}, u_2 = 5, \\ 9\tfrac{1}{2} \leq u_3 \leq 11\tfrac{1}{2}, u_1 + u_3 = 28 \end{array} \right\}.
$$

This set is depicted in Figure 15.1. The interpretation is that, if the profits are divided according to any vector $[u_1 \; u_2 \; u_3]^{\mathsf{T}} \in C(1,1,1)$, no group of farmers is better off by not participating in the grand coalition. If, on the other hand, the profits are divided according to a vector that is not in $C(1,1,1)$, then there is some group of farmers that is not satisfied, and which will not want to be part of the grand coalition.

Since the number of constraints in the definition of $C(\mathbf{a})$ may be very large when the number of farmers is large, it is in general not at all straightforward to find the core, or even a point in the core. In the remainder of this section, we will use the dual of model (GP) to determine a so-called Owen point of the core, and show that it is in fact a point in the core.

For any (sub)coalition vector $\mathbf{a} \in \mathbb{N}^n$, the dual of model (GP) reads:

$$
\begin{aligned}
v(\mathbf{a}) = \min \; & \mathbf{a}^{\mathsf{T}} \mathbf{P}^{\mathsf{T}} \mathbf{y} \\
\text{s.t. } & \mathbf{A}^{\mathsf{T}} \mathbf{y} \geq \mathbf{c} \\
& \mathbf{y} \geq \mathbf{0}.
\end{aligned}
\qquad\qquad \text{(GD)}
$$

Both the primal and the dual model are in standard form; see Section 1.2.1.

Theorem 15.3.1.
Model (GP) and (GD) are both feasible.

Proof. Model (GP) is feasible, because $\mathbf{x} = \mathbf{0}$ is a feasible point. To see that model (GD) is feasible as well, notice first that $\mathbf{A} \geq \mathbf{0}$ and $\mathbf{c} \geq \mathbf{0}$. We will show that the vector $\mathbf{y}$, of which the entries satisfy:

$$y_i = \max_{\ell=1,\ldots,r} \left\{ \frac{c_\ell}{a_{i\ell}} \,\middle|\, a_{i\ell} > 0 \right\}$$

is a feasible point of model (GD). Note that y_i in the above expression is well-defined, because we assumed that no row of $\mathbf{A}$ consists of all zeroes, and hence there exists at least one ℓ such that $a_{i\ell} > 0$ in the expression. To prove that $\mathbf{y}$ is indeed a feasible point of model (GD), take any $j \in \{1,\ldots,r\}$. Let $I = \{i \mid a_{ij} > 0\}$. Since none of the columns of $\mathbf{A}$ consist of all zeroes, it follows that $I \neq \emptyset$. By the definition of y_i, we have that $y_i \geq \frac{c_j}{a_{ij}}$ for $i \in I$. Therefore, we have that:

$$(\mathbf{A}^\mathsf{T}\mathbf{y})_j = \sum_{i=1}^{m} a_{ij}y_i = \sum_{i \in I} a_{ij}y_i \geq \sum_{i \in I} a_{ij}\frac{c_j}{a_{ij}} = \sum_{i \in I} c_j \geq c_j.$$

This proves that model (GD) is indeed feasible. $\square$

Because model (GP) and (GD) are both feasible, they have the same optimal objective value. Now suppose that $\mathbf{y}^*$ is an optimal solution of model (GD). Define the vector $\mathbf{u}^* = \begin{bmatrix} u_1^* & \ldots & u_n^* \end{bmatrix}^\mathsf{T}$ by:

$$\mathbf{u}^* = \mathbf{P}^\mathsf{T}\mathbf{y}^*.$$

The vector $\mathbf{u}^*$ is called an *Owen point* of the game (with coalition vector $\mathbf{a}$). The following theorem shows that Owen points are points in the core of the game.

Theorem 15.3.2.
For any Owen point $\mathbf{u}^*$ of a linear production game with coalition vector $\mathbf{a}$, we have that $\mathbf{u}^* \in C(\mathbf{a})$.

Proof. Let $\mathbf{y}^*$ be any optimal solution of model (GD), and let $\mathbf{u}^* = \mathbf{P}^\mathsf{T}\mathbf{y}^*$. Note that $v(\mathbf{a})$ is the optimal objective value of both model (GP) and model (GD). Therefore, $v(\mathbf{a}) = \mathbf{a}^\mathsf{T}\mathbf{P}^\mathsf{T}\mathbf{y}^* = \mathbf{a}^\mathsf{T}\mathbf{u}^*$. So, to show that $\mathbf{u}^* \in C(\mathbf{a})$, it suffices to show that $(\mathbf{a}')^\mathsf{T}\mathbf{u}^* \geq v(\mathbf{a}')$ for every $\mathbf{a}' \geq \mathbf{0}$ with $\mathbf{a}' \subseteq \mathbf{a}$. Replace $\mathbf{a}$ by $\mathbf{a}'$ in model (GD). Since $\mathbf{y}^*$ is a feasible (but not necessarily optimal) solution of the new model, it follows that $(\mathbf{a}')^\mathsf{T}\mathbf{u}^* = (\mathbf{a}')^\mathsf{T}\mathbf{P}^\mathsf{T}\mathbf{y}^* \geq v(\mathbf{a}')$. This proves the theorem. $\square$

For the case of the grand coalition of three farmers, an Owen point is easily calculated from the following model:

$$
\begin{aligned}
v(1,1,1) = \min\ & 6y_1 + 14y_2 + 6y_3 \\
\text{s.t.}\quad & 2y_1 + 2y_2 + y_3 \ge 5 \\
& \qquad\quad 4y_2 \qquad\ \ge 2 \\
& \ y_1 + 3y_2 + y_3 \ge 7 \\
& \ y_1, y_2, y_3 \ge 0.
\end{aligned}
$$

The vector $\mathbf{y}^* = \left[\frac{1}{4}\ 2\frac{1}{4}\ 0\right]^{\mathsf{T}}$ is an optimal solution of this model, with $v(1,1,1) = 33$. Hence, in the case with one farmer of each type, we obtain the Owen point:

$$
\mathbf{u}^* = \left[18\tfrac{1}{2}\ \ 5\ \ 9\tfrac{1}{2}\right]^{\mathsf{T}}.
$$

This is the point $\mathbf{u}^1$ in Figure 15.1. Thus, when the total profit is divided so that the farmers of Type 1, Type 2, and Type 3 receive $18\frac{1}{2}$, 5, and $9\frac{1}{2}$ ($\times€\,1{,}000$), respectively, then no farmer can make a higher profit by forming other coalitions. The above model for $v(1,1,1)$ has a unique solution, and so the game has a unique Owen point. This does not mean that there are no other points in the core for which, for instance, the Type 3 farmer obtains a larger share; see Figure 15.1. So, in general, not every (extreme) point of the core corresponds to an Owen point. However, it may happen that a game has more than one Owen point. This happens if model (GD) has multiple optimal solutions; for example the model corresponding to $\mathbf{a}^2$ in Figure 15.2 of the next section. Also, note that Owen points are of particular interest if the number of farmers is large, because for Owen points not all (exponentially many!) coalition values have to be determined. In Exercise 15.6.11, an example is given of a linear production game with several Owen points. Actually, in this example the core is equal to the set of Owen points.

15.4 Profit distribution for arbitrary numbers of farmers

The problem of this case study is to determine a distribution of the total profit in such a way that all farmers are satisfied with their respective shares. A complicating factor is that only the number of the types of farmers (namely, three) is known, not the actual number of farmers that will participate. In this section, we will analyze how the set of Owen points depends on the numbers of farmers of each type.

We first make the following important observation. Let p be a positive integer, and $\mathbf{a}$ an n-vector. A game with coalition vector $p\mathbf{a} = \left[pa_1\ \dots\ pa_n\right]^{\mathsf{T}}$ is called a *p-replica game*. Hence, for a p-replica game with coalition vector $p\mathbf{a}$, an Owen point is calculated from the following model:

$$
\begin{aligned}
v(p\mathbf{a}) = \min\ & (p\mathbf{a})^{\mathsf{T}}\mathbf{P}^{\mathsf{T}}\mathbf{y} \\
\text{s.t.}\ & \mathbf{A}^{\mathsf{T}}\mathbf{y} \ge \mathbf{c} \qquad\qquad\qquad (\text{pGD}) \\
& \mathbf{y} \ge \mathbf{0}.
\end{aligned}
$$

Clearly, $v(p\mathbf{a}) = pv(\mathbf{a})$, and model (GD) and model (pGD) have the same set of optimal solutions. Hence, model (GD) and model (pGD) produce the same set of Owen points. Moreover, any Owen point $\mathbf{u}^* = \mathbf{P}^\mathsf{T}\mathbf{y}^*$ of (GD) is an Owen point of (pGD) for each $p \geq 1$. This means that the set of Owen points of a game depends only on the *proportions* of Type 1, Type 2, and Type 3 farmers, and not on the total number of farmers. Hence, in order to gain insight in the set of Owen points as a function of the coalition vector, we may restrict our attention to fractional coalition vectors $[a_1\ a_2\ a_3]^\mathsf{T}$ satisfying $a_1 \geq 0$, $a_2 \geq 0$, $a_3 \geq 0$, and $a_1 + a_2 + a_3 = 1$. The values of a_1, a_2, and a_3 represent the fraction of farmers that are of Type 1, Type 2, and Type 3. By the above discussion, we have that the set of Owen points of any coalition vector $[a_1\ a_2\ a_3]^\mathsf{T}$ can be determined by solving (pGD) applied to the fractional coalition vector:

$$\left[\frac{a_1}{a_1 + a_2 + a_3} \quad \frac{a_2}{a_1 + a_2 + a_3} \quad \frac{a_3}{a_1 + a_2 + a_3} \right]^\mathsf{T}.$$

Hence, we will restrict our attention to only such fractional coalition vectors.

15.4.1 The profit distribution

In order to determine how the set of Owen points depends on the fractional coalition vector $\mathbf{a}$, we solved model (GD) for various such vectors. We found the following four Owen points, denoted by ①, ②, ③, and ④, respectively:

$$① = \begin{bmatrix} 18\frac{1}{2} & 5 & 9\frac{1}{2} \end{bmatrix}^\mathsf{T} \text{ corresponding to } \mathbf{y}_1^* = \begin{bmatrix} \frac{1}{4} & 2\frac{1}{4} & 0 \end{bmatrix}^\mathsf{T},$$

$$② = \begin{bmatrix} 18 & 5 & 11 \end{bmatrix}^\mathsf{T} \text{ corresponding to } \mathbf{y}_2^* = \begin{bmatrix} 0 & 2 & 1 \end{bmatrix}^\mathsf{T},$$

$$③ = \begin{bmatrix} 15 & 12 & 13 \end{bmatrix}^\mathsf{T} \text{ corresponding to } \mathbf{y}_3^* = \begin{bmatrix} 5\frac{1}{2} & \frac{1}{2} & 0 \end{bmatrix}^\mathsf{T},$$

$$④ = \begin{bmatrix} 15 & 6\frac{1}{2} & 18\frac{1}{2} \end{bmatrix}^\mathsf{T} \text{ corresponding to } \mathbf{y}_4^* = \begin{bmatrix} 0 & \frac{1}{2} & 5\frac{1}{2} \end{bmatrix}^\mathsf{T}.$$

It is possible to analytically derive the values of a_1, a_2, a_3 for which each of these points is an Owen point; see Section 15.4.2. We have summarized these derivations in Figure 15.2. Each point $[a_1\ a_2]^\mathsf{T}$ in the graph corresponds to the coalition vector $[a_1\ a_2\ 1-a_1-a_2]^\mathsf{T}$. Thus, the midpoint of the triangle represents the coalition vector $\begin{bmatrix} \frac{1}{3} & \frac{1}{3} & \frac{1}{3} \end{bmatrix}^\mathsf{T}$, which represents all coalitions with equal numbers of Type 1, Type 2, and Type 3. The Owen point we find for such coalitions is ①. This corresponds to payouts of € 18,000 to each Type 1 farmer, € 5,000 to each Type 2 farmer, and € 9,500 to each Type 3 farmer.

The point $\mathbf{a}^2 = \begin{bmatrix} \frac{3}{5} & \frac{1}{5} \end{bmatrix}^\mathsf{T}$ represents all coalitions in which 60% of the farmers are of Type 1, 20% of are of Type 2, and 20% of Type 3. For such a coalition, all four vectors $\mathbf{y}_1^*, \mathbf{y}_2^*, \mathbf{y}_3^*$, and $\mathbf{y}_4^*$ are optimal for model (GD), and so all points of the set $\mathrm{conv}\{\mathbf{y}_1^*, \mathbf{y}_2^*, \mathbf{y}_3^*, \mathbf{y}_4^*\}$ are optimal as well. In Exercise 15.6.4 the reader is asked to draw the feasible region of model (GD) in this case, and to show that the optimal set is a facet (see Appendix D) of the feasible region. Observe also that, in Figure 15.2, each region for which a particular Owen point

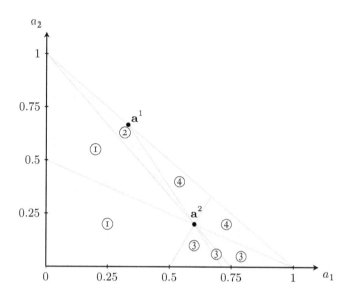

Figure 15.2: Owen points for different distributions of farmers of Type 1, Type 2, and Type 3. Each point $[a_1 \ a_2]^T$ with $a_1 \geq 0$, $a_2 \geq 0$, $a_1 + a_2 \leq 1$, corresponds to a coalition vector $[a_1 \ a_2 \ a_3]^T$ with $a_3 = 1 - a_1 - a_2$. The marked points are $\mathbf{a}^1 = \left[\frac{1}{3} \ \frac{2}{3} \right]^T$ and $\mathbf{a}^2 = \left[\frac{3}{5} \ \frac{1}{5} \right]^T$.

occurs is convex. In Exercise 15.6.5, the reader is asked to prove that this observation holds in general.

From Figure 15.2, we can draw the following conclusions. In the case study of this chapter, we are considering the situation in which, at the beginning of the season, three types of farmers simultaneously decide to sell their products to a factory where these products are used to produce other goods with a certain market value. It is not known in advance how many farmers offer products and what share of the total profit (which will only become known at the end of the season) they can ask. These uncertainties imply that at the beginning of the season, the optimal proportions of shares can only be estimated. We have solved the problem by determining for every possible coalition a reasonable profit distribution scheme in which not too many farmers are (not too) dissatisfied with their shares.

The proportion of each type of farmer may be estimated from historical data. For example, if we know from earlier years that the number of farmers is roughly the same for each type, then we know that the fractional coalition vector will be 'close' to $\left[\frac{1}{3} \ \frac{1}{3} \ \frac{1}{3} \right]^T$, and hence it is safe to agree to divide the profit according to the proportions corresponding to Owen point ①, which means paying out € 18,500 to each Type 1 farmer, € 5,000 to each Type 2 farmer, and € 9,500 to each Type 3 farmer.

However, if the proportions of farmers do not correspond to a point in the area marked ① in Figure 15.2, then Owen point ① is not in the core. In that case the share of one of the farmer types is too high, and the other types can obtain a large share of the total profits by forming a subcoalition.

15.4.2 Deriving conditions such that a given point is an Owen point

We would like to determine for which fractional coalition vectors $\mathbf{a}$ the vector $\mathbf{y}_i^*$ is an optimal solution of (GD). To do so, we will determine for which vectors $\mathbf{a}$ the corresponding feasible basic solution is optimal. Consider model (GD) in standard form with slack variables:

$$
\begin{aligned}
v(\mathbf{a}) = - \max\ & (-\mathbf{Pa})^{\mathsf{T}} \mathbf{y} \\
\text{s.t.}\ & -\mathbf{A}^{\mathsf{T}} \mathbf{y} + \mathbf{y}_s = -\mathbf{c} \\
& \mathbf{y} \geq \mathbf{0}, \mathbf{y}_s \geq \mathbf{0}.
\end{aligned}
\tag{15.4}
$$

We will write $\mathbf{y} = \begin{bmatrix} y_1\ y_2\ y_3 \end{bmatrix}^{\mathsf{T}}$ and $\mathbf{y}_s = \begin{bmatrix} y_4\ y_5\ y_6 \end{bmatrix}^{\mathsf{T}}$. The vector of objective coefficients of model (15.4) is given by the vector $-\mathbf{Pa}$. For convenience, define:

$$
\bar{\mathbf{c}} = -\mathbf{Pa} = \begin{bmatrix}
-2a_1 - 2a_2 - 2a_3 \\
-8a_1 - 2a_2 - 4a_3 \\
-2a_1 - a_2 - 3a_3 \\
0 \\
0 \\
0
\end{bmatrix}.
$$

Moreover, the technology matrix of model (15.4) is:

$$
\begin{bmatrix} -\mathbf{A}^{\mathsf{T}}\ \mathbf{I}_3 \end{bmatrix} = \begin{bmatrix}
-2 & -2 & -1 & 1 & 0 & 0 \\
0 & -4 & 0 & 0 & 1 & 0 \\
-1 & -3 & -1 & 0 & 0 & 1
\end{bmatrix}.
$$

To illustrate the calculation process, let $i = 1$ and consider the dual solution $\mathbf{y}_1^* = \begin{bmatrix} \frac{1}{4}\ 2\frac{1}{4}\ 0 \end{bmatrix}^{\mathsf{T}}$. The corresponding optimal values of the slack variables are $y_4^* = y_6^* = 0$, and $y_5^* = 7$. Since the model has three constraints, any feasible basic solution has three basic variables. Thus, the basic variables corresponding to $\mathbf{y}_1^*$ are y_1, y_2, and y_5. We therefore have that $BI = \{1, 2, 5\}$ and $NI = \{3, 4, 6\}$. The basis matrix $\mathbf{B}$ and the matrix $\mathbf{N}$ corresponding to this choice of basic variables are:

$$
\mathbf{B} = \begin{bmatrix}
-2 & -2 & 0 \\
0 & -4 & 1 \\
-1 & -3 & 0
\end{bmatrix}, \text{ and } \mathbf{N} = \begin{bmatrix}
-1 & 1 & 0 \\
0 & 0 & 0 \\
-1 & 0 & 1
\end{bmatrix}.
$$

Recall that a feasible basic solution of a maximizing LO-model is optimal if and only if it is feasible and the current objective coefficients are nonpositive; see also Section 5.1. Since the constraints of (GD) do not depend on $\mathbf{a}$, we have that $\mathbf{y}_i^*$ is a feasible solution of (GD) for all $\mathbf{a}$. In other words, the feasibility conditions do not restrict the set of vectors $\mathbf{a}$ for which $\mathbf{y}_i^*$ is optimal. This means that we only have to look at the current objective coefficients.

These are given by:

$$\bar{\mathbf{c}}_{NI}^\mathsf{T} - \bar{\mathbf{c}}_{BI}^\mathsf{T} \mathbf{B}^{-1}\mathbf{N}$$

$$= \begin{bmatrix} -2a_1 - a_2 - 3a_3 \\ 0 \\ 0 \end{bmatrix}^\mathsf{T} - \begin{bmatrix} -2a_1 - 2a_2 - 2a_3 \\ -8a_1 - 2a_2 - 4a_3 \\ 0 \end{bmatrix}^\mathsf{T} \begin{bmatrix} -\frac{3}{4} & 0 & \frac{1}{2} \\ \frac{1}{4} & 0 & -\frac{1}{2} \\ 1 & 1 & -2 \end{bmatrix} \begin{bmatrix} -1 & 1 & 0 \\ 0 & 0 & 0 \\ -1 & 0 & 1 \end{bmatrix}$$

$$= \begin{bmatrix} \frac{1}{2}a_1 - \frac{3}{2}a_3 & \frac{1}{2}a_1 - a_2 - \frac{1}{3}a_3 & -3a_1 - a_3 \end{bmatrix},$$

where the second line is obtained by straightforward (but tedious) calculations. Since the current objective coefficients need to be nonpositive, we conclude from these calculations that $\mathbf{y}_1^*$ is an optimal solution of (GD) if and only if:

$$a_1 \leq 3a_3, \quad a_1 \leq 2a_2 + a_3, \quad 3a_1 + a_3 \geq 0.$$

The third inequality may be dropped, because it is implied by $a_1 \geq 0$ and $a_3 \geq 0$. To draw these inequalities in a two-dimensional figure, we substitute the equation $a_3 = 1 - a_1 - a_2$. We then find that, for any given nonnegative vector $\mathbf{a} = \begin{bmatrix} a_1 & a_2 & 1-a_1-a_2 \end{bmatrix}$, $\mathbf{y}_1^*$ is an optimal solution of (GD) if and only if:

$$4a_1 + 3a_2 \leq 3, \quad 2a_1 \leq 1 + a_2.$$

Together with $a_1 \geq 0$ and $a_2 \geq 0$, these are the inequalities that bound the region for ① in Figure 15.2. We have carried out the same calculations for $\mathbf{y}_2^*$, $\mathbf{y}_3^*$, and $\mathbf{y}_4^*$. The results are listed below.

	Solution	Basic variables	Optimality conditions	
①	$\mathbf{y}_1^*$	y_1, y_2, y_5	$4a_1 + 3a_2 \leq 3$	$2a_1 \leq 1 + a_2$
②	$\mathbf{y}_2^*$	y_2, y_3, y_5	$4a_1 + 3a_2 \geq 3$	$7a_1 + 4a_2 \leq 5$
③	$\mathbf{y}_3^*$	y_1, y_2, y_4	$a_1 + 2a_2 \leq 1$	$2a_1 \geq 1 + a_2$
④	$\mathbf{y}_4^*$	y_2, y_3, y_4	$a_1 + 2a_2 \geq 1$	$7a_1 + 4a_2 \geq 5$

15.5 Sensitivity analysis

In this last section we will investigate how the profit distribution, in particular the core of the game, is sensitive to changes of the market prices of the end products and of the production process. In order to avoid extensive calculations, we will restrict ourselves to the case of three farmers and the grand coalition, and where the profit is divided according to the Owen point. We leave it to the reader to carry out calculations for other coalitions and more farmers.

The market prices of the products Bis1, Bis2, and Bis3 are given by the vector $\mathbf{c} = \begin{bmatrix} 5 & 2 & 7 \end{bmatrix}^\mathsf{T}$. In Table 15.3(a), we list the Owen points and the corresponding total profit $v(\mathbf{1})$ for $\Delta = 0, 1, \ldots, 13$, with $\mathbf{c}(\Delta) = \begin{bmatrix} \Delta & 2 & 7 \end{bmatrix}^\mathsf{T}$. So only the first coordinate entry of $\mathbf{c}$ is subject to change.

Δ	Owen point			$v(1)$
0	[$18\frac{2}{3}$	$4\frac{2}{3}$	$9\frac{1}{3}$]	$32\frac{2}{3}$
$\vdots$		$\vdots$		$\vdots$
4	[$18\frac{2}{3}$	$4\frac{2}{3}$	$9\frac{1}{3}$]	$32\frac{2}{3}$
5	[$18\frac{1}{2}$	5	$9\frac{1}{2}$]	33
6	[18	6	10]	34
7	[$17\frac{1}{2}$	7	$10\frac{1}{2}$]	35
8	[17	8	11]	36
9	[$16\frac{1}{2}$	9	$11\frac{1}{2}$]	37
10	[16	10	12]	38
11	[$15\frac{1}{2}$	11	$12\frac{1}{2}$]	39
12	[15	12	13]	40
13	[16	13	14]	43
14	[17	14	15]	46

(a) Changing the price of Bis1.

Δ	Owen point			$v(1)$
0	[18	8	23]	49
1	[17	$7\frac{1}{2}$	$21\frac{1}{2}$]	46
2	[16	7	20]	43
3	[$18\frac{1}{2}$	5	$9\frac{1}{2}$]	33
4	[14	5	8]	27
5	[$11\frac{3}{4}$	5	$7\frac{1}{4}$]	24
6	[$10\frac{2}{5}$	5	$6\frac{4}{5}$]	$22\frac{1}{5}$
7	[$9\frac{1}{2}$	5	$6\frac{1}{2}$]	21
8	[$8\frac{6}{7}$	5	$6\frac{2}{7}$]	$20\frac{1}{7}$
9	[$8\frac{3}{8}$	5	$6\frac{1}{8}$]	$19\frac{1}{2}$
10	[8	5	6]	19
11	[8	5	6]	19
$\vdots$		$\vdots$		$\vdots$

(b) Changing the amount of flour for Bis3.

Table 15.3: Sensitivity analysis.

In Table 15.3(a) we may observe the interesting fact that for $\Delta \geq 12$ it holds that for any subcoalition S it does not make a difference whether or not a farmer who is not in S joins S: the value of S plus this farmer is equal to the value of S plus the value of the coalition consisting of this single farmer. So the addition of this farmer has no added value for the coalition S. In such situations we call this farmer a dummy player.

In general, a farmer of Type $k \in \{1, \ldots, n\}$ is called a *dummy player* with respect to the coalition vector **a** if it holds that

$$v(\mathbf{a}' + a_k\mathbf{e}_k) = v(\mathbf{a}') + v(a_k\mathbf{e}_k) \text{ for each } \mathbf{a}' \subseteq \mathbf{a} - a_k\mathbf{e}_k.$$

Note that $v(a_k\mathbf{e}_k) = a_k v(\mathbf{e}_k)$. We will show if $\Delta = 12$, i.e., the market price of Bis1 is € 12, then all three farmers are in fact dummy players. It is left to the reader to carry out the various calculations; we only present the results. We start with the optimal values of all subcoalitions of the grand coalition.

$$v(1,1,1) = 40, \; v(1,1,0) = 27, \; v(1,0,1) = 28, \; v(0,1,1) = 25,$$
$$v(1,0,0) = 15, \; v(0,1,0) = 12, \; v(0,0,1) = 13.$$

For the Type 1 farmer, it holds that:

$$v(1,1,0) - v(0,1,0) = 27 - 12 = 15 = v(1,0,0),$$
$$v(1,0,1) - v(0,0,1) = 28 - 13 = 15 = v(1,0,0),$$
$$v(1,1,1) - v(0,1,1) = 40 - 25 = 15 = v(1,0,0).$$

For the Type 2 farmer, we find that:

$$
\begin{aligned}
v(1,1,0) - v(1,0,0) &= 27 - 15 = 12 = v(0,1,0),\\
v(0,1,1) - v(0,0,1) &= 25 - 13 = 12 = v(0,1,0),\\
v(1,1,1) - v(1,0,1) &= 40 - 28 = 12 = v(0,1,0).
\end{aligned}
$$

For the Type 3 farmer, we have that:

$$
\begin{aligned}
v(1,0,1) - v(1,0,0) &= 28 - 15 = 13 = v(0,0,1),\\
v(0,1,1) - v(0,1,0) &= 25 - 12 = 13 = v(0,0,1),\\
v(1,1,1) - v(1,1,0) &= 40 - 27 = 13 = v(0,0,1).
\end{aligned}
$$

Hence, all three farmers are in fact dummy players for $\Delta = 12$. In Exercise 15.6.8, the reader is asked to show that this is true for every $\Delta \geq 12$. This means that for $\Delta \geq 12$ no farmer is better off when forming coalitions with other farmers. For $0 \leq \Delta \leq 4$, all the grand coalitions have the same value and the same corresponding point in the core, but no farmer is dummy; see Exercise 15.6.8.

By varying the market price of Bis2 ($\mathbf{c} = \begin{bmatrix} 5 & \Delta & 7 \end{bmatrix}^{\mathsf{T}}$), it can be shown that for $0 \leq \Delta \leq 9$ the total profit satisfies $v(1,1,1) = 33$ and that the Owen point is $\begin{bmatrix} 18\frac{1}{2} & 5 & 9\frac{1}{2} \end{bmatrix}^{\mathsf{T}}$. For $\Delta \geq 10$ all three farmers are dummy players; see Exercise 15.6.9.

By varying the market price of Bis3 ($\mathbf{c} = \begin{bmatrix} 5 & 2 & \Delta \end{bmatrix}^{\mathsf{T}}$), it can be shown that, for $0 \leq \Delta \leq 3\frac{1}{2}$, the Owen point is $\begin{bmatrix} 8 & 5 & 6 \end{bmatrix}^{\mathsf{T}}$ and the total profit satisfies $v(1,1,1) = 19$. For $\Delta \geq 8$, none of the farmers become dummy, and the Owen points are of the form $\alpha \begin{bmatrix} 4 & 1 & 2 \end{bmatrix}^{\mathsf{T}}$ with $\alpha > 0$.

Finally, we investigate how a change of the production process influences the Owen point. For instance, what happens if the proportions of the input factors needed for the production of Bis3 are changed? Consider the matrix:

$$
\mathbf{A}(\Delta) = \begin{bmatrix} 2 & 0 & 1 \\ 2 & 4 & \Delta \\ 1 & 0 & 1 \end{bmatrix}.
$$

Table 15.3(b) presents for $\Delta \geq 0$ the corresponding Owen points and the values $v(1,1,1)$ of the total profit. It can be shown that for $\Delta \geq 10$ all the farmers are dummy players. For $0 \leq \Delta \leq 3$, there are multiple Owen points. In Exercise 15.6.9 the reader is asked to design similar tables for the other entries of the technology matrix $\mathbf{A}$.

In all cases considered above, there is a certain value of Δ for which all values, either larger or smaller than Δ, give rise to a situation in which all the farmers are dummy players. In these situations the shares assigned are the same for all subcoalitions; cooperation does not make sense from a profit distribution point of view. We leave it to the reader to analyze the case when there are four or more types of farmers.

15.6 Exercises

Exercise 15.6.1. Consider model (GP) in Section 15.2. Let $\mathbf{P} \geq \mathbf{0}$ (meaning that each entry of $\mathbf{P}$ is nonnegative). Show that, for each two coalition vectors $\mathbf{a}' \geq \mathbf{0}$ and $\mathbf{a}'' \geq \mathbf{0}$ with $\mathbf{a}' \subseteq \mathbf{a}''$, it holds that $v(\mathbf{a}') \leq v(\mathbf{a}'')$. Show that the maximum total profit is attained by the grand coalition.

Exercise 15.6.2. Let $\mathbf{a} \in \mathbb{N}^n$. Let $C(\mathbf{a})$ be the core corresponding to $\mathbf{a}$; see Section 15.3. Show that $C(\mathbf{a})$ is a polyhedral set with dimension at most $n-1$.

Exercise 15.6.3. Let $\mathbf{a} \in \mathbb{N}^n$, and let $S(\mathbf{a}) = \{k \mid a_k > 0, k = 1, \ldots, n\}$, called the *support* of the vector $\mathbf{a}$. Define for each $k \in S(\mathbf{a})$:

$$M_k(\mathbf{a}) = v(\mathbf{a}) - v(\mathbf{a} - a_k \mathbf{e}_k), \text{ and}$$

$$m_k(\mathbf{a}) = \max\left\{ v(\mathbf{a}') - \sum_{l \in S(a')\setminus\{k\}} M_l(\mathbf{a}) \;\middle|\; \mathbf{a}' \subseteq \mathbf{a}, k \in S(\mathbf{a}') \right\}.$$

(a) Show that $M_k(\mathbf{a})/a_k$ is the highest possible share a farmer of Type k can ask without risking the withdrawal of another farmer, and $m_k(\mathbf{a})/a_k$ is the lowest possible share with which a farmer of Type k is satisfied.

The *Tijs point*[1] $\boldsymbol{\tau}(\mathbf{a})$ of a linear production game is an n-vector whose entries $\tau_k(\mathbf{a})$, $k = 1, \ldots, n$, are defined by:

$$\tau_k(\mathbf{a}) = (1 - \alpha)m_k(\mathbf{a})/a_k + \alpha M_k(\mathbf{a})/a_k,$$

where $\alpha \in [0, 1]$ is chosen such that $\mathbf{a}^\mathsf{T}\boldsymbol{\tau}(\mathbf{a}) = v(\mathbf{a})$.

(b) Show that $C(\mathbf{a}) \subseteq \{\mathbf{u} \in \mathbb{N}^n \mid m_k(\mathbf{a}) \leq a_k u_k \leq M_k(\mathbf{a}) \text{ for } k = 1, \ldots, n\}$.

(c) Determine $\boldsymbol{\tau}(1, 1, 1)$ for the case study described in this chapter, and verify that $\boldsymbol{\tau}(1, 1, 1) \in C(1, 1, 1)$.

Exercise 15.6.4. Consider the feasible region of model (GD) for the data used in the case study in this chapter.

(a) Draw the feasible region. Determine the five extreme points, the three extreme rays, and the two bounded facets of this feasible region.

(b) Show that the vectors $\mathbf{y}_1^*, \mathbf{y}_2^*, \mathbf{y}_3^*, \mathbf{y}_4^*$ (see Section 15.4) correspond to Owen points of the grand coalition (i.e., $a_1, a_2, a_3 > 0$), and that the vector $\begin{bmatrix} 0 & 2\frac{1}{2} & 0 \end{bmatrix}^\mathsf{T}$ does not.

---

[1] In honor of STEF H. TIJS (born 1937) who introduced the concept in 1981, although he called it the *compromise value*.

Exercise 15.6.5. Assume that the matrix $\mathbf{P}$ in model (GD) in Section 15.3 is nonsingular. Show that if the coalition vectors $\mathbf{a}'$ and $\mathbf{a}''$ correspond to the same point $\mathbf{u}$ in the core, then all convex combinations of $\mathbf{a}'$ and $\mathbf{a}''$ correspond to $\mathbf{u}$.

Exercise 15.6.6. Consider Figure 15.2. Let the total profit be distributed according to the proportions $18\frac{1}{2} : 5 : 9\frac{1}{2}$, for a Type 1 farmer, a Type 2 farmer, and a Type 3 farmer, respectively. Calculate in all cases for which ① is not an Owen point: the core, the values of $m_k(\mathbf{a})$ and $M_k(\mathbf{a})$ for all $k = 1, \ldots, n$ (see Exercise 15.6.3), the Tijs point (see Exercise 15.6.3), and the deviations from the satisfactory shares.

Exercise 15.6.7. Let $\mathbf{a}$ be a coalition vector with $\mathbf{a} \neq \mathbf{0}$. Let $\mathbf{y}^*$ be an optimal solution of model (GD) (see Section 15.3) for each subcoalition vector $\mathbf{a}'$ of $\mathbf{a}$. Show that, in this case, each farmer of Type k is a dummy player with respect to $\mathbf{a}$.

Exercise 15.6.8. Consider Table 15.3(a) in Section 15.5.

(a) Draw the perturbation function for $2y_1 + 2y_2 + y_3 = \Delta$ for $\delta = 0, \ldots, 14$ (see Section 15.3).

(b) Describe the relationships between the perturbation function from (a) and the values of Table 15.3(a).

(c) Show that all farmers are dummy players with respect to the grand coalition for $\Delta \geq 12$. (Hint: use a drawing of the feasible region and Exercise 15.6.7.)

(d) Why are the Owen points equal for $0 \leq \Delta \leq 4$? Are there dummy players?

(e) Explain why the values of $v(\mathbf{1})$ increase, and why the Owen points increase lexicographically with Δ.

Exercise 15.6.9. Calculate tables similar to Table 15.3(a) for the cases $\mathbf{c} = \begin{bmatrix} 5 & \Delta & 7 \end{bmatrix}^{\mathsf{T}}$ and $\mathbf{c} = \begin{bmatrix} 5 & 2 & \Delta \end{bmatrix}^{\mathsf{T}}$ with $\Delta \in \mathbb{R}$, and answer similar questions as the ones in Exercise 15.6.8; see Section 15.5. Determine and analyze similar tables when the entries of the production matrix $\mathbf{A}$ are subject to change; compare in this respect Table 15.3(b) and the corresponding remarks in Section 15.5. Pay special attention to the situations where the values of the Owen points do not increase or decrease monotonically with Δ. See for example Table 15.3(b): the values of the Owen points decrease for $\Delta = 1, 2, 3$, increase for $\Delta = 3, 4$, and decrease again for $\Delta \geq 3$.

Exercise 15.6.10. Consider a linear production game with $\mathbf{A} = \begin{bmatrix} 12 & 1 & 3 \\ 40 & 10 & 90 \end{bmatrix}$, $\mathbf{P} = \begin{bmatrix} 100 & 0 \\ 240 & 480 \end{bmatrix}$, and $\mathbf{c} = \begin{bmatrix} 40 & 6 & 30 \end{bmatrix}^{\mathsf{T}}$.

(a) Solve the LO-model corresponding to this linear production game; see Section 15.2, model (GP).

(b) Determine all optimal solutions of the dual model and the set of Owen points; see Section 15.3, model (GD).

(c) Determine the core for the grand coalition of this game, and show that the set of Owen points is a proper subset of the core.

Exercise 15.6.11.   Show that, for the linear production game defined by $\mathbf{A} = \begin{bmatrix} 1 & 0 & 1 \\ 0 & 1 & 1 \end{bmatrix}$, $\mathbf{P} = \begin{bmatrix} 1 & 0 \\ 0 & 1 \end{bmatrix}$, and $\mathbf{c} = \begin{bmatrix} 296 & 0 & 488 \end{bmatrix}^{\mathsf{T}}$, the set of Owen points is equal to the core.

CHAPTER 16

Minimizing trimloss when cutting
cardboard

Overview

Cutting stock problems arise, among others, when materials such as paper, cardboard, and textiles are manufactured in rolls of large widths. These rolls have to be cut into subrolls of smaller widths. It is not always possible to cut the rolls without leftovers. These leftovers are called trimloss. In this section we will discuss a cutting stock problem where rolls of cardboard need to be cut such that the trimloss is minimized. This type of trimloss problem is one of the oldest industrial applications of Operations Research. The solution algorithm of PAUL C. GILMORE (born 1925) and RALPH E. GOMORY (born 1929), published in 1963, will be discussed here.

16.1 Formulating the problem

The cardboard factory BGH manufactures and cuts cardboard. The cutting department receives the cardboard on standard rolls with a width of 380 cm, which have to be cut into subrolls with smaller widths. The total length of a standard roll is 1,000 m. The customers order subrolls of cardboard in various widths. An example of an order package is given in Table 16.1. Although customers usually order full-length subrolls, we assume in this chapter that, as long as the customers receive subrolls of the correct total length, they do not mind if the subrolls are shipped in parts (e.g., two subrolls of 500 m instead of one subroll of 1,000 m).

The question is how to cut the standard rolls such that the demand of all customers is satisfied, and the amount of wasted cardboard, the so-called *trimloss*, is as small as possible. The management of BGH wants to use a computer package that generates for each order package the best cutting pattern with a minimum amount of trimloss. We will show that the cutting problem can be formulated as an LO-model.

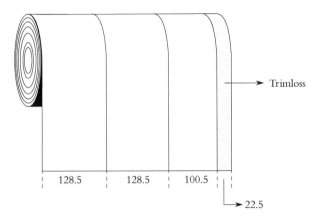

128.5   128.5   100.5

Trimloss

22.5

Figure 16.1: A cutting pattern.

Standard rolls are cut as follows. First, the knives of the cutting machine are set up in a certain position. Next, the cardboard goes through the cutting machine as it is unrolled. This generates a *cutting pattern* of the cardboard in the longitudinal direction. Once the subrolls have attained the desired length, the standard roll is cut off in the perpendicular direction. An example is depicted in Figure 16.1.

Since the knives of the cutting machine can be arranged in any position, standard rolls can be cut into subrolls in many different ways. We consider two cutting patterns to be equal if they both consist of the same combination of demanded subrolls. Table 16.2 lists thirteen cutting patterns for the case described in Table 16.1. So, for example, cutting pattern 7 consists of cutting a roll into two subrolls of 128.5 cm and one subroll of 100.5 cm (resulting in 22.5 cm of trimloss); this cutting pattern 7 is the pattern depicted in Figure 16.1. The orders are labeled $i = 1, \ldots, m$, and the cutting patterns are labeled $j = 1, \ldots, n$. In the case of Table 16.2, we have that $m = 6$; the number n of cutting patterns is certainly more than the thirteen listed in this table.

For $i = 1, \ldots, m$, and $j = 1, \ldots, n$, define:

x_j = the number of standard rolls for which cutting pattern j is used;
b_i = the demand for subrolls of order i;

Order label	# of subrolls	Width (cm)
1	20	128.5
2	9	126.0
3	22	100.5
4	18	86.5
5	56	82.0
6	23	70.5

Table 16.1: Order package.

	Width		Cutting pattern												
			1	2	3	4	5	6	7	8	9	10	11	12	13	...
1	128.5	1	0	0	0	0	0	2	2	2	2	1	1	0	...
2	126.0	0	2	0	0	0	0	0	0	0	0	1	1	0	...
3	100.5	0	0	1	0	0	0	1	0	0	0	1	0	0	...
4	86.5	0	0	0	1	0	0	0	1	0	0	0	0	0	...
5	82.5	0	0	0	0	1	0	0	0	1	0	0	0	0	...
6	70.5	0	0	0	0	0	1	0	0	0	1	0	0	5	...

Table 16.2: A number of cutting patterns.

W_i = the width of order i;
a_{ij} = the number of widths i in cutting pattern j.

The values of the entries a_{ij} are given in the body of Table 16.2, and the values of the b_i's are the entries in the second column of Table 16.1. The a_{ij}'s are integers, and the b_i's are not necessarily integers. The columns $\mathbf{A}_1, \ldots, \mathbf{A}_n$ ($\in \mathbb{R}^m$) of the matrix $\mathbf{A} = \{a_{ij}\}$ refer to the various feasible cutting patterns. Define $\mathbf{b} = \begin{bmatrix} b_1 & \ldots & b_m \end{bmatrix}^\mathsf{T}$.

The general LO-model can now be formulated as follows:

$$
\begin{aligned}
\min \quad & \sum_{j=1}^{n} x_j \\
\text{s.t.} \quad & \sum_{j=1}^{n} \mathbf{A}_j x_j = \mathbf{b} \\
& x_j \geq 0 \qquad \text{for } j = 1, \ldots, n.
\end{aligned}
\tag{16.1}
$$

The objective function denotes the total number of standard rolls needed to satisfy the demand, and has to be minimized. The m constraints (written in matrix notation in (16.1)) express the fact that the number of demanded subrolls has to be equal to the number of produced subrolls. One might wonder why b_i and x_j have not been taken to be integer-valued in (16.1), since they refer to a certain number of standard rolls. The reason is that standard rolls are first unrolled before being cut. The cutting of a certain pattern can be stopped before the end of the roll has been reached. In that case the machine that rotates the roll is stopped, and the knives are readjusted for the next cutting pattern. As said, the customers usually do not demand full subrolls for which the total cardboard length is precisely the length of a standard roll, since the cardboard on the subrolls will usually be cut into sheets of cardboard. So, the values of b_i and x_j do not need to be integers. For example, a demand of three subrolls of 128.5 cm and two subrolls of 100.5 cm can be obtained by using pattern 7 on one-and-a-half standard roll and pattern 3 on half a standard roll; see Figure 16.2.

The *trimloss* of an order package is the difference between the number of used standard rolls and the number of the demanded standard rolls, i.e.,

$$
380 z^* - \sum_{i=1}^{m} W_i b_i,
$$

with z^* the optimal objective value of model (16.1). In this definition the trimloss is expressed in cm units; the company expresses trimloss in tons of cardboard. Since

$$
\min \left(380 \sum_{j=1}^{n} x_j - \sum_{i=1}^{m} W_i b_i \right) = 380 \left(\min \sum_{j=1}^{n} x_j \right) - \sum_{i=1}^{m} W_i b_i,
$$

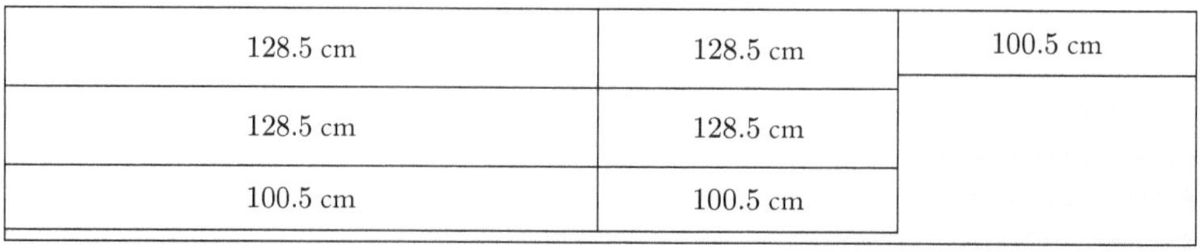

128.5 cm	128.5 cm	100.5 cm
128.5 cm	128.5 cm	
100.5 cm	100.5 cm	

Figure 16.2: A noninteger 'solution'.

minimizing the trimloss is equivalent to minimizing the number of standard rolls needed for the demanded production. Note that any cardboard that remains on the roll, i.e., that is not cut, is not counted as trimloss, because this remaining cardboard may still be used for a future order.

16.2 Gilmore-Gomory's solution algorithm

For a small order package of six orders it is possible to calculate all cutting patterns (provided that the values of the W_i are not too small compared to the total width of the roll). For example, for the data in the previous section, there are 135 different patterns. However, for more extensive orders, the number of cutting patterns can easily increase to hundreds of thousands or millions. We will show how to solve model (16.1) without using all cutting patterns, and hence without taking into account all decision variables $x_1, \ldots, x_n$. In fact, the matrix $\mathbf{A}$ will only exist virtually; it is never constructed explicitly. Instead, only a (hopefully small) submatrix of $\mathbf{A}$ is generated. Each iteration step of the Gilmore-Gomory algorithm consists of two phases. In the first phase, an LO-model is solved, and in the second phase, a knapsack model (see Section 7.2.3) is solved, which either establishes optimality, or leads to a cutting pattern that is added to the submatrix of $\mathbf{A}$. We will first formulate the algorithm.

Algorithm 16.2.1. (*Gilmore-Gomory algorithm*)

Input: Model (16.1), with an order package, containing the amount of demanded subrolls with the corresponding widths.

Output: An optimal solution of model (16.1).

Step 0: *Initialization.* Choose an initial full row rank matrix $\mathbf{A}^{(1)}$, of which the columns correspond to cutting patterns. For instance, take $\mathbf{A}^{(1)} = \mathbf{I}_m$. Go to Step 1.

Step 1: *Simplex algorithm step.* Let $\mathbf{A}^{(k)}$, $k \geq 1$, be the current technology matrix (after k iterations of the Gilmore-Gomory algorithm), of which the columns correspond to cutting patterns; let $J(k)$ be the index set of the

columns of $\mathbf{A}^{(k)}$. Solve the LO-model:

$$\min \sum_{j \in J(k)} x_j$$

$$\text{s.t.} \quad \sum_{j \in J(k)} \mathbf{A}_j^{(k)} x_j = \mathbf{b} \qquad (\mathrm{P}_k)$$

$$x_j \geq 0 \text{ for } j \in J(k),$$

with $\mathbf{A}_j^{(k)}$ the j'th column of $\mathbf{A}^{(k)}$. Let $y_1^{(k)}, \ldots, y_m^{(k)}$ be the values of an optimal dual solution, corresponding to the current optimal basis matrix of (P_k). Go to Step 2.

Step 2: *Column generation.* Solve the knapsack model:

$$\max \sum_{i=1}^{m} y_i^{(k)} u_i$$

$$\sum_{i=1}^{m} W_i u_i \leq 380 \qquad (\mathrm{K}_k)$$

$$u_1, \ldots, u_m \geq 0.$$

Let $\mathbf{u}^{(k)} = \begin{bmatrix} u_1^{(k)} & \ldots & u_m^{(k)} \end{bmatrix}^{\mathsf{T}}$ be an optimal solution of (K_k), and let α_k be the optimal objective value of (K_k). Go to Step 3.

Step 3: *Optimality test and stopping rule.* If $\alpha_k > 1$, then let $\mathbf{A}^{(k+1)} = \begin{bmatrix} \mathbf{A}^{(k)} & \mathbf{u}^{(k)} \end{bmatrix}$ and return to Step 1. If $\alpha_k \leq 1$, then stop: the pair $\{\mathbf{u}^{(k)}, \alpha_k\}$ is an optimal solution of model (16.1).

We will give an explanation of the various steps of the Gilmore-Gomory algorithm.

Step 0. The initial matrix can always be chosen such that it contains a nonsingular (m, m)-submatrix, for instance $\mathbf{I}_m$. This takes care of the feasibility of model (P_k) for $k = 1$. In order to speed up the calculations, $\mathbf{A}^{(1)}$ can be chosen such that it contains many good cutting patterns.

Step 1. In this step, the original model (16.1) is solved for the submatrix $\mathbf{A}^{(k)}$ of the 'cutting pattern' matrix $\mathbf{A}$. Recall that $\mathbf{A}$ exists only virtually; the number of columns (i.e., cutting patterns) is generally very large. Any basis matrix $\mathbf{B}^{(k)}$, corresponding to an optimal solution of (P_k), is a basis matrix in $\mathbf{A}$, but need not be an optimal basis matrix of $\mathbf{A}$. Let $\mathbf{A} \equiv \begin{bmatrix} \mathbf{B}^{(k)} & \mathbf{N}^{(k)} \end{bmatrix}$. The objective coefficients of the current simplex tableau corresponding to $\mathbf{B}^{(k)}$ are zero (by definition), and the objective coefficients vector corresponding to $\mathbf{N}^{(k)}$ is:

$$\mathbf{1}^{\mathsf{T}} - \mathbf{1}^{\mathsf{T}} (\mathbf{B}^{(k)})^{-1} \mathbf{N}^{(k)}$$

(see Section 3.8; $\mathbf{c}_{NI} = \mathbf{1} \in \mathbb{R}^{|NI|}$ and $\mathbf{c}_{BI} = \mathbf{1} \in \mathbb{R}^{|BI|}$, with BI corresponding to $\mathbf{B}^{(k)}$ and NI to $\mathbf{N}^{(k)}$). Therefore, the current objective coefficient in position $\sigma \ (\in NI)$

is denoted and defined as follows:

$$\bar{c}_\sigma = 1 - \sum_{i=1}^{m}(\mathbf{1}^\mathsf{T}(\mathbf{B}^{(k)})^{-1})_i a_{i\sigma}.$$

Since $\mathbf{B}^{(k)}$ is optimal for $\mathbf{A}^{(k)}$, it follows that, for each $\sigma \in BI$, the current objective coefficient $\bar{c}_\sigma$ corresponding to $\mathbf{A}^{(k)}$ is nonnegative, i.e., $\bar{c}_\sigma = 0$. If $\bar{c}_\sigma \geq 0$ for all $\sigma \in NI$, then $\mathbf{B}^{(k)}$ is optimal for (16.1). We try to find, among the possibly millions of $\bar{c}_\sigma$'s with $\sigma \in NI$, one with a negative value. To that end, we determine:

$$\min_{\sigma \in NI} \bar{c}_\sigma = \min\left(1 - \sum_{i=1}^{m}(\mathbf{1}^\mathsf{T}(\mathbf{B}^{(k)})^{-1})_i a_{i\sigma}\right) = 1 - \max_{\sigma \in NI} \mathbf{1}^\mathsf{T}(\mathbf{B}^{(k)})^{-1}\mathbf{A}_\sigma,$$

with $\mathbf{A}_\sigma$ the σ'th (cutting pattern) column of $\mathbf{A}$. So the problem is to determine a cutting pattern $\mathbf{A}_\sigma$ with a smallest current objective coefficient $\bar{c}_\sigma$. Hence, the entries of $\mathbf{A}_\sigma$ have to be considered as decision variables, say $\mathbf{u} = \mathbf{A}_\sigma$, in the following model:

$$\begin{aligned} \max \quad & \mathbf{1}^\mathsf{T}(\mathbf{B}^{(k)})^{-1}\mathbf{u} \\ \text{s.t.} \quad & \mathbf{L}^\mathsf{T}\mathbf{u} \leq 380 \\ & \mathbf{u} \geq \mathbf{0}, \text{ and integer.} \end{aligned} \qquad (\text{K})$$

This model is a knapsack model (see Section 7.2.3) with objective coefficients vector $((\mathbf{B}^{(k)})^{-1})^\mathsf{T}\mathbf{1}$. An optimal solution $\mathbf{u}^{(k)}$ of (K) is certainly one of the many cutting patterns, since $\mathbf{L}^\mathsf{T}\mathbf{u}^{(k)} \leq 380$ and $\mathbf{u}^{(k)} \geq \mathbf{0}$, and all entries of $\mathbf{u}^{(k)}$ are integers.

An optimal dual solution $y_1^{(k)}, \ldots, y_m^{(k)}$ of (P_k) can be calculated by means of a computer package that solves (P_k); in the case of nondegeneracy, the $y_i^{(k)}$'s are precisely the shadow prices of the m constraints of (P_k). Let $\mathbf{y} = \left[y_1^{(k)} \ \ldots \ y_m^{(k)}\right]^\mathsf{T}$. Theorem 4.2.4 implies that (take $\mathbf{c}_{BI} = \mathbf{1}$):

$$\mathbf{y}^\mathsf{T} = \mathbf{1}^\mathsf{T}(\mathbf{B}^{(k)})^{-1}$$

is an optimal solution of the dual model $\max\left\{\mathbf{b}^\mathsf{T}\mathbf{y} \mid \mathbf{A}^{(k)}\mathbf{y} \leq \mathbf{1}\right\}$ of (P_k).

Step 2. Substituting $\mathbf{1}^\mathsf{T}(\mathbf{B}^{(k)})^{-1} = \mathbf{y}^\mathsf{T}$ in (K), the model (K_k) follows. It now follows immediately from the remarks in Ad Step 1 that an optimal solution of the knapsack model (K_k) is some column of the matrix $\mathbf{A}$. This column is denoted by $\mathbf{u}^{(k)}$, and the optimal objective value of (K_k) is denoted by α_k.

Step 3. The smallest current nonbasic objective coefficient is $1 - \alpha_k$. Hence, if $1 - \alpha_k \geq 0$, then all current nonbasic objective coefficients are nonnegative, and so an optimal solution of (16.1) has been reached; see Theorem 3.5.1. If $1 - \alpha_k < 0$, then the current matrix $\mathbf{A}^{(k)}$ is augmented with the column $\mathbf{u}^{(k)}$, and the process is repeated for $\mathbf{A}^{(k+1)} = \left[\mathbf{A}^{(k)} \ \mathbf{u}^{(k)}\right]$ instead of $\mathbf{A}^{(k)}$. Obviously, $\mathbf{u}^{(k)}$ is not a column of $\mathbf{A}^{(k)}$, otherwise the model (P_k) was not solved to optimality.

The knapsack model (K_k) is an extension of the (binary) knapsack problem as formulated in Section 7.2.3. The binary knapsack problem formulated in Section 7.2.3 has $\{0, 1\}$-variables, whereas model (K_k) has nonnegative integer variables. Such a problem is called a *bounded knapsack problem*. The general form of a bounded knapsack problem is:

$$\begin{aligned}
\max \quad & c_1 x_1 + \ldots + c_n x_n \\
\text{s.t.} \quad & a_1 x_1 + \ldots + a_n x_n \leq b \\
& x_i \leq u_i \quad \text{for } i = 1, \ldots, n \\
& x_1, \ldots, x_n \text{ integer,}
\end{aligned}$$

with n the number of objects, c_i (≥ 0) the value obtained if object i is chosen, b (≥ 0) the amount of an available resource, a_i (≥ 0) the amount of the available resource used by object i, and u_i (≥ 0) the upper bound on the number of objects i that are allowed to be chosen; $i = 1, \ldots, n$. Clearly, the binary knapsack problem is a special case of the bounded knapsack problem, namely the case where $u_i = 1$ for $i = 1, \ldots, n$. A branch-and-bound algorithm for solving $\{0, 1\}$-knapsack problems has been discussed in Section 7.2.3. In Exercise 16.4.2, the reader is asked to describe a branch-and-bound algorithm for the bounded knapsack problem.

16.3 Calculating an optimal solution

In this section we will show how an optimal solution of the cutting stock problem can be calculated for the data in Table 16.1. One could start the calculations by applying the simplex algorithm for a reasonable number of cutting patterns. Instead of the first six columns of Table 16.2, it is better to use the diagonal matrix with the vector $\begin{bmatrix} 2 & 3 & 3 & 4 & 4 & 5 \end{bmatrix}$ as its main diagonal, because these patterns are clearly more efficient than patterns 1–6. We augment the first matrix $\mathbf{A}^{(1)}$ with the remaining seven columns of Table 16.2.

We give the details of the first iteration:

Step 1. Solve the LO-model:

$$\begin{aligned}
\min \quad & x_1 + \ldots + x_{13} \\
\text{s.t.} \quad & 2x_1 + 2x_7 + 2x_8 + 2x_9 + 2x_{10} + x_{11} + x_{12} = 20 \\
& 3x_2 + x_{11} + x_{12} = 9 \\
& 3x_3 + x_7 + x_{11} = 22 \\
& 4x_4 + x_8 = 18 \\
& 4x_5 + x_9 = 56 \\
& 5x_6 + x_{10} + 5x_{13} = 23 \\
& x_1, \ldots, x_{13} \geq 0.
\end{aligned}$$

An optimal solution can be calculated using a computer package: $x_1 = x_8 = x_9 = x_{10} = x_{11} = x_{12} = x_{13} = 0$, $x_2 = 3$, $x_3 = 4$, $x_4 = 4.5$, $x_5 = 14$, $x_6 = 4.6$, $x_7 = 10$, with optimal objective value $z = 40.1$. The knapsack problem to be solved in the next step uses as objective coefficients the optimal dual values of the current constraints: $y_1 = y_2 = y_3 = 0.3333$, $y_4 = y_5 = 0.25$, $y_6 = 0.20$. (Usually, these optimal dual values are reported in the output of a linear optimization computer package.)

Order	Optimal solution						Optimal dual value
	3.027	4.676	14.973	4.216	9.000	0.351	
	Cutting pattern						
1	0	1	1	0	0	1	0.3514
2	0	0	0	0	1	0	0.3243
3	0	0	0	3	1	1	0.2703
4	1	0	1	0	0	0	0.2162
5	1	3	2	0	1	0	0.2162
6	3	0	0	1	1	2	0.1892

Table 16.3: Optimal solution.

Step 2. Solve the knapsack problem:

$$\max \; 0.3333u_1 + 0.3333u_2 + 0.3333u_3 + 0.25u_4 + 0.25u_5 + 0.20u_6$$
$$\text{s.t.} \quad 128.5u_1 + 126.0u_2 + 100.5u_3 + 86.5u_4 + 82.0u_5 + 70.5u_6 \le 380$$
$$u_1, \ldots, u_6 \ge 0, \text{ and integer.}$$

The optimal solution (generated by a computer package) is: $u_1 = u_2 = u_4 = u_5 = 0$, $u_3 = 3$, $u_6 = 1$. The optimal objective value satisfies $\alpha_1 = 1.2$. Since $\alpha_1 > 1$, optimality has not yet been reached.

Step 3. Construct matrix $\mathbf{A}^{(2)}$ from $\mathbf{A}^{(1)}$ by adding the column $\begin{bmatrix} 0 & 0 & 3 & 0 & 0 & 1 \end{bmatrix}^{\mathsf{T}}$.

In the next iteration, Step 1 and Step 2 are repeated for the matrix $\mathbf{A}^{(2)}$. The algorithm stops at iteration k if the optimal objective value α_k of model (K_k) is ≤ 1. It is left to the reader to carry out the remaining calculations leading to an optimal solution. With our calculations, the Gilmore-Gomory algorithm performed ten iterations. The resulting optimal solution is listed in Table 16.3.

The order in which the six cutting patterns (the six columns in Table 16.3) are actually carried out is left to the planning department of the company; they can, for instance, take into account the minimization of the number of adjustments of the knives. The first decision variable in Table 16.3 has the value 3.027. If this is the first pattern that will be cut from standard rolls, then the number 3.027 means that three standard rolls are needed plus a 0.027 fraction of a standard roll. This last part is very small, and may not be delivered to the customer. The reason for this is the following. On average, standard rolls contain about 1,000 meters of cardboard. This means that this 0.027 fraction is approximately 27 meters. The margin in an order is usually about 0.5%, i.e., the customer may expect either 0.5% less or 0.5% more of his order. So, BGH may decide to round down the number of rolls cut according to the first pattern to 3, and round up the number of rolls cut according to the third pattern to 15. Whether or not this change is acceptable depends on how much 27 meters is compared to the size of the affected orders. The sizes of the affected orders are:

order 1: $20 \times 1.285 \times 1,000 = 25,700 \text{ m}^2$ cardboard,

order 5: $56 \times 0.820 \times 1,000 = 45,920 \text{ m}^2$ cardboard, and order 6: $23 \times 0.705 \times 1,000 = 16,215$ m^2 cardboard.

If the change is implemented, then:

order 1 is increased by $1 \times 1.285 \times 27 = 34.695$ m^2 (0.13% of 25,700 m^2),

order 5 is decreased by $1 \times 0.820 \times 27 = 22.140$ m^2 (0.05% of 45,920 m^2), and

order 6 is decreased by $3 \times 0.705 \times 27 = 57.105$ m^2 (0.35% of 16,215 m^2).

For all three orders, the 27 meters is certainly less than 0.5%, and so the customers of orders 1, 5, and 6 can expect to receive a slightly different amount than what they ordered. The price of the order is of course adjusted to the quantity that is actually delivered.

Note that the optimal solution is nondegenerate. Hence, the optimal dual values in the rightmost column of Table 16.3 are the shadow prices for the six constraints. It may be noted that all shadow prices are positive. This means that if the number of demanded subrolls in an order increases by one, then the number of needed standard rolls increases with the value of the shadow price. In Exercise 16.4.1, the reader is asked to show that the shadow prices are nonnegative.

The company BGH may use shadow prices to deviate (within the set margin of 0.5%) from the demanded orders. Consider for instance order 4 in which eighteen subrolls of width 86.5 cm are demanded. The shadow price of this order is 0.2162. This means that if one more subroll is produced, then a 0.2162 fraction of a standard roll will be needed (assuming that the other orders remain the same). This a 0.2162 fraction of a standard roll is $0.2162 \times 3.80 \times 1,000 = 821.56$ m^2. However, one subroll of width 86.5 cm consists of $0.865 \times 1,000 = 865.00$ m^2 of cardboard. So producing more of order 4 is profitable, and the customer of order 4 may expect 0.5% more cardboard. In case this additional 0.5% is produced and delivered, the order package has actually been changed, and so the model has to be solved for the changed order package (with 18.09 instead of 18 for order 4). The shadow prices will change, and based on these new shadow prices it can be decided whether or not another order will be decreased or increased. In practice, these additional calculations are not carried out; based on the shadow prices corresponding to the original order package some orders are slightly decreased and some increased if that is profitable for the company.

16.4 Exercises

Exercise 16.4.1. Consider the cutting stock problem as described in this chapter.

(a) Show that model (16.1) is equivalent to a model where the '=' signs of the restrictions are replaced by '≥' signs.

(b) Show that the constraints of model (16.1) correspond to nonnegative optimal dual values.

Exercise 16.4.2. Consider the bounded knapsack problem.

(a) Generalize Theorem 7.2.1 to the case of the bounded knapsack problem.

(b) Explain how the branch-and-bound algorithm of Section 7.2.3 can be generalized to the case of the bounded knapsack problem.

(c) Use the branch-and-bound algorithm described in (b) to solve the following model:

$$\begin{aligned}
\max \quad & 2x_1 + 3x_2 + 8x_3 + x_4 + 6x_5 \\
\text{s.t.} \quad & 3x_1 + 7x_2 + 12x_3 + 2x_4 + 6x_5 \leq 26 \\
& 0 \leq x_i \leq 3 \text{ and integer, for } i = 1, \ldots, 5.
\end{aligned}$$

Exercise 16.4.3. Use a computer package to solve the model in Exercise 16.4.2(c).

Exercise 16.4.4. The factory PAPYRS produces paper in standard rolls of width 380 cm. For a certain day, the order package is as follows:

Customer	Number of demanded subrolls	Width (in cm)
1	43	150
2	31	110
3	10	70

These orders need to be cut from the standard rolls with a minimum amount of trimloss.

(a) Design an LO-model that describes this problem, and determine all possible cutting patterns.

(b) Solve the problem by means of the Gilmore-Gomory algorithm.

(c) What are the shadow prices of the demand constraints? One of the shadow prices turns out to be zero; what does this mean? How much can the corresponding order be increased without increasing the number of subrolls needed?

(d) Suppose that the customers want subrolls with the same diameter as the standard rolls. Therefore, the optimal solution needs to be integer-valued. Solve this problem by means of an ILO-package. Are all cutting patterns necessary to determine this integer optimal solution? Show that the problem has multiple optimal solutions.

Exercise 16.4.5. For the armament of concrete, the following is needed: 24 iron bars with a length of 9.50 m, 18 bars of 8 m, 44 bars of 6.20 m, 60 bars of 5.40 m, and 180 bars of 1.20 m. All bars have the same diameter. The company that produces the bars has a large inventory; all these bars have a standard length of 12 m. The problem is how to cut the demanded bars out of the standard ones, such that the waste is minimized.

(a) Formulate this problem in terms of a mathematical model.

(b) Discuss the differences with the cutting stock problem discussed in this chapter.

(c) Determine an optimal solution, and the percentage of trimloss.

CHAPTER 17

Off-shore helicopter routing

Overview

This case study deals with the problem of determining a flight schedule for helicopters to off-shore platform locations for exchanging crew people employed on these platforms. The model is solved by means of an LO-model in combination with a column generation technique. Since the final solution needs to be integer-valued, we have chosen a round-off procedure to obtain an integer solution.

17.1 Problem description

In 1962, the countries around the North Sea started to drill for gas in the North Sea motivated by the discovery of a large gas field in the northern part of The Netherlands. Since then, a lot of production platforms have been built there. This study concerns the off-shore locations on the Dutch continental shelf of the North Sea; see Figure 17.1.

The transportation of people to these platforms is carried out by helicopters which have as their base an airport near Amsterdam. The transportation of goods is usually done by boats, but this is not considered in this study. The purpose of the present study is to develop a transportation schedule for the demanded crew exchanges on the various platforms such that the total cost of the transportation is as low as possible.

There are two types of platforms, namely drill platforms (so-called *drill rigs*), and production platforms where the actual production takes place. Moreover, there are two types of production platforms, namely the large ones with 20 to 60 employees, and the so-called satellites served and controlled by computers from the large platforms. A survey of the locations of 51 Dutch platforms, where people are working, is depicted in Figure 17.2. We consider only these platforms. The coordinates of the 51 platforms are listed in Table 17.1. The first coordinate is the longitude and the second one the latitude; they are calculated with the airport as the origin.

#	Coord.		#	Coord.		#	Coord.	
1	1	35	18	76	20	35	61	32
2	2	36	19	73	21	36	65	32
3	21	10	20	77	23	37	69	35
4	43	21	21	76	24	38	71	28
5	46	29	22	84	25	39	79	32
6	27	35	23	81	27	40	35	23
7	29	37	24	63	18	41	37	24
8	58	20	25	70	18	42	38	22
9	57	29	26	66	19	43	68	10
10	58	33	27	68	19	44	65	52
11	57	41	28	65	20	45	63	53
12	68	49	29	67	20	46	75	7
13	72	47	30	63	21	47	73	8
14	73	47	31	64	21	48	75	12
15	71	49	32	67	22	49	55	55
16	75	51	33	69	28	50	60	55
17	72	20	34	68	30	51	58	59

Figure 17.1: Continental shelves of the North Sea.   **Table 17.1:** Coordinates of the platforms (in units of 4.5 km).

There are two types of crews working on the platforms, namely regular crew which are employees that have their regular jobs on the platforms, and irregular crew such as surveyors, repair people, or physicians. The regular crew on the Dutch owned platforms work every other week.

The helicopters that fly from the airport to the platforms transport new people, and helicopters that leave the platforms or return to the airport transport the leaving people. The number of people leaving the platform does have to be equal to the number of newly arriving people. However, in this case study, we will assume that each arriving person always replaces a leaving person, and so the number of people in a helicopter is always the same during a flight. A crew exchange is the exchange of an arriving person by a leaving person.

In the present study, the transportation is carried out by helicopters with 27 seats; the rental price is about $6,000 per week. We assume that there are enough helicopters available to carry out all demanded crew exchanges.

17.2   Vehicle routing problems

Problems that deal with the planning of cargo on vehicles and the routing of vehicles are called *vehicle routing problems* (VRPs). To be more precise, a VRP can be characterized by the following features:

  a number of depots with known locations;

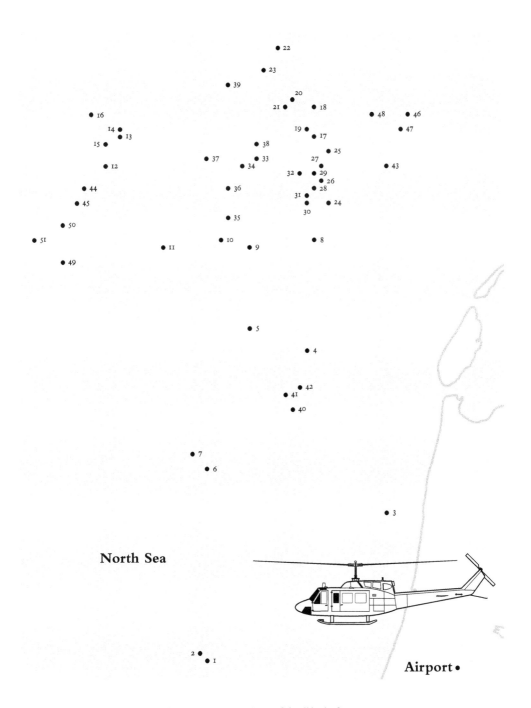

Figure 17.2: Locations of the 51 platforms.

a number of customers with known locations;

a given demand for commodities;

a number of vehicles with known capacities;

limited driving times;

deliveries take place within certain time windows.

In terms of the present off-shore transportation problem, these features become:

there is a single depot, the airport;

the customers are the platforms;

the demand is the demand for crew exchanges;

the vehicles are the helicopters;

the limited driving time is the radius of action, called the *range*, of the helicopter, determined by the maximum amount of fuel that can be carried;

the time windows are the usual daily working hours of one week without the week end.

The helicopters are not allowed to fly at night, and the amount of fuel that can be carried limits the range of the helicopter. It is assumed that the range is large enough to reach any platform and to return to the airport. One major difference from the usual VRP is that the demanded crew exchanges of a platform need not be carried out by one helicopter. This is an example of the so-called *split delivery VRP*, in which the demand of one customer can be delivered in multiple parts. VRPs face the same computational difficulty as the traveling salesman problem (see Section 4.8), and so optimal solutions are usually not obtainable within a reasonable amount of time. We will design a linear optimization model that can be solved by means of so-called column generation; the obtained solution is then rounded off to an integer solution, which is in general not optimal but which may be good enough for practical purposes.

17.3   Problem formulation

Let $\mathcal{P} = \{P_1, \ldots, P_N\}$ be the set of platforms, and let $\mathcal{D} = \{D_1, \ldots, D_N\}$ be the set of numbers of demanded crew exchanges at the platforms. We may assume that $D_i \geq 1$ for all $i = 1, \ldots, N$, because if $D_i = 0$ for some i, then we can just delete platform P_i from the problem. There are known distances between any pair of platforms, and between the airport and any platform. These distances are given by $d(A, B)$ for each $A, B \in \mathcal{P} \cup \{\text{Airport}\}$. In our particular case study, all distances are expressed in units of 4.5 km, which is about 2.415 nautical miles. Throughout this chapter, we assume that the distances satisfy the *triangle inequality*:

$$d(A, C) \leq d(A, B) + d(B, C) \qquad \text{for each } A, B, C \in \mathcal{P} \cup \{\text{Airport}\}.$$

Informally, this inequality states that flying straight from A to C is shorter (or, at least, not longer) than flying from A to B, and then to C. Clearly, this inequality holds in the case of helicopter routing.

The helicopters have a *capacity* C, which is the number of seats (excluding the pilot) in the helicopter, and a *range* R, which is the maximum distance that the helicopter can travel in one flight. We assume that all helicopters have the same capacity and the same range. In the case study, for the capacity C of all helicopters we take $C = 23$, so that four seats are always left available in case of emergencies or for small cargo. For the helicopter range, we take $R = 200$ ($\times 4.5$ km).

A *flight* f is defined as a vector $\mathbf{w}_f = \begin{bmatrix} w_{1f} & \cdots & w_{Nf} \end{bmatrix}^{\mathsf{T}}$, where the value of w_{if} indicates the number of crew exchanges at platform P_i to be performed by flight f. If $w_{if} = 0$, this means that flight f does not visit platform P_i. According to this definition, two flights visiting the same set of platforms but which have different crew exchange patterns are considered different flights. The total traveled distance d_f during flight f is the length of a shortest traveling salesman tour that includes the airport and all platforms P_i that satisfy $w_{if} > 0$. A *feasible flight* is a flight such that the range R is not exceeded, and the number of crew exchanges during that flight does not exceed the capacity C. To be precise, a flight f is feasible if:

$$d_f \leq R \qquad \text{and} \qquad \sum_{i=1}^{N} w_{if} \leq C.$$

Let $\mathcal{F} = \{f_1, \ldots, f_F\}$ be the set of all feasible flights. Clearly, for a set of locations together with the airport there are usually several possibilities of feasible flights, and even several possibilities of shortest feasible flights. In fact, the number F of feasible flights grows exponentially with the number of platform locations, and the range and capacity of the helicopter (see Exercise 17.10.2).

Let $j \in \{1, \ldots, F\}$. With slight abuse of notation, we will write w_{ij} instead of w_{if_j} for each platform P_i and each flight f_j. Similarly, we will write d_j instead of d_{f_j}.

A *feasible flight schedule* is a finite set $\mathcal{S}$ of feasible flights such that the demand for crew exchanges on all platforms is satisfied by the flights in $\mathcal{S}$. To be precise, $\mathcal{S}$ is a feasible flight schedule if:

$$\sum_{f_j \in \mathcal{S}} w_{ij} = D_i \qquad \text{for } i = 1, \ldots, N.$$

It may be necessary to use the same flight multiple times, so that it is possible that the same flight f_j appears multiple times in $\mathcal{S}$. Such a set in which elements are allowed to appear multiple times is also called a *multiset*.

Example 17.3.1. *Let* $\mathcal{P} = \{P_1, P_2, P_3, P_4\}$, $\mathcal{W} = \{15, 9, 3, 8\}$, *and* $C = 10$. *Let,* $\mathcal{S} = \{f_1, f_2\}$, *where flight* f_1 *visits the platforms* $P_1(10)$, $P_2(2)$, $P_3(3)$, *and* f_2 *visits the* platforms $P_1(5)$, $P_2(7)$, $P_4(8)$. The numbers in parentheses refer to the number of crew exchanges during that flight. That is, we have $\mathbf{w}_1 = \begin{bmatrix} w_{11} & w_{21} & w_{31} & w_{41} \end{bmatrix}^\top = \begin{bmatrix} 10 & 2 & 3 & 0 \end{bmatrix}^\top$ and $\mathbf{w}_2 = \begin{bmatrix} w_{12} & w_{22} & w_{32} & w_{42} \end{bmatrix}^\top = \begin{bmatrix} 5 & 7 & 0 & 8 \end{bmatrix}^\top$. Then $\mathcal{S}$ is a feasible flight schedule. Notice that $\mathcal{F}$ is much larger than $\mathcal{S}$, because, for example, it contains all flights consisting of the same platforms as f_1, but with smaller numbers of crew exchanges. It can be calculated that, in this case, $F = |\mathcal{F}| = 785$, assuming there is no range restriction, i.e., $R = \infty$ (see also Exercise 17.10.2).

It is assumed that the cost of operating a helicopter is a linear function of the total traveled distance. Therefore, we are looking to solve the problem of finding a feasible flight schedule which minimizes the total traveled distance of the flights.

The number of demanded crew exchanges on a certain platform may exceed the capacity of the helicopter. In that case a helicopter could visit such a platform as many times as needed with all seats occupied until the demand is less than the capacity. However, in general, this strategy does not lead to an optimal feasible flight schedule, since the helicopters have a limited range. This will become clear from the following simple example.

Example 17.3.2. *Consider three platforms: P_1 with demand 1, P_2 with demand 18, and P_3 with demand 1; see Figure 17.3. Suppose that the capacity of the helicopter is 10. If the helicopter first flies to P_2 and performs 10 crew exchanges there, this leaves 8 crew members to be exchanged at P_2, and 1 on each of P_1 and P_2. Thus, the flight schedule becomes (see the left diagram in Figure 17.3):*

$$Airport \rightarrow P_2(10) \rightarrow Airport,$$
$$Airport \rightarrow P_3(1) \rightarrow P_2(8) \rightarrow P_1(1) \rightarrow Airport.$$

The numbers given in parentheses are the numbers of crew exchanges that are performed at the corresponding platforms. Now suppose that the pairwise distances between each of P_1, P_2, P_3, and the airport is 1, and the range of the helicopter is 3. Then, the flight $Airport \rightarrow P_3 \rightarrow P_2 \rightarrow P_1 \rightarrow Airport$ exceeds the range of the helicopter. Therefore, the helicopter needs multiple flights to exchange the crews remaining after the flight $Airport \rightarrow P_2 \rightarrow Airport$. The flight schedule becomes (see the middle diagram in Figure 17.3):

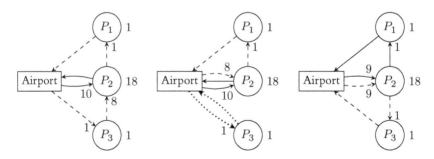

Figure 17.3: The limited range of the helicopters. The numbers next to the nodes are the demands. The different flights are drawn using the solid, dashed, and dotted arrows. The number of crew exchanges are next to each arc.

$$Airport \rightarrow P_2(10) \rightarrow Airport,$$
$$Airport \rightarrow P_2(8) \rightarrow P_2(1) \rightarrow Airport,$$
$$Airport \rightarrow P_1(3) \rightarrow Airport.$$

There is, however, a better schedule that only takes two flights (see the right diagram in Figure 17.3):

$$Airport \rightarrow P_2(9) \rightarrow P_1(1) \rightarrow Airport,$$
$$Airport \rightarrow P_2(9) \rightarrow P_3(1) \rightarrow Airport.$$

The situation in which the number of demanded crew exchanges on a platform exceeds the capacity of the helicopter appears quite often in practice. Clearly, such a platform has to be visited more than once, but the model should also take into account the limited range of the helicopters.

Because of the computational difficulty of the off-shore transportation problem (and VRPs in general), we cannot hope to design an algorithm that solves the problem within a reasonable amount of time. We therefore resort to a heuristic algorithm that (hopefully) produces a solution whose objective value is reasonably close to the optimal objective value, and which runs in an acceptable amount of computing time. In the remainder of this chapter, we will formulate an integer linear optimization model that solves the off-shore transportation problem. We will describe an algorithm to solve the LO-relaxation of this model, and show how an optimal solution of the LO-relaxation can be rounded off to find a feasible solution for the off-shore transportation problem. We will then show some computational results to assess the quality of the solutions produced by this procedure.

17.4 ILO formulation

The present off-shore transportation problem can be formulated as an integer linear optimization model. In Section 17.3, we introduced the following notation:

N = the number of platforms;
i = the platform location index, with $i \in \{1, \ldots, N\}$;
F = the number of feasible helicopter flights;
$\mathcal{F}$ = the set of all feasible helicopter flights $\{f_1, \ldots, f_F\}$;
j = the flight index, with $j \in \{1, \ldots, F\}$;
D_i = the number of demanded crew exchanges at platform P_i;
w_{ij} = the number of crew exchanges on platform P_i during flight f_j;
d_j = the total traveled distance during flight f_j;
C = the capacity, i.e., the number of available seats, of the helicopters;
R = the range of the helicopters.

In order to formulate the problem as an ILO-model, we introduce the following decision variables:

x_j = the number of times flight f_j is performed ($j = 1, \ldots, F$).

The ILO-model can be formulated as follows:

$$\min \sum_{j=1}^{F} d_j x_j \tag{FF}$$

$$\text{s.t.} \sum_{j=1}^{F} w_{ij} x_j = D_i \qquad\qquad \text{for } i = 1, \ldots, N$$

$$x_j \geq 0, \text{ and integer} \qquad\qquad \text{for } j = 1, \ldots, F.$$

The objective function is the sum of $d_j x_j$ over all feasible flights $f_j \in \mathcal{F}$, where $d_j x_j$ is the total length of the flight f_j (any flight f_j may be carried out several times). The total number of passengers with destination platform P_i on all helicopter flights in the schedule has to be equal to the number D_i of demanded crew exchanges of platform P_i. The technology matrix $\mathbf{W} = \{w_{ij}\}$ of the model has N rows and F columns; one can easily check that $\mathbf{W}$ contains the identity matrix $\mathbf{I}_N$, and hence it has full row rank.

Although the formulation (FF) may look harmless, it is in fact a hard model to solve. Not only are we dealing with an ILO-model, but since the value of F is in general very large, the number of columns of the matrix $\mathbf{W}$ is very large. In fact, $\mathbf{W}$ may be too large to fit in the memory of a computer. This means that the usual simplex algorithm cannot be applied to solve even the LO-relaxation of model (FF) in which the variables do not need to be integer-valued. Even if the matrix $\mathbf{W}$ fits in the memory of a computer, finding an entering column for the current basis may take an exorbitant amount of computer time.

We will take the following approach to solving model (FF). We will first show how to solve the relaxation of model (FF), which we denote by (RFF), using a *column generation procedure*, as in Chapter 16. This procedure starts with a feasible basic solution of (RFF) and in each iteration determines an 'entering column' without explicitly knowing the matrix $\mathbf{W}$. In Section 17.7 a round-off procedure is formulated, which derives integer solutions from the solutions of model (RFF).

17.5   Column generation

An effective method for solving LO-models with a large number of columns is the *column generation procedure* (CG-procedure); see also Chapter 16. This is an iterative procedure which works just like the simplex algorithm. The main difference lies in the procedure that determines the column that enters the basis in each iteration. Because the technology matrix is very large, we cannot simply go through all the nonbasic columns, calculate their current objective coefficients, and choose a column with a negative current objective coefficient. This would, if even possible, take an exorbitant amount of time. Instead, we will assume that the technology matrix is only virtually known and use a different method to 'generate an entering column'.

We will explain the method by means of model (RFF). For this model, finding an initial feasible basic solution is straightforward. Indeed, an initial feasible basic solution requires N basic variables; for $i = 1, \ldots, N$, we just take the flight f_{j_i} that performs $\min\{C, D_i\}$ crew exchanges at platform P_i and immediately flies back to the airport. Let $\mathbf{W}^{(1)}$ be the basis matrix corresponding to this feasible basic solution. Note that $\mathbf{W}^{(1)}$ is a diagonal matrix with $\min\{C, D_i\}$ $(i = 1, \ldots, N)$ as the diagonal entries, and hence it is invertible. Note also that $\mathbf{W}^{(1)}$ is a submatrix of $\mathbf{W} = \{w_{ij}\}$.

Next consider the k'th iteration step of the procedure $(k = 1, 2, \ldots)$. At the beginning of the k'th iteration, we have a submatrix $\mathbf{W}^{(k)}$ of $\mathbf{W}$, which is – during the next iteration step – augmented with a new column of $\mathbf{W}$. This entering column is generated by means of solving a knapsack problem plus a traveling salesman problem.

Let $\mathbf{B}$ be an optimal basis matrix of $\mathbf{W}^{(k)}$ in model (RFF), with $\mathbf{W}$ replaced by $\mathbf{W}^{(k)}$, and let $\mathbf{y} = (\mathbf{B}^{-1})^\mathsf{T} \mathbf{d}_{BI}$ be an optimal dual solution corresponding to $\mathbf{B}$ of this model; see Theorem 4.2.4. Notice that $\mathbf{B}$ is also a basis matrix for model (RFF). Therefore, we may write, as usual, $\mathbf{W} \equiv [\mathbf{B} \ \mathbf{N}]$. Now consider any nonbasic variable x_j (corresponding to flight f_j). The current objective coefficient corresponding to x_j is $d_j - \sum_{i=1}^{N} w_{ij} y_i$; see Exercise 17.10.1. Suppose that we can determine a column with the smallest current objective coefficient, c^*. If $c^* < 0$, then we have determined a column corresponding to a variable x_j that has a negative current objective coefficient and we augment $\mathbf{W}^{(k)}$ with this column. If, however $c^* \geq 0$, then we conclude that no such column exists, and hence an optimal solution has been found. We can indeed determine such a column with the smallest current objective coefficient, by solving the following model:

$$c^* = \min \ d_{S(\mathbf{w})} - \sum_{i=1}^{N} y_i w_i$$

$$\text{s.t.} \ \sum_{i=1}^{N} w_i \leq C$$

$$w_i \ \leq D_i \qquad \text{for } i = 1, \ldots, N \qquad \text{(CG)}$$

$$d_{S(\mathbf{w})} \ \leq R,$$

$$w_i \geq 0 \text{ and integer} \qquad \text{for } i = 1, \ldots, N,$$

where $S(\mathbf{w}) = \{i \mid w_i > 0\}$ and $d_{S(\mathbf{w})}$ is the length of a shortest route of a flight visiting platforms P_i in $S(\mathbf{w})$, starting and ending at the airport. In this model, the decision variables are the w_i's (i.e., the number of crew exchanges to be performed at each platform that is visited). Recall that the y_i's are known optimal dual values obtained from the basis matrix $\mathbf{B}$, and they are therefore parameters of model (CG). This model is a well-defined optimization model, but it is by no means a linear optimization model or even an integer linear optimization model. In fact, model (CG) is a nonlinear model, since it contains the term $d_{S(\mathbf{w})}$ which is defined as the shortest total traveled distance of a flight from the airport to all platforms in S and back to the airport. This means that $d_S(\mathbf{w})$ is the optimal objective value of a traveling salesman problem.

Assume for the time being that we can solve model (CG). The model determines a column $\begin{bmatrix} w_1^* & \ldots & w_N^* \end{bmatrix}^\mathsf{T}$ of $\mathbf{W}$. The optimal objective value c^* is the current objective coefficient of this column. If $c^* < 0$, then this vector is not a column of $\mathbf{W}^{(k)}$, because otherwise $\mathbf{B}$ would have been an optimal basis matrix in model (RFF) with $\mathbf{W}$ replaced by $\mathbf{W}^{(k)}$. This means that we find a column of $\mathbf{W}$ that is not in $\mathbf{W}^{(k)}$. We therefore construct $\mathbf{W}^{(k+1)}$ by augmenting $\mathbf{W}^{(k)}$ with the column vector $\begin{bmatrix} w_1^* & \ldots & w_N^* \end{bmatrix}^\mathsf{T}$. Notice that if the corresponding column index (i.e., the flight index) is denoted by j^*, then $w_{ij^*} = w_i^*$.

In model (CG), the minimum of all current objective coefficients is determined, including the ones that correspond to basic variables. Since the current objective coefficients corresponding to the basic variables all have value zero, the fact that the minimum over of all current objective coefficients is determined does not influence the solution. Recall that, if all current objective coefficients $d_j - \sum_i w_{ij} y_i$ in the current simplex tableau are nonnegative, which means that the optimal objective value of (CG) is nonnegative, then an optimal solution in (RFF) has been reached. Therefore, if $c^* \geq 0$, then no column with negative current objective coefficient exists, and the current feasible basic solution is optimal.

The question that remains to be answered is: how should model (CG) be solved? Before we can solve model (CG), note that the model can be thought of a two-step optimization model. In the first step, we consider all possible subsets S of $\mathcal{P}$ and compute, for every such subset S, the corresponding optimal values of the w_i's, and the corresponding optimal objective value $c(S)$. Next, we choose a set S for which the value of $c(S)$ is minimized (subject to the constraint $d_S \leq R$). To be precise, for each subset S of $\mathcal{P}$, we define the following model:

$$
\begin{aligned}
c(S) = d_S \;-\; & \max \sum_{i=1}^{N} y_i w_i \\
\text{s.t.} \;\; & \sum_{i=1}^{N} w_i \leq C && (CG_S) \\
& w_i \;\; \leq D_i && \text{for } i = 1, \ldots, N \\
& w_i \;\; = 0 && \text{if } P_i \notin S \text{ for } i = 1, \ldots, N \\
& w_i \geq 0 \text{ and integer} && \text{for } i = 1, \ldots, N.
\end{aligned}
$$

Now, (CG) is equivalent to solving the following model:

$$
\begin{aligned}
c^* = \min_{S \subseteq \mathcal{P} \setminus \{\emptyset\}} \;\; & c(S) \\
\text{s.t.} \;\; & d_S \leq R,
\end{aligned}
\qquad (CG^*)
$$

where the minimum is taken over all nonempty subsets S of $\mathcal{P}$. In order to solve (CG^*), we may just enumerate all nonempty subsets of $\mathcal{P}$, solve model (CG_S) for each S, and choose, among all sets S that satisfy $d_S \leq R$, the one that minimizes $c(S)$. In fact, it turns out that we do not need to consider all subsets S. Thus, to solve model (CG), we distinguish the following procedures:

(A) For each nonempty subset S of the set $\mathcal{P}$ of platforms, we formulate and solve a traveling salesman problem and a knapsack problem that solves model (CG_S).

(B) We design a procedure for generating subsets S of $\mathcal{P}$, while discarding – as much as possible – subsets S that only delay the solution procedure.

We will discuss (A) and (B) in the next two subsections.

17.5.1 Traveling salesman and knapsack problems

Let S be a nonempty subset of $\mathcal{P}$. Model (CG_S) is a simple knapsack problem, in the following way. Let d_S be the total traveled distance on a flight from the airport to the platforms in S and back to the airport, to be calculated by using Table 17.1. Since S is now fixed, the term d_S is a constant, and nothing more than the length of a shortest traveling salesman tour (see Section 4.8) including the airport and the platforms in S. The length of such a shortest tour can be determined by simply calculating all possible tours and selecting a smallest one. As long as the number of platforms in S is not too large, this procedure works fast enough. Since in the present situation, S will almost never contain more than five platforms (in which case there are $\frac{5!}{2} = 60$ possible tours), we use this procedure to calculate shortest traveling salesman tours.

If $d_S > R$, then S does not allow a feasible flight. So we may concentrate on the case in which $d_S \leq R$. By eliminating the variables w_i with $i \notin S$ (which are restricted to have value zero), the maximization problem in model (CG_S) can be written as the following knapsack problem (KP_S):

$$
\begin{aligned}
\max \ & \sum_{i \in S} y_i w_i \\
\text{s.t.} \ & \sum_{i \in S} w_i \leq C && (KP_S) \\
& w_i \leq D_i && \text{for } i = 1, \ldots, N \\
& w_i \geq 0 \text{ and integer} && \text{for } i = 1, \ldots, N.
\end{aligned}
$$

An optimal solution w_i^* $(i \in S)$ of (KP_S) can easily be calculated in the following way. First, sort the values of the y_i's with $i \in S$ in nonincreasing order: say $y_{j_1} \geq y_{j_2} \geq \ldots \geq y_{j_N} \geq 0$. Then, take $w_{j_1} = \min\{C, D_{j_1}\}$. If $C \leq D_{j_1}$, then take $w_{j_1} = C$, and $w_{j_2} = \ldots = w_{j_N} = 0$. Otherwise, take $w_{j_1} = D_{j_1}$ and $w_{j_2} = \min\{C - D_{j_1}, D_{j_2}\}$, and proceed in a way similar to the calculation of w_{j_1}. In terms of a knapsack problem, this means that the knapsack is filled with as much as possible of the remaining item with the highest profit. See also Section 7.2.3. Note that this optimal solution of (KP_S) is integer-valued.

Model (CG) can be solved by individually considering each possible subset S, and solving for this fixed S the knapsack problem (KP_S) (and therefore model (CG_S)). This is certainly a time-consuming procedure, since the number of nonempty subsets S of $\mathcal{P}$ is $2^N - 1$. Since $N = 51$ in the present case study, this amounts to roughly 2.2×10^{15} subsets. In Section

17.5.2, we will present a more clever method, that excludes a large number of subsets S from consideration when solving (KP_S).

17.5.2 Generating 'clever' subsets of platforms

First, all platforms are sorted according to nonincreasing values of y_i, and relabeled. Thus, we now have that $y_1 \geq y_2 \geq \ldots \geq y_N$. Fixing this ordering during the column generation step has the advantage that the y_i's do not need to be sorted for every platform subset S when (KP_S) is solved. Moreover, we generate all platform subsets S according to the lexicographical ordering, $S = \{P_1\}$ being the first set, and $S = \{P_N\}$ being the last. For example, if $N = 3$ and $\mathcal{P} = \{P_1, P_2, P_3\}$, then the subsets of $\mathcal{P}$ are ordered as follows: $\{P_1\}, \{P_1, P_2\}, \{P_1, P_2, P_3\}, \{P_1, P_3\}, \{P_2\}, \{P_2, P_3\}, \{P_3\}$. Since the platforms are ordered according to their optimal dual values, subsets with high optimal dual values are generated first.

We call a platform subset S_2 a *lex-superset* of a platform subset S_1, if $S_1 \subset S_2$ with $S_1 \neq S_2$ and S_2 appears later in the lexicographical order than S_1. For instance, $\{P_1, P_2, P_3\}$ is a lex-superset of $\{P_1\}$, but not of $\{P_2\}$.

We will formulate two procedures which enable us to exclude subsets S, together with all their lex-supersets, from consideration.

(B1) *Exceeding the range.*

   If a platform subset S satisfies $d_S > R$, then any S' such that $S \subseteq S'$ satisfies $d_{S'} > R$ (due to the triangle inequality) and hence all supersets of S may be excluded from consideration. Since we are considering the sets in the lexicographical ordering, all supersets of S that are not lex-supersets have already been considered, so in fact we only exclude the lex-supersets from consideration.

(B2) *Exceeding the capacity.*

   Suppose that the current S does not contain all platforms, i.e., $S \neq \mathcal{P}$. If $\sum_{i \in S} D_i \geq C$, then all lex-supersets of S can be excluded from consideration. The reason is that, when adding a new platform, say P_t, to the current S, then the optimal value of w_t is zero, because the value of y_t is less than the values of all y_i's with $P_i \in S$. Hence, P_t is not visited during the flight corresponding to $S \cup \{t\}$. By excluding all lex-supersets of S, we have created the advantage that only platform subsets are considered for which the optimal values of the w_i's in (KP_S) are strictly positive, and so the traveling salesman tour for such sets does not visit platforms that have zero demand for crew exchanges.

If $c(S) \geq 0$ for all S, then an optimal solution of model (RFF) with technology matrix $\mathbf{W}^{(k)}$ is also an optimal solution of the original model (RFF). It turns out that the calculations can be speeded up, when the current matrix $\mathbf{W}^{(k)}$ is augmented with several new columns for which $c(S) < 0$ at once.

The solution procedure of model (RFF) can now be formulated as follows. For the submatrices $\mathbf{W}^{(k)}$ of $\mathbf{W}$ ($k = 1, 2, \ldots$), define J_k as the set of flight indices corresponding to the columns of $\mathbf{W}^{(k)}$.

Algorithm 17.5.1. (*Simplex algorithm with column generation*)

Input: Models (RFF) and (CG); coordinates of platforms, the vector of demanded numbers of crew exchanges $\begin{bmatrix} D_1 & \ldots & D_N \end{bmatrix}^{\mathsf{T}}$, the range R, and the capacity C.

Output: An optimal solution of model (RFF).

Step 0:   *Initialization.*
Let $\mathbf{W}^{(1)} = \mathbf{A}_W$, with $\mathbf{A}_W = \{a_{ij}\}$ the diagonal matrix with $a_{ii} = \min\{D_i, C\}$. Calculate d_f for each $f \in J_1$. Go to Step 1.

Step 1:   *Simplex algorithm step.*
Let $\mathbf{W}^{(k)} = \{w_{ij}^{(k)}\}$, $k \geq 1$, be the current 'flight' matrix (a submatrix of the virtually known matrix $\mathbf{W}$). Solve the LO-model:

$$
\begin{aligned}
\min \quad & \sum_{j \in J_k} d_j x_j \\
\text{s.t.} \quad & \sum_{j \in J_k} w_{ij}^{(k)} x_j = D_i && \text{for } i = 1, \ldots, N && (FF_k) \\
& x_j \geq 0 && \text{for } j \in J_k.
\end{aligned}
$$

Let $\mathbf{y}^{(k)} = \begin{bmatrix} y_1^{(k)} & \ldots & y_N^{(k)} \end{bmatrix}^{\mathsf{T}}$ be an optimal dual solution corresponding to the current optimal basis matrix of model (FF_k). Go to Step 2.

Step 2:   *Column generation step.*
Label the platforms from $1, \ldots, N$ according to nonincreasing values of $y_i^{(k)}$. Using this labeling, order all platform subsets S according to the lexicographical ordering.

Determine the optimal objective value $c(S^*) = d_{S^*} - \sum_{i \in S^*} y_i^{(k)} w_i^*$ of model (CG) by calculating, for successively (according to the lexicographical ordering) all platform subsets S, feasible solutions of (CG), thereby taking into account the procedures (B1) and (B2). Let $\mathbf{w}^* = \{w_i\}$, with $w_i = w_{i^*}$ for $i \in S^*$ and $w_i = 0$ otherwise.

If $c(S^*) \geq 0$, then an optimal solution has been reached.

If $c(S^*) < 0$, then define:

$$
\mathbf{W}^{(k+1)} \equiv \begin{bmatrix} \mathbf{W}^{(k)} & \mathbf{w}^* \end{bmatrix},
$$

and return to Step 1.

| Number of platforms: 51 |
| Coordinates of the platforms: see Table 17.1 |
| Capacity of the helicopter: 23 |
| Range of the helicopter: 200 (units of 4.5 km) |

#	Exch.	#	Exch.	#	Exch.	#	Exch.
1	0	14	8	27	6	40	40
2	20	15	8	28	4	41	40
3	3	16	8	29	15	42	16
4	12	17	10	30	19	43	40
5	14	18	10	31	16	44	5
6	4	19	45	32	14	45	20
7	8	20	12	33	12	46	20
8	8	21	30	34	11	47	5
9	30	22	3	35	14	48	4
10	8	23	2	36	5	49	20
11	8	24	10	37	5	50	4
12	8	25	12	38	0	51	4
13	8	26	9	39	13		

Table 17.2: Demanded crew exchanges. (# = Platform index, Exch. = Number of demanded crew exchanges)

Note that this algorithm always returns an optimal solution, because a feasible solution is constructed in Step 0 (meaning that the model is feasible), and the model cannot be unbounded because the optimal objective value cannot be negative.

It has been mentioned already that the traveling salesman problem in Step 2 can be solved by considering all tours and picking a shortest one. If the sets S become large, then more advanced techniques are needed, since the number of tours becomes too large. Recall that if $|S| = n$, the number of tours is $\frac{1}{2}n!$. Also, recall that the calculations are speeded up when, in each iteration, multiple columns for which $c(S) < 0$ are added to $\mathbf{W}^{(k)}$.

17.6   Dual values as price indicators for crew exchanges

The solution of the linear optimization model (RFF) in combination with the column generation model (CG) is in general not integer-valued. Using a standard LO-package for the simplex algorithm, we calculated a (noninteger-valued) optimal solution for the input instance of Table 17.2. The package also generates optimal dual values. In the current section, we will use these optimal dual values as indicators for the price of crew exchanges. After 217 simplex iterations, we found an optimal solution of model (RFF) for the input instances of Table 17.2; see Table 17.3. The 'j' column of Table 17.3 contains labels for the nonzero-valued optimal decision variables x_j^* of model (RFF); the zero-valued variables are omitted. We have labeled these flight numbers from 1 through 41. Note that some values of x_j^* are already integers. The third column of Table 17.3 contains the platforms that are

j	x_j^*	Platform visits	Crew exchanges	Distance ($\times 4.5$ km)
1*	1.000	2	20	72.111
2	0.172	3, 40, 7, 6	3, 8, 8, 4	104.628
3	0.250	4, 42	12, 11	96.862
4	1.000	4, 5	9, 14	110.776
5	0.182	8, 30	4, 19	132.858
6	0.909	8, 30	8, 15	132.858
7	0.828	3, 42, 40	3, 16, 4	89.111
8	1.000	9, 35, 10	1, 14, 8	138.846
9	1.261	9	23	127.906
10	0.217	11, 49	8, 15	162.138
11	0.783	11, 44, 45	8, 5, 10	168.380
12	0.739	12, 15, 13	8, 7, 8	175.034
13	0.298	13, 14, 16	7, 8, 8	182.152
14	0.298	14, 16, 12	8, 8, 7	182.398
15	0.404	14, 16, 15	8, 8, 7	182.033
16	1.000	17, 19	10, 13	152.101
17	1.000	18, 20, 19	10, 12, 1	162.182
18	1.348	19	23	151.921
19	1.304	21	23	159.399
20	0.800	24, 31	10, 13	136.041
21	0.200	24, 26, 28	10, 9, 4	138.105
22	1.000	26, 27, 25	5, 6, 12	145.194
23	0.550	26, 29, 28	4, 15, 4	140.102
24	0.250	28, 32, 31	4, 14, 5	141.355
25	0.750	29, 32	9, 14	142.441
26	0.272	30, 31	7, 16	134.765
27	0.522	33, 34, 36	7, 11, 5	152.756
28	0.217	33, 34	12, 11	151.024
29	0.478	33, 34, 36	12, 6, 5	152.765
30	1.000	37, 39, 23, 22	5, 13, 2, 3	184.442
31	0.828	40, 7, 6	11, 8, 4	104.145
32	1.139	40	23	83.762
33	1.739	41	23	88.204
34	1.739	43	23	137.463
35	0.217	44, 45	5, 18	167.805
36	0.435	45, 50	19, 4	167.328
37	0.261	46, 47	18, 5	150.999
38	0.261	46, 48	19, 4	156.280
39	0.739	48, 46, 47	4, 14, 5	156.627
40	0.435	49, 51	19, 4	165.516
41	0.565	49, 51, 50	15, 4, 4	168.648

The total traveled distance is 3,903.62 ($\times 4.5$ km)

Table 17.3: Noninteger optimal solution of model (RFF). Flight 1 (marked with an asterisk) is the only one that does not use the full capacity of the helicopter.

Pl.	Dual	Pl.	Dual	Pl.	Dual	Pl.	Dual
1	–	14	7.723	27	6.282	40	3.642
2	3.606	15	7.587	28	6.095	41	3.835
3	3.803	16	8.393	29	6.105	42	3.946
4	4.455	17	6.623	30	5.776	43	5.977
5	5.049	18	7.130	31	5.896	44	7.711
6	4.857	19	6.605	32	6.250	45	7.181
7	5.582	20	7.023	33	6.566	46	6.550
8	5.776	21	6.930	34	6.566	47	6.620
9	5.561	22	11.377	35	6.103	48	7.957
10	5.981	23	7.217	36	6.913	49	6.941
11	7.252	24	5.940	37	7.662	50	7.724
12	7.637	25	6.443	38	–	51	8.408
13	7.604	26	6.036	39	7.505		

Table 17.4: Dual optimal values (# = Platform number, Dual = dual value).

visited during the corresponding flight, while the fourth column contains the performed crew exchanges. The last column contains the total traveled distance during that flight. Platform 4, for instance, is visited by flight $f = 3$ and by flight $f = 4$; the demanded number of crew exchanges satisfies $12x_3^* + 9x_4^* = 12$ (see Table 17.2).

Table 17.4 lists the values of our optimal dual solution. A dual value gives an indication of the price in distance units per crew exchange on the corresponding platform. For instance, the crew exchanges for platform 22 have the highest optimal dual value, namely 11.377 km per crew exchange. This means that a crew exchange for platform 22 is relatively the most expensive one. This is not very surprising, since platform 22 is far from the airport, and the number of demanded crew exchanges is relatively low (namely, three); compare in this respect flight number 7 in Table 17.3. Note that the platforms 1 and 38 do not demand crew exchanges.

The optimal dual values of Table 17.4 give only an indication for the relative 'price' per crew exchange, because they refer to the LO-relaxation, and not to an optimal integer solution of the model.

17.7　A round-off procedure for determining an integer solution

An integer solution can be calculated by means of a round-off procedure, while taking into account the limited range of the helicopters; see also Section 7.1.4. Below, we have formulated a round-off algorithm that generates integer-valued solutions of model (FF).

Algorithm 17.7.1. *(Rounding-off algorithm)*

Input: The coordinates of the platforms, the demanded crew exchanges, the helicopter range and the capacity.

Output: A feasible flight schedule.

The values of the right hand side vector $\mathbf{d} = \begin{bmatrix} D_1 & \dots & D_N \end{bmatrix}^{\mathsf{T}}$ of (RFF) are updated in each iteration. Let (RFF$\mathbf{d}$) denote the model with right hand side values given by $\mathbf{d}$.

Step 1: Solve model (RFF$\mathbf{d}$) by means of the 'simplex algorithm with column generation'. If the solution of (RFF$\mathbf{d}$) is integer-valued, then stop. Otherwise, go to Step 2.

Step 2: Choose, arbitrarily, a variable of (RFF$\mathbf{d}$) with a positive optimal value, say $x_j > 0$.

If $x_j < 1$, then set $x_j := \lceil x_j \rceil = 1$; if $x_j > 1$, then set $x_j := \lfloor x_j \rfloor$. The flight f_j is carried out either $\lceil x_j \rceil$ or $\lfloor x_j \rfloor$ times, and the current right hand side values D_i are updated by subtracting the number of crew exchanges carried out by flight f_j. (The new LO-model now contains a number of D_i's whose values are smaller than in the previous model.)

If the new right hand side values D_i are all 0, then stop; all crew exchanges are performed. Otherwise, remove all columns from $\mathbf{W}$ that do not correspond to feasible flights anymore; moreover, extend $\mathbf{W}$ with the diagonal matrix $\mathbf{A}_W$ with $a_{ii} = \min\{D_i, C\}$ on its main diagonal.

Return to Step 1.

In Step 2, a certain positive noninteger entry is rounded-off to an integer value. It is not known *a priori* which variable should be taken. The reader may try several choices. When a random number generator is used, then each time the round-off algorithm is performed, a different integer solution can be expected.

Example 17.7.1. *Choose flight f_3 (i.e., choose $j = 3$) in Table 17.3 to be rounded-off. Then x_3^* becomes equal to 1, and so flight f_3 visits P_4 and P_{42} with 12 and 11 crew exchanges, respectively. However, P_4 and P_{42} are also visited during $j = 4$ and $j = 7$, respectively. So, if 0.250 is rounded-off to 1, then D_4 becomes $12 - 12 = 0$, and D_{42} becomes $16 - 11 = 5$. Platform P_{40} is visited during the flights f_j with $j = 2, 7, 31, 32$. Note that $0.172 \times 8 + 0.828 \times 4 + 0.828 \times 11 + 1.139 \times 23 = 40$, which is precisely the number of demanded crew exchanges on P_{40}. When flight f_{32} is selected to be rounded-off from 1.139 to 1, then the value of D_{40} becomes $40 - 23 = 17$ in the next iteration.*

Table 17.5 lists the results of the round-off algorithm. For instance, flight f_1 in Table 17.5 (this is flight 27 in Table 17.3) is Airport $\to P_{33}(7) \to P_{34}(11) \to P_{36}(5) \to$ Airport; the distance traveled is 152.7564 ($\times 4.5$ km).

Flight number	Platform visits	Crew exchanges	Distance ($\times 4.5$ km)
1	33, 34, 36	7, 11, 5	152.7564
2	41, 42	7, 16	90.2472
3	8, 30	8, 15	132.8583
4*	2	20	72.1110
5	3, 40, 7, 6	3, 8, 8, 4	104.6284
6	24, 31, 30	10, 9, 4	136.0911
7	11, 45	8, 15	165.9590
8	21	23	159.3989
9	19	23	151.9210
10	28, 32, 31	4, 14, 5	141.3553
11	4, 5	9, 14	110.7763
12	13, 16, 15	8, 8, 7	181.7217
13	9	23	127.9062
14	4, 9, 35, 10	3, 7, 5, 8	138.8715
15	40	23	83.7616
16	20, 22, 23, 39	5, 3, 2, 13	181.8675
17	18, 20, 21	10, 6, 7	162.8635
18	12, 44, 45, 50, 49	8, 5, 5, 4, 1	176.6813
19	17, 19, 20, 33	10, 7, 1, 5	164.5113
20	41	23	88.2043
21	46, 47	18, 5	150.9991
22	25, 27, 29, 31	12, 6, 3, 2	146.4471
23	26, 29, 41	9, 12, 2	144.4623
24*	40, 41	9, 8	88.2190
25	19, 48, 46, 43	15, 4, 2, 2	166.5272
26	15, 14, 37, 35	1, 8, 5, 9	179.1725
27	43	23	137.4627
28*	43	15	137.4627
29	49, 51	19, 4	165.5163

The total traveled distance is 4,040.7605 ($\times 4.5$ km)

Table 17.5: A feasible flight schedule. The flights marked with an asterisk do not use the full capacity of the helicopter.

The flight schedule in Table 17.5 is feasible: no flight distance exceeds the range of the helicopters and its capacities. Moreover, the capacity of the helicopters is not always completely used. For instance, flight 4 visits only platform 2, where all 20 demanded crew exchanges are performed, so three seats are not used.

17.8 Computational experiments

The solution method developed in the previous sections will be applied to a number of input instances in order to obtain an idea of the quality of the method. To that end, we have taken the 51 platforms and fifteen input instances with different numbers of crew

Input	Lower bound	Best	Dev. (in %)	Average	Dev. (in %)	Worst	Dev. (in %)
1	3,903.62	4,040.98	3.6	4,078.27	4.5	4,103.22	5.2
2	4,286.56	4,408.24	2.9	4,471.11	4.4	4,602.74	7.4
3	4,020.77	4,188.79	4.2	4,231.99	5.3	4,259.74	6.0
4	3,452.46	3,526.14	2.2	3,595.79	4.2	3,653.31	5.9
5	3,639.30	3,826.51	5.2	3,885.54	6.8	3,961.94	8.9
6	4,064.77	4,230.52	4.1	4,261.91	4.9	4,293.35	5.7
7	4,157.05	4,297.44	3.4	4,345.64	4.6	4,396.77	5.8
8	3,356.93	3,457.80	3.1	3,502.42	4.4	3,572.05	6.5
9	3,864.65	3,958.66	2.5	4,006.40	3.7	4,049.87	4.8
10	3,893.02	3,962.48	1.8	4,007.01	3.0	4,169.03	7.1
11	3,590.70	3,677.64	2.5	3,712.40	3.4	3,822.41	6.5
12	3,576.03	3,715.27	3.9	3,804.32	6.4	3,948.79	10.5
13	4,147.14	4,312.30	4.0	4,350.88	5.0	4,403.13	6.2
14	3,781.28	3,882.50	2.7	3,928.02	3.9	4,085.97	8.1
15	4,243.10	4,419.23	4.2	4,467.56	5.3	4,592.53	8.3

Table 17.6: Comparing fifteen input instances for the round-off algorithm. Each instance was solved sixteen times using the column generation method and the round-off procedure. The lower bound is given by the optimal objective value of the LO-relaxation (RFF). The column 'Dev.' contains the relative deviation from the lower bound.

exchanges. We do not specify these instances here, but they are available online (see `http://www.lio.yoriz.co.uk/`). The C++ code that we used is also available on that website.

The helicopter range remains 200, and the capacity 23 seats. The results are listed in Table 17.6. The first column contains the labels of the 15 input instances. The second column contains the lower bounds, which are the optimal solutions of model (RFF) in combination with model (CG). For each instance we have applied the round-off algorithm sixteen times. Columns three and four contain the best results ('Best') and the corresponding deviations (in %) from the lower bound ('Dev.'). Similarly, the fifth and the sixth columns contain the average optimal objective values ('Average') and the corresponding deviations from the lower bound ('Dev.'). Finally, the seventh and the eighth columns contain the worst of the sixteen results ('Worst') and the corresponding deviations from the lower bound ('Dev.').

It can be seen from Table 17.6 that the average integer solution deviates about 4 to 6% from the lower bound noninteger solution, while all integer solutions are within 11% of this lower bound. On the other hand the gap between the worst and the best solution is at most 7%. The conclusion is that the method developed in this section generates solutions that are practically useful, in spite of the fact that they are usually not optimal. The fact that the column generation procedure and the round-off procedure require a lot of computer time makes the method suitable only when there is enough time available. In the case of short term planning, caused for instance by sudden changes in the crew exchange demand, a faster method is needed, and it may be necessary to resort to heuristics.

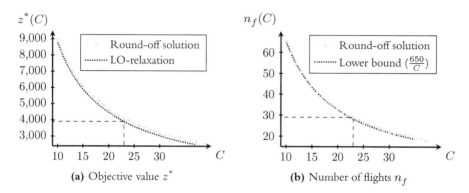

Figure 17.4: Sensitivity with respect to the helicopter capacity C.

17.9 Sensitivity analysis

In this last section we will investigate how the optimal objective value and the number of flights change when the values of the helicopter capacity C and the range R change. The initial helicopter range remains 200, and the initial capacity remains 23 seats. We are considering again the demands given in Table 17.2.

We first consider what happens when the value of the helicopter capacity C changes. Note that C does not just appear as a right hand side value or as an objective coefficient in model (RFF). Also note that model (FF) is an integer optimization model. As a result, the sensitivity analysis methods introduced in Chapter 5 do not apply. We therefore have to resort to solving the models repeatedly for different values of C. The values for model (RFF) in this section were computed by running the column generation procedure for each of the chosen values of C. For the rounded solutions, we applied, for each of the chosen values of C, the round-off procedure four times and averaged the result.

Figure 17.4(a) shows how the optimal objective value $z^*(C)$ of the LO-relaxation (RFF) and the optimal objective value of the rounded solutions of (FF) change as the value of C changes. Using a least-squares procedure, we find that the relationship between the optimal objective value of (FF) and the helicopter capacity roughly C satisfies:

$$z^*(C) \approx 295 + \tfrac{87,126}{C}. \tag{17.1}$$

Notice that choosing $C = R = \infty$ gives a lower bound on the optimal objective value $z^*(\cdot)$. This case, however, is equivalent to finding an optimal traveling salesman tour through all platforms, starting and finishing at the airport. We have used a TSP solver to compute such an optimal tour. This tour, which has length 334.7 ($\times 4.5$ km), is shown in Figure 17.5. Since this is a lower bound on the optimal objective value $z^*(C')$, we have that $z^*(C) \geq 334.7$ for all C. This means that relationship (17.1) is not valid for large values of C.

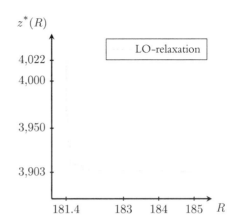

Figure 17.5: An optimal traveling salesman tour that visits each platform exactly once, starting and finishing at the airport.

Figure 17.6: Sensitivity with respect to the helicopter range R.

We are also interested in how the number of flights that need to be carried out depends on C. Figure 17.4(b) shows this dependence. The total demand is 650. Since each flight has capacity 23, this means that at least $\left\lceil \frac{650}{23} \right\rceil = 29$ flights need to be carried out. Note that the solution given in Table 17.5 consists of exactly 29 flights. In general, a lower bound on the number of flights to be carried out is $\left\lceil \frac{650}{C} \right\rceil$. The dotted line in Figure 17.4(b) shows this lower bound. It is clear from the figure that, for this particular instance, the number of flights is usually close to the lower bound. Thus, the capacity of the helicopter is crucial for the optimal objective value.

Now consider the sensitivity of the optimal solution with respect to the range of the helicopters. We first consider increasing the value of the range R, starting at $R = 200$. Let $z^*(R)$ denote the optimal objective value of the LO-relaxation (RFF) when setting the helicopter range to R. When increasing the value of R, the set of all possible flights $\mathcal{F}$ becomes larger. So, we expect that $z^*(R)$ is nonincreasing in R. Our calculations show that, in fact, $z^*(R)$ is constant for $R \geq 200$, i.e., the optimal objective value of model (RFF) does not change when increasing the value of R. Because round-off solutions for (FF) depend on the optimal solution of (RFF), this means that increasing the value of R has no effect on the round-off solutions of model (FF) either. This makes sense, because all flights except flights that are carried out in the solutions are significantly shorter than 200 ($\times 4.5$ km); see Table 17.3 and Table 17.5. This indicates that most flights are individually restricted not by the limited range of the helicopters, but by their limited capacity. Therefore, it makes sense that increasing the helicopter range will not enable us to choose longer flights, and hence increasing the value of R has no impact on the schedule.

Next, we consider decreasing the value of the range R, again starting at $R = 200$. Note that platform P_{16} has coordinates $(75, 51)$, which means that the distance from the airport to this platform is $\sqrt{75^2 + 51^2} \approx 90.697$ ($\times 4.5$ km). Since every flight that carries out a crew exchange at platform P_{22} has to travel at least twice this distance, it follows that for $R < 2 \times \sqrt{75^2 + 51^2} \approx 181.395$ the model is infeasible. On the other hand, consider the solution of the LO-relaxation given in Table 17.3. From this table, we see that the longest flight is flight f_{30}, which has distance 184.442 ($\times 4.5$ km). Therefore, this solution is an optimal solution for model (RFF) for all values of R with $184.442 \leq R \leq 200$. This means that decreasing the value of R, starting from 200, has no impact on the schedule, except when $181.395 \leq R \leq 184.442$. Figure 17.6 shows how the optimal objective value depends on the value of R in this range. For $181.395 \leq R \leq 181.51$, there is a sharp drop of the optimal objective value, starting at $z^*(181.395) = 4{,}022.62$, down to $z^*(181.51) = 3{,}934.92$. For $R \geq 181.51$, the value of $z^*(R)$ slowly decreases to $3{,}903.62$.

17.10 Exercises

Exercise 17.10.1. Consider the relaxation (RFF) of model (FF) in Section 17.4. Show that $d_j - \sum_{i=1}^{N} w_{ij} y_i$ is the j'th current nonbasic objective coefficient corresponding to some basis matrix $\mathbf{B}$ in $\mathbf{W}$ with $\mathbf{y} = (\mathbf{B}^{-1})^{\mathsf{T}} \mathbf{d}_{BI}$, and $\mathbf{W} \equiv [\mathbf{B} \ \mathbf{N}]$.

Exercise 17.10.2. Show that the number F of feasible flights may grow exponentially with each of the following parameters individually:

(a) The number of platform locations.

(b) The helicopter range R.

(c) The helicopter capacity C.

Exercise 17.10.3. Consider the column generation procedure described in Section 17.5. Let (RFFk) be model (RFF), but with $\mathbf{W}$ replaced by $\mathbf{W}^{(k)}$. Show that any optimal basis matrix of model (RFFk) is an optimal basis matrix of model (RFF).

Exercise 17.10.4. Calculate the percentage of platform subsets that is excluded by the procedures (B1) and (B2) in Section 17.5.2.

Exercise 17.10.5. Apply the following heuristics to determine feasible flight schedules for the input data of Section 17.1.

(a) *Nearest neighbor heuristic:*

Start at the airport and go to the platform that is closest to the airport. Then go to the next closest platform, and continue until either the capacity of the helicopter has been reached or its range (taking into account that the helicopter has to return to the airport). The heuristic stops when all demanded crew exchanges have been performed.

(b) *Sweep heuristic*:

A needle sweeps around, with the airport as its center. The initial position of the needle can be either platform 1 (the sweep is clockwise), or platform 46 (the sweep is counterclockwise). In the case of a clockwise sweep, the first visited platform is 1, then 2, 6, 7, and so on. For each newly encountered platform, a shortest route is calculated for this new platform together with the previously encountered platforms. A flight is formed when either the capacity of the helicopter has been reached or its range. The sweep of the needle is continued starting with the last assigned platform. Note that it may happen that there are some demanded crew exchanges left on this last platform. The heuristic stops when all demanded crew exchanges have been performed.

Compare the solutions with the one derived in Table 17.5.

CHAPTER 18

The catering service problem

Overview

The catering service problem is a problem that arises when the services of a caterer are to be scheduled. It is based on a classical paper by WILLIAM PRAGER (1903–1980); see also Prager (1956). The version discussed in this section will be formulated as a transshipment model (see Section 8.3.1). It will be solved by means of the network simplex algorithm.

18.1 Formulation of the problem

A catering company provides meals on customers' order. The caterer has booked its services for seven consecutive days; these seven days form the planning period. The company provides a clean napkin with each meal. The demanded number of meals and napkins during the planning period is known. The number of napkins demanded is listed in Table 18.1. After a napkin has been used by a customer, it is put into the basket of dirty napkins. The dirty napkins are sent to the laundry. The laundry provides a fast service and a slow service. Cleaning napkins via fast service takes two days, and via slow service four days. The fast laundering cost is 0.75 per napkin, and the slow laundering cost is 0.50 per napkin. Dirty napkins need not be sent to the laundry immediately after use; the caterer can decide to carry them over to the next day, or even later. As a consequence, the basket of dirty napkins on day j ($j = 1, \ldots, 7$) may consist of napkins used on day j and of napkins used earlier.

When the number of napkins that return from the laundry on a certain day is not enough to meet the demand of napkins on that day, the caterer can decide to buy new napkins; the price of a new napkin is 3.

At the beginning of each day, the company has to decide how many new napkins should be purchased in order to meet the demand, taking into account the napkins that return from

Day	Napkins demanded
1	23
2	14
3	19
4	21
5	18
6	14
7	15

Table 18.1: Number of demanded napkins.

the laundry. At the end of each day, the company has to make four decisions concerning the stock of dirty napkins:

How many dirty napkins should be sent to the slow laundry service?

How many dirty napkins should be sent to the fast laundry service?

How many dirty napkins should be carried over to the next day?

How many new napkins should be purchased?

The objective is to find a sequence of decisions such that the total cost is minimized. We assume that the caterer has no napkins (clean or dirty) on the first day of the planning period. Moreover, all demanded napkins can be purchased. A feasible sequence of decisions is summarized in Table 18.2.

The figures in the 'Laundered' column of Table 18.2 refer to the number of napkins returning from the laundry. For example, on day 5 there are four napkins returning from the laundry; they were sent to the slow laundry service on day 1 (see column 'Slow'). The fourteen napkins that were sent to the fast laundry service on day 3 (see column 'Fast') also return on day 5. Hence, there are eighteen napkins returning from the laundry on day 5. The column 'Purchased' indicates the number of purchased napkins, while the 'Carry over' column shows how many dirty napkins are carried over to the next day. For example, at the beginning of day 5 the company has a stock of seven dirty napkins since they were held on day 4; during that day, eighteen napkins are used. So, at the end of day 5, the company has twenty-five dirty napkins. The 'Fast' column shows that ten of these dirty napkins are sent to the fast laundry service, while the 'Carry over' column indicates that fifteen dirty napkins are carried over to the next day.

The cost associated with the decision sequence of Table 18.2 is $3 \times 44 = \$132$ for buying new napkins, $\$0.75 \times 71 = \$53.25$ for fast cleaning, and $\$0.50 \times 9 = \$4.50$ for slow cleaning. So the total cost is $\$189.75$. Note that at the end of the planning period the caterer has 44 dirty napkins.

| Day | Supply of laundered napkins | | Laundering option | | |
	Laundered	Purchased	Fast	Slow	Carry over
1	–	23	19	4	–
2	–	14	14	–	–
3	19	–	14	5	–
4	14	7	14	–	7
5	18	–	10	–	15
6	14	–	–	–	29
7	15	–	–	–	44

Table 18.2: A feasible decision sequence.

18.2 The transshipment problem formulation

Define for each day j ($j = 1, \ldots, 7$) of the planning period, the following variables:

p_j = the number of napkins purchased on day j;
s_j = the number of dirty napkins sent to the slow laundry service on day j;
f_j = the number of dirty napkins sent to the fast laundry service on day j;
h_j = the number of dirty napkins that is held until day $j + 1$.

Furthermore, define the set $\{P_0, \ldots, P_7\}$ of supply nodes or *sources* of napkins; these nodes have the property that the number of 'leaving' napkins exceeds the number of 'entering' napkins. We define P_0 as the source where enough new napkins, say 125, are available during the whole planning period; these napkins can be supplied by P_0 any day of the planning period. For $j \in \{1, \ldots, 7\}$, P_j may be thought of as the basket containing the dirty napkins at the end of day j, and P_0 as the company where the new napkins are purchased. When the caterer decides to send f_j napkins to the fast laundry service on day j, then we can think of it as source P_j supplying these f_j napkins as clean napkins on day $j + 2$; when on day j the caterer sends s_j napkins to the slow laundry service, then P_j supplies s_j clean napkins on day $j + 4$. Moreover, the caterer can decide to carry over h_j napkins to the next day; source P_j then transfers dirty napkins to P_{j+1}. So for example, the source P_1 can supply clean napkins either on day 3 (if fast service is used), or on day 5 (if slow service is used), or can transfer dirty napkins to P_2 (if some dirty napkins are carried over to the next day).

In addition to the sources, we define the set $\{Q_0, Q_1, \ldots, Q_7\}$ of demand nodes or *sinks*. Sinks are defined as nodes where the number of 'entering' napkins exceeds the number of 'leaving' napkins. For $j = 1, \ldots, 7$, let Q_j denote the sink where the napkins are demanded for day j. For reasons that will become clear later on, we introduce the sink Q_0. At the end of the planning period all napkins in the network are transported to Q_0 at zero cost.

We will now present a transshipment problem (see Section 8.3.1) formulation of the catering service problem. Let $G = (\mathcal{V}, \mathcal{A})$ be the network with node set $\mathcal{V} = \{P_0, P_1, \ldots, P_7, Q_0, Q_1, \ldots, Q_7\}$ (the P_i's are the sources and the Q_i's the sinks), and arc set $\mathcal{A}$ joining pairs of nodes in $\mathcal{V}$; see Figure 18.1. Hence, $|\mathcal{V}| = 16$, and $|\mathcal{A}| = 23$.

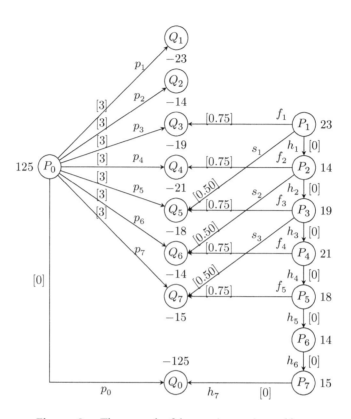

Figure 18.1: The network of the catering service problem.

In Figure 18.1, the horizontal arcs correspond to fast laundry service: source P_{j-2} can supply to sink Q_j ($j = 2, \ldots, 7$). The slanted arcs to the right of the column of sinks, represent slow laundry services: source P_{j-4} can supply to sink Q_j ($j = 4, \ldots, 7$). The vertical arcs, between the sources, represent the option of carrying over dirty napkins to the next day. So, for sink Q_j the required napkins can be delivered by either source P_0 (buying napkins), or by source P_{j-2} (dirty napkins cleaned by the fast laundry service), or by source P_{j-4} (dirty napkins cleaned by the slow laundry service).

The supply-demand vector $\mathbf{b} = \begin{bmatrix} b_0 & \ldots & b_{15} \end{bmatrix}^\mathsf{T}$ is defined as follows. Let $i, j \in \mathcal{V}$. If node i is a source, then define b_i as the supply of napkins at node i; for node j being a sink, define b_j as the negative demand of napkins at node j. The introduction of the sink Q_0 yields that $\sum_{i \in \mathcal{V}} b_i = 0$. Transportation at the end of day 7 to sink Q_0 is free, and so the introduction of this node does not increase the total cost. In Figure 18.1, the following supply-demand vector $\mathbf{b} = \begin{bmatrix} b_0 & b_1 & \ldots & b_7 & b_8 & \ldots & b_{14} & b_{15} \end{bmatrix}^\mathsf{T}$ corresponds to the set of nodes $\{P_0, P_1, \ldots, P_7, Q_0, Q_1, \ldots, Q_7\}$:

$$\mathbf{b} = [125\ 23\ 14\ 19\ 21\ 18\ 14\ 15\ -125\ -23\ -14\ -19\ -21\ -18\ -14\ -15]^\mathsf{T}.$$

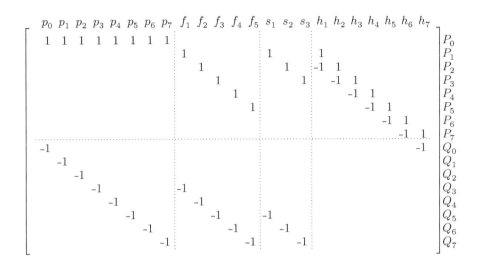

Figure 18.2: The transshipment matrix $\mathbf{A}$ corresponding to the data of Table 18.1. (The zeros have been omitted for readability.)

The first eight entries of $\mathbf{b}$ correspond to the sources, while the last eight entries of $\mathbf{b}$ correspond to the sinks in the network of Figure 18.1.

The flow of napkins through G is represented by the variables p_j, f_j, s_j, and h_j. In Figure 18.1, these flow variables are written next to the arcs. Between square brackets are denoted the costs of shipping one napkin through the corresponding arc. In the case of Figure 18.1, we have that:

$$\mathbf{c} = [0\ 3\ 3\ 3\ 3\ 3\ 3\ 3\ 0.75\ 0.75\ 0.75\ 0.75\ 0.75\ 0.50\ 0.50\ 0.50\ 0\ 0\ 0\ 0\ 0\ 0\ 0]^{\mathsf{T}}$$

$$\mathbf{x} = [p_0\ p_1\ p_2\ p_3\ p_4\ p_5\ p_6\ p_7\ f_1\ f_2\ f_3\ f_4\ f_5\ s_1\ s_2\ s_3\ h_1\ h_2\ h_3\ h_4\ h_5\ h_6\ h_7]^{\mathsf{T}}.$$

The variables f_6, f_7, s_4, s_5, s_6, and s_7 are not present in the vector $\mathbf{x}$ for reasons mentioned below. Note that there is a difference between the graph of Figure 18.1 and the graph of the transportation problem of Section 8.2.1. In a transportation problem, goods can only be supplied by sources and delivered to sinks; there are no shipments of goods between sources and between sinks, nor are there 'intermediate' nodes between sources and sinks. However, in Figure 18.1, source P_1 can, for instance, deliver to source P_2. The network presented in Figure 18.1 is in fact the network of a transshipment problem; see Section 8.3.1. Note that the network of Figure 18.1 has no intermediate or transshipment nodes.

The network of Figure 18.1 contains no arcs that correspond to sending dirty napkins to the slow laundry service after day 3, and no arcs that correspond to sending dirty napkins to the fast laundry service after day 5. The reason for this phenomenon is that napkins sent to the slow laundry service after day 3 do not return within the planning period. Also napkins sent to the fast laundry service after day 5 do not return within the planning period.

The transshipment matrix $\mathbf{A}$ corresponding to the data of Table 18.1 is shown in Figure 18.2. The ILO-model corresponding to the data of the catering service problem of Figure 18.1, can now be formulated as follows:

$$\min 3p_1 + 3p_2 + 3p_3 + 3p_4 + 3p_5 + 3p_6 + 3p_7 + 0.75f_1 + 0.75f_2$$
$$+ 0.75f_3 + 0.75f_4 + 0.75f_5 + 0.50s_1 + 0.50s_2 + 0.50s_3$$

$$\begin{aligned}
\text{s.t.} \quad p_0 + p_1 + p_2 + p_3 + p_4 + p_5 + p_6 + p_7 &= 125 \\
f_1 + s_1 + h_1 &= 23 \\
f_2 + s_2 - h_1 + h_2 &= 14 \\
f_3 + s_3 - h_2 + h_3 &= 19 \\
f_4 - h_3 + h_4 &= 21 \\
f_5 - h_4 + h_5 &= 18 \\
-h_5 + h_6 &= 14 \\
-h_6 + h_7 &= 15 \\
-p_0 - h_7 &= -125 \\
-p_1 &= -23 \\
-p_2 &= -14 \\
-p_3 - f_1 &= -19 \\
-p_4 - f_2 &= -21 \\
-p_5 - f_3 - s_1 &= -18 \\
-p_6 - f_4 - s_2 &= -14 \\
-p_7 - f_5 - s_3 &= -15 \\
\end{aligned}$$
$$p_0, p_1, \ldots, p_7, f_1, \ldots, f_5, s_1, s_2, s_3, h_1, \ldots, h_7 \geq 0, \text{ and integer.}$$

The technology matrix $\mathbf{A}$ is totally unimodular, and so − since $\mathbf{b}$ is an integer vector − this model has an integer optimal solution; see Theorem 4.2.4. (Why is the model feasible?) In the next section we will solve the model.

18.3 Applying the network simplex algorithm

In this section, the network simplex algorithm is applied to the catering service problem as formulated in Section 18.2.

Initialization. An initial feasible tree solution is easily found. Namely, buy all demanded napkins, and keep the dirty ones until the end of the planning period. Obviously, this solution is feasible, but not optimal. The corresponding values of the basic variables are:

$$\begin{array}{llll}
p_0 = 1 & p_1 = 23 & p_2 = 14 & p_3 = 19 \\
p_4 = 21 & p_5 = 18 & p_6 = 14 & p_7 = 15 \\
h_1 = 23 & h_2 = 37 & h_3 = 56 & h_4 = 77 \\
h_5 = 95 & h_6 = 109 & h_7 = 124.
\end{array}$$

The cost of this solution is $124 \times \$3 = \$372$. The spanning tree associated with this feasible tree solution is shown in Figure 18.3.

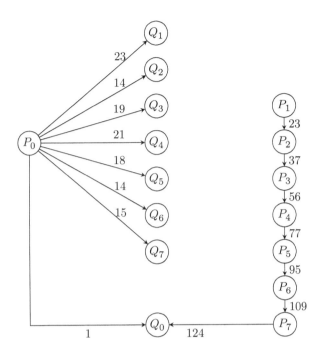

Figure 18.3: Initial feasible tree solution.

Iteration 1. Let $\begin{bmatrix} y_{P_0} & y_{P_1} & \cdots & y_{P_7} & y_{Q_0} & y_{Q_1} & \cdots & y_{Q_7} \end{bmatrix}$ be the vector of dual decision variables. Select node P_0 as the root of the spanning tree, and take $y_{P_0} = 0$. The remaining dual variables are determined according to the procedure formulated in Step 1 of the network simplex algorithm. Namely, first determine the values for the nodes for which the path to the root consists of only one arc. These are the nodes $Q_0, Q_1, \ldots, Q_7$. From node Q_0 we work our way up to node P_1, resulting in the following dual values:

$$
\begin{array}{llll}
y_{P_0} = 0 & y_{P_1} = 0 & y_{P_2} = 0 & y_{P_3} = 0 \\
y_{P_4} = 0 & y_{P_5} = 0 & y_{P_6} = 0 & y_{P_7} = 0 \\
y_{Q_0} = 0 & y_{Q_1} = -3 & y_{Q_2} = -3 & y_{Q_3} = -3 \\
y_{Q_4} = -3 & y_{Q_5} = -3 & y_{Q_6} = -3 & y_{Q_7} = -3.
\end{array}
$$

The nonbasic variables are f_1, f_2, f_3, f_4, f_5, s_1, s_2, s_3. The dual costs of the arcs corresponding to these nonbasic variables are:

$$
\begin{bmatrix} \bar{c}_{f_1} & \bar{c}_{f_2} & \bar{c}_{f_3} & \bar{c}_{f_4} & \bar{c}_{f_5} & \bar{c}_{s_1} & \bar{c}_{s_2} & \bar{c}_{s_3} \end{bmatrix}
$$
$$
= \begin{bmatrix} -2.25 & -2.25 & -2.25 & -2.25 & -2.25 & -2.50 & -2.50 & -2.50 \end{bmatrix}.
$$

All these costs are negative, and so we can arbitrarily select an entering arc with the smallest value. We choose arc (P_1, Q_5), corresponding to the flow variable s_1, as the entering arc. The resulting cycle is Q_5-P_0-Q_0-P_7-P_6-P_5-P_4-P_3-P_2-P_1. Arc (P_0, Q_0), corresponding to the flow variable p_0, is the only arc that points in the direction of the entering arc (P_1, Q_5). So, the flow on (P_1, Q_5) is increased by a certain amount, while

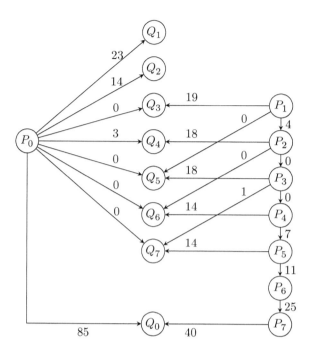

Figure 18.4: Optimal solution.

the flow on the other arcs of the cycle is decreased by the same amount. We change the flow on the cycle until the flow on one of the arcs becomes zero. The first arc with a zero flow is (P_0, Q_5). Hence arc (P_0, Q_5) leaves the set of basic variables. The new basic variables have the following values:

$$
\begin{array}{llll}
p_0 = 19 & p_1 = 23 & p_2 = 14 & p_3 = 19 \\
p_4 = 21 & p_6 = 14 & p_7 = 15 & s_1 = 18 \\
h_1 = 5 & h_2 = 19 & h_3 = 38 & h_4 = 59 \\
h_5 = 77 & h_6 = 91 & h_7 = 106.
\end{array}
$$

The cost of this solution is: $106 \times \$3 = \$318$ for buying new napkins, and $18 \times \$0.50 = \$9.00$ for cleaning eighteen napkins via the slow laundry service. So the total cost is $\$327.00$.

Iteration 2. Subtracting the dual cost of the entering arc from the current value of y_{Q_5}, the following new dual values are obtained:

$$
\begin{array}{llll}
y_{P_0} = 0 & y_{P_1} = 0 & y_{P_2} = 0 & y_{P_3} = 0 \\
y_{P_4} = 0 & y_{P_5} = 0 & y_{P_6} = 0 & y_{P_7} = 0 \\
y_{Q_0} = 0 & y_{Q_1} = -3 & y_{Q_2} = -3 & y_{Q_3} = -3 \\
y_{Q_4} = -3 & y_{Q_5} = -\frac{1}{2} & y_{Q_6} = -3 & y_{Q_7} = -3.
\end{array}
$$

Day	Supply of napkins		Laundering option		
---	Laundered	Purchased	Fast	Slow	Carry over
1	–	23	19	–	4
2	–	14	18	–	–
3	19	–	18	1	–
4	18	3	14	–	7
5	18	–	14	–	11
6	14	–	–	–	24
7	15	–	–	–	–

Table 18.3: An optimal schedule.

The dual costs of the arcs that are not in the current tree are:

$$\begin{bmatrix} \bar{c}_{f_1} & \bar{c}_{f_2} & \bar{c}_{f_3} & \bar{c}_{f_4} & \bar{c}_{f_5} & \bar{c}_{s_1} & \bar{c}_{s_2} & \bar{c}_{s_3} \end{bmatrix}$$
$$= \begin{bmatrix} 2.50 & -2.25 & -2.25 & 0.25 & -2.25 & -2.25 & -2.50 & -2.50 \end{bmatrix}.$$

We choose s_2 as the entering basic variable. The flow along the corresponding arc is 14. One can easily check that the arc corresponding to p_6 leaves the set of basic variables.

Iteration 10. It is left to reader to check that the network simplex algorithm terminates after ten iterations. The spanning tree with optimal flow is shown in Figure 18.4. The corresponding optimal basic variables satisfy:

$$\begin{array}{llll} p_0 = 85 & p_1 = 23 & p_2 = 14 & p_4 = 3 \\ f_1 = 19 & f_2 = 18 & f_3 = 18 & f_4 = 14 \\ f_5 = 14 & s_3 = 1 & h_1 = 4 & h_4 = 7 \\ h_5 = 11 & h_6 = 25 & h_7 = 40. \end{array}$$

The cost of this optimal solution consists of the following parts: $40 \times \$3 = \$120$ for buying new napkins, $83 \times \$0.75 = \$62.50$ for cleaning napkins by the fast cleaning service, and $\$0.50$ for using the slow cleaning service. The total cost is $\$182.75$. Table 18.3 lists the corresponding optimal decision schedule.

The fact that on days 1 and 2 the demanded napkins are all purchased is not quite unexpected, since it is assumed that the caterer has no napkins at the beginning of the planning period and the fast laundry service takes already two days. It is also clear that if the caterer already has a number of napkins at the beginning of day 1 (but not more than 37), then this number can be subtracted from the number of purchased napkins on days 1 and 2.

18.4 Sensitivity analysis

The ILO-model of the catering service problem with technology matrix $\mathbf{A}$, right hand side vector $\mathbf{b}$ and objective vector $\mathbf{c}$ (see Section 18.2) has only equality constraints. This means that it does not make sense to change only one of the right hand side values at a time, because that would make the model infeasible. On the other hand, when we are interested

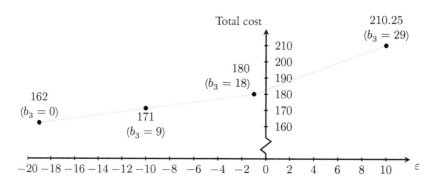

Figure 18.5: Perturbation function for b_3 ($b_3 = 19 + \varepsilon$).

in changes of the demand b_i, then two right hand side values need to be changed, namely both b_i and $-b_i$.

18.4.1 Changing the inventory of the napkin supplier

First we will consider the number of napkins available at the napkin supplier. Currently, the number of available napkins is 125. We may wonder how small the inventory of napkins may be chosen while maintaining the same optimal schedule. The minimum number turns out to be 40, and the reasons are the following. On day 1 and day 2, a total of $23 + 14 = 37$ napkins need to be purchased. One can easily check that, even if all napkins are fast-laundered, then three napkins need to be purchased on day 4. Only if the prices of fast and slow cleaning are increased, or the price of new napkins goes down, then the caterer might consider buying more than 40 napkins; see Section 18.4.3. In the extreme case, when all napkins are purchased, the napkin supplier needs a stock of $(23 + 14 + 19 + 21 + 18 + 14 + 15 =)$ 124 napkins.

18.4.2 Changing the demand

Consider first the case that the number of demanded napkins on one of the first two days changes. We take day 1. A change of the value of b_1 immediately influences the value of p_1, the number of napkins purchased on day 1. Let $p_i(b_1)$ be the number of napkins purchased on day i, when the number of demanded napkins on day 1 is b_1. Clearly, $p_1(b_1) = b_1$ for $b_1 \geq 0$; $p_2(b_1) = 14$ for $b_1 \geq 0$; $p_3(b_1) = 19 - b_1$ for $b_1 = 0, \ldots, 19$, $p_3(b_1) = 0$ for $b_1 \geq 19$; $p_4(b_1) = 7$ for $b_1 = 0, 1, \ldots, 19$, $p_4(b_1) = 26 - b_1$ for $b_1 = 19, \ldots, 26$, and $p_4(b_1) = 0$ for $b_1 \geq 26$. The table below lists the sequences of napkins to be purchased in the optimal solutions.

b_1	0	1	...	18	19	20	21	...	25	26	...
Day 1	0	1	...	18	19	20	21	...	25	26	...
Day 2	14	14	...	14	14	14	14	...	14	14	...
Day 3	19	18	...	1	0	0	0	...	0	0	...
Day 4	7	7	...	7	7	6	5	...	1	0	...

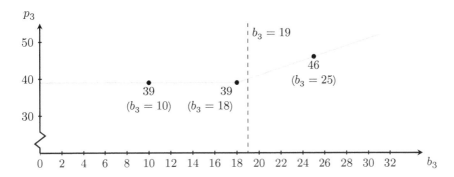

Figure 18.6: Number of purchased versus demanded napkins.

No napkins need be purchased on day 3 when $b_1 \geq 19$, and no napkins need be purchased on days 3 and 4 when $b_1 \geq 26$. In Exercise 18.6.2, the reader is asked to draw the perturbation function for b_1, i.e., the total costs as a function of b_1.

Now consider day 3. The perturbation function for b_3, the number of demanded napkins on day 3, is depicted in Figure 18.5. For $b_3 \geq 18$ (i.e., $\varepsilon \geq -1$) the total cost increases more than for $0 \leq b_3 \leq 18$ (i.e., $-19 \leq \varepsilon \leq -1$). This is caused by the fact that for $b_3 > 18$ more napkins need to be purchased, while for $0 \leq b_3 \leq 18$ the number of purchased napkins is constant (namely, equal to 39); see Figure 18.6.

18.4.3 Changing the price of napkins and the laundering costs

Figure 18.7 shows the tolerance region of the costs of fast laundry and slow laundry. It denotes the possible changes of the costs of fast laundering (F) and of slow laundering (S), such that the optimal decision schedule for $F = 0.75$ and $S = 0.50$ does not change. So within the shaded area of Figure 18.7, the feasible tree solution of Figure 18.4 is optimal.

To determine this tolerance region, we need to determine the values of F and S for which the current optimal tree solution remains optimal. Varying the values of F and S affects only the cost vector $\mathbf{c}$ of the LO-model. Let $\mathbf{A}$ be the technology matrix of the LO-model (see Section 18.2), and construct $\tilde{\mathbf{A}}$ from $\mathbf{A}$ by deleting the last row. Moreover, let $\mathbf{B}$ be the $(15, 15)$-submatrix of $\tilde{\mathbf{A}}$ that is the basis matrix for the optimal tree solution, i.e., $\mathbf{B}$ contains the columns of $\mathbf{A}$ corresponding to the variables $p_0, p_1, p_2, p_4, f_1, \ldots, f_5, s_3, h_1, h_2, h_3, h_4, h_7$. Let $\mathbf{N}$ consist of the columns of $\tilde{\mathbf{A}}$ corresponding to the nonbasic variables. Then, $\tilde{\mathbf{A}} \equiv [\mathbf{B}\ \mathbf{N}]$. Let $\mathbf{c}_{BI}(F, S)$ and $\mathbf{c}_{NI}(F, S)$ be the new cost vector as a function of F and S. Thus, we have that:

$$\mathbf{c}_{NI}(F, S) = \begin{bmatrix} 3 & 3 & 3 & 3 & S & S & 0 & 0 \end{bmatrix}^\mathsf{T}, \text{ and}$$

$$\mathbf{c}_{BI}(F, S) = \begin{bmatrix} 3 & 3 & 3 & 0 & F & F & F & F & F & S & 0 & 0 & 0 & 0 & 0 \end{bmatrix}^\mathsf{T}.$$

Using Theorem 3.5.1, the current tree solution is optimal if and only if

$$\left(\mathbf{c}_{NI}^\mathsf{T}(F, S) - \mathbf{c}_{BI}^\mathsf{T}(F, S) \mathbf{B}^{-1} \mathbf{N} \right)_\alpha \geq 0 \qquad \text{for } \alpha \in NI.$$

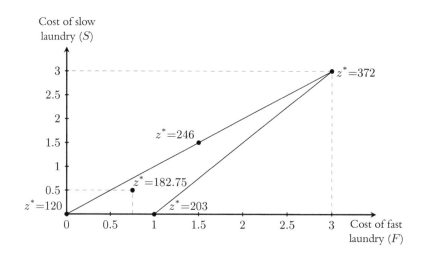

Figure 18.7: Tolerance region for the costs of fast and slow laundry.

Note that the vector in parentheses in this expression contains the current objective coefficients of the nonbasic variables p_3, p_5, p_6, p_7, s_1, s_2, h_5, and h_6. It can be checked that this system of inequalities is equivalent to

$$F \leq S \leq \tfrac{3}{2}(F-1),$$

which corresponds precisely to the shaded area in Figure 18.7.

It follows from Figure 18.7 that the management of the laundry can increase the prices for fast laundering up to $1.33, and for slow laundering up to $0.75, without taking the risk that the caterer will buy more new napkins than the 40 that are already purchased.

Figure 18.8 shows the perturbation function for the price of buying a napkin. It is clear from this figure that the price of $3.00 for one napkin is relatively high, because the caterer buys as few napkins as possible: increasing the price of a napkin does not change the decision schedule.

The caterer starts buying more napkins when the price drops below $1.25. This, however, is not profitable for the napkin supplier, because then the revenue of $(40 \times \$3.00 =)$ $120 drops drastically. For instance, if the price of a napkin is $1.25, then the total revenue is $(23 + 14 + 14 + 21 + 0 + 0 + 0 \times \$1.25 =)$ $67.50. In Figure 18.8, we have shown the sales figures in the corresponding price intervals.

18.5 GMPL model code

```
1       T;
2       M;
3       p_napkin;
```

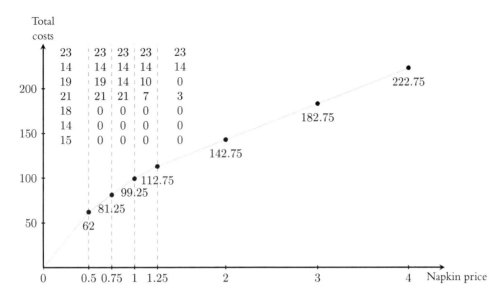

Figure 18.8: Perturbation function for the napkin price.

```
4         p_fast;
5         p_slow;
6        d{t      1..T} >= 0;
7
8      p{t      0..T}   >= 0;
9      f{t      1..T-2} >= 0;
10     s{t      1..T-4} >= 0;
11     h{t      1..T}   >= 0;
12
13        costs:
14     {t      1..T} p_napkin * p[t] +    {t      1..T-2} p_fast * f[t]
15                                    +    {t      1..T-4} p_slow * s[t];
16        P0:
17     {t      0..T} p[t] = M;
18        P {t      1..T}:
19  (    t <= T-2        f[t]      0) + (    t <= T-4       s[t]      0)
20                                   -  (    t >= 2        h[t-1]      0)
21                                   +  h[t] = d[t];
22        Q0:
23  -p[0] - h[7] = -M;
24        Q {t      1..T}:
25  -p[t] - (    t >= 3       f[t-2]      0)
26        - (    t >= 5       s[t-4]      0) = -d[t];
27
28     ;
29
30     M := 125;
31     T := 7;
32     p_napkin := 3;
33     p_fast := 0.75;
34     p_slow := 0.5;
```

```
35          d :=
36      1  23
37      2  14
38      3  19
39      4  21
40      5  18
41      6  14
42      7  15;
43      ;
```

18.6 Exercises

Exercise 18.6.1.   Consider the catering service problem of this chapter. Determine the perturbation functions of:

(a)   the demand b_1 on day 1, and

(b)   the demand b_2 on day 2.

Give for each interval, between kink points of the graphs, the schedule of amounts of napkins to be purchased.

Exercise 18.6.2.   Draw the perturbation function for b_1, i.e., the total costs as a function of b_1.

Exercise 18.6.3.   Complete the graph of Figure 18.7 by considering also the nonoptimal fast/slow laundering combinations.

Exercise 18.6.4.   Consider the catering service problem of this chapter.

(a)   What would be the price of a napkin if the caterer buys all its napkins?

(b)   Calculate prices for fast and slow cleaning such that the caterer buys all the napkins.

Exercise 18.6.5.   In Section 18.3 (at the end of the section), what would happen if more than 37 napkins were available at the beginning of day 1?

Appendices

Appendix A

Mathematical proofs

A mathematical *proof* is a deductive argument for a mathematical statement such as a *theorem* or a *lemma*. Any such mathematical statement consists of a set of *assumptions*, and one or more *conclusions*. A proof of a given statement is an argument that (hopefully) convinces the reader that the conclusions of the statement logically follow from its assumptions.

A proof demonstrates that a statement is always true, and so it is not sufficient to just enumerate a large number of confirmatory cases. Instead, a proof consists of sentences that logically follow from the assumptions, definitions, earlier sentences in the proof, and other previously established theorems or lemmas. A proof can in principle be traced back to self-evident or assumed statements that are known as *axioms* or *postulates*.

There are usually multiple, very different, ways of proving a statement, and we therefore usually refer to 'a proof' of a given statement, rather than 'the proof' (unless, of course, we refer to a *particular* proof). Often, there is one proof that is regarded as 'more elegant' than all others. The famous mathematician PAUL ERDŐS (1913–1996) spoke of The Book, a visualization of a book in which God has written down the most elegant proof of every theorem. Using this metaphor, the job of a mathematician is to 'discover' the pages of The Book; see also Aigner and Ziegler (2010).

Although proofs use logical and mathematical symbols, they usually consist of a considerable amount of natural language (e.g., English). This means that any proof usually admits a certain degree of ambiguity. It is certainly possible to write proofs that are completely unambiguous. Such proofs are considered in so-called *proof theory*, and they are usually written in a symbolic language without natural language. These proofs are suitable for verification with the help of a computer, but they are usually very hard to read for humans. The interested reader is referred to, e.g., von Plato, J. (2008).

So, providing a proof of a theorem is usually a matter of balancing the right level of detail and the right degree of readability. On the one hand, purely formal are usually hard to read for humans, because they are too detailed. On the other hand, a proof that contains too few details or skips important details is hard to understand because it leaves out critical insights. Note that this also depends on the intended audience. An extreme case of 'too few details' is a so-called *proof by intimidation*. Wikipedia gives the following definition:

> *"A proof by intimidation is a jocular phrase used mainly in mathematics to a style of presenting a purported mathematical proof by giving an argument loaded with jargon and appeal to obscure results, so that the audience is simply obliged to accept it, lest they have to admit their ignorance and lack of understanding."*

The phrase is used, for example, when the author is an authority in a field, presenting a proof to people who *a priori* respect the author's insistence that the proof is valid, or when the author claims that a statement is true because it is trivial, or simply because he or she says so.

In this appendix, we give an overview of the three most often used forms of (serious) mathematical proofs, and illustrate these with examples.

A.1 Direct proof

A *direct proof* starts with the assumptions of the theorem and the definitions of the relevant concepts, and then deduces, step by step, new true statements, finally leading to the conclusion of the theorem. In the process, the assumptions, axioms, definitions, and earlier theorems may be used. For example, the fact that the sum of two even integers is always even can be shown through a direct proof:

Theorem A.1.1.
Let x and y be even integers. Then, $x + y$ is even.

Proof. Since x and y are even, there exist integers a and b such that $x = 2a$ and $y = 2b$. Hence, we have that $x + y = 2a + 2b = 2(a + b)$. Therefore, $x + y$ has 2 as a factor, so that $x + y$ is even. This proves the theorem. □

Most proofs in this book are direct proofs. Note that the proof of Theorem A.1.1 is considered a proof because (1) each sentence is a precise statement that logically follows from the assumptions of the theorem and the preceding sentences (notice the words "hence" and "therefore"), and (2) the final sentence of the proof is exactly the conclusion of the theorem.

A.2 Proof by contradiction

A *proof by contradiction* (also called a *reductio ad absurdum*, which is Latin for *reduction to absurdity*) deduces the conclusion "if P, then Q" from its *contrapositive* "if not Q, then not P". In a proof by contradiction, we assume that the conclusion that we would like to prove is false, and then logically deduce that one or more assumptions are not true. A proof by contradiction usually ends with the sentence "this contradicts the fact that . . .", "contrary to the assumptions of the theorem", or plainly "a contradiction". For example, a proof by contradiction can be used to prove the following theorem:

> **Theorem A.2.1.**
> There is no largest even integer.

Proof. Suppose for a contradiction that the statement of the theorem is false, i.e., suppose that there does exist a largest even integer. Let M be that largest even integer. Define $N = M + 2$. Then, N is an even integer, and $N > M$. Hence, N is an even integer that is larger than M. This contradicts the fact that M is the largest even integer. This proves the theorem. □

The proof of Theorem 2.1.1 is another example of a proof by contradiction.

A.3 Mathematical induction

Mathematical induction is a method of proof that is used to prove a statement that depends on a natural number n. The principle of mathematical induction can be compared to dominoes. The process starts by pushing the first domino. Once the process has started, the process then continues by itself: the first domino hits the second domino, so that the second domino falls. In turn, the second domino hits the third domino, and so on. The end result is that all dominoes fall. The main components of this process are the facts that (1) the first domino falls by pushing it manually, and (2) once domino n falls, domino $n + 1$ falls as well ($n \geq 1$).

Mathematical induction consists of two elements in an analogous way. The first element is the so-called *base case*, which is an (often obvious) choice for the lowest number n_0 for which the statement should be proved. The second element is the so-called *induction step*, which asserts that if the statement holds for n, then it also holds for $n + 1$. The assumption that the statement holds for n is called the *induction hypothesis*. Once the validity of these two elements has been proved, the principle of mathematical induction states that the statement holds for all natural numbers $n \geq n_0$.

The general structure of a proof by mathematical induction is as follows. For $n \in \mathbb{N}$, let $P(n)$ be a mathematical statement involving n ($\in \mathbb{N}$). Usually, $P(0)$ or $P(1)$ is used as the induction hypothesis, and hence the proof starts by showing that $P(0)$ or $P(1)$ is a true statement. The proof then proceeds by taking any $n \in \mathbb{N}$, and showing the induction step that if $P(n)$ is true, then $P(n+1)$ is true, i.e., that the induction hypothesis "$P(n)$ is true" implies "$P(n+1)$ is true". Once these two elements have been proved, we may conclude that $P(n)$ is true for all natural numbers $n \geq 1$.

As an example, we prove the following theorem using mathematical induction.

Theorem A.3.1.
For any $n \geq 1$, it holds that $1 + 2 + \ldots + n = \frac{n(n+1)}{2}$.

Proof. For $n \geq 1$, let $P(n)$ be the statement "$1 + 2 + \ldots + n = n(n+1)/2$". We first prove the base case $P(1)$. $P(1)$ is the statement "$1 = (1 \cdot 2)/2$", which is obviously true. Thus we have shown that the base case $P(1)$ is indeed true. Next, we prove the induction step, i.e., we will prove that, for $n \geq 1$, if $P(n)$ holds, then $P(n+1)$ holds. Let $n \geq 1$ be given and assume that $P(n)$ is true (this is the induction hypothesis). Since $P(n)$ is true, we know that the statement "$1 + 2 + \ldots + n = n(n+1)/2$" is true. It remains to prove that $P(n+1)$ is true, i.e., that the statement "$1 + 2 + \ldots + (n+1) = (n+1)(n+2)/2$" is true. The proof is as follows:

$$1 + 2 + \ldots + (n+1) = (1 + 2 + \ldots + n) + (n+1)$$

$$= \frac{n(n+1)}{2} + (n+1) \qquad \text{(by the induction hypothesis)}$$

$$= \frac{n(n+1) + 2(n+1)}{2}$$

$$= \frac{(n+2)(n+1)}{2}.$$

Thus, $P(n+1)$ is indeed true. Since we have proved the base case and the induction step, the principle of mathematical induction implies that $P(n)$ is true for all $n \geq 1$. $\qquad \square$

Another example of mathematical induction is the proof of Theorem 8.1.4.

The principle of mathematical induction allows us to prove a statement for infinitely many integers n. It is important to note, however, that this does not necessarily mean that the conclusion holds for $n = \infty$. The interested read may use mathematical induction to prove that "for any integer $n \geq 0$, the number n is finite". Clearly, this statement is false for $n = \infty$.

Linear algebra

In this appendix we review some basic results from vector and matrix algebra. For a more extensive treatment of linear algebra, the reader is referred to, e.g., Strang (2009).

B.1 Vectors

The set of real numbers is denoted by $\mathbb{R}$. For any positive integer n, the elements of the space $\mathbb{R}^n$ are column arrays of size n, called *vectors* of size n, also called *n-vectors*; n is called the *dimension* of the space. Vectors are denoted by lower case boldface letters. For example, $\mathbf{a} = \begin{bmatrix} 3 & 5 & 1 & 0 & 0 \end{bmatrix}^\mathsf{T}$ is a vector of size 5; the superscript 'T' means that the row vector $\begin{bmatrix} 3 & 5 & 1 & 0 & 0 \end{bmatrix}$ is *transposed* to the column vector

$$\begin{bmatrix} 3 \\ 5 \\ 1 \\ 0 \\ 0 \end{bmatrix};$$

see also Section B.2 on 'matrices'. In $\mathbb{R}^2$ and $\mathbb{R}^3$ vectors can be represented by an arrow pointing from the origin to the point with coordinates the entries of the vector. Vectors in $\mathbb{R}^n$ are also called *points*. The *all-zero vector*, denoted by $\mathbf{0}$, is a vector with all entries equal to zero. For any positive integer i, the *i'th unit vector*, denoted by $\mathbf{e}_i$, is a vector in which all entries are zero except for a one in the i'th position. For example, the third unit vector in $\mathbb{R}^5$ is $\mathbf{e}_3 = \begin{bmatrix} 0 & 0 & 1 & 0 & 0 \end{bmatrix}^\mathsf{T}$. The *all-ones vector*, denoted by $\mathbf{1}$, is a vector with each entry equal to 1. The sizes of the vectors $\mathbf{0}$, $\mathbf{e}_i$, and $\mathbf{1}$ will be clear from the context.

Vectors of the same size can be added. For instance, let $\mathbf{a}_1$ and $\mathbf{a}_2$ be two vectors in $\mathbb{R}^n$, denoted by $\mathbf{a}_1 = \begin{bmatrix} a_{11} & a_{21} & \ldots & a_{n1} \end{bmatrix}^\mathsf{T}$, and $\mathbf{a}_2 = \begin{bmatrix} a_{12} & a_{22} & \ldots & a_{n2} \end{bmatrix}^\mathsf{T}$. Then the *addition* or the *sum* of $\mathbf{a}_1$ and $\mathbf{a}_2$ is denoted and defined by:

$$\mathbf{a}_1 + \mathbf{a}_2 = \begin{bmatrix} a_{11} + a_{12} & a_{21} + a_{22} & \dots & a_{n1} + a_{n2} \end{bmatrix}^{\mathsf{T}}.$$

Let k be any real number. Then the *scalar product* of the vector $\mathbf{a} = \begin{bmatrix} a_1 & a_2 & \dots & a_n \end{bmatrix}^{\mathsf{T}}$ with k is denoted and defined by:

$$k\mathbf{a} = \begin{bmatrix} ka_1 & ka_1 & \dots & ka_n \end{bmatrix}^{\mathsf{T}}.$$

The number k in this scalar product is called the *scalar*. Note that addition, as well as scalar multiplication are performed componentwise. For a given $k \neq 0$ and $\mathbf{a} \neq \mathbf{0}$, the vectors $k\mathbf{a}$ and $-k\mathbf{a}$ are called *opposite*.

The *inner product* of the vectors $\mathbf{a} = \begin{bmatrix} a_1 & \dots & a_n \end{bmatrix}^{\mathsf{T}}$ and $\mathbf{b} = \begin{bmatrix} b_1 & \dots & b_n \end{bmatrix}^{\mathsf{T}}$ is denoted and defined by:

$$\mathbf{a}^{\mathsf{T}}\mathbf{b} = a_1 b_1 + \dots + a_n b_n = \sum_{k=1}^{n} a_k b_k.$$

For example, if $\mathbf{x} = \begin{bmatrix} 0 & 1 & -5 \end{bmatrix}^{\mathsf{T}}$ and $\mathbf{y} = \begin{bmatrix} 6 & 4 & -2 \end{bmatrix}^{\mathsf{T}}$, then $\mathbf{x}^{\mathsf{T}}\mathbf{y} = 0 \times 6 + 1 \times 4 + (-5) \times (-2) = 14$.

The *Euclidean norm*[1] of the vector $\mathbf{a} = \begin{bmatrix} a_1 & \dots & a_n \end{bmatrix}^{\mathsf{T}}$ is denoted and defined by:

$$\|\mathbf{a}\| = \sqrt{a_1^2 + \dots + a_n^2} = \left(\sum_{k=1}^{n} a_k^2 \right)^{\frac{1}{2}}.$$

It can be easily checked that $\|\mathbf{a}+\mathbf{b}\|^2 = \|\mathbf{a}\|^2 + \|\mathbf{b}\|^2 + 2\mathbf{a}^{\mathsf{T}}\mathbf{b}$. The *angle* α ($0 \leq \alpha < 2\pi$) between two nonzero vectors $\mathbf{a}$ and $\mathbf{b}$ is defined by:

$$\mathbf{a}^{\mathsf{T}}\mathbf{b} = \|\mathbf{a}\| \, \|\mathbf{b}\| \cos \alpha,$$

which is the *cosine rule* for $\mathbf{a}$ and $\mathbf{b}$. For any two vectors $\mathbf{a}$ and $\mathbf{b}$ of the same size, the following inequality holds:

$$|\mathbf{a}^{\mathsf{T}}\mathbf{b}| \leq \|\mathbf{a}\| \, \|\mathbf{b}\|,$$

where $|\cdot|$ is the absolute value. This inequality is called the *Cauchy-Schwarz inequality*[2]. The inequality is a direct consequence of the cosine-rule for the vectors $\mathbf{a}$ and $\mathbf{b}$. Two vectors $\mathbf{a}$ and $\mathbf{b}$ of the same size are called *orthogonal* or *perpendicular* if $\mathbf{a}^{\mathsf{T}}\mathbf{b} = 0$. Note that the cosine-rule for orthogonal vectors $\mathbf{a}$ and $\mathbf{b}$ implies that $\cos \alpha = 0$, and so $\alpha = \frac{1}{2}\pi$, i.e., the angle between $\mathbf{a}$ and $\mathbf{b}$ is in fact $90°$. Note that for each i and j, with $i \neq j$, it holds

---

[1]Named after the Greek mathematician EUCLID OF ALEXANDRIA, who lived from the mid-4th century to the mid-3rd century B.C.

[2]Also known as the Cauchy-Bunyakovsky-Schwarz inequality, in honor of the French mathematician AUGUSTIN-LOUIS CAUCHY (1789–1857), the Ukrainian mathematician VIKTOR Y. BUNYAKOVSKY (1804–1889), and the German mathematician HERMANN A. SCHWARZ (1843–1921).

that $\mathbf{e}_i$ and $\mathbf{e}_j$ are perpendicular, where $\mathbf{e}_i$ and $\mathbf{e}_j$ are the i'th and j'th unit vectors of the same size, respectively.

The set $\mathbb{R}^n$, equipped with the above defined 'addition', 'scalar multiplication', 'Euclidean norm', and 'inner product', is called the n-dimensional *Euclidean vector space*. From now on we refer to $\mathbb{R}^n$ as the Euclidean vector space of dimension n. Note that $\mathbb{R}^0 = \{\mathbf{0}\}$.

Let $\mathbf{a}_1, \ldots, \mathbf{a}_k$, and $\mathbf{b}$ be vectors of the same size. The vector $\mathbf{b}$ is called a *linear combination* of the vectors $\mathbf{a}_1, \ldots, \mathbf{a}_k$ if there are scalars $\lambda_1, \ldots, \lambda_k$ such that:

$$\mathbf{b} = \lambda_1 \mathbf{a}_1 + \ldots + \lambda_k \mathbf{a}_k = \sum_{j=1}^{k} \lambda_j \mathbf{a}_j;$$

if, in addition, $\lambda_1 + \ldots + \lambda_k = 1$, then $\mathbf{b}$ is called an *affine combination* of $\mathbf{a}_1, \ldots, \mathbf{a}_k$. For example, the set of all linear combinations – called the *linear hull* – of the vectors $\mathbf{e}_1$ and $\mathbf{e}_2$ in $\mathbb{R}^3$ is the set $\left\{ \begin{bmatrix} x_1 & x_2 & x_3 \end{bmatrix}^\mathsf{T} \in \mathbb{R}^3 \,\middle|\, x_3 = 0 \right\}$, whereas the set of all affine combinations – called the *affine hull* – of $\mathbf{e}_1$ and $\mathbf{e}_2$ is $\left\{ \begin{bmatrix} x_1 & x_2 & x_3 \end{bmatrix}^\mathsf{T} \,\middle|\, x_1 + x_2 = 1, x_3 = 0 \right\}$; see also Appendix D, Section D.1.

The collection of vectors $\{\mathbf{a}_1, \ldots, \mathbf{a}_k\}$ in $\mathbb{R}^n$ with $k \geq 1$ is called *linearly independent* if:

$$\lambda_1 \mathbf{a}_1 + \ldots + \lambda_k \mathbf{a}_k = \mathbf{0} \text{ implies that } \lambda_1 = \ldots = \lambda_k = 0.$$

On the other hand, the collection of vectors $\{\mathbf{a}_1, \ldots, \mathbf{a}_k\} \subset \mathbb{R}^n$ is called *linearly dependent* if it is not linearly independent, i.e., if there exist scalars $\lambda_1, \ldots, \lambda_k$, not all zero, such that:

$$\lambda_1 \mathbf{a}_1 + \ldots + \lambda_k \mathbf{a}_k = \mathbf{0}.$$

For example, the set $\{\mathbf{e}_i, \mathbf{e}_j\}$, with $i \neq j$, is a linearly independent collection of vectors, while the set $\left\{ \begin{bmatrix} 1 & 4 & 1 \end{bmatrix}^\mathsf{T}, \begin{bmatrix} 2 & -3 & 1 \end{bmatrix}^\mathsf{T}, \begin{bmatrix} 3 & 1 & 2 \end{bmatrix}^\mathsf{T} \right\}$ is a linearly dependent collection of vectors, because $1 \times \begin{bmatrix} 1 & 4 & 1 \end{bmatrix}^\mathsf{T} + 1 \times \begin{bmatrix} 2 & -3 & 1 \end{bmatrix}^\mathsf{T} + (-1) \times \begin{bmatrix} 3 & 1 & 2 \end{bmatrix}^\mathsf{T} = \begin{bmatrix} 0 & 0 & 0 \end{bmatrix}^\mathsf{T}$. Note that any set of vectors containing $\mathbf{0}$ is linearly dependent. Although linear independence is a property of a collection of vectors rather than its individual vectors, for brevity, we often say that the vectors $\mathbf{a}_1, \ldots, \mathbf{a}_k$ are linearly independent (and similarly for linear dependence).

It can be shown that the maximum number of linearly independent vectors in $\mathbb{R}^n$ is equal to n. A set of n linearly independent vectors in $\mathbb{R}^n$ is called a *basis* of $\mathbb{R}^n$. For example, the set $\{\mathbf{e}_1, \ldots, \mathbf{e}_n\}$ is a basis. The reason of using the term 'basis' of $\mathbb{R}^n$ is that any vector $\mathbf{x}$ in $\mathbb{R}^n$ can be written as a linear combination of n basis vectors. For instance, for any vector $\mathbf{x} = \begin{bmatrix} x_1 & \ldots & x_n \end{bmatrix}^\mathsf{T} \in \mathbb{R}^n$, it holds that:

$$\mathbf{x} = x_1 \mathbf{e}_1 + \ldots + x_n \mathbf{e}_n.$$

In fact, for any basis $\mathbf{a}_1, \ldots, \mathbf{a}_n$ of $\mathbb{R}^n$ and any $\mathbf{x} \in \mathbb{R}^n$, the representation $\mathbf{x} = \lambda_1 \mathbf{a}_1 + \ldots + \lambda_n \mathbf{a}_n$ is unique. Indeed, suppose that it also holds that $\mathbf{x} = \lambda_1' \mathbf{a}_1 + \ldots + \lambda_n' \mathbf{a}_n$, then $\mathbf{0} = \mathbf{x} - \mathbf{x} = (\lambda_1 - \lambda_1')\mathbf{a}_1 + \ldots + (\lambda_n - \lambda_n')\mathbf{a}_n$, and so in fact $\lambda_1 - \lambda_1' = \ldots = \lambda_n - \lambda_n' = 0$.

The following theorem shows that, if we are given a basis of $\mathbb{R}^n$ and an additional nonzero vector, then another basis can be constructed by exchanging one of the vectors of the basis with the additional vector.

Theorem B.1.1.
Let $n \geq 1$. Let $\mathbf{a}_1, \ldots, \mathbf{a}_n \in \mathbb{R}^n$ be linearly independent vectors, and let $\mathbf{b} \in \mathbb{R}^n$ be any nonzero vector. Then, there exists $k \in \{1, \ldots, n\}$ such that $\mathbf{a}_1, \ldots, \mathbf{a}_{k-1}, \mathbf{b}, \mathbf{a}_{k+1}, \ldots, \mathbf{a}_n$ are linearly independent.

Proof. Because $\mathbf{a}_1, \ldots, \mathbf{a}_n$ are linearly independent vectors, there exist scalars $\lambda_1, \ldots, \lambda_n$ such that $\mathbf{b} = \lambda_1 \mathbf{a}_1 + \ldots + \lambda_n \mathbf{a}_n$. Since $\mathbf{b} \neq \mathbf{0}$, there exists an integer $k \in \{1, \ldots, n\}$ such that $\lambda_k \neq 0$. Without loss of generality, we may assume that $k = 1$. We will show that $\mathbf{b}, \mathbf{a}_2, \ldots, \mathbf{a}_n$ are linearly independent. To that end, let $\mu_1, \ldots, \mu_n$ be such that $\mu_1 \mathbf{b} + \mu_2 \mathbf{a}_2 + \ldots + \mu_n \mathbf{a}_n = \mathbf{0}$. Substituting $\mathbf{b} = \lambda_1 \mathbf{a}_1 + \ldots + \lambda_n \mathbf{a}_n$, we obtain that:

$$\mu_1 \lambda_1 \mathbf{a}_1 + (\mu_1 \lambda_2 + \mu_2)\mathbf{a}_2 + \ldots + (\mu_1 \lambda_n + \mu_n)\mathbf{a}_n = \mathbf{0}.$$

Since $\mathbf{a}_1, \ldots, \mathbf{a}_n$ are linearly independent, it follows that $\mu_1 \lambda_1 = 0$, $\mu_1 \lambda_2 + \mu_2 = 0$, $\ldots$, $\mu_1 \lambda_n + \mu_n = 0$. Since $\lambda_1 \neq 0$, we have that $\mu_1 = 0$, and this implies that $\mu_2 = \ldots = \mu_n = 0$. Hence, $\mathbf{b}, \mathbf{a}_2, \ldots, \mathbf{a}_n$ are linearly independent. $\qquad\square$

B.2 Matrices

Let $m, n \geq 1$. A *matrix* is a rectangular array of real numbers, arranged in m rows and n columns; it is called an (m, n)-matrix. The set of all (m, n)-matrices is denoted by $\mathbb{R}^{m \times n}$. Matrices are denoted by capital boldface letters. Two matrices are said to be *of the same size* if they have the same number of rows and the same number of columns. The entry in row i and column j, called *position* (i, j), of the matrix $\mathbf{A}$ is denoted by a_{ij}; we also write $\mathbf{A} = \{a_{ij}\}$. An $(m, 1)$-matrix is a column vector in $\mathbb{R}^m$, and an $(1, n)$-matrix is a row vector whose transpose is a vector in $\mathbb{R}^n$. If $m = n$, then the matrix is called a *square matrix*. The vector $\begin{bmatrix} a_{11} & \ldots & a_{mm} \end{bmatrix}^\mathsf{T}$ of the (m, n)-matrix $\mathbf{A} = \{a_{ij}\}$ with $m \leq n$ is called the *main diagonal* of $\mathbf{A}$; if $m \geq n$, then the main diagonal is $\begin{bmatrix} a_{11} & \ldots & a_{nn} \end{bmatrix}^\mathsf{T}$.

A *submatrix* of the matrix $\mathbf{A}$ is obtained from $\mathbf{A}$ by deleting a number of rows and/or columns from $\mathbf{A}$. The *all-zero matrix* in $\mathbb{R}^{m \times n}$, denoted by $\mathbf{0}$, is the matrix with only zero entries. The *identity matrix* in $\mathbb{R}^{n \times n}$, denoted by $\mathbf{I}_n$, is a square matrix with entries equal to 1 on the main diagonal and 0 entries elsewhere; i.e., if $\mathbf{I}_n = \{\iota_{ij}\}$, then $\iota_{ii} = 1$ for $i = 1, \ldots, n$, and $\iota_{ij} = 0$ for $i, j = 1, \ldots, n$ and $i \neq j$. An (m, n)-matrix $\mathbf{A} = \{a_{ij}\}$ with $m \geq n$ is called *upper triangular* if the nonzero entries are in the triangle on and above the positions $(1, 1), \ldots, (m, m)$; i.e., $a_{ij} = 0$ for all $i, j \in \{1, \ldots, m\}$ with $i > j$.

Similarly, an (m, n)-matrix $\mathbf{A} = \{a_{ij}\}$ with $m \leq n$ is called *lower triangular* if the nonzero entries are in the triangle on and below the positions $(1, 1), \ldots, (m, m)$; i.e., $a_{ij} = 0$ for all $i, j \in \{1, \ldots, m\}$ with $i < j$. A *diagonal matrix* is a square matrix which is upper as well as lower triangular.

The *transpose* of the matrix $\mathbf{A} = \{a_{ij}\}$ is denoted and defined by:

$$\mathbf{A}^\mathsf{T} = \{a_{ji}\}.$$

Notice that if $\mathbf{A}$ is an (m, n)-matrix, then $\mathbf{A}^\mathsf{T}$ is an (n, m)-matrix. Note also that $(\mathbf{A}^\mathsf{T})^\mathsf{T} = \mathbf{A}$, and that for any diagonal matrix $\mathbf{A}$, it holds that $\mathbf{A}^\mathsf{T} = \mathbf{A}$. A matrix $\mathbf{A}$ is called *symmetric* if $\mathbf{A}^\mathsf{T} = \mathbf{A}$.

The *addition* or *sum* of two matrices $\mathbf{A} = \{a_{ij}\}$ and $\mathbf{B} = \{b_{ij}\}$ of the same size is denoted and defined by:

$$\mathbf{A} + \mathbf{B} = \{a_{ij} + b_{ij}\}.$$

The *scalar product* of the matrix $\mathbf{A} = \{a_{ij}\}$ with the scalar k is denoted and defined by:

$$k\mathbf{A} = \{ka_{ij}\}.$$

Let $\mathbf{A} = \{a_{ij}\}$ be an (m, p)-matrix, and $\mathbf{B} = \{b_{ij}\}$ an (p, n)-matrix. Then the *matrix product* of $\mathbf{A}$ and $\mathbf{B}$ is an (m, n)-matrix, which is denoted and defined by:

$$\mathbf{A}\mathbf{B} = \left\{ \sum_{k=1}^{n} a_{ik}b_{kj} \right\};$$

i.e., the (i, j)'th entry of $\mathbf{A}\mathbf{B}$ is equal to the inner product of the i'th row of $\mathbf{A}$ and the j'th column of $\mathbf{B}$. Note that $(\mathbf{A}\mathbf{B})^\mathsf{T} = \mathbf{B}^\mathsf{T}\mathbf{A}^\mathsf{T}$.

B.3 Matrix partitions

Let $\mathbf{A}$ be an (m, n)-matrix. A *partition* of $\mathbf{A}$ is a selection of submatrices of $\mathbf{A}$ such that the submatrices form rows and columns in $\mathbf{A}$. The $(5, 6)$-matrix:

$$\mathbf{A} = \begin{bmatrix} a_{11} & a_{12} & a_{13} & a_{14} & a_{15} & a_{16} \\ a_{21} & a_{22} & a_{23} & a_{24} & a_{25} & a_{26} \\ a_{31} & a_{32} & a_{33} & a_{34} & a_{35} & a_{36} \\ a_{41} & a_{42} & a_{43} & a_{44} & a_{45} & a_{46} \\ a_{51} & a_{52} & a_{53} & a_{54} & a_{55} & a_{56} \end{bmatrix}$$

can, for example, be partitioned into the submatrices

$$\mathbf{A}_{11} = \begin{bmatrix} a_{11} & a_{12} & a_{13} \\ a_{21} & a_{22} & a_{23} \\ a_{31} & a_{32} & a_{33} \end{bmatrix}, \qquad \mathbf{A}_{12} = \begin{bmatrix} a_{14} & a_{15} \\ a_{24} & a_{25} \\ a_{34} & a_{35} \end{bmatrix}, \qquad \mathbf{A}_{13} = \begin{bmatrix} a_{16} \\ a_{26} \\ a_{36} \end{bmatrix},$$

$$\mathbf{A}_{21} = \begin{bmatrix} a_{41} & a_{42} & a_{43} \\ a_{51} & a_{52} & a_{53} \end{bmatrix}, \qquad \mathbf{A}_{22} = \begin{bmatrix} a_{44} & a_{45} \\ a_{54} & a_{55} \end{bmatrix}, \qquad \mathbf{A}_{23} = \begin{bmatrix} a_{46} \\ a_{56} \end{bmatrix}.$$

The original matrix is then written as:

$$\mathbf{A} = \left[\begin{array}{c:c:c} \mathbf{A}_{11} & \mathbf{A}_{12} & \mathbf{A}_{13} \\ \hdashline \mathbf{A}_{21} & \mathbf{A}_{22} & \mathbf{A}_{23} \end{array} \right].$$

Partitioned matrices can be multiplied if the various submatrices in the partition have appropriate sizes. This can be made clear by means of the following example. Let $\mathbf{A}$ and $\mathbf{B}$ be partitioned as follows:

$$\mathbf{A} = \left[\begin{array}{c:c} \mathbf{A}_{11} & \mathbf{A}_{12} \\ \hdashline \mathbf{A}_{21} & \mathbf{A}_{22} \end{array} \right], \text{ and } \mathbf{B} = \left[\begin{array}{c:c:c} \mathbf{B}_{11} & \mathbf{B}_{12} & \mathbf{B}_{13} \\ \hdashline \mathbf{B}_{21} & \mathbf{B}_{22} & \mathbf{B}_{23} \end{array} \right],$$

where $\mathbf{A}_{11}$ is an (m_1, n_1) matrix, $\mathbf{A}_{12}$ is an (m_1, n_2) matrix, $\mathbf{A}_{21}$ is an (m_2, n_1) matrix, $\mathbf{A}_{22}$ is an (m_2, n_2) matrix, $\mathbf{B}_{11}$ is a (p_1, q_1) matrix, $\mathbf{B}_{12}$ is a (p_1, q_2) matrix, $\mathbf{B}_{13}$ is a (p_1, q_3) matrix, $\mathbf{B}_{21}$ is a (p_2, q_1) matrix, $\mathbf{B}_{22}$ is a (p_2, q_2) matrix, and $\mathbf{B}_{23}$ is a (p_2, q_3) matrix, with $n_1 = p_1$ and $n_2 = p_2$. Then the product $\mathbf{AB}$ can be calculated as follows:

$$\mathbf{AB} = \left[\begin{array}{c:c:c} \mathbf{A}_{11}\mathbf{B}_{11} + \mathbf{A}_{12}\mathbf{B}_{21} & \mathbf{A}_{11}\mathbf{B}_{12} + \mathbf{A}_{12}\mathbf{B}_{22} & \mathbf{A}_{11}\mathbf{B}_{13} + \mathbf{A}_{12}\mathbf{B}_{23} \\ \hdashline \mathbf{A}_{21}\mathbf{B}_{11} + \mathbf{A}_{22}\mathbf{B}_{21} & \mathbf{A}_{21}\mathbf{B}_{12} + \mathbf{A}_{22}\mathbf{B}_{22} & \mathbf{A}_{21}\mathbf{B}_{13} + \mathbf{A}_{22}\mathbf{B}_{23} \end{array} \right].$$

A matrix is called *block diagonal* if it can be written as:

$$\mathbf{A} = \begin{bmatrix} \mathbf{A}_1 & \mathbf{0} & \cdots & \mathbf{0} \\ \mathbf{0} & \mathbf{A}_2 & \cdots & \mathbf{0} \\ \vdots & \vdots & \ddots & \vdots \\ \mathbf{0} & \mathbf{0} & \cdots & \mathbf{A}_k \end{bmatrix},$$

where $\mathbf{A}_1, \ldots, \mathbf{A}_k$ are square matrices. The matrices $\mathbf{A}_1, \ldots, \mathbf{A}_k$ are called the *blocks* of the matrix $\mathbf{A}$.

B.4 Elementary row/column operations; Gaussian elimination

An *elementary row operation* on a matrix $\mathbf{A}$ is one of the following operations:

(R1) Row i and row j of $\mathbf{A}$ are interchanged;

(R2) Row i of $\mathbf{A}$ is multiplied by a nonzero scalar k;

(R3) Row i is replaced by 'row i plus k times row j'.

Similarly, an *elementary column operation* on $\mathbf{A}$ is one of the following operations:

(C1) Column i and column j of $\mathbf{A}$ are interchanged;

(C2) Column i of $\mathbf{A}$ is multiplied by a nonzero scalar k;

(C3) Column i is replaced by 'column i plus k times column j'.

Note that an elementary column operation on $\mathbf{A}$ is an elementary row operation on $\mathbf{A}^\mathsf{T}$. The three rules (C1)–(C3) are obtained from the rules (R1)–(R3), by replacing 'row' by 'column'. Two matrices $\mathbf{A}$ and $\mathbf{B}$ that differ in a sequence of elementary row and/or column operations are called *equivalent*, and this equivalence is written as $\mathbf{A} \sim \mathbf{B}$.

Gaussian elimination[3] is the process of transforming a nonzero entry into a zero entry using a fixed nonzero entry as *pivot entry*, whereas the transformation is carried out by means of elementary row and/or column operations. More precisely, Gaussian elimination can be described as follows. Let a_{ij} be the selected (nonzero!) pivot entry. Gaussian elimination allows us to create nonzero entries in row i as well as in column j as follows. Suppose that $a_{kj} \neq 0$, and we want a zero entry in position $(k, j); k \neq i$. The transformation that accomplishes this is:

(GR) Subtract (a_{kj}/a_{ij}) times row i from row k, and take the result as the new row k.

Similarly, we can apply Gaussian elimination to the columns. Suppose we want a zero in position $(i, l); l \neq j$. Then rule (GR) is applied on the transposed matrix. Hence, the rule for Gaussian elimination on columns is:

(GC) Subtract (a_{il}/a_{ij}) times column j from column l, and take the result as the new column l.

The reader may check that rule (GR) uses (R2) and (R3), while rule (GC) uses (C2) and (C3). Hence, the matrices that appear after applying Gaussian elimination are equivalent to the original matrix.

We will now show how matrices are transformed into equivalent matrices for which all entries below the main diagonal are zero. Such an equivalent matrix is called a *Gauss form* of the original matrix, and the transformation that leads to a Gauss form is called a *Gaussian reduction*. The procedure uses the entries on the main diagonal as pivot entries. The procedure can be repeated for the transposed matrix. This leads to an equivalent matrix with all nonzero entries on the main diagonal. In a further step, all nonzero main diagonal entries can be made equal to one. A *Gauss-Jordan form*[4] of a matrix is an equivalent 0-1 *matrix* (a matrix consisting of only zeros and/or ones) with all ones on the main diagonal.

---

[3]In honor of the German mathematician CARL F. GAUSS (1777–1855), although the method was known to the Chinese as early as 179 A.D.

[4]Named after the German mathematicians CARL F. GAUSS (1777–1855) and WILHELM JORDAN (1842–1899).

Example B.4.1. *Consider the matrix:*

$$\mathbf{A} = \begin{bmatrix} 0 & 1 & 2 & 4 & 0 \\ 2 & 1 & 1 & 0 & 1 \\ 1 & 0 & 1 & 2 & 4 \end{bmatrix}.$$

We start by applying Gaussian elimination with pivot entries the entries in the positions $(1,1)$, $(2,2)$, respectively. Note that the entry in position $(1,1)$ is 0, and so it cannot be used as pivot entry. Therefore, row 1 and row 2 can be interchanged, so obtaining a nonzero entry in position $(1,1)$. In the next step, the entry in position $(3,1)$ is made zero, by using pivot entry 2 in position $(1,1)$. This is achieved by subtracting $\frac{1}{2}$ times row 1 from row 3. Since the third row contains nonintegers, we multiply row 3 by two. The entry in position $(3,2)$ can be made zero by adding row 2 and row 3. We obtain the following sequence of equivalent matrices:

$$\mathbf{A} \overset{1}{\sim} \begin{bmatrix} 2 & 1 & 1 & 0 & 1 \\ 0 & 1 & 2 & 4 & 0 \\ 1 & 0 & 1 & 2 & 4 \end{bmatrix} \overset{2}{\sim} \begin{bmatrix} 2 & 1 & 1 & 0 & 1 \\ 0 & 1 & 2 & 4 & 0 \\ 0 & -1 & 1 & 4 & 7 \end{bmatrix} \overset{3}{\sim} \begin{bmatrix} 2 & 1 & 1 & 0 & 1 \\ 0 & 1 & 2 & 4 & 0 \\ 0 & 0 & 3 & 8 & 7 \end{bmatrix} \overset{4}{\sim} \begin{bmatrix} 2 & 0 & 0 \\ 1 & 1 & 0 \\ 1 & 2 & 3 \\ 0 & 4 & 8 \\ 1 & 0 & 7 \end{bmatrix}^{\mathsf{T}}$$

$$\overset{5}{\sim} \begin{bmatrix} 2 & 0 & 0 \\ 0 & 1 & 0 \\ 0 & 2 & 3 \\ 0 & 4 & 8 \\ 0 & 0 & 7 \end{bmatrix}^{\mathsf{T}} \overset{6}{\sim} \begin{bmatrix} 2 & 0 & 0 \\ 0 & 1 & 0 \\ 0 & 0 & 3 \\ 0 & 0 & 0 \\ 0 & 0 & 0 \end{bmatrix}^{\mathsf{T}} \overset{7}{\sim} \begin{bmatrix} 2 & 0 & 0 & 0 & 0 \\ 0 & 1 & 0 & 0 & 0 \\ 0 & 0 & 3 & 0 & 0 \end{bmatrix} \overset{8}{\sim} \begin{bmatrix} 1 & 0 & 0 & 0 & 0 \\ 0 & 1 & 0 & 0 & 0 \\ 0 & 0 & 1 & 0 & 0 \end{bmatrix}.$$

Step 1. *Row 1 and row 2 are interchanged. Notation: $r_1 \leftrightarrow r_2$.*

Step 2. *Row 1 is subtracted from two times row 3. Notation: $r_3 := 2r_3 - r_1$.*

Step 3. *$r_3 := r_3 + r_2$. Note that the matrix is now in Gauss form.*

Step 4. *$\mathbf{A}$ is transposed to $\mathbf{A}^{\mathsf{T}}$.*

Step 5. *$\frac{1}{2}$ times column 1 of $\mathbf{A}$ is subtracted from column 2 of $\mathbf{A}$; this is the same as: $\frac{1}{2}$ times row 1 of $\mathbf{A}^{\mathsf{T}}$ is subtracted from row 2 of $\mathbf{A}^{\mathsf{T}}$. Notation: $k_2 := k_2 - \frac{1}{2}k_1$. Moreover, $k_3 := k_3 - \frac{1}{2}k_1$, $k_5 := k_5 - \frac{1}{2}k_1$.*

Step 6. *$k_3 := k_3 - 2k_2$ and $k_4 := k_4 - 4k_2$.*

Step 7. *$k_4 := k_4 - \frac{8}{3}k_3$ and $k_5 := k_5 - \frac{7}{3}k_3$.*

Step 8. *$\mathbf{A}^{\mathsf{T}}$ is transposed back to $\mathbf{A}$.*

Step 9. *Row 1 is divided by 2. Notation: $r_1 := r_1/2$. Moreover, $r_3 := r_3/2$. Note that the matrix is now in Gauss-Jordan form.*

In general, it holds that any (m, n)-matrix $\mathbf{A}$, with $m \leq n$, is equivalent to a Gauss-Jordan form, being a $\{0, 1\}$ diagonal (m, m) matrix $\mathbf{D}_A$ extended with an all-zero $(m, n - m)$ matrix $\mathbf{0}$, i.e.:

$$\mathbf{A} \sim \begin{bmatrix} \mathbf{D}_A & \mathbf{0} \end{bmatrix}.$$

In case $\mathbf{A}$ is an (m, n)-matrix with $m \geq n$, the Gauss-Jordan form becomes:

$$\mathbf{A} \sim \begin{bmatrix} \mathbf{D}_A \\ \mathbf{0} \end{bmatrix},$$

with $\mathbf{D}_A$ a diagonal (n, n)-matrix with only entries equal to 0 and 1 on its main diagonal. The procedure of Gaussian reduction can formally be described by means of the following algorithm.

Algorithm B.4.1. (*Gaussian reduction*)

Input: Values for the entries of the (m, n)-matrix $\mathbf{A} = \{a_{ij}\}$.

Output: A Gauss form of the matrix $\mathbf{A}$.

The algorithm successively uses the current entries in the positions $(1, 1), \ldots, (m, m)$ as pivot entries; if $m \leq n$, then only the positions $(1, 1), \ldots, (m - 1, m - 1)$ need to be used. Suppose the algorithm has already considered the columns 1 to $k - 1$. So the current column is k $(1 \leq k \leq m)$.

Step 1. If all entries below position (k, k) in the current column k are zero, then if $k < m$ go to Step 4, and if $k = m$ then stop. Otherwise, go to Step 2.

Step 2. Select a nonzero entry below position (k, k) in the current column k, say in position (i, k) with $i > k$. Then interchange (elementary row operation (R1)) row k with row i in the current matrix. This yields that the entry in position (k, k) becomes nonzero. Go to Step 3.

Step 3. Apply Gaussian elimination on the rows below row k, by using the current entry in position (k, k) as pivot entry. The result is that all entries in column k, below position (k, k), become equal to zero. Go to Step 4.

Step 4. As long as $k < m$, return to Step 1 with $k := k + 1$. If $k = m$, then stop.

The *rank* of a matrix $\mathbf{A}$, denoted by $\mathrm{rank}(\mathbf{A})$, is defined as the maximum number of linearly independent columns of $\mathbf{A}$. It can be shown that the rank of a matrix is also equal to the maximum number of linearly independent rows, and so $\mathbf{A}$ and $\mathbf{A}^\mathsf{T}$ have the same rank.

Although a Gauss-Jordan form of a matrix is not uniquely determined, the number of entries that is equal to 1 is fixed. Based on this fact, it can be shown that the rank of a matrix is equal to the number of ones in any Gauss-Jordan form of $\mathbf{A}$.

If $m \leq n$ and $\mathrm{rank}(\mathbf{A}) = m$, then $\mathbf{A}$ is said to have *full row rank*. Similarly, the matrix $\mathbf{A}$, with $m \geq n$, has *full column rank* if $\mathrm{rank}(\mathbf{A}) = n$. A square matrix $\mathbf{A}$ with full row (column) rank is called *nonsingular* or *regular*, whereas $\mathbf{A}$ is called *singular* otherwise. Note that any Gauss-Jordan form of a singular matrix with $m \leq n$ contains an all-zero row, whereas any Gauss-Jordan form of a regular matrix is (equivalent to) the identity matrix.

Theorem B.4.1.
Let $\mathbf{A}$ be an (m, n)-matrix with $m \leq n$, and $\text{rank}(\mathbf{A}) = m$. Then, $\mathbf{A}\mathbf{A}^\mathsf{T}$ is square, symmetric, and nonsingular.

Proof. We leave it to the reader to show that $\mathbf{A}\mathbf{A}^\mathsf{T}$ is square and symmetric. The fact that $\mathbf{A}\mathbf{A}^\mathsf{T}$ is nonsingular, is shown by proving that the columns of $\mathbf{A}\mathbf{A}^\mathsf{T}$ are linearly independent. Let the entries of the m-vector $\mathbf{y}$ form a linear combination of the columns of $\mathbf{A}\mathbf{A}^\mathsf{T}$ such that $\mathbf{A}\mathbf{A}^\mathsf{T}\mathbf{y} = \mathbf{0}$. (The reader may check that $\mathbf{A}\mathbf{A}^\mathsf{T}\mathbf{y}$ is in fact a shorthand notation for a linear combination of the columns of $\mathbf{A}\mathbf{A}^\mathsf{T}$ with as 'weights' the entries of $\mathbf{y}$.) Multiplying $\mathbf{A}\mathbf{A}^\mathsf{T}\mathbf{y} = \mathbf{0}$ by $\mathbf{y}^\mathsf{T}$ gives: $\mathbf{0} = \mathbf{y}^\mathsf{T}\mathbf{A}\mathbf{A}^\mathsf{T}\mathbf{y} = (\mathbf{A}^\mathsf{T}\mathbf{y})^\mathsf{T}(\mathbf{A}^\mathsf{T}\mathbf{y}) = (\mathbf{A}^\mathsf{T}\mathbf{y})^2$. Hence, $\mathbf{A}^\mathsf{T}\mathbf{y} = \mathbf{0}$. Since the rows of $\mathbf{A}$ are linearly independent, it follows that $\mathbf{y} = \mathbf{0}$. Hence, $\mathbf{A}\mathbf{A}^\mathsf{T}$ has independent columns, and this means that $\mathbf{A}\mathbf{A}^\mathsf{T}$ is nonsingular. $\qquad\square$

B.5 Solving sets of linear equalities

Let m and n be positive integers. A system of m *linear equalities* in n variables $x_1, \ldots, x_n$ is any system of the form:

$$
\begin{aligned}
a_{11}x_1 + \ldots + a_{1n}x_n &= b_1 \\
\vdots \qquad\qquad \vdots \quad\;\; &\;\; \vdots \\
a_{m1}x_1 + \ldots + a_{mn}x_n &= b_m.
\end{aligned}
$$

Such a system can be written in the following compact form:

$$\mathbf{A}\mathbf{x} = \mathbf{b},$$

where $\mathbf{A} = \begin{bmatrix} a_{11} & \cdots & a_{1n} \\ \vdots & & \vdots \\ a_{m1} & \cdots & a_{mn} \end{bmatrix}$, $\mathbf{b} = \begin{bmatrix} b_1 & \cdots & b_m \end{bmatrix}^\mathsf{T}$, and $\mathbf{x} = \begin{bmatrix} x_1 & \cdots & x_n \end{bmatrix}^\mathsf{T}$.

By solving a system of linear equations $\mathbf{A}\mathbf{x} = \mathbf{b}$, we mean determining the set:

$$X = \{\mathbf{x} \in \mathbb{R}^n \mid \mathbf{A}\mathbf{x} = \mathbf{b}\}.$$

The set X is called the *solution space* of the system of equalities. If $X \neq \emptyset$ then the system $\mathbf{A}\mathbf{x} = \mathbf{b}$ is called *solvable*, or *consistent*; otherwise, $\mathbf{A}\mathbf{x} = \mathbf{b}$ is called *nonsolvable*, or *inconsistent*. It can be shown that if $\begin{bmatrix} \mathbf{A}_1 & \mathbf{b}_1 \end{bmatrix}$ and $\begin{bmatrix} \mathbf{A}_2 & \mathbf{b}_2 \end{bmatrix}$ are row-equivalent matrices, which means that $\begin{bmatrix} \mathbf{A}_1 & \mathbf{b}_1 \end{bmatrix}$ and $\begin{bmatrix} \mathbf{A}_2 & \mathbf{b}_2 \end{bmatrix}$ can be transformed into each other by means of elementary row operations, then the systems of linear equalities $\mathbf{A}_1\mathbf{x} = \mathbf{b}_1$ and $\mathbf{A}_2\mathbf{x} = \mathbf{b}_2$ have the same solution space. Based on this fact, the solution set can be determined by means of elementary row operations on the matrix $\begin{bmatrix} \mathbf{A} & \mathbf{b} \end{bmatrix}$. Actually, we will apply Algorithm B.4.1. The following examples illustrate the procedure. In the first example the solution space is nonempty, and in the second one it is empty.

Example B.5.1. *Consider the following system of linear equalities ($m = 3, n = 4$):*

$$\begin{array}{rrrrr} x_1 & - & x_2 & + & x_3 & - & x_4 & = & 6, \\ -2x_1 & + & x_2 & - & x_3 & + & x_4 & = & 3, \\ -x_1 & + & 2x_2 & - & 3x_3 & & & = & 3. \end{array}$$

The calculations are carried out on the extended matrix $\begin{bmatrix} \mathbf{A} & \mathbf{b} \end{bmatrix}$. Applying Algorithm B.4.1, leads to the following sequence of row-equivalent matrices:

$$\left[\begin{array}{rrrr|r} 1 & -1 & 1 & -1 & 6 \\ -2 & 1 & -1 & 1 & 3 \\ -1 & 2 & -3 & 0 & 3 \end{array}\right] \sim \left[\begin{array}{rrrr|r} 1 & -1 & 1 & -1 & 6 \\ 0 & -1 & 1 & -1 & 15 \\ 0 & 1 & -2 & -1 & 11 \end{array}\right] \sim \left[\begin{array}{rrrr|r} 1 & 0 & 0 & 0 & -9 \\ 0 & 1 & -1 & 1 & -15 \\ 0 & 0 & -1 & -2 & 26 \end{array}\right].$$

Rewriting the last matrix as a system of linear equalities gives:

$$\begin{array}{rrrrr} x_1 & & & & & = & -9, \\ & x_2 & - & x_3 & + & x_4 & = & -15, \\ & & & -x_3 & - & 2x_4 & = & 29. \end{array}$$

For any fixed value of x_4, the values of x_1, x_2, and x_3 are fixed as well. Take $x_4 = \alpha$. Then, $x_3 = -26 - 2\alpha$, $x_2 = -15 + (-26 - 2\alpha) + \alpha = -41 - 3\alpha$, and $x_1 = -9$. The reader may check that these values for x_1, x_2, x_3, and x_4 satisfy the original system of equalities. The solution space can be written as:

$$X = \left\{ \left. \begin{bmatrix} -9 \\ -41 \\ -26 \\ 0 \end{bmatrix} + \alpha \begin{bmatrix} 0 \\ -3 \\ -2 \\ 1 \end{bmatrix} \right| \alpha \in \mathbb{R} \right\}.$$

The solution space is nonempty and one-dimensional.

In the following example, the solution space is empty.

Example B.5.2. *Consider the following system of linear equalities:*

$$\begin{array}{rrrrr} x_1 & + & x_2 & + & x_3 & + & x_4 & = & 1, \\ 2x_1 & & & & & + & 2x_4 & = & 4, \\ & & x_2 & + & x_3 & & & = & 1. \end{array}$$

Applying Algorithm B.4.1, we find the following sequence of row-equivalent matrices:

$$\left[\begin{array}{rrrr|r} 1 & 1 & 1 & 1 & 1 \\ 2 & 0 & 0 & 2 & 4 \\ 0 & 1 & 1 & 0 & 1 \end{array}\right] \sim \left[\begin{array}{rrrr|r} 1 & 1 & 1 & 1 & 1 \\ 0 & -2 & -2 & 0 & 2 \\ 0 & 1 & 1 & 0 & 1 \end{array}\right] \sim \left[\begin{array}{rrrr|r} 1 & 1 & 1 & 1 & 1 \\ 0 & -2 & -2 & 0 & 2 \\ 0 & 0 & 0 & 0 & 2 \end{array}\right].$$

The last matrix corresponds to the following system of linear equalities:

$$
\begin{aligned}
x_1 + \quad x_2 + \; x_3 + x_4 &= 1, \\
-2x_2 - 2x_3 \quad\;\; &= 2, \\
0 \qquad\qquad\qquad &= 2.
\end{aligned}
$$

This system of linear equalities is obviously inconsistent, and so the original system has no solution, i.e., its solution space is empty.

B.6 The inverse of a matrix

Let $\mathbf{A}$ be any (n, n)-matrix, $n \geq 1$. It can be shown that if there exists an (n, n)-matrix $\mathbf{X}$ such that:

$$\mathbf{XA} = \mathbf{AX} = \mathbf{I}_n, \tag{B.1}$$

then $\mathbf{X}$ is uniquely determined by (B.1); $\mathbf{X}$ is then called the *inverse* of $\mathbf{A}$, and is denoted by $\mathbf{A}^{-1}$. If we write $\mathbf{X} = \begin{bmatrix} \mathbf{x}_1 & \ldots & \mathbf{x}_n \end{bmatrix}$, and $\mathbf{I}_n = \begin{bmatrix} \mathbf{e}_1 & \ldots & \mathbf{e}_n \end{bmatrix}$, then the expression $\mathbf{AX} = \mathbf{I}_n$ can be written as:

$$\mathbf{Ax}_1 = \mathbf{e}_1, \ldots, \mathbf{Ax}_n = \mathbf{e}_n.$$

These n equalities can be solved by means of Gaussian elimination, applied to the matrices:

$$\begin{bmatrix} \mathbf{A} \mid \mathbf{e}_1 \end{bmatrix}, \ldots, \begin{bmatrix} \mathbf{A} \mid \mathbf{e}_n \end{bmatrix},$$

which can be compactly written as: $\begin{bmatrix} \mathbf{A} \mid \mathbf{I}_n \end{bmatrix}$. If $\mathbf{A}$ is nonsingular, it is possible to transform $\mathbf{A}$ to the identity matrix by means of elementary row operations. In other words, the following transformation is carried out:

$$\begin{bmatrix} \mathbf{A} \mid \mathbf{I}_n \end{bmatrix} \sim \begin{bmatrix} \mathbf{I}_n \mid \mathbf{A}^{-1} \end{bmatrix}.$$

Since the inverse exists only for nonsingular matrices, nonsingular matrices are also called *invertible*. The following example illustrates the procedure.

Example B.6.1. *We will determine the inverse of* $\mathbf{A} = \begin{bmatrix} 1 & 3 & 3 \\ 1 & 4 & 3 \\ 1 & 3 & 4 \end{bmatrix}$. *The matrix* $\mathbf{A}$ *is extended with the identity matrix* $\mathbf{I}_3$. *Applying elementary row operations, we obtain the following sequence of equivalent matrices:*

$$
\left[\begin{array}{ccc|ccc}
1 & 3 & 3 & 1 & 0 & 0 \\
1 & 4 & 3 & 0 & 1 & 0 \\
1 & 3 & 5 & 0 & 0 & 1
\end{array}\right]
\sim
\left[\begin{array}{ccc|ccc}
1 & 3 & 3 & 1 & 0 & 0 \\
0 & 1 & 0 & -1 & 1 & 0 \\
0 & 0 & 2 & -1 & 0 & 1
\end{array}\right]
$$

$$
\sim
\left[\begin{array}{ccc|ccc}
1 & 0 & 53 & 4 & -3 & 0 \\
0 & 1 & 0 & -1 & 1 & 0 \\
0 & 0 & 2 & -1 & 0 & 1
\end{array}\right]
\sim
\left[\begin{array}{ccc|ccc}
2 & 0 & 0 & 11 & -6 & -3 \\
0 & 1 & 0 & -1 & 1 & 0 \\
0 & 0 & 2 & -1 & 0 & 1
\end{array}\right]
$$

$$\sim \begin{bmatrix} 1 & 0 & 0 \\ 0 & 1 & 0 \\ 0 & 0 & 1 \end{bmatrix} \left| \begin{matrix} 5\frac{1}{2} & -3 & -1\frac{1}{2} \\ -1 & 1 & 0 \\ -\frac{1}{2} & 0 & \frac{1}{2} \end{matrix} \right] = \left[\mathbf{I}_3 \big| \mathbf{A}^{-1} \right].$$

Therefore, the inverse of $\mathbf{A}$ *is* $\mathbf{A}^{-1} = \begin{bmatrix} 5\frac{1}{2} & -3 & -1\frac{1}{2} \\ -1 & 1 & 0 \\ -\frac{1}{2} & 0 & \frac{1}{2} \end{bmatrix}$. *The reader may check that* $\mathbf{AA}^{-1} =$ $\mathbf{I}_3 = \mathbf{A}^{-1}\mathbf{A}$.

In the following example the inverse does not exist.

Example B.6.2. *Let* $\mathbf{A} = \begin{bmatrix} 1 & 6 & 4 \\ -1 & 3 & 5 \\ 2 & 3 & -1 \end{bmatrix}$. *We find the following sequence of equivalent matrices:*

$$\left[\mathbf{A} \big| \mathbf{I}_3 \right] \sim \left[\begin{matrix} 1 & 6 & 4 \\ -1 & 3 & 5 \\ 2 & 3 & -1 \end{matrix} \right| \left. \begin{matrix} 1 & 0 & 0 \\ 0 & 1 & 0 \\ 0 & 0 & 1 \end{matrix} \right] \sim \left[\begin{matrix} 1 & 6 & 4 \\ 0 & 9 & 9 \\ 0 & -9 & -9 \end{matrix} \right| \left. \begin{matrix} 1 & 0 & 0 \\ 1 & 1 & 0 \\ -2 & 0 & 1 \end{matrix} \right]$$

$$\sim \left[\begin{matrix} 1 & 6 & 4 \\ 0 & 9 & 9 \\ 0 & 0 & 0 \end{matrix} \right| \left. \begin{matrix} 1 & 0 & 0 \\ 1 & 1 & 0 \\ -1 & 1 & 1 \end{matrix} \right].$$

Here, the elimination process stops, since the first three entries of the third row of the last matrix consists of all zeros. Hence, this process does not lead to the inverse of $\mathbf{A}$. *In fact,* $\mathbf{A}$ *has no inverse since it is a singular matrix.*

B.7 The determinant of a matrix

The *determinant* of a square matrix is a real number that is nonzero if the matrix is nonsingular, and equal to zero if the matrix is singular. So the determinant of a matrix can be seen as a criterion for the (non)singularity of the matrix. The actual definition of the determinant of the matrix $\mathbf{A} = \{a_{ij}\}$, denoted by $\det(\mathbf{A})$, is given in the following iterative manner. First, if $\mathbf{A} = [a_{11}]$ is a $(1,1)$-matrix, then $\det(\mathbf{A}) = \det([a_{11}]) = a_{11}$. For any $n \geq 1$, we define:

$$\det(\mathbf{A}) = \sum_{i=1}^{n} a_{i1}(-1)^{i+1} \det(\mathbf{A})_{\{1,\dots,n\}\setminus\{i\},\{1,\dots,n\}\setminus\{1\}}.$$

In this expression, $(\mathbf{A})_{I,J}$ is the submatrix of $\mathbf{A}$ that consists of the rows with indices in I, and the columns with indices in J, with $I, J \subseteq \{1,\dots,n\}$. The right hand side in the expression above is called a *Laplace expansion*[5] of the determinant. The *cofactor* of the matrix entry a_{ij} of the (n,n)-matrix $\mathbf{A}$ is denoted and defined by:

$$\mathrm{cof}(a_{ij}) = (-1)^{i+j} \det(\mathbf{A})_{\{1,\dots,n\}\setminus\{i\},\{1,\dots,n\}\setminus\{j\}}.$$

---

[5]Named after the French mathematician and astronomer Pierre-Simon Laplace (1749–1827).

So the cofactor of a_{ij} in the (n, n)-matrix $\mathbf{A}$ is equal to $(-1)^{i+j}$ times the determinant of the matrix $\mathbf{A}$ without row i and column j.

In the above definition of determinant, we have taken the cofactors of the entries in the first column of $\mathbf{A}$. It can be shown that the Laplace expansion can be done along any column or row of $\mathbf{A}$.

Example B.7.1.   *We will calculate the determinant of a $(2, 2)$ and a $(3, 3)$-matrix. It is left to the reader to check the results for Laplace expansions for other rows and columns. For* $\mathbf{A} = \begin{bmatrix} a_{11} & a_{12} \\ a_{21} & a_{22} \end{bmatrix}$, *it holds that:*

$$\det(\mathbf{A}) = a_{11}(-1)^{1+1}\det(a_{22}) + a_{21}(-1)^{2+1}\det(a_{12}) = a_{11}a_{22} - a_{12}a_{21}.$$

For $\mathbf{A} = \begin{bmatrix} a_{11} & a_{12} & a_{13} \\ a_{21} & a_{22} & a_{23} \\ a_{31} & a_{32} & a_{33} \end{bmatrix}$, *it holds that:*

$$\det(\mathbf{A}) = a_{11}(-1)^{1+1}\det\left(\begin{bmatrix} a_{22} & a_{23} \\ a_{32} & a_{33} \end{bmatrix}\right) + a_{21}(-1)^{2+1}\det\left(\begin{bmatrix} a_{12} & a_{13} \\ a_{32} & a_{33} \end{bmatrix}\right)$$

$$+ a_{31}(-1)^{3+1}\det\left(\begin{bmatrix} a_{12} & a_{13} \\ a_{22} & a_{23} \end{bmatrix}\right)$$

$$= a_{11}(a_{22}a_{33} - a_{23}a_{32}) - a_{21}(a_{12}a_{33} - a_{13}a_{32}) + a_{31}(a_{12}a_{23} - a_{13}a_{22})$$

$$= a_{11}a_{22}a_{33} + a_{12}a_{23}a_{31} + a_{13}a_{21}a_{32}$$

$$- a_{13}a_{22}a_{31} - a_{12}a_{21}a_{33} - a_{11}a_{32}a_{23}.$$

Without proofs, we mention the following properties of determinants. Let $\mathbf{A}_1, \ldots, \mathbf{A}_n$ be the column vectors of the (n, n)-matrix $\mathbf{A}$. Let α be a scalar. Then, for each $i, j \in \{1, \ldots, n\}$ with $i \neq j$, the following properties hold:

   $\det(\mathbf{A}_1 \cdots \alpha\mathbf{A}_i \cdots \mathbf{A}_n) = \alpha\det(\mathbf{A})$, and $\det(\alpha\mathbf{A}) = \alpha^n\det(\mathbf{A})$.

   $\det(\mathbf{A}_1 \cdots \mathbf{A}_i \cdots \mathbf{A}_j \cdots \mathbf{A}_n) = \det(\mathbf{A}_1 \cdots \mathbf{A}_i + \alpha\mathbf{A}_j \cdots \mathbf{A}_j \cdots \mathbf{A}_n)$.

   $\det(\mathbf{I}_n) = 1$.

   $\det(\mathbf{A}_1 \cdots \mathbf{A}_i \cdots \mathbf{A}_j \cdots \mathbf{A}_n) = -\det(\mathbf{A}_1 \cdots \mathbf{A}_j \cdots \mathbf{A}_i \cdots \mathbf{A}_n)$.

   $\mathbf{A}$ is nonsingular if and only if $\det(\mathbf{A}) \neq 0$.

   $\det(\mathbf{A}_1 \cdots \mathbf{A}_i' + \mathbf{A}_i'' \cdots \mathbf{A}_n) = \det(\mathbf{A}_1 \cdots \mathbf{A}_i' \cdots \mathbf{A}_n) + \det(\mathbf{A}_1 \cdots \mathbf{A}_i'' \cdots \mathbf{A}_n)$.

   $\det(\mathbf{A}^\top) = \det(\mathbf{A})$.

   If $\mathbf{A}$ is nonsingular, then $\det(\mathbf{A}^{-1}) = 1/\det(\mathbf{A})$.

   If $\mathbf{A}$ is either an upper or a lower triangular matrix, then $\det(\mathbf{A})$ is equal to the product of the main diagonal entries of $\mathbf{A}$.

   For any two (n, n)-matrices $\mathbf{A}$ and $\mathbf{B}$, it holds that $\det(\mathbf{AB}) = \det(\mathbf{A}) \times \det(\mathbf{B})$.

If $\mathbf{A} = \begin{bmatrix} \mathbf{B} & \mathbf{0} \\ \mathbf{C} & \mathbf{D} \end{bmatrix}$, where $\mathbf{B}$ and $\mathbf{D}$ are square matrices, then $\det(\mathbf{A}) = \det(\mathbf{B}) \times \det(\mathbf{D})$.

If $\mathbf{A}$ is a block diagonal matrix (see Section B.3) with blocks $\mathbf{A}_1, \ldots, \mathbf{A}_k$, then $\det(\mathbf{A}) = \det(\mathbf{A}_1) \times \cdots \times \det(\mathbf{A}_k)$.

(*Cramer's rule*[6]) Consider the system $\mathbf{Ax} = \mathbf{b}$, with $\mathbf{A}$ a nonsingular (n, n)-matrix, $\mathbf{b} \in \mathbb{R}^n$, and $\mathbf{x}$ a vector consisting of n variables. Let $\mathbf{A}(i; \mathbf{b})$ be the matrix consisting of the columns of $\mathbf{A}$ except for column i which is exchanged by the vector $\mathbf{b}$. Then the unique solution to this system is given by:

$$x_i = \frac{\det(\mathbf{A}(i; \mathbf{b}))}{\det(\mathbf{A})} \text{ for } i = 1, \ldots, n.$$

From the fact that $\det(\mathbf{A}^\mathsf{T}) = \det(\mathbf{A})$, it follows that all properties concerning columns hold for rows as well.

Example B.7.2. *We will solve a system of equalities by means of Cramer's rule. Consider the following system of three linear equalities in three variables:*

$$\begin{aligned} 4x_2 + x_3 &= 2, \\ -2x_1 + 5x_2 - 2x_3 &= 1, \\ 3x_1 + 4x_2 + 5x_3 &= 6. \end{aligned}$$

We first check that the coefficient matrix is nonsingular: $\det \begin{bmatrix} 0 & 4 & 1 \\ -2 & 5 & -2 \\ 3 & 4 & 5 \end{bmatrix} = -7 \neq 0$, *and so this matrix is in fact nonsingular. Using Cramer's rule, we find that:*

$$x_1 = -\tfrac{1}{7} \det \begin{bmatrix} 2 & 4 & 1 \\ 1 & 5 & -2 \\ 6 & 4 & 5 \end{bmatrix} = 4, x_2 = -\tfrac{1}{7} \det \begin{bmatrix} 0 & 2 & 1 \\ -2 & 1 & -2 \\ 3 & 6 & 5 \end{bmatrix} = 1, \text{ and}$$

$$x_3 = -\tfrac{1}{7} \det \begin{bmatrix} 0 & 4 & 2 \\ -2 & 5 & 1 \\ 3 & 4 & 6 \end{bmatrix} = -2.$$

By substituting these values into the initial system, we find that this is in fact a solution.

It is left to the reader to show that for any nonsingular (n, n)-matrix $\mathbf{A}$ and any vector $\mathbf{b} \in \mathbb{R}^n$, the system $\mathbf{Ax} = \mathbf{b}$ has the unique solution $\mathbf{x} = \mathbf{A}^{-1}\mathbf{b}$.

The *adjoint* of the (n, n)-matrix $\mathbf{A} = \{a_{ij}\}$ is denoted and defined by:

$$\text{adj}(\mathbf{A}) = \text{adj}(\{a_{ij}\}) = \{\text{cof}(a_{ji})\}.$$

---

[6]Named after the Swiss mathematician GABRIEL CRAMER (1704–1752).

It can be shown that for any nonsingular matrix $\mathbf{A}$, it holds that:

$$\mathbf{A}^{-1} = \frac{1}{\det(\mathbf{A})}\,\mathrm{adj}(\mathbf{A}).$$

This formula enables us to calculate the inverse of a nonsingular matrix by means of the adjoint. We illustrate this with the following example.

Example B.7.3. *Let* $\mathbf{A} = \begin{bmatrix} 1 & 2 & 3 \\ 2 & 3 & 0 \\ 0 & 1 & 2 \end{bmatrix}$. *The determinant of* $\mathbf{A}$ *is:*

$$\det(\mathbf{A}) = 1\times3\times2+2\times0\times0+3\times2\times1-3\times3\times0-2\times2\times2-1\times1\times0 = 6+6-8 = 4.$$

Since $\det(\mathbf{A}) \neq 0$, *it follows that the inverse exists. The adjoint of* $\mathbf{A}$ *satisfies*

$$\mathrm{adj}(\mathbf{A}) = \begin{bmatrix} 6 & -1 & -9 \\ -4 & 2 & 6 \\ 2 & -1 & -1 \end{bmatrix}, \text{ and hence } \mathbf{A}^{-1} = \tfrac{1}{4}\begin{bmatrix} 6 & -1 & -9 \\ -4 & 2 & 6 \\ 2 & -1 & -1 \end{bmatrix}.$$

B.8 Linear and affine spaces

A *linear subspace* S of the n-dimensional Euclidean vector space $\mathbb{R}^n$ $(n \geq 0)$ is a nonempty subset that satisfies the following two conditions:

(i) If $\mathbf{x}, \mathbf{y} \in S$, then $\mathbf{x} + \mathbf{y} \in S$;

(ii) If $\mathbf{x} \in S$ and $\lambda \in \mathbb{R}$, then $\lambda\mathbf{x} \in S$.

Examples of linear subspaces of $\mathbb{R}^3$ are:

$\mathbb{R}^3$ itself.

The singleton set $\{\mathbf{0}\}$.

The set $\left\{ [x_1, x_2, x_3]^\top \ \middle| \ 2x_1 + 3x_2 = 0 \right\}$.

Note that linear subspaces always contain the all-zero vector.

Let S be a linear subspace of $\mathbb{R}^n$. A set of vectors $\mathbf{v}_1, \ldots, \mathbf{v}_k$ is said to *span* S if every vector $\mathbf{x}$ in S can be written as a linear combination of $\mathbf{v}_1, \ldots, \mathbf{v}_k$. For any nontrivial linear subspace S of $\mathbb{R}^n$, there is a unique smallest positive integer d – called the *dimension* of S; notation: $\dim(S)$ – such that every vector in S can be expressed as a linear combination of a fixed set of d independent vectors in S. Any such set of d vectors in S is said to form a *basis* for S. Although the vectors that form a basis are not unique, they must be linearly independent. It can be shown that for any positive integer d, a set of d linearly independent vectors is a basis for the linear subspace spanned by the d vectors. The linear subspace consisting of the all-zero vector only is called the *trivial subspace*. Also note that

$\dim(\mathbb{R}^n) = n$, so that $\dim(\{\mathbf{0}\}) = 0$. The fact that the dimension d of a subspace is uniquely defined is proved in the following theorem.

Theorem B.8.1.
There exists a unique smallest positive integer d such that every vector in S can be expressed as a linear combination of a fixed set of d independent vectors in S.

Proof. We will show that if $\mathbf{v}_1, \ldots, \mathbf{v}_r$ and $\mathbf{w}_1, \ldots, \mathbf{w}_s$ are both bases for S, then $r = s$. Suppose, to the contrary, that $r < s$. Since $\mathbf{v}_1, \ldots, \mathbf{v}_r$ is a basis for $\mathbf{S}$, it follows for each $j = 1, \ldots, s$ that there are scalars $a_{1j}, \ldots, a_{rj}$ such that:

$$\mathbf{w}_j = a_{1j}\mathbf{v}_1 + \ldots + a_{rj}\mathbf{v}_r. \tag{B.2}$$

Let $\mathbf{W}$ be the matrix with columns $\mathbf{w}_1, \ldots, \mathbf{w}_s$, let $\mathbf{V}$ be the matrix with columns $\mathbf{v}_1, \ldots, \mathbf{v}_r$, and let $\mathbf{A} = \{a_{ij}\}$ be an (r, s)-matrix. Then (B.2) is equivalent to: $\mathbf{W} = \mathbf{VA}$. Since $r < s$, the system $\mathbf{Ax} = \mathbf{0}$ must have a nonzero solution $\mathbf{x}_0 \neq \mathbf{0}$ (this can be seen by writing $\mathbf{Ax} = \mathbf{0}$ in Gauss-Jordan form). Hence, $\mathbf{Wx}_0 = \mathbf{V}(\mathbf{Ax}_0) = \mathbf{V0} = \mathbf{0}$. This means that the columns of $\mathbf{W}$ are linearly dependent, which contradicts the fact that the columns form a basis. Thus, we have that $r = s$. For $r > s$, a similar argument holds. $\qquad \square$

The *row space* $\mathcal{R}(\mathbf{A})$ of the (m, n)-matrix $\mathbf{A}$ consists of all linear combinations of the rows of $\mathbf{A}$; i.e., $\mathcal{R}(\mathbf{A}) = \{\mathbf{A}^\top \mathbf{y} \mid \mathbf{y} \in \mathbb{R}^m\}$. The reader may check that $\mathcal{R}(\mathbf{A})$ is a linear subspace of $\mathbb{R}^n$. Similarly, the *column space* of $\mathbf{A}$ consists of all linear combinations of the columns of $\mathbf{A}$. Note that the column space of $\mathbf{A}$ is a linear subspace of $\mathbb{R}^m$, and that it is equal to $\mathcal{R}(\mathbf{A}^\top)$. Also note that if $\dim \mathcal{R}(\mathbf{A}) = m$, then $\mathbf{A}$ has full row rank and the rows of $\mathbf{A}$ form a basis for $R(\mathbf{A})$; if $\dim \mathcal{R}(\mathbf{A}^\top) = n$, then $\mathbf{A}$ has full column rank and the columns of $\mathbf{A}$ form a basis for $\mathcal{R}(\mathbf{A}^\top)$.

The *null space* of the (m, n)-matrix $\mathbf{A}$ is denoted and defined by:

$$\mathcal{N}(\mathbf{A}) = \{\mathbf{x} \in \mathbb{R}^n \mid \mathbf{Ax} = \mathbf{0}\}.$$

The reader may check that $\mathcal{N}(\mathbf{A})$ is a linear subspace of $\mathbb{R}^n$. For any (m, n)-matrix $\mathbf{A}$ the following properties hold:

(1) $\mathcal{R}(\mathbf{A}) \cap \mathcal{N}(\mathbf{A}) = \{\mathbf{0}\}$.

(2) $\mathcal{R}(\mathbf{A}) \perp \mathcal{N}(\mathbf{A})$.

(3) $\dim \mathcal{R}(\mathbf{A}) + \dim \mathcal{N}(\mathbf{A}) = n$.

Here, $\mathcal{R}(\mathbf{A}) \perp \mathcal{N}(\mathbf{A})$ means that the linear spaces $\mathcal{R}(\mathbf{A})$ and $\mathcal{N}(\mathbf{A})$ are *perpendicular*, i.e., for every $\mathbf{r} \in \mathcal{R}(\mathbf{A})$ and every $\mathbf{s} \in \mathcal{N}(\mathbf{A})$, we have that $\mathbf{r}^\top \mathbf{s} = 0$.

Property (1) is a direct consequence of property (2). For the proof of property (2), let $\mathbf{r} \in \mathcal{R}(\mathbf{A})$ and $\mathbf{s} \in \mathcal{N}(\mathbf{A})$. Because $\mathbf{r} \in \mathcal{R}(\mathbf{A})$, there exists a vector $\mathbf{y} \in \mathbb{R}^m$ such that $\mathbf{r} = \mathbf{A}^\mathsf{T}\mathbf{y}$. Hence, we have that $\mathbf{r}^\mathsf{T}\mathbf{s} = (\mathbf{A}^\mathsf{T}\mathbf{y})^\mathsf{T}\mathbf{s} = \mathbf{y}^\mathsf{T}\mathbf{A}\mathbf{s} = \mathbf{y}^\mathsf{T}\mathbf{0} = 0$. Property (3) can be shown as follows. Let $\dim \mathbb{R}(\mathbf{A}) = p \leq n$. Writing the system $\mathbf{A}\mathbf{x} = \mathbf{0}$ in Gauss-Jordan form, we easily see that $n - p$ variables of $x_1, \ldots, x_n$ can be given an arbitrary value. Hence, $\dim \mathcal{N}(\mathbf{A}) = n - p$.

It follows from properties (1), (2), and (3) that any vector $\mathbf{x} \in \mathbb{R}^n$ can be written as:

$$\mathbf{x} = \mathbf{x}_R + \mathbf{x}_N \text{ with } \mathbf{x}_R \in \mathcal{R}(\mathbf{A}), \ \mathbf{x}_N \in \mathcal{N}(\mathbf{A}), \text{ and } \mathbf{x}_R^\mathsf{T}\mathbf{x}_N = 0.$$

Example B.8.1.   *Consider the matrix* $\mathbf{A} = \begin{bmatrix} 1 & 0 & -1 \\ 0 & 1 & 2 \end{bmatrix}$. *Then,*

$$\mathcal{R}(\mathbf{A}) = \{\mathbf{A}^\mathsf{T}\mathbf{y} \mid \mathbf{y} \in \mathbb{R}^2\} = \left\{ y_1 \begin{bmatrix} 1 \\ 0 \\ -1 \end{bmatrix} + y_2 \begin{bmatrix} 0 \\ 1 \\ 2 \end{bmatrix} \ \middle| \ y_1, y_2 \in \mathbb{R} \right\},$$

and $\mathcal{N}(\mathbf{A}) = \left\{ \alpha \begin{bmatrix} 1 \\ -2 \\ 1 \end{bmatrix} \ \middle| \ \alpha \in \mathbb{R} \right\}.$

Therefore, we have that $\dim \mathcal{R}(\mathbf{A}) = 2$ *and* $\dim \mathcal{N}(\mathbf{A}) = 1$. *Note that property* (3) *holds. Property* (2) *also holds:*

$$\left(\alpha_1 \begin{bmatrix} 1 & 0 & -1 \end{bmatrix} + \alpha_2 \begin{bmatrix} 0 & 1 & 2 \end{bmatrix} \right) \left(\alpha \begin{bmatrix} 1 \\ -2 \\ 1 \end{bmatrix} \right)$$

$$= \underbrace{\alpha_1 \alpha \begin{bmatrix} 1 & 0 & -1 \end{bmatrix} \begin{bmatrix} 1 \\ -2 \\ 1 \end{bmatrix}}_{=0} + \underbrace{\alpha_2 \alpha \begin{bmatrix} 0 & 1 & 2 \end{bmatrix} \begin{bmatrix} 1 \\ -2 \\ 1 \end{bmatrix}}_{=0} = 0.$$

Property (1) *now holds trivially. One can easily check that the vectors* $\begin{bmatrix} 1 & 0 & -1 \end{bmatrix}^\mathsf{T}$, $\begin{bmatrix} 1 & 1 & 1 \end{bmatrix}^\mathsf{T}$, *and* $\begin{bmatrix} 1 & -2 & 1 \end{bmatrix}^\mathsf{T}$ *are linearly independent and span* $\mathbb{R}^3$.

An *affine subspace* of $\mathbb{R}^n$ is any subset of the form $\mathbf{a} + L$, with $\mathbf{a} \in \mathbb{R}^n$ and L a linear subspace of $\mathbb{R}^n$. By definition, $\dim(\mathbf{a} + L) = \dim L$. Examples of affine subspaces are all linear subspaces ($\mathbf{a} = \mathbf{0}$). Affine subspaces of $\mathbb{R}^n$ can be written in the form $\{\mathbf{x} \in \mathbb{R}^n \mid \mathbf{A}\mathbf{x} = \mathbf{b}\}$ with $\mathbf{A} \in \mathbb{R}^{m \times n}$, $\mathbf{b} \in \mathbb{R}^m$, and the system $\mathbf{A}\mathbf{x} = \mathbf{b}$ is consistent. Clearly, the system $\mathbf{A}\mathbf{x} = \mathbf{b}$ is consistent if and only if the vector $\mathbf{b}$ can be expressed as a linear combination of the columns of $\mathbf{A}$. Namely, if $\mathbf{a}_1, \ldots, \mathbf{a}_n$ are the columns of $\mathbf{A}$ and $\mathbf{x} = \begin{bmatrix} x_1 & \ldots & x_n \end{bmatrix}^\mathsf{T}$, then $\mathbf{A}\mathbf{x} = \mathbf{b}$ is equivalent to:

$$\mathbf{b} = x_1 \mathbf{a}_1 + \ldots + x_n \mathbf{a}_n.$$

APPENDIX C

Graph theory

In this appendix some basic terminology and results from graph theory are introduced. For a more detailed treatment of graph theory, we refer to Bondy and Murty (1976) and West (2001).

C.1 Undirected graphs

A *graph* G is defined as a nonempty finite set $V(G)$ of elements called *nodes*, a finite set $\mathcal{E}(G)$ of elements called *edges*, and an incidence relation which associates with each edge either one or two nodes called *end nodes*. Two nodes that are incident with a common edge are called *adjacent*, as are two edges which are incident with a common node. We will write $G = (V, \mathcal{E})$. An edge is called a *loop* if the number of its end nodes is one. Two or more edges with the same set of end nodes are said to be *parallel*. A graph without loops and parallel edges is called a *simple graph*.

It is customary to represent a graph by means of a two-dimensional picture. The nodes are drawn as dots and the edges as curves joining the dots corresponding to the end nodes. It may happen that two edge-curves intersect at some point different from the node-dots; such edge-crossings are ignored as they do not represent anything in the structure of the graph. A graph can be drawn in different ways. Two graphs are said to be *isomorphic* if one of them can be obtained by relabeling the nodes of the other. For example, the graphs in Figure C.1 are isomorphic.

Let $n, m \geq 1$ be integers. The *complete graph* K_n is the simple graph with n nodes and (all) $\binom{n}{2} = \frac{1}{2}n(n-1)$ edges. A *bipartite graph* is a graph $G = (V, \mathcal{E})$ of which the node set V can be partitioned into two sets V_1 and V_2 with $V_1 \cup V_2 = V$, $V_1 \cap V_2 = \emptyset$ and each edge has one end in V_1 and one in V_2. The *complete bipartite graph* $K_{m,n}$ is the simple bipartite graph with $|V_1| = m$ and $|V_2| = n$ such that there is an edge between each node in V_1 and each node in V_2. The *degree* $d(x)$ of a node x of a graph $G = (V, \mathcal{E})$ is the number

585

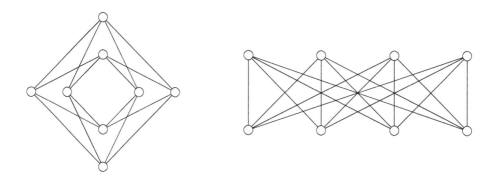

Figure C.1: Isomorphic graphs.

of edges incident with x, loops being counted twice. An *isolated node* is a node of degree 0. Counting the number of edges in each node, we obtain the formula:

$$\sum_{x \in \mathcal{V}(G)} d(x) = 2|\mathcal{E}|.$$

It can be shown that in any graph the number of nodes of odd degree is even. A graph in which each node has the same degree is called *regular*. Hence, for each $n \geq 1$, the complete graph K_n is regular of degree $n - 1$ (see Figure C.2), and the complete bipartite graph $K_{n,n}$ is regular of degree n (see Figure C.3).

C.2 Connectivity and subgraphs

A graph G with node set $\mathcal{V}$ is called *disconnected* if there exist two nonempty subsets $\mathcal{V}_1$ and $\mathcal{V}_2$ of $\mathcal{V}$ with $\mathcal{V}_1 \cap \mathcal{V}_2 = \emptyset$ and $\mathcal{V}_1 \cup \mathcal{V}_2 = \mathcal{V}$, such that there are no edges between $\mathcal{V}_1$ and $\mathcal{V}_2$; otherwise, G is called *connected*.

A graph H is called a *subgraph* of a graph G if $\mathcal{V}(H) \subseteq \mathcal{V}(G)$, $\mathcal{E}(H) \subseteq \mathcal{E}(G)$, and each edge of H has the same end nodes in H as in G. Note that every graph is a subgraph of itself. A subgraph that is not equal to G itself is called a *proper subgraph* of G. A *spanning subgraph* of G is a subgraph H of G such that $\mathcal{V}(H) = \mathcal{V}(G)$ and $\mathcal{E}(H) \subseteq \mathcal{E}(G)$.

A *path* between distinct nodes v_0 and v_k in a graph G is a connected subgraph P of G such that $\mathcal{V}(P) = \{v_0, \ldots, v_k\}$, $|\mathcal{E}(P)| = k$, the nodes v_0 and v_k have degree one, and the nodes $v_1, \ldots, v_{k-1}$ have degree two in P. We will write $P = v_0\text{-}v_1\text{-}\cdots\text{-}v_k$. The integer k is called the *length* of the path and is equal to the number of edges in P. Starting at v_0 and finishing at v_k, the path P is said to *traverse* the nodes in $\mathcal{V}(P)$ and the edges in $\mathcal{E}(P)$. A *cycle of length* k (≥ 1) in a graph G is a connected subgraph C of G such that $\mathcal{V}(C) = \{v_1, \ldots, v_k\}$, $|\mathcal{E}(C)| = k$, and all nodes in $\mathcal{V}(C)$ have degree two in C. We will write $C = v_1\text{-}v_2\text{-}\cdots\text{-}v_k\text{-}v_1$. A cycle is called *odd* if k is odd and *even* if k is even. It can be shown that if every node of G has degree at least two, then G contains a cycle. The

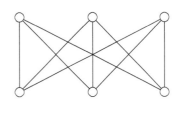

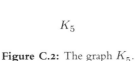

K_5 $\qquad\qquad\qquad\qquad\qquad\qquad\qquad\qquad\quad$ $K_{3,3}$

Figure C.2: The graph K_5. $\qquad\qquad\qquad$ **Figure C.3:** The graph $K_{3,3}$.

relationship between the concepts of connectivity and path are expressed in the following theorem.

Theorem C.2.1.
For any graph $G = (\mathcal{V}, \mathcal{E})$, the following assertions are equivalent:

(i) $\quad G$ is connected;

(ii) $\quad$ For each $u, v \in \mathcal{V}$ with $u \neq v$, there is a path between u and v.

Proof of Theorem C.2.1. (i) $\Rightarrow$ (ii): Take any $u, v \in \mathcal{V}$. Let $\mathcal{V}_1$ consist of all nodes $w \in \mathcal{V}$ such there is a path between u and w. We have to show that $v \in \mathcal{V}_1$. Assume to the contrary that $v \notin \mathcal{V}_1$. Define $\mathcal{V}_2 = \mathcal{V} \setminus \mathcal{V}_1$. Since G is connected, there exist adjacent $u_1 \in \mathcal{V}_1$ and $v_1 \in \mathcal{V}_2$. Hence, there is a path from u to v_1 that traverses v. But this means that $v \in \mathcal{V}_1$, which is a contradiction. Hence, there is a path between u and v.

(ii) $\Rightarrow$ (i): Assume to the contrary that G is not connected. Then there are nodes u and v and sets $\mathcal{V}_1, \mathcal{V}_2 \subset \mathcal{V}$ with $u \in \mathcal{V}_1, v \in \mathcal{V}_2, \mathcal{V}_1 \cup \mathcal{V}_2 = \mathcal{V}, \mathcal{V}_1 \neq \emptyset, \mathcal{V}_2 \neq \emptyset, \mathcal{V}_1 \cap \mathcal{V}_2 = \emptyset$ and no edges between $\mathcal{V}_1$ and $\mathcal{V}_2$. However, since there is a path between u and v, there has to be an edge between $\mathcal{V}_1$ and $\mathcal{V}_2$ as well. So we arrive at a contradiction, and therefore G is connected. $\quad\square$

Using the concept of cycle, the following characterization of bipartite graphs can be formulated.

Theorem C.2.2.
A graph is bipartite if and only if it contains no odd cycles.

Proof of Theorem C.2.2. First suppose that $G = (\mathcal{V}, \mathcal{E})$ is bipartite with bipartition $(\mathcal{V}_1, \mathcal{V}_2)$. Let $C = v_1\text{-}\cdots\text{-}v_k\text{-}v_1$ be a cycle in G (so $\{v_1, v_2\}, \ldots, \{v_{k-1}, v_k\}, \{v_k, v_1\} \in \mathcal{E}(C)$). Without loss of generality, we may assume that $v_1 \in \mathcal{V}_1$. Since G is bipartite, it follows that $v_2 \in \mathcal{V}_2$. Similarly, $v_3 \in \mathcal{V}_1$, and – in general – $v_{2i} \in \mathcal{V}_2$ and $v_{2i+1} \in \mathcal{V}_1$. Clearly, $v_1 \in \mathcal{V}_1$ and $v_k \in \mathcal{V}_2$.

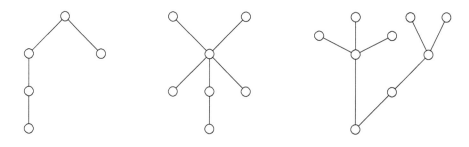

Figure C.4: Trees.

Hence, $k = 2i$ for some integer i. Hence, the number of edges of C is even.

For the proof of the converse, we may obviously restrict ourselves to connected graphs. Thus, let $G = (V, \mathcal{E})$ be a connected graph without odd cycles. Choose an arbitrary node u and define a partition (V_1, V_2) of V by setting:

$$V_1 = \{x \in V \mid d(u, x) \text{ is even}\} \text{ and } V_2 = \{y \in V \mid d(u, y) \text{ is odd}\},$$

where $d(u, v)$ is as the minimum length of a path between the nodes u and v. We will show that (V_1, V_2) is a bipartition of V. To see this, let any two nodes v and w in V_1 be given. Let P be a shortest path (in terms of number of edges) between u and v, and let Q be a shortest path between u and w. Hence, the lengths of both P and Q are even. Now let $\bar{u}$ be the last common node of P and Q, when traversing the paths starting at u. The $(u, \bar{u})$-sections of both P and Q (which may be different) are shortest paths between u and $\bar{u}$, and must therefore have the same length. Call P_1 be the $(\bar{u}, v)$-section of P, and Q_1 the $(\bar{u}, w)$-section of Q. Clearly, P_1 and Q_1 have either both even length or both odd length. Hence, the path v-P_1-$\bar{u}$-Q_1-w between v and w is of even length. If v and w were adjacent, then (P_1, Q_1, w, v) would be a cycle of odd length. This contradicts the assumption. Therefore, no two nodes in V_1 are adjacent. Similarly, no two nodes in V_2 are adjacent. Hence, G is bipartite. □

C.3   **Trees and spanning trees**

A *tree* is a connected graph that contains no cycles. Examples of trees are depicted in Figure C.4. We give the following obvious properties of trees without proof.

Theorem C.3.1.
Let $G = (V, \mathcal{E})$ be a simple graph. Then the following assertions hold.

(1) G is a tree if and only if there is exactly one path between any two nodes.

(2) If G is a tree, then $|\mathcal{E}| = |V| - 1$.

(3) Every tree T in G with $|V(T)| \geq 2$ has at least two nodes of degree one.

Notice that it follows from Theorem C.3.1 that the addition of a new edge (between nodes that are in the tree) leads to a unique cycle in that tree. A *spanning tree* of a graph G is a spanning subgraph of G that is a tree. So a spanning tree of G is a tree in G that contains all nodes of G.

C.4 Eulerian tours, Hamiltonian cycles, and Hamiltonian paths

In the earliest known paper on graph theory, the Swiss mathematician LEONHARD EULER (1707–1783) showed in 1736 that it was impossible to cross each of the seven bridges of Kaliningrad once and only once during a walk through the town. A plan of Kaliningrad with the river Pregel and the corresponding graph are shown in Figure C.5.

An *Eulerian tour* in a graph G is a cycle in G that traverses each edge of G exactly once. A graph is called *Eulerian* if it contains a Eulerian tour. (The terms 'tour' and 'cycle' are interchangeable.)

Theorem C.4.1.
A graph G is Eulerian if and only if:

(a) G is connected, and

(b) the degree of each node is even.

Proof of Theorem C.4.1. The necessity of the conditions (a) and (b) is obvious. So, we need to show that (a) and (b) are sufficient for a graph being Eulerian. We first decompose G into cycles $C_1, \ldots, C_k$. Construct a cycle C in G as follows. Start at an arbitrary non-isolated node v_1. Since v_1 is not isolated, it has a neighbor v_2. Because the degree of v_2 is even and at least one (because v_2 is adjacent to v_1), v_2 has degree at least two. So, v_2 has a neighbor v_3. By repeating this argument, we can continue this process until we reach a node that has been visited before. This must happen, because the graph has a finite number of nodes. We thus construct a path v_1-v_2-$\ldots$-v_q. Let p be such that $v_p = v_q$. Then, $C = v_p$-v_{p+1}-$\ldots$-v_q is a

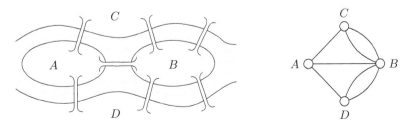

Figure C.5: The bridges of Kaliningrad and the corresponding graph.

Figure C.6: The Hamiltonian cycles of K_4.

cycle in G. Now remove the edges of C from G, and repeat this procedure until there are no remaining edges. Note that removing these edges decreases the degree of each of the nodes v_p, $v_{p+1}, \ldots, v_{q-1}$ by two and therefore, after removing them, the degree of each node remains even, and the procedure can be repeated until all nodes have zero degree.

The above process yields cycles $C_1, \ldots, C_k$ such that $E(G) = \bigcup_{i=1}^{k} E(C_i)$, and no two cycles share an edge. Order $C_1, \ldots, C_k$ by keeping a current (ordered) set of cycles, starting with the set $\{C_1\}$, and iteratively adding a cycle that has a node in common with at least one of the cycles in the current set of cycles. If, in iteration t, we cannot add a new cycle, then the cycles are partitioned into $\{C_{i_1}, \ldots, C_{i_t}\}$ and $\{C_{i_{t+1}}, \ldots, C_{i_k}\}$ such that no cycle of the former set has a node in common with the latter set, which implies that G is disconnected, contrary to the assumptions of the theorem. So, we may assume that $C_1, \ldots, C_k$ are ordered such that, for $j = 1, \ldots, k$, the cycle C_j has a node in common with at least one of $C_1, \ldots, C_{j-1}$.

Now, iteratively construct new cycles $C_1', \ldots, C_k'$ as follows. Let $C_1' = C_1$. For $i = 2, \ldots, k$, write $C_{i-1}' = v_1\text{-}v_2\text{-}\ldots\text{-}v_r\text{-}\ldots\text{-}v_{l-1}\text{-}v_1$ and $C_i = v_r\text{-}u_1\text{-}\ldots\text{-}u_s\text{-}v_r$, where v_r is a common node of C_{i-1}' and C_i, and define $C_i' = v_1\text{-}v_2\text{-}\ldots\text{-}v_r\text{-}u_1\text{-}\ldots\text{-}u_s\text{-}v_r\text{-}\ldots\text{-}v_{l-1}\text{-}v_1$. After k iterations, C_k' is an Eulerian cycle.   □

Using Theorem C.4.1, it is now obvious that the graph of Figure C.6 is not Eulerian, and so it is not possible to make a walk through Kaliningrad by crossing the seven bridges precisely once.

A *Hamiltonian cycle*[1] in $G = (\mathcal{V}, \mathcal{E})$ is a cycle in G of length $|\mathcal{V}|$ that traverses each node of G precisely once. The three Hamiltonian cycles of K_4 are depicted in Figure C.6.

A graph is said to be *Hamiltonian* if it contains a Hamiltonian cycle. It can be shown that the number of Hamiltonian cycles in K_n is $\frac{1}{2}(n-1)!$. A *Hamiltonian path* in $G = (\mathcal{V}, \mathcal{E})$ is a path in G of length $|\mathcal{V}| - 1$ that traverses all nodes of G precisely once. Note that a graph may contain a Hamiltonian path but not a Hamiltonian cycle (for example the leftmost graph in Figure C.4). On the other hand, a graph that has a Hamiltonian cycle certainly has a Hamiltonian path.

---

[1]Named after the Irish mathematician SIR WILLIAM ROWAN HAMILTON (1805–1865).

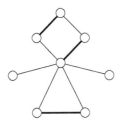

Figure C.7: A maximum matching.

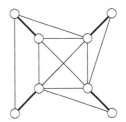

Figure C.8: A perfect matching.

In contrast with the case of Eulerian graphs, no nontrivial necessary and sufficient condition for a graph to be Hamiltonian is known; actually, the problem of finding such a condition is one of the main unsolved problems of graph theory.

C.5 **Matchings and coverings**

A subset M of $\mathcal{E}$ is called a *matching* in the graph $G = (\mathcal{V}, \mathcal{E})$ if no two edges of M are adjacent in G. If every node of G is an end of an edge of the matching M, then M is called a *perfect matching*, also called a *1-factor*. Note that a perfect matching is a spanning subgraph of that graph and exists only if the number of nodes of the graph is even. A matching M in G is called a *maximum matching* if G has no matching M' with $|M'| > |M|$. The number of edges in a maximum matching of G is denoted by $\nu(G)$. Clearly, every perfect matching is maximal. Figure C.7 depicts a maximum matching, and Figure C.8 depicts a perfect matching. Notice that, based on the definition, the empty set is a matching in any graph.

A *node covering* of a graph G is a subset K of $\mathcal{V}(G)$ such that every edge of G has at least one end in K. A node-covering K is called a *minimum node covering* if G has no node-covering K' with $|K'| < |K|$; see Figure C.9 and Figure C.10. The number of elements in a minimum node-covering of G is denoted by $\tau(G)$.

The following theorem shows that there is an interesting relationship between $\nu(G)$ and $\tau(G)$. The first part of the theorem asserts that the size of a maximum matching is at most the size of a minimum node covering. The second part asserts that, if G is bipartite, then the two are in fact equal. The latter is known as Kőnig's theorem[2]. The proof is omitted here.

Theorem C.5.1.
For any graph G, it holds that $\nu(G) \leq \tau(G)$. Moreover, equality holds when G is bipartite.

---

[2]Named after the Hungarian mathematician DÉNES KŐNIG (1884–1944).

Figure C.9: A node-covering.

Figure C.10: A minimum node-covering.

Proof of Theorem C.5.1. The first assertion of the theorem follows directly from the fact that edges in a matching M of a graph G are disjoint, whereas a node-covering contains at least one node of each edge of the matching. For a proof of the second statement, we refer to, e.g., West (2001).   □

For example, for G being the *unit-cube graph* corresponding to the unit cube in $\mathbb{R}^3$; see Figure C.11. Because this graph is bipartite, it follows that $\nu(G) = \tau(G)$. Also, it can easily be shown that $\nu(K_n) = \lfloor \frac{n}{2} \rfloor$, and that $\tau(K_n) = n$ ($n \geq 1$). Notice that if a graph G is not bipartite, this does not necessarily mean that $\nu(G) = \tau(G)$. See Figure C.12 for an example.

C.6 Directed graphs

A *directed graph* $D = (\mathcal{V}, \mathcal{A})$ is defined as a finite set $\mathcal{V}$ of elements called *nodes*, a finite set $\mathcal{A}$ of elements called *arcs* or *directed edges*, and two functions $h\colon \mathcal{A} \to \mathcal{V}$ and $t\colon \mathcal{A} \to \mathcal{V}$. For each $a \in \mathcal{A}$, $h(a)$ is called the *head node* of arc a, and $t(a)$ is called the *tail node* of arc a. Any arc a is said to be directed *from* $t(a)$ *to* $h(a)$; it is a *loop-arc* if its head node and tail node coincide. Arcs are also denoted as ordered pairs: $a = (x, y) \in \mathcal{A}$ means $x = t(a)$ and $y = h(a)$.

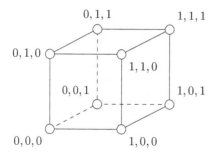

Figure C.11: The unit-cube graph in $\mathbb{R}^3$.

Figure C.12: Nonbipartite graph G that satisfies $\nu(G) = \tau(G)$.

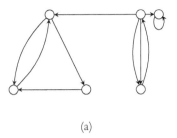 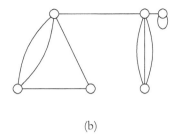

(a)   (b)

Figure C.13: (a) A digraph and (b) its underlying graph.

For convenience, we shall abbreviate 'directed graph' to *digraph*. A digraph is sometimes also called a *network*. A digraph D' is a *subdigraph* of D if $\mathcal{V}(D') \subset \mathcal{V}(D)$, $\mathcal{A}(D') \subset \mathcal{A}(D)$, and each arc of D' has the same head node and tail node in D' as in D.

With every digraph D, we associate an undirected graph G on the same node set but for which the arc-directions are discarded. To be precise, $\mathcal{V}(G) = \mathcal{V}(D)$ and for each arc $a \in \mathcal{A}$ with head node u and tail node v, there is an edge of G with end nodes u and v. This graph G is called the *underlying graph* of D; see Figure C.13.

Just as with graphs, digraphs (and also *mixed graphs*, in which some edges are directed and some edges are undirected) have a simple pictorial representation. A digraph is represented by a diagram of its underlying graph together with arrows on its edges, each arrow pointing towards the head node of the corresponding arc.

Every concept that is valid for graphs, automatically applies to digraphs as well. Thus the digraph in Figure C.13 is connected, because the underlying graph has this property. However, there are many notions that involve the orientation, and these apply only to digraphs.

A *directed path* in D is a nonempty sequence $W = (v_0, a_1, v_1, \ldots, a_k, v_k)$, whose terms are alternately distinct nodes and arcs, such that for $i = 1, \ldots, k$ the arc a_i has head node v_i and tail node v_{i-1}. As can be done in graphs, a directed path can often be represented by its node sequence $(v_0, v_1, \ldots, v_k)$; the *length* of this path is k.

Two nodes u and v are *diconnected* in D if there exist both a directed path from u to v and a directed path from v to u. A digraph D is called *strongly connected* if each two nodes in D are diconnected. The digraph in Figure C.14 is not strongly connected.

The *indegree* $d_D^-(v)$ of a node v in D is the number of arcs with head node v; the *outdegree* $d_D^+(v)$ of v is the number of arcs with tail node v. In Figure C.14, e.g., $d^-(v_1) = 2$, $d^+(v_2) = 0$, $d^+(v_3) = 2$, and $d^-(v_4) = 1$. The subscript D will usually be omitted from indegree and outdegree expressions.

Directed cycles are defined similar to (undirected) cycles. A *directed Hamiltonian path* (*cycle*) is a directed path (cycle) that includes every node of D. A *directed Eulerian tour* of D is a directed

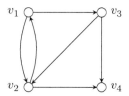

Figure C.14: Digraph that is not strongly connected.

tour that traverses each arc of d exactly once. The proof of the following theorem is left to the reader.

Theorem C.6.1.
A digraph $D = (\mathcal{V}, \mathcal{A})$ contains a directed Eulerian tour if and only if D is strongly connected and $d^+(v) = d^-(v)$ for each $v \in \mathcal{V}$.

C.7 Adjacency matrices and incidence matrices

Graphs can conveniently be represented by means of matrices. Given a simple graph $G = (\mathcal{V}, \mathcal{E})$, its *adjacency matrix* $\mathbf{A}(G) = \{a_{ij}\}$ is defined by:

$$a_{ij} = \begin{cases} 1 & \text{if } \{i, j\} \in \mathcal{E}, \\ 0 & \text{otherwise.} \end{cases}$$

In the case of a digraph, the sum of all the entries in row i gives the outdegree of node i, and the sum of the entries in column j gives the indegree of node j. In case the graph is undirected, the adjacency matrix is symmetric; the sum of the entries of column i gives twice the degree of node i and the same holds for sum of the entries of row i.

The *node-edge incidence matrix* $\mathbf{M}_{VE} = \{m_{ie}\}$ of a graph $G = (\mathcal{V}, \mathcal{E})$ is defined as follows:

$$m_{ie} = \begin{cases} 1 & \text{if } i \in \mathcal{V}, e \in \mathcal{E} \text{ and } e \text{ is incident with } i \\ 0 & \text{otherwise.} \end{cases}$$

Note that the row sums of node-edge incidence matrices are the degrees of the nodes, and that each column contains precisely two entries equal to 1. See also Section 4.3.4.

The *node-arc incidence matrix* $\mathbf{M}_{VA} = \{m_{ia}\}$ associated with the digraph $G = (\mathcal{V}, \mathcal{A})$ is defined as follows:

$$m_{ie} = \begin{cases} 1 & \text{if } i \in \mathcal{V}, a \in \mathcal{A}, \text{ and } i = t(a) \\ -1 & \text{if } i \in \mathcal{V}, a \in \mathcal{A}, \text{ and } i = h(a) \\ 0 & \text{otherwise.} \end{cases}$$

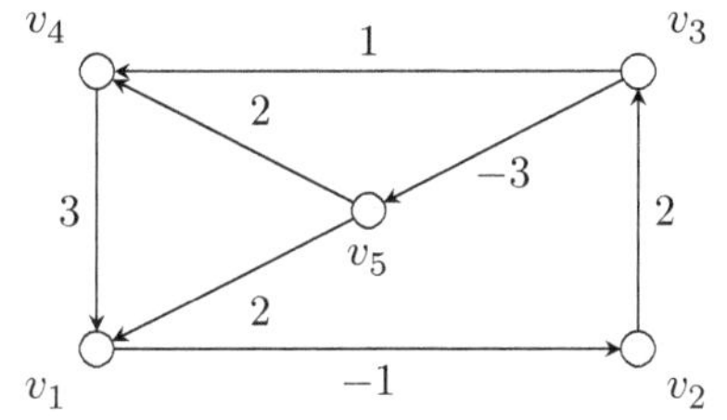

Figure C.15: Weighted digraph.

Note that, in each column of a node-arc incidence matrix, exactly one entry is $+1$, exactly one entry is -1, and all other entries are 0. See also Section 4.3.4.

C.8 Network optimization models

Let $D = (\mathcal{V}, \mathcal{A})$ be a digraph with node set $\mathcal{V}$ and arc set $\mathcal{A}$, and $\ell : \mathcal{A} \to \mathbb{R}$ a *weight* or *length function* on $\mathcal{A}$. For each $(v, w) \in \mathcal{A}$, $\ell(v, w)$ is called the *weight* or *length* of (v, w). If no arc (v, w) exists in D, we define $\ell(v, w) = \infty$; moreover, $\ell(v, v) = 0$ for each $v \in \mathcal{V}$. Without loss of generality, it can therefore be assumed that the graphs are complete, meaning that all arcs are present.

Example C.8.1. *Consider the weighted digraph of Figure C.15 (arcs with weights ∞ and 0 are omitted). The weight function ℓ in this example is defined by $\ell(v_1, v_2) = -1$, $\ell(v_2, v_3) = 2$, $\ell(v_3, v_4) = 1$, $\ell(v_4, v_1) = 3$, $\ell(v_5, v_1) = 2$, $\ell(v_3, v_5) = -3$, and $\ell(v_5, v_4) = 2$. Note that some weights are negative.*

The *weight* of a path or cycle is defined as the sum of the weights of the individual arcs comprising the path or cycle.

We will give the formulation of the most common problems in network optimization theory. For the algorithms that solve these problems, we only mention the names of the most common ones. The interested reader is referred to, e.g., Ahuja *et al.* (1993) and Cook *et al.* (1997).

The *shortest path problem* is the problem of determining a minimum weight (also called a shortest) path between two given nodes in a weighted graph. The most famous algorithm that solves the shortest path problem is due to EDSGER W. DIJKSTRA (1930–2002). This so-called *Dijkstra's shortest path algorithm* (1959) finds a shortest path from a specified node s (the source) to all other nodes, provided that the arc weights are nonnegative. In the case where arc weights are allowed to be negative, it may happen that there is a cycle of negative length in the graph. In that case, finite shortest paths may not exist for all pairs of nodes. An algorithm due to ROBERT W. FLOYD (1936–2001) and STEPHEN WARSHALL (1935–2006), the so-called *Floyd-Warshall algorithm* (1962), finds a shortest path between all pairs of nodes and stops whenever a cycle of negative length occurs.

The *minimum spanning tree problem* is the problem of finding a spanning tree of minimum length in a weighted connected graph. *Prim's minimum spanning tree algorithm* (1957), due to ROBERT C. PRIM (born 1921), solves the problem for undirected graphs. In case the graph is directed, the spanning trees are usually so-called spanning *arborescences*, which are trees in which no two arcs are directed into the same node. The *minimum arborescence algorithm* (1968), due to JACK R. EDMONDS (born 1934), solves this problem.

The *maximum flow problem* (see Section 8.2.5) is the problem of determining the largest possible amount of flow that can be sent through a digraph from a source to a sink taking into account certain capacities on the arcs. The *minimum cost flow problem* is the problem of sending a given amount of flow through a weighted digraph against minimum costs. These problems may be solved with the *flow algorithms* (1956) discovered by LESTER R. FORD (born 1927) and DELBERT R. FULKERSON (1924–1976).

The *maximum cardinality matching problem* is the problem of finding a matching in a graph with maximum cardinality. The *minimum weight matching problem* is the problem of determining a perfect matching in a weighted graph with minimum weight. These problems can be solved by means of JACK R. EDMONDS' *matching algorithm* (1965).

The *postman problem*, also called the *Chinese postman problem*, is the problem of finding a shortest tour in a weighted graph such that all edges and/or arcs are traversed at least once and the begin and end node of the tour coincide. In case the graph is either directed or undirected, then postman problems can be solved by means of algorithms designed in 1973 by JACK R. EDMONDS and ELLIS L. JOHNSON (born 1937). In the case of a mixed graph an optimal postman route – if one exists – is much more difficult to determine. Actually, there are cases in which no efficient optimal solution procedure is currently available.

The *traveling salesman problem* is the problem of finding a shortest Hamiltonian cycle in a weighted graph. Also for this problem no efficient solution technique is available. On the other hand, an extensive number of *heuristics* and *approximation algorithms* (efficient algorithms that generate feasible solutions) have been designed which can usually solve practical problems satisfactorily. For further discussion, we refer to, e.g., Applegate *et al.* (2006). WILLIAM J. COOK (born 1957) published an interesting book describing the history of the traveling salesmen problem, and the techniques that are used to solve this problem; see Cook (2012). See also Section 9.4.

Appendix D

Convexity

This appendix contains basic notions and results from the theory of convexity. We start with developing some properties of sets, and formulating the celebrated theorem of Karl Weierstrass (1815–1897). For a more extensive account of convexity, we refer to Grünbaum (2003).

D.1 Sets, continuous functions, Weierstrass' theorem

For $n \geq 1$, let $S \subseteq \mathbb{R}^n$. The *affine hull* of the set S is denoted and defined by:

$$\text{aff}(S) = \left\{ \mathbf{x} \in \mathbb{R}^n \;\middle|\; \begin{array}{l} \text{there exist points } \mathbf{s}_1, \ldots, \mathbf{s}_k \in S \text{ such that} \\ \mathbf{x} = \sum_{i=1}^{k} \lambda_i \mathbf{s}_i \text{ with } \sum_{i=1}^{k} \lambda_i = 1 \text{ and } \lambda_1, \ldots, \lambda_k \in \mathbb{R} \end{array} \right\}.$$

For example, $\text{aff}(\{\mathbf{v}\}) = \{\mathbf{v}\}$ for any $\mathbf{v} \in \mathbb{R}^n$. Moreover, if $\mathbf{a}$ and $\mathbf{b}$ are different points in $\mathbb{R}^2$, then $\text{aff}(\{\mathbf{a}, \mathbf{b}\}) = \{\lambda\mathbf{a} + (1 - \lambda)\mathbf{b} \mid \lambda \in \mathbb{R}\}$ is the line through $\mathbf{a}$ and $\mathbf{b}$. So, the affine hull of a set can be considered as the smallest linear extension of that set. See also Appendix B, Section B.1.

For any $\mathbf{a} \in \mathbb{R}^n$ and any $\varepsilon > 0$, the set:

$$B_\varepsilon(\mathbf{a}) = \{\mathbf{x} \in \mathbb{R}^n \mid \|\mathbf{x} - \mathbf{a}\| < \varepsilon\}$$

is called the *ε-ball* around $\mathbf{x}$. Let $S \subset \mathbb{R}^n$. The *closure* of S is denoted and defined by:

$$\text{cl}(S) = \{\mathbf{x} \in \mathbb{R}^n \mid S \cap B_\varepsilon(\mathbf{x}) \neq \emptyset \text{ for all } \varepsilon > 0\}.$$

The *interior* of S is denoted and defined by:

$$\text{int}(S) = \{\mathbf{x} \in \mathbb{R}^n \mid B_\varepsilon(\mathbf{x}) \subset S \text{ for some } \varepsilon > 0\}.$$

Note that $\text{int}(\{\lambda[1\ 0]^{\mathsf{T}} + (1-\lambda)[3\ 0]^{\mathsf{T}} \mid 0 \leq \lambda \leq 1\}) = \emptyset$. The *relative interior* of a set S in $\mathbb{R}^n$ is defined and denoted by:

$$\text{relint}(S) = \{\mathbf{x} \in \mathbb{R}^n \mid B_\varepsilon(\mathbf{x}) \cap \text{aff}(S) \subset S \text{ for some } \varepsilon > 0\}.$$

Note that the set

$$\text{relint}\left(\left\{\lambda[1\ 0]^{\mathsf{T}} + (1-\lambda)[3\ 0]^{\mathsf{T}} \,\middle|\, 0 \leq \lambda \leq 1\right\}\right)$$
$$= \text{conv}([1\ 0]^{\mathsf{T}}, [3\ 0]^{\mathsf{T}}) \setminus \left\{[1\ 0]^{\mathsf{T}}, [3\ 0]^{\mathsf{T}}\right\}$$

is nonempty. If $S = \text{int}(S)$, then the set S is called *open*; if $S = \text{cl}(S)$, then S is called *closed*. The point $\mathbf{a}$ is said to be in the *boundary* of S, if for each $\varepsilon > 0$, $B_\varepsilon(\mathbf{a})$ contains at least one point in S and at least one point not in S. The set S is called *bounded* if there exists $\mathbf{x} \in \mathbb{R}^n$ and a number $r > 0$ such that $S \subset B_r(\mathbf{x})$.

A sequence $(\mathbf{x}_1, \mathbf{x}_2, \ldots)$ of points is called *convergent* with *limit point* $\mathbf{x}$ if and only if for each $\varepsilon > 0$ there exists an integer M such that $\mathbf{x}_k \in B_\varepsilon(\mathbf{x})$ for each $k \geq M$. We use the notation $\lim_{k\to\infty} \mathbf{x}_k = \mathbf{x}$. The following theorem holds for bounded sets. Any sequence $(\mathbf{x}_1, \mathbf{x}_2, \ldots)$ of points in a bounded set S contains a convergent subsequence $(\mathbf{x}_{i(1)}, \mathbf{x}_{i(2)}, \ldots)$.

It can be shown that S is closed if and only if S contains all its boundary points. Another characterization is the following one. The set S is closed if and only if for any convergent sequence $(\mathbf{x}_1, \mathbf{x}_2, \ldots)$ of points in S, it holds that $\lim_{k\to\infty} \mathbf{x}_k \in S$. Moreover, S is open if and only if S does not contain any of its boundary points. A set may be neither open nor closed, and the only sets in $\mathbb{R}^n$ that are both open and closed are $\emptyset$ and $\mathbb{R}^n$ (why?).

A *(real-valued) function* f, defined on a subset S of $\mathbb{R}^n$, associates with each point $\mathbf{x}$ in S a real number $f(\mathbf{x})$; the notation $f: S \to \mathbb{R}$ means that the *domain* of f is S and that its *range* $f(S)$ is a subset of $\mathbb{R}$.

Let S be a nonempty subset of $\mathbb{R}^n$. The point $\mathbf{a} \in \mathbb{R}^n$ is called a *minimizer* of the function $f: S \to \mathbb{R}$, if $f(\mathbf{a}) \leq f(\mathbf{x})$ for every $\mathbf{x} \in S$. The set of all minimizers of f is denoted by $\text{argmin}\{f(\mathbf{x}) \mid \mathbf{x} \in S\}$. Similarly, $\mathbf{a}$ is called a *maximizer* of the function $f: S \to \mathbb{R}$, if $f(\mathbf{a}) \geq f(\mathbf{x})$ for every $\mathbf{x} \in S$. The set of all maximizers of f is denoted by $\text{argmax}\{f(\mathbf{x}) \mid \mathbf{x} \in S\}$. The function $f: S \to \mathbb{R}$ is called *bounded from below* (*bounded from above*) on S, if there is a real number α (β) such that $f(\mathbf{x}) \geq \alpha$ $(f(\mathbf{x}) \leq \beta)$ for each $\mathbf{x} \in S$. The *largest lower bound*, also called the *infimum*, of the function $f: S \to \mathbb{R}$ is defined as the largest real number L (if it exists) such that $L \leq f(\mathbf{x})$ for each $\mathbf{x} \in S$. We use the notation $L = \inf\{f(\mathbf{x}) \mid \mathbf{x} \in S\}$. If no such largest real number exists, then this is either because $f(\mathbf{x})$ takes on arbitrarily large negative values, or because S is the empty set. In the former case, $\inf\{f(\mathbf{x}) \mid \mathbf{x} \in S\}$ is defined to be equal to $-\infty$; in the latter case, it is defined to be equal to ∞. Similarly, the *least upper bound*, also called the *supremum*, of $f: S \to \mathbb{R}$ is defined as the smallest real number U (if it exists) such that

$f(\mathbf{x}) \le U$ for each $\mathbf{x} \in S$. We use the notation $U = \sup\{f(\mathbf{x}) \mid \mathbf{x} \in S\}$. If $f(\mathbf{x})$ takes on arbitrarily large (positive) values, then $\sup\{f(\mathbf{x}) \mid \mathbf{x} \in S\} = \infty$; if S is the empty set, then $\sup\{f(\mathbf{x}) \mid \mathbf{x} \in S\} = -\infty$.

Let $\mathbf{a} \in S$. The function $f : S \to \mathbb{R}$ is called *continuous* at $\mathbf{a}$, if for each $\varepsilon > 0$ there is a $\delta > 0$ such that for each $\mathbf{x} \in B_\delta(\mathbf{a})$, it holds that $|f(\mathbf{x}) - f(\mathbf{a})| < \varepsilon$. The function f is called *continuous on S* if f is continuous at every point of S. It can be shown that, if f is a continuous function on S, then for any convergent sequence $(\mathbf{x}_1, \mathbf{x}_2, \ldots)$ in S with limit point $\mathbf{x}$, it holds that $\lim_{k \to \infty} f(\mathbf{x}_k) = f(\mathbf{x})$.

With these definitions, we can formulate and prove the following famous theorem of KARL WEIERSTRASS (1815–1897).

Theorem D.1.1. (*Weierstrass' theorem*)
Let S be a nonempty closed bounded subset of $\mathbb{R}^n$, and $f : S \to \mathbb{R}$ a continuous function on S. Then, the function f attains a minimum and a maximum on S (meaning that $\operatorname{argmin}\{f(\mathbf{x}) \mid \mathbf{x} \in S\} \neq \emptyset \neq \operatorname{argmax}\{f(\mathbf{x}) \mid \mathbf{x} \in S\}$).

Proof of Theorem D.1.1. We only show that f attains a minimum on S. First note that, since f is a continuous function on the closed and bounded set S, it follows that f is bounded below on S. Moreover, since S is nonempty, the largest lower bound of f on S exists, say $\alpha = \inf\{f(\mathbf{x}) \mid \mathbf{x} \in S\}$. For any positive integer k and $0 < \varepsilon < 1$, define $S(\varepsilon, k) = \left\{\mathbf{x} \in S \mid \alpha \le f(\mathbf{x}) \le \alpha + \varepsilon^k\right\}$. Since α is the infimum of f on S, it follows that $S(\varepsilon, k) \neq \emptyset$. For each $k = 1, 2, \ldots$, select a point $\mathbf{x}_k \in S(\varepsilon, k)$. Since S is bounded, the sequence $(\mathbf{x}_1, \mathbf{x}_2, \ldots)$ contains a convergent subsequence, say $(\mathbf{x}_{i(1)}, \mathbf{x}_{i(2)}, \ldots)$; let $\hat{\mathbf{x}}$ be its limit point. The closedness of S implies that $\hat{\mathbf{x}} \in S$, and the continuity of f implies that $\alpha = \lim_{k \to \infty} f(\mathbf{x}_{i(k)}) = f(\hat{\mathbf{x}})$. Actually, we have shown that there is a point $\hat{\mathbf{x}} \in S$ such that $f(\hat{\mathbf{x}}) = \alpha = \inf\{f(\mathbf{x}) \mid \mathbf{x} \in S\}$. Hence, $\hat{\mathbf{x}} \in \operatorname{argmin}\{f(\mathbf{x}) \mid \mathbf{x} \in S\}$. $\qquad\square$

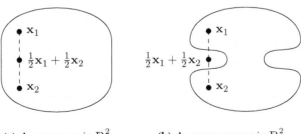

(a) A convex set in $\mathbb{R}^2$.   (b) A nonconvex set in $\mathbb{R}^2$.

Figure D.1: Examples of a convex and a nonconvex set in $\mathbb{R}^2$.

D.2   Convexity, polyhedra, polytopes, and cones

We use the following notation for *line segments* in $\mathbb{R}$, also called *intervals*. Let a and b be real numbers with $a < b$.

$$[a, b] = \{\lambda a + (1 - \lambda)b \mid 0 \leq \lambda \leq 1\},$$
$$(a, b] = \{\lambda a + (1 - \lambda)b \mid 0 \leq \lambda < 1\}.$$

The intervals $[a, b)$ and (a, b) are defined analogously. We also define:

$$(-\infty, a] = \{x \in \mathbb{R} \mid x \leq a\}, \text{ and } [a, \infty) = \{x \in \mathbb{R} \mid x \geq a\}.$$

Again, (∞, a) and (a, ∞) are defined analogously. Notice that $(-\infty, \infty) = \mathbb{R}$.

A set S in the Euclidean vector space (see Appendix B) $\mathbb{R}^n$ is said to be *convex* if for any two points $\mathbf{x}_1$ and $\mathbf{x}_2$ in S also the line segment $[\mathbf{x}_1, \mathbf{x}_2] = \{\lambda \mathbf{x}_1 + (1 - \lambda)\mathbf{x}_2 \mid 0 \leq \lambda \leq 1\}$, joining $\mathbf{x}_1$ and $\mathbf{x}_2$, belongs to S. Thus, S is convex if and only if $\mathbf{x}_1, \mathbf{x}_2 \in S \implies [\mathbf{x}_1, \mathbf{x}_2] \subset S$. Figure D.1(a) shows an example of a convex set; Figure D.1(b) shows an example of a nonconvex set.

Let $k \geq 1$, and let $\mathbf{x}_1, \ldots, \mathbf{x}_k$ be points in $\mathbb{R}^n$. For any set of scalars $\{\lambda_1, \ldots, \lambda_k\}$ with $\lambda_1, \ldots, \lambda_k \geq 0$ and $\lambda_1 + \ldots + \lambda_k = 1$, $\lambda_1 \mathbf{x}_1 + \ldots + \lambda_k \mathbf{x}_k$ is called a *convex combination* of $\mathbf{x}_1, \ldots, \mathbf{x}_k$. Moreover, the *convex hull* of the points $\mathbf{x}_1, \ldots, \mathbf{x}_k$ is denoted and defined by:

$$\text{conv}(\{\mathbf{x}_1, \ldots, \mathbf{x}_k\}) = \left\{ \sum_{i=1}^{k} \lambda_i \mathbf{x}_i \,\middle|\, \lambda_i \geq 0 \text{ for } i = 1, \ldots, k, \text{ and } \sum_{i=1}^{k} \lambda_i = 1 \right\};$$

i.e., the convex hull of a set S is the set of all convex combinations of the points of S. Instead of $\text{conv}(\{\cdot\})$, we will also write $\text{conv}(\cdot)$. Note that *singletons* (sets consisting of one point), as well as the whole space $\mathbb{R}^n$ are convex. The empty set $\emptyset$ is convex by definition. Some more convex sets are introduced below.

The set $H = \{\mathbf{x} \in \mathbb{R}^n \mid \mathbf{a}^\mathsf{T} \mathbf{x} = b\}$ with $\mathbf{a} \neq \mathbf{0}$ is called a *hyperplane* in $\mathbb{R}^n$; the vector $\mathbf{a}$ is called the *normal vector* of H. Note that for any two points $\mathbf{x}_1$ and $\mathbf{x}_2$ in the hyperplane H, it holds that $\mathbf{a}^\mathsf{T}(\mathbf{x}_1 - \mathbf{x}_2) = b - b = 0$, and so $\mathbf{a}$ is perpendicular to H; see Appendix B. A set of hyperplanes is called *linearly dependent* (*linearly independent*, respectively) if the corresponding normal vectors are linearly dependent (independent, respectively). With each hyperplane $H = \{\mathbf{x} \in \mathbb{R}^n \mid \mathbf{a}^\mathsf{T} \mathbf{x} = b\}$ are associated two (closed) *halfspaces*:

$$H^+ = \{\mathbf{x} \in \mathbb{R}^n \mid \mathbf{a}^\mathsf{T} \mathbf{x} \leq b\} \text{ and } H^- = \{\mathbf{x} \in \mathbb{R}^n \mid \mathbf{a}^\mathsf{T} \mathbf{x} \geq b\}.$$

Note that $H^+ \cap H^- = H$, and that $H^+ \cup H^- = \mathbb{R}^n$. See also Section 2.1.1.

The intersection of a finite number of halfspaces is called a *polyhedral set*, or a *polyhedron*. A nonnegative polyhedral set F in $\mathbb{R}^n$ ($n \geq 1$) can be written as:

$$F = \{\mathbf{x} \in \mathbb{R}^n \mid \mathbf{A}\mathbf{x} \leq \mathbf{b}, \mathbf{x} \geq \mathbf{0}\},$$

with $\mathbf{A}$ an (m, n) matrix, and $\mathbf{b} \in \mathbb{R}^m$. For $m = 0$, we define $F = \mathbb{R}^n$. In the context of linear optimization, this is called a *feasible region* (see Section 1.2). The convex hull of a finite number of points is called a *polytope*. It can be shown that a polytope is a bounded polyhedral set.

A nonempty set K in $\mathbb{R}^n$ is called a *cone* with apex $\mathbf{0}$, if $\mathbf{x} \in K$ implies that $\lambda\mathbf{x} \in K$ for all $\lambda \geq 0$; if, in addition, K is convex, then K is called a *convex cone*. A cone K is *pointed* if $K \cap (-K) = \{\mathbf{0}\}$. For any set of vectors $\mathbf{a}_1, \ldots, \mathbf{a}_p$ in $\mathbb{R}^n$, we define:

$$\text{cone}(\{\mathbf{a}_1, \ldots, \mathbf{a}_p\}) = \left\{\sum_{i=1}^{p} \lambda_i \mathbf{a}_i \,\middle|\, \lambda_i \geq 0 \text{ for } i = 1, \ldots, p\right\}.$$

A *polyhedral cone* in $\mathbb{R}^n$ is a set of the form $\{\mathbf{x} \in \mathbb{R}^n \mid \mathbf{A}\mathbf{x} \leq \mathbf{0}\}$, with $\mathbf{A}$ an (m, n)-matrix; $m, n \geq 1$. The *polar cone* of any nonempty set S in $\mathbb{R}^n$ is denoted and defined by:

$$S^0 = \{\mathbf{x} \in \mathbb{R}^n \mid \mathbf{x}^\top\mathbf{y} \leq 0 \text{ for all } \mathbf{y} \in S\};$$

if S is empty, then by definition $S^0 = \mathbb{R}^n$. Using the cosine-rule (see Appendix B), it follows that for any $\mathbf{x} \in S^0$ and any $\mathbf{y} \in S$, it holds that $\mathbf{x}^\top\mathbf{y} = \|\mathbf{x}\| \|\mathbf{y}\| \cos \alpha \leq 0$, where α is the angle between $\mathbf{x}$ and $\mathbf{y}$. Since $\|\mathbf{x}\| \geq 0$ and $\|\mathbf{y}\| \geq 0$, it follows that $\cos \alpha \leq 0$, which means that the angle between $\mathbf{x}$ and $\mathbf{y}$ satisfies $\frac{1}{2}\pi \leq \alpha \leq 1\frac{1}{2}\pi$.

It is left to the reader to show that hyperplanes, halfspaces, polyhedral sets and cones, polytopes, convex cones, and polar cones are in fact all convex sets. Moreover, if S is convex, then so are $\text{int}(S)$ and $\text{cl}(S)$.

Let X and Y be two sets in $\mathbb{R}^n$. A hyperplane $\{\mathbf{x} \in \mathbb{R}^n \mid \mathbf{w}^\top\mathbf{x} + b = 0\}$ is called a *separating hyperplane* for X and Y if $\mathbf{w}^\top\mathbf{x} + b > 0$ for all $\mathbf{x} \in X$ and $\mathbf{w}^\top\mathbf{y} + b < 0$ for all $\mathbf{y} \in Y$ (or vice versa). The following theorem shows that if X and Y are disjoint closed bounded nonempty convex sets, such a separating hyperplane always exists.

Theorem D.2.1.
Let $X, Y \subseteq \mathbb{R}^n$ be closed bounded nonempty convex sets such that $X \cap Y \neq \emptyset$. Then, there exist $\mathbf{w} \in \mathbb{R}^n$ and $b \in \mathbb{R}$ such that $\mathbf{w}^\top\mathbf{x} + b > 0$ for all $\mathbf{x} \in X$ and $\mathbf{w}^\top\mathbf{y} + b < 0$ for all $\mathbf{y} \in Y$.

Proof. Because X and Y are closed and bounded nonempty sets, the Cartesian product (see Section D.4) $X \times Y = \{(\mathbf{x}, \mathbf{y}) \mid \mathbf{x} \in X, \mathbf{y} \in Y\}$ is closed, bounded, and nonempty as well. Define $f: X \times Y \to \mathbb{R}$ by $f(\mathbf{x}, \mathbf{y}) = \|\mathbf{x} - \mathbf{y}\|$. Then, by Weierstrass' Theorem (Theorem

D.1.1), there exist $\mathbf{x}^* \in X$ and $\mathbf{y}^* \in Y$ such that $f(\mathbf{x}^*, \mathbf{y}^*) = \|\mathbf{x}^* - \mathbf{y}^*\|$ is minimized. If $\|\mathbf{x}^* - \mathbf{y}^*\| = 0$, then $\mathbf{x}^* = \mathbf{y}^*$, contrary to the fact that $X \cap Y \neq \emptyset$. Therefore, $\|\mathbf{x}^* - \mathbf{y}^*\| > 0$. Define $\mathbf{w} = 2(\mathbf{x}^* - \mathbf{y}^*)$ and $b = (\mathbf{y}^*)^\mathsf{T}\mathbf{y}^* - (\mathbf{x}^*)^\mathsf{T}\mathbf{x}^*$. By substituting these expressions into $\mathbf{w}^\mathsf{T}\mathbf{x}^* + b$, we find that:

$$\begin{aligned}\mathbf{w}^\mathsf{T}\mathbf{x}^* + b &= (\mathbf{x}^*)^\mathsf{T}\mathbf{x}^* + (\mathbf{y}^*)^\mathsf{T}\mathbf{y}^* - 2(\mathbf{y}^*)^\mathsf{T}\mathbf{x}^* \\ &= (\mathbf{x}^* - \mathbf{y}^*)^\mathsf{T}(\mathbf{x}^* - \mathbf{y}^*) = \|\mathbf{x}^* - \mathbf{y}^*\| > 0.\end{aligned}$$

Now, suppose for a contradiction that there exists $\mathbf{x}' \in X$ with $\mathbf{w}^\mathsf{T}\mathbf{x}' + b \leq 0$. For $\varepsilon \in (0,1)$, define $\mathbf{x}''(\varepsilon) = \mathbf{x}^* + \varepsilon\mathbf{d}$ with $\mathbf{d} = \mathbf{x}' - \mathbf{x}^*$. Observe that, because X is convex, $\mathbf{x}''(\varepsilon) \in X$ for all $\varepsilon \in (0,1)$. We have that:

$$\begin{aligned}(\mathbf{x}''(\varepsilon) - \mathbf{y}^*)^\mathsf{T}(\mathbf{x}''(\varepsilon) - \mathbf{y}^*) &= (\mathbf{x}^* - \mathbf{y}^* + \varepsilon\mathbf{d})^\mathsf{T}(\mathbf{x}^* - \mathbf{y}^* + \varepsilon\mathbf{d}) &&\text{(D.1)}\\ &= (\mathbf{x}^* - \mathbf{y}^*)^\mathsf{T}(\mathbf{x}^* - \mathbf{y}^*) + 2\varepsilon(\mathbf{x}^* - \mathbf{y}^*)^\mathsf{T}\mathbf{d} + \varepsilon^2\mathbf{d}^\mathsf{T}\mathbf{d}.\end{aligned}$$

From the facts that $\mathbf{w}^\mathsf{T}\mathbf{x}^* + b > 0$ and $\mathbf{w}^\mathsf{T}\mathbf{x}' + b \leq 0$, it follows that $(\mathbf{x}^* - \mathbf{y}^*)^\mathsf{T}\mathbf{d} = \mathbf{w}^\mathsf{T}(\mathbf{x}' - \mathbf{x}^*) < 0$. Therefore, for $\varepsilon > 0$ small enough, (D.1) implies that

$$(\mathbf{x}''(\varepsilon) - \mathbf{y}^*)^\mathsf{T}(\mathbf{x}''(\varepsilon) - \mathbf{y}^*) < (\mathbf{x}^* - \mathbf{y}^*)^\mathsf{T}(\mathbf{x}^* - \mathbf{y}^*),$$

contrary to the choice of $\mathbf{x}^*$ and $\mathbf{y}^*$. This proves that $\mathbf{w}^\mathsf{T}\mathbf{x} + b > 0$ for all $\mathbf{x} \in X$. The proof that $\mathbf{w}^\mathsf{T}\mathbf{y} + b < 0$ for all $\mathbf{y} \in Y$ is similar and is left to the reader. □

The requirement in Theorem D.2.1 that X and Y are closed and bounded is necessary. Indeed, the sets

$$X = \left\{\begin{bmatrix} x_1 \\ x_2 \end{bmatrix} \in \mathbb{R}^2 \;\middle|\; x_1 \leq 0\right\} \text{ and } Y = \left\{\begin{bmatrix} x_1 \\ x_2 \end{bmatrix} \in \mathbb{R}^2 \;\middle|\; x_2 > 0, x_1 \geq \frac{1}{x_2}\right\}$$

are both closed and convex (but not bounded), but there does not exist a separating hyperplane for X and Y. It can be shown, however, that it suffices that one of the sets X, Y is closed and bounded.

D.3 Faces, extreme points, and adjacency

Let C be a convex set in $\mathbb{R}^n$ of dimension d ($0 \leq d \leq n$), and let k satisfy $0 \leq k \leq d-1$. A subset S (of dimension k) in C is called a $(k\text{-})face$ of C if there exists a hyperplane H in $\mathbb{R}^n$ such that $C \subset H^+$ and $C \cap H = S$, where H^+ is the halfspace corresponding to H; an $(n-1)$-face is called a *facet*. The vector $\mathbf{v}$ is called an *extreme point* or *vertex* of the convex set C if $\mathbf{v}$ is a 0-face of C. Two vertices $\mathbf{u}$ and $\mathbf{v}$ of a convex set C are called *adjacent* on C if $\mathrm{conv}(\mathbf{u}, \mathbf{v})$ is a 1-face of C.

In Section 2.1.2, a different definition of extreme point is used. In Theorem D.3.1, we give three equivalent definitions of extreme point.

Theorem D.3.1.

Let $H_1, \ldots, H_p$ be hyperplanes in $\mathbb{R}^n$. Let $p \geq n$, and let $F = \bigcap_{i=1}^{p} H_i^+ (\neq \emptyset)$. For each $\mathbf{v} \in F$, the following assertions are equivalent:

(i) $\mathbf{v}$ is a 0-face of F;

(ii) $\mathbf{v}$ is not a convex combination of two other points of F;

(iii) There exist n independent hyperplanes $H_{i(1)}, \ldots, H_{i(n)}$ (among the hyperplanes $H_1, \ldots, H_p$) corresponding to n halfspaces $H_{i(1)}^+, \ldots, H_{i(n)}^+$, respectively, such that

$$\{\mathbf{v}\} = \bigcap_{k=1}^{n} H_{i(k)}.$$

$H_{i(1)}, \ldots, H_{i(n)}$ are called hyperplanes that *determine* $\mathbf{v}$.

Proof of Theorem D.3.1. (i) $\implies$ (ii): Assume to the contrary that there are points $\mathbf{u}$ and $\mathbf{w}$ in F, both different from $\mathbf{v}$, such that $\mathbf{v} \in \text{conv}(\mathbf{u}, \mathbf{w})$. We may choose $\mathbf{u}$ and $\mathbf{w}$ such that $\mathbf{v} = \frac{1}{2}(\mathbf{u} + \mathbf{w})$. Since $\mathbf{v}$ is a 0-face of F, there is a hyperplane H such that $F \subset H^+$ and $H \cap F = \{\mathbf{v}\}$. Let $H^+ = \left\{ \mathbf{x} \mid \mathbf{a}^\mathsf{T}\mathbf{x} \leq b \right\}$. Assume that $\mathbf{u} \notin H$. If $\mathbf{a}^\mathsf{T}\mathbf{u} < b$, then $\mathbf{a}^\mathsf{T}\mathbf{v} = \mathbf{a}^\mathsf{T}(\frac{1}{2}(\mathbf{u} + \mathbf{v})) = \frac{1}{2}\mathbf{a}^\mathsf{T}\mathbf{u} + \frac{1}{2}\mathbf{a}^\mathsf{T}\mathbf{v} < \frac{1}{2}b + \frac{1}{2}b = b$. This contradicts the fact that $\mathbf{v} \in H$. For $\mathbf{a}^\mathsf{T}\mathbf{u} > b$, the same contradiction is obtained. Hence, $\mathbf{u} \in H$. But this contradicts the fact that $\mathbf{v}$ is the only point in $H \cap F$.

(ii) $\implies$ (iii): Let $F = \{\mathbf{x} \mid \mathbf{A}\mathbf{x} \leq \mathbf{b}\}$, with the rows of $\mathbf{A}$ being the normal vectors of the hyperplanes $H_1, \ldots, H_p$. Let $\mathbf{A}_{(v)}$ be the submatrix of $\mathbf{A}$ consisting of the rows of $\mathbf{A}$ that correspond to the hyperplanes H_i that are binding at $\mathbf{v}$; i.e., if $\mathbf{a}_i$ is a row of $\mathbf{A}_{(v)}$ then $\mathbf{a}_i^\mathsf{T}\mathbf{v} = b_i$, and if $\mathbf{a}_i$ is a row of $\mathbf{A}$ that is not a row of $\mathbf{A}_{(v)}$ then $\mathbf{a}_i^\mathsf{T}\mathbf{v} < b_i$, where $H_i = \left\{ \mathbf{x} \mid \mathbf{a}_i^\mathsf{T}\mathbf{x} = b_i \right\}$ is the corresponding hyperplane. Let I be the index set of the rows of $\mathbf{A}$ that are in $\mathbf{A}_{(v)}$, and J the index set of the rows of $\mathbf{A}$ that are not in $\mathbf{A}_{(v)}$. Note that $|I \cup J| = p$. We will show that $\text{rank}(\mathbf{A}_{(v)}) = n$. Assume, to the contrary, that $\text{rank}(\mathbf{A}_{(v)}) \leq n - 1$. This means that the columns of $\mathbf{A}_{(v)}$ are linearly dependent; see Appendix B. Hence, there exists a vector $\boldsymbol{\lambda} \neq \mathbf{0}$ such that $\mathbf{A}_{(v)}\boldsymbol{\lambda} = \mathbf{0}$. Since $\mathbf{a}_i^\mathsf{T}\mathbf{v} < b_i$ for each $i \in J$, there is a (small enough) $\varepsilon > 0$ such that $\mathbf{a}_i^\mathsf{T}(\mathbf{v} + \varepsilon\boldsymbol{\lambda}) \leq b_i$ and $\mathbf{a}_i^\mathsf{T}(\mathbf{v} - \varepsilon\boldsymbol{\lambda}) \leq b_i$ for $i \in J$. On the other hand, since $\mathbf{a}_i\boldsymbol{\lambda} = 0$ for each $i \in I$, it follows that $\mathbf{a}_i^\mathsf{T}(\mathbf{v} + \varepsilon\boldsymbol{\lambda}) \leq b_i$ and $\mathbf{a}_i^\mathsf{T}(\mathbf{v} - \varepsilon\boldsymbol{\lambda}) \leq b_i$ for $i \in I$. Hence, $\mathbf{A}(\mathbf{v} + \varepsilon\boldsymbol{\lambda}) \leq \mathbf{b}$ and $\mathbf{A}(\mathbf{v} + \varepsilon\boldsymbol{\lambda}) \leq \mathbf{b}$, and so $\mathbf{v} + \varepsilon\boldsymbol{\lambda} \in F$ and $\mathbf{v} - \varepsilon\boldsymbol{\lambda} \in F$. Since $\mathbf{v}$ is a convex combination of $\mathbf{v} + \varepsilon\boldsymbol{\lambda}$ and $\mathbf{v} - \varepsilon\boldsymbol{\lambda}$, we have obtained a contradiction with (ii). Therefore, $\text{rank}(\mathbf{A}_{(v)}) = n$. Let $H_{i(1)}, \ldots, H_{i(n)}$ be hyperplanes corresponding to n independent rows of $\mathbf{A}_{(v)}$. Since n independent hyperplanes in an n-dimensional space can intersect in one point, and we know that $\mathbf{v} \in H_{i(1)}, \ldots, \mathbf{v} \in H_{i(n)}$, it follows that $H_{i(1)}, \ldots, H_{i(n)}$ determine $\mathbf{v}$. This proves (iii).

(iii) $\implies$ (i): Let $\mathbf{a}_1, \ldots, \mathbf{a}_n$ be the linearly independent normals of $H_{i(1)}, \ldots, H_{i(n)}$, respectively. Choose a vector $\mathbf{c}$ in the interior of cone$\{\mathbf{a}_1, \ldots, \mathbf{a}_n\}$ and scalars $\lambda_1, \ldots, \lambda_n > 0$ with $\lambda_1 + \ldots + \lambda_n = 1$, and such that $\mathbf{c} = \lambda_1\mathbf{a}_1 + \ldots + \lambda_n\mathbf{a}_n$. Define $H = \left\{\mathbf{x} \in \mathbb{R}^n \mid \mathbf{c}^\mathsf{T}\mathbf{x} = \mathbf{c}^\mathsf{T}\mathbf{v}\right\}$. We will show that $F \cap H = \{\mathbf{v}\}$. Take any $\mathbf{v}' \in F \cap H$. Then $\mathbf{c}^\mathsf{T}\mathbf{v}' = \mathbf{c}^\mathsf{T}\mathbf{v}$, and so $0 = \mathbf{c}^\mathsf{T}(\mathbf{v}' - \mathbf{v}) = (\lambda_1\mathbf{a}_1 + \ldots + \lambda_n\mathbf{a}_m)^\mathsf{T}(\mathbf{v}' - \mathbf{v}) = \lambda_1(\mathbf{a}_1^\mathsf{T}(\mathbf{v}' - \mathbf{v})) + \ldots + \lambda_n(\mathbf{a}_n^\mathsf{T}(\mathbf{v}' - \mathbf{v}))$. Since $\mathbf{v}' \in F$, it follows that $\mathbf{a}_j^\mathsf{T}\mathbf{v}' \leq b_j$, and hence $\mathbf{a}_j^\mathsf{T}(\mathbf{v}' - \mathbf{v}) \leq b_j - b_j = 0$ for all $j = 1, \ldots, n$. Moreover, since $\lambda_1, \ldots, \lambda_n > 0$, it follows that $\mathbf{a}_j^\mathsf{T}(\mathbf{v}' - \mathbf{v}) = 0$ for all $j = 1, \ldots, n$. Hence, both $\mathbf{v}'$ and $\mathbf{v}$ are in $H_{i(1)}, \ldots, H_{i(n)}$. But this implies that conv$(\mathbf{v}', \mathbf{v}) \subset \bigcap_{k=1}^n H_{i(k)}$. Since $\dim(\bigcap_{k=1}^n H_{i(k)}) = 0$, it follows that $\mathbf{v}' = \mathbf{v}$. This proves the theorem. $\square$

It follows from Theorem D.1.1 that $\mathbf{v}$ is an extreme point of $F = \{\mathbf{x} \in \mathbb{R}^n \mid \mathbf{A}\mathbf{x} \leq \mathbf{b}\}$ if and only if the system $\mathbf{A}\mathbf{x} = \mathbf{b}$ contains a subsystem $\mathbf{A}'\mathbf{x} = \mathbf{b}'$ that has $\mathbf{v}$ as its unique solution. In Section 2.2.5, a definition for adjacency has been given that is different from the above one. Theorem D.3.1 shows that both definitions are equivalent.

Theorem D.3.2.
Let $H_1, \ldots, H_p$ be hyperplanes in $\mathbb{R}^n$. Let $p \geq n + 1$, and let $F = \bigcap_{i=1}^p H_i^+ (\neq \emptyset)$. For each $\mathbf{u}, \mathbf{v} \in F$, the following assertions are equivalent:

(i) $\mathbf{u}$ and $\mathbf{v}$ are adjacent extreme points of F;

(ii) There exist $n+1$ hyperplanes $H_{i(1)}, H_{i(2)}, \ldots, H_{i(n)}, H_{i(n+1)}$, among $H_1, \ldots, H_p$, such that $H_{i(1)}, \ldots, H_{i(n)}$ determine $\mathbf{u}$, and $H_{i(2)}, \ldots, H_{i(n+1)}$ determine $\mathbf{v}$.

Proof of Theorem D.3.2. (i) $\Rightarrow$ (ii): As in the proof of Theorem D.3.1, let $\mathbf{A}$ be the matrix corresponding to the hyperplanes $H_1, \ldots, H_p$. Let $\mathbf{A}_{(u,v)}$ be the submatrix of $\mathbf{A}$ consisting of the rows of A that correspond to the hyperplanes that are binding at both $\mathbf{u}$ and $\mathbf{v}$. Similar to the proof of (ii) $\Rightarrow$ (iii) of Theorem D.3.1, it follows that rank$(\mathbf{A}_{(u,v)}) = n - 1$. So, there are $n - 1$ independent hyperplanes $H_{i(2)}, \ldots, H_{i(n)}$ among $H_1, \ldots, H_p$ that are binding at both $\mathbf{u}$ and $\mathbf{v}$. Since $\mathbf{u}$ and $\mathbf{v}$ are different extreme points of F, two more hyperplanes exist: one to determine $\mathbf{u}$, and one to determine $\mathbf{v}$; say $H_{i(1)}$ and $H_{i(n+1)}$, respectively. I.e., the set $\{H_{i(1)}, \ldots, H_{i(n)}\}$ determines $\mathbf{u}$, and the set $\{H_{i(2)}, \ldots, H_{i(n+1)}\}$ determines $\mathbf{v}$.

(ii) $\Rightarrow$ (i): This proof is similar to the proof of (iii) $\Rightarrow$ (i) in Theorem D.3.1. The vector $\mathbf{c}$ is chosen as follows. Let $\mathbf{a}_2, \ldots, \mathbf{a}_n$ be the normals of the $n-1$ independent hyperplanes that are binding at both $\mathbf{u}$ and $\mathbf{v}$. Choose $\mathbf{c}$ in the interior of cone$(\mathbf{a}_2, \ldots, \mathbf{a}_n)$. It then can be shown that the hyperplane $H = \left\{\mathbf{x} \mid \mathbf{c}^\mathsf{T}\mathbf{x} = \mathbf{c}^\mathsf{T}\mathbf{u} = \mathbf{c}^\mathsf{T}\mathbf{v}\right\}$ satisfies $F \cap H = \text{conv}(\mathbf{u}, \mathbf{v})$. $\square$

In Section 2.1.3, we stated that a face of a face of the feasible region of an LO-model is a face of the feasible region itself. The following theorem makes this statement precise.

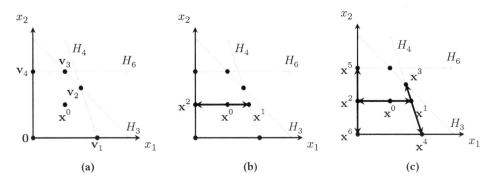

Figure D.2: Writing the vertex $\mathbf{x}^0$ as a convex combination of vertices of F.

Theorem D.3.3.
Let F be a polyhedral set in $\mathbb{R}^n$. Then any face F_J of F is a polyhedral set, and any face of F_J is also a face of F.

Proof. Let $\mathcal{H}^+ = \{H_1, \ldots, H_p\}$ be the set of halfspaces that determine the polyhedral set F, and let $\mathcal{H}$ be the set of corresponding hyperplanes. Let F_J be a face of F. Moreover, let $\mathcal{H}_J \subseteq \mathcal{H}$ be the set of hyperplanes that determine F_J in F, meaning that for each $H \in \mathcal{H}_J$, it holds that $F_J \subseteq H$. Hence, $F_J = F \cap (\bigcap \mathcal{H}_J)$. Let $\mathcal{H}_J^+$ be the set of halfspaces H^+ corresponding to $H \in \mathcal{H}_J$, and let $\mathcal{H}_J^-$ be the set of halfspaces H^- corresponding to $H \in \mathcal{H}_J$. Since each hyperplane in $\mathcal{H}_J$ is the intersection of its two corresponding halfspaces, we have that $F_J = (\bigcap \mathcal{H}^+) \cap (\bigcap \mathcal{H}_J^+) \cap (\bigcap \mathcal{H}_J^-)$. Therefore, F_J is a polyhedral set. Moreover, the set $\mathcal{H}$ contains all hyperplanes corresponding to the halfspaces that determine F_J.

Now assume that F_I is a face of the polyhedral set F_J. Let $\mathcal{H}_I$ be the set of hyperplanes with $\mathcal{H}_I \subseteq \mathcal{H}$ that determine F_I in F_J. Hence, $F_I = F_J \cap (\bigcap \mathcal{H}_I)$. Define $\mathcal{H}_{I \cup J} = \mathcal{H}_I \cup \mathcal{H}_J$. Then, we have that $F_I = (F \cap (\bigcap \mathcal{H}_J)) \cap (\bigcap \mathcal{H}_I) = F \cap (\bigcap \mathcal{H}_{I \cup J})$, so that F_I is in fact a face of F. □

In the remainder of this section, we give a proof of Theorem 2.1.2 for the case when the feasible region F is bounded. This is the 'if' part of Theorem 2.1.6. The following example illustrates the main idea behind the proof.

Example D.3.1. *Consider the feasible region F of Model Dovetail, and suppose that we want to write the point $\mathbf{x}^0 = \begin{bmatrix} 3 & 3 \end{bmatrix}^\mathsf{T}$ as a convex combination of the vertices of F; see Figure D.2(a). We may choose any vector $\mathbf{z}^0 \in \mathbb{R}^2 \setminus \{\mathbf{0}\}$, and move away from $\mathbf{x}^0$ in the direction $\mathbf{z}^0$, i.e., we move along the halfline $\mathbf{x}^0 + \alpha \mathbf{z}^0$ with $\alpha \geq 0$. For instance, let $\mathbf{z}^0 = \begin{bmatrix} 1 & 0 \end{bmatrix}^\mathsf{T}$. We move away from $\mathbf{x}^0$ in the direction $\mathbf{z}^0$ as far as possible, while staying inside the feasible region F. Doing that, we arrive at the point $\mathbf{x}^1 = \mathbf{x}^0 + 2\mathbf{z}^0 = \begin{bmatrix} 5 & 3 \end{bmatrix}^\mathsf{T}$, where we 'hit' the hyperplane H_4 and we cannot move* *any further. Note that $\mathbf{x}^1$ is a point of the face $F_{\{4\}}$, which is the line segment $\mathbf{v}_1\mathbf{v}_2$. The point $\mathbf{x}^1$ along with the vector $2\mathbf{z}^0$ originating from $\mathbf{x}^0$ is shown in Figure D.2(b). Similarly, we can move away as far as possible from $\mathbf{x}^0$ in the direction $-\mathbf{z}^0 = \begin{bmatrix} -1 & 0 \end{bmatrix}^{\mathsf{T}}$, until we hit the hyperplane H_2, arriving at point $\mathbf{x}^2 = \mathbf{x}^0 - 3\mathbf{z}^0 = \begin{bmatrix} 0 & 3 \end{bmatrix}^{\mathsf{T}}$. Note that $\mathbf{x}^2$ is a point of the face $F_{\{2\}}$, which is the line segment $\mathbf{0}\mathbf{v}_4$.*

Now we have two new points $\mathbf{x}^1$ and $\mathbf{x}^2$, with the property that $\mathbf{x}^0$, the original point of interest, lies on the line segment $\mathbf{x}^1\mathbf{x}^2$. To be precise, since $\mathbf{x}^1 = \mathbf{x}^0 + 2\mathbf{z}^0$ and $\mathbf{x}^2 = \mathbf{x}^0 - 3\mathbf{z}^0$, it follows by eliminating $\mathbf{z}^0$ that $3\mathbf{x}^1 + 2\mathbf{x}^2 = 5\mathbf{x}^0$ or, equivalently, that $\mathbf{x}^0 = \frac{3}{5}\mathbf{x}^1 + \frac{2}{5}\mathbf{x}^2$, which means that $\mathbf{x}^0$ is in fact a convex combination of $\mathbf{x}^1$ and $\mathbf{x}^2$. The next step is to repeat this process for the two points $\mathbf{x}^1$ and $\mathbf{x}^2$.

So consider $\mathbf{x}^1$. Again, we choose a vector $\mathbf{z}_1 \in \mathbb{R}^2 \setminus \{\mathbf{0}\}$ and move away from $\mathbf{x}^1$ in the direction $\mathbf{z}^1$. Now, however, since $\mathbf{x}^1$ is a point of the face $F_{\{2\}}$, we will make sure that, as we move in the direction $\mathbf{z}^1$, we stay inside the face $F_{\{2\}}$. Since every point in this face satisfies the equation $3x_1 + x_2 = 18$, this means that $\mathbf{z}^1 = \begin{bmatrix} z_1^1 & z_2^1 \end{bmatrix}^{\mathsf{T}}$ should satisfy:

$$3(x_1^1 + z_1^1) + (x_2^1 + z_2^1) = 18 \text{ or, equivalently, } 3z_1^1 + z_2^1 = 0.$$

Therefore, we choose $\mathbf{z}^1 = \begin{bmatrix} -1 & 3 \end{bmatrix}^{\mathsf{T}}$. As we move from $\mathbf{x}^1$ in the direction $\mathbf{z}^1$, we hit the hyperplane H_3, arriving at the point $\mathbf{x}^3 = \begin{bmatrix} 4\frac{1}{2} & 4\frac{1}{2} \end{bmatrix}^{\mathsf{T}}$; see Figure D.2(c). Similarly, moving from $\mathbf{x}^1$ in the direction $-\mathbf{z}^1$, we hit the hyperplane H_1, arriving at the point $\mathbf{x}^4 = \begin{bmatrix} 6 & 0 \end{bmatrix}^{\mathsf{T}}$. The point $\mathbf{x}^1$ lies on the line segment $\mathbf{x}^3\mathbf{x}^4$, so that $\mathbf{x}^1$ can be written as a convex combination of $\mathbf{x}^3$ and $\mathbf{x}^4$. Note that $\mathbf{x}^3$ is the unique point in $F_{\{3,4\}}$, and hence $\mathbf{x}^3$ is a vertex of $F_{\{3,4\}}$. From Theorem D.3.3, it follows that $\mathbf{x}^3$ is also a vertex of F. Similarly, $\mathbf{x}^4$ is the unique point in $F_{\{1,4\}}$ and hence $\mathbf{x}^4$ is a vertex of F. Therefore, $\mathbf{x}^1$ is a convex combination of two vertices of F. In a similar way, $\mathbf{x}^2$ is a convex combination of two vertices of F, namely $\mathbf{0}$ and $\mathbf{v}_4$.

So we have established that $\mathbf{x}^1$ and $\mathbf{x}^2$ are convex combinations of vertices of F. Because $\mathbf{x}^0$ is a convex combination of $\mathbf{x}^1$ and $\mathbf{x}^2$, it is straightforward to check that $\mathbf{x}^0$ is a convex combination of vertices of F. In Exercise 2.3.12, the reader is asked to work out the values of $\lambda_1, \ldots, \lambda_5 \geq 0$ with $\sum_{i=1}^{5} \lambda_i$ such that $\mathbf{x}^0 = \lambda\mathbf{v}_1 + \ldots + \lambda_4\mathbf{v}_4 + \lambda_5\mathbf{0}$.

We are now ready to give a proof of Theorem 2.1.2. This proof generalizes the construction method described in the example above.

Proof of Theorem 2.1.2. (Bounded case)   As in Section 2.1.3, we may write the feasible region of any standard LO-model as:

$$F = \left\{ \mathbf{x} \in \mathbb{R}^n \mid \mathbf{a}_i^{\mathsf{T}}\mathbf{x} \leq b_i, i = 1, \ldots, n+m \right\}.$$

The assumption that F is bounded implies that F_J is bounded for all $J \subseteq \{1, \ldots, n+m\}$. We will prove by induction on $|J|$ that for any $J \subseteq \{1, \ldots, n+m\}$:

Any $\hat{\mathbf{x}} \in F_J$ can be written as a convex combination of vertices of F_J. $(\star)$

Note that the choice $J = \emptyset$ is equivalent to the statement of the theorem.

We use induction to prove ($\star$). As the base case we take $J = \{1, \ldots, n+m\}$. Because F is the feasible region of a standard LO-model, we have that:

$$F_{\{1,\ldots,n+m\}} = \left\{ \mathbf{x} \in \mathbb{R}^n \;\middle|\; \mathbf{a}_i^\mathsf{T} \mathbf{x} = b_i, i = 1, \ldots, n+m \right\}$$
$$= \left\{ \mathbf{x} \in \mathbb{R}^n \;\middle|\; \mathbf{Ax} = \mathbf{b}, \mathbf{x} = \mathbf{0} \right\}.$$

Therefore, either $F_{\{1,\ldots,n+m\}} = \{\mathbf{0}\}$, or $F_{\{1,\ldots,n+m\}} = \emptyset$. In either case, ($\star$) holds trivially.

For the general case, let $J \subseteq \{1, \ldots, n+m\}$ and assume that ($\star$) holds for all J' with $|J'| > |J|$. Let $\hat{\mathbf{x}} \in F_J$. If F_J is the empty set or a singleton, then ($\star$) holds trivially and we are done. So we may assume that F_J contains at least two distinct points, $\mathbf{u}$ and $\mathbf{u}'$, say. Let $\mathbf{d} = \mathbf{u} - \mathbf{u}'$. We have that $\mathbf{d} \neq \mathbf{0}$ and $\mathbf{a}_j^\mathsf{T} \mathbf{d} = \mathbf{a}_j^\mathsf{T}(\mathbf{u} - \mathbf{u}') = b_j - b_j = 0$ for each $j \in J$. Consider moving away as far as possible from $\hat{\mathbf{x}}$ in the directions $\mathbf{d}$ and $-\mathbf{d}$, while staying inside F_J. Formally, define:

$$\alpha_1 = \max\left\{ \alpha \in \mathbb{R} \;\middle|\; \begin{array}{l} \mathbf{a}_j^\mathsf{T}(\hat{\mathbf{x}} + \alpha\mathbf{d}) = b_j \text{ for } j \in J \\ \mathbf{a}_j^\mathsf{T}(\hat{\mathbf{x}} + \alpha\mathbf{d}) \leq b_j \text{ for } j \in \bar{J} \end{array} \right\}, \text{ and}$$

$$\alpha_2 = \max\left\{ \alpha \in \mathbb{R} \;\middle|\; \begin{array}{l} \mathbf{a}_j^\mathsf{T}(\hat{\mathbf{x}} - \alpha\mathbf{d}) = b_j \text{ for } j \in J \\ \mathbf{a}_j^\mathsf{T}(\hat{\mathbf{x}} - \alpha\mathbf{d}) \leq b_j \text{ for } j \in \bar{J} \end{array} \right\}.$$

(The vector $\alpha_1 \mathbf{d}$ can informally be described as the furthest we can move away from $\hat{\mathbf{x}}$ in the direction $\mathbf{d}$ without leaving F_J, and similarly for $-\alpha_2 \mathbf{d}$ in the direction $-\mathbf{d}$. Since $\mathbf{a}_j^\mathsf{T} \mathbf{d} = 0$ for each $j \in J$, we have that $\mathbf{a}_j^\mathsf{T}(\hat{\mathbf{x}} + \alpha\mathbf{d}) = \mathbf{a}_j^\mathsf{T}\hat{\mathbf{x}} = b_j$ for all $j \in J$. So, all of the equality constraints defined by $\mathbf{a}_j^\mathsf{T}\mathbf{x} = b_j$ are still satisfied as we move in the direction $\mathbf{d}$ or $-\mathbf{d}$.)

Because $\hat{\mathbf{x}} \in F_J$ and F_J is bounded, we have that α_1 and α_2 are nonnegative and finite. By the choice of α_1, there exists an index $j \in \bar{J}$ such that $\mathbf{a}_j^\mathsf{T}(\hat{\mathbf{x}} + \alpha_1\mathbf{d}) = b_j$. Hence, $\hat{\mathbf{x}} + \alpha_1\mathbf{d} \in F_{J \cup \{j\}}$, and it follows from the induction hypothesis that $\hat{\mathbf{x}} + \alpha_1\mathbf{d}$ can be written as a convex combination of vertices of $F_{J \cup \{j\}}$. Since, by Theorem D.3.3, the vertices of $F_{J \cup \{j\}}$ are also vertices of F_J, it follows that $\hat{\mathbf{x}} + \alpha_1\mathbf{d}$ can be written as a convex combination of the vertices $\mathbf{v}_1, \ldots, \mathbf{v}_k$, say, of F_J. Similarly, $\hat{\mathbf{x}} - \alpha_2\mathbf{d}$ can be written as a convex combination of vertices of F_J. Therefore, we have that:

$$\hat{\mathbf{x}} + \alpha_1\mathbf{d} = \sum_{i=1}^{k} \lambda_i \mathbf{v}_i, \text{ and } \hat{\mathbf{x}} - \alpha_2\mathbf{d} = \sum_{i=1}^{k} \lambda_i' \mathbf{v}_i,$$

with $\sum_{i=1}^{k} \lambda_i = 1$ and $\sum_{i=1}^{k} \lambda_i' = 1$. Observe that $\hat{\mathbf{x}}$ can be written as a convex combination of $\hat{\mathbf{x}} + \alpha_1\mathbf{d}$ and $\hat{\mathbf{x}} - \alpha_2\mathbf{d}$, namely:

$$\hat{\mathbf{x}} = \nu(\hat{\mathbf{x}} + \alpha_1\mathbf{d}) + (1-\nu)(\hat{\mathbf{x}} - \alpha_2\mathbf{d}), \text{ with } \nu = \frac{\alpha_2}{\alpha_1 + \alpha_2} \quad (\in [0,1]).$$

Therefore we may write $\hat{\mathbf{x}}$ as follows:

$$\hat{\mathbf{x}} = \nu \sum_{i=1}^{k} \lambda_i \mathbf{v}_i + (1-\nu) \sum_{i=1}^{k} \lambda_i' \mathbf{v}_i = \sum_{i=1}^{k} (\nu\lambda_i + (1-\nu)\lambda_i') \mathbf{v}_i.$$

Since $\sum_{i=1}^{k}(\nu\lambda_i + (1-\nu)\lambda_i') = \nu\sum_{i=1}^{k}\lambda_i + (1-\nu)\sum_{i=1}^{k}\lambda_i' = 1$, this proves that $\hat{\mathbf{x}}$ can be written as a convex combination of vertices of F_J. $\qquad\qquad\square$

D.4   Convex and concave functions

Let C be a convex set in $\mathbb{R}^n$. The function $f\colon C \to \mathbb{R}$ is called a *convex function* if for all $\mathbf{x}_1$ and $\mathbf{x}_2$ in C and all $\lambda \in \mathbb{R}$ with $0 < \lambda < 1$, it holds that:

$$f(\lambda\mathbf{x}_1 + (1-\lambda)\mathbf{x}_2) \le \lambda f(\mathbf{x}_1) + (1-\lambda)f(\mathbf{x}_2). \qquad (*)$$

The function $f\colon C \to \mathbb{R}$ is called *concave* if the '$\ge$' sign holds in $(*)$. The function f is called a *strictly convex function* if the strict inequality sign '$<$' holds in $(*)$; for a *strictly concave function*, the '$>$' sign holds in $(*)$.

Examples of convex functions are $f(x) = x^2$ with $x \in \mathbb{R}$, $f(x_1, x_2) = 3x_1^2 + 2x_2^2$ with $[x_1 \ x_2]^{\mathsf{T}} \in \mathbb{R}^2$. Note that for any two points $[\mathbf{a} \ f(\mathbf{a})]^{\mathsf{T}}$ and $[\mathbf{b} \ f(\mathbf{b})]^{\mathsf{T}}$, the line segment connecting these two points is 'above' the graph of the convex function f and the region 'above' the graph is convex. This last property can be formulated as follows. Let C be a convex set in $\mathbb{R}^n$, and $f\colon C \to \mathbb{R}$. The *epigraph* of f on the set C is denoted and defined by:

$$\mathrm{epi}(f) = \{(\mathbf{x}, \lambda) \in C \times \mathbb{R} \mid f(\mathbf{x}) \le \lambda\}.$$

It can be shown that $\mathrm{epi}(f)$ on C is convex if and only if f is convex. Moreover, it can be shown that, for any concave function f, it holds that $\mathrm{epi}(-f)$ is convex.

The expression

$$\prod_{k=1}^{p} D_k = D_1 \times \ldots \times D_p = \left\{ [\mathbf{x}_1 \ \ldots \ \mathbf{x}_p]^{\mathsf{T}} \mid \mathbf{x}_k \in D_k \text{ for } k = 1, \ldots, p \right\}$$

is called the *Cartesian product*[1] of the sets $D_1, \ldots, D_k$.

Theorem D.4.1.
For each $k = 1, \ldots, p$, let $f_k\colon D_k \to \mathbb{R}$ be a (strictly) convex (concave) function. Then the function $\sum_{k=1}^{p} f_k\colon \prod_{k=1}^{p} D_k \to \mathbb{R}$, defined by

$$\left(\sum_{k=1}^{p} f_k\right)(\mathbf{x}_1, \ldots, \mathbf{x}_p) = \sum_{k=1}^{p} f_k(\mathbf{x}_k),$$

with $\mathbf{x}_i \in D_i$ for $i = 1, \ldots, p$, is a (strictly) convex (concave) function.

---

[1] Named after the French philosopher, mathematician, and writer RENÉ DESCARTES (1596–1650).

Proof of Theorem D.4.1. We give a proof for convex functions and $p = 2$; all other cases can be proved similarly, and they are left to the reader. So, let $f_1(\mathbf{x})$ and $f_2(\mathbf{y})$ be convex functions; $\mathbf{x} \in D_1$ and $\mathbf{y} \in D_2$. Note that D_1 and D_2 may have different dimensions. Also note that $(f_1 + f_2)(\mathbf{x} + \mathbf{y}) = f_1(\mathbf{x}) + f_2(\mathbf{y})$. Let $0 \leq \lambda \leq 1, \mathbf{x}_1, \mathbf{x}_2 \in D_1$, and $\mathbf{y}_1, \mathbf{y}_2 \in D_2$. Then, $f_1(\lambda \mathbf{x}_1 + (1-\lambda)\mathbf{x}_2) + f_2(\lambda \mathbf{y}_1 + (1-\lambda)\mathbf{y}_2) \leq \lambda(\mathbf{f}_1(\mathbf{x}_1) + f_2(\mathbf{y}_1)) + (1-\lambda)(f_1(\mathbf{x}_2) + f_2(\mathbf{y}_2))$.

$\square$

The proof of Theorem D.4.2 is based on the following fact, which we state without proof. Let F be a nonempty convex set in $\mathbb{R}^n$, and let $f : F \to \mathbb{R}$ be a convex function. Then f is continuous on the relative interior of F.

Theorem D.4.2.
Let F be a nonempty bounded subset of $\mathbb{R}^n$, and let $f : F \to \mathbb{R}$ be a strictly convex function on F. Then, the function f has a unique minimizer on F.

Proof of Theorem D.4.2. Since f is continuous on $\text{relint}(S)$, it follows from Weierstrass' theorem (Theorem D.1.1) that f attains a minimum on $\text{relint}(S)$. The fact that there is a unique minimizer is a direct consequence of the fact that f is strictly convex.

$\square$

APPENDIX E

Nonlinear optimization

Interior point algorithms (see Chapter 6) rely strongly on optimization methods from non-linear constrained optimization. For that purpose, in this appendix, we will present some of the theory concerning nonlinear constrained optimization, and we will show how linear optimization can be viewed as a special case of nonlinear optimization.

E.1 Basics

Consider the function $f : \mathbb{R}^n \to \mathbb{R}^m$ $(n, m \geq 1)$. We say that f is *continuous* at the point $\hat{\mathbf{x}}$ $(\in \mathbb{R}^n)$ if, for any sequence of points $\mathbf{x}_1, \mathbf{x}_2, \ldots$ in $\mathbb{R}^n$ that converges to $\hat{\mathbf{x}}$, we have that $\lim_{k \to \infty} f(\mathbf{x}_k) = f(\hat{\mathbf{x}})$. We write $f(\mathbf{x}) = f\left(\begin{bmatrix} x_1 & \ldots & x_n \end{bmatrix}^\mathsf{T}\right)$. Whenever this does not cause any confusion, we write $f(x_1, \ldots, x_n)$ instead of $f\left(\begin{bmatrix} x_1 & \ldots & x_n \end{bmatrix}^\mathsf{T}\right)$.

If $m = 1$, then f is called a *scalar function*. We are mainly interested in scalar functions. So, from now on we assume that $m = 1$, i.e., we consider the function $f : \mathbb{R}^n \to \mathbb{R}$ $(n \geq 1)$. The function f is *differentiable* at $\hat{\mathbf{x}}$ $(\in \mathbb{R}^n)$ if there exists a vector $\nabla f(\hat{\mathbf{x}})$ $(\in \mathbb{R}^n)$ such that, for any sequence of points $\mathbf{x}_1, \mathbf{x}_2, \ldots$ in $\mathbb{R}^n$ that converges to $\hat{\mathbf{x}}$, we have that $\lim_{k \to \infty} \frac{f(\mathbf{x}_k) - h(\mathbf{x}_k)}{\|\hat{\mathbf{x}} - \mathbf{x}_k\|} = 0$ where $h(\mathbf{x}) = f(\hat{\mathbf{x}}) + \nabla f(\hat{\mathbf{x}})^\mathsf{T}(\mathbf{x} - \hat{\mathbf{x}})$. (Note that $\|\mathbf{x}\|$ is the Euclidean norm of the vector $\mathbf{x}$; see Appendix B.1.) Intuitively, the function f is differentiable at $\hat{\mathbf{x}}$ if the function $h(\mathbf{x})$ is a 'good approximation' of f near $\hat{\mathbf{x}}$. See also Appendix D.1.

If f is differentiable at $\hat{\mathbf{x}}$, then the *partial derivative of f with respect x_i* at the point $\hat{\mathbf{x}}$ is given by:

$$\frac{\partial f}{\partial x_i}(\hat{\mathbf{x}}) = \lim_{\varepsilon \to 0} \frac{f(\hat{\mathbf{x}} + \varepsilon \mathbf{e}_i) - f(\hat{\mathbf{x}})}{\varepsilon},$$

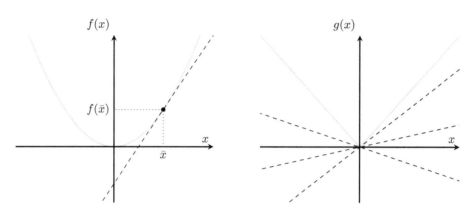

Figure E.1: The functions $f(x) = x^2$ (solid line), and $h(x) = \hat{x}^2 + 2\hat{x}(x - \hat{x})$ (dashed line).

Figure E.2: The functions $f(x) = |x|$ (solid line), and $h(x) = |\hat{x}| + \alpha(x - \hat{x})$ for different choices of α (dashed lines).

where $\mathbf{e}_i$ is the i'th unit vector in $\mathbb{R}^n$, and $i = 1, \ldots, n$. The vector $\nabla f(\hat{\mathbf{x}})$ (if it exists) in the definition above is called the *gradient* of f at $\hat{\mathbf{x}}$. It satisfies the following equation:

$$\nabla f(\hat{\mathbf{x}}) = \begin{bmatrix} \dfrac{\partial f}{\partial x_1}(\hat{\mathbf{x}}) \\ \vdots \\ \dfrac{\partial f}{\partial x_n}(\hat{\mathbf{x}}) \end{bmatrix}.$$

A function f is *continuously differentiable* at $\hat{\mathbf{x}}$ if f is differentiable at $\hat{\mathbf{x}}$ and the gradient of f is continuous at $\hat{\mathbf{x}}$. If $n = 1$, then there is only one partial derivative, and this partial derivative is then called the *derivative* of f.

The following two examples provide differentiable functions.

Example E.1.1.   *Let $f : \mathbb{R} \to \mathbb{R}$ be defined by $f(x) = x^2$. We will show that f is differentiable at every point $\hat{x} \in \mathbb{R}$, and that $\nabla f(x) = 2x$. Let $\hat{x} \in \mathbb{R}$ be given. In the notation introduced above, we have that $h(x) = f(\hat{x}) + 2\hat{x}(x - \hat{x})$; see also Figure E.1. Let $x_1, x_2, \ldots$ be any sequence in $\mathbb{R}$ that converges to $\hat{x}$. Then, we have that:*

$$\lim_{k \to \infty} \frac{f(x_k) - h(x_k)}{|\hat{x} - x_k|} = \lim_{k \to \infty} \frac{x_k^2 - \hat{x}^2 - 2\hat{x}(x_k - \hat{x})}{|\hat{x} - x_k|} = \lim_{k \to \infty} \frac{x_k^2 - 2\hat{x}x_k + \hat{x}^2}{|\hat{x} - x_k|}$$

$$= \lim_{k \to \infty} \frac{(\hat{x} - x_k)^2}{|\hat{x} - x_k|} = \lim_{k \to \infty} |\hat{x} - x_k| = 0.$$

Hence, f is differentiable at any point $\hat{x} \in \mathbb{R}$, and the derivative of f at $\hat{x}$ is $2\hat{x}$.

Example E.1.2.   *Consider the function $f : \mathbb{R}^2 \to \mathbb{R}$ defined by $f(y, z) = \alpha y^2 + \beta yz + \gamma z^2$. This function is continuous at all points in $\mathbb{R}^2$ because, for any point $\begin{bmatrix} \hat{y} & \hat{z} \end{bmatrix}^\mathsf{T}$ and for any sequence of* *points $[\hat{y}_1 \ \hat{z}_1]^\mathsf{T}$, $[\hat{y}_2 \ \hat{z}_2]^\mathsf{T}$, ... in $\mathbb{R}^2$ that converge to $[\hat{y} \ \hat{z}]^\mathsf{T}$, we have that:*

$$\lim_{k\to\infty} f(y_k, z_k) = \lim_{k\to\infty} \alpha y_k^2 + \beta y_k z_k + \gamma z_k^2$$

$$= \lim_{k\to\infty} \alpha y_k^2 + \lim_{k\to\infty} \beta y_k z_k + \lim_{k\to\infty} \gamma z_k^2 = \alpha \hat{y}^2 + \beta \hat{y}\hat{z} + \gamma \hat{z}^2.$$

The partial derivatives of f with respect to y and z at the point $[\hat{y} \ \hat{z}]^\mathsf{T}$ are given by:

$$\frac{\partial f}{\partial y}(\hat{y}, \hat{z}) = 2\alpha \hat{y} + \beta \hat{z}, \ and \ \frac{\partial f}{\partial z}(\hat{y}, \hat{z}) = \beta \hat{y} + 2\gamma \hat{z}.$$

Hence, the gradient of f is given by:

$$\nabla f(\hat{y}, \hat{z}) = \begin{bmatrix} 2\alpha \hat{y} + \beta \hat{z} \\ \beta \hat{y} + 2\gamma \hat{z} \end{bmatrix}.$$

The following example provides a function that is not differentiable.

Example E.1.3. *Consider the function $f : \mathbb{R} \to \mathbb{R}$ defined by $f(x) = |x|$. We will show that f is not differentiable at $\hat{x} = 0$. Suppose for a contradiction that f is differentiable at 0. Then, it has a gradient, say $\nabla f(0) = \alpha$ for some $\alpha \in \mathbb{R}$. Define $h(x) = f(0) + \alpha(x - 0) = \alpha x$; see also Figure E.2. Because we assumed that f is differentiable at 0, we must have that:*

$$\alpha = \nabla f(0) = \lim_{k\to\infty} \frac{f(x_k) - h(x_k)}{|\hat{x} - x_k|} = \lim_{k\to\infty} \frac{|x_k| - \alpha x_k}{|x_k|} = \lim_{k\to\infty} \left(1 - \frac{\alpha x_k}{|x|}\right) \quad \text{(E.1)}$$

for every sequence $x_1, x_2, \ldots$ in $\mathbb{R}$ that converges to 0. However, this is not true. For example, choose the sequence $x_k = \frac{(-1)^k}{k}$, i.e., $x_1 = -1$, $x_2 = \frac{1}{2}$, $x_3 = -\frac{1}{3}$, $x_4 = -\frac{1}{4}$, ... This sequences converges to 0, but we have that:

$$1 - \frac{\alpha x_k}{|x|} = 1 - \alpha \frac{(-1)^k/k}{|(-1)^k/k|} = 1 - \alpha \frac{(-1)^k/k}{1/k} = 1 - \alpha(-1)^k = \begin{cases} 1 + \alpha & \text{if } k \text{ is odd} \\ 1 - \alpha & \text{if } k \text{ is even}. \end{cases}$$

We now distinguish two cases:

$\alpha \neq 0$. *Then, $\lim_{k\to\infty} \left(1 - \frac{\alpha x_k}{|x|}\right)$ does not exist, and hence (E.1) does not hold.*

$\alpha = 0$. *Then, (E.1) does not hold either, because the left hand side equals $(\alpha =) \ 0$, and the right hand side equals $\lim_{k\to\infty} \left(1 - \frac{\alpha x_k}{|x|}\right) = 1$.*

Hence, the function $f(x) = |x|$ is not differentiable at $\hat{x} = 0$.

The gradient has two important geometric interpretations. First, $\nabla f(\hat{\mathbf{x}})$ is a vector that is perpendicular to the so-called *level set* $\{\mathbf{x} \in \mathbb{R}^n \mid f(\mathbf{x}) = f(\hat{\mathbf{x}})\}$. Second, $\nabla f(\mathbf{x})$ points in the direction in which the function f increases. In fact, $\nabla f(\mathbf{x})$ is a direction in which f increases fastest; it is therefore also called the *direction of steepest ascent*.

For $\mathbf{p} \in \mathbb{R}^n$, the *directional derivative of f in the direction* $\mathbf{p}$ (if it exists) is denoted and defined as:

$$D(f(\mathbf{x}); \mathbf{p}) = \lim_{\varepsilon \to 0} \frac{f(\mathbf{x} + \varepsilon \mathbf{p}) - f(\mathbf{x})}{\varepsilon}.$$

So the i'th partial derivative is just $D(f(\mathbf{x}); \mathbf{e}_i)$ $(i = 1, \ldots, n)$.

The *Hessian*[1] of f at $\hat{\mathbf{x}}$ (if it exists) is the matrix $\nabla^2 f(\hat{\mathbf{x}})$, defined as:

$$\nabla^2 f(\hat{\mathbf{x}}) = \begin{bmatrix} \dfrac{\partial^2 f}{\partial x_1 \partial x_1}(\hat{\mathbf{x}}) & \cdots & \dfrac{\partial^2 f}{\partial x_1 \partial x_n}(\hat{\mathbf{x}}) \\ \vdots & & \vdots \\ \dfrac{\partial^2 f}{\partial x_n \partial x_1}(\hat{\mathbf{x}}) & \cdots & \dfrac{\partial^2 f}{\partial x_n \partial x_n}(\hat{\mathbf{x}}) \end{bmatrix}.$$

E.2 Nonlinear optimization; local and global minimizers

This appendix deals with optimization models that are not necessarily linear optimization models. The most general form of an optimization model is:

$$\begin{aligned} \max\ & f(\mathbf{x}) \\ \text{s.t.}\ & \mathbf{x} \in F, \end{aligned} \tag{E.2}$$

where $f : F \to \mathbb{R}$ is the *objective function*, F is the *feasible region*, and $\mathbf{x}$ is the vector of *decision variables*. We will assume that $F \subseteq \mathbb{R}^n$ for some integer $n \geq 1$, and therefore $\mathbf{x}$ is a vector of n real numbers.

We assume that the set F can be described by a set of equality constraints $g_j(\mathbf{x}) = 0$ $(j = 1, \ldots, p)$ and inequality constraints $h_k(\mathbf{x}) \geq 0$ $(k = 1, \ldots, q)$. We also assume that the functions f, g_j, and h_k $(j = 1, \ldots, p, \ k = 1, \ldots, q)$ are continuously differentiable. The set F is then given by:

$$F = \left\{ \mathbf{x} \in \mathbb{R}^n \ \middle|\ \begin{array}{l} g_j(\mathbf{x}) = 0 \ \text{for } j = 1, \ldots, p, \\ h_k(\mathbf{x}) \leq 0 \ \text{for } k = 1, \ldots, q \end{array} \right\}.$$

Consequently, the models that we deal with in this appendix can be written in the form:

$$\begin{aligned} \max\ & f(\mathbf{x}) \\ \text{s.t.}\ & g_j(\mathbf{x}) = 0 \quad \text{for } j = 1, \ldots, p \\ & h_k(\mathbf{x}) \leq 0 \quad \text{for } k = 1, \ldots, q. \end{aligned} \tag{E.3}$$

Example E.2.1.   *In linear optimization, the function f is linear, i.e., it can be written as $f(\mathbf{x}) = \mathbf{c}^\mathsf{T} \mathbf{x}$ with $\mathbf{c} \in \mathbb{R}^n$. The feasible region F is the intersection of finitely many halfspaces; see Section 2.1.1. The requirement that $\mathbf{x} \in F$ can then be written as $h_k(\mathbf{x}) = \mathbf{a}_k^\mathsf{T} \mathbf{x} - b_k$, for $k = 1, \ldots, q$.*

---

[1] Named after the German mathematician LUDWIG OTTO HESSE (1811–1874).

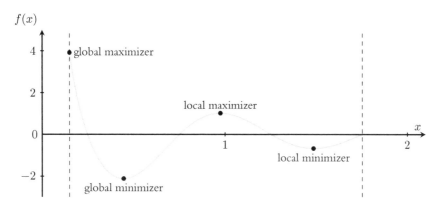

Figure E.3: Global and local minimizers and maximizers of the function $f(x) = \dfrac{\cos(2\pi x)}{x}$ for $0.15 \leq x \leq 1.75$.

Any point $\mathbf{x} \in F$ is called a *feasible point* (or *feasible solution*). The goal of optimization is to find a feasible point $\mathbf{x}^*$ for which the value of $f(\mathbf{x}^*)$ is as large as possible. A *global maximizer* for model (E.2) is a feasible point $\mathbf{x}^* \in F$ such that, for all $\mathbf{x} \in F$, we have that $f(\mathbf{x}^*) \geq f(\mathbf{x})$.

Although the goal of optimization is to find a global maximizer, in many cases in nonlinear optimization, this is too much to ask for. In such cases, we are already satisfied with a so-called local maximizer. Intuitively, a local maximizer is a feasible point $\hat{\mathbf{x}}$ such that no feasible point 'close' to it has a larger objective value than $f(\hat{\mathbf{x}})$. To make this intuition precise, first define, for $\hat{\mathbf{x}} \in \mathbb{R}^n$ and for $\varepsilon > 0$, the (restricted) ε-*ball* (around $\hat{\mathbf{x}}$) as $B_\varepsilon(\hat{\mathbf{x}}) = \{\mathbf{x} \in F \mid \|\mathbf{x} - \hat{\mathbf{x}}\| \leq \varepsilon\}$; see also Section D.1. We say that the point $\hat{\mathbf{x}} \in F$ is a *local maximizer* if there exists an $\varepsilon > 0$, such that for all $\mathbf{x} \in B_\varepsilon(\hat{\mathbf{x}})$, we have that $f(\hat{\mathbf{x}}) \geq f(\mathbf{x})$. A *global minimizer* and a *local minimizer* are defined similarly. Figure E.3 illustrates these concepts.

E.3 Equality constraints; Lagrange multiplier method

Let $n, p \geq 1$. For $j = 1, \ldots, p$, let $f \colon \mathbb{R}^n \to \mathbb{R}$ and $g_j \colon \mathbb{R}^n \to \mathbb{R}$ be differentiable functions. Let $\mathbf{x} = \begin{bmatrix} x_1 & \ldots & x_n \end{bmatrix}^{\mathsf{T}} \in \mathbb{R}^n$. Consider the following constrained optimization model (E.4):

$$\begin{aligned} \max\ & f(\mathbf{x}) \\ \text{s.t.}\ & g_j(\mathbf{x}) = 0 \ \text{ for } j = 1, \ldots, p. \end{aligned} \tag{E.4}$$

In general, it is very difficult to solve constrained optimization models and many practical techniques work only in special cases. The *Lagrange multiplier method* for constrained optimization models is one such technique. The method was developed by the French mathematician JOSEPH LOUIS LAGRANGE (1736–1813). We will briefly review the technique in this section. We start with an example that gives some intuition behind the method.

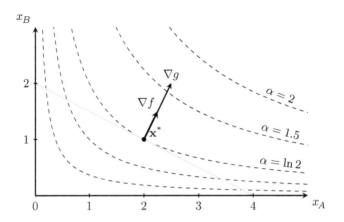

Figure E.4: Consumer choice model with parameters $p_A = 1$, $p_B = 2$, and $b = 4$. The solid line is the budget constraint $p_A x_A + p_B x_B = b$. The level sets
$$\left\{ \begin{bmatrix} x_A \\ x_B \end{bmatrix} \,\middle|\, U(x_A, x_B) = \ln(x_A) + \ln(x_B) = \alpha \right\}$$ of the objective function, for different values of α, are drawn as dashed curves.

Example E.3.1.   *Consider the following consumer choice model. A consumer can buy two different products, say products A and B. The prices of the products are p_A and p_B (in dollars), respectively, and the consumer has a budget of b dollars. Suppose that the consumer receives a certain amount of 'happiness' or 'utility' $U(x_A, x_B)$ from buying amounts x_A and x_B of products A and B, respectively. The consumer can spend all her money on just one product, but she is also allowed to spend some of her money on product A, and some on product B. The optimization model she faces is:*

$$\max\ U(x_A, x_B)$$
$$\text{s.t.}\quad p_A x_A + p_B x_B = b. \tag{E.5}$$

In this example, we choose $U(x_A, x_B) = \ln(x_A) + \ln(x_B)$. Note that we need $x_A > 0$ and $x_B > 0$ in order for $U(x_A, x_B)$ to make sense, but we will ignore these constraints in this example. In the notation of model (E.3), we have that $f(x_A, x_B) = \ln(x_A) + \ln(x_B)$, and $g_1(x_A, x_B) = b - p_A x_A - p_B x_B$.

Figure E.4 shows the constraint $g_1(x_A, x_B) = 0$, together with some of the level sets of f. Let $\alpha > 0$. Recall that the level set corresponding to α is given by the equation $f(x_A, x_B) = \alpha$, where α is a constant. It is not hard to work out that the level sets of f satisfy the equation $x_A x_B = e^{\alpha}$ for $\alpha > 0$.

To solve the optimization model, suppose we 'move along' the constraint $p_A x_A + p_B x_B = b$. As can be seen from Figure E.4, different level sets cross the constraint, which means that at different points on the line, the objective functions takes on different values. If, at a given point $\begin{bmatrix} x_A & x_B \end{bmatrix}^{\top}$ on this line, the level set of f crosses the line $p_A x_A + p_B x_B = b$, then we can move along the constraint to decrease the objective value. The only situation in which this is not possible, is when the level set of f is tangent to the constraint, i.e., the level set of f does not cross the constraint. Thus, if $\mathbf{x}^*$ $\left(= \begin{bmatrix} x_A^* & x_B^* \end{bmatrix}^{\top} \right)$ is an optimal point, then the level set of the objective function at $\mathbf{x}^*$ is tangent to the constraint at $\mathbf{x}^*$. Since the gradient of a function is perpendicular to its level set, this is equivalent to requiring that the gradients of f and g are parallel at $\mathbf{x}^*$, i.e., $\nabla f(x_A^*, x_B^*) = \lambda \nabla g(x_A^*, x_B^*)$ for some $\lambda \in \mathbb{R}$. Therefore, any optimal point $\begin{bmatrix} x_A^* & x_B^* \end{bmatrix}^\top$ satisfies

$$\begin{bmatrix} 1/x_A^* \\ 1/x_B^* \end{bmatrix} = \lambda \begin{bmatrix} p_A \\ p_B \end{bmatrix}, \text{ and}$$

$$g(x_A^*, x_B^*) = b - p_A x_A^* - p_B x_B^* = 0.$$

Note that λ is unrestricted in sign. This is only a necessary condition for $\mathbf{x}^*$ being an optimal point; it is in general not sufficient. We can summarize the system of (three) equations by defining a convenient function, which is called the Lagrange function associated with the optimization model:

$$\mathcal{L}(x_A, x_B, \lambda) = \nabla f(x_A, x_B) - \lambda \nabla g(x_A, x_B)$$
$$= -\ln(x_A) - \ln(x_B) - \lambda(b - p_A x_A - p_B x_B).$$

The beauty of this is that it suffices to equate the gradient of this Lagrange function to the all-zero vector. Taking derivatives with respect to x_A, x_B, and λ, and equation them to zero, we obtain the following system of equations:

$$\nabla \mathcal{L}(x_A^*, x_B^*, \lambda^*) = \begin{bmatrix} 1/x_A^* - p_A \lambda^* \\ 1/x_B^* - p_B \lambda^* \\ p_A x_A^* + p_B x_B^* - b \end{bmatrix} = \begin{bmatrix} 0 \\ 0 \\ 0 \end{bmatrix}.$$

Solving the first two equations of this system yields that $x_A^* = \frac{1}{p_A \lambda^*}$, $x_B = \frac{1}{p_B \lambda^*}$. The third equations reads $p_A x_A^* + p_B x_B^* - b = 0$. Substituting the values for x_A^* and x_B^* yields:

$$p_A x_A^* + p_B x_B^* - b = p_A \frac{1}{p_A \lambda^*} + p_B \frac{1}{p_B \lambda^*} - b = \frac{2}{\lambda^*} - b = 0.$$

This implies that $\lambda^* = 2/b$, and therefore, the optimal solution satisfies:

$$x_A^* = \frac{b}{2p_A}, \ x_B^* = \frac{b}{2p_B}, \ \lambda^* = 2/b.$$

The utility value corresponding to this optimal solution is:

$$\ln\left(\frac{b}{2p_A}\right) + \ln\left(\frac{b}{2p_B}\right).$$

Notice that the rate of change of the objective value with respect to b is:

$$\frac{\partial}{\partial b}\left[\ln\left(\frac{b}{2p_A}\right) + \ln\left(\frac{b}{2p_B}\right)\right] = \frac{2p_A}{b} \times \frac{1}{2p_A} + \frac{2p_A}{b} \times \frac{1}{2p_A} = \frac{2}{b},$$

which corresponds exactly to the optimal value of λ^*. This is no coincidence: λ can in fact be viewed as a variable of the dual model of (E.5). Duality theory for nonlinear optimization lies beyond the scope of this book. The interested reader is referred to, e.g., Bazaraa et al. (1993), or Boyd and Vandenberghe (2004). Duality theory for linear optimization is the subject of Chapter 4.

The Lagrange multiplier method can be formulated as follows. Let $\mathbf{x} = \begin{bmatrix} x_1 & \dots & x_n \end{bmatrix}^{\mathsf{T}} \in \mathbb{R}^n$, and $\boldsymbol{\lambda} = \begin{bmatrix} \lambda_1 & \dots & \lambda_p \end{bmatrix}^{\mathsf{T}} \in \mathbb{R}^p$. The *Lagrange function* (also called the *Lagrangian*) $\mathcal{L} \colon \mathbb{R}^n \times \mathbb{R}^p \to \mathbb{R}$ is defined by:

$$\mathcal{L}(\mathbf{x}, \boldsymbol{\lambda}) = f(\mathbf{x}) - \sum_{j=1}^{p} \lambda_j g_j(\mathbf{x}).$$

A *constrained stationary point* of model (E.4) is any feasible point $\hat{\mathbf{x}}$ ($\in \mathbb{R}^n$) for which there exists a vector $\hat{\boldsymbol{\lambda}}$ ($\in \mathbb{R}^p$) such that $\begin{bmatrix} \hat{\mathbf{x}} \\ \hat{\boldsymbol{\lambda}} \end{bmatrix}$ is a stationary point of the function $\mathcal{L}$, meaning that $\begin{bmatrix} \hat{\mathbf{x}} \\ \hat{\boldsymbol{\lambda}} \end{bmatrix}$ is a solution of the following set of equations:

$$\frac{\partial f}{\partial x_i}(\mathbf{x}) - \sum_{j=1}^{p} \lambda_j \frac{\partial g_j}{\partial x_i}(\mathbf{x}) = 0 \qquad \text{for } i = 1, \dots, n, \qquad \text{(stationarity)}$$

$$g_j(\mathbf{x}) = 0 \qquad \text{for } j = 1, \dots, p. \qquad \text{(feasibility)}$$

Note that the first set of equations is equivalent to $\frac{\partial \mathcal{L}}{\partial x_i}(\mathbf{x}, \boldsymbol{\lambda}) = 0$ for $i = 1, \dots, n$. It can be written in vector notation as $\nabla f(\mathbf{x}) = \sum_{j=1}^{p} \lambda_j \nabla g_j(\mathbf{x})$, which states that the gradient of f at $\hat{\mathbf{x}}$ should be a linear combination of the gradients of the constraints. The second set is equivalent to $\frac{\partial \mathcal{L}}{\partial \lambda_j}(\mathbf{x}, \boldsymbol{\lambda}) = 0$ for $j = 1, \dots, p$.

The variables $\lambda_1, \dots, \lambda_p$ are called the *Lagrange multipliers* of (E.4). The Lagrange multiplier method for solving (E.4) can now be described as follows. By a 'point at infinity' we mean the limit of a direction; i.e., if $\mathbf{a}$ is any nonzero point in $\mathbb{R}^n$, then $\lim_{\lambda \to \infty} \lambda \mathbf{a}$ is the *point at infinity* corresponding to $\mathbf{a}$.

Algorithm E.3.1. (*Lagrange multiplier algorithm*)

Input: Values for the parameters of model (E.4).

Output: Either the message 'the model has no optimal solution', or the set of all optimal solutions of the model.

Step 1: Construct the Lagrange function:

$$\mathcal{L}(\mathbf{x}, \boldsymbol{\lambda}) = f(\mathbf{x}) - \sum_{j=1}^{p} \lambda_j g_j(\mathbf{x}).$$

Step 2: Find the set X of all constrained stationary points, by solving the following set of $n + p$ equations:

$$\frac{\partial \mathcal{L}}{\partial x_i}(\mathbf{x}) = 0 \text{ for } i = 1, \dots, n, \text{ and } g_j(\mathbf{x}) = 0 \text{ for } j = 1, \dots, p.$$

If this set of constraints is inconsistent, then stop: model (E.4) has no optimal solution. Otherwise, continue.

Step 3: Calculate the objective value $f(\mathbf{x})$ at each $\mathbf{x} \in X$, and – where appropriate – at points at infinity satisfying the constraints.

Step 4: Let z^* be the largest finite (if it exists) objective value $f(\mathbf{x})$ among all $\mathbf{x} \in X$. Let X^* be the set of points $\mathbf{x} \in X$ such that $f(\mathbf{x}) = z^*$. If z^* is at least as large as the largest objective value of the points at infinity, then stop: X^* is the set of all optimal points of (E.4). Otherwise, model (E.4) has no optimal solution.

In the case of a minimizing model, the same algorithm can be used, except for Step 4, where smallest values have to be selected. The following example illustrates the Lagrange multiplier method.

Example E.3.2. *Consider the model*

$$\begin{aligned} \min \ & x_1^2 + x_2^2 + x_3^2 \\ \text{s.t.} \ & x_1 + x_2 + x_3 = 0 \\ & x_1 + 2x_2 + 3x_3 = 1. \end{aligned}$$

Thus, in the notation of (E.4), we have that $f(\mathbf{x}) = -x_1^2 - x_2^2 - x_3^2$, $g_1(\mathbf{x}) = x_1 + x_2 + x_3$, and $g_2(\mathbf{x}) = x_1 + 2x_2 + 3x_3 - 1$. The Lagrange function reads:

$$\mathcal{L}(\mathbf{x}, \boldsymbol{\lambda}) = -x_1^2 - x_2^2 - x_3^2 - \lambda_1(x_1 + x_2 + x_3) - \lambda_2(x_1 + 2x_2 + 3x_3 - 1).$$

Any constrained stationary point $\begin{bmatrix} \mathbf{x} \\ \boldsymbol{\lambda} \end{bmatrix}$ satisfies:

$$\frac{\partial \mathcal{L}}{\partial x_1}(\mathbf{x}, \boldsymbol{\lambda}) = -2x_1 - \lambda_1 - \lambda_2 = 0, \quad \frac{\partial \mathcal{L}}{\partial \lambda_1}(\mathbf{x}, \boldsymbol{\lambda}) = -x_1 - x_2 - x_3 = 0,$$

$$\frac{\partial \mathcal{L}}{\partial x_2}(\mathbf{x}, \boldsymbol{\lambda}) = -2x_2 - \lambda_1 - 2\lambda_2 = 0, \quad \frac{\partial \mathcal{L}}{\partial \lambda_2}(\mathbf{x}, \boldsymbol{\lambda}) = -x_1 - 2x_2 - 3x_3 + 1 = 0,$$

$$\frac{\partial \mathcal{L}}{\partial x_3}(\mathbf{x}, \boldsymbol{\lambda}) = -2x_3 - \lambda_1 - 3\lambda_2 = 0.$$

It follows that $x_1 = -\frac{1}{2}(\lambda_1 + \lambda_2)$, $x_2 = -\frac{1}{2}(\lambda_1 + 2\lambda_2)$, and $x_3 = -\frac{1}{2}(\lambda_1 + 3\lambda_2)$. Substituting these values into the constraint equations, and solving for λ_1 and λ_2 yields that $\lambda_1 = 2$ and $\lambda_2 = -1$. Substituting these values back into the expressions found for x_1, x_2, and x_3, yields that the unique constrained stationary point is $\begin{bmatrix} -\frac{1}{2} & 0 & \frac{1}{2} \end{bmatrix}^\mathsf{T}$. The corresponding objective value is $(f(-\frac{1}{2}, 0, \frac{1}{2}) =) \frac{1}{2}$. The feasible region contains points at infinity. However, in all directions to infinity the value of the objective function tends to ∞. The conclusion is that the optimal solution reads: $x_1^* = -\frac{1}{2}$, $x_2^* = 0$, and $x_3^* = \frac{1}{2}$, and the corresponding optimal objective value is $\frac{1}{2}$.

E.4 Models with equality and inequality constraints; Karush-Kuhn-Tucker conditions

The previous section dealt with optimization models that have only equality constraints. Now consider a more general optimization model, with both equality and inequality constraints, of the form:

$$
\begin{aligned}
\max\ & f(\mathbf{x}) \\
\text{s.t.}\ & g_j(\mathbf{x}) = 0 && \text{for } j = 1, \ldots, p \\
& h_k(\mathbf{x}) \leq 0 && \text{for } k = 1, \ldots, q,
\end{aligned}
\tag{E.6}
$$

where $f\colon \mathbb{R}^n \to \mathbb{R}$, $g_j\colon \mathbb{R}^n \to \mathbb{R}$ (for $j = 1, \ldots, p$) and $h_k\colon \mathbb{R}^n \to \mathbb{R}$ (for $k = 1, \ldots, q$) are all real-valued differentiable functions. The *Lagrange function* for this model is defined as:

$$
\mathcal{L}(\mathbf{x}, \boldsymbol{\mu}, \boldsymbol{\lambda}) = f(\mathbf{x}) - \sum_{j=1}^{p} \lambda_j g_j(\mathbf{x}) - \sum_{k=1}^{q} \mu_k h_k(\mathbf{x}),
$$

where $\boldsymbol{\lambda} = \begin{bmatrix} \lambda_1 & \ldots & \lambda_p \end{bmatrix}^{\mathsf{T}} \in \mathbb{R}^p$ and $\boldsymbol{\mu} = \begin{bmatrix} \mu_1 & \ldots & \mu_q \end{bmatrix}^{\mathsf{T}} \in \mathbb{R}^q$. The stationarity conditions corresponding to this model are known as the *Karush-Kuhn-Tucker conditions*[2] (or *KKT conditions*). They are the following $n + p + 3q$ equations and inequalities:

Stationarity:
$$
\frac{\partial f}{\partial x_i}(\mathbf{x}) - \sum_{j=1}^{p} \lambda_j \frac{\partial g_j}{\partial x_i}(\mathbf{x}) - \sum_{k=1}^{q} \mu_k \frac{\partial h_k}{\partial x_i}(\mathbf{x}) = 0 \qquad \text{for } i = 1, \ldots, n,
$$

Primal feasibility:
$$
\begin{aligned}
g_j(\mathbf{x}) &= 0 && \text{for } j = 1, \ldots, p, \\
h_k(\mathbf{x}) &\leq 0 && \text{for } k = 1, \ldots, q,
\end{aligned}
$$

Dual feasibility:
$$
\mu_k \geq 0 \qquad \text{for } k = 1, \ldots, q,
$$

Complementary slackness relations:
$$
\mu_k h_k(\mathbf{x}) = 0 \qquad \text{for } k = 1, \ldots, q.
$$

Note that the first two sets of equations are equivalent to $\nabla \mathcal{L}(\mathbf{x}, \boldsymbol{\mu}, \boldsymbol{\lambda}) = \mathbf{0}$.

The following theorem states that, under certain regularity conditions, the KKT conditions hold at any local maximum. Although these regularity conditions are satisfied in many cases in practice, they do not hold in all situations. The regularity conditions are usually referred to as the *constraint qualifications*. There are several such constraint qualifications, each of which, if satisfied, guarantees that the KKT conditions hold. Two commonly used constraint qualifications are:

---

[2]Named after the American mathematicians WILLIAM KARUSH (1917–1997) and HAROLD WILLIAM KUHN (1925–2014), and the Canadian mathematician ALBERT WILLIAM TUCKER (1905–1995).

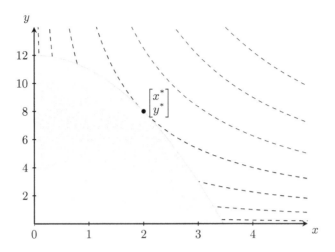

Figure E.5: The feasible region of the model in Example E.4.1. The level sets $\left\{\begin{bmatrix} x \\ y \end{bmatrix} \;\middle|\; xy = \alpha \right\}$ of the objective function, for different values of α, are drawn as dashed curves.

Linearity constraint qualification. For $j = 1, \ldots, p$ and $k = 1, \ldots, q$, the functions $g_j(\mathbf{x})$ and $h_k(\mathbf{x})$ are affine. (A function $t \colon \mathbb{R}^n \rightarrow \mathbb{R}$ is called *affine* if it can be written as $t(\mathbf{x}) = \mathbf{a}^\mathsf{T} \mathbf{x} + b$, where $\mathbf{a} \in \mathbb{R}^n$ and $b \in \mathbb{R}$.)

Slater's condition. Slater's condition requires that (a) the function $g_j(\mathbf{x})$ is affine for all $j = 1, \ldots, p$, (b) the function $h_k(\mathbf{x})$ is convex for all $k = 1, \ldots, q$, and (c) there exists a feasible point $\hat{\mathbf{x}}$ such that $h_k(\hat{\mathbf{x}}) < 0$ for all $k = 1, \ldots, q$. Conditions (a) and (b) guarantee that the feasible region of the model is a convex set (why is this the case?). Condition (c) states that the relative interior (see Section D.1) of the feasible region is nonempty.

Clearly, the linearity constraint qualification holds in the case when model (E.6) is an LO-model, so that the KKT conditions apply to linear optimization.

Theorem E.4.1. (*Necessity of the Karush-Kuhn-Tucker conditions*)
For $j = 1, \ldots, p$, $k = 1, \ldots, q$, let $f, g_j, h_k \colon \mathbb{R}^n \rightarrow \mathbb{R}$. Let $\mathbf{x}^*$ be a local maximizer of model (E.6). If one of the constraint qualifications holds, then there exist $\boldsymbol{\lambda}^* \in \mathbb{R}^p$ and $\boldsymbol{\mu}^* \in \mathbb{R}^q$ such that $\mathbf{x}^*$, $\boldsymbol{\lambda}^*$, and $\boldsymbol{\mu}^*$ satisfy the KKT conditions.

The KKT conditions can, in principle, be used to find an optimal solution to any model (for which one of the constraint qualifications is satisfied). However, the equations and inequalities that constitute the KKT conditions are in general very hard to solve, and they can usually only be solved for very small models. Even the relatively simple model in the following example requires tedious calculations.

Example E.4.1.   *Consider the nonlinear optimization model:*

$$\max xy$$
$$\text{s.t.} \quad x^2 + y \le 12$$
$$x, y \ge 0.$$

Figure E.5 illustrates the feasible region and the levels sets of the objective function of this model. Since the objective function is continuous and the feasible region $\left\{ \begin{bmatrix} x \\ y \end{bmatrix} \;\middle|\; x^2 + y \le 12, \; x, y \ge 0 \right\}$ is a closed bounded subset of $\mathbb{R}^2$, Theorem D.1.1 guarantees the existence of an optimal point of this model. Putting the model into the form of (E.6), we have that $f(x, y) = xy$, $p = 0$, $q = 3$, $h_1(x, y) = x^2 + y - 12$, $h_2(x, y) = -x$, and $h_3(x, y) = -y$. The corresponding Lagrange function is:

$$\mathcal{L}(x, y, \mu_1, \mu_2, \mu_3) = xy - \mu_1(x^2 - y + 12) + \mu_2 x + \mu_3 y.$$

The KKT conditions are:

$y - 2\mu_1 x + \mu_2 = 0$	(i)	$\mu_1 \ge 0$	(vi)
$x - \mu_1 + \mu_3 = 0$	(ii)	$\mu_2 \ge 0$	(vii)
$x^2 + y - 12 \le 0$	(iii)	$\mu_3 \ge 0$	(viii)
$x \ge 0$	(iv)	$\mu_1(x^2 + y - 12) = 0$	(ix)
$y \ge 0$	(v)	$-\mu_2 x = 0$	(x)
		$-\mu_3 y = 0.$	(xi)

This system of equations and inequalities can be solved by considering a few different cases. First, suppose that $\mu_1 = 0$. Then (i) implies that $y = -\mu_2$, and hence (v) and (vii) together imply that $y = \mu_2 = 0$. Similarly, (ii) implies that $x = -\mu_3$, and hence (iv) and (viii) together imply that $x = \mu_3 = 0$. It is straightforward to verify that the solution $x = y = \mu_1 = \mu_2 = \mu_3 = 0$ indeed satisfies the conditions (i)–(xi). The corresponding objective value is 0.

Second, suppose that $\mu_1 > 0$. By (ix), we have that $x^2 + y = 12$, and hence $x = \sqrt{12 - y}$. We consider three subcases:

> $x = 0$. *Then, we have that $y = 12 > 0$. Equation (xi) implies that $\mu_3 = 0$, and hence (ii) implies that $\mu_1 = x = 0$, contradicting the fact that $\mu_1 > 0$.*

> $y = 0$. *Then, we have that $x = \sqrt{12} > 0$. Equation (x) implies that $\mu_2 = 0$, and hence (i) implies that $\mu_1 = \frac{y}{2} = 0$, contradicting the fact that $\mu_1 > 0$.*

> $x > 0$ *and* $y > 0$. *Then, equations (x) and (xi) imply that $\mu_2 = \mu_3 = 0$. Hence, using (i) and (ii), we have that $x = \mu_1$ and $y = 2\mu_1 x = 2x^2$. So, we have that $12 = x^2 + y = 3x^2$, which implies that either $x = -2$ (which we may ignore because we have that $x \ge 0$) or $x = 2$. So, we must have $x = 2$, and therefore $y = 8$. Thus, we find the following solution of the conditions (i)–(xi):*

$$x = 2, y = 8, \mu_1 = 2, \mu_2 = 0, \mu_3 = 0.$$

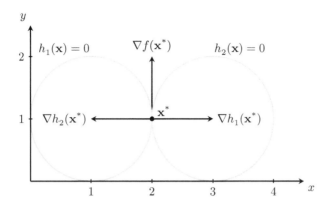

Figure E.6: Nonlinear optimization model for which the KKT conditions fail.

The corresponding objective value is $(2 \times 8 =)$ 16.

We have found the two points of the KKT conditions (i)–(xi)*. Clearly, the first one (the all-zero solution) is not optimal, because the corresponding objective value is* 0*, and we know that there exists a point with corresponding objective value* 16*. Since we also know that an optimal point exists, it must be the second solution, i.e.,* $x^* = 2$, $y^* = 8$*.*

Note that the first KKT conditions can be written in vector form as follows:

$$\nabla f(\mathbf{x}) = \sum_{j=1}^{p} \lambda_j \nabla g_j(\mathbf{x}) + \sum_{k=1}^{q} \mu_k \nabla h_k(\mathbf{x}). \tag{E.8}$$

This means that, assuming that some constraint qualification condition holds, any optimal point satisfies the property that the gradient of the objective function at that point can be written as a linear combination of the gradients of the constraints at that point. The weights in this linear combination are precisely the corresponding Lagrange multipliers. The following example shows that there are cases in which the KKT conditions fail, although an optimal solution exists. The reader may check that the constraint qualifications listed above Theorem E.4.1 do not hold.

Example E.4.2. *Consider the following optimization model with decision variables* x *and* y:

$$\max y$$
$$\text{s.t.} \quad (x-1)^2 + (y-1)^2 \le 1$$
$$(x-3)^2 + (y-1)^2 \le 1.$$

The feasible region and the constraints of this problem are drawn in Figure E.6. The first constraint restricts the values of x and y to lie inside a unit circle (i.e., with radius 1) centered at the point $\begin{bmatrix} 1 & 1 \end{bmatrix}^{\mathsf{T}}$*; the second constraint restricts the values to lie inside a unit circle centered at* $\begin{bmatrix} 3 & 1 \end{bmatrix}^{\mathsf{T}}$*. From the figure, it should be clear that the only feasible, and hence optimal, point of the model is the point* $\begin{bmatrix} 2 & 1 \end{bmatrix}^{\mathsf{T}}$*. However, consider the KKT conditions. Setting* $f(x,y) = y$, $h_1(x,y) = (x-1)^2 + (y-1)^2 - 1$, and $h_2(x, y) = (x - 3)^2 + (y - 1)^2 - 1$, *we have that, at the point* $\begin{bmatrix} 2 & 1 \end{bmatrix}^\mathsf{T}$:

$$\nabla f(x, y) = \begin{bmatrix} 0 \\ 1 \end{bmatrix},$$

$$\nabla h_1(x, y) = \begin{bmatrix} 2(x - 1) \\ 2(y - 1) \end{bmatrix} = \begin{bmatrix} 2 \\ 0 \end{bmatrix},$$

$$\nabla h_2(x, y) = \begin{bmatrix} 2(x - 3) \\ 2(y - 1) \end{bmatrix} = \begin{bmatrix} -2 \\ 0 \end{bmatrix}.$$

Hence, at the unique optimal point, the gradient of the objective function cannot be written as a linear combination of the gradients of the constraints at that point, and therefore the KKT conditions have no solution. This can also seen from the figure: the arrow pointing up is the gradient vector of the objective function at the optimal point, and the arrows pointing left and right are the gradient vectors of the constraints at the optimal point.

E.5   Karush-Kuhn-Tucker conditions for linear optimization

The KKT conditions can directly be applied to LO-models. Consider the standard LO-model from Section 1.2.1:

$$\max\{\mathbf{c}^\mathsf{T}\mathbf{x} \mid \mathbf{A}\mathbf{x} \leq \mathbf{b}, \mathbf{x} \geq \mathbf{0}\},$$

where $\mathbf{A}$ is an (m, n)-matrix, $\mathbf{x} \in \mathbb{R}^n$, $\mathbf{c} \in \mathbb{R}^n$, and $\mathbf{b} \in \mathbb{R}^m$ ($m, n \geq 1$). In the notation of model (E.6), we have that:

$$f(\mathbf{x}) = \mathbf{c}^\mathsf{T}\mathbf{x}, \quad \text{and} \quad \begin{bmatrix} h_1(\mathbf{x}) \\ \vdots \\ h_m(\mathbf{x}) \\ h_{m+1}(\mathbf{x}) \\ \vdots \\ h_{m+n}(\mathbf{x}) \end{bmatrix} = \begin{bmatrix} \mathbf{A}\mathbf{x} - \mathbf{b} \\ -\mathbf{x} \end{bmatrix}.$$

The corresponding Lagrange function is $\mathcal{L}(\mathbf{x}, \mathbf{y}, \boldsymbol{\lambda}) = \mathbf{c}^\mathsf{T}\mathbf{x} - \mathbf{y}^\mathsf{T}(\mathbf{A}\mathbf{x} - \mathbf{b}) + \boldsymbol{\lambda}^\mathsf{T}\mathbf{x}$. Therefore, the KKT conditions for this model are:

$$\mathbf{c} - \mathbf{A}^\mathsf{T}\mathbf{y} + \boldsymbol{\lambda} = \mathbf{0} \qquad \text{(stationarity)}$$

$$\mathbf{A}\mathbf{x} - \mathbf{b} \leq \mathbf{0}, -\mathbf{x} \leq \mathbf{0} \qquad \text{(primal feasibility)}$$

$$\mathbf{y} \geq \mathbf{0}, \boldsymbol{\lambda} \geq \mathbf{0} \qquad \text{(dual feasibility)}$$

$$\mathbf{y}^\mathsf{T}(\mathbf{A}\mathbf{x} - \mathbf{b}) = \mathbf{0}, \quad \boldsymbol{\lambda}^\mathsf{T}(-\mathbf{x}) = \mathbf{0}. \qquad \text{(complementary slackness)}$$

The stationarity equation states that $\lambda = \mathbf{A}^\mathsf{T}\mathbf{y} - \mathbf{c}$. Substituting $\mathbf{A}^\mathsf{T}\mathbf{y} - \mathbf{c}$ for λ and rearranging the terms gives that the KKT conditions are equivalent to:

$$\mathbf{Ax} \leq \mathbf{b}, \mathbf{x} \geq \mathbf{0} \qquad\qquad \text{(primal feasibility)}$$
$$\mathbf{y} \geq \mathbf{0}, \mathbf{A}^\mathsf{T}\mathbf{y} \geq \mathbf{c} \qquad\qquad \text{(dual feasibility)}$$
$$\mathbf{y}^\mathsf{T}(\mathbf{b} - \mathbf{Ax}) = \mathbf{0}, \quad (\mathbf{A}^\mathsf{T}\mathbf{y} - \mathbf{c})^\mathsf{T}\mathbf{x} = \mathbf{0} \qquad \text{(complementary slackness)}$$

Recall that Theorem E.4.1 states that any optimal point of a standard LO-model satisfies this system of equations and inequalities, i.e., they form a necessary condition for optimality. Compare in this respect Theorem 4.3.1, which states that the KKT conditions are in fact necessary and sufficient.

APPENDIX F

Writing LO-models in GNU
MathProg (GMPL)

In this appendix, the *GNU MathProg Modeling Language* (or *GMPL*) is described. The GMPL language is a subset of the AMPL language, which is one of a few common modeling languages for mathematical optimization, including linear optimization. Similar languages include AIMMS, AMPL, GAMS, and Xpress-Mosel. These languages provide a tool for describing and solving complex optimization models. One of the advantages of using such a language is the fact that models can be expressed in a notation that is very similar to the mathematical notation that is used for expressing optimization models. This allows for concise and readable definitions of the models, and avoids having to explicitly construct the abstract ingredients of LO-models, namely the vector objective coefficients, the technology matrix, and the vector of right hand side values.

We chose to describe GMPL because software for it is available on the internet at no cost. We think that most modeling languages are similar to GMPL, so once one knows how to write models in this language, the transition to other modeling languages is relatively easy.

GMPL models may be solved using, for instance, the GLPK package, which is available for download at `http://www.gnu.org/software/glpk/`. For small models, the reader is invited to use the online solver on this book's website: `http://www.lio.yoriz.co.uk/`. This online solver provides a user-friendly interface for solving GMPL models. Moreover, the code for the listings in this book are available online.

F.1 Model Dovetail

The following listing shows how to write Model Dovetail in GMPL. The line numbers are not part of the model code; they have been added to make it easy to refer to particular lines.

```
 1  /* Decision variables */
 2      x1 >= 0;                       # number of boxes (x 100,000) of long matches
 3      x2 >= 0;                       # number of boxes (x 100,000) of short matches
 4
 5  /* Objective function */
 6          z:       3*x1 + 2*x2;
 7
 8  /* Constraints */
 9              c11:    x1 +    x2 <=  9;  # machine capacity (1.1)
10              c12: 3*x1 +    x2 <= 18;  # wood (1.2)
11              c13:    x1         <=  7;  # boxes for long matches (1.3)
12              c14:           x2 <=  6;  # boxes for short matches (1.4)
13
14      ;
```

Listing F.1: Model Dovetail.

Line 1 is a comment. Anything between /* and */ is regarded as a comment. Comments are ignored by the computer package and are only there for clarification. Similarly, anything on a line that appears after the # sign is regarded as a comment.

Lines 2–3 define the decision variables x1 and x2. Any variable definition starts with the keyword '   '. They are defined to be nonnegative decision variables by the '>= 0' sign. To declare the nonpositive decision variable x, we would insert the definition '   x <= 0'. In fact, the 0 can be replaced by any value. For instance, to declare the variable x that has to have value at least 15, one can insert '   x >= 15'. It is also possible to specify both upper and lower bounds. For instance: x >= 0, <= 20 defines the nonnegative decision variable x whose value should be at most 20. Notice that the variable definitions, like every definition in GMPL, end with a semicolon. Note also that GMPL is case sensitive, i.e., the expressions 'y1' and 'Y1' are seen as two different variables.

Line 6 defines the objective. The word '   ' indicates that we are dealing with a max-imizing LO-model. We could have used '   ' to indicate a minimizing LO-model. The word '   ' is followed by 'z', which is the name of the objective. The following colon separates the name of the objective from the objective function. The objective function ($3x_1 + 2x_2$ in this case) is written using the customary arithmetic signs +, −, *, and /. The semicolon marks the end of the objective definition.

Lines 9–12 define the constraints. Each constraint starts with the keywords '   '. As with the objective, these keywords are followed by the name of the constraint, a colon, the constraint expression, and a closing semicolon. The names of the constraints ('c11', 'c12', 'c13', 'c14') must be chosen to be unique. The sign '<=' means that we are defining a '$\leq$' constraint. A '$\geq$' constraint is defined using the sign '>=', and an equality constraint using '='.

Finally, line 14 marks the end of the model.

F.2 Separating model and data

It is good practice to separate the model formulation (i.e., the equations) from the data for the model (i.e., the numbers). For instance, we may rewrite Model Dovetail as follows. Define:

$\mathcal{P}$ = the set of end products, i.e., long matches and short matches.

$\mathcal{I}$ = the set of input products required for the production of the products, i.e., machine time, wood, boxes for long matches, and boxes for short matches.

p_j = the profit from selling one unit of end product j $(j \in \mathcal{P})$.

a_{ij} = the number of units of input product i that is required for the production of one unit of end product j $(i \in \mathcal{I}, j \in \mathcal{P})$.

c_i = the amount of input product available for production $(i \in \mathcal{I})$.

Then, Model Dovetail may be rewritten as:

$$\max \sum_{j \in \mathcal{P}} p_j x_j$$

$$\text{s.t.} \sum_{j \in \mathcal{P}} a_{ij} x_j \leq c_i \qquad\qquad \text{for } i \in \mathcal{I} \qquad\qquad (\text{F.1})$$

$$x_j \geq 0 \qquad\qquad \text{for } j \in \mathcal{P}.$$

This model has one variable for each element of $\mathcal{P}$, and one constraint for each element of $\mathcal{I}$. In GMPL, this model is written as follows:

```
1  /* Model parameters */
2     P;                            # set of products
3     I;                            # input products
4     p{P};                         # profit per unit of product
5     c{I};                         # availability of each input product
6     a{I, P};                      # number of units of input required to produce
7                                   # one unit of product
8
9  /* Decision variables */
10    x{P} >= 0;                    # number of boxes (x 100,000) of each product
11
12 /* Objective function */
13        z:
14     {j    P} p[j] * x[j];
15
16 /* Constraints */
17            input {i    I}:       # one constraint for each input product
18        {j    P} a[i, j] * x[j] <= c[i];
19
20 /* Model data */
21    ;
22
23    P := long short;
24    I := machine wood boxlong boxshort;
```

```
25
26          p :=
27    long     3
28    short    2;
29
30          c :=
31    machine      9
32    wood        18
33    boxlong      7
34    boxshort     6;
35
36        a : long short :=
37    machine    1    1
38    wood       3    1
39    boxlong    1    0
40    boxshort   0    1;
41
42      ;
```

Listing F.2: Model Dovetail (separating model and data).

This GMPL listing consists of two parts. The first part (lines 1–18) concerns the definition of model (F.1), and there is no reference to the particular data of the model (i.e., the precise elements of the sets $\mathcal{P}$ and $\mathcal{I}$, the numerical values of parameters a_{ij}, c_i, and p_j). The values of the sets and parameters are specified in the second part (lines 20–41). An LO-model can only be solved if all these numerical values of the sets and parameters have been specified. This separation has a number of advantages: (1) it is easier to read the listing when model and data are specified separately, (2) solving the same model with a new data set is a matter of simply replacing the second part of the listing, and (3) especially when the model has many sets and parameters, and these sets and parameters appear in different places in the model, it saves a lot of repetition in the code.

Lines 2–3 define the two sets $\mathcal{P}$ and $\mathcal{I}$, without specifying the elements of the sets. Lines 4–6 define the different parameters of the model, again without specifying their numerical values. The parameters are defined with index sets. For example, line 4 defines the parameter p, which has index set P. This means that, for every element j in P, a scalar parameter p[j] is defined. Similarly, lines 5–6 define, for each element i in I and for each element j in P, the parameters c[i] and a[i,j].

Line 10 defines the decision variables of the model. It defines one nonnegative variable x[j] for each j in P.

Lines 13–14 define the objective. As in Listing F.1, the objective is to maximize the total profit. Since it is not known at this stage how many terms will be involved in this expression, we take the sum over all elements j of P of the total profit due to selling product j, i.e., p[j]*x[j].

Lines 17–18 compactly summarize the constraints of the model. In contrast to the previous listing, this expression defines multiple constraints at once: the expression '{i   I}' means that we are defining one constraint for each element i of I. The left hand side of each of the constraints consists of a summation, and the right hand side contains a parameter.

Line 21 signals to the solver package that the model has been specified, and that the remainder of the listing will contain data specifications. Lines 23–24 specify the entries of the sets P and I. The entries consist of plain words separated by spaces. So, P contains two entries, and I contains four entries.

Lines 26–28 specify the parameter p. Recall that p is indexed by the set of products P. Thus, since P has two entries, we need to specify two values, namely p[long] and p[short]. This is exactly what these lines do: they state that p[long] $= 3$ and p[short] $= 2$. The parameter c is specified in lines 30–34 in a similar manner.

Lines 36–41 specify the two-dimensional parameter a in table form. Line 36 announces that the parameter a will be specified. The next line contains the elements of the first index set, I. These elements can follow in any order, as long as the numerical values are consistently listed in the same order. Lines 38–41 start with the name of an element j of the set P, followed by the values of a[i,j].

If all sets and model parameters have been given numerical values, the solver package has enough information to construct an LO-model: it derives the number of decision variables from the number of elements of the set P, the number of constraints from the number of elements of the set I, the vector of objective coefficients from the definition of the objective function, and the entries of the technology matrix and the right hand side vector from the constraint definitions. The solver uses an algorithm (i.e., the simplex algorithm) to solve the resulting LO-model, and then, if an optimal solution exists, translates the optimal solution back into the variables names defined in the model.

F.3 Built-in operators and functions

The GMPL language has a number of built-in operators and functions. Some commonly used ones, along with their interpretations, are:

x*y	$x \times y$;
x/y	$\frac{x}{y}$;
x**y, x^y	x^y;
x y	integer quotient x/y (i.e., x/y without the fractional part);
x y	remainder of x divided by y;
(x)	absolute value $\|x\|$;
(x)	arctan x (in radians);
(y, x)	arctan y/x (in radians);
(S)	cardinality $\|S\|$ of the set S;
(x)	x rounded up;

(x)	$\cos x$ (in radians);
(x)	e^x;
(x)	x rounded down;
(x)	natural logarithm $\ln x$;
(x)	base-10 logarithm $\log_{10} x$;
(x1,...,xn)	the maximum of the values $x_1, x_2, \ldots, x_n$;
(x1,...,xn)	the minimum of the values $x_1, x_2, \ldots, x_n$;
(x)	x rounded to the nearest integer;
(x, n)	x rounded to the nearest number with n decimals;
(x)	$\sin x$ (in radians);
(x)	$\sqrt{x}$.

Recall that the solver constructs an LO-model (i.e., the technology matrix **A**, and the vectors **b** and **c**) from the model code. Because the operators and functions listed above are all nonlinear, they cannot appear in the constructed LO-model. Instead, any of these operators and functions is evaluated when the solver constructs the LO-model. Hence, they may only involve parameter values (which are known when the LO-model is constructed), and they cannot involve any decision variables.

Another special expression in GMPL is the 'if, then, else' expression. We will explain this expression by means of an example. Suppose that we have an LO-model with decision variables $x_1, \ldots, x_T$, and $s_1, \ldots, s_T$, where x_t represents the account balance (in dollars) of a bank account at the end of year t, and s_t denotes the (net) earnings during year t ($t = 1, \ldots, T$). Let x_0 be the initial account balance. Let r be the annual interest rate. For simplicity, assume that the interest is paid at the beginning of the year, based on the balance at that time point. The evolution of the account balance can be expressed as the following set of constraints:

$$x_t = (1 + r)x_{t-1} + s_t, \qquad \text{for } t = 1, \ldots, T. \tag{F.2}$$

It may be tempting to write these constraints as follows in GMPL code:

```
1      r;
2      T;
3    x{1..T} >= 0;
4    s{1..T} >= 0;
5        balance{t    1..T}:
6    x[t] = (1+r) * x[t−1] + s[t];
```

However, this code is invalid, because x[0] is not defined. The problem lies in the fact that $x_1, \ldots, x_T$ are variables of the model, whereas x_0 is a parameter of the model. It is not possible to define x[0] as a parameter. Instead, a new parameter, x0, say, has to be defined. This means that, in constraint (F.2) with $t = 1$, x0 should be used instead of x[0] and, for $t \neq 1$, x[t] should be used. This can be written in GMPL code as follows:

```
1      x0;
```

```
2                  balance{t    1..T}:
3     x[t] = (1+r) * (   t == 1      x0     x[t−1]) + s[t];
```

The expression ' t == 1   x0   x[t−1]' is evaluated to x0 if t = 1, and to x[t-1] otherwise. So, the constraint is interpreted as:

$$x[t] = (1+r) * x0 + s[t] \qquad \text{for } t = 1, \text{ and}$$
$$x[t] = (1+r) * x[t−1] + s[t] \qquad \text{for } t = 2, \dots, T.$$

The general form of the 'if, then, else' expression in GMPL is:

condition   *value-if-true*   *value-if-false*.

If *condition* is true, then the expression is equal to *value-if-true*; otherwise it is equal to *value-if-false*. Note that, as with the operators and functions above, any 'if, then, else' expression is evaluated when the GMPL solver constructs the technology matrix $\mathbf{A}$, and the vectors $\mathbf{b}$ and $\mathbf{c}$. This means that the *condition* may only depend on the values of the model parameters, and not on the values of the decision variables. However, *value-if-true* and *value-if-false* may contain decision variables.

It is important to note that the 'if, then, else' expression discussed here is different from the conditional relationships that are discussed in Section 7.3. The latter are conditional relationships between decision variables. Such relationships generally require the use of mixed integer optimization algorithms and are computationally harder to deal with; see Chapter 7. See also Section 18.5 for an application that uses 'if, then, else' expressions.

F.4 Generating all subsets of a set

In Section 7.2.4, we formulated Model 7.2.2. This model has a constraint for each subset of the ground set $X = \{0, \dots, n\}$ $(n \geq 0)$. GMPL has no built-in functionality to generate all subsets of a given set. A simple way to generate all subsets of X is to write each number k $(\in \{0, \dots, 2^{n+1}-1\})$ as a *binary number* $d_n d_{n-1} \cdots d_2 d_1 d_0$, with $d_i = d_i(k) \in \{0,1\}$ and $k = \sum_{i=0}^{n} 2^i d_i(k)$. For example, for $n = 3$, the numbers $0, \dots, 15$ in binary notation are:

$$0000, 0001, 0010, 0011, 0100, 0101, 0110, 0111,$$
$$1000, 1001, 1010, 1011, 1100, 1101, 1110, 1111.$$

So, for instance the number 13 can be written as 1101, i.e., $d_3 = 1, d_2 = 1, d_1 = 0, d_0 = 1$. The subsets of X can now be constructed by considering each $k \in \{0, \dots, 2^{n+1}-1\}$, and defining the set S_k by letting $i \in S_k$ if and only if $d_i(k) = 1$. Doing this for the binary numbers listed above, we obtain the following subsets $S_0, \dots, S_{15}$ of $\{0, 1, 2, 3\}$:

$$\emptyset, \{0\}, \{1\}, \{0,1\}, \{2\}, \{0,2\}, \{1,2\}, \{0,1,2\},$$
$$\{3\}, \{0,3\}, \{1,3\}, \{0,1,3\}, \{2,3\}, \{0,2,3\}, \{1,2,3\}, \{0,1,2,3\}.$$

For $k = 0, \ldots, 2^{n+1}-1$ and $i = 0, \ldots, n$, let $d_i(k)$ be the i'th binary digit of the number k. It can be checked that:

$$d_i(k) = \lfloor 2^{-i}k \rfloor \pmod{2} = \begin{cases} 0 & \text{if } \lfloor 2^{-i}k \rfloor \text{ if even} \\ 1 & \text{if } \lfloor 2^{-i}k \rfloor \text{ if odd}. \end{cases}$$

In this expression, $x \pmod{2}$ denotes the remainder after dividing x by 2. So, for instance $12 \pmod{2} = 0$, and $13 \pmod{2} = 1$. For $k = 0, \ldots, 2^{n+1}-1$, we have that the subset S_k of X satisfies:

$$S_k = \left\{ i \mid \lfloor 2^{-i}k \rfloor = 1 \pmod{2}, i = 0, \ldots, n \right\}.$$

This can be written in GMPL code as follows:

```
1       n;                              # largest number in the set X
2    X := 0 .. n;                       # the set X
3    SI := 0 .. 2^(n+1)-1;              # index set of all subsets of X
4    S {k    SI} :=                     # define the set S[k], with k in SI
5        {i   X :      (k/2^i)    2 = 1} i;
```

Listing F.3: Generating all subsets of a set.

Note that, in Model 7.2.2, there are no subtour elimination constraints for $S = \emptyset$ and $S = J$. So, to implement this model in GMPL, we should define SI to be the numbers $1, \ldots, 2^{n+1}-2$.

1	vector containing all ones
A	typical notation for a matrix
$\{a_{ij}\}$	matrix with entries a_{ij} $(i = 1, \ldots, m,\ j = 1, \ldots, n)$
$\mathbf{A}^{-1}$	inverse matrix of **A** (see Appendix B.6)
$\mathbf{A}^{\mathsf{T}}$	transpose matrix of **A** (see Appendix B.1)
$(\mathbf{A})_{I,J}$	submatrix of the matrix **A** consisting of the rows with indices in I and the columns with indices in J (see Section 2.2.1)
$(\mathbf{A})_{\star,J}$	submatrix of the matrix **A** consisting of the columns with indices in J (see Section 2.2.1)
$(\mathbf{A})_{I,\star}$	submatrix of the matrix **A** consisting of the columns with indices in J (see Section 2.2.1)
a	typical notation for a (column) vector
a_i	i'th entry of the vector **a**
$\mathrm{adj}(\mathbf{A})$	adjoint matrix of the matrix **A** (see Appendix B.7)
$\mathrm{aff}(S)$	affine hull of a set (see Appendix D.1)
$\mathbf{A}(G)$	adjacency matrix of the graph G (see Appendix C.7)
$\mathrm{argmin}\{f(\mathbf{x})\}$	set of vectors **x** for which f is minimal (see Appendix D.1)
$\mathrm{argmax}\{f(\mathbf{x})\}$	set of vectors **x** for which f is maximal (see Appendix D.1)
BI	typical notation for the set of indices of the current basic variables
$\mathcal{A}(G)$	arc set of the (directed) graph G (see Appendix C.6)
$|a|$	absolute value of the real number a
$\|\mathbf{a}\|$	the Euclidean norm of the vector **a** (see Appendix B.1)
$[a, b]$	closed interval with end points a and b (see Appendix D.2)
(a, b)	open interval with end points a and b (see Appendix D.2)
$\{a_1, \ldots, a_n\}$	set with specified elements $a_1, \ldots, a_n$

$\overline{\mathbf{B}}$	complementary dual matrix of $\mathbf{B}$ (see Section 2.2.3)
$B_\varepsilon(\mathbf{a})$	open ball with radius ε around $\mathbf{a}$ (see Appendix D.1)
$\mathrm{cl}(S)$	closure of the set S (see Appendix D.1)
$\mathrm{cof}(a_{ij})$	cofactor of the entry a_{ij} in the matrix $\{a_{ij}\}$ (see Appendix B)
$\mathrm{cone}(\mathbf{a}_1,\dots,\mathbf{a}_k)$	cone generated by the vectors $\mathbf{a}_1,\dots,\mathbf{a}_k$ (see Appendix D.2)
$\mathrm{conv}(S)$	convex hull of the set S (see Appendix D.2)
$\det(\mathbf{A})$	determinant of the matrix $\mathbf{A}$ (see Appendix B.7)
$\dim(S)$	dimension of the set S (see Appendix D.3)
$d(v)$	degree of vertex v (see Appendix C.1)
$d^-(v)$	indegree of vertex v (see Appendix C.6)
$d^+(v)$	outdegree of vertex v (see Appendix C.6)
δ_{ij}	typical notation for a binary variable (see Chapter 7)
$\dfrac{\partial f(x_1,\dots,x_n)}{\partial x_i}$	partial derivative of f with respect to x_i
$\mathbf{e}_i$	i'th unit vector, $\mathbf{e}_i = \begin{bmatrix} 0 & \dots & 0 & 1 & 0 & \dots & 0 \end{bmatrix}^{\mathsf{T}}$, with '1' in the i'th position
$\mathcal{E}(G)$	edge set of the graph G (see Appendix C.1)
$\mathrm{epi}(f)$	epigraph of the function f (see Appendix D.4)
∇f	gradient of the function f (see Appendix E.1)
$G = (\mathcal{N}, \mathcal{A})$	notation for a directed graph G with node set $\mathcal{N}$, arc set $\mathcal{A}$ (see Appendix C.6)
$G = (\mathcal{V}, \mathcal{E})$	notation for an undirected graph G with node set $\mathcal{V}$, edge set $\mathcal{E}$ (see Appendix C.1)
$h(a)$	head node of the arc a (see Appendix C.6)
H	typical notation for a hyperplane (see Section 2.1.1)
H^+, H^-	typical notation for halfspaces (see Section 2.1.1)
$I + k$	$\{i + k \mid i \in I\}$ for any integer k (see Section 2.2.1)
$I - k$	$\{i - k \mid i \in I\}$ for any integer k (see Section 2.2.1)
$\mathbf{I}_m$	identity matrix with m rows and m columns
$\inf f(x)$	infimum of the function f (see Appendix D.1)
$\bar{I}$	complement of the set I
K_n	complete graph on n nodes (see Appendix C.1)
$K_{m,n}$	complete bipartite graph on $m + n$ nodes (see Appendix C.1)
lexmin	lexicographic minimum (see Section 14.6.1)
lim	limit
M	typical notation for a large constant
$\mathbf{M}_{VA}$	node-arc incidence matrix of the digraph $G = (\mathcal{V}, \mathcal{A})$ (see Appendix C.7)
$\mathbf{M}_{VE}$	node-edge incidence matrix of the graph $G = (\mathcal{V}, \mathcal{E})$ (see Appendix C.7)
(m, n)-matrix	matrix with m rows and n columns
max	maximize
min	minimize

NI	typical notation for the set of indices of the current basic variables
$\mathcal{N}(\mathbf{A})$	null space of the matrix $\mathbf{A}$ (see Appendix B)
$\mathbb{N}$	set of positive integers
$n!$	n factorial
$\binom{n}{k}$	binomial coefficient 'n choose k' ($= \frac{n!}{k!(n-k)!}$)
$O(\cdot)$	big-O notation (see Chapter 9)
$\mathcal{R}(\mathbf{A})$	row space of the matrix $\mathbf{A}$ (see Appendix B)
relint	(see Appendix D.1)
$\mathbb{R}$	set of all real numbers
$\mathbb{R}^n$	set of all n-dimensional real-valued vectors
$\mathbb{R}^{m \times n}$	set of all real-valued (m, n)-matrices
s.t.	subject to
sup	supremum (see Appendix D.1)
$|S|$	number of elements in the set S
$t(a)$	tail node of the arc a (see Appendix C.6)
$\mathcal{V}(G)$	node set of the graph G (see Appendix C.1)
$\mathbf{x}$	typical notation for the vector of (primal) decision variables
$\mathbf{x}_s$	typical notation for the vector of (primal) slack variables
x_i	typical notation for a (primal) decision variable
$\begin{bmatrix} x_1 & \cdots & x_n \end{bmatrix}^\mathsf{T}$	typical notation for a column vector with specified entries
$\{x \mid P(x)\}$	set of all elements x satisfying the property P
$\mathbf{y}$	typical notation for the vector of dual decision variables
$\mathbf{y}_s$	typical notation for the vector of dual slack variables
y_i	typical notation for a dual decision variable
z	typical notation for the objective value
z^*	typical notation for the optimal objective value
$\mathbb{Z}$	set of integers (i.e., the positive and negative integers, and 0)
$=$	is equal to
$\neq$	is not equal to
$\approx$	is approximately equal to
$\equiv$	is equivalent to, up to an appropriate permutation of the rows and/or columns
$:=$	assign to
$<$	less than
$\leq$	less than or equal; if both sides of the inequality are vectors, then the inequality is understood to be entry-wise
$>$	greater than
$\ll$	is much smaller than
$\gg$	is much greater than

$\geq$	greater than or equal; if both sides of the inequality are vectors, then the inequality is understood to be entry-wise
$\sim$	is equivalent to
$\emptyset$	empty set
$\backslash$	set difference
$\in$	is an element of
$\notin$	is not an element of
$\subset$	is a subset of (but not equal to)
$\subseteq$	is a subset of (or equal to)
$\cup$	set union
$\cap$	set intersection
$\bigcap_{i=1}^{n} S_i$	intersection of the sets $S_1, \ldots, S_n$
$\bigcup_{i=1}^{n} S_i$	union of the sets $S_1, \ldots, S_n$
$\perp$	is perpendicular to
∞	infinity
$\vee$	logical inclusive disjunction (see Section 7.3.2)
$\underline{\vee}$	logical exclusive disjunction (see Section 7.3.2)
$\wedge$	logical conjunction (see Section 7.3.2)
$\Rightarrow$	logical implication (see Section 7.3.2)
$\Leftrightarrow$	logical equivalence (see Section 7.3.2)
$\neg$	logical negation (see Section 7.3.2)

Bibliography

Ahuja, R. K., Magnanti, T. L., and Orlin, J. B. (1993), *Network Flows: Theory, Algorithms, and Applications*, Prentice Hall, Englewood Cliffs, New Jersey.

Aigner, M., and Ziegler, G. M. (2010), *Proofs from The Book*, Fourth edition, Springer.

Albers, D. J., Reid, C., and Dantzig, G. B. (1986), 'An Interview with George B. Dantzig: The Father of Linear Programming', *The College Mathematics Journal* **17**(4), 292–314.

Alpaydin, E. (2004), *Introduction to Machine Learning*, Second edition, MIT Press.

Altier, W. J. (1999), *The Thinking Manager's Toolbox: Effective Processes for Problem Solving and Decision Making*, Oxford University Press.

Applegate, D. L., Bixby, R. E., Chvátal, V., and Cook, W. J. (2006), *The Traveling Salesman Problem: A Computational Study*, Princeton University Press.

Arora, S., and Barak, B. (2009), *Computational Complexity: A Modern Approach*, Cambridge University Press, New York.

Arsham, H., and Oblak, M. (1990), 'Perturbation Analysis of General LP Models: A Unified Approach to Sensitivity, Parametric, Tolerance, and More-for-Less Analysis', *Mathematical and Computer Modelling* **13**(8), 79–102.

Barnett, S. (1990), *Matrices. Methods and Applications*, Oxford Applied Mathematics and Computing Science Series, Clarendon Press, Oxford.

Bazaraa, M. S., Jarvis, J. J., and Sherali, H. D. (1990), *Linear Programming and Network Flows*, Second edition, John Wiley & Sons, Inc., New York.

Bazaraa, M. S., Sherali, H. D., and Shetty, C. M. (1993), *Nonlinear Programming: Theory and Algorithms*, John Wiley & Sons, Inc., New York.

Beasley, J. E. (1996), *Advances in Linear and Integer Programming*, Oxford University Press.

Bertsimas, D., and Tsitsiklis, J. N. (1997), *Introduction to Linear Optimization*, Athena Scientific, Belmont, Massachusetts.

Bixby, R. E., Fenelon, M., Gu, Z., Rothberg, E., and Wunderling, R. (2000), 'MIP: Theory and Practice – Closing the Gap', *System Modelling and Optimization: Methods, Theory, and Applications* **174**, 19–49.

Bondy, J. A., and Murty, U. S. R. (1976), *Graph Theory with Applications*, MacMillan, London.

Boyd, S. P., and Vandenberghe, L. (2004), *Convex Optimization*, Cambridge University Press.

Bröring, L. (1996), Linear Production Games, Master's thesis, University of Groningen, The Netherlands.

Chartrand, G., Lesniak, L., and Zhang, P. (2010), *Graphs & Digraphs*, Fifth edition, CRC Press.

Checkland, P. (1999), *Systems Thinking, Systems Practice*, John Wiley & Sons, Inc., New York.

Chen, D. S., Batson, R. G., and Dang, Y. (2011), *Applied Integer Programming: Modeling and Solution*, John Wiley & Sons, Inc., New York.

Chvátal, V. (1983), *Linear Programming*, W.H. Freeman and Company, New York.

Ciriani, T. A., and Leachman, R. C. (1994), *Optimization in Industry 2: Mathematical Programming and Modeling Techniques in Practice*, John Wiley & Sons, Inc., New York.

Cook, W. J. (2012), *In Pursuit of the Traveling Salesman: Mathematics at the Limits of Computation*, Princeton University Press, Princeton and Oxford.

Cook, W. J., Cunningham, W. H., Pulleyblank, W. R., and Schrijver, A. (1997), *Combinatorial Optimization*, John Wiley & Sons, Inc., New York.

Cooper, W. W., Seiford, L. M., and Zhu, J., eds (2011), *Handbook on Data Envelopment Analysis*, Second edition, Springer.

Coppersmith, D., and Winograd, S. (1990), 'Matrix Multiplication via Arithmetic Progressions', *Journal of Symbolic Computation* **9**(3), 251–280.

Cornuéjols, G. (2008), 'Valid Inequalities for Mixed Integer Linear Programs', *Mathematical Programming* **112**(1), 3–44.

Cristianini, N., and Shawe-Taylor, J. (2000), *An Introduction to Support Vector Machines and Other Kernel-Based Learning Methods*, Cambridge University Press.

Cunningham, W. H., and Klincewicz, J. G. (1983), 'On Cycling in the Network Simplex Method', *Mathematical Programming* **26**(2), 182–189.

Dantzig, G. B. (1963), *Linear Programming and Extensions*, Princeton University Press, Princeton, New Jersey.

Dantzig, G. B. (1982), 'Reminiscences about the Origins of Linear Programming', *Operations Research Letters* **1**(2), 43–48.

Dantzig, G. B., and Thapa, M. N. (1997), *Linear Programming 1: Introduction*, Springer.

Fang, S.-C., and Puthenpura, S. (1993), *Linear Optimization and Extensions: Theory and Algorithms*, Prentice Hall, Inc.

Ferris, M. C., Mangasarian, O. L., and Wright, S. J. (2007), *Linear Programming with Matlab*, MPS-SIAM Series on Optimization.

Fletcher, R. (2000), *Practical Methods of Optimization*, John Wiley & Sons, Inc., New York.

Friedman, J. W. (1991), *Game Theory with Applications to Economics*, Second edition, Oxford University Press.

Gal, T. (1986), 'Shadow Prices and Sensitivity Analysis in Linear Programming under Degeneracy', *OR Spectrum* **8**(2), 59–71.

Gal, T. (1995), *Postoptimal Analyses, Parametric Programming and Related Topics*, Walter de Gruyter, Berlin.

Gill, P. E., Murray, W., and Wright, M. H. (1982), *Practical Optimization*, Academic Press, Inc., London.

Gilmore, P. C., and Gomory, R. E. (1963), 'A Linear Programming Approach to the Cutting Stock Problem – Part II', *Operations Research* **11**(6), 863–888.

Goldfarb, D., and Todd, M. J. (1989), 'Linear Programming', *Optimization* **1**, 73–170.

Greenberg, H. J. (1986), 'An Analysis of Degeneracy', *Naval Research Logistics Quarterly* **33**(4), 635–655.

Greenberg, H. J. (2010), Myths and Counterexamples in Mathematical Programming, Technical Report, `http://glossary.computing.society.informs.org/myths/CurrentVersion/myths.pdf`.

Griffel, D. H. (1989*a*), *Linear Algebra and its Applications. Vol. 1: A First Course*, Halsted Press.

Griffel, D. H. (1989*b*), *Linear Algebra and its Applications. Vol. 2: More Advanced*, Halsted Press.

Grünbaum, B. (2003), *Convex Polytopes*, Springer Verlag.

Guéret, C., Prins, C., and Sevaux, M. (2002), *Applications of Optimisation with Xpress-MP*, Dash Optimization.

Haimovich, M. (1983), *The Simplex Algorithm is Very Good! On the Expected Number of Pivot Steps and Related Properties of Random Linear Programs*, Columbia University Press.

Hartsfield, N., and Ringel, G. (2003), *Pearls in Graph Theory: A Comprehensive Introduction*, Dover Publications.

den Hertog, D. (1994), *Interior Point Approach to Linear, Quadratic and Convex Programming: Algorithms and Complexity*, Kluwer Academic Publisher.

Horn, R. A., and Johnson, C. R. (1990), *Matrix Analysis*, Cambridge University Press.

Ignizio, J. P. (1985), *Introduction to Linear Goal Programming*, Sage Publications.

Ignizio, J. P. (1991), *Introduction to Expert Systems: the Development and Implementation of Rule-Based Expert Systems*, McGraw-Hill, New York.

Jeffrey, P., and Seaton, R. (1995), 'The Use of Operational Research Tools: A Survey of Operational Research Practitioners in the UK', *Journal of the Operational Research Society* **46**, 797–808.

Jenkins, L. (1990), 'Parametric Methods in Integer Linear Programming', *Annals of Operations Research* **27**(1), 77–96.

Johnson, E. L., and Nemhauser, G. L. (1992), 'Recent Developments and Future Directions in Mathematical Programming', *IBM Systems Journal* **31**(1), 79–93.

Johnson, L. W., Riess, R. D., and Arnold, J. T. (2011), *Introduction to Linear Algebra*, Sixth edition, Pearson.

Karmarkar, N. (1984), 'A New Polynomial-Time Algorithm for Linear Programming', *Combinatorica* **4**, 373–395.

Keys, P. (1995), *Understanding the Process of Operational Research: Collected Readings*, John Wiley & Sons, Inc., New York.

Khachiyan, L. G. (1979), 'A Polynomial Algorithm in Linear Programming', *Doklady Akademia Nauk SSR* **224**(5), 1093–1096.

Klee, V., and Minty, G. J. (1972), How Good is the Simplex Algorithm?, *in* O. Shisha, ed., 'Inequalities III', Academic Press, New York and London, 159–175.

Lawler, E. L., Lenstra, J. K., Rinnooy Kan, A. H. G., and Shmoys, D. B., eds (1985), *The Traveling Salesman Problem: A Guided Tour of Combinatorial Optimization*, John Wiley & Sons, Inc., New York.

Lay, D. C. (2012), *Linear Algebra and its Applications*, Fourth edition, Pearson.

Lee, J. (2004), *A First Course in Combinatorial Optimization*, Cambridge University Press.

Lenstra, A. K., Rinnooy Kan, A. H. G., and Schrijver, A. (1991), *History of Mathematical Programming – A Collection of Personal Reminiscences*, North-Holland.

Littlechild, S. C., and Shutler, M. F., eds (1991), *Operations Research in Management*, Prentice Hall International, London.

Liu, J. S., Lu, L. Y. Y., Lu, W. M., and Lin, B. J. Y. (2013), 'Data Envelopment Analysis 1978–2010: A Citation-Based Literature Survey', *Omega* **41**(1), 3–15.

Luenberger, D. G., and Ye, Y. (2010), *Linear and Nonlinear Programming*, Springer.

Martello, S., and Toth, P. (1990), *Knapsack Problems: Algorithms and Computer Implementations*, John Wiley & Sons, Inc., New York.

Martin, R. K. (1999), *Large Scale Linear and Integer Optimization: A Unified Approach*, Springer.

Matoušek, J., and Gärtner, B. (2007), *Understanding and Using Linear Programming*, Springer.

McMahan, H. B., Holt, G., Sculley, D., Young, M., Ebner, D., Grady, J., Nie, L., Phillips, T., Davydov, E., Golovin, D. *et al.* (2013), Ad Click Prediction: A View from the Trenches, *in* 'Proceedings of the 19th ACM SIGKDD International Conference on Knowledge Discovery and Data Mining', ACM, 1222–1230.

Michalewicz, Z., and Fogel, D. B. (2004), *How to Solve It: Modern Heuristics*, Springer.

Mirchandani, P. B., and Francis, R. L., eds (1990), *Discrete Location Theory*, John Wiley & Sons, Inc., New York.

Mitra, G., Lucas, C., Moody, S., and Hadjiconstantinou, E. (1994), 'Tools for Reformulating Logical Forms into Zero-One Mixed Integer Programs', *European Journal of Operational Research* **72**(2), 262–276.

Müller-Merbach, H. (1981), 'Heuristics and their Design: A Survey', *European Journal of Operational Research* **8**(1), 1–23.

Nemhauser, G. L., and Wolsey, L. A. (1999), *Integer and Combinatorial Optimization*, John Wiley & Sons, Inc., New York.

Nering, E. D., and Tucker, A. W. (1993), *Linear Programs and Related Problems*, Academic Press, London.

Padberg, M. W. (1999), *Linear Optimization and Extensions*, Springer Verlag.

Papadimitriou, C. H., and Steiglitz, K. (1998), *Combinatorial Optimization: Algorithms and Complexity*, Dover Publications.

Picard, J.-C. (1976), 'Maximal Closure of a Graph and Applications to Combinatorial Problems', *Management Science* **22**(11), 1268–1272.

Pinedo, M. L. (2012), *Scheduling: Theory, Algorithms, and Systems*, Springer.

von Plato, J. (2008), The Development of Proof Theory, Technical Report, `http://plato.stanford.edu/entries/proof-theory-development/`.

Powell, S. G., and Baker, K. R. (2004), *The Art of Modeling with Spreadsheets*, John Wiley & Sons, Inc., New York.

Prager, W. (1956), 'On the Caterer Problem', *Management Science* **3**(1), 15–23.

Ribeiro, C. C., and Urrutia, S. (2004), 'OR on the Ball: Applications in Sports Scheduling and Management', *OR/MS Today* **31**, 50–54.

Rockafellar, R. T. (1997), *Convex Analysis*, Princeton University Press.

Roos, C., Terlaky, T., and Vial, J.-Ph. (2006), *Interior Point Methods for Linear Optimization*, Second edition, Springer.

Schrage, L., and Wolsey, L. (1985), 'Sensitivity Analysis for Branch and Bound Integer Programming', *Operations Research* **33**(5), 1008–1023.

Schrijver, A. (1998), *Theory of Linear and Integer Programming*, John Wiley & Sons, Inc., New York.

Schrijver, A. (2003), *Combinatorial Optimization*, Springer.

Schweigman, C. (1979), *Doing Mathematics in a Developing Country: Linear Programming with Applications in Tanzania*, Tanzania Publishing House.

Sierksma, G., and Ghosh, D. (2010), *Networks in Action: Text and Computer Exercises in Network Optimization*, Springer.

Sierksma, G., and Tijssen, G. A. (1998), 'Routing Helicopters for Crew Exchanges on Off-Shore Locations', *Annals of Operations Research* **76**, 261–286.

Sierksma, G., and Tijssen, G. A. (2003), 'Degeneracy Degrees of Constraint Collections', *Mathematical Methods of Operations Research* **57**(3), 437–448.

Sierksma, G., and Tijssen, G. A. (2006), 'Simplex Adjacency Graphs in Linear Optimization', *Algorithmic Operations Research* **1**(1).

Strang, G. (2009), *Introduction to Linear Algebra*, Fourth edition, Wellesley-Cambridge Press.

Strassen, V. (1969), 'Gaussian Elimination is not Optimal', *Numerische Mathematik* **13**(4), 354–356.

Terlaky, T., and Zhang, S. (1993), 'Pivot Rules for Linear Programming: A Survey on Recent Theoretical Developments', *Annals of Operations Research* **46–47**(1), 203–233.

Thurston, W. P. (1998), On Proof and Progress in Mathematics, *in* T. Tymoczko, ed., 'New Directions in the Philosophy of Mathematics', Princeton University Press, 337–55.

Tijs, S. H., and Otten, G. J. (1993), 'Compromise Values in Cooperative Game Theory', *TOP* **1**(1), 1–36.

Tijssen, G. A., and Sierksma, G. (1998), 'Balinski-Tucker Simplex Tableaus: Dimensions, Degeneracy Degrees, and Interior Points of Optimal Faces', *Mathematical Programming* **81**(3), 349–372.

Truemper, K. (1990), 'A Decomposition Theory for Matroids. V. Testing of Matrix Total Unimodularity', *Journal of Combinatorial Theory, Series B* **49**(2), 241–281.

Vanderbei, R. J. (2014), *Linear Programming: Foundations and Extensions*, Fourth edition, Springer.

van de Vel, M. L. J. (1993), *Theory of Convex Structures*, North Holland.

Ward, J. E., and Wendell, R. E. (1990), 'Approaches to Sensitivity Analysis in Linear Programming', *Annals of Operations Research* **27**(1), 3–38.

West, D. B. (2001), *Introduction to Graph Theory*, Second edition, Prentice Hall, Upper Saddle River.

Williams, H. P. (2013), *Model Building in Mathematical Programming*, Fifth edition, Wiley.

Williams, V. V. (2012), Multiplying Matrices Faster than Coppersmith-Winograd, *in* 'STOC '12 Proceedings of the 44th Annual ACM Symposium on Theory of Computing', 887–898.

Williamson, D. P., and Shmoys, D. B. (2011), *The Design of Approximation Algorithms*, Cambridge University Press.

Wolsey, L. A. (1998), *Integer Programming*, John Wiley & Sons, Inc., New York.

Wright, M. B. (2009), '50 Years of OR in Sport', *Journal of the Operational Research Society* **60**, 161–168.

Zhang, S. (1991), 'On Anti-Cycling Pivoting Rules for the Simplex Method', *Operations Research Letters* **10**(4), 189–192.

Zimmermann, H. J. (2001), *Fuzzy Set Theory – and its Applications*, Springer.

Zwols, Y., and Sierksma, G. (2009), 'OR Practice – Training Optimization for the Decathlon', *Operations Research* **57**(4), 812–822.

Author index